Absolute Value Function

$$f(x) = |x|$$

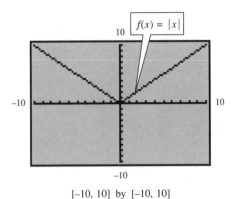

$f(x) = |x|$

[−10, 10] by [−10, 10]
Xscl = 1   Yscl = 1

Rational Function

$$f(x) = \frac{1}{x}$$

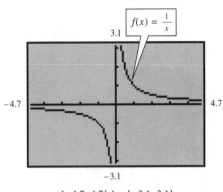

$f(x) = \frac{1}{x}$

[−4.7, 4.7] by [−3.1, 3.1]
Xscl = 1   Yscl = 1

Exponential Function $a > 1$

$$f(x) = a^x,\ a > 1$$

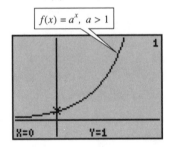

$f(x) = a^x,\ a > 1$

Exponential Function $a < 1$

$$f(x) = a^x,\ a < 1$$

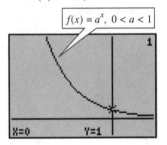

$f(x) = a^x,\ 0 < a < 1$

Logarithmic Function $a > 1$

$$f(x) = \log_a x,\ a > 1$$

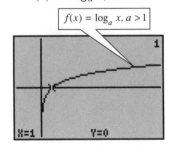

$f(x) = \log_a x,\ a > 1$

Logarithmic Function $a < 1$

$$f(x) = \log_a x,\ a < 1$$

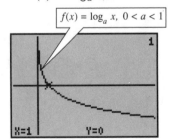

$f(x) = \log_a x,\ 0 < a < 1$

A GRAPHICAL APPROACH TO COLLEGE ALGEBRA

SECOND EDITION

John Hornsby
UNIVERSITY OF NEW ORLEANS

Margaret L. Lial
AMERICAN RIVER COLLEGE

ADDISON-WESLEY

An imprint of Addison Wesley Longman, Inc.

Reading, Massachusetts • Menlo Park, California • New York • Harlow, England
Don Mills, Ontario • Sydney • Mexico City • Madrid • Amsterdam

Sponsoring Editor: Bill Poole

Editorial Project Manager: Christine O'Brien

Assistant Editor: Rachel S. Reeve

Managing Editor: Karen Guardino

Production Supervisor: Rebecca Malone

Project Coordination: Elm Street Publishing Services, Inc.

Marketing Manager: Brenda L. Bravener

Marketing Coordinator: Stephanie Baldock

Prepress Services Buyer: Caroline Fell

Manufacturing Manager: Ralph Mattivello

Manufacturing Buyer: Evelyn Beaton

Text Design: Cynthia Crampton

Cover Design: Barbara T. Atkinson

Cover Photography: © SuperStock

Photo Credits

p. 1 © Stephen Poulin/SuperStock   **p. 121** © 1997 PhotoDisc, Inc.   **p. 203** © 1997 PhotoDisc, Inc.   **p. 315** © 1997 PhotoDisc, Inc.   **p. 395** © 1997 PhotoDisc, Inc. **p. 466** © 1997 PhotoDisc, Inc.   **p. 516** © 1997 PhotoDisc, Inc.   **p. 609** © 1997 PhotoDisc, Inc.

Library of Congress Cataloging-in-Publication Data

Hornsby, E. John.
    A graphical approach to college algebra / John Hornsby, Margaret L. Lial—2nd ed.
        p.   cm.
    Includes index.
    ISBN 0-321-02847-3
    1. Algebra.   I. Lial, Margaret L.   II. Title.
QA152.2.H68   1999
512.9—dc21   98-18058
                                                                                        CIP

1 2 3 4 5 6 7 8 9 10—RNT—01009998

To the memory of Jack Hornsby

Contents

CHAPTER 3   *Polynomial Functions*   *203*

CHAPTER 4   *Rational, Root, and Inverse Functions*   *315*

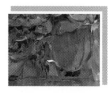

CHAPTER 5 *Exponential and Logarithmic Functions* *395*

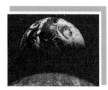

CHAPTER 6 *The Conic Sections and Parametric Equations* *466*

CHAPTER 7 *Matrices and Systems of Equations and Inequalities* *516*

CHAPTER 8   *Further Topics in Algebra*   *609*

CHAPTER R   *Reference: Basic Algebraic Concepts*   *674*

Preface

In writing the previous edition of this text, we pursued the project with the firm commitment that it would not be merely an adaptation of our traditional text with only lip service paid to the use of technology. In so doing, we realized that a completely new approach would be necessary, based on the premise that all students would have graphing calculators on the first day of class and use them throughout the course.

The first edition was prepared and class-tested at the University of New Orleans from 1993 to 1995. It was published in 1996 after thousands of hours of work, not only by the authors but by reviewers, answer-checkers, editors, class-testers, and students.

This second edition reflects our combined years of experience as classroom teachers, emphasizes our enthusiasm for teaching with graphing calculators, and provides refinement of the first edition of the text through incorporation of the many helpful suggestions of both teachers and students.

PHILOSOPHY OF OUR APPROACH

Throughout the first five chapters, we present the various classes of functions studied in a standard college algebra text. Chapter 1 introduces functions and relations, using the linear function as the basis for the presentation. In this chapter we introduce the following approach used throughout the next four chapters.

After introducing a class of function:

- ◆ We first examine the nature of its graph.
- ◆ Next we discuss the analytic solution of equations based on that function.
- ◆ We then show how to provide graphical support for that solution using a graphing calculator.
- ◆ Having established these two methods of solving equations, we move on to the analytic methods of solving the associated inequalities.
- ◆ We then show how the analytic solutions of these inequalities can also be supported graphically. We use two approaches to graphical methods of solving equations and inequalities: the *x-intercept method* and the *intersection-of-graphs method*. We continually review and reinforce these methods throughout.
- ◆ Finally, once the student has a feel for the particular class of function under consideration, we use our analytic and graphical methods to solve interesting applications involving that function.

By consistently using this approach with all the different classes of functions, students become aware that we are always following the same general procedure and applying that procedure to a new kind of function.

THE APPROACH TO TECHNOLOGY

We wrote this text with the idea in mind that technology can be used to help us better understand the mathematical concepts. We continually emphasize that it is essential to understand the mathematical concepts and apply them hand-in-hand with the calculator. Our good friend, Peg Crider of Tomball College, says it best: *"Your brain is the most powerful tool in the whole process."*

Because technology seems to be ever-changing, we feel strongly that this text should not attempt to teach the student how to use a particular model of calculator. However, we do use actual graphing calculator–generated screens in addition to the traditional art found in textbooks. All screens in the text can be duplicated by the TI-83 graphing calculator, manufactured by Texas Instruments. TI-Graph Link software was used to render the calculator screens. In addition, the *Graphing Calculator Manual* by Stuart Moskowitz that accompanies this text provides students with keystroke operations for many of the more popular graphing calculators.

OVERVIEW OF THE CONTENT

We have found that the function concept is frequently a difficult one for students to grasp. Rather than present (and possibly confuse) the student with a variety of functions at the outset, we begin with the linear function in Chapter 1, analyzing its graph, solving linear equations and inequalities, and then solving applications dealing exclusively with linear functions. In Chapter 1, we also immediately begin to explore the capabilities of graphing calculators to help students better understand algebraic concepts.

In Chapter 2 we examine the graphs of the basic algebraic functions and their associated symmetries, transformations, and operations. Here we use the absolute value function to extend the concepts presented in Chapter 1, again using the graph/equation/inequality/application approach that we feature throughout the text. In Chapter 3 we present polynomial functions, focusing first on quadratic functions and then expanding the discussion to higher degree functions. Chapter 4 covers the rational and root functions, using the same approach, and concludes with a section on inverse functions that leads naturally into Chapter 5 on exponential and logarithmic functions. In Chapter 6 we introduce the conic sections and parametric equations. Chapter 7 covers the various methods of solving linear and nonlinear systems and includes matrix methods for solving linear systems. The appropriate use of graphing calculators to help explain concepts and confirm solutions continues to be stressed in these two chapters. Chapter 8 covers various other topics in algebra, and Chapter R, a "reference" chapter for basic algebraic concepts, provides examples and exercises for review and reference.

NEW AND ENHANCED FEATURES

We have been very pleased with the response to the first edition of this text, and at the request of those who have used the book and our reviewers, we have included the following new or enhanced features.

Meaningful Applications of Mathematics   With the assistance of Gary Rockswold of Mankato State University, we have provided more than 600 new applied examples and exercises that focus on real-life applications of mathematics. To further supplement the material, we also open each chapter with an interesting application that can be solved using the methods introduced in that chapter. Additionally, all applications are titled, and an index of applications can be found on pages I-1–I-4.

Increased Emphasis on Modeling We have included a large number of applications that provide data, often in tabular form. These exercises provide opportunity for the students to construct and analyze mathematical models. Section 1.7 has been newly written to focus on linear models and regression. Here we first introduce the concept of modeling, and we then continue to feature data for other types of models throughout the subsequent chapters.

Increased Emphasis on Using Tables When we wrote the first edition of this text, tables were not found on some models of graphing calculators. This is no longer the case, and we have included table use in both examples and exercises in this edition.

Reference Chapter on Basic Algebraic Concepts The reference chapter has been updated and now includes exercises that test each of the concepts. Answers to the odd-numbered exercises in this chapter appear in the answer section.

New Quick Reference Guide This tear-out card, bound into the back of the text, serves as a handy guide to the Reference Chapter. For each section of the text, it suggests sections to review in the Reference Chapter before undertaking the study of the content in that section.

New Chapter Projects Each chapter concludes with a project that can be used as either an individual or collaborative learning activity. The project provides an opportunity for students to see how the material in the chapter they have just studied can be applied. The projects were written by Stuart Moskowitz of Humboldt State University, and we greatly appreciate the excellent contributions that he has made to both editions of this text.

New Chapter Summaries The chapter summaries are now provided in an easy-to-read grid format. They provide a section-by-section summary of important concepts that should assist students in reviewing and preparing for examinations.

New Chapter Tests We now offer a carefully written chapter test for each chapter. Students can use these to prepare for examinations, and instructors may wish to pattern their classroom tests after them.

New Sections At the request of users of the text and reviewers, we have included sections on Linear Models (Section 1.7) and Partial Fractions (Section 7.8) in this edition.

New Analytic and Graphical Solution Identification Many examples within the text highlight both ANALYTIC and GRAPHICAL solutions. This feature provides strong support for a multirepresentational approach to problem solving and shows students the value of solving analytically and supporting graphically.

New "What Went Wrong" Feature Using graphing technology to study mathematics opens up a whole new area of error analysis. In anticipation of typical student errors, we have included this feature that allows students and instructors to discuss such errors. This feature was suggested some time ago by a reviewer whose name we cannot remember, but we wish to thank that reviewer for this excellent suggestion.

New Web Site A new Web site has been established—designed to increase student success in the course by offering section-by-section tutorial help, enhancement of text chapter projects, downloadable programs for TI graphing calculators, and author tips. This icon ⬥ alerts students at times when this site would be helpful. The site will also be useful to instructors by providing dynamic resources for use in their classes. http://hepg.awl.com Keyword: Hornsby

CONTINUING FEATURES

The following features from the first edition have been retained.

Technology Notes   Notes in the margin provide tips to students on how to use graphing calculators more effectively.

Cautions and Notes   These warn of common errors and misconceptions.

For Group Discussion   This feature appears within the exposition and offers material for instructors and students to discuss in a classroom setting.

Relating Concepts Exercises   These groups of exercises tie together different topics and highlight the connections among various concepts and skills. By working the entire group in sequence, the student can appreciate the relationship among topics that earlier may have seemed unrelated.

Writing and Conceptual Exercises   In addition to exercises that test concepts and skills or that present the mathematical concepts in a real-world applied setting, we have also included many writing (marked with a 🗎) and conceptual exercises. These are designed to help students reach a deeper level of understanding of the mathematical ideas being considered and to get them more actively involved in their own learning.

SUPPLEMENTS

FOR THE STUDENT

PRINTED SUPPLEMENTS

Student's Solutions Manual, ISBN 0-321-03945-9, *Norma James, New Mexico State University*

- ◆ Detailed solutions to odd-numbered Section Exercises, all Relating Concepts Exercises, odd-numbered Review Exercises and all Chapter Test Items.
- ◆ Ask your bookstore about ordering.

Graphing Calculator Manual, ISBN 0-321-03948-3, *Stuart Moskowitz, Humboldt State University*

- ◆ Graphing calculator usage instruction
- ◆ Keystroke operations for the following calculator models: TI-82®, TI-83®, TI-85®, TI-86®, Casio9850 Plus®, and HP38G®
- ◆ Worked-out examples taken directly from the text

MEDIA SUPPLEMENTS
Web Site

- ◆ Includes section-by-section tutorial help, enhanced Chapter Projects from the main text, study tips from the authors, downloadable TI-83® graphing calculator programs, and links to other sites
  http://hepg.awl.com Keyword: Hornsby

InterAct Math Tutorial Software, Windows ISBN 0-321-03547-X
Macintosh ISBN 0-321-03548-8

Throughout the text, this icon 🖳 indicates when this software would be helpful to students.

Interact Math Tutorial Software has been developed and designed by professional software engineers working closely with a team of experienced math educators. Interact Math Tutorial Software includes exercises that are linked with every objective in the textbook and require the same computational and problem-solving skills as their companion exercises in the text. Each exercise has an example and an interactive guided solution that are designed to involve students in the solution process and to help them identify precisely where they are having trouble. The software recognizes common student errors and provides students with appropriate customized feedback. With its sophisticated answer recognition capabilities, Interact Math Tutorial Software recognizes appropriate forms of the same answer for any kind of input. It also tracks student activity and scores for each section which can then be printed out. The software is free to qualifying adopters or can be bundled with books for sale to students.

Videotape Series, ISBN 0-321-03951-3

Throughout the text, this icon indicates when these videotapes would be helpful to students.

- Keyed specifically to text
- An engaging team of lecturers provide comprehensive coverage of each section.
- Selected odd-numbered exercises from the text presented
- Opportunity is given to solve a problem before the solution to the problem is given.
- Can be ordered by mathematics instructors or departments

FOR THE INSTRUCTOR

PRINTED SUPPLEMENTS

Instructor's Solutions Manual, ISBN 0-321-03943-2, *Norma James, New Mexico State University*

- Detailed solutions to all Section Exercises, Relating Concepts Exercises, Chapter Review Exercises, Chapter Tests, and Chapter Projects
- Free to instructors with textbook adoption

Instructor's Testing Manual, ISBN 0-321-03949-1

- Contains four tests per chapter modeled on the chapter tests found in the text as well as the answers to all of the test questions included in the manual
- Free to instructors with textbook adoption

MEDIA SUPPLEMENTS

TestGen-EQ with QuizMaster-EQ, Windows ISBN 0-321-03541-0
                                   Macintosh ISBN 0-321-03542-9

TestGen-EQ is a computerized test generator with algorithmically defined problems organized specifically for this textbook. Its user-friendly graphical interface enables instructors to select, view, edit and add test items, then print tests in a variety of fonts and forms. Seven question types are available, and search and sort features let the instructor quickly locate questions and arrange them in a preferred order. A built-in question editor gives the user the power to create graphs, import graphics, insert mathematical symbols and templates, and insert variable numbers or text. An "Export to HTML" feature lets instructors create practice tests that can be posted to a Web site. Tests created with TestGen-EQ can be used with QuizMaster-EQ, which enables students to take exams on a computer network. QuizMaster-EQ automatically grades the exams, stores results on disk, and allows the instructor to view or print a variety of reports for individual students, classes, or courses. This program is available in Windows and Macintosh formats and is free to adopters of the text.

InterAct Math Plus Software   Interact Math Plus combines course management and on-line testing with the features of the basic Interact Math Tutorial Software to create an invaluable teaching resource. Consult your Addison-Wesley representative for details.

ACKNOWLEDGMENTS

We wish to thank the many teachers and students who have given us valuable suggestions that have made this a better book. It has been deeply gratifying to have had some of you say "You've done it just the way it should be done." Such comments remain some of the most cherished memories of our professional careers. We wish to thank Anne Kelly, formerly of HarperCollins College Publishers, for signing the book and believing in it. Thanks also go out to those individuals who provided input into the first edition.

William A. Armstrong, *Lakeland Community College* ◆ Janis Cimperman, *St. Cloud State University* ◆ Susan Danielson, *University of New Orleans* ◆ Gerry Fitch, *Louisiana State University* ◆ Bill Hebert, *De La Salle High School (New Orleans, LA)* ◆ Norma James, *New Mexico State University* ◆ Michael Shafferkötter, *University of New Orleans*

Thanks also to these reviewers of the first edition.

John Baldwin, *University of Illinois–Chicago* ◆ Jim Birdsall, *Santa Fe Community College* ◆ Dick J. Clark, *Portland Community College* ◆ William L. Grimes, *Central Missouri State University* ◆ Bruce Hoelter, *Raritan Valley Community College* ◆ Dick Little, *Baldwin-Wallace University* ◆ Dan Loprieno, *Harper College* ◆ Virginia E. Lund, *Pensacola Junior College* ◆ Karen Mitchell, *Rowan-Cabarrus Community College* ◆ Shelle A. Palaski, *Northeast Missouri State University* ◆ Richard Schori, *Oregon State University* ◆ Kathy Soderbom, *Massasoit Community College* ◆ John P. Thomas, *College of Lake County* ◆ Mahbobeh Vezvaei, *Kent State University* ◆ Tom Williams, *Rowan-Cabarrus Community College* ◆ Karl M. Zilm, *Lewis and Clark Community College*

The following reviewers of the second edition provided many valuable suggestions, ideas, and criticisms.

William Armstrong, *Lakeland Community College* ◆ Brian Balman, *Johnson County Community College* ◆ Beth Beno, *South Suburban College* ◆ Randall Brian, *Vincennes University* ◆ Hongwei Chen, *Christopher Newport University* ◆ Donald Clayton, *Madisonville Community College* ◆ John Collado, *South Suburban College* ◆ Al Coons, *Pima Community College* ◆ Michael Dauzat, *Louisiana State University* ◆ Marie Dupuis, *Milwaukee Area Technical College* ◆ David Ebert, *Peninsula College* ◆ Jane Ellett, *Northeastern Louisiana University* ◆ Eunice Everett, *Seminole Community College* ◆ Odene Forsythe, *Westark Community College* ◆ Madelyn Gould, *DeKalb College* ◆ William Grimes, *Central Missouri State University* ◆ Heidi Howard, *Florida Community College* ◆ Miles Hubbard, *St. Cloud State University* ◆ Rebecca Isaac-Fahey, *Lexington Community College* ◆ Judith Jones, *Valencia Community College* ◆ John Khoury, *Brevard Community College* ◆ Michael Kirby, *Tidewater Com-*

munity College ◆ Helen Kolman, *Central Piedmont Community College* ◆ Frank Lombardo, *Daytona Beach Community College* ◆ Patricia Mower, *Washburn University* ◆ Nancy Olson, *Johnson County Community College* ◆ Linda Parrish, *Brevard Community College* ◆ Kathy Rodgers, *University of Southern Indiana* ◆ Deirdre Smith, *University of Arizona*

We thank Otis Taylor, of Addison Wesley Longman, for his contributions to the success of the first edition. To our good friend, Charlie Dawkins, we wish the happiest of retirements. Kitty Pellissier provided her usual excellent help in checking answers, and Becky Troutman assisted in preparing the Index of Applications. Terry McGinnis continues to be the unsung hero of our textbook production process. We thank Bob Martin, of Tarrant County Junior College–Northeast, for accuracy checking all the examples in the text. Norma James and Stuart Moskowitz continue the outstanding contributions they began with the first edition. Thanks also go out to Barbara Atkinson, Donna Bagdasarian, Stephanie Baldock, Brenda Bravener, Karen Guardino, Becky Malone, Christine O'Brien, Bill Poole, Rachel Reeve, and Greg Tobin of Addison Wesley Longman for their input. Ann Sargent of Elm Street Publishing Services assisted in the production and did her usual excellent job. Paul Van Erden continues to prepare the most accurate indexes we have seen.

A FINAL WORD

In the conclusion to the preface for the first edition, we wrote:

> We hope that this book begins to make a difference in the manner in which algebra is presented and learned as we move into the twenty-first century. We ask that both instructors and students pursue its contents with an open mind, ready to teach and learn in the manner that only now, after so many thousands of years, is possible. We, like Newton, can do so only because we "have stood on the shoulders of giants."

Judging from the many positive comments we have received since its publication, we feel that we are indeed on our way to making that difference. This is a most special project for us, and we are grateful to those who have contributed to its acceptance.

John Hornsby
Margaret L. Lial

Rectangular Coordinates, Functions, and Analysis of Linear Functions

*A*pproximating data with linear relations and functions is one of the most important and fundamental mathematical techniques we use today. Although most real-world applications are nonlinear, we can often use linear approximations to give accurate estimations. For example, the shape of Earth is round, not flat. Yet, when a building is constructed, the curvature of Earth's surface is seldom taken into account. Instead, it is assumed that the surface is level over the relatively small distance covered by the building. In this case, we use a linear approximation to accurately solve a nonlinear problem. However, when freeways were built across the United States, the curvature of Earth's surface had to be taken into account. If the distance or interval is small, linear approximations can lead to accurate estimations. Their advantage is that they are simple and easy to compute. On the other hand, if the distance or interval is large, then a linear approximation may lead to incorrect results.

In the late 1920s the famous observational astronomer Edwin P. Hubble (1889–1953) determined by careful measurement both the distances to several galaxies and the velocities at which they were moving away from Earth. Four galaxies with their distances in megaparsecs and velocities in kilometers per second are listed in the table. (One megaparsec is approximately 1.9×10^{19} miles.)

Is there any relationship between the data that could be used to predict the distance from Earth to the galaxy Hydra? Could the age of the universe be estimated using these data? Edwin Hubble made one of the most important discoveries in astronomy when he determined that a linear relationship existed between the distance and velocity of a galaxy. His important finding resulted in Hubble's Law. Because of this significant contribution to the understanding of our expanding universe, the Hubble Space Telescope was named after him. For galaxies relatively close to Earth, Hubble's linear relationship has been

Galaxy	Distance	Velocity
Virgo	15	1600
Ursa Minor	200	15,000
Corona Borealis	290	24,000
Bootes	520	40,000
Hydra	?	60,000

shown to be accurate. How far into deep space this linear relationship holds remains uncertain. Before we can calculate the distance to the galaxy Hydra or approximate the age of the universe, we must first understand relations and functions. Using relations and functions to approximate real data, we will be able to answer these and other important questions.*

1.1   REAL NUMBERS, LOGIC, AND COORDINATE SYSTEMS

Sets of Real Numbers   ◆   Roots   ◆   Logic   ◆   Coordinate Systems   ◆   Viewing Windows

SETS OF REAL NUMBERS

The idea of counting goes back to the beginning of our civilization. When people first counted they used only the **natural numbers,** written in set notation as

$$\{1, 2, 3, 4, 5, \ldots\}.$$

Much more recent is the idea of counting *no* object—that is, the idea of the number 0. Including 0 with the set of natural numbers gives the set of **whole numbers.**

$$\{0, 1, 2, 3, 4, 5, \ldots\}$$

(These and other sets of numbers are summarized later in this section.)

As the need for other kinds of numbers arose, additional sets of numbers were developed. Though the need for negative numbers may seem obvious to us today, they are a relatively recent development in the history of mathematics. The negatives of the natural numbers, included with the set of whole numbers, gives the very useful set of **integers,**

$$\{\ldots, -4, -3, -2, -1, 0, 1, 2, 3, \ldots\}.$$

Integers can be shown pictorially with a **number line.** (A number line is similar to a thermometer on its side.) As an example, the elements of the set $\{-3, -1, 0, 1, 3, 5\}$ are located on the number line in Figure 1.

$$-5 \quad -4 \quad -3 \quad -2 \quad -1 \quad 0 \quad 1 \quad 2 \quad 3 \quad 4 \quad 5$$

FIGURE 1

The result of dividing two integers, with a nonzero divisor, is called a *rational number*. By definition, the **rational numbers** are the elements of the set

$$\left\{\frac{p}{q} \;\middle|\; p, q \text{ are integers and } q \neq 0 \right\}.$$

This definition, which is given in *set-builder notation*, is read "the set of all elements p/q such that p and q are integers and $q \neq 0$." Examples of rational numbers include $\frac{3}{4}$, $-\frac{5}{8}$, $\frac{7}{2}$, and $-\frac{14}{9}$. All integers are rational numbers, since any integer can be written as the quotient of itself and 1.

Rational numbers can be located on a number line by a process of subdivision. For example, $\frac{5}{8}$ can be located by dividing the interval from 0 to 1 into 8 equal parts, then

---

Sources: Acker, A., and C. Jaschek, *Astronomical Methods and Calculations*, John Wiley & Sons, 1986.

Sharov, A., and I. Novikov, *Edwin Hubble, The Discoverer of the Big Bang Universe*, Cambridge University Press, 1993.

labeling the fifth part $\frac{5}{8}$. Several rational numbers are located on the number line in Figure 2.

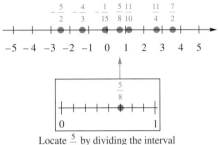

Locate $\frac{5}{8}$ by dividing the interval
from 0 to 1 into 8 equal parts.
FIGURE 2

The set of all numbers that correspond to points on a number line is called the set of **real numbers.** The set of real numbers is shown in Figure 3.

FIGURE 3

A real number that is not rational is called an **irrational number.** The set of irrational numbers includes $\sqrt{3}$ and $\sqrt{5}$ but not $\sqrt{1}$, $\sqrt{4}$, $\sqrt{9}$, . . . , which equal 1, 2, 3, . . . , and hence are rational numbers. Another irrational number is π, which is approximately equal to 3.14159. The numbers in the set $\left\{-\frac{2}{3}, 0, \sqrt{2}, \sqrt{5}, \pi, 4\right\}$ can be located on a number line as shown in Figure 4. (Only $\sqrt{2}$, $\sqrt{5}$, and π are irrational here. The others are rational.)

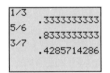

These displays are correct to 10 decimal places. The final 6 in the decimal for $\frac{3}{7}$ is obtained by rounding off (see Section 1.5).

FIGURE 4

Real numbers can also be defined in another way, in terms of decimals. Using repeated subdivisions, any real number can be located (at least in theory) as a point on a number line. By this process, the set of real numbers can be defined as the set of all decimals. Every rational number has a decimal representation that either terminates (comes to an end) or repeats in a fixed "block" of digits. Here are some examples.

Rational Numbers Whose Decimals Terminate	Rational Numbers Whose Decimals Repeat
$\frac{1}{4} = .25$	$\frac{1}{3} = .3333\ldots$
$\frac{3}{8} = .375$	$\frac{5}{6} = .8333\ldots$
$\frac{7}{4} = 1.75$	$\frac{3}{7} = .428571428571\ldots$

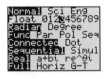

The second line in the top screen indicates that displays will be correct to 3 decimal places. Compare the bottom screen to the one given above.

The three dots at the end of the repeating decimals indicate that the pattern of digits established continues indefinitely. Another way to indicate the repeating digits is to place a bar over the part that repeats. Thus, we would write the following.

$$\frac{1}{3} = .\overline{3} \qquad \frac{5}{6} = .8\overline{3} \qquad \text{and} \qquad \frac{3}{7} = .\overline{428571}$$

If at any time we use an approximation for a rational number, we use the $\approx$ symbol to indicate "is approximately equal to." Thus, it would technically be incorrect to write $\frac{2}{3} = .67$; we should write $\frac{2}{3} \approx .67$ if an approximation is warranted. We call .67 an *approximation of $\frac{2}{3}$ to the nearest hundredth,* while $.\overline{6}$ is the *exact decimal representation for $\frac{2}{3}$.*

Note   In this text we will often make distinctions about whether an approximation or an exact value is required. As we progress in our work, more will be said about this.

The decimal representation of an irrational number will neither terminate nor repeat. The locations of $\sqrt{2}$, $\sqrt{5}$, and π on the number line in Figure 4 were determined by observing these calculator approximations:

$$\sqrt{2} \approx 1.414213562 \qquad \sqrt{5} \approx 2.236067977 \qquad \pi \approx 3.141592654.$$

EXAMPLE 1   *Identifying Elements of Subsets of the Real Numbers*   Let set $A = \left\{ -8, -6, -.75, 0, .\overline{09}, \sqrt{2}, \sqrt{5}, 6, \frac{107}{4} \right\}$. List the elements from set A that belong to each of the following sets: (a) real numbers, (b) integers, (c) rational numbers, (d) irrational numbers, (e) whole numbers, and (f) natural numbers.

SOLUTION

(a) Because every element of A can be represented by a point on a number line, all elements are real numbers.

(b) The integers are -8, -6, 0, and 6.

(c) The rational numbers are -8, -6, $-.75$, 0, $.\overline{09}$, 6, and $\frac{107}{4}$.

(d) The irrational numbers are $\sqrt{2}$ and $\sqrt{5}$.

(e) The whole numbers are 0 and 6.

(f) The only natural number in the set is 6.   ◆

The relationships among the various subsets of the real numbers, along with examples in the sets, are shown in Figure 5.

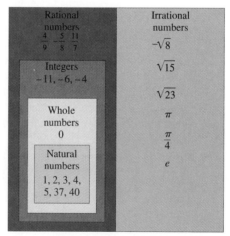

FIGURE 5 The Real Numbers

Note   This text was written with the assumption that the student has access to a graphing calculator at all times. For this reason, the term "calculator" is understood to mean a *graphing* calculator.

ROOTS

The most common irrational numbers that we will encounter in this course are roots— **square roots, cube roots,** and so forth. A formal discussion of roots will follow in

Chapter 4. For now, you should be able to use your calculator to find roots. Calculators usually have dedicated keys for square and cube roots, and have functions that allow for other roots. You should consult your owner's manual to see how to find roots on your particular model.

EXAMPLE 2   *Finding Roots on a Calculator*   Use a calculator to find approximations for the following roots. (Note: It is often convenient to use the fact that $\sqrt[n]{a} = a^{1/n}$ for appropriate values for a and n to find roots.)

(a) $\sqrt{23}$ **(b)** $\sqrt[3]{87}$ **(c)** $\sqrt[4]{12}$

SOLUTION

(a)   The screen in Figure 6(a) shows that an approximation for $\sqrt{23}$ is 4.795831523. It is displayed twice, once for $\sqrt{23}$ and once for $23^{1/2}$.

(b) $\sqrt[3]{87} \approx 4.431047622$. See Figure 6(b).

(c)   Figure 6(c) indicates $\sqrt[4]{12} \approx 1.861209718$ in three different ways.   ◆

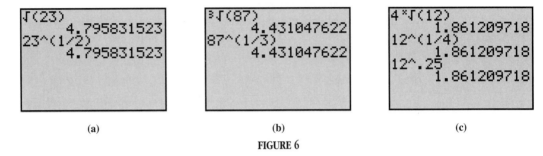

(a)   (b)   (c)

FIGURE 6

LOGIC

A statement in mathematics is defined to be an assertion that is either true or false.

> **Graphing calculators are designed to return
> a 1 for a true statement and a 0 for a false statement.**

Figure 7 on the following page shows how a graphing calculator responds to a true statement and to a false statement.

When two simple statements are joined with the word *and,* the resulting statement is true only when both of the simple statements it consists of are true. For example,

$$4 + 1 = 5 \quad \text{and} \quad 3 - 2 = 1$$

is a true statement because $4 + 1 = 5$ is true and $3 - 2 = 1$ is true. However,

$$4 + 1 = 5 \quad \text{and} \quad 3 - 2 = 5$$

is false, since the second simple statement is false. Figure 8 illustrates the calculator responses to these statements.

Two simple statements joined with the word *or* form a true statement if at least one of the simple statements is true. Thus,

$$3 + 2 = 5 \text{ or } 3 - 2 = 5 \text{ is true,} \quad \text{but} \quad 3 + 2 = 6 \text{ or } 3 - 2 = 5 \text{ is false,}$$

as Figure 9 shows.

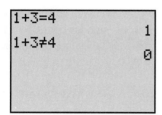

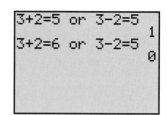

The first statement is true (1) and the second is false (0).

FIGURE 7

The first statement is true (1) and the second is false (0).

FIGURE 8

FIGURE 9

COORDINATE SYSTEMS

Figure 10 shows a number line with the points corresponding to several different numbers marked on the line. A number that corresponds to a particular point on a line is called the **coordinate** of the point. For example, the leftmost marked point in Figure 10 has coordinate -4. The correspondence between points on a line and the real numbers is called a **coordinate system** for the line. (The phrase "the point on a number line with coordinate a" will be abbreviated as "the point with coordinate a," or simply "the point a.")

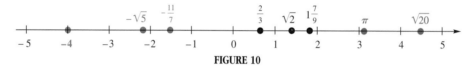

FIGURE 10

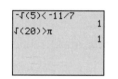

Both inequalities are true.

If the real number a is to the left of the real number b on a number line, then **a is less than b,** written $a < b$. If a is to the right of b, then **a is greater than b,** written $a > b$. For example, in Figure 10, $-\sqrt{5}$ is to the left of $-\frac{11}{7}$ on the number line, so $-\sqrt{5} < -\frac{11}{7}$, while $\sqrt{20}$ is to the right of π, indicating $\sqrt{20} > \pi$.

Note   Remember that the "point" of the inequality symbol goes toward the smaller number.

As an alternative to this geometric definition of "is less than" or "is greater than," there is an algebraic definition: if a and b are two real numbers and if the difference $a - b$ is positive, then $a > b$. If $a - b$ is negative, then $a < b$. The geometric and algebraic statements of order are summarized as follows.

Statement	Geometric Form	Algebraic Form
$a > b$	a is to the right of b.	$a - b$ is positive.
$a < b$	a is to the left of b.	$a - b$ is negative.

The symbols $<$ and $>$ can be combined with the symbol for equality. The statement $a \leq b$ means "a is less than or equal to b" and is true if either $a < b$ or $a = b$. Similarly, $a \geq b$ means "a is greater than or equal to b" and is true if either $a > b$ or $a = b$. Statements involving $<$ or $>$ are called **strict** inequalities, while those involving $\leq$ or $\geq$ are called **nonstrict** inequalities. We can negate any of these symbols by using a slash bar ($/$).

EXAMPLE 3 *Showing Why Inequality Statements Are True* The list below shows several statements and the reason why each is true.

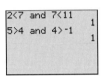

Compare to the first two entries of the chart in Example 3.

Statement	Reason
$8 \leq 10$	$8 < 10$
$8 \leq 8$	$8 = 8$
$-9 \geq -14$	$-9 > -14$
$-8 \not> -2$	$-8 < -2$
$4 \not< 2$	$4 > 2$

◆

The inequality $a < b < c$ says that b is *between* a and c, since

$$a < b < c$$

means

$$a < b \quad \text{and} \quad b < c.$$

In the same way,

$$a \leq b \leq c$$

means

$$a \leq b \quad \text{and} \quad b \leq c.$$

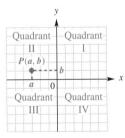

See the Caution statement. $2 < 7 < 11$ and $5 > 4 > -1$ are entered as "and" statements.

Caution When writing these "between" statements, make sure that both inequality symbols point in the same direction, toward the smallest number. For example,

both $2 < 7 < 11$ and $5 > 4 > -1$

are true statements, but $3 < 5 > 2$ is meaningless. Generally, it is best to rewrite statements such as $5 > 4 > -1$ as $-1 < 4 < 5$, which is the order of these numbers on the number line.

A number line is an example of a one-dimensional coordinate system, and it is sufficient to graph real numbers. If we place two number lines at right angles, intersecting at their origins, we obtain a two-dimensional **rectangular coordinate system.** It is customary to have one of these lines vertical and the other horizontal. They intersect at the **origin** of the system, designated 0. The horizontal line is called the **x-axis,** and the vertical line is called the **y-axis.** On the x-axis, positive numbers are located to the right of the origin, while negative numbers are located to the left. On the y-axis, positive numbers are located above the origin, negative numbers below.

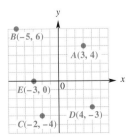

FIGURE 11

This rectangular coordinate system is also called the **Cartesian coordinate system,** named after Rene Descartes (1596–1650). The plane into which the coordinate system is introduced is the **coordinate plane,** or **xy-plane.** The x-axis and y-axis divide the plane into four regions, or **quadrants,** labeled as shown in Figure 11. The points on the x-axis and y-axis belong to no quadrant.

Each point P in the xy-plane corresponds to a unique ordered pair (a, b) of real numbers. The numbers a and b are the **coordinates** of point P. We call a the **x-coordinate** and b the **y-coordinate.** To locate on the xy-plane the point corresponding to the ordered pair $(3, 4)$, for example, draw a vertical line through 3 on the x-axis and a horizontal line through 4 on the y-axis. These two lines intersect at point A in Figure 12. Point A corresponds to the ordered pair $(3, 4)$. Also in Figure 12, B corresponds to the ordered pair $(-5, 6)$, C to $(-2, -4)$, D to $(4, -3)$, and E to $(-3, 0)$. The point P corresponding to the ordered pair (a, b) often is written as $P(a, b)$ as in Figure 11 and referred to as "the point (a, b)."

FIGURE 12

TECHNOLOGY NOTE

You should read your owner's manual to see how to alter the viewing window on your screen. Remember that different settings will result in different views of graphs. When you adjust the settings on cameras, telescopes, and binoculars, different views of the subject are obtained; the same goes for graphs generated by calculators.

VIEWING WINDOWS

The characteristic that distinguishes this text from traditional algebra texts is that it features full integration of modern-day graphing calculators. A graphing calculator differs from a typical scientific calculator in many ways, the most obvious of which is that it allows the user to plot points and a variety of graphs at the touch of keys.

The rectangular (Cartesian) coordinate system theoretically extends indefinitely in all directions. We are limited to illustrating only a portion of such a system in a text figure. Similar limitations are found in portraying coordinate systems on calculator screens. For this reason, the student should become familiar with the key on the calculator that sets the limits for x- and y-coordinates. The most common term used to refer to these limits is "window." (There also may be other designations.) Figure 13 shows a calculator screen that has been set to have a minimum x-value of -10, a maximum x-value of 10, a minimum y-value of -10, and a maximum y-value of 10. Additionally, the tick marks on the axes have been set to be 1 unit apart. Throughout this book, this window will be called the *standard viewing window.*

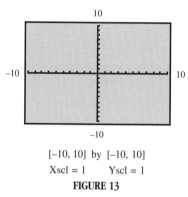

[−10, 10] by [−10, 10]

Xscl = 1   Yscl = 1

FIGURE 13

In order to convey important information about a viewing window, in this text we will use the following abbreviations:

Xmin: minimum value of x Ymin: minimum value of y

Xmax: maximum value of x Ymax: maximum value of y

Xscl: scale (distance between   Yscl: scale (distance between
tick marks) on the x-axis   tick marks) on the y-axis

To further condense this information, we will often use the following symbolism:

Xmin   Xmax   Ymin   Ymax

[−10, 10]   by   [−10, 10]

Xscl = 1   Yscl = 1

The symbols above indicate the viewing window information for the window in Figure 13.

All calculators have a standard viewing window. Viewing windows may be changed by manually entering the information, or by using the zoom feature of the calculator. The graphing screen is made up of pixels, which are small areas that, when illuminated, will represent points in the plane. The coordinates of the pixels may be found by using the trace feature of the calculator. When we begin our study of graphs later in this chapter, we will say more about the zoom and trace features.

Figure 14 shows several other viewing windows, with the important information. Notice that (b) and (c) look exactly alike, and unless we are told what the settings are, we have no way of distinguishing between them. Paying careful attention to window settings will be an important part of our work in this text.

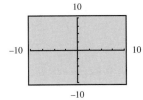

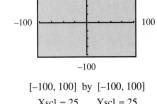

[−4.7, 4.7] by [−3.1, 3.1]
Xscl = 1   Yscl = 1

(a)

[−10, 10] by [−10, 10]
Xscl = 2.5   Yscl = 2.5

(b)

[−100, 100] by [−100, 100]
Xscl = 25   Yscl = 25

(c)

FIGURE 14

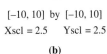   A student learning how to use a graphing calculator
could not understand why the axes on the graph were so "thick," as seen in Figure A,
while those on a friend's calculator were not, as seen in Figure B.

1. What went wrong in Figure A?
2. How can the student correct the problem so that the axes look like those in Figure B?

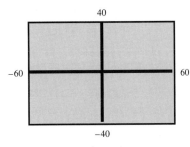

[−60, 60] by [−40, 40]
Xscl = 1   Yscl = 1

A

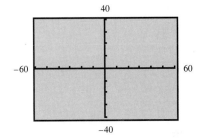

[−60, 60] by [−40, 40]
Xscl = 10   Yscl = 10

B

1.1   EXERCISES   **Tape 1**

For each of the following sets, list all elements that belong to (a) natural numbers, (b) whole numbers, (c) integers, (d) rational numbers, (e) irrational numbers, and (f) real numbers.

1. $\left\{-6, -\dfrac{12}{4}, -\dfrac{5}{8}, -\sqrt{3}, 0, .31, .\overline{3}, 2\pi, 10, \sqrt{17}\right\}$

2. $\left\{-8, -\dfrac{14}{7}, -.245, 0, \dfrac{6}{2}, 8, \sqrt{81}, \sqrt{12}\right\}$

3. $\left\{-\sqrt{100}, -\dfrac{13}{6}, -1, 5.23, 9.\overline{14}, 3.14, \dfrac{22}{7}\right\}$

4. $\left\{-\sqrt{49}, -.405, -.\overline{3}, .1, 3, 18, 6\pi, 56\right\}$

Graph each set of numbers on a number line.

5. $\{-4, -3, -2, -1, 0, 1\}$

6. $\{-6, -5, -4, -3, -2\}$

7. $\left\{-.5, .75, \dfrac{5}{3}, 3.5\right\}$

8. $\left\{-.6, \dfrac{9}{8}, 2.5, \dfrac{13}{4}\right\}$

Each rational number written in common fraction form in Exercises 9–16 has its decimal equivalent appearing in the columns on the right. Without using a calculator, if possible, match the fraction with its decimal equivalent.

9. $\dfrac{1}{5}$ **10.** $\dfrac{2}{3}$ 　　　　　　　**A.** .5   **B.** .67

11. $\dfrac{67}{100}$ **12.** $\dfrac{12}{10}$ 　　　　　　**C.** .75   **D.** .2

13. $\dfrac{3}{4}$ **14.** $\dfrac{50}{100}$ 　　　　　　**E.** $.\overline{6}$ **F.** .125

15. $\dfrac{3}{11}$ **16.** $\dfrac{1}{8}$ 　　　　　　　**G.** $.\overline{27}$ **H.** 1.2

Each rational number in Exercises 17–24 has a decimal equivalent that repeats. Use the bar symbolism to write the decimal. Use a calculator.

17. $\dfrac{5}{6}$ 　　　　**18.** $\dfrac{1}{9}$ 　　　　**19.** $-\dfrac{13}{3}$ 　　　　**20.** $-\dfrac{9}{11}$

21. $\dfrac{6}{27}$ 　　　　**22.** $\dfrac{5}{33}$ 　　　　**23.** $\dfrac{9}{110}$ 　　　　**24.** $\dfrac{77}{990}$

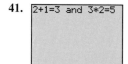 **25.** Explain the difference between the rational numbers .87 and $.\overline{87}$.

26. A student, using her powerful new calculator, found the decimal 1.414213562 when she evaluated $\sqrt{2}$. Is this decimal the exact value of $\sqrt{2}$, or just an approximation? Should she write $\sqrt{2} = 1.414213562$ or $\sqrt{2} \approx 1.414213562$?

Use a calculator to find a decimal approximation of each root or power. Give as many decimal places as your calculator shows.

27. $\sqrt{58}$ 　　　　**28.** $\sqrt{97}$ 　　　　**29.** $\sqrt[3]{33}$ 　　　　**30.** $\sqrt[3]{91}$

31. $\sqrt[4]{86}$ 　　　　**32.** $\sqrt[4]{123}$ 　　　　**33.** $19^{1/2}$ 　　　　**34.** $29^{1/3}$

Decide which of the following symbols may be placed in the blank to make a true statement: $<$, $\le$, $>$, $\ge$. *There may be more than one correct answer.*

35. -5 _____ -4 　　　　**36.** -1.3 _____ $-.6$ 　　　　**37.** 8 _____ 4

38. 9 _____ 8.9 　　　　**39.** -6 _____ -6 　　　　**40.** 2 _____ 2

Decide whether the calculator will return a 0 (for false) or a 1 (for true) for the indicated screen.

41. `2+1=3 and 3*2=5`

42. `8>9 and 9<4`

43. `5-6=⁻1 or 5-6=1`

44. 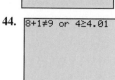 `8+1≠9 or 4≥4.01`

Locate the following points on a rectangular coordinate system. Identify the quadrant, if any, in which each point lies.

45. $(2, 3)$ **46.** $(-1, 2)$ **47.** $(-3, -2)$ **48.** $(1, -4)$ **49.** $(0, 5)$

50. $(-2, -4)$ **51.** $(-2, 4)$ **52.** $(3, 0)$ **53.** $(-2, 0)$ **54.** $(3, -3)$

Recall from elementary algebra that the product of two numbers with the same signs is positive and the product of two numbers with different signs is negative. A similar rule holds for quotients. Name the possible quadrants in which the point (x, y) can lie if the given condition is true.

55. $xy > 0$ **56.** $xy < 0$ **57.** $\dfrac{x}{y} < 0$ **58.** $\dfrac{x}{y} > 0$

59. If the x-coordinate of a point is 0, the point must lie on which axis?

60. If the y-coordinate of a point is 0, the point must lie on which axis?

It is important to become familiar with your graphing calculator so that you can operate it efficiently. Answer the following questions about your particular calculator, using a complete sentence or sentences.

 61. How do you set the screen in order to obtain the standard viewing window?

62. What are the minimum and maximum values of x and y in your standard viewing window?

Give the values of Xmin, Xmax, Ymin, and Ymax for the screen shown, given the values for Xscl and Yscl. Use the notation described in this section.

63.

Xscl = 1, Yscl = 5

64.

Xscl = 5, Yscl = 1

65.

Xscl = 10, Yscl = 50

66.

Xscl = 50, Yscl = 10

67.

Xscl = 100, Yscl = 100

68.

Xscl = 75, Yscl = 75

Using the notation described in the text, set the viewing window of your calculator to the following specifications.

69. $[-10, 10]$ by $[-10, 10]$
Xscl = 1   Yscl = 1

70. $[-40, 40]$ by $[-30, 30]$
Xscl = 5   Yscl = 5

71. $[-5, 10]$ by $[-5, 10]$
Xscl = 3   Yscl = 3

72. $[-3.5, 3.5]$ by $[-4, 10]$
Xscl = 1   Yscl = 1

73. $[-100, 100]$ by $[-50, 50]$
Xscl = 20   Yscl = 25

74. $[-4.7, 4.7]$ by $[-3.1, 3.1]$
Xscl = .5   Yscl = .5

75. Set your viewing window to $[-10, 10]$ by $[-10, 10]$ and then set Xscl to 0 and Yscl to 0. Do you notice any tick marks on the axes? Make a conjecture as to how to set a screen with no tick marks on the axes.

76. Set your viewing window to $[-50, 50]$ by $[-50, 50]$ and then set Xscl to 1 and Yscl to 1. Observe this screen and describe the appearance of the axes as compared to those seen in the standard window. Why do you think they appear this way? How can you change your scale settings so that this "problem" is alleviated?

1.2   INTRODUCTION TO RELATIONS AND FUNCTIONS

Set-Builder and Interval Notation and Their Number Line Graphs ◆ Relations, Domain, and Range ◆ Functions ◆ Tables and Lists ◆ Function Notation

In this section we introduce some of the most important concepts in the study of mathematics: relation, function, domain, and range. In order to make our work simpler, various types of set notation are useful. We begin by discussing two types: set-builder and interval notation.

SET-BUILDER AND INTERVAL NOTATION AND THEIR NUMBER LINE GRAPHS

Inequalities and variables can be used to specify sets of real numbers. Suppose we wish to symbolize the set of real numbers greater than -2. One way to symbolize this is $\{x \mid x > -2\}$, read "the set of all x such that x is greater than -2." This is called **set-builder notation,** since the variable x is used to "build" the set. On a number line, we show the elements of this set (the set of all real numbers to the right of -2) by drawing a line from -2 to the right. We use a parenthesis at -2 since -2 is not an element of the given set. The result, shown in Figure 15, is called the **graph** of the set $\{x \mid x > -2\}$.

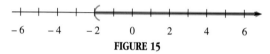

FIGURE 15

The set of numbers greater than -2 is an example of an **interval** on the number line. A simplified notation, called **interval notation,** is used for writing intervals. For example, using this notation, the interval of all numbers greater than -2 is written as $(-2, \infty)$. The **infinity symbol** ∞ does not indicate a number; it is used to show that the interval includes all real numbers greater than -2. The left parenthesis indicates that -2 is not included. A parenthesis is always used next to the infinity symbol in interval notation. The set of all real numbers is written in interval notation as $(-\infty, \infty)$.

A chart summarizing the names of various types of intervals is on the next page.

Caution   Notice how the interval notation for the open interval (2, 5) looks exactly like the notation for the ordered pair (2, 5). While this does not usually cause confusion, as the interpretation is determined by the context of the use, we will, when the need arises, distinguish between them by referring to "the interval (2, 5)" or "the point (2, 5)."

RELATIONS, DOMAIN, AND RANGE

The bar graph shown in Figure 16 is typical of the kinds of graphs found in magazines and newspapers. Notice that it shows the number of visitors, in millions, to national parks in the United States for selected years between 1950 and 1990. Every year is paired with a number of visitors, and we may depict this information in ordered-pair form as well, agreeing that the first component represents the year and the second component represents the number of visitors in millions:

$$(1950, 14), (1960, 28), (1970, 46), (1980, 47), (1990, 57).$$

Later in this section we will return to this set of ordered pairs.

Such a set of ordered pairs is called a *relation.*

EXAMPLE 1 *Determining Domains and Ranges from Graphs* Give the domain and the range of each graph in

(a) Figure 19(a) **(b)** Figure 19(b) **(c)** Figure 19(c).

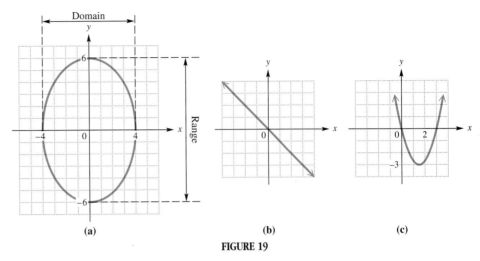

(a)   (b)   (c)

FIGURE 19

SOLUTION

(a) In Figure 19(a), the *x*-values of the points on the graph include all numbers between −4 and 4, inclusive. The *y*-values include all numbers between −6 and 6, inclusive. Using interval notation, the domain is [−4, 4], and the range is [−6, 6].

(b) In Figure 19(b), the arrowheads indicate that the line extends indefinitely left and right, as well as up and down. Therefore, both the domain and the range are the set of all real numbers, written (−∞, ∞).

(c) In Figure 19(c), the arrowheads indicate that the graph extends indefinitely left and right, as well as upward. The domain is (−∞, ∞). Because there is a least *y*-value, −3, the range includes all numbers greater than or equal to −3, written [−3, ∞). ◆

EXAMPLE 2 *Finding Domain and Range from a Calculator Window* Figure 20 shows a graph on a screen with viewing window [−5, 5] by [−5, 5], with Xscl = 1 and Yscl = 1. By observation, give the domain and the range of this relation.

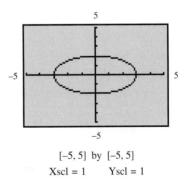

[−5, 5] by [−5, 5]

Xscl = 1   Yscl = 1

FIGURE 20

TECHNOLOGY NOTE

In Figure 20 we see a calculator-generated graph that is formed by a rather jagged curve. These are sometimes called *jaggies* and are typically found on low-resolution graphers, such as graphing calculators. In general, we should remember that most curves we will study in this book are smooth, and the jaggies are just a part of the limitations of technology.

SOLUTION Since the scales on both axes are 1, we see that the graph *appears* to have a minimum *x*-value of −3, a maximum *x*-value of 3, a minimum *y*-value of −2, and a maximum *y*-value of 2. Therefore, observation leads us to conclude that the domain is [−3, 3] and the range is [−2, 2]. ◆

Caution   As we shall see many times while discussing calculator-generated graphs, simple observation is not enough to guarantee accuracy in determining domains and ranges. Calculators have capabilities which allow us to improve our accuracy, but even then, it is essential to understand the mathematical concepts behind graphing before we can be certain that our observations are correct. This is why in this book, we will study concepts and technological capabilities in an integrated fashion.

FUNCTIONS

In Figure 16, we noticed that every year depicted corresponded to one number of visitors. Also, looking back at the relations F, G, and H introduced earlier in this section, we notice that in F and G, each x-value appears only once in the relation, while in H, the x-value -2 appears twice: it is paired with 1 in one ordered pair, while it is paired with 0 in another. The national park information and relations F and G are simple examples of a very important kind of relation, known as a *function*.

DEFINITION OF FUNCTION

A **function** is a relation in which each element in the domain corresponds to exactly one element in the range.*

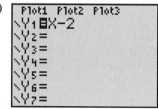

The function defined by $y = 9x - 5$ is defined by the user as $Y_1 = 9X - 5$. (Notice that the calculator requires an input that is a function.)

Suppose a group of students gets together each Monday evening to study algebra (and perhaps watch football). A number giving the student's weight to the nearest kilogram can be associated with each member of this set of students. Since each student has only one weight at a given time, the relationship between the students and their weights is a function. The domain is the set of all students in the group, while the range is the set of all the weights of the students.

If x represents any element in the domain, x is called the **independent variable.** If y represents any element in the range, y is called the **dependent variable,** because the value of y *depends on* the value of x. That is, in the example above, each weight depends on the student associated with it.

In most mathematical applications of functions, the correspondence between the domain and range elements is defined with an equation, like $y = 9x - 5$. The equation is usually solved for y, as it is here, because y is the dependent variable. As we choose values from the domain for x, we can easily determine the corresponding y-values of the ordered pairs of the function. (These equations need not use only x and y as variables; any appropriate letters may be used. In physics, for example, t is often used to represent the independent variable *time*.)

EXAMPLE 3   *Deciding Whether a Relation Is a Function*   Decide whether the following relations are functions. Give the domain and range of each relation.

(a) $\{(1, 2), (3, 4), (5, 6), (7, 8), (9, 10)\}$ **(b)** $\{(1, 1), (1, 2), (1, 3), (2, 4)\}$

(c)

X	Y₁
-5	2
-4	2
-3	2
-2	2
-1	2
0	2
1	2

X = -5

(d)

Plot1 Plot2 Plot3
\Y₁ ⊟ X−2
\Y₂ =
\Y₃ =
\Y₄ =
\Y₅ =
\Y₆ =
\Y₇ =

---

*An alternative definition of function based on the idea of correspondence is given later in the section.

SOLUTION

(a) The domain is the set {1, 3, 5, 7, 9}, and the range is {2, 4, 6, 8, 10}. Since each element in the domain corresponds to just one element in the range, this set is a function. The correspondence is shown below using D for the domain and R for the range.

$$D = \{1, 3, 5, 7, 9\}$$
$$R = \{2, 4, 6, 8, 10\}$$

(b) The domain here is {1, 2}, and the range is {1, 2, 3, 4}. As shown in the correspondence below, one element in the domain, 1, has been assigned three different elements from the range, so this relation is not a function.

$$D = \{1, 2\}$$
$$R = \{1, 2, 3, 4\}$$

(c) This is a table of ordered pairs generated by a graphing calculator. Here, the domain is {−5, −4, −3, −2, −1, 0, 1}, and the range is {2}. Although every element in the domain corresponds to the same range element, this is a function because each element in the domain has exactly one range element assigned to it.

(d) Since Y_1 is always found by subtracting 2 from x, each x corresponds to just one value of Y_1, so this relation is a function. Any number can be used for x, and each x will give a number 2 smaller for Y_1; thus, both the domain and the range are the set of real numbers, or in interval notation, $(-\infty, \infty)$.   ◆

There is a quick way to tell whether a given graph is the graph of a function. Figure 21 shows two graphs. In the graph for part (a), each value of x leads to only one value of y, so that this is the graph of a function. On the other hand, the graph in part (b) is not the graph of a function. For example, if $x = x_1$, the vertical line through x_1 intersects the graph at two points, showing that there are two values of y that correspond to this x-value. This idea is known as the *vertical line test* for a function.

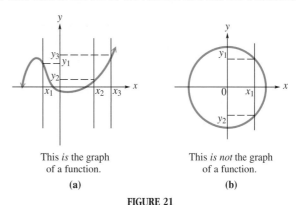

This *is* the graph   This *is not* the graph
of a function.   of a function.
      (a)   **(b)**

FIGURE 21

VERTICAL LINE TEST

If each vertical line intersects a graph in no more than one point, the graph is the graph of a function.

EXAMPLE 4   *Using the Vertical Line Test and Determining Domain and Range*

(a) Is the graph in Figure 22 the graph of a function? Specify the domain and the range using interval notation.

(b)   Assuming the graph in Figure 23 extends left and right indefinitely and upward indefinitely, does it appear to be the graph of a function? What are the domain and the range if Xscl = 1, Yscl = 1 (use observation)?

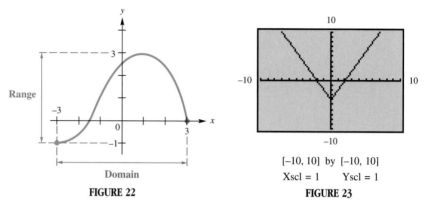

FIGURE 22

[−10, 10] by [−10, 10]
Xscl = 1   Yscl = 1
FIGURE 23

SOLUTION

(a)   The graph satisfies the vertical line test, and is therefore the graph of a function. As indicated by the annotations at the left and below the graph, the domain appears to be [−3, 3] and the range appears to be [−1, 3].

(b)   It appears that no vertical line will intersect the graph more than once, so we may conclude that it is the graph of a function. Since we are told that it extends left and right indefinitely, the domain is (−∞, ∞). It appears that the lowest point on the graph has the ordered pair (0, −4), and since we know that the graph extends upward indefinitely, the range appears to be the interval [−4, ∞).

(Graphs that are generated by graphing calculators will not exhibit arrowheads, and thus we will need to be aware of the type of function we are observing in order to determine the domain and the range.)   ◆

While the concept of function is crucial to the study of mathematics, the definition of function may vary in wording from text to text. We now give an alternative definition of function that will be helpful in understanding the function notation that follows.

ALTERNATIVE DEFINITION OF FUNCTION

A function is a correspondence in which each element x from a set called the domain is paired with one and only one element y from a set called the range.

This idea of correspondence, or mapping, can be illustrated as shown in Figure 24, where the function f consists of the ordered pairs in the national parks example.

Function f

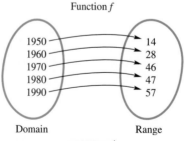

Domain   Range

FIGURE 24

In this example, function *f* pairs 1950 with 14, 1960 with 28, and so on. Each domain value is paired with one and only one range value.

TABLES AND LISTS

A convenient way to specify ordered pairs in a function is by a **table**. An equation such as $y = 9x - 5$ can be used to describe a function. If we choose *x*-values to be 0, 1, 2, . . . , 6, then the corresponding *y*-values are

$$y = 9(0) - 5 = -5 \qquad y = 9(4) - 5 = 31$$

$$y = 9(1) - 5 = 4 \qquad y = 9(5) - 5 = 40$$

$$y = 9(2) - 5 = 13 \qquad y = 9(6) - 5 = 49$$

$$y = 9(3) - 5 = 22$$

These ordered pairs $(0, -5)$, $(1, 4)$, . . . , $(6, 49)$ can be organized in a table. A graphing calculator can do this in seconds as well. The result is shown in Figure 25.

x	y
0	-5
1	4
2	13
3	22
4	31
5	40
6	49

FIGURE 25

Note In Figure 25, the function is denoted Y_1. Occasionally there might be a slight discrepancy between the way we represent a function in the text and the way our illustrative screen represents it (in this case, *y* and Y_1). This is not something to be overly concerned about: the equations $y = 9x - 5$ and $Y_1 = 9X - 5$ are satisfied by the same ordered pairs, and thus define the same function.

This screen indicates that the table will *start* with 0 and have an *increment* of 1. Both variables will appear *automatically*.

In this text we will use the symbol TblStart to represent the initial value of the independent variable (0 in the example above) and ΔTbl to represent the difference, or increment, between successive values of the independent variable (1 in the example above, because $1 - 0 = 1$, $2 - 1 = 1$, $3 - 2 = 1$, and so on).

Graphing calculators also allow us to define **lists** of numbers. In the national parks illustration (Figure 16), we can define the set of years as list $L_1 = \{1950, 1960, 1970, 1980, 1990\}$, and the set of visitor numbers as list $L_2 = \{14, 28, 46, 47, 57\}$. Notice that in both lists the variables appear in the same order as they appear in the set of ordered pairs. Figure 26 shows these lists. If we set a window of [1950, 1990] by [−10, 80] with

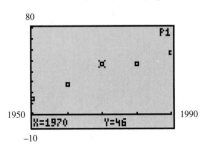

[1950, 1990] by [−10, 80]
Xscl = 10   Yscl = 10

FIGURE 26   **FIGURE 27**

Xscl $= 10$ and Yscl $= 10$, we can plot the points (x, y), where each x comes from L_1 and each corresponding y comes from L_2. Notice also that any point, such as (1970, 46), can be highlighted using an appropriate feature of the calculator. See Figure 27 on the previous page.

FUNCTION NOTATION

To say that y is a function of x means that for each value of x from the domain of the function, there is exactly one value of y. To emphasize that y *is a function of* x, or that y depends on x, it is common to write

$$y = f(x),$$

with $f(x)$ read "f of x." This notation is called **function notation.** For the function f illustrated earlier in Figure 24,

$$f(1950) = 14 \qquad \text{because (1950, 14) belongs to the correspondence,}$$
$$f(1960) = 28 \qquad \text{because (1960, 28) belongs to the correspondence,}$$

and so on.

Function notation is used frequently when functions are defined by equations. As an example, for the function defined by the equation $y = 9x - 5$, we may name this function f and write

$$f(x) = 9x - 5.$$

Note that $f(x)$ is simply another name for y. In this function f, if $x = 2$, then we find y, or $f(2)$, by replacing x with 2.

$$f(2) = 9 \cdot 2 - 5$$
$$= 18 - 5$$
$$= 13.$$

The statement "if $x = 2$, then $y = 13$" is abbreviated with function notation as

$$f(2) = 13.$$

Also, $f(0) = 9 \cdot 0 - 5 = -5$, and $f(-3) = -32$.

These ideas and the symbols used to represent them can be explained as follows.

Name of the function

Defining expression

$$y = \widehat{(f(x))} = \overbrace{9x - 5}$$

Value of the function

Name of the independent variable

Caution   The symbol f(x) *does not* indicate "f times x," but represents the y-value for the indicated x-value. As shown above, f(2) is the y-value that corresponds to the x-value 2.

Function notation is also available on many models of graphing calculators. For example, suppose a function is defined by $Y_1 = 9X - 5$ as seen in Figure 28(a). This represents the same ordered pairs defined by $f(x) = 9x - 5$ just discussed. The function notation $Y_1(2)$, for example, means that 2 is substituted for X, yielding $9(2) - 5 = 13$. Figure 28(b) illustrates this example, as well as examples for domain values -2 and -3.

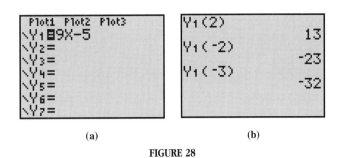

(a)   **(b)**

FIGURE 28

EXAMPLE 5   *Using Function Notation*   In each of the following, find $f(3)$.

(a) $f(x) = 3x - 7$

(b)   the function f depicted in Figure 29

(c)   the function g graphed in Figure 30

(d)   the function $Y_1 = f(x)$ defined on the screen in Figure 31

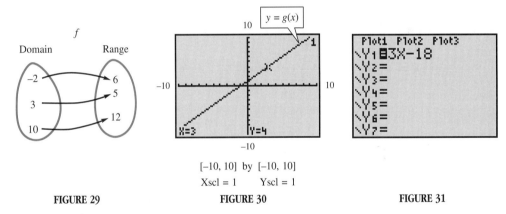

FIGURE 29   **FIGURE 30**   **FIGURE 31**

$[-10, 10]$ by $[-10, 10]$
Xscl = 1   Yscl = 1

SOLUTION

(a)   Replace x with 3 to get

$$f(3) = 3(3) - 7$$
$$= 9 - 7$$
$$= 2.$$

This result means that the ordered pair $(3, 2)$ belongs to the function, and this ordered pair lies on the graph of this function.

(b)   In the correspondence shown, 3 in the domain is paired with 5 in the range, so $f(3) = 5$.

(c)   From the information displayed at the bottom of the screen, when X = 3, Y = 4, so $g(3) = 4$.

(d)   Using a calculator with function notation capability, we find that $Y_1(3) = -9$. See Figure 32 on the next page.   ◆

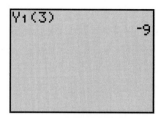

FIGURE 32

The algebraic operations used in the following example are covered in beginning and intermediate algebra courses. You should be able to perform them in this course. Refer to the reference Chapter R that appears in this book if necessary.

EXAMPLE 6   *Using Function Notation*   If $f(x) = x^2 + 3x - 4$, find and simplify each of the following.

(a) $f(x + h)$

(b) $f(x + h) - f(x)$

(c) $\dfrac{f(x + h) - f(x)}{h}, \quad h \neq 0$

SOLUTION

(a)   To find $f(x + h)$, replace x with $x + h$ in the rule for f, and simplify.

$$f(x + h) = (x + h)^2 + 3(x + h) - 4$$
$$= x^2 + 2xh + h^2 + 3x + 3h - 4$$

(b) $f(x + h) - f(x) = (x^2 + 2xh + h^2 + 3x + 3h - 4) - (x^2 + 3x - 4)$
$$= x^2 + 2xh + h^2 + 3x + 3h - 4 - x^2 - 3x + 4$$
$$= 2xh + h^2 + 3h$$

(c) $\dfrac{f(x + h) - f(x)}{h} = \dfrac{2xh + h^2 + 3h}{h}$
$$= \dfrac{h(2x + h + 3)}{h}$$
$$= 2x + h + 3 \qquad \blacklozenge$$

1.2   EXERCISES   **Tape 1**

Write each of the following using interval notation, and then graph each set on the real number line.

1. $\{x \mid -1 < x < 4\}$ **2.** $\{x \mid x \geq -3\}$ **3.** $\{x \mid x < 0\}$

4. $\{x \mid 8 > x > 3\}$ **5.** $\{x \mid 1 \leq x < 2\}$ **6.** $\{x \mid -5 < x \leq -4\}$

Using the variable x, write each of the following using set-builder notation.

7. $(-4, 3)$ **8.** $[2, 7)$ **9.** $(-\infty, -1]$ **10.** $(3, \infty)$

11.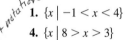

12.

13.

14.

 15. Explain how to determine whether a parenthesis or a square bracket is used when graphing an inequality on a number line.

16. The three-part inequality $a < x < b$ means "a is less than x and x is less than b." Which one of the following inequalities is not satisfied by some real number x?

   (a) $-3 < x < 5$ **(b)** $0 < x < 4$ **(c)** $-3 < x < -2$ **(d)** $-7 < x < -10$

Determine the domain and the range of each relation, and tell whether the relation is a function. If it is calculator-generated, assume that the graph extends indefinitely.

17. $\{(5, 1), (3, 2), (4, 9), (7, 6)\}$

18. $\{(8, 0), (5, 4), (9, 3), (3, 8)\}$

19.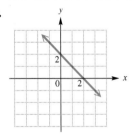

20.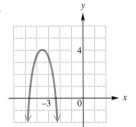

21.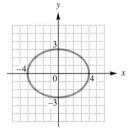

22.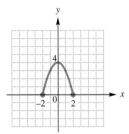

23.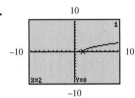

24.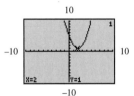

25.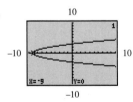

26.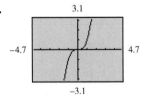

27.

X	Y₁
-5	-1
-2	2
-1	3
-.5	3.5
0	4
1.75	5.75
3.5	7.5

X= -5

28.

X	Y₁
-2	5
-1	0
0	-3
5	12
9	60
10	77
13	140

X= -2

29. $x \in L_1, y \in L_2$

L1	L2	---- 1
0	100	------
10	150	
10	175	
20	200	
30	250	
40	300	
50	400	

L1 ={0,10,10,20,...

30. $x \in L_1, y \in L_2$

L1	L2	---- 1
1940	100	------
1950	200	
1960	300	
1970	400	
1980	500	
1980	600	
1990	700	

L1 ={1940,1950,1...

31.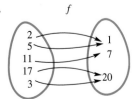

32.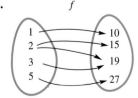

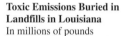

33. Explain in your own words each of the following terms.

(**a**) relation   (**b**) function   (**c**) domain of a function   (**d**) range of a function

Toxic Emissions in Landfills   *The graph shown here gives the number, in millions of pounds, of toxic emissions that were buried in landfills in Louisiana between 1987 and 1994. Suppose that f defines the number of pounds of emissions as a function of the year. Find each of the following values.*

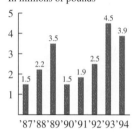

Toxic Emissions Buried in Landfills in Louisiana
In millions of pounds

(*Source*: Environmental Protection Agency, Louisiana Department of Environmental Quality.)

34. $f(1988)$ **35.** $f(1990)$ **36.** $f(1991)$

37. For what two values of x is $f(x)$ equal?

38. What is the domain of f? What is the range?

Coast-Down Time   *According to an article in the December 1994 issue of* Scientific American, *the coast-down time for a typical 1993 car as it drops 10 miles per hour from an initial speed depends on variations from the standard condition (automobile in neutral, average drag, and tire pressure). The accompanying graph illustrates some of these conditions with coast-down time in seconds graphed as a function of initial speed in miles per hour.*

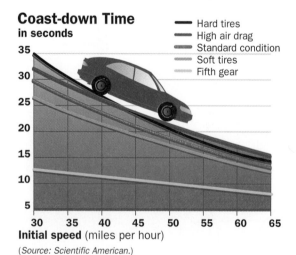

(*Source: Scientific American.*)

Use the graph to answer the following questions.

39. What is the approximate coast-down time in fifth gear if the initial speed is 40 miles per hour?

40. What is the approximate coast-down time with soft tires if the initial speed is 35 miles per hour?

41. For what standard-condition initial speed is the coast-down time 27 seconds?

42. For what initial speed is the coast-down time the same for the conditions of high air drag and hard tires?

Suppose that Y_1, Y_2, *and* Y_3 *define functions as shown in this screen.*

Find each of the following values. First calculate by hand, and then use your calculator to support your answer.

43. $Y_1(9)$ **44.** $Y_1(-4)$ **45.** $Y_2(-2)$ **46.** $Y_2(3)$ **47.** $Y_3(16)$

48. $Y_3(100)$ **49.** $Y_1(.5)$ **50.** $Y_2(0)$ **51.** $Y_3(0)$ **52.** $Y_2(.5)$

Suppose that Y_1 *and* Y_2 *define functions that generate the following table.*

X	Y1	Y2
0	-5	6
1	-4	7
4	-1	8
9	4	9
16	11	10
25	20	11
36	31	12
X=0		

Use the table to find each of the following values.

53. $Y_1(0)$ **54.** $Y_2(0)$ **55.** $Y_2(16)$ **56.** $Y_1(16)$ **57.** $Y_1(36)$ **58.** $Y_2(36)$

Suppose that f is a function defined by the set of ordered pairs (x, y) such that x is an element of list L_1 *and y is the corresponding element of list* L_2.

L1	L2	-----	1
3	2		
5	4		
7	6		
9	8		
11	10		
13	12		
L1(1)=3			

Answer each of the following.

59. What is $f(7)$?   **60.** What is $f(13)$?   Ans. 12

61. For what x is $f(x) = 10$ (if any)?

62. For what x is $f(x) = 9$ (if any)?

Find each of the following function values.

63. $f(3)$, if $f(x) = -2x + 9$ **64.** $f(6)$, if $f(x) = -2x + 8$

65. $f(11)$, for the function f in Exercise 31

66. $f(5)$, for the function f in Exercise 31

67. $f(4)$

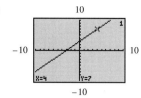

68. $f(8)$

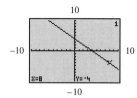

69. $f(3)$

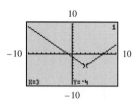

70. $f(-2)$

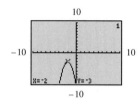

71. If f is the function graphed in Exercise 20, and $f(x) = 4$, what is x?

72. If f is the function graphed in Exercise 22, and $f(x) = 0$, what are the values of x?

In Exercises 73–80, a function is given. Find the simplified form of the function value specified.

73. $f(x) = 3x - 7$; find $f(a + 2)$

74. $f(x) = -9x + 2$; find $f(t - 3)$

75. $g(x) = 2x^2 - 3x + 4$; find $g(r + 1)$

76. $g(x) = -3x^2 + 4x - 9$; find $g(s - 3)$

77. $F(x) = x$; find $F(7p + 2s - 1)$

78. $F(x) = -x$; find $F(\sqrt{2})$

79. $f(x) = x^2$; find $f(x + h) - f(x)$

80. $f(x) = x^3$; find $f(x + h) - f(x)$

1.3   LINEAR FUNCTIONS

Functions Defined by Linear Equations   ◆   Locating a Point on a Graph with a Calculator   ◆   Slope of a Line

FUNCTIONS DEFINED BY LINEAR EQUATIONS

A variable, a numeral, or a product of numerals and variables is called a **term,** or a **monomial.** Examples of terms, or monomials, are

$$x, \quad -yz, \quad 3, \quad 5x^2y, \quad \text{and} \quad -2x^3.$$

The numerical factor in a term is called the **numerical coefficient,** or simply **coefficient,** of the term. The coefficients of the terms above are 1 (understood), -1 (understood, indicated by the $-$ sign), 3, 5, and -2. The degree of a term in a single variable is simply the exponent of the variable. Thus, the degree of the term x is 1, since $x = x^1$, and the degree of $-2x^3$ is 3. A nonzero numeral, or **constant,** is defined to have degree 0. The number 0 has no degree.

Suppose that we add a term of degree 1 in x to a constant—for example,

$$3x + 6.$$

This is an example of a **binomial,** and its degree is the largest of the degrees of all of its terms. Therefore, $3x + 6$ is a binomial of degree 1, and is said to be **linear.** (The word *linear* refers to degree 1.) In this section, we will examine functions defined by linear binomials. For example, the function f defined by $3x + 6$ may be symbolized using function notation as

$$f(x) = 3x + 6.$$

This function f is called a linear function.

DEFINITION OF LINEAR FUNCTION

A function f defined by $f(x) = ax + b$, where a and b are real numbers, is called a linear function.

Similarly, an equation such as $y = 3x + 6$ is called a linear equation. A **solution** of such an equation is an ordered pair (x, y) that makes the equation true. Verify that $(0, 6)$, $(-1, 3)$, $(-2, 0)$, and $(1, 9)$ are all solutions of $y = 3x + 6$.

The traditional method of graphing linear equations involves plotting points whose coordinates are solutions of the equation, and then joining them with a straight line. Figure 33(a) shows the ordered pairs just mentioned for the linear equation $y = 3x + 6$. It is accompanied by a *table of values*. Some graphing calculators will generate such tables. Notice that the points appear to lie in a straight line; that is indeed the case. Since we may substitute *any* real number for x, we join these points with a line to obtain the graph of the function, as shown in Figure 33(b). This line is actually the graph of the relation $\{(x, y) \mid y = 3x + 6\}$.

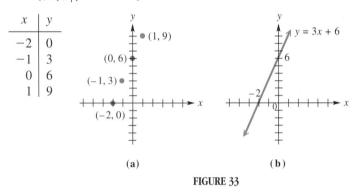

x	y
-2	0
-1	3
0	6
1	9

(a)   (b)

FIGURE 33

Using function notation, we can also say that the graph in Figure 33(b) is the graph of the linear function $f(x) = 3x + 6$.

GRAPH OF A LINEAR FUNCTION

The graph of the linear function $f(x) = ax + b$ is the same as the graph of the line whose equation is $y = ax + b$.

The graph of a linear function can be created on a calculator window. Most graphing calculators allow the user to graph several functions in the same window. These are sometimes entered by using subscripted y variables: y_1, y_2, y_3, and so on. (In some cases, upper case is used for the y-variable.) To graph the function $f(x) = 3x + 6$ on a calculator, enter $3x + 6$ for one of the y-variables. If we choose $[-10, 10]$ by $[-10, 10]$ as our viewing window, with Xscl = Yscl = 1, we get the graph shown in Figure 34(a) on the next page.

A graphing calculator will also give a table of selected points, as shown in Figure 34(b). And some of the latest models allow the user to view both the graph and the table using a split screen. See Figures 34(c) and 34(d).

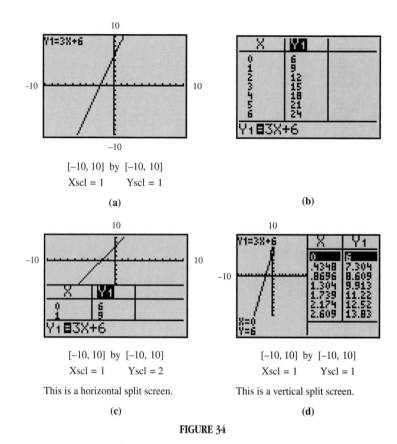

FIGURE 34

Calculators generate graphs such as the one shown in Figure 34(a) by plotting a large number of points in a very short amount of time. The choice of the viewing window will give drastically different views of a graph, and we will often devote discussion to choosing windows appropriate to the problem at hand. Figure 35 shows three different views of the graph of $f(x) = 3x + 6$, with the viewing windows noted. Notice in particular the one in Figure 35(c), since it is graphed in a "square" viewing window. You should consult your calculator manual to see if your calculator has this capability, and how to obtain such a square window.

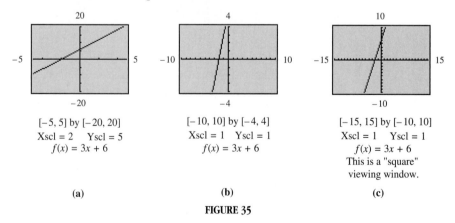

FIGURE 35

From geometry we know that two distinct points determine a line. Therefore, if we know the coordinates of a minimum of two points on a line, we can graph the line. For the equation $y = 3x + 6$, suppose we let $x = 0$ and find y: $y = 3x + 6 = 3(0) + 6 = 6.$

Now, let $y = 0$ and solve for x.

$$y = 3x + 6$$
$$0 = 3x + 6$$
$$-6 = 3x \quad \text{or} \quad x = -2$$

We have found that the points $(0, 6)$ and $(-2, 0)$ lie on the graph of $y = 3x + 6$, and this is sufficient for obtaining the graph in Figure 33(b). The numbers 6 and -2 are called the **y- and x-intercepts** of the line.

---

x- AND y-INTERCEPTS

To find the y-intercept of the graph of $y = ax + b$, let $x = 0$ and solve for y. To find the x-intercept, let $y = 0$ and solve for x (assuming $a \neq 0$).

The x-intercept of the graph of a linear function is a value that makes $f(x) = 0$ a true statement; that is, it causes the function value to become *zero*. In general, such a number is called a *zero* of the function.

ZERO OF A FUNCTION

Let f be a function. Then any number c for which $f(c) = 0$ is true is called a **zero** of the function f. (From a graphical viewpoint, c is an x-intercept of the graph of f.)

EXAMPLE 1   *Graphing a Line Using Intercepts*   Without using a calculator, graph the function $f(x) = -2x + 5$. Then support the answer with a calculator-generated graph. What is the zero of f?

SOLUTION   ANALYTIC   This is the same as the graph of $y = -2x + 5$. The table below shows the x- and y-intercepts.

	x	y	
	0	5	← y-intercept
x-intercept →	2.5	0	

The graph of $f(x) = -2x + 5$ is shown in Figure 36(a).

GRAPHICAL   A calculator-generated graph is shown in Figure 36(b). Because $f(2.5) = 0$, 2.5 is the zero of f.   ◆

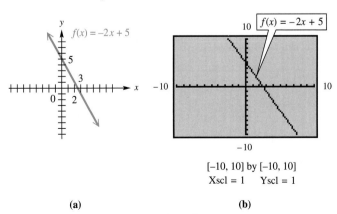

$[-10, 10]$ by $[-10, 10]$
Xscl = 1   Yscl = 1

(a)   **(b)**

FIGURE 36

Since it is possible to obtain infinitely many viewing windows for a graph, we will often be interested in choosing a window that shows the most important features of a particular graph. We will call such a graph a **comprehensive graph.** Keep in mind that the choice of window for a comprehensive graph is not unique—there will be many acceptable ones.*

Each time we introduce a new kind of graph, we will state the requirements for a comprehensive graph. For a line, we have the following.

COMPREHENSIVE GRAPH OF A LINE

A comprehensive graph of a line will show all intercepts of the line.

EXAMPLE 2   *Finding a Comprehensive Graph of a Line*   Find a comprehensive graph of the function $g(x) = -.75x + 12.5$.

SOLUTION   There are many ways that we could choose a viewing window for this comprehensive graph. The window $[-10, 10]$ by $[-10, 10]$ of Figure 37(a) does not show either intercept, so it will not do. We must increase Xmax and Ymax to show them, so if we use $[-10, 20]$ by $[-10, 20]$, for example, a comprehensive graph is obtained. See Figure 37(b).   ◆

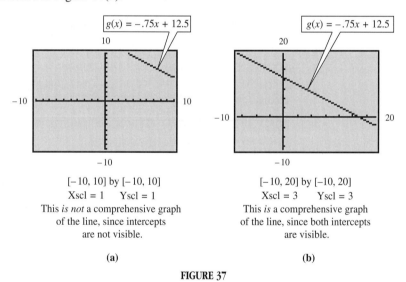

$[-10, 10]$ by $[-10, 10]$
Xscl = 1   Yscl = 1
This *is not* a comprehensive graph
of the line, since intercepts
are not visible.

(a)

$[-10, 20]$ by $[-10, 20]$
Xscl = 3   Yscl = 3
This *is* a comprehensive graph
of the line, since both intercepts
are visible.

(b)

FIGURE 37

Suppose that for a linear function $f(x) = ax + b$, we have $a = 0$. Then our function becomes

$$f(x) = b,$$

where b is some real number. Its graph is a special kind of straight line.

EXAMPLE 3   *Sketching the Graph of f(x) = b*   Consider the function $f(x) = -3$.

(a)   Sketch its graph on a rectangular coordinate system.

(b)   Plot a comprehensive graph in an appropriate window of a calculator.

---

*The term *comprehensive graph* was coined by Shoko Aogaichi Brant and Edward A. Zeidman in the text *Intermediate Algebra: A Functional Approach* (HarperCollins College Publishers, 1996), with the assistance of Professor Brant's daughter, Jennifer. The authors thank them for permission to use the terminology in this text.

SOLUTION

(a)   Since y always equals -3, the value of y can never be 0. This means that the graph has no x-intercept. The only way a straight line can have no x-intercept is for it to be parallel to the x-axis, as shown in Figure 38(a).

(b)   Using the viewing window $[-5, 5]$ by $[-5, 2]$, we find the same horizontal line shown in part (a). See Figure 38(b) and compare it to Figure 38(a).   ◆

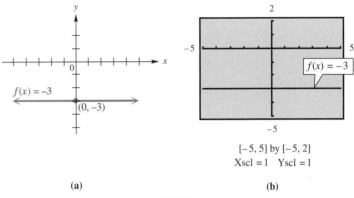

$[-5, 5]$ by $[-5, 2]$
Xscl $= 1$ Yscl $= 1$

(a)   (b)

FIGURE 38

The function considered in Example 3 is an example of a constant function.

DEFINITION OF CONSTANT FUNCTION

A function of the form $f(x) = b$, where b is a real number, is called a **constant function.** Its graph is a horizontal line with y-intercept b. For $b \neq 0$, it has no x-intercept.

We will agree that unless otherwise specified, the domain of a linear function will be the set of all real numbers. The range of a nonconstant linear function is also the set of all real numbers. For the constant function $f(x) = b$, the range is $\{b\}$.

LOCATING A POINT ON A GRAPH WITH A CALCULATOR

A graphing calculator allows the user to locate a point on a graph while displaying the coordinates at the same time. This can be done by tracing along the graph or giving the appropriate command by entering the x-coordinate of the point. Figure 39 shows some typical screens with designated points identified for the graph of $y = 3x$. Notice that in each case the y-value is three times the x-value, since this is how y is defined.

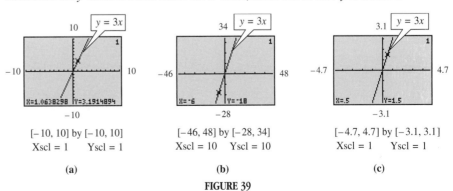

$[-10, 10]$ by $[-10, 10]$ $[-46, 48]$ by $[-28, 34]$ $[-4.7, 4.7]$ by $[-3.1, 3.1]$
Xscl $= 1$ Yscl $= 1$ Xscl $= 10$ Yscl $= 10$ Xscl $= 1$ Yscl $= 1$

(a)   (b)   (c)

FIGURE 39

Graphing calculators also provide the user with the option of using decimal or integer increments for *x*-values when tracing. Figure 40 shows screens for $y = 2x + 5$, using decimal increments and integer increments for *x*. Note that we have included the coordinates of a point on the graph at the bottom of the screen.

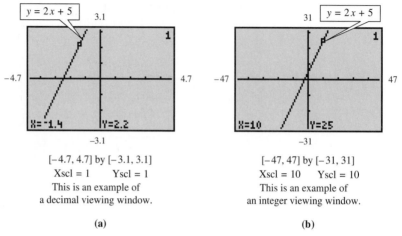

[−4.7, 4.7] by [−3.1, 3.1]
Xscl = 1   Yscl = 1
This is an example of
a decimal viewing window.

(a)

[−47, 47] by [−31, 31]
Xscl = 10   Yscl = 10
This is an example of
an integer viewing window.

(b)

Note: Different models of calculators use different screens
for decimal and integer windows. Make a note
of what your particular model uses.

FIGURE 40

EXAMPLE 4   *Using a Calculator to Support a Function Value on a Graph*
For the function $f(x) = 3x - 5$, find $f(2.1)$ analytically, and then support the answer by using an appropriate graphing feature of a calculator.

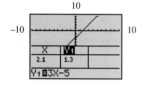

[−10, 10] by [−10, 10]
Xscl = 1   Yscl = 1
This is another way that a calculator can display the result required in Example 4.

SOLUTION   ANALYTIC   Recall from Section 1.2 that to find $f(2.1)$, we substitute 2.1 for *x*.

$$f(2.1) = 3(2.1) - 5$$
$$= 6.3 - 5$$
$$= 1.3$$

Since $f(2.1) = 1.3$, the point $(2.1, 1.3)$ must lie on the graph of the line $y = 3x - 5$.

GRAPHICAL   This is supported in Figure 41, where the screen indicates that when $x = 2.1, y = 1.3$.   ◆

TECHNOLOGY NOTE

To duplicate Figure 41, you must learn how your particular model will plot a point with a designated *x*-value. This is an excellent example of the function concept: you input an *x*-value, and the calculator determines the *y*-value and shows you the point on the graph of the function.

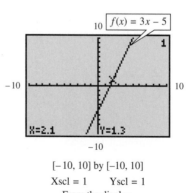

[−10, 10] by [−10, 10]
Xscl = 1   Yscl = 1
From the display,
$f(2.1) = 1.3$.
FIGURE 41

SLOPE OF A LINE

According to statistics provided by the College Board, in 1981 the average annual cost for tuition and fees at private four-year colleges was $4113. By 1993, this cost had risen to $11,025. Figure 42 shows this in graphical form, much like the kinds of graphs used in magazines and newspapers. The graphed line is actually somewhat misleading, since it indicates that the rise in cost was the same from year to year. However, we can use the graph to determine the average yearly rise in cost. Over the 12-year span, the cost increased $6912. Therefore, the average yearly rise was

$$\frac{\$6912}{12} = \$576.$$

The number 12 was obtained by subtracting $1993 - 1981$, and $6912 was found by subtracting $11,025 - $4113. The quotient shown above is an illustration of the *slope* of the line joining (1981, 4113) and (1993, 11,025). The slope concept is of major importance in the study of linear functions.

Annual Cost for Tuition and Fees at Private Four-Year Colleges

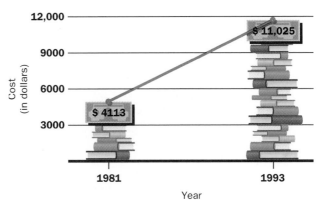

(*Source:* The College Board.)

FIGURE 42

The graph of the line $y = 3x + 1$ is shown in Figure 43. A table of selected points on the graph follows.

x	y
-2	-5
-1	-2
0	1
1	4
2	7

x-increase is 1.   *y*-increase is 3.

(2, 7)
(1, 4)
(0, 1)
(−1, −2)
(−2, −5)

FIGURE 43

Notice that for each increase of 1 for the x-value, the y-value increases by 3. For example, when x increases from 1 to 2, y increases from 4 to 7. This idea is basic to the concept of slope of a line. Geometrically, the slope is a numerical measure of the steepness of the line. (This may be interpreted as the ratio of *rise* to *run*.) To find this measure, start with the line through the two distinct points (x_1, y_1) and (x_2, y_2), as shown in Figure 44, where $x_1 \neq x_2$. The difference

$$x_2 - x_1$$

is called the **change in x** and denoted by Δx (read "delta x"), where Δ is the Greek letter *delta*. In the same way, the **change in y** can be written

$$\Delta y = y_2 - y_1.$$

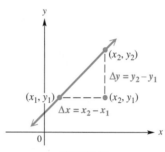

FIGURE 44

The *slope* of a nonvertical line is defined as the quotient of the change in y and the change in x, as follows.

SLOPE

The **slope** m of the line through the points (x_1, y_1) and (x_2, y_2) is

$$m = \frac{\Delta y}{\Delta x} = \frac{y_2 - y_1}{x_2 - x_1},$$

where $\Delta x \neq 0$.

Caution   When using the slope formula, be sure that it is applied correctly. It makes no difference which point is (x_1, y_1) or (x_2, y_2); however, it is important to be consistent. Start with the x- and y-values of one point (either one) and subtract the corresponding values of the *other* point.

EXAMPLE 5   *Finding Slope Using the Slope Formula*   The table of points shown in Figure 45 was generated by a graphing calculator with Y_1 defined by a linear function. Use any two points in the table to determine the slope of the line that is the graph of Y_1. Sketch the graph by hand.

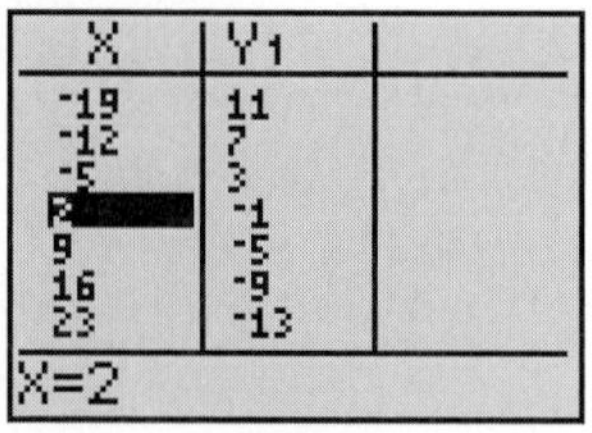

X	Y1	
-19	11	
-12	7	
-5	3	
2	-1	
9	-5	
16	-9	
23	-13	

X=2

FIGURE 45

SOLUTION   Because the slope of a line is the same regardless of the two points chosen, we can choose any two points and apply the slope formula. Suppose we choose $(2, -1)$ and $(-5, 3)$. Then, if

$$(2, -1) = (x_1, y_1) \text{ and } (-5, 3) = (x_2, y_2),$$

$$m = \frac{y_2 - y_1}{x_2 - x_1} = \frac{3 - (-1)}{-5 - (2)} = \frac{4}{-7} = -\frac{4}{7}.$$

See Figure 46. On the other hand, if $(2, -1) = (x_2, y_2)$ and $(-5, 3) = (x_1, y_1)$, the slope would be

$$m = \frac{-1 - 3}{2 - (-5)} = \frac{-4}{7} = -\frac{4}{7},$$

the same answer. This example illustrates the earlier statement that the slope is the same no matter which point is considered first. Also, using similar triangles from geometry, it can be shown that the slope is the same no matter which two different points on the line are chosen.   ◆

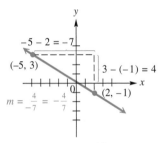

FIGURE 46

WHAT WENT WRONG ⍰   A student just starting to use a graphing calculator was attempting to graph $y = \frac{1}{2}x + 15$. He knew that the graph should be a line with slope $\frac{1}{2}$ and y-intercept 15. However, he obtained the blank screen shown here.

1. What went wrong?
2. How can he obtain a comprehensive graph of this linear function?

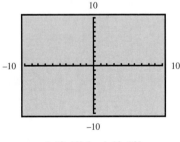

[-10, 10] by [-10, 10]
Xscl = 1   Yscl = 1

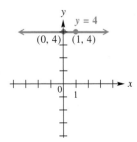

FIGURE 47

EXAMPLE 6   *Finding Slope Using the Slope Formula*   The line $y = 4$ is graphed in Figure 47. What is the slope of this line?

SOLUTION   We must choose any two points that lie on the line. For any choice of x, we have $y = 4$. Choose, for example, $(0, 4)$ and $(1, 4)$. Then, by the slope formula,

$$m = \frac{4 - 4}{1 - 0} = \frac{0}{1} = 0.$$

The slope of this line is 0.   ◆

In Figure 43, a line with *positive* slope is graphed. In Figure 46, we show a line with *negative* slope, and in Figure 47, a line with slope 0 is given. Notice that the line with positive slope *rises* from left to right, the line with negative slope *falls* from left to right, and the line with slope 0 is *horizontal*. In general, we have the following.

GEOMETRIC ORIENTATION BASED ON SLOPE

For a line with slope m, if $m > 0$, the line rises from left to right. If $m < 0$, it falls from left to right. If $m = 0$, the line is horizontal.

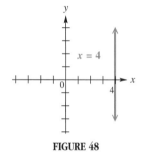

FIGURE 48

In the formula for slope, we have the condition $\Delta x = x_2 - x_1 \neq 0$. This means that $x_2 \neq x_1$. If we graph a line with two points having the same x-values, we get a vertical line. For example, the line with equation $x = 4$ is graphed in Figure 48. Notice that this is *not* the graph of a function, since 4 appears as the first number in more than one ordered pair. If we were to apply the slope formula, the denominator would be 0. As a result, the slope of such a line is *undefined*.

VERTICAL LINE

A vertical line with x-intercept a has an equation of the form

$$x = a.$$

Its slope is undefined.

TECHNOLOGY NOTE

It is essential that your calculator be in function mode (rather than parametric, polar, sequence, etc.) in order to study the material here. The other modes are covered in later chapters.

Note   It is not possible to *graph* a vertical line with a calculator in function mode. It is possible, however, to *draw* a vertical line with some models. Figure 49 shows the line $x = 4$ drawn, using a Texas Instruments TI-83 calculator. It is not possible to trace along this line with the calculator. (A vertical line can be *graphed* using parametric equations, covered later in the text.)

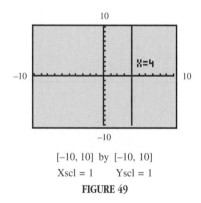

[−10, 10] by [−10, 10]
Xscl = 1   Yscl = 1
FIGURE 49

If we go back to Figure 43, we see that the slope of the line, 3, is the same as the coefficient of x in its equation ($y = 3x + 1$). Also, the graph of $y = 4$, which may also be written $y = 0x + 4$, has slope 0, which is the coefficient of x. (See Figure 47.) Consider the linear function $f(x) = ax + b$. Suppose that we increase the value of x by 1, to get $x + 1$. Then, by the slope formula,

$$
\begin{aligned}
m &= \frac{y_2 - y_1}{x_2 - x_1} = \frac{f(x + 1) - f(x)}{(x + 1) - x} \\
&= \frac{[a(x + 1) + b] - (ax + b)}{x + 1 - x} \\
&= \frac{ax + a + b - ax - b}{1} \\
&= \frac{a}{1} \\
&= a.
\end{aligned}
$$

Therefore, for each increase of 1 for the x-value, the y-value increases by a. This is consistent with our earlier observations, and leads us to a very important result concerning linear functions.

SLOPE AND y-INTERCEPT FOR A LINEAR FUNCTION

The slope of the graph of the linear function $f(x) = ax + b$ is a. The y-intercept of the graph is $f(0) = b$.

Because the slope of the graph of $f(x) = ax + b$ is a, it is often convenient to use m rather than a in the general form of the equation. Therefore, we will sometimes write

$$f(x) = mx + b \quad \text{or} \quad y = mx + b$$

to indicate a linear function. The slope is m and the y-intercept is b. This is generally called the *slope-intercept form* of the equation of a line.

SLOPE-INTERCEPT FORM

The slope-intercept form of the equation of a line is $y = mx + b$, where m is the slope and b is the y-intercept. (Linear functions are often written in the form $f(x) = ax + b$, where a is the slope and b is the y-intercept of the graph.)

EXAMPLE 7   *Matching a Graph with an Equation*   In Figure 50, we have four calculator-generated lines. Their equations are

$$y = 2x + 3, \quad y = -2x + 3, \quad y = 2x - 3, \quad \text{and} \quad y = -2x - 3,$$

but not necessarily in this order. Match each equation with its graph.

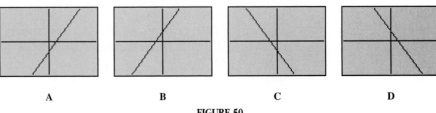

| A | B | C | D |

FIGURE 50

SOLUTION   Keep in mind that the sign of m determines whether the graph rises or falls from left to right. Also, if $b > 0$, the y-intercept is *above* the x-axis, while if $b < 0$, the y-intercept is *below* the x-axis. Therefore,

$y = 2x + 3$ is shown in B, since the graph rises from left to right, and the y-intercept is positive;

$y = -2x + 3$ is shown in D, since the graph falls from left to right, and the y-intercept is positive.

Verify that $y = 2x - 3$ is shown in A, and $y = -2x - 3$ is shown in C.   ◆

EXAMPLE 8   *Interpreting Slope in an Application of U.S. Auto Industry Employees*   The graph shown in Figure 51 is approximately linear. Discuss the slope of this "line," and explain how the slope relates to the actual employee numbers from 1988 to 1994.

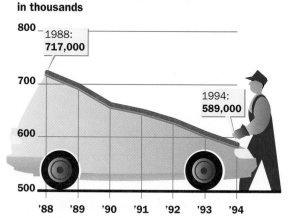

U.S. Auto Industry Employees
in thousands

1988: **717,000**

1994: **589,000**

'88 '89 '90 '91 '92 '93 '94

(*Source: USA Today,* Sec. B, p. 1, 6/29/94.)

FIGURE 51

SOLUTION   Because the approximate line has a negative slope, we can interpret this as meaning that, from 1988 to 1994, employee numbers decreased, or were on the decline. We could also find an average annual decline or the decline between any two years during the given period.   ◆

For Group Discussion   With your calculator set for the standard viewing window, graph these four lines on the same screen.

$$y_1 = .5x + 1 \qquad y_2 = x + 1 \qquad y_3 = 2x + 1 \qquad y_4 = 4x + 1$$

Each line has a positive slope. As the slope m becomes larger, what do you notice about the steepness of the line?

Now repeat the experiment, but change each coefficient of x to its negative. As the absolute value of the slope becomes larger, what do you notice about the steepness of the line?

The result of the "For Group Discussion" exercise should lead to the following.

> STEEPNESS OF A LINE
>
> For the line $y = mx + b$, as $|m|$ becomes larger, the line becomes steeper.

The final example shows how the geometric interpretation of slope can be used to graph a line by hand if we know the slope of the line and a point that lies on the line.

EXAMPLE 9 *Using the Slope and a Point to Graph a Line* Graph the line through (2, 1) that has slope $-\frac{4}{3}$.

SOLUTION Start by locating the point (2, 1) on the graph. Find a second point on the line by using the definition of slope.

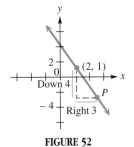

FIGURE 52

$$\text{slope} = \frac{\text{change in } y}{\text{change in } x} = \frac{-4}{3}$$

Move *down* 4 units from (2, 1) and then 3 units to the *right*. Draw a line through this second point P and (2, 1), as shown in Figure 52. The slope also could be written as

$$\frac{\text{change in } y}{\text{change in } x} = \frac{4}{-3}.$$

In that case, the second point is located *up* 4 units and 3 units to the *left*. Verify that this approach produces the same line. ◆

The method of Example 9 can be adapted to graphing the linear function $f(x) = ax + b$. Since the y-intercept is b, we know that one point on the graph is $(0, b)$. Since a is the slope, we can determine another point by interpreting a as "rise over run." For example, to graph $f(x) = 1.5x + 3$, start at $(0, 3)$, and since $1.5 = 1\frac{5}{10} = \frac{15}{10} = \frac{3}{2}$, move 3 units up and 2 units to the right to find a second point on the line, $(2, 6)$. Join the points with a straight line to obtain the graph, as shown in Figure 53.

TECHNOLOGY NOTE

If you wish to graph the function shown in Figure 53 with a calculator, it may be to your advantage to use the decimal form of the slope, 1.5, rather than the common fraction form, $\frac{3}{2}$. If you use the common fraction form, it will be necessary to use parentheses around it so that the calculator will perform the operations in the correct order. This is one advantage of using decimal numerals rather than fractions when working with graphing calculators.

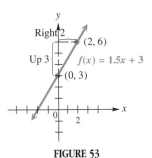

FIGURE 53

1.3 EXERCISES 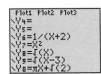 Tape 1

The two screens show functions Y_1 through Y_0 defined on a graphing calculator. Determine whether the function is linear by observing the screens. Do not actually graph. If the function is linear, give the slope m and the y-intercept b.

1. Y_1 2. Y_2 3. Y_3

4. Y_4 5. Y_5 6. Y_6

7. Y_7 8. Y_8 9. Y_9

10. Y_0

Graph each of the following linear functions by hand. You may wish to support your answer by graphing on a calculator. Also, give (a) the x-intercept, (b) the y-intercept, (c) the domain, (d) the range, and (e) the slope of the line.

11. $f(x) = x - 4$ **12.** $f(x) = -x + 4$ **13.** $f(x) = 3x - 6$ **14.** $f(x) = \dfrac{2}{3}x - 2$

15. $f(x) = -\dfrac{2}{5}x + 2$ **16.** $f(x) = \dfrac{4}{3}x - 3$ **17.** $f(x) = 3x$ **18.** $f(x) = -.5x$

19. Based on the graphs of the functions in Exercises 17 and 18, what conclusion can you make about one particular point that *must* lie on the graph of the line $y = ax$ (where $b = 0$)?

20. Using the geometric definition of slope and your answer to Exercise 19, give the equation of the line whose graph is shown here.

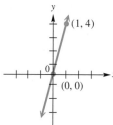

Graph each of the following by hand. For Exercises 21, 22, and 24, you may wish to support your answer by graphing on a calculator. Also, give (a) the x-intercept (if any), (b) the y-inter-cept (if any), (c) the domain, (d) the range, and (e) the slope of the line (if defined).

21. $f(x) = -3$ **22.** $f(x) = 5$ **23.** $x = -1.5$ **24.** $f(x) = \dfrac{5}{4}$ **25.** $x = 2$ **26.** $x = -3$

27. What special name is given to the functions found in Exercises 21, 22, and 24?

28. Give the equation of the line illustrated.

(a)

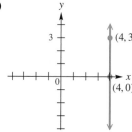

(b)

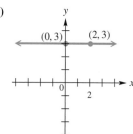

29. What is the equation of the *x*-axis?

30. What is the equation of the *y*-axis?

Graph each linear function on a calculator, using the two different windows given. One window gives a comprehensive graph (as defined in this section), while the other does not. State which window gives the comprehensive graph.

31. $f(x) = 4x + 20$
  Window A: $[-10, 10]$ by $[-10, 10]$
  Window B: $[-10, 10]$ by $[-5, 25]$

32. $f(x) = -5x + 30$
  Window A: $[-10, 10]$ by $[-10, 40]$
  Window B: $[-5, 5]$ by $[-5, 40]$

33. $f(x) = 3x + 10$
  Window A: $[-3, 3]$ by $[-5, 5]$
  Window B: $[-5, 5]$ by $[-10, 14]$

34. $f(x) = -6$
  Window A: $[-5, 5]$ by $[-5, 5]$
  Window B: $[-10, 10]$ by $[-10, 10]$

For each function in Exercises 35–38, find the function value indicated by analytic methods. Then, use the capabilities of your calculator to locate the point with the specified x-value on a graph, and show that the y-value corresponds to the one you found analytically.

35. $f(x) = -3x + .25; \quad f(4.3)$ **36.** $f(x) = 4x - 1.3; \quad f(-1.4)$

37. $f(x) = 2.9x + 10;\quad f(-1.3)$ **38.** $f(x) = 5;\quad f(6.7)$

Find the slope of the line that passes through the given points.

39. $(-2, 1)$ and $(3, 2)$ **40.** $(-2, 3)$ and $(-1, 2)$ **41.** $(8, 4)$ and $(-1, -3)$

42. $(-4, -3)$ and $(5, 0)$ **43.** $(-6, 5)$ and $(12, 5)$ **44.** $(3, 6)$ and $(3, 1)$

45. Based on your answer to Exercise 43, what special kind of straight line passes through these points? Sketch its graph. Is this the graph of a function?

46. Based on your answer to Exercise 44, what special kind of straight line passes through these points? Sketch its graph. Is this the graph of a function?

Find the slope of the line defined by Y_1, *based on the given table of points.*

47.
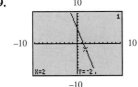

48.
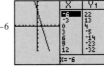

Find the slope of the line graphed.

49.
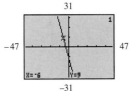
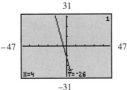

50.
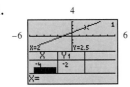
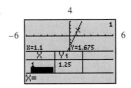

51.

52.

A linear function is defined as Y_1 *and a table of points is generated. Find the slope m of the line that is the graph of* Y_1. *Then use the table to find the y-intercept b. Finally, give the equation.*

53.

54.

Exercises 55 and 56 deal with linear "models." We will investigate such models in more detail in Section 1.7.

55. *5000-Meter Run in the Olympic Games*   The graph shows the winning times (in minutes) at the Olympic Games for the 5000-meter run together with a linear approximation of the data.

Olympic Time for 5000 Meter Run
in minutes

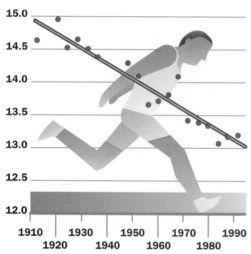

(*Source:* United States Olympic Committee.)

(a) The equation for the linear approximation is $y = -.0221x + 57.14$. What does the slope of this line represent? Why is the slope negative?

(b) Can you think of any reason why there are no data points for the years 1940 and 1944?

(c) Do you see a limitation of this linear model?

56. *U.S. Radio Stations on the Air*   The graph shows the number of U.S. radio stations on the air along with a linear function that models the data.

(a) Discuss the accuracy of the linear function.

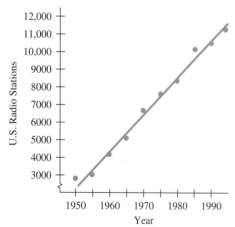

(*Source*: National Association of Broadcasters.)

(b) Use the two data points (1950, 2773) and (1994, 11,600) to find the approximate slope of the line shown. Interpret this number.

57. Match each equation with the line that would most closely resemble its graph.

(a) $y = 2$ **(b)** $y = -2$

(c) $x = 2$ **(d)** $x = -2$

A

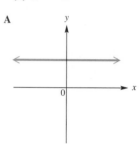

B

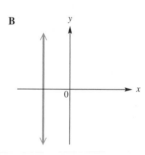

C

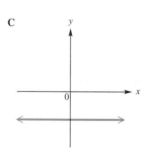

D

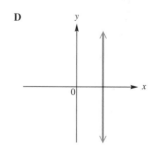

58. Match each equation with its calculator-generated graph.

    (a) $y = 3x + 2$ **(b)** $y = -3x + 2$ **(c)** $y = 3x - 2$ **(d)** $y = -3x - 2$

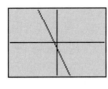

A

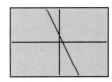

B

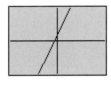

C

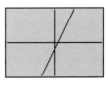

D

59. Without actually plotting points, sketch by hand a line on rectangular coordinate axes that would resemble the graph of $y = mx + b$ if

    (a) $m > 0, b > 0$ **(b)** $m > 0, b < 0$ **(c)** $m < 0, b > 0$ **(d)** $m < 0, b < 0$.

60. Explain why the function defined by Y_1 cannot be a linear function, based on the given table of points.

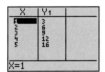

In Exercises 61–66, two lines are graphed on the same screen. Decide whether y_1 or y_2 has a slope of larger absolute value. Also, determine whether the slope of each line is positive, negative, or zero.

61.

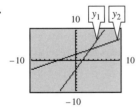

62.

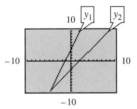

63.

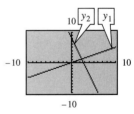

64.

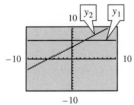

65.

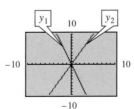

66.

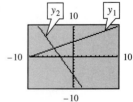

Sketch by hand the graph of the line passing through the given point and having the given slope. Indicate two points on the line.

67. Through $(-1, 3)$, $m = \dfrac{3}{2}$ **68.** Through $(-2, 8)$, $m = -1$ **69.** Through $(3, -4)$, $m = -\dfrac{1}{3}$

70. Through $(-2, -3)$, $m = -\dfrac{3}{4}$ **71.** Through $(-1, 4)$, $m = 0$ **72.** Through $\left(\dfrac{9}{4}, 2\right)$, undefined slope

73. Through $(0, -4)$, $m = \dfrac{3}{4}$ **74.** Through $(0, 5)$, $m = -2.5$

75. Give the equation of the line described in Exercise 73.

76. Give the equation of the line described in Exercise 74.

77. Refer to Example 7 and accompanying Figure 50. Explain why the graph of $y = 2x - 3$ must be the line shown in A.

78. Repeat Exercise 77 for $y = -2x - 3$ and the line shown in C.

1.4 EQUATIONS OF LINES AND GEOMETRIC CONSIDERATIONS

Point-Slope Form of the Equation of a Line ◆ Other Forms of the Equation of a Line ◆ The Pythagorean Theorem, the Distance Formula, and the Midpoint Formula ◆ Parallel and Perpendicular Lines

POINT-SLOPE FORM OF THE EQUATION OF A LINE

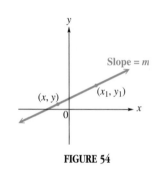

FIGURE 54

The equation of a line can be found if we know a point on the line and the slope of the line. Figure 54 shows a line passing through the fixed point (x_1, y_1) and having slope m. (Assuming that the line has a slope guarantees that it is not vertical.) Let (x, y) be any other point on the line. By the definition of slope, the slope of the line is

$$\frac{y - y_1}{x - x_1}.$$

Since the slope of the line is m,

$$\frac{y - y_1}{x - x_1} = m.$$

Multiplying both sides by $x - x_1$ gives

$$y - y_1 = m(x - x_1).$$

This result, called the *point-slope form* of the equation of a line, identifies points on a given line: a point (x, y) lies on the line through (x_1, y_1) with slope m if and only if

$$y - y_1 = m(x - x_1).$$

POINT-SLOPE FORM

The line with slope m passing through the point (x_1, y_1) has an equation

$$y - y_1 = m(x - x_1).$$

This is called the **point-slope form** of the equation of a line.

EXAMPLE 1   *Using the Point-Slope Form*   The screens in Figures 55(a) and 55(b) show two points and their coordinates at the bottoms of the screens. Use the point-slope form to find the equation of the line joining the two points.

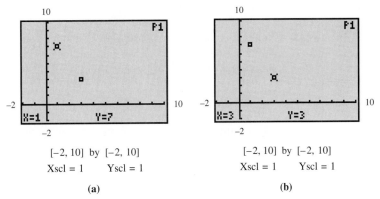

[−2, 10] by [−2, 10]
Xscl = 1   Yscl = 1

(a)

[−2, 10] by [−2, 10]
Xscl = 1   Yscl = 1

(b)

FIGURE 55

SOLUTION   As indicated, the points are (1, 7) and (3, 3). First, we find the slope of the line using the slope formula.

$$m = \frac{7 - 3}{1 - 3} = \frac{4}{-2} = -2$$

Now, using either point, say (1, 7), apply the point-slope form.

$$y - y_1 = m(x - x_1)$$
$$y - 7 = -2(x - 1) \qquad x_1 = 1, m = -2, y_1 = 7$$
$$y - 7 = -2x + 2 \qquad \text{Distributive property}$$
$$y = -2x + 9 \qquad \text{Add 7.} \qquad \blacklozenge$$

EXAMPLE 2   *Using the Point-Slope Form (Given Two Points)*   The table of points shown in Figure 56 indicates at the bottom that the equation of the line containing the points is $y = 2x - 1$. Suppose that the equation were not displayed. Use two points from the table to show how the equation can be determined.

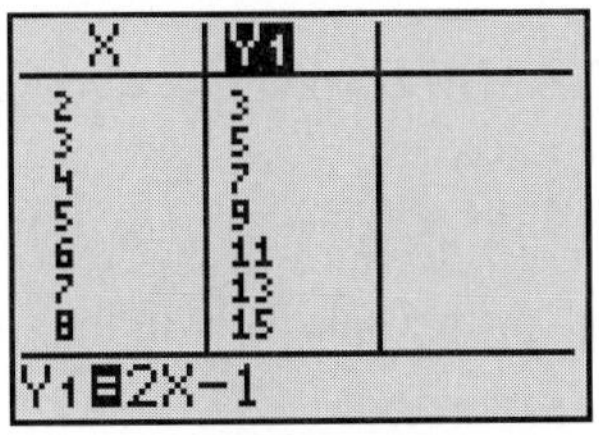

FIGURE 56

SOLUTION   Find the slope of the line first. We may choose any two points, so let us choose (2, 3) and (6, 11).

$$m = \frac{11 - 3}{6 - 2} = \frac{8}{4} = 2$$

Now, use either point, say (2, 3), with $m = 2$ in the point-slope form.

$$y - 3 = 2(x - 2)$$
$$y - 3 = 2x - 4$$
$$y = 2x - 1$$

This final equation agrees with the displayed equation at the bottom of the table.

$\blacklozenge$

OTHER FORMS OF THE EQUATION OF A LINE

Notice that the equations found in Examples 1 and 2 were given in $y = mx + b$ form. As mentioned earlier, this very important form of the equation of a line is called the *slope-intercept form.* We repeat it here.

SLOPE-INTERCEPT FORM

The **slope-intercept form** of the equation of a line with slope m and y-intercept b is

$$y = mx + b.$$

This form is probably the most useful form of the equation of a line, since at a single glance we can determine the slope and the y-intercept. In addition, for the purpose of graphing lines on a graphing calculator, this is the form required for input, as we saw in Section 1.3.

Other forms of the equation of a line include

$$Ax + By = C \quad \text{and} \quad Ax + By + C = 0,$$

such as $3x + 2y = 6$ and $3x + 2y - 6 = 0$. Texts often refer to these as **standard form.** One advantage of standard form is that it allows quick calculation for both intercepts. For example, if we begin with $3x + 2y = 6$, we can find the x-intercept by letting $y = 0$ and the y-intercept by letting $x = 0$.

$$x\text{-intercept: } 3x + 2(0) = 6 \qquad y\text{-intercept: } 3(0) + 2y = 6$$
$$3x = 6 \qquad\qquad\qquad 2y = 6$$
$$x = 2 \qquad\qquad\qquad y = 3$$

Having this information is particularly useful if we wish to sketch the graph of the line by hand.

EXAMPLE 3 *Graphing an Equation in $Ax + By = C$ Form* Graph $3x + 2y = 6$

(a) by hand.

(b) with a calculator in the standard viewing window.

SOLUTION

(a) ANALYTIC As seen above, the points $(2, 0)$ and $(0, 3)$ lie on the graph. Plot these two points and join them with a straight line. See Figure 57(a).

(b) GRAPHICAL To begin, we must first solve the equation for y.

$$3x + 2y = 6 \qquad \text{Given}$$
$$2y = -3x + 6 \qquad \text{Subtract } 3x.$$
$$y = -1.5x + 3 \qquad \text{Divide by 2.}$$

Using this last equation, we obtain the graph in the desired viewing window as shown in Figure 57(b). ◆

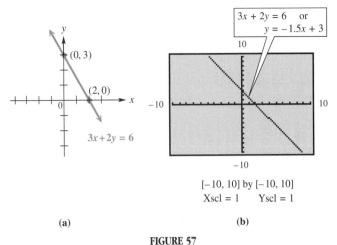

FIGURE 57

Note Because of the usefulness of the slope-intercept form over the so-called standard form in the context of the graphing calculator approach of this text, we will emphasize the slope-intercept form in most of our work.

THE PYTHAGOREAN THEOREM, THE DISTANCE FORMULA, AND THE MIDPOINT FORMULA

The Pythagorean theorem from geometry, which gives a relationship concerning the lengths of the sides of a right triangle, is the basis for many results in mathematics. In a right triangle, the sides that form the right angle are called *legs* and the side opposite the right angle (the longest side) is called the *hypotenuse*.

PYTHAGOREAN THEOREM

In a right triangle, the sum of the squares of the lengths of the legs is equal to the square of the length of the hypotenuse.

$$a^2 + b^2 = c^2$$

An outline of a proof of this theorem is found in Exercise 67.

EXAMPLE 4 *Using the Pythagorean Theorem* Use the Pythagorean theorem to find the length of the hypotenuse in the triangle in Figure 58.

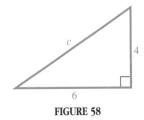

FIGURE 58

SOLUTION   By the Pythagorean theorem, the length of the hypotenuse is

$$c = \sqrt{a^2 + b^2} \qquad \text{Take square roots.}$$
$$= \sqrt{4^2 + 6^2} \qquad \text{Let } a = 4 \text{ and } b = 6.$$
$$= \sqrt{16 + 36}$$
$$= \sqrt{52} = \sqrt{4 \cdot 13}$$
$$= \sqrt{4} \cdot \sqrt{13}$$
$$= 2\sqrt{13}. \qquad \blacklozenge$$

Caution   When using the equation $c^2 = a^2 + b^2$, be sure that the length of the hypotenuse is substituted for c, and that the lengths of the legs are substituted for a and b. Errors often occur because values are substituted incorrectly.

The *converse* of the Pythagorean theorem is also true. That is, if a, b, and c are lengths of the sides of a triangle and $a^2 + b^2 = c^2$, then the triangle is a right triangle with hypotenuse c.

An important formula that is derived using the Pythagorean theorem is the formula that allows us to find the distance between two points in a plane. Figure 59 shows the points $P(-4, 3)$ and $R(8, -2)$.

To find the distance between these two points, complete a right triangle as shown in the figure. This right triangle has its right angle at $(8, 3)$. The horizontal side of the triangle has length

$$|8 - (-4)| = 12,$$

where absolute value is used to make sure that the distance is not negative. The vertical side of the triangle has length

$$|3 - (-2)| = 5.$$

By the Pythagorean theorem, the length of the remaining side of the triangle is

$$\sqrt{12^2 + 5^2} = \sqrt{144 + 25} = \sqrt{169} = 13.$$

The distance between $(-4, 3)$ and $(8, -2)$ is 13.

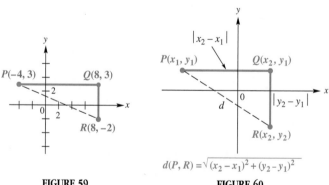

FIGURE 59   **FIGURE 60**

To obtain a general formula for the distance between two points in a coordinate plane, let $P(x_1, y_1)$ and $R(x_2, y_2)$ be any two distinct points in a plane, as shown in Figure 60. Complete a triangle by locating point Q with coordinates (x_2, y_1). Using the Pythagorean theorem gives the distance between P and R, written $d(P, R)$, as

$$d(P, R) = \sqrt{(x_2 - x_1)^2 + (y_2 - y_1)^2}.$$

Note   The use of absolute value bars is not necessary in this formula, since for all real numbers a and b, $|a - b|^2 = (a - b)^2$.

The distance formula can be summarized as follows.

> ## DISTANCE FORMULA
>
> Suppose that $P(x_1, y_1)$ and $R(x_2, y_2)$ are two points in a coordinate plane. Then the distance between P and R, written $d(P, R)$, is given by the **distance formula,**
>
> $$d(P, R) = \sqrt{(x_2 - x_1)^2 + (y_2 - y_1)^2}.$$

Although the figure used in the proof of the distance formula assumes that P and R are not on a horizontal or vertical line, the result is true for any two points.

EXAMPLE 5 *Using the Distance Formula* The segment drawn on the screen in Figure 61 joins the points $P(-8, 4)$ and $Q(3, -2)$. Use the distance formula to find the length of this segment (or, equivalently, the distance $d(P, Q)$).

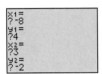

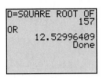

A graphing calculator can be programmed to apply the distance formula. Compare with Example 5.

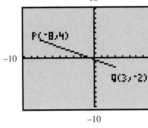

[−10, 10] by [−10, 10]
Xscl = 1   Yscl = 1
FIGURE 61

SOLUTION The length of the segment, or $d(P, Q)$, is found as follows.

$$d(P, Q) = \sqrt{[3 - (-8)]^2 + (-2 - 4)^2} \qquad x_1 = -8,\ y_1 = 4,$$
$$= \sqrt{11^2 + (-6)^2} \qquad\qquad x_2 = 3,\ y_2 = -2$$
$$= \sqrt{121 + 36} = \sqrt{157}. \qquad \blacklozenge$$

Note As shown in Example 5, it is customary to leave the distance between two points in radical form rather than approximating it with a calculator (unless, of course, it is otherwise specified).

Given the coordinates of the endpoints of a line segment, it is possible to find the coordinates of the *midpoint* of the segment; that is, the point on the segment that lies the same distance from both endpoints. Point M in Figure 62 is the midpoint of the segment joining (x_1, y_1) and (x_2, y_2).

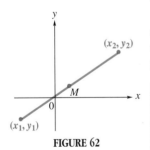

FIGURE 62

> ## MIDPOINT FORMULA
>
> The midpoint of the line segment with endpoints (x_1, y_1) and (x_2, y_2) is
>
> $$\left(\frac{x_1 + x_2}{2}, \frac{y_1 + y_2}{2} \right).$$

In other words, the midpoint formula says that the coordinates of the midpoint of a segment are found by calculating the *average* of the x-coordinates and the average of the y-coordinates of the endpoints of the segment. In Exercise 68, you are asked to verify that the coordinates above satisfy the definition of midpoint.

EXAMPLE 6 *Using the Midpoint Formula* Find the midpoint M of the segment with endpoints $(8, -4)$ and $(-9, 6)$.

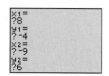

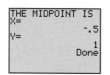

A graphing calculator can be programmed to apply the midpoint formula. Compare with Example 6.

SOLUTION   Use the midpoint formula to find that the coordinates of *M* are

$$\left(\frac{8 + (-9)}{2}, \frac{-4 + 6}{2}\right) = \left(-\frac{1}{2}, 1\right). \qquad \blacklozenge$$

EXAMPLE 7   *Using the Midpoint Formula to Predict Welfare Payment Data*
The graph in Figure 63 shows an idealized linear relationship for the average monthly payment to families with dependent children in 1994 dollars. Use this information and apply the midpoint formula to approximate the average payment in 1987.

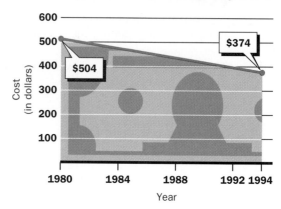

Average Monthly Family Payment

(*Sources:* Office of Financial Management, Administration for Children and Families.)

FIGURE 63

SOLUTION   Notice that the year 1987 lies exactly halfway between 1980 and 1994. Therefore, we can use the midpoint formula to find the coordinates of the midpoint of the segment joining (1980, 504) and (1994, 374). The *y*-coordinate will give the approximate average payment in 1987.

$$\text{midpoint} = \left(\frac{1980 + 1994}{2}, \frac{504 + 374}{2}\right) = (1987, 439)$$

Thus, in 1987, the payment was about $439. $\blacklozenge$

PARALLEL AND PERPENDICULAR LINES

For Group Discussion   In the standard viewing window of your calculator, graph all four of the following lines.

$$y_1 = 2x - 6 \qquad y_2 = 2x - 2 \qquad y_3 = 2x \qquad y_4 = 2x + 4$$

What is the slope of each line? What geometric term seems to describe the set of lines?

Two lines in a plane are *parallel* if they do not intersect. Although the exercise in the "For Group Discussion" box above does not actually prove the result that follows, it provides good visual support.

PARALLEL LINES

Two distinct nonvertical lines are parallel if and only if they have the same slope.

EXAMPLE 8   *Using the Slope Relationship for Parallel Lines*   Find the equation of the line that passes through the point (3, 5) and is parallel to the line with the equation $2x + 5y = 4$. Then graph both lines in the window $[-10, 10]$ by $[-10, 10]$ to provide visual support for your answer.

SOLUTION   ANALYTIC   Since it is given that the point (3, 5) is on the line, we need only find the slope to use the point-slope form. Find the slope by writing the equation of the given line in slope-intercept form. (That is, solve for y.)

$$2x + 5y = 4$$
$$y = -\frac{2}{5}x + \frac{4}{5}$$

The slope is $-\frac{2}{5}$. Since the lines are parallel, $-\frac{2}{5}$ is also the slope of the line whose equation is to be found. Substituting $m = -\frac{2}{5}$, $x_1 = 3$, and $y_1 = 5$ into the point-slope form gives

$$y - y_1 = m(x - x_1)$$
$$y - 5 = -\frac{2}{5}(x - 3)$$
$$5(y - 5) = -2(x - 3)$$
$$5y - 25 = -2x + 6$$
$$5y = -2x + 31$$
$$y = -\frac{2}{5}x + \frac{31}{5}.\qquad \text{Slope-intercept form of the desired line}$$

GRAPHICAL   To provide visual support, we now graph the given line and the line just determined. They are $y = -\frac{2}{5}x + \frac{4}{5}$ (the slope-intercept form of the given equation) and $y = -\frac{2}{5}x + \frac{31}{5}$. It is often easier to enter decimal forms of fractions rather than $\frac{a}{b}$ (common fraction) forms since the latter often require use of parentheses to indicate grouping. Therefore, we will enter these two equations as

$$y_1 = -.4x + .8 \qquad \text{(the given line)}$$

and $\qquad y_2 = -.4x + 6.2 \qquad$ (the desired line).

As seen in Figure 64, the lines *seem* to be parallel, providing visual support for our result.   ◆

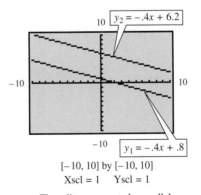

$[-10, 10]$ by $[-10, 10]$
Xscl = 1   Yscl = 1

These lines appear to be parallel.
Because the slopes are equal, they
are indeed parallel.

FIGURE 64

Caution   When using graphing calculator technology, we must always be aware that visual support (as seen in Example 8) does not necessarily prove our result. For example, Figure 65 shows the graphs of $y_1 = -.5x + 4$ and $y_2 = -.5001x + 2$. Although they *seem* to be parallel by visual inspection, they are *not* parallel because the slope of y_1 is $-.5$ and the slope of y_2 is $-.5001$, and these are unequal slopes. The fact that they are very close to each other causes the lines to *appear* to be parallel on the viewing screen. This discussion provides an excellent illustration as to why we can never completely rely on graphical methods; analytic methods *must* be studied hand-in-hand with graphical methods.

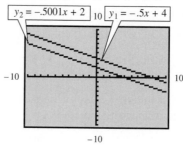

$$[-10, 10] \text{ by } [-10, 10]$$
$$\text{Xscl} = 1 \quad \text{Yscl} = 1$$

These lines appear to be parallel.
However, because the slopes are
not equal, they *are not* truly parallel.

FIGURE 65

For Group Discussion   With your calculator viewing window set for a "square" window, graph each pair of lines. Do each group separately, clearing the screen of one group before proceeding to the next.

I	II	III	IV
$y_1 = 4x + 1$	$y_1 = -\dfrac{2}{3}x + 3$	$y_1 = 6x - 3$	$y_1 = \dfrac{13}{7}x - 3$
$y_2 = -.25x + 3$	$y_2 = 1.5x - 4$	$y_2 = -\dfrac{1}{6}x + 4$	$y_2 = -\dfrac{7}{13}x + 4$

What geometric term applies to each pair of lines? What is the product of the slopes in each pair of lines?

Again, as in the earlier "For Group Discussion" activity, while we have not proved the result stated below, we have provided good visual support for it.

PERPENDICULAR LINES

Two lines, neither of which is vertical, are perpendicular if and only if their slopes have a product of -1.

TECHNOLOGY NOTE

Read your owner's manual to determine the number of pixels your screen displays horizontally and vertically. This will determine the aspect ratio of the screen.

For example, if the slope of a line is $-\frac{3}{4}$, the slope of any line perpendicular to it is $\frac{4}{3}$, since $\left(-\frac{3}{4}\right)\left(\frac{4}{3}\right) = -1$. We often refer to numbers like $-\frac{3}{4}$ and $\frac{4}{3}$ as "negative reciprocals." A proof of this result is outlined in Exercise 69.

Because the standard viewing windows for graphing calculators are usually designed with an aspect ratio of approximately 3 to 2 (that is, the number of pixels across is about $\frac{3}{2}$ times the number of pixels up and down), in order to give visual support for perpendicularity, a "square" setting is necessary. Figure 66 illustrates this.

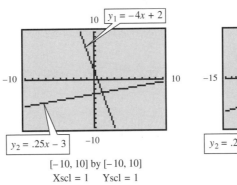

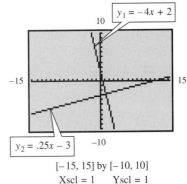

[−10, 10] by [−10, 10]
Xscl = 1   Yscl = 1

Although the graphs of these lines are perpendicular, they do not seem to be when graphed in a standard viewing window.

(a)

[−15, 15] by [−10, 10]
Xscl = 1   Yscl = 1

Visual support for perpendicularity is more obvious using a "square" viewing window.

(b)

FIGURE 66

EXAMPLE 9 *Using the Slope Relationship for Perpendicular Lines* Find the equation of the line that passes through the point (3, 5) and is perpendicular to the line with the equation $2x + 5y = 4$. Then graph both lines in a square viewing window to provide visual support for your answer.

SOLUTION ANALYTIC In Example 8, we found that the slope of the given line is $-\frac{2}{5}$, so the slope of any line perpendicular to it is $\frac{5}{2}$. Therefore, use $m = \frac{5}{2}$, $x_1 = 3$, and $y_1 = 5$ in the point-slope form.

$$y - 5 = \frac{5}{2}(x - 3)$$
$$2(y - 5) = 5(x - 3)$$
$$2y - 10 = 5x - 15$$
$$2y = 5x - 5$$
$$y = 2.5x - 2.5$$

GRAPHICAL Graphing $y_1 = -.4x + .8$ (the point-slope form of the given equation) and $y_2 = 2.5x - 2.5$ (the point-slope form of the equation just determined) in a *square* viewing window provides support for our answer. See Figure 67. ◆

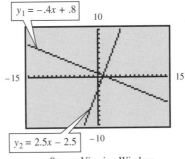

Square Viewing Window

FIGURE 67

1.4   EXERCISES      Tape 1

Write the equation of the line through the given point and having the indicated slope. Give the equation in slope-intercept form, if possible.

1. Through $(1, 3)$, $m = -2$

2. Through $(2, 4)$, $m = -1$

3. Through $(-5, 4)$, $m = -1.5$

4. Through $(-4, 3)$, $m = .75$

5. Through $(-8, 1)$, $m = -.5$

6. Through $(6, 1)$, $m = 0$

In Exercises 7–14, find the equation of the line described. Give the equation in slope-intercept form, if possible.

7. the line joining the two points identified in the screens

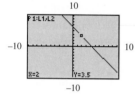

8. the line joining the two points identified in the screens

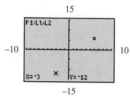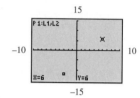

9. the line shown in the two views

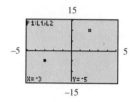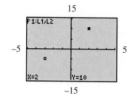

10. the line shown in the two views

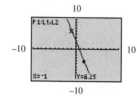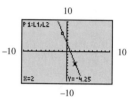

11. the line shown in this split screen

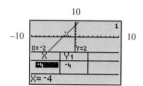

12. the line shown in this split screen

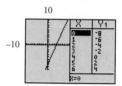

13. the line defined by Y_1 that yields the given table of points

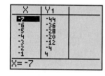

14. the line defined by Y_1 that yields the given table of points

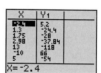

15. Give the equation of the line shown in the screen.

(a)

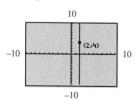

(b)

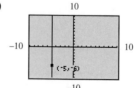

(c)

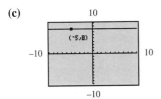

(d)

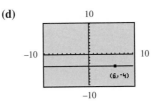

16. Fill in each blank with the appropriate response: The line $x + 2 = 0$ has x-intercept _____. It (does/does not) have a y-intercept. The slope of this line is (zero/undefined). The line $4y = 2$ has y-intercept _____. It (does/does not) have an x-intercept. The slope of this line is (zero/undefined).

17. What is true about y for every point on the x-axis?

18. What is true about x for every point on the y-axis?

Graph each of the following lines by hand, finding intercepts to determine two points on the line, as explained in Example 3(a).

19. $x - y = 4$ **20.** $x + y = 4$ **21.** $3x - y = 6$ **22.** $2x - 3y = 6$ **23.** $2x + 5y = 10$ **24.** $4x - 3y = 9$

A line having an equation of the form $y = kx$, where k is a real number, $k \neq 0$, will always pass through the origin. To graph such an equation by hand, we must determine a second point and then join the origin and that second point with a straight line. Use this method to graph each of the following.

25. $y = 3x$ **26.** $y = -2x$ **27.** $y = -.75x$ **28.** $y = 1.5x$

Write each equation in the form $y = mx + b$. Then, using your calculator, graph the line in the window indicated.

29. $5x + 3y = 15$
  $[-10, 10]$ by $[-10, 10]$

30. $6x + 5y = 9$
  $[-10, 10]$ by $[-10, 10]$

31. $-2x + 7y = 4$
  $[-5, 5]$ by $[-5, 5]$

32. $-.23x - .46y = .82$
  $[-5, 5]$ by $[-5, 5]$

33. $1.2x + 1.6y = 5.0$
  $[-6, 6]$ by $[-4, 4]$

34. $2y - 5x = 0$
  $[-10, 10]$ by $[-10, 10]$

Use the Pythagorean theorem to find the length of the unknown side of the right triangle. In each case, a and b represent the lengths of the legs and c represents the length of the hypotenuse.

35. $a = 8, b = 15$; find c

36. $a = 7, b = 24$; find c

37. $a = 13, c = 85$; find b

38. $a = 14, c = 50$; find b

39. $a = 5, b = 8$; find c

40. $a = 9, b = 10$; find c

41. $a = \sqrt{13}, c = \sqrt{29}$; find b

42. $a = \sqrt{7}, c = \sqrt{11}$; find b

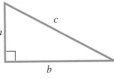

Typical Labeling

Find (a) the distance between P and Q and (b) the coordinates of the midpoint of the segment joining P and Q.

43. $P(5, 7), Q(13, -1)$ **44.** $P(-2, 5), Q(4, -3)$ **45.** $P(-8, -2), Q(-3, -5)$ **46.** $P(-6, -10), Q(6, 5)$

47.

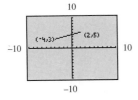

$P = (-4, 3), Q = (2, 5)$

48.

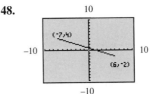

$P = (-7, 4), Q = (6, -2)$

49. *Cost of Private College Education*   During the period 1983–1991, the average annual cost (in dollars) of tuition and fees at private four-year colleges rose in an approximately linear fashion. The graph depicts this growth using a line segment. Use the midpoint formula to approximate the cost during the year 1987. (*Source:* The College Board.)

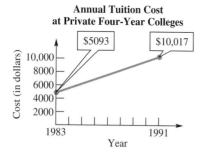

Annual Tuition Cost at Private Four-Year Colleges

to approximate the federal debt in 1987. (*Source:* U.S. Office of Management and Budget.)

Federal Debt

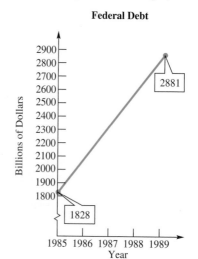

50. *Federal Debt*   From 1985 to 1989, the federal debt (in billions of dollars) rose in an approximately linear fashion. See the graph. Use the information from the graph

Poverty Level Income Cutoffs   *The table lists how poverty level income cutoffs (in dollars) for a family of four have changed over time. Use this information for Exercises 51–53.*

Year	Income
1960	3022
1970	3968
1975	5500
1980	8414
1985	10,989
1990	13,359

(*Source:* U.S. Census Bureau.)

51. Use the midpoint formula to approximate the poverty level cutoff in 1965.

52. The midpoint formula will give an exact answer if the data have what type of relationship?

53. Do the income cutoffs have a linear relationship with time? (*Hint:* Consider the data for 1970 and 1980.)

54. Suppose that segment PQ has midpoint M. The coordinates of P are $(-3, 6)$ and the coordinates of M are $(5, 8)$. Find the coordinates of Q.

Find the equation of the line satisfying the given conditions, giving it in slope-intercept form, if possible.

55. Through $(-1, 4)$, parallel to $x + 3y = 5$

56. Through $(3, -2)$, parallel to $2x - y = 5$

57. Through $(1, 6)$, perpendicular to $3x + 5y = 1$

58. Through $(-2, 0)$, perpendicular to $8x - 3y = 7$

59. Through $(-5, 7)$, perpendicular to $y = -2$

60. Through $(1, -4)$, perpendicular to $x = 4$

61. Through $(-5, 8)$, parallel to $y = -.2x + 6$

62. Through $(-4, -7)$, parallel to $x + y = 5$

63. Through the origin, perpendicular to $2x + y = 6$

64. Through the origin, parallel to $y = -3.5x + 7.4$

65. The figure shows the graphs of $y_1 = 2.3x + .57$ and $y_2 = 2.3001x - 4.8$ in a viewing window $[-10, 10]$ by $[-10, 10]$. The student unfamiliar with the concepts presented in this section may conclude that these two lines are parallel. Write a short paragraph explaining why they are or are not parallel.

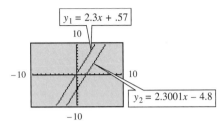

$[-10, 10]$ by $[-10, 10]$
Xscl = 1   Yscl = 1

66. The figure shows the graphs of $y_1 = -3x + 4$ and $y_2 = \frac{1}{3}x - 4$ in a typical standard viewing window. The student unfamiliar with the concepts presented in this section may conclude that these two lines are not perpendicular. Write a short paragraph explaining why they are or are not perpendicular. If they *are* perpendicular lines, explain how to set a graphing calculator so that this result may be more easily supported.

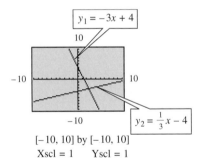

$[-10, 10]$ by $[-10, 10]$
Xscl = 1   Yscl = 1

67. The figure shown is a square made up of four right triangles and a smaller square. By use of the method of *equal areas*, the Pythagorean theorem may be proved. Fill in the blanks with the missing information.

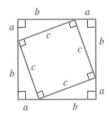

(a) The length of a side of the large square is _____, so its area is (_____)² or _____.

(b) The area of the large square may also be found by obtaining the sum of the areas of the four right triangles and the smaller square. The area of each

right triangle is _____, so the sum of the areas of the four right triangles is _____. The area of the smaller square is _____.

(c) The sum of the areas of the four right triangles and the smaller square is _____.

(d) Since the areas in (a) and (c) represent the area of the same figure, the expressions there must be equal. Setting them equal to each other we obtain _____ = _____.

(e) Subtract $2ab$ from each side of the equation in (d) to obtain the desired result _____ = _____.

68. Suppose that the endpoints of a line segment have coordinates (x_1, y_1) and (x_2, y_2).

(a) Show that the distance between (x_1, y_1) and $\left(\frac{x_1 + x_2}{2}, \frac{y_1 + y_2}{2}\right)$ is the same as the distance between (x_2, y_2) and $\left(\frac{x_1 + x_2}{2}, \frac{y_1 + y_2}{2}\right)$.

(b) Show that the sum of the distances between (x_1, y_1) and $\left(\frac{x_1 + x_2}{2}, \frac{y_1 + y_2}{2}\right)$ and (x_2, y_2) and $\left(\frac{x_1 + x_2}{2}, \frac{y_1 + y_2}{2}\right)$ is equal to the distance between (x_1, y_1) and (x_2, y_2).

(c) From the results of parts (a) and (b), what conclusion can be made?

69. Refer to the given figure and complete parts (a)–(h) to prove that if two lines are perpendicular, and neither line is parallel to an axis, then the lines have slopes whose product is -1.

(a) In triangle OPQ, angle POQ is a right angle if and only if

$$[d(O, P)]^2 + [d(O, Q)]^2 = [d(P, Q)]^2.$$

What theorem from geometry assures us of this?

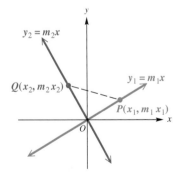

(b) Find an expression for the distance $d(O, P)$.

(c) Find an expression for the distance $d(O, Q)$.

(d) Find an expression for the distance $d(P, Q)$.

(e) Use your results from parts (b)–(d), and substitute into the equation in part (a). Simplify to show that this leads to the equation $-2m_1m_2x_1x_2 - 2x_1x_2 = 0$.

(f) Factor $-2x_1x_2$ from the final form of the equation in part (e).

(g) Use the zero-product property from intermediate algebra to solve the equation in part (f) to show that $m_1m_2 = -1$.

(h) State your conclusion based on parts (a)–(g).

70. The following tables show ordered pairs for linear functions defined in Y_1 and Y_2. Determine whether the lines are parallel, perpendicular, or neither parallel nor perpendicular.

(a)

X	Y₁	Y₂
0	-3	4
1	1	3.75
2	5	3.5
3	9	3.25
4	13	3
5	17	2.75
6	21	2.5

X=0

(b)

X	Y₁	Y₂
0	-3	5.2
1	2	5.4
2	7	5.6
3	12	5.8
4	17	6
5	22	6.2
6	27	

X=0

(c)

X	Y₁	Y₂
0	-3	12
1	2	17
2	7	22
3	12	27
4	17	32
5	22	37
6	27	42

X=0

Beginning in this section, we will occasionally include groups of exercises under the heading "Relating Concepts." These exercise groups are designed to be worked in sequential order. Their purpose is to show connections and relationships among various topics that appear in the section and perhaps in earlier sections.

RELATING CONCEPTS (EXERCISES 71–80)

The accompanying table was generated for a linear function by using a graphing calculator. It identifies seven points on the graph of the function. Work Exercises 71–80 in order, to see connections between the slope formula, the distance formula, the midpoint formula, and linear functions.

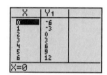

X	Y₁
0	-6
1	-3
2	0
3	3
4	6
5	9
6	12

X=0

71. Use the first two points in the table to find the slope of the line. (Section 1.3)

72. Use the second and third points in the table to find the slope of the line. (Section 1.3)

73. Make a conjecture by filling in the blanks: If we use any two points on a line to find its slope, we find that the slope is _____ in all cases. (Section 1.3)

74. Use the distance formula to find the distance between the first two points in the table. (Section 1.4)

75. Use the distance formula to find the distance between the second and fourth points in the table. (Section 1.4)

76. Use the distance formula to find the distance between the first and fourth points in the table. (Section 1.4)

77. Add the results in Exercises 74 and 75, and compare the sum to the answer you found in Exercise 76. What do you notice?

78. Fill in the blanks, basing your answers on your observations in Exercises 74–77: If points A, B, and C lie on a line in that order, then the distance between A and B added to the distance between _____ and _____ is equal to the distance between _____ and _____.

79. Use the midpoint formula to find the midpoint of the segment joining $(0, -6)$ and $(6, 12)$. Compare your answer to the middle entry in the table. What do you notice? (Section 1.4)

80. If the table were set up to show an x-value of 4.5, what would be the corresponding y-value?

1.5 SOLUTION OF LINEAR EQUATIONS; ANALYTIC METHOD AND GRAPHICAL SUPPORT

Solving Linear Equations Analytically ◆ Graphical Support for Solutions of Linear Equations ◆ Identities and Contradictions

One of the primary topics in the study of algebra is the process of solving equations. The simplest type of equation, the *linear equation*, is discussed in this section.

SOLVING LINEAR EQUATIONS ANALYTICALLY

In this text, we will use two distinct approaches to equation solving. The **analytic approach** is the method you have probably seen in previous courses, where paper and pencil are used to transform complicated equations into simpler ones. In so doing, mathematical concepts are applied, and there is no reliance upon graphical representation. Most of the equations that we will encounter in this text are solvable by strictly analytic methods, and the student must realize that this approach *is not to be downplayed* just because the text is based on graphing calculator use.

The **graphical approach** is the method that distinguishes this text from most of the texts you have previously encountered. We will often *support* our analytic solutions by using graphical techniques. Occasionally, we will encounter equations that are very difficult or even impossible to solve using analytic methods, and for those we may choose to present a solution that is strictly graphical in nature. It is important to realize that both analytic and graphical approaches will be used, and part of becoming a good mathematics student is learning when to use and when not to use each approach.

An **equation** is a statement that two expressions are equal. To **solve** an equation means to find all numbers that make the equation a true statement. Such numbers are called **solutions** or **roots** of the equation. A number that is a solution of an equation is said to **satisfy** the equation, and the solutions of an equation make up its **solution set.**

In this section, we will concentrate on solving equations that have the largest power of the variable equal to 1. These are called *linear equations*.

LINEAR EQUATION IN ONE VARIABLE

A **linear equation** in one variable is an equation that can be written in the form

$$ax + b = 0,$$

where $a \neq 0$.

One way to solve an equation is to rewrite it as a series of simpler equations, each of which has the same solution set as the original one. Such equations are said to be **equivalent equations.** These simpler equations are obtained by using the addition and multiplication properties of equality.

ADDITION AND MULTIPLICATION PROPERTIES OF EQUALITY

For real numbers a, b, and c:

$a = b$ and $a + c = b + c$ are **equivalent.** (*The same number may be added to both sides of an equation without changing the solution set.*)

If $c \neq 0$, then $a = b$ and $ac = bc$ are **equivalent.** (*Both sides of an equation may be multiplied by the same nonzero number without changing the solution set.*)

Extending the addition property of equality allows us to subtract the same number from both sides. Similarly, extending the multiplication property of equality allows us to divide both sides of an equation by the same nonzero number.

EXAMPLE 1   *Solving a Linear Equation Analytically*   Solve $10 + 3(2x - 4) = 17 - (x + 5)$.

SOLUTION   Use the distributive property and then collect like terms to get the following series of simpler equivalent equations.

$$10 + 3(2x - 4) = 17 - (x + 5)$$

$10 + 6x - 12 = 17 - x - 5$	Distributive property
$-2 + 7x = 12$	Add x to each side; combine terms.
$7x = 14$	Add 2 to each side.
$x = 2$	Multiply both sides by $\frac{1}{7}$.

An *analytic* check to determine whether 2 is indeed the solution of this equation requires that we substitute 2 for x in the original equation to see if a true statement is obtained.

$10 + 3(2x - 4) = 17 - (x + 5)$		Original equation
$10 + 3(2 \cdot 2 - 4) = 17 - (2 + 5)$	?	Let $x = 2$.
$10 + 3(4 - 4) = 17 - 7$	?	
$10 = 10$		True

Since replacing x with 2 results in a true statement, 2 is the solution of the given equation. The solution set is therefore $\{2\}$.   ◆

When fractions or decimals appear in an equation, our work can be made simpler by multiplying both sides by the least common denominator of all the fractions in the equation. Examples 2 and 3 illustrate these types of equations.

EXAMPLE 2   *Solving a Linear Equation with Fractional Coefficients Analytically*   Solve $\dfrac{x + 7}{6} + \dfrac{2x - 8}{2} = -4$.

SOLUTION   Start by eliminating the fractions. Multiply both sides by 6.

$$6\left[\frac{x + 7}{6} + \frac{2x - 8}{2}\right] = 6 \cdot (-4)$$

$6\left(\dfrac{x + 7}{6}\right) + 6\left(\dfrac{2x - 8}{2}\right) = 6(-4)$	Distributive property
$x + 7 + 3(2x - 8) = -24$	
$x + 7 + 6x - 24 = -24$	Distributive property
$7x - 17 = -24$	Combine terms.
$7x = -7$	Add 17.
$x = -1$	Divide by 7.

Analytic check:

$$\frac{x + 7}{6} + \frac{2x - 8}{2} = -4 \qquad \text{Original equation}$$

$$\frac{(-1) + 7}{6} + \frac{2(-1) - 8}{2} = -4 \qquad ? \quad \text{Let } x = -1.$$

$$\frac{6}{6} + \frac{-10}{2} = -4 \qquad ?$$

$$1 + (-5) = -4 \qquad ?$$

$$-4 = -4 \qquad \text{True}$$

Our analytic check indicates that $\{-1\}$ is the solution set.   ◆

EXAMPLE 3   *Solving a Linear Equation with Decimal Coefficients Analytically*   Solve $.06x + .09(15 - x) = .07(15)$.

SOLUTION   Since each decimal number is given in hundredths, multiply both sides of the equation by 100. (This is done by moving the decimal points two places to the right.)

$$.06x + .09(15 - x) = .07(15)$$

$$6x + 9(15 - x) = 7(15) \qquad \text{Multiply by 100.}$$

$$6x + 9(15) - 9x = 105 \qquad \text{Distributive property}$$

$$-3x + 135 = 105 \qquad \text{Combine like terms.}$$

$$-3x = -30 \qquad \text{Subtract 135.}$$

$$x = 10 \qquad \text{Divide by } -3.$$

Analytic check:

$$.06x + .09(15 - x) = .07(15) \qquad \text{Original equation}$$

$$.06(10) + .09(15 - 10) = .07(15) \qquad ? \quad \text{Let } x = 10.$$

$$.6 + .09(5) = 1.05 \qquad ?$$

$$.6 + .45 = 1.05 \qquad ?$$

$$1.05 = 1.05 \qquad \text{True}$$

The solution set is $\{10\}$.   ◆

The equations solved in Examples 1–3 each have a single solution. Such equations are called **conditional equations.** Later in this section, we will see that equations may have no solutions or infinitely many solutions.

GRAPHICAL SUPPORT FOR SOLUTIONS OF LINEAR EQUATIONS

Let us go back to the linear equation solved in Example 1. We found that the number 2 makes the equation $10 + 3(2x - 4) = 17 - (x + 5)$ a true statement. We can also look at this situation from a standpoint of functions. The statement

"Solve $10 + 3(2x - 4) = 17 - (x + 5)$"

can be re-worded as follows:

> "Find the value(s) in the domain of the functions
>
> $$f(x) = 10 + 3(2x - 4)$$
>
> and $g(x) = 17 - (x + 5)$
>
> that give the same function value(s) (i.e., range values)."

From a graphing perspective, this can be interpreted as follows:

> "Find the *x*-value(s) of the point(s) of intersection of the graphs of
> $$f(x) = 10 + 3(2x - 4) \text{ and } g(x) = 17 - (x + 5)."$$

Since the graphs of both functions are straight lines (because they are linear functions), they will intersect in a single point, no point, or infinitely many points. See Figure 68.

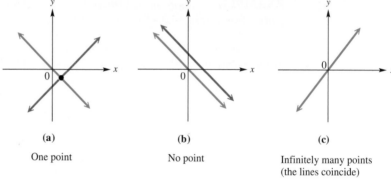

(a)	(b)	(c)
One point	No point	Infinitely many points (the lines coincide)

FIGURE 68

We are considering only conditional equations *for now*, so we need only consider the first of these situations, as seen in Figure 68(a). To support graphically our analytic work in Example 1, we graph the lines $y_1 = 10 + 3(2x - 4)$ and $y_2 = 17 - (x + 5)$ in the same viewing window, and use the capabilities of the calculator to find the point of intersection. We see in Figure 69 that the point of intersection is (2, 10). The *x*-coordinate here, 2, is the solution of the equation, while the *y*-coordinate, 10, is the value that we get when we substitute 2 for *x* in both of the original expressions. (See the analytic check in Example 1.)

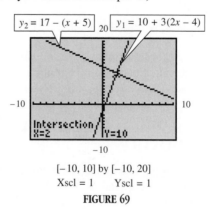

[−10, 10] by [−10, 20]
Xscl = 1   Yscl = 1

FIGURE 69

EXAMPLE 4   *Providing Visual Support for an Analytic Solution*   Support the result of Example 2: that is,

$$\frac{x + 7}{6} + \frac{2x - 8}{2} = -4$$

has solution set $\{-1\}$, using the graphical method described above.

SOLUTION   Enter $y_1 = \frac{x + 7}{6} + \frac{2x - 8}{2}$ and $y_2 = -4$. See Figure 70. The two straight lines intersect at $(-1, -4)$, providing support that $\{-1\}$ is the solution set. Remember, if $(-1, -4)$ is the point of intersection, -1 is the solution, and -4 is the value obtained when -1 is substituted into both expressions.   ◆

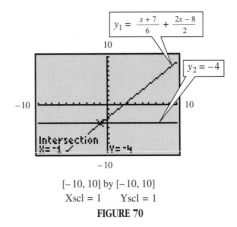

[-10, 10] by [-10, 10]
Xscl = 1   Yscl = 1
FIGURE 70

We now state the first of two methods of equation solving by graphical methods that we will use in this book.

INTERSECTION-OF-GRAPHS METHOD OF GRAPHICAL SOLUTION

To solve the equation

$$f(x) = g(x)$$

graphically, graph $y_1 = f(x)$ and $y_2 = g(x)$. The x-coordinate of any point of intersection of the two graphs is a solution of the equation.

TECHNOLOGY NOTE

The words *root, solution,* and *zero* all refer to the same basic concept. If there is one thing that you should remember when studying methods of solving equations, it is this: *The real solutions (or roots or zeros) of an equation of the form* $f(x) = 0$ *correspond to the x-intercepts of the graph of* $y = f(x)$. For this reason, modern graphing calculators are programmed so that x-intercepts (roots, zeros) can be located.

At this point, we must make two important observations. First, since we are studying *linear* functions in this chapter, our examples and exercises will consist of functions whose graphs are straight lines. The intersection-of-graphs method of solving equations will be applied to linear functions now, but we will, in later chapters, apply it to many other kinds of functions as well.

Second, we should realize that graphing calculators have the capability of determining the coordinates of the point of intersection of two lines to a great degree of accuracy, and will, in many cases, give the exact decimal values. Of course, if a coordinate is an irrational number, the decimal shown will be only an approximation.

There is a second method of graphical support for solving equations. Suppose that we once again wish to solve

$$f(x) = g(x).$$

By subtracting $g(x)$ from both sides, we obtain

$$f(x) - g(x) = 0.$$

Notice that $f(x) - g(x)$ is simply a function itself. Let us call it $F(x)$. Then we only need to solve

$$F(x) = 0$$

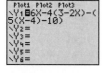

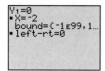

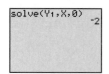

to obtain the solution set of the original equation. In Section 1.3, we learned that any number that satisfies this equation is an x-intercept of the graph of $y = F(x)$. It is also called a *zero* of F. Using this idea, we now state another method of solving an equation graphically.

---

x-INTERCEPT METHOD OF GRAPHICAL SOLUTION

To solve the equation

$$f(x) = g(x)$$

graphically, graph $y = f(x) - g(x) = F(x)$. Any x-intercept of the graph of $y = F(x)$ (or zero of the function F) is a solution of the equation.

The x-intercept method is used in the next example.

EXAMPLE 5   *Providing Visual Support for an Analytic Solution*   It can be shown analytically that the solution set of

$$6x - 4(3 - 2x) = 5(x - 4) - 10$$

is $\{-2\}$. Use the x-intercept method of graphical solution to support this result.

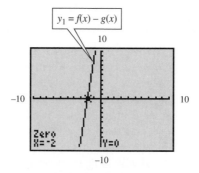

SOLUTION   Begin by letting

$$f(x) = 6x - 4(3 - 2x) \quad \text{and} \quad g(x) = 5(x - 4) - 10.$$

Then, find $f(x) - g(x)$ and enter it as y_1 in a calculator.

$$y_1 = f(x) - g(x)$$
$$= 6x - 4(3 - 2x) - (5(x - 4) - 10)$$

Graph this function to get the straight line shown in Figure 71. By choosing a standard viewing window, we see that the x-intercept is -2, supporting the information given in the statement of the problem.   ◆

In Example 5, we use a graphical method. Graphing calculators also have *solver* features that allow the user to define a function Y_1 (see the top screen) and solve the equation $0 = Y_1$ (see the second screen). The two bottom screens show two different solver features, both giving the correct solution, -2, as seen in Example 5. Solver routines often require the user to provide a "guess."

$[-10, 10]$ by $[-10, 10]$

Xscl = 1   Yscl = 1

The x-intercept of the graph is the **zero** of the function defined by $f(x) - g(x)$. It is also the solution of the equation $f(x) = g(x)$.

FIGURE 71

Caution   When using the *x*-intercept method of graphical solution, as in Example 5, it is a common error to forget to enter symbols of inclusion around the expression that is being subtracted. Notice how we used parentheses around $g(x) = 5(x - 4) - 10$ when we determined the expression for y_1. Graphing calculator technology provides us a new respect for the need to use symbols of inclusion correctly!

For Group Discussion   Repeat Example 5, but this time, rather than graphing
$$y_1 = f(x) - g(x),$$
graph
$$y_2 = g(x) - f(x).$$
Observe the graph of y_2. Does it have the same *x*-intercept as y_1? Can you make a conjecture concerning the order in which the two functions are subtracted when using the *x*-intercept method of solution?

As stated earlier, we can use graphical methods to find approximate solutions of equations that, for one reason or another, are difficult to solve algebraically. In this book, we will occasionally require approximate solutions. We now give rules for rounding decimal numbers to a particular place value.

RULES FOR ROUNDING

To round a number to a place value to the right of the decimal point:
Step 1   Locate the **place** to which the number is being rounded.
Step 2   Look at the next **digit to the right** of the place to which the number is being rounded.
Step 3A If this digit is **less than 5,** drop all digits to the right of the place to which the number is being rounded. Do *not change* the digit in the place to which the number is being rounded.
Step 3B If this digit is **5 or greater,** drop all digits to the right of the place to which the number is being rounded. *Add 1* to the digit in the place to which the number is being rounded.

To round a number to a place value to the left of the decimal point:

Step 1   Same as above.
Step 2   Same as above.
Step 3A If this digit is **less than 5,** do *not change* the digit, and replace the digits to the right with zeros.
Step 3B If this digit is **5 or greater,** *add 1* to it, and replace all digits to the right with zeros.

If either situation requires that 1 be added to 9, replace the 9 with a 0 and add 1 to the digit to the left of 9.

EXAMPLE 6   *Approximating a Solution of a Linear Equation Graphically*
Use the intersection-of-graphs method to approximate the solution of

$$.51(\sqrt{2} + 3x) - .21(\pi x + 6.1) = 7$$

to the nearest hundredth.

SOLUTION   Let $y_1 = .51(\sqrt{2} + 3x) - .21(\pi x + 6.1)$ and let $y_2 = 7$. Since we do not have any idea of what the solution might be, graph them both in the standard viewing

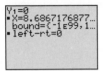

The result of Example 6 can also be found using a solver routine.

window. As seen in Figure 72, the point of intersection does indeed lie in the standard window. Using the capabilities of the calculator, we can find the coordinates of the point of intersection, as seen at the bottom of the screen. If we round the *x*-coordinate to the nearest hundredth, we have {8.69} as the solution set of the equation. ◆

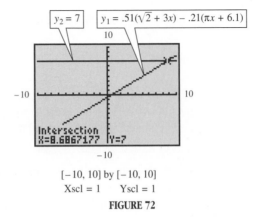

[−10, 10] by [−10, 10]
Xscl = 1   Yscl = 1
FIGURE 72

IDENTITIES AND CONTRADICTIONS

Every equation solved thus far in this section has been a conditional equation. However, since the graphs of two linear functions may not intersect or may overlap (as shown in Figures 68(b) and (c)), there are two other situations that may occur when solving linear equations.

A **contradiction** is an equation that has no solution, as seen in the next example.

EXAMPLE 7   *Identifying a Contradiction Analytically and Graphically*
Consider the equation

$$-2x + 5x - 9 = 3(x - 4) - 5.$$

(a)   Solve it analytically.

(b)   Support the result of part (a) graphically.

SOLUTION

(a)   ANALYTIC $-2x + 5x - 9 = 3(x - 4) - 5$ Given equation

$3x - 9 = 3x - 12 - 5$ Combine like terms and use the distributive property.

$-9 = -17$ Subtract 3x and combine terms.

Notice that this final equation is obviously false. When this happens, the equation is a contradiction, and the solution set is the *empty* or *null set*, symbolized ∅.

(b)   GRAPHICAL   Let

$$y_1 = -2x + 5x - 9$$

and $y_2 = 3(x - 4) - 5.$

Graphing both in the viewing window [−5, 10] by [−20, 10] gives us the picture shown in Figure 73. Notice that the lines *appear* to be parallel, supporting our analytic work.

Remember again that this graph is only support, and not an actual proof that there are no solutions. ◆

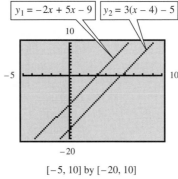

$$y_1 = -2x + 5x - 9 \qquad y_2 = 3(x - 4) - 5$$

$[-5, 10]$ by $[-20, 10]$
Xscl = 1   Yscl = 1
y_1 and y_2 appear to be
parallel, and indeed are.
FIGURE 73

An **identity** is an equation that is true for all values in the domain of its variable. Example 8 illustrates an identity.

EXAMPLE 8   *Identifying an Identity Analytically and Graphically*   Consider the equation

$$9x + 4 - 3x = 2(3x + 4) - 4.$$

(a)   Solve it analytically.

(b)   Support the result of part (a) graphically.

SOLUTION

(a)   ANALYTIC $9x + 4 - 3x = 2(3x + 4) - 4$ Given equation

$6x + 4 = 6x + 8 - 4$ Combine like terms and use the distributive property.

$4 = 4$ Subtract 6x and combine 8 − 4 to get 4.

This final equation is obviously true. When this happens, the equation is an identity, and for linear equations, the solution set is {all real numbers}, or $(-\infty, \infty)$. (In later chapters, we will see that some identities will have certain values excluded from their solution sets, but will still have infinitely many solutions.)

(b)   GRAPHICAL   Let

$$y_1 = 9x + 4 - 3x$$

and $$y_2 = 2(3x + 4) - 4.$$

Proceeding as we did in Example 7, we graph both using a window that will give us a comprehensive graph. As seen in Figure 74 on the next page, a standard window appears to give us only one line, indicating an overlap of the graphs. This supports our conclusion in part (a) that all real numbers are solutions. Again, this is not a proof—only visual support. ◆

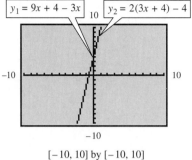

$$\boxed{y_1 = 9x + 4 - 3x} \quad \boxed{y_2 = 2(3x + 4) - 4}$$

$[-10, 10]$ by $[-10, 10]$
Xscl = 1   Yscl = 1

y_1 and y_2 appear to
coincide, and indeed do.

FIGURE 74

For Group Discussion   Duplicate the screens shown in Figures 75(a), (b), and (c), using your telephone number in Figure 75(b).

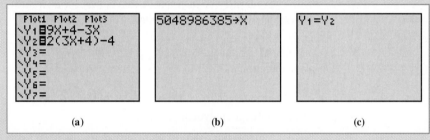

(a)   (b)   (c)

FIGURE 75

1. Based on the result in Example 8, should the calculator return a 0 (for false) or a 1 (for true)?
2. Change the number stored in X to any other number you wish. Repeat Item 1.
3. Discuss the results. Why does the calculator return the value that it does?

1.5   EXERCISES   Tape 1

*A **zero** of a function f is a number c that satisfies $f(c) = 0$. For example, 4 is a zero of $f(x) = 2x - 8$, since $f(4) = 2(4) - 8 = 8 - 8 = 0$. Use an analytic method to determine the zero of the linear function f by solving $f(x) = 0$.*

1. $f(x) = -3x - 12$ 2. $f(x) = 5x - 30$ 3. $f(x) = 5x$

4. $f(x) = -2x$ 5. $f(x) = 2(3x - 5) + 8(4x + 7)$ 6. $f(x) = -4(2x - 3) + 8(2x + 1)$

Find the zero of the linear function shown in the screen. Use analytic methods. You may wish to support your answer by graphing the function and using the capability of your calculator.

7. Y_1

8. Y_2

9. Y_3

10. If c is a zero of the linear function $f(x) = mx + b, m \neq 0$, then the point at which the graph intersects the x-axis has coordinates (_____, _____).

In Exercises 11–16, two linear functions, y_1 and y_2, are graphed in a viewing window with the point of intersection of the graphs given in the display at the bottom. Using the intersection-of-graphs method of graphical solution, give the solution set of $y_1 = y_2$.

11.

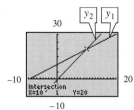

12.

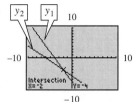

13.

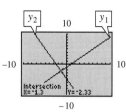

14.

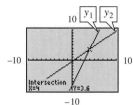

15.

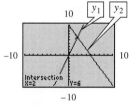

16.

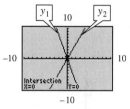

17. Give the interpretation of the y-value shown at the bottom of the display in Exercises 11–16 above.

18. If $y_1 = f(x)$ and $y_2 = g(x)$, and the solution set of $y_1 = y_2$ is $\{-4\}$, what is the value of $f(-4) - g(-4)$? How does your answer relate to the x-intercept method of graphical solution of equations?

In Exercises 19–22, linear functions y_1 and y_2 have been defined, and then the graph of $y_1 - y_2$ has been graphed in an appropriate viewing window. Use the x-intercept method of graphical solution to solve the equation $y_1 = y_2$.

19.

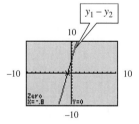

20.

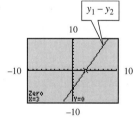

21.

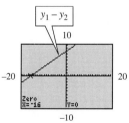

22.

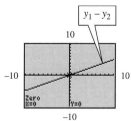

23. If the x-intercept method of graphical solution leads to the graph of a horizontal line above or below the x-axis, what is the solution set of the equation? What special name is this kind of equation given?

24. If the x-intercept method of graphical solution leads to a horizontal line that coincides with the x-axis, what is the solution set of the equation? What special name is this kind of equation given?

Solve each equation analytically. Check it analytically by direct substitution, and then support your solution graphically.

25. $2x - 5 = x + 7$

26. $9x - 17 = 2x + 4$

27. $.01x + 3.1 = 2.03x - 2.96$

28. $.04x + 2.1 = .02x + 1.92$

29. $-(x + 5) - (2 + 5x) + 8x = 3x - 5$

30. $-(8 + 3x) + 5 = 2x + 3$

31. $\dfrac{2x + 1}{3} + \dfrac{x - 1}{4} = \dfrac{13}{2}$

32. $\dfrac{x - 2}{4} + \dfrac{x + 1}{2} = 1$

33. $.40x + .60(100 - x) = .45(100)$

34. $1.30x + .90(.50 - x) = 1.00(50)$

35. $2[x - (4 + 2x) + 3] = 2x + 2$

36. $6[x - (2 - 3x) + 1] = 4x - 6$

37. $\dfrac{5}{6}x - 2x + \dfrac{1}{3} = \dfrac{1}{3}$

38. $\dfrac{3}{4} + \dfrac{1}{5}x - \dfrac{1}{2} = \dfrac{4}{5}x$

39. $5x - (8 - x) = 2[-4 - (3 + 5x - 13)]$

40. $-[x - (4x + 2)] = 2 + (2x + 7)$

The table shows selected ordered pairs for two linear functions Y_1 and Y_2. Use the table to solve the given equation.

41. $Y_1 = Y_2$

X	Y1	Y2
0	4	16
1	5	14
2	7	12
3	7	10
4	8	8
5	9	6
6	10	4

X=0

42. $Y_1 = Y_2$

X	Y1	Y2
-9	-35	-23
-8	-32	-22
-7	-29	-21
-6	-26	-20
-5	-23	-19
-4	-20	-18
-3	-17	-17

X=-9

43. $Y_1 - Y_2 = 0$

X	Y1	Y2
0	0	3
.5	1.5	3.5
1	3	4
1.5	4.5	4.5
2	6	5
2.5	7.5	5.5
3	9	6

X=0

44. $Y_1 - Y_2 = 0$

X	Y1	Y2
5.5	8	8
6	9	8.5
6.5	10	9
7	11	9.5
7.5	12	10
8	13	10.5
8.5	14	11

X=5.5

Use the approach of Example 6 to find the approximate solution, to the nearest hundredth, of each equation in Exercises 45–50. Use the intersection-of-graphs method.

45. $4(.23x + \sqrt{5}) = \sqrt{2}x + 1$

46. $9(-.84x + \sqrt{17}) = \sqrt{6}x - 4$

47. $2\pi x + \sqrt[3]{4} = .5\pi x - \sqrt{28}$

48. $3\pi x - \sqrt[4]{3} = .75\pi x + \sqrt{19}$

49. $.23(\sqrt{3} + 4x) - .82(\pi x + 2.3) = 5$

50. $-.15(6 + \sqrt{2}x) + 1.4(2\pi x - 6.1) = 10$

Repeat Exercises 45–50, but use the x-intercept method of graphical solution.

51. Exercise 45

52. Exercise 46

53. Exercise 47

54. Exercise 48

55. Exercise 49

56. Exercise 50

Each equation in Exercises 57–62 is either an identity or a contradiction. Determine which of these it is using the analytic method, and then support your answer graphically, as explained in Examples 7(b) and 8(b).

57. $6(2x + 1) = 4x + 8\left(x + \dfrac{3}{4}\right)$

58. $3(x + 2) - 5(x + 2) = -2x - 4$

59. $-4[6 - (-2 + 3x)] = 21 + 12x$

60. $4[6 - (1 + 2x)] + 10x = 2(10 - 3x) + 8x$

61. $7[2 - (3 + 4x)] - 2x = 9 + 2(1 - 15x)$

62. $-3[-5 - (-9 + 2x)] = 2(3x - 1)$

63. Verify the result in Example 3 graphically.

64. The screens show two views of graphical support for the solution of the equation

$$f(x) = 3,$$

where $y_1 = f(x)$ and $y_2 = 3$, using the intersection-of-graphs method.

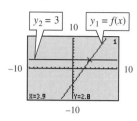

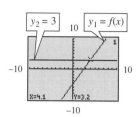

| First View | Second View |

Which one of these choices could be a possible solution for this equation? Why?

(a) 3.89   (b) 3.892   (c) 2.95   (d) 4.0

65. The figures show two views of a linear function $y = f(x)$.

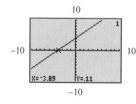

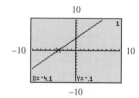

| First View | Second View |

Which one of these choices could be a possible solution for $f(x) = 0$? Why?

(a) -4.05 (b) .11   (c) $-.1$ (d) 0

66. If you had your choice to use the intersection-of-graphs method or the *x*-intercept method of graphical solution to approximate the solution of

$$\sqrt{1.7x} - 2[3x + \pi] - 4.79x = \pi[2.3x - (2x + 4)],$$

which method would you prefer? (There is no right or wrong answer here—it's just a matter of preference.)

67. If you were asked to solve

$$2x + 3 = 4x - 12$$

by the *x*-intercept method of graphical solution, why would you *not* get the correct answer by graphing $y_1 = 2x + 3 - 4x - 12$?

68. Try to solve each of these equations by the *x*-intercept method of graphical solution. What happens? What is the solution set in each case?

(a) $4x + 8 = 2(2x + 3)$ (b) $4x + 8 = 2(2x + 4)$

Use an equation-solving routine to find the solution set of each equation. Give your answer correct to the nearest hundredth.

69. $\dfrac{1}{.38}x - \sqrt{2}\,(x + \pi) = 0$

70. $-\dfrac{5}{.183}x - \sqrt[3]{4}\,(2 - x) = 0$

71. $3\pi x - \sqrt[5]{7}\,(9 - 8x) = (\pi^2)x - 3$

72. $\dfrac{1}{\pi}x + \sqrt[6]{9}\,(3 - 2x) = (\pi^3)x - 1$

1.6 SOLUTION OF LINEAR INEQUALITIES; ANALYTIC METHOD AND GRAPHICAL SUPPORT

Using Line Graphs to Illustrate Inequality ◆ Solving Linear Inequalities Analytically ◆ Graphical Support for Solutions of Linear Inequalities ◆ Three-Part Linear Inequalities

USING LINE GRAPHS TO ILLUSTRATE INEQUALITY

The graphs in Figure 76 depict how the emissions of carbon into the atmosphere rose in both Western Europe (yellow graph) and in Eastern Europe and the former USSR (red graph) from 1950 to 1990. Such emissions are important to scientists, since carbon combines with oxygen to form carbon dioxide, which is believed to cause the "greenhouse effect" and contribute to global warming.

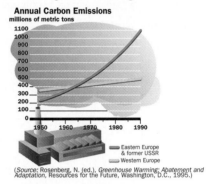

(*Source:* Rosenberg, N. (ed.), *Greenhouse Warming: Abatement and Adaptation,* Resources for the Future, Washington, D.C., 1995.)

FIGURE 76

We can use this graph to illustrate some concepts introduced thus far. For example, we see that the graphs intersect at approximately (1963, 400), meaning that in 1963, emissions were about *equal* in Western and Eastern Europe (400 million metric tons). This observation can be compared to the intersection-of-graphs method of solving equations presented in Section 1.5. Prior to 1963, the emissions in Eastern Europe and the former USSR were *less* than those of Western Europe, indicated by the fact that the red graph lies *below* the yellow graph to the left of 1963. On the other hand, emissions in Eastern Europe and the former USSR were *greater* than those of Western Europe after 1963, because for those years the red graph lies *above* the yellow graph to the right of 1963.

For Group Discussion   Both the red and yellow graphs in Figure 76 are graphs of functions. We will refer to them as $y = \text{red}(x)$ and $y = \text{yellow}(x)$ in the discussion questions that follow. Use the graphs to approximate your answers in Items 1–4.

1. What is the value of $y = \text{red}(1950)$? Interpret this in terms of annual carbon emissions.

2. What is the value of $y = \text{yellow}(1950)$?

3. For what value of x is $\text{red}(x)$ equal to 1100?

4. For what value of x is $\text{yellow}(x) = 500$?

5. Use your answers in Items 3 and 4 to support the statement that after 1963, annual carbon emissions in Eastern Europe and the former USSR were greater than those in Western Europe.

The ideas just discussed can be generalized to provide graphical support for solutions of linear inequalities.

SOLVING LINEAR INEQUALITIES ANALYTICALLY

An equation says that two expressions are equal, while an **inequality** says that one expression is greater than, greater than or equal to, less than, or less than or equal to another. As with equations, a value of the variable for which the inequality is true is a solution of the inequality, and the set of all such solutions is the solution set of the inequality. Two inequalities with the same solution set are **equivalent inequalities.**

Inequalities are solved with the following properties of inequality.

PROPERTIES OF INEQUALITY

For real numbers a, b, and c:

a. $a < b$ and $a + c < b + c$ are equivalent.
(The same number may be added to both sides of an inequality without changing the solution set.)

b. If $c > 0$, then $a < b$ and $ac < bc$ are equivalent.
(Both sides of an inequality may be multiplied by the same positive number without changing the solution set.)

c. If $c < 0$, then $a < b$ and $ac > bc$ are equivalent.
(Both sides of an inequality may be multiplied by the same negative number without changing the solution set, as long as the direction of the inequality symbol is reversed.)

Similar properties exist for $>$, $\leq$, and $\geq$.

Note Because division is defined in terms of multiplication, the word "multiplied" may be replaced by "divided" in parts (b) and (c) of the properties of inequality. Similarly, in part (a) the words "added to" may be replaced with "subtracted from."

Pay careful attention to part (c); if both sides of an inequality are multiplied by a negative number, the direction of the inequality symbol must be reversed. For example, starting with the true statement $-3 < 5$ and multiplying both sides by the positive number 2 gives

$$-3 \cdot 2 < 5 \cdot 2$$
$$-6 < 10,$$

still a true statement. On the other hand, starting with $-3 < 5$ and multiplying both sides by the *negative* number -2 gives a true result only if the direction of the inequality symbol is reversed:

$$-3(-2) > 5(-2)$$
$$6 > -10.$$

A similar situation exists when dividing both sides by a negative number. In summary, the following statement can be made.

When multiplying or dividing both sides of an inequality by a negative number, we must *reverse* the direction of the inequality symbol to obtain an equivalent inequality.

A linear inequality in one variable is defined in a way similar to a linear equation in one variable.

LINEAR INEQUALITY IN ONE VARIABLE

A **linear inequality** in one variable is an inequality that can be written in one of the following forms, where $a \neq 0$:

$$ax + b > 0 \qquad ax + b < 0 \qquad ax + b \geq 0 \qquad ax + b \leq 0.$$

We solve a linear inequality analytically using the same steps as those used to solve a linear equation.

EXAMPLE 1   *Solving a Linear Inequality Analytically*   Solve the inequality $3x - 2(2x + 4) \leq 2x + 1$. Express the solution set using interval notation.

SOLUTION

$$3x - 2(2x + 4) \leq 2x + 1$$

$$3x - 4x - 8 \leq 2x + 1 \qquad \text{Distributive property}$$

$$-x - 8 \leq 2x + 1 \qquad \text{Combine like terms.}$$

$$-3x \leq 9 \qquad \text{Subtract } 2x \text{ and add 8.}$$

$$x \geq -3 \qquad \text{Divide by } -3 \text{ and reverse the direction of the inequality symbol.}$$

The solution set is $[-3, \infty)$.   ◆

If a linear inequality involves fractions or decimals as coefficients, we use the same procedure as described in Section 1.5 to clear them: multiply both sides by the least common denominator (which, in the case of decimals, will be a power of 10).

For Group Discussion   With the class divided into two groups, have each group solve one of the following inequalities analytically.

$$\text{I} \qquad\qquad\qquad \text{II}$$

$$\frac{x + 7}{6} + \frac{2x - 8}{2} > -4 \qquad .06x + .09(15 - x) > .07(15)$$

After deciding on the correct solution for each, compare the solution set of inequality I with that in Example 2 of Section 1.5, and the solution set of inequality II with that in Example 3 of Section 1.5. What is the same in each case? How do they differ in each case?

From the result of this group discussion, we should conclude that solving a linear inequality will give us an interval whose endpoint (included if the symbol is $\leq$ or $\geq$, excluded if the symbol is $<$ or $>$) is the solution of the corresponding linear equation.

GRAPHICAL SUPPORT FOR SOLUTIONS OF LINEAR INEQUALITIES

In Section 1.5, we learned two methods of graphical support for solutions of equations. We will now extend these methods to solutions of inequalities. Suppose that two linear functions f and g are graphed, as shown in Figure 77, and the equation $f(x) = g(x)$ is conditional. Then according to the figure, the solution set is $\{x_1\}$ since, by applying the intersection-of-graphs method, it consists of the x-coordinate of the point of intersection of the two lines.

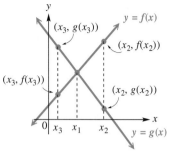

FIGURE 77

Notice that if we choose any x value *less than* x_1, such as x_3 in the figure, the point $(x_3, f(x_3))$ is *below* the point $(x_3, g(x_3))$ and on the same vertical line, indicating that $f(x_3) < g(x_3)$. Similarly, if we choose any x-value *greater than* x_1, such as x_2 in the figure, the point $(x_2, f(x_2))$ is *above* the point $(x_2, g(x_2))$, indicating that $f(x_2) > g(x_2)$. This discussion leads to the following extension of the intersection-of-graphs method of solution of equations.

INTERSECTION-OF-GRAPHS METHOD OF SOLUTION OF LINEAR INEQUALITIES

Suppose that f and g are linear functions. The solution set of $f(x) > g(x)$ is the set of all real numbers x such that the graph of f is *above* the graph of g. The solution set of $f(x) < g(x)$ is the set of all real numbers x such that the graph of f is *below* the graph of g.

If an inequality involves one of the symbols $\geq$ or $\leq$, the same method is applied, with the solution of the corresponding equation included in the solution set. This is summarized as follows.

SPECIFYING INTERVALS OF SOLUTION FOR LINEAR INEQUALITIES

If f and g are linear functions, and $f(x) = g(x)$ has a single solution, k, the solution set of

$$f(x) > g(x) \quad \text{or} \quad f(x) < g(x)$$

will be of the form (k, ∞) or $(-\infty, k)$, with the endpoint of the interval not included. On the other hand, the solution set of

$$f(x) \geq g(x) \quad \text{or} \quad f(x) \leq g(x)$$

will be of the form $[k, \infty)$ or $(-\infty, k]$, with the endpoint of the interval included.

EXAMPLE 2   *Providing Graphical Support for an Analytic Solution*   The inequality

$$3x - 2(2x + 4) \leq 2x + 1,$$

solved in Example 1, has solution set $[-3, \infty)$. Support this result graphically.

SOLUTION   Start by entering the left side as y_1 and the right side as y_2.

$$y_1 = 3x - 2(2x + 4)$$
$$y_2 = 2x + 1$$

The graph, shown in Figure 78, indicates that the point of intersection of the two lines is $(-3, -5)$. The x-coordinate, -3, gives the included endpoint of the solution set of the inequality. Because the graph of y_1 is *below* the graph of y_2 when x is *greater than* -3, the solution set of $[-3, \infty)$ is supported.   ◆

TECHNOLOGY NOTE

When supporting the solution set in Example 2 using graphs, the calculator will not determine whether the endpoint of the interval is included or excluded. This must be done by looking at the inequality symbol in the given inequality.

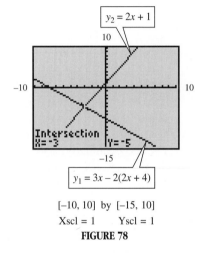

$[-10, 10]$ by $[-15, 10]$
Xscl = 1   Yscl = 1
FIGURE 78

Since there is another method of providing graphical support for the solution of a linear equation (the x-intercept method), it too can be extended to linear inequalities. Suppose that we wish to solve

$$f(x) > g(x).$$

Subtracting $g(x)$ from both sides gives

$$f(x) - g(x) > 0.$$

If we call the expression on the left side $F(x)$, we are interested in solving

$$F(x) > 0.$$

We know that $F(x) = 0$ has as its solution the x-intercept of the graph of $y = F(x)$. Therefore, all solutions of $F(x) > 0$ will be the x-values of the points *above* the point at which the graph intersects the x-axis. Similarly, all solutions of $F(x) < 0$ will be the x-values of the points *below* the point at which the graph intersects the x-axis.

TECHNOLOGY NOTE

If two functions defined by Y_1 and Y_2 are already entered into your calculator, you can enter Y_3 as $Y_2 - Y_1$. Then, if you direct the calculator to graph Y_3 only, you can solve the equation $Y_1 = Y_2$ by finding the x-intercept of Y_3. Consult your owner's manual to see how this is accomplished. It will save you a lot of time and effort.

> ### x-INTERCEPT METHOD OF SOLUTION OF LINEAR INEQUALITIES
>
> The solution set of $F(x) > 0$ is the set of all real numbers x such that the graph of F is *above* the x-axis. The solution set of $F(x) < 0$ is the set of all real numbers x such that the graph of F is *below* the x-axis.

Figure 79 illustrates this discussion, and summarizes the solution sets for the appropriate inequalities.

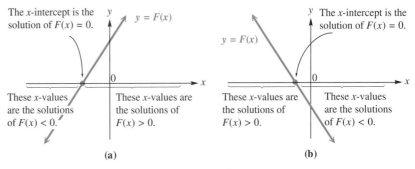

FIGURE 79

EXAMPLE 3   *Providing Graphical Support for an Analytic Solution*   Let $f(x) = -2(3x + 1)$ and $g(x) = 4(x + 2)$. If we let $y_1 = f(x)$ and $y_2 = g(x)$ and graph $y_1 - y_2$, we obtain the graph shown in Figure 80.

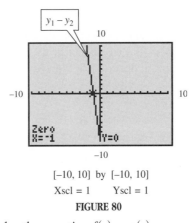

[−10, 10] by [−10, 10]

Xscl = 1   Yscl = 1

FIGURE 80

(a)   Use the graph to solve the *equation* $f(x) = g(x)$.

(b)   Use the graph to solve the *inequality* $f(x) < g(x)$.

(c)   Use the graph to solve the *inequality* $f(x) > g(x)$.

(d)   Use the results of parts (b) and (c) to express the solution sets of $f(x) \le g(x)$ and $f(x) \ge g(x)$.

SOLUTION

(a)   The *x*-intercept (identified as "Zero" on the screen) of $y_1 - y_2$ is -1, so the solution set of this *equation* is $\{-1\}$.

(b)   Because the symbol is $<$, we want to find the *x*-values of the points *below* the *x*-axis. All such points are to the *right* of $x = -1$, leading to the solution set $(-1, \infty)$.

(c)   Because the symbol is $>$, this time we choose the *x*-values of the points *above* the *x*-axis. All such points are to the *left* of $x = -1$, leading to the solution set $(-\infty, -1)$.

(d)   Notice here that the only difference is that we now have the symbols that include equality. Therefore, the solution sets are $[-1, \infty)$ and $(-\infty, -1]$, respectively.

◆

Either of the graphical methods of supporting the analytic solution of inequalities may be used to approximate the solutions of an inequality that involves complicated expressions or "messy" numbers. In Example 6 of Section 1.5, we solved such an equation using the intersection-of-graphs method. Now we will apply the *x*-intercept method to one of the related inequalities.

EXAMPLE 4   *Approximating a Solution of a Linear Inequality Graphically*
Use the *x*-intercept method to find the solution set of the inequality below. Express the endpoint correct to the nearest hundredth.

$$.51(\sqrt{2} + 3x) - .21(\pi x + 6.1) > 7$$

SOLUTION   If we let y_1 represent the left side of the inequality and y_2 represent the right side, we have

$$y_1 > y_2$$
$$y_1 - y_2 > 0.$$

We now graph $y_1 - y_2$ and look for *x*-values of the points that lie *above* the *x*-axis. Figure 81 shows this graph after using a calculator to obtain the *x*-intercept. As we would expect, the *x*-intercept of the graph is, to the nearest hundredth, 8.69, confirming our result in Section 1.5. Since the graph lies *above* the *x*-axis for *x*-values *to the right* of 8.69, we express the solution set as the interval (8.69, ∞). ◆

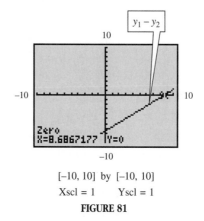

[–10, 10] by [–10, 10]
Xscl = 1   Yscl = 1
FIGURE 81

In Example 4, the endpoint we gave in the solution set was an approximation, to the nearest hundredth. In solving inequalities graphically, it may happen that, due to the approximation process, the appropriate symbol for the endpoint, a parenthesis or a bracket, may not actually be valid. However, we will state the following agreement, to be used throughout this text.

AGREEMENT ON INCLUSION OR EXCLUSION OF ENDPOINTS FOR APPROXIMATIONS

When an approximation is used for an endpoint in specifying an interval, we will continue to use parentheses in specifying inequalities involving < or >, and square brackets in specifying inequalities involving ≤ or ≥.

THREE-PART LINEAR INEQUALITIES

Let us now consider a linear inequality that is actually made up of two separate linear inequalities:

$$-2 < 5 + 3x < 20.$$

The solution set of this three-part inequality consists of all real numbers that make $5 + 3x$ lie in the interval $(-2, 20)$. Therefore, we are seeking the numbers that are solutions of the two inequalities

$$-2 < 5 + 3x \qquad \text{and} \qquad 5 + 3x < 20$$

at the same time. Rather than solve each one separately and then take the intersection of their solution sets, we may find the solution set more quickly by working with all three expressions at the same time, as shown in the following example.

EXAMPLE 5   *Solving a Three-Part Inequality Analytically and Supporting the Solution Graphically*   Solve the three-part inequality

$$-2 < 5 + 3x < 20$$

analytically. Then, support the solution graphically.

SOLUTION   ANALYTIC   To begin, work will all three expressions at the same time.

$-2 < 5 + 3x < 20$	Given inequality
$-7 < 3x < 15$	Subtract 5 from each expression.
$\dfrac{-7}{3} < x < 5$	Divide each expression by 3.

The open interval $\left(-\frac{7}{3}, 5\right)$ is the solution set of the inequality, as determined by analytic methods.

GRAPHICAL   To support this solution graphically, we graph

$$y_1 = -2,$$
$$y_2 = 5 + 3x,$$

and

$$y_3 = 20$$

in the viewing window $[-20, 6]$ by $[-20, 25]$. See the two views in Figure 82. The x-values of the points of intersection are $-\frac{7}{3} = -2.\overline{3}$, and 5, confirming that our analytic work is correct. Notice how the slanted line, y_2, lies *between* the graphs of $y_1 = -2$ and $y_3 = 20$ for x-values between $-\frac{7}{3}$ and 5.   ◆

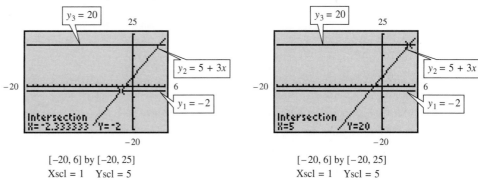

First View	Second View
$[-20, 6]$ by $[-20, 25]$	$[-20, 6]$ by $[-20, 25]$
Xscl = 1 Yscl = 5	Xscl = 1 Yscl = 5

FIGURE 82

EXAMPLE 6 *Using Parentheses and Brackets Appropriately* Use the analytic solution in Example 5 to express the solution set of each three-part inequality using interval notation.

(a) $-2 \le 5 + 3x < 20$

(b) $-2 \le 5 + 3x \le 20$

SOLUTION

(a) Because the symbol at the left is $\le$, we use a square bracket, and because the symbol at the right is $<$, we use a parenthesis. The solution set is the half-open interval $\left[-\frac{7}{3}, 5\right)$.

(b) The solution set is the closed interval $\left[-\frac{7}{3}, 5\right]$. ◆

1.6 EXERCISES   Tape 2

Beverage Can Production The line graphs depict the annual beverage can production in the United States from 1965 to 1990. Use interval notation to represent the years that satisfy each description. Include both endpoints of the interval.

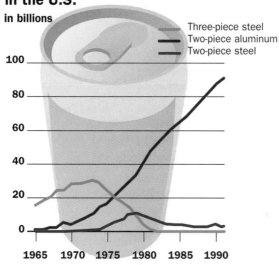

Annual Beverage Can Production in the U.S.

(*Source: Scientific American*, September 1994.)

1. Production of three-piece steel cans is greater than or equal to production of two-piece aluminum cans.

2. Production of three-piece steel cans is less than or equal to production of two-piece aluminum cans.

3. Production of two-piece steel cans is less than or equal to production of three-piece steel cans.

4. Production of two-piece steel cans is greater than or equal to production of three-piece steel cans.

5. Production of two-piece steel cans is less than or equal to production of two-piece aluminum cans.

6. Production of two-piece steel cans is greater than or equal to production of two-piece aluminum cans.

Refer to the graphs of the linear functions $y_1 = f(x)$ and $y_2 = g(x)$ in the figure to find the solution set of each equation or inequality in Exercises 7–18. (Remember that the solution sets consist of x-values.)

7. $f(x) = g(x)$

8. $y_1 = y_2$

9. $f(x) > g(x)$

10. $y_1 > y_2$

11. $y_1 < y_2$

12. $f(x) < g(x)$

13. $g(x) \geq f(x)$

14. $y_2 \geq y_1$

15. $f(x) - g(x) = 0$

16. $y_1 - y_2 = 0$

17. $g(x) - f(x) = 0$

18. $y_2 - y_1 = 0$

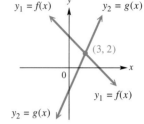

 19. Explain why the solution sets in Exercises 7 and 15 are the same.

 20. Explain why the solution sets in Exercises 16 and 18 are the same.

In Exercises 21–24, refer to the graph of the linear function $y = f(x)$ to solve the inequalities specified in parts (a)–(d). Express solution sets in interval notation.

21. (a) $f(x) > 0$

    (b) $f(x) < 0$

    (c) $f(x) \geq 0$

    (d) $f(x) \leq 0$

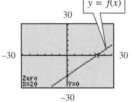

22. (a) $f(x) < 0$

    (b) $f(x) \leq 0$

    (c) $f(x) \geq 0$

    (d) $f(x) > 0$

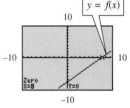

23. (a) $f(x) \leq 0$

    (b) $f(x) > 0$

    (c) $f(x) < 0$

    (d) $f(x) \geq 0$

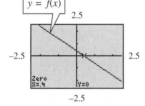

24. (a) $f(x) < 0$

    (b) $f(x) > 0$

    (c) $f(x) \leq 0$

    (d) $f(x) \geq 0$

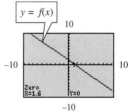

In Exercises 25 and 26, f and g are linear functions.

25. If the solution set of $f(x) \geq g(x)$ is $[4, \infty)$, what is the solution set of each?

    (a) $f(x) = g(x)$ (b) $f(x) > g(x)$ (c) $f(x) < g(x)$

26. If the solution set of $f(x) < g(x)$ is $(-\infty, 3)$, what is the solution set of each?

    (a) $f(x) = g(x)$ (b) $f(x) \geq g(x)$ (c) $f(x) \leq g(x)$

In the given screen, Y_1 and Y_2 are defined as linear functions. Give the solution set of the inequality based on the table of values shown. (Remember that both Y_1 and Y_2 have domain $(-\infty, \infty)$.)

27. (a) $Y_1 > Y_2$

    (b) $Y_1 \geq Y_2$

    (c) $Y_1 < Y_2$

    (d) $Y_1 \leq Y_2$

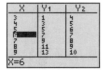

28. (a) $Y_1 > Y_2$

    (b) $Y_1 \geq Y_2$

    (c) $Y_1 < Y_2$

    (d) $Y_1 \leq Y_2$

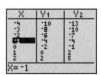

Solve each inequality analytically, giving the solution set using interval notation. Support your answer graphically. (Hint: Once part (a) is done, the answer to part (b) follows from the answer to part (a).)

29. (a) $10x + 5 - 7x \geq 8(x + 2) + 4$

    (b) $10x + 5 - 7x < 8(x + 2) + 4$

30. (a) $6x + 2 + 10x > -2(2x + 4) + 10$

    (b) $6x + 2 + 10x \leq -2(2x + 4) + 10$

31. (a) $x + 2(-x + 4) - 3(x + 5) < -4$

    (b) $x + 2(-x + 4) - 3(x + 5) \geq -4$

32. (a) $-11x - (6x - 4) + 5 - 3x \leq 1$

    (b) $-11x - (6x - 4) + 5 - 3x > 1$

33. (a) $\frac{1}{3}x - \frac{1}{5}x \leq 2$

    (b) $\frac{1}{3}x - \frac{1}{5}x > 2$

34. (a) $\frac{3x}{2} + \frac{4x}{7} \geq -5$

    (b) $\frac{3x}{2} + \frac{4x}{7} < -5$

35. (a) $\frac{x - 2}{2} - \frac{x + 6}{3} > -4$

    (b) $\frac{x - 2}{2} - \frac{x + 6}{3} \leq -4$

36. (a) $\frac{2x + 3}{5} - \frac{3x - 1}{2} < \frac{4x + 7}{2}$

    (b) $\frac{2x + 3}{5} - \frac{3x - 1}{2} \geq \frac{4x + 7}{2}$

37. (a) $.6x - 2(.5x + .2) \leq .4 - .3x$

    (b) $.6x - 2(.5x + .2) > .4 - .3x$

38. (a) $-.9x - (.5 + .1x) \leq -.3x - .5$

    (b) $-.9x - (.5 + .1x) > -.3x - .5$

39. (a) $-\frac{1}{2}x + .7x > 5$

    (b) $-\frac{1}{2}x + .7x \leq 5$

40. (a) $\frac{3}{4}x - .2x > 6$

    (b) $\frac{3}{4}x - .2x \leq 6$

Use the approach of Example 4 to find the solution set of each inequality in Exercises 41–46. Give the solution set using interval notation, with endpoint rounded to the nearest hundredth. Use the x-intercept method.

41. $4(.28x + \sqrt{6}) - \sqrt{2}x > 1$

42. $9(-.78x + \sqrt{12}) - \sqrt{7}x > -4$

43. $2.3\pi + \sqrt[3]{7} \leq .6\pi x - \sqrt{21}$

44. $\frac{3}{7}\pi - \sqrt[3]{9} \leq \frac{2}{7}\pi x - \sqrt{23}$

45. $.29(\sqrt{6} + 5x) < .74(\pi x + 4.1) - 6$

46. $-.28(\sqrt{7} + 3x) < .89(2\pi x + 4.2) + 10$

Refer to the graphs of y_1, y_2, and y_3 in the two views of the figure shown here. Assume that the lines for y_1 and y_2 are parallel. Give the solution set for each equation or inequality in Exercises 47–55. (Remember that solution sets consist of x-values.)

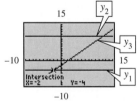 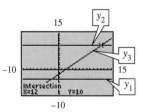

First View   Second View

47. $y_3 < y_2$ **48.** $y_1 < y_3$ **49.** $y_3 \geq y_2$ **50.** $y_1 \geq y_3$ **51.** $y_1 < y_3 < y_2$ **52.** $y_2 = y_3$

53. $y_2 = y_1$ (The graphs are parallel lines.)

54. $y_2 > y_1$

55. $y_2 \leq y_1$

56. Draw a sketch indicating the following: "The functions $y_1 = f(x)$ and $y_2 = g(x)$ are both linear, and the solution set of $f(x) < g(x)$ is $(-2, \infty)$. Furthermore, $f(-2) = g(-2) = 6$."

Solve each of the following three-part inequalities analytically. Support your answer graphically, as explained in Example 5.

57. $4 \leq 2x + 2 \leq 10$

58. $-4 \leq 2x - 1 \leq 5$

59. $-10 > 3x + 2 > -16$

60. $4 > 6x + 5 > -1$

61. $-3 \leq \dfrac{x-4}{-5} < 4$

62. $1 < \dfrac{4x-5}{-2} < 9$

If two linear functions y_1 and y_2 do not have a single point in common, their graphs will either be parallel (Figure A) or will coincide (Figure B). In these cases, the solution set of an inequality of one of the forms

$$y_1 < y_2 \qquad y_1 \leq y_2 \qquad y_1 > y_2 \qquad y_1 \geq y_2$$

will either be $\emptyset$ or $(-\infty, \infty)$.

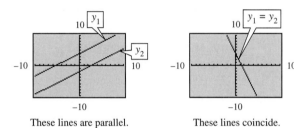

These lines are parallel.   These lines coincide.

A   B

Use Figure A to find the solution set of each inequality in Exercises 63–66.

63. $y_1 > y_2$

64. $y_1 \geq y_2$

65. $y_1 \leq y_2$

66. $y_1 < y_2$

Use Figure B to find the solution set of each inequality in Exercises 67–70.

67. $y_1 > y_2$

68. $y_1 \geq y_2$

69. $y_1 \leq y_2$

70. $y_1 < y_2$

71. Graph the function $f(x) = 3x - 18$ so that a comprehensive graph is shown, and then use the capabilities of your calculator to locate the x-intercept.

    (a) What is the x-intercept? What is the solution set of $f(x) = 0$?

    (b) Now trace to the right of the x-intercept. What do you notice about every y-value (in comparison to 0)? What is the solution set of $f(x) > 0$?

    (c) Return to the x-intercept, and then trace to the left of it. What do you notice about every y-value (in comparison to 0)? What is the solution set of $f(x) < 0$?

72. Sketch the graph of a linear function $y = f(x)$ satisfying these conditions: the x-intercept is 3, the graph passes through the point $(4, 7)$. Now, use the graph to solve each of the following.

    (a) $f(x) = 0$

    (b) $f(x) \geq 0$

    (c) $f(x) \leq 0$

    (d) $f(x) = 12$ (*Hint:* You will have to use the methods of Section 1.4 to find the equation that defines f.)

1.7   LINEAR MODELS

An Example of a Linear Model   ◆   Finding a Least Squares Regression Line   ◆   The Correlation Coefficient   ◆   The Celsius/Fahrenheit Relationship

AN EXAMPLE OF A LINEAR MODEL

The advent of graphing calculators provides us with opportunities to study many kinds of interesting problems that were previously too difficult to solve in college algebra. One of the most important types of problems that we can now study is that of fitting graphs to data. This is an example of a topic in statistics called **regression.** In this section, we will see how it is possible to fit lines to some sets of data. This is called **linear modeling.**

Based on statistics from a *USA Today* article, the table gives the total amounts (in millions of dollars) of government-guaranteed student loans from 1986 through 1994. See also Figure 83.

Year	Amount
1986	8.6
1987	9.8
1988	11.8
1989	12.5
1990	12.3
1991	13.5
1992	14.7
1993	16.5
1994	18.2

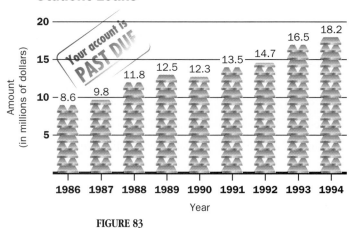

FIGURE 83

Now, let us agree that $x = 0$ will represent the year 1986, $x = 1$ will represent 1987, and so on, up to $x = 8$ representing 1994. Then the data can be represented by the following function:

$$\{(0, 8.6), (1, 9.8), (2, 11.8), (3, 12.5), (4, 12.3), (5, 13.5), (6, 14.7),$$
$$(7, 16.5), (8, 18.2)\}.$$

This function can be plotted by entering the domain values in list L_1, the corresponding range values in list L_2, and graphing in the window $[-1, 10]$ by $[-1, 20]$. See Figure 84.

These data lie approximately in a line. We can find the equation of a line that fits these data fairly well by choosing two representative points and using the methods presented earlier to find the equation.

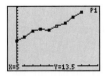

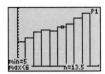

The TI-83 calculator allows the user to plot data using *discrete points* as in Figure 84, with *line segments* joining the points (see top screen), or as a *histogram* (see bottom screen).

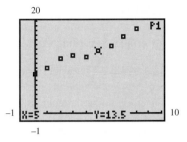

[−1, 10] by [−1, 20]
Xscl = 1   Yscl = 1
The display at the bottom indicates that for X = 5 (the year 1991), Y=13.5 (the amount $13.5 million).

FIGURE 84

EXAMPLE 1   *Finding a Line to Fit Student Loan Data by Choosing Two Data Points*   Use the data points (1, 9.8) and (7, 16.5) from the student loan example to find the equation of a line that approximates these data. Graph the line and the data points together using a graphing calculator.

SOLUTION   First find the slope, using the slope formula.

$$m = \frac{16.5 - 9.8}{7 - 1} = \frac{6.7}{6} = \frac{67}{60}$$

Now, use this slope with the point (1, 9.8).

$$y - 9.8 = \frac{67}{60}(x - 1)$$

$$y - 9.8 = \frac{67}{60}x - \frac{67}{60}$$

$$y = \frac{67}{60}x + \frac{521}{60}$$

$$y = 1.11\overline{6}x + 8.68\overline{3}$$

Enter this equation and graph along with the data points to get the screen shown in Figure 85. Notice that the line provides a fairly accurate fit.   ◆

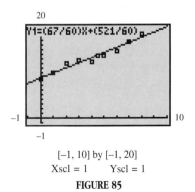

[−1, 10] by [−1, 20]
Xscl = 1   Yscl = 1

FIGURE 85

EXAMPLE 2   *Using a Linear Model to Predict a Student Loan Amount*   Use the linear model found in Example 1 to approximate the amount (in millions of dollars) of government-guaranteed student loans in 1991.

SOLUTION   Using the model equation $y = 1.11\overline{6}x + 8.68\overline{3}$, we substitute 5 for x (since $x = 5$ corresponds to 1991) to get $y \approx 14.3$. In 1991, the amount was actually 13.5 million dollars, meaning that this model is off by about .8 million dollars.   ◆

FINDING A LEAST SQUARES REGRESSION LINE

The method used to find the model in Example 1 used only simple algebraic concepts presented earlier in this chapter. It is reasonable to expect that a method exists to find the line of "best fit." Graphing calculators are capable of finding this regression line. This method uses a technique from statistics known as **least squares regression.** In the formulas stated below, the Greek letter sigma (Σ) represents the summation of all terms following it; for example, Σx represents the sum of all x-values, Σxy represents the sum of the products of the corresponding x- and y-values, and so on.

REGRESSION COEFFICIENT FORMULAS

The **least squares line** $y = ax + b$ that provides the best fit to the data points $(x_1, y_1), (x_2, y_2), \ldots, (x_n, y_n)$ has

$$a = \frac{n(\Sigma xy) - (\Sigma x)(\Sigma y)}{n(\Sigma x^2) - (\Sigma x)^2}$$

and

$$b = \frac{\Sigma y - a(\Sigma x)}{n}.$$

EXAMPLE 3   *Finding the Least Squares Line*

(a)   Use the regression coefficient formulas to find the line of best fit to the data points just preceding Example 1.

(b)   Graph this line and the data points using a calculator with the same window as the one shown in Figure 84.

(c)   Use a graphing calculator with regression line capability to support your result from part (a).

SOLUTION

(a)   ANALYTIC   The chart that follows summarizes the necessary information, as well as the sums.

x	y	xy	x^2
0	8.6	0	0
1	9.8	9.8	1
2	11.8	23.6	4
3	12.5	37.5	9
4	12.3	49.2	16
5	13.5	67.5	25
6	14.7	88.2	36
7	16.5	115.5	49
8	18.2	145.6	64
Sums: 36	117.9	536.9	204

From the chart,

$$\Sigma x = 36, \qquad \Sigma y = 117.9, \qquad \Sigma xy = 536.9, \qquad \text{and} \qquad \Sigma x^2 = 204.$$

Also, $n = 9$ since there are 9 pairs of values. Now find a with the formula given above.

$$a = \frac{9(536.9) - 36(117.9)}{9(204) - 36^2}$$
$$= 1.088\overline{3}$$

Finally, use this value of a to find b.

$$b = \frac{117.9 - 1.088\overline{3}(36)}{9}$$
$$= 8.74\overline{6}$$

Therefore, the line of best fit has the equation $y \approx 1.08833x + 8.7467$.

(b)   Figure 86 shows the line of best fit and the data points plotted together.

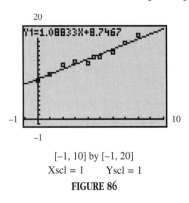

[−1, 10] by [−1, 20]
Xscl = 1   Yscl = 1
FIGURE 86

(c)   GRAPHICAL   Figure 87 shows how a TI-83 calculator computes the same regression line as the one found in part (a). Except for inputting the information and pressing the correct keys, the calculator does all the work.   ◆

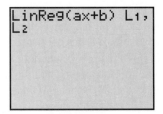
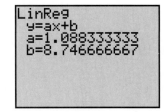

FIGURE 87

The slope of the regression line, $1.088\overline{3}$, represents the average rate of change, meaning that the amount of guaranteed student loans increased about 1.09 million dollars per year (on the *average*). The *y*-intercept, $8.74\overline{6}$, indicates the approximate amount for 1986, where $x = 0$. Compare this to the actual amount, 8.6, and you will see how close this is.

For Group Discussion

1. Use the least squares regression line found in Example 3(a) to approximate the amount of guaranteed student loans in 1992. Do this by entering the equation as a *y*-variable, and use various methods such as function notation, table construction, and locating a point on the line. (*Hint:* In all cases, $x = 6$.)

2. Suppose we make the assumption that the model continues to hold in 1995. What would be the amount of loans that year?

3. Discuss a drawback of the prediction made in Item 2.

THE CORRELATION COEFFICIENT

Once an equation for the line of best fit (the least squares line) has been found, it is reasonable to ask, "Just how good is this line for predictive purposes?" If the points already observed fit the line quite closely, then future pairs of points can be expected to do so. If the points are widely scattered about even the "best-fitting" line, then predictions are not likely to be accurate. In general, the closer the data points lie to the least squares line, the more likely it is that the entire *population* of (x, y) points really do form a line, that is, that *x* and *y* really are related linearly. Also, the better the fit, the more confidence we can have that our least squares line is a good estimator of the true population line.

One common measure of the strength of the linear relationship in the sample is called the **correlation coefficient,** denoted *r*. It is calculated from the data according to the following formula.

CORRELATION COEFFICIENT FORMULA

In linear regression, the strength of the linear relationship is measured by the correlation coefficient

$$r = \frac{n(\Sigma xy) - (\Sigma x)(\Sigma y)}{\sqrt{n(\Sigma x^2) - (\Sigma x)^2} \cdot \sqrt{n(\Sigma y^2) - (\Sigma y)^2}}.$$

The value of *r* is always between -1 and 1, or perhaps equal to -1 or 1. Values of exactly 1 or -1 indicate that the least squares line goes exactly through all the data points. If *r* is close to 1 or -1, but not exactly equal, then the line comes "close" to fitting through all the data points, and we say that the linear correlation between *x* and *y* is "strong." If *r* is equal, or nearly equal, to 0, there is no linear correlation, or the correlation is weak. This would mean that the points are a totally disordered conglomeration. Or it may be that the points form an ordered pattern but one which is not linear. Other types of curve fitting will be discussed in later chapters.

Figure 88 shows examples of data points and interpretations of correlation.

EXAMPLE 4   *Finding the Correlation Coefficient*

(a)   Use the correlation coefficient formula to find *r* for the data used in Example 1.

(b)   Use a graphing calculator to support the result in part (a).

SOLUTION

(a)   ANALYTIC   Almost all values needed to find *r* were computed in Example 3.

$$n = 9 \qquad \Sigma x^2 = 204 \qquad \Sigma xy = 536.9$$
$$\Sigma y = 117.9 \qquad \Sigma x = 36$$

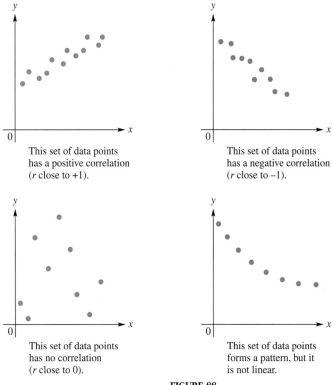

FIGURE 88

The only missing value is Σy^2. Squaring each y in the original data and adding the squares gives

$$\Sigma y^2 = 1618.61.$$

Now use the formula to find r.

$$r = \frac{9(536.9) - (36)(117.9)}{\sqrt{9(204) - 36^2} \cdot \sqrt{9(1618.61) - (117.9)^2}}$$

$$r \approx .9791964961$$

This value of r indicates a high positive correlation, because it is so close to $+1$.

(b)   GRAPHICAL   Figure 89 shows how a TI-83 calculator displays the correlation coefficient r. It supports the result in part (a).   ◆

TECHNOLOGY NOTE

The involved calculations in Examples 3 and 4 were shown primarily to illustrate the complexities of the computation. Students should learn how their individual calculators compute regression lines and correlation coefficients. See the owner's manual.

FIGURE 89

THE CELSIUS/FAHRENHEIT RELATIONSHIP

A classic, well-known example of a linear model is found in the equation that relates the Celsius and Fahrenheit temperature scales. In the following example, we derive the linear function, or model, that relates these two temperature scales.

EXAMPLE 5   *Determining the Linear Function (Model) that Relates Celsius and Fahrenheit*   There is a linear relationship between the Celsius and Fahrenheit temperature scales. When C = 0°, F = 32°, and when C = 100°, F = 212°. Use this information to express F as a function of C.

SOLUTION   Think of ordered pairs of temperatures (C, F), where C and F represent corresponding Celsius and Fahrenheit temperatures. The equation that relates the two scales has a straight-line graph that contains the points (0, 32) and (100, 212). The slope of this line can be found by using the slope formula.

$$m = \frac{212 - 32}{100 - 0} = \frac{180}{100} = \frac{9}{5}$$

Now, think of the point-slope form of the equation in terms of C and F, where C replaces x and F replaces y. Use $m = \frac{9}{5}$, and $(C_1, F_1) = (0, 32)$.

$$F - F_1 = m(C - C_1)$$

$$F - 32 = \frac{9}{5}(C - 0) \qquad F_1 = 32, \, m = \frac{9}{5}, \, C_1 = 0$$

$$F - 32 = \frac{9}{5}C$$

$$F = \frac{9}{5}C + 32 \qquad \text{Solve for F.}$$

This final equation expresses F as a function of C.   ◆

EXAMPLE 6   *Graphing the Celsius/Fahrenheit Equation*   Graph the function $y = \frac{9}{5}x + 32$ on a graphing calculator in the window $[-50, 50]$ by $[-35, 35]$. Let x take on the values -5, 35, and 0, and interpret the displays.

SOLUTION   Enter $\frac{9}{5}$ as 1.8. Here y represents the Fahrenheit temperature that corresponds to the Celsius temperature x, and the various displays shown in Figure 90 are interpreted as follows:

(a)   When the Celsius temperature is $-5°$, the Fahrenheit temperature is 23°.

(b)   When Celsius is 35°, Fahrenheit is 95°.

(c)   When Celsius is 0°, Fahrenheit is 32°.

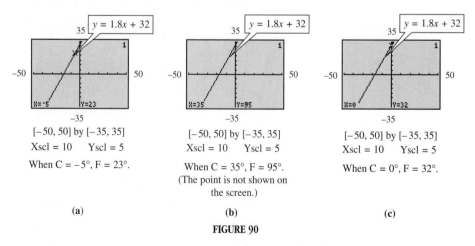

[−50, 50] by [−35, 35]
Xscl = 10   Yscl = 5
When C = −5°, F = 23°.

[−50, 50] by [−35, 35]
Xscl = 10   Yscl = 5
When C = 35°, F = 95°.
(The point is not shown on the screen.)

[−50, 50] by [−35, 35]
Xscl = 10   Yscl = 5
When C = 0°, F = 32°.

(a)   (b)   (c)

FIGURE 90

If you were to locate the point where $x = 100$, what would be the value of y?   ◆

1.7   EXERCISES    Tape 2

In the screens in Exercises 1–6, a set of data is plotted. Describe the data as (a) having a positive correlation coefficient close to +1, (b) having a negative correlation coefficient close to −1, or (c) having no linear correlation.

1.
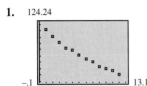

2.
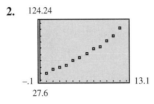

3.
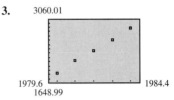

4.
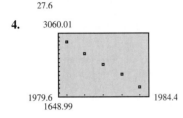

5.
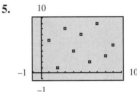

6.
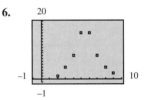

7. *Income of Residents in a Village*   Suppose that a sociologist gathers data on ten residents of a small village in a remote country in order to get an idea of how annual income in U.S. dollars relates to age in that village. She obtains the following information.

Resident	Age	Annual Income
A	19	2150
B	23	2550
C	27	3250
D	31	3150
E	36	4250
F	40	4200
G	44	4350
H	49	5000
I	52	4950
J	54	5650

(a) Plot by hand the set of data points, where x represents age and y represents income.

(b) Use your data points in part (a) to sketch an estimated regression line.

(c) Make a chart similar to the one found in Example 3(a) to organize the values of x, y, xy, and x^2. Find the sums as well.

(d) Use the regression coefficient formulas to find a and b, and give the equation of the least squares line.

(e) Use the result of part (d) to predict the income of a village resident who is 35 years old.

(f) Find Σy^2, and then use the correlation coefficient formula to find the value of r. Discuss the meaning of this value.

8. *Corn Production and Fertilization*   In a study to determine the linear relationship between the length (in decimeters) of an ear of corn (y) and the amount (in tons per acre) of fertilizer used (x), the following data were collected.

$$n = 10 \qquad \Sigma xy = 75$$
$$\Sigma x = 30 \qquad \Sigma x^2 = 100$$
$$\Sigma y = 24 \qquad \Sigma y^2 = 80$$

(a) Find an equation for the least squares line.

(b) Find the coefficient of correlation.

(c) If 3 tons per acre of fertilizer are used, what length (in decimeters) would the equation in (a) predict for an ear of corn?

9. *Celsius and Fahrenheit Temperatures*   In an experiment to determine the linear relationship between temperatures on the Celsius scale (y) and on the Fahrenheit scale (x), a student got the following results.

$$n = 5 \qquad \Sigma xy = 28{,}050$$
$$\Sigma x = 376 \qquad \Sigma x^2 = 62{,}522$$
$$\Sigma y = 120 \qquad \Sigma y^2 = 13{,}450$$

(a) Find an equation for the least squares line.

(b) Find the reading on the Celsius scale that corresponds to a reading of 120° Fahrenheit, using the equation of part (a).

(c) Find the coefficient of correlation.

10. *Heights and Weights of Men*   A sample of 10 adult men gave the following data on their heights and weights.

Height (inches) (x)	Weight (pounds) (y)
62	120
62	140
63	130
65	150
66	142
67	130
68	135
68	175
70	149
72	168

(a) Find the equation of the least squares line.

(b) Using the results of part (a), predict the weight of a man whose height is 60 inches.

(c) What would be the predicted weight of a man whose height is 70 inches?

(d) Compute the coefficient of correlation.

11. *Reading Ability and IQ Scores*   The table below gives reading ability scores and IQs for a group of 10 individuals.

Reading (x)	IQ (y)
83	120
76	104
75	98
85	115
74	87
90	127
75	90
78	110
95	134
80	119

(a) Find the equation of the least squares line.

(b) Use the regression line equation to estimate the IQ of a person with a reading score of 65.

(c) Find the correlation coefficient.

12. *Sales of a Company*   Sales, in thousands of dollars, of a certain company are shown here.

Year (x)	Sales (y)
0	48
1	59
2	66
3	75
4	80
5	90

(a) Find the equation of the least squares line.

(b) Find the coefficient of correlation.

(c) Predict the sales after 6 years.

13. *Cost of Private College Education*   The table lists the average annual cost (in dollars) of tuition and fees at private four-year colleges for selected years.

Year	Tuition and Fees
1981	4113
1983	5093
1985	6121
1987	7116
1989	8446
1991	10,017
1993	11,025

(*Source:* The College Board.)

(a) Determine a linear function defined by $f(x) = mx + b$ that models the data, using the points $(1, 4113)$ and $(13, 11{,}025)$. Graph f and the data on the same coordinate axes. What does the slope of the graph of f indicate?

(b) Use this function to approximate the tuition and fees in the year 1990. Compare it with the true value of $9340.

(c) Discuss the accuracy of using f to estimate the cost of private colleges in the years 1970 and 2010.

(d) Use a calculator to find the least squares regression line and correlation coefficient.

(e) Use the result in part (d) to predict the cost in 2001.

14. *Federal Debt*   The table lists the total federal debt (in billions of dollars) from 1985 to 1989.

Year	Federal Debt
1985	1828
1986	2130
1987	2354
1988	2615
1989	2881

(*Source:* U.S. Office of Management and Budget.)

(a) Plot the data by letting $x = 0$ correspond to 1985. Discuss any trends of the federal debt over this time period.

(b) Find a linear function defined by $f(x) = mx + b$ that approximates the data, using the points $(0, 1828)$ and $(4, 2881)$. What does the slope of the graph of f represent? Graph f and the data on the same coordinate axes.

(c) Use f to estimate the federal debt in the years 1984 and 1990. Compare your results to the true values of 1577 and 3191 billion dollars.

(d) Now use f to estimate the federal debt in the years 1980 and 1994. Compare your results to the true values of 914 and 4690 billion dollars.

(e) Use a calculator to find the least squares regression line and correlation coefficient.

(f) Use the result in part (e) to estimate the federal debt in the year 1991.

15. *Cost of Public Education*   The table lists the average annual cost (in dollars) of tuition and fees at public four-year colleges for selected years.

Year	Tuition and Fees
1981	909
1983	1148
1985	1318
1987	1537
1989	1781
1991	2137
1993	2527

(*Source:* The College Board.)

 (a) Plot the cost of public colleges data, letting $x = 0$ correspond to 1980. Is the data *exactly* linear? Could the data be *approximated* by a linear function?

(b) Determine a linear function f defined by $f(x) = mx + b$ that models the data, using the points (1, 909) and (11, 2137). Graph f and the data on the same coordinate axes. What does the slope of the graph of f indicate?

(c) Use this function to approximate the tuition and fees in the year 1984. Compare it with the true value of $1228.

(d) Use a calculator to find the least squares regression line and correlation coefficient.

(e) Use the result in part (d) to predict the tuition and fees in the year 2001.

16. *Distances and Velocities of Galaxies*   The table lists the distances (in megaparsecs) and velocities (in kilo-

meters per second) of four galaxies moving rapidly away from Earth.

Galaxy	Distance	Velocity
Virgo	15	1600
Ursa Minor	200	15,000
Corona Borealis	290	24,000
Bootes	520	40,000

(*Sources:* Acker, A., and C. Jaschek, *Astronomical Methods and Calculations,* John Wiley & Sons, 1986. Karttunen, H. (editor), *Fundamental Astronomy*, Springer-Verlag, 1994.)

(a) Plot the data using distance for the *x*-values and velocity for the *y*-values. What type of relationship seems to hold between the data?

(b) Find a linear equation in the form $y = mx$ that models these data, using the points (520, 40,000) and (0, 0). Graph your equation with the data on the same coordinate axes.

(c) The galaxy Hydra has a velocity of 60,000 km/sec. How far away is it?

(d) The value of m is called the **Hubble constant.** The Hubble constant can be used to estimate the age of the universe A (in years) using the formula $A = \dfrac{9.5 \times 10^{11}}{m}$. Approximate A using your value of m.

(e) Astronomers currently place the value of the Hubble constant between 50 and 100. What is the range for the age of the universe A?

(f) Use the data points determined by the information in the table and your calculator to find the least squares regression line.

The data in Exercises 17–20 were adapted from the 1995 Information Please Almanac. In each case, obtain the least squares regression line and the correlation coefficient. Make a statement about the correlation.

17. *City Sizes in the United States and Mayors' Salaries*

Rank	City	Population (x)	Mayor's Salary (y)
1	New York	7,311,966	$130,000
2	Los Angeles	3,489,779	123,778
3	Chicago	2,768,483	115,000
4	Houston	1,690,180	133,000
5	Philadelphia	1,552,572	110,000
6	San Diego	1,148,851	65,300
7	Dallas	1,022,497	2600
8	Phoenix	1,012,230	37,500
9	Detroit	1,012,110	117,000
10	San Antonio	966,437	3000

18. *City Sizes in the World (Population and Area)*

Rank	City	Population, x (in thousands)	Area, y (in square miles)
1	Tokyo-Yokohama, Japan	28,447	1089
2	Mexico City, Mexico	23,913	522
3	Saõ Paulo, Brazil	21,539	451
4	Seoul, South Korea	19,065	342
5	New York, United States	14,638	1274
6	Osaka-Kobe-Kyoto, Japan	14,060	495
7	Bombay, India	13,532	95
8	Calcutta, India	12,885	209
9	Rio de Janeiro, Brazil	12,788	260
10	Buenos Aires, Argentina	12,232	535

19. *Gestation Period and Life Span of Animals*

Animal	Average Gestation or Incubation Period, x (days)	Record Life Span, y (years)
Cat	63	26
Dog	63	24
Duck	28	15
Elephant	624	71
Goat	151	17
Guinea pig	68	6
Hippopotamus	240	49
Horse	336	50
Lion	108	29
Parakeet	18	12
Pig	115	22
Rabbit	31	15
Sheep	151	16

20. *Revolution, Rotation, and Distance of the Planets*   In this case, relate **(a)** period of revolution (y) to mean distance from sun (x), and **(b)** period of rotation (y) to mean distance from sun (x).

Planet	Period of Revolution (days)	Period of Rotation (hours)	Mean Distance From Sun (millions of miles)
Mercury	88	1416	36
Venus	225	5832	67
Earth	365	24	93
Mars	687	25	142
Jupiter	4329	10	484
Saturn	10,753	11	887
Uranus	30,660	17	1784
Neptune	60,225	16	2796
Pluto	90,520	153	3666

RELATING CONCEPTS (EXERCISES 21–26)

Work the following problems in sequence. They should relate some of the concepts presented so far in this chapter.

21. Consider the points $A(2, 8)$ and $B(6, 16)$. Find the slope m of line AB. (Section 1.3)

22. Use the point-slope form of the equation of a line to find the equation of line AB. Then write it in slope-intercept form. (Section 1.4)

23. Use the midpoint formula to find the midpoint of segment AB. (Section 1.4)

24. Use the x-value you found in Exercise 23, and substitute it into the equation you found in Exercise 22. What is the corresponding y-value? Does it agree with the one you found in Exercise 23 for the midpoint?

25. Use your calculator's linear regression capability to find the equation of the least squares line for the data points A and B. Does it agree with the equation you found in Exercise 22? (Section 1.7)

26. Use the statistical plot function of your calculator to plot points A and B along with the regression line. Then, show that the midpoint you found in Exercise 23 actually lies on this line. (Section 1.7)

Celsius and Fahrenheit Temperatures   Use the result of Examples 5 and 6 to answer the following questions.

27. When the Celsius temperature is 50°, what is the Fahrenheit temperature?

28. When the Celsius temperature is −40°, what is the Fahrenheit temperature?

29. When the Fahrenheit temperature is 90°, what is the Celsius temperature?

30. When the Fahrenheit temperature is −40°, what is the Celsius temperature?

1.8 OTHER APPLICATIONS OF LINEAR FUNCTIONS

Formulas ◆ Problem-Solving Strategies ◆ Applications of Linear Equations ◆ Direct Variation ◆ Break-Even Analysis ◆ Applications of Linear Inequalities ◆ Applications Involving Real Data

We will now investigate a variety of ways that linear functions, equations, and inequalities can be used to solve problems.

FORMULAS

In many applications, we use formulas that give a general relationship among several quantities in a problem situation. For example, the formula

$$A = \frac{1}{2}h(b_1 + b_2)$$

gives the area A of a trapezoid in terms of its height (h) and its two parallel bases (b_1 and b_2). (See Figure 91.) Notice that A is alone on one side of the equation. Suppose, however, that we want to have the formula arranged in such a way that it is solved for b_1. The methods of solving linear equations analytically can be adapted so that this goal is accomplished.

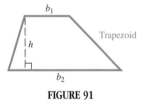

FIGURE 91

EXAMPLE 1   *Solving the Trapezoid Area Formula for a Specified Linear Variable*   Solve the formula $A = \frac{1}{2}h(b_1 + b_2)$ for b_1.

SOLUTION   We treat the equation as if b_1 were the only variable and all other variables are constants.

$$A = \frac{1}{2}h(b_1 + b_2) \qquad \text{Given formula}$$

$$2A = h(b_1 + b_2) \qquad \text{Multiply by 2.}$$

$$\frac{2A}{h} = b_1 + b_2 \qquad \text{Multiply by } \tfrac{1}{h}.$$

$$\frac{2A}{h} - b_2 = b_1 \qquad \text{Subtract } b_2.$$

$$b_1 = \frac{2A}{h} - b_2$$

An alternative method can also be applied after multiplying both sides by 2.

$$2A = h(b_1 + b_2)$$

$$2A = b_1h + b_2h \qquad \text{Distributive property; commutative property}$$

$$2A - b_2h = b_1h \qquad \text{Subtract } b_2h.$$

$$\frac{2A - b_2h}{h} = b_1 \qquad \text{Multiply by } \frac{1}{h}.$$

This is equivalent to the result found by the first method.   ◆

A list of some of the most useful *geometric* formulas can be found on the inside covers of this book. Three common formulas are given below.

SIMPLE INTEREST   MOTION   PERCENTAGE

Simple Interest   Interest = Principal × Rate × Time ($I = PRT$)
Motion   Distance = Rate × Time ($D = RT$)
Percentage   Percentage = Base × Rate ($P = BR$)

PROBLEM-SOLVING STRATEGIES

Probably the most famous study of problem-solving techniques was developed by George Polya (1888–1985), among whose many publications was the modern classic *How to Solve It*. In this book, Polya proposed a four-step process for problem solving.

POLYA'S FOUR-STEP PROCESS FOR PROBLEM SOLVING

1. **Understand the problem.** You cannot solve a problem if you do not understand what you are asked to find. The problem must be read and analyzed carefully. You will probably need to read it several times. After you have done so, ask yourself, "What must I find?"
2. **Devise a plan.** There are many ways to attack a problem and decide what plan is appropriate for the particular problem you are solving. (In this text, the plan will usually be to solve an equation or an inequality.)
3. **Carry out the plan.** Once you know how to approach the problem, carry out your plan. You may run into "dead ends" and unforeseen roadblocks, but be persistent. If you are able to solve a problem without a struggle, it isn't much of a problem, is it?
4. **Look back and check.** Check your answer to see that it is reasonable. Does it satisfy the conditions of the problem? Have you answered all the questions the problem asks? Can you solve the problem a different way and come up with the same answer?

A tool that we have to help us solve problems that George Polya did not is the technology of graphing calculators. We will use calculators whenever possible in our work, but we must remember that they will not *solve* problems for us—*we* must solve the problems by using our own ingenuity and skills, and let our calculators support our results. (Refer to the preface of this book, and read the quote from Peg Crider.)

APPLICATIONS OF LINEAR EQUATIONS

The next example illustrates an application of ratio.

EXAMPLE 2   *Determining the Dimensions of a Television Screen*   The Panasonic CinemaVision Projection television is one of a new generation of televisions that boasts a 16:9 aspect ratio technology. This means that the length of its rectangular screen is $\frac{16}{9}$ times its width. If the perimeter of the screen is 136 inches, find the length and the width of the screen.

SOLUTION   ANALYTIC   If we let x represent the width of the screen, then $\frac{16}{9}x$ can represent the length. See Figure 92. The formula for the perimeter of a rectangle is $P = 2L + 2W$. Let $P = 136$, $L = \frac{16}{9}x$, and $W = x$ in the formula, and solve the equation analytically.

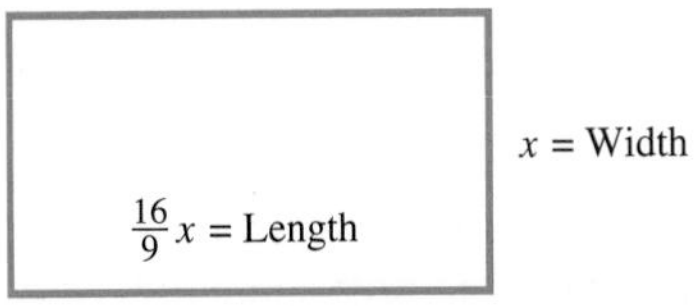

$x =$ Width

$\frac{16}{9}x =$ Length

FIGURE 92

$$136 = 2\left(\frac{16}{9}x\right) + 2x$$

$$136 = \frac{32}{9}x + 2x$$

$$136 = \frac{50}{9}x \qquad \text{Add like terms.}$$

$$x = 24.48 \qquad \text{Multiply by } \tfrac{9}{50}.$$

Since x represents the width, the width of the screen is 24.48 inches. The length is $\frac{16}{9}(24.48) = 43.52$ inches.

To check our answer analytically, find the sum of the lengths of the sides (which is the meaning of *perimeter*):

$$24.48 + 43.52 + 24.48 + 43.52 = 136 \text{ inches.}$$

GRAPHICAL   To check our answer graphically, we can use the intersection-of-graphs method, with $y_1 = 2\left(\frac{16}{9}\right)x + 2x$ and $y_2 = 136$. As seen in Figure 93, the point of intersection of the graphs is (24.48, 136). The x-coordinate supports our answer of 24.48 inches for the width.   ◆

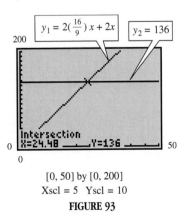

$[0, 50]$ by $[0, 200]$
Xscl = 5   Yscl = 10
FIGURE 93

Note   Our checks provide support only for our graphical and analytic work. If the problem is not set up correctly, a check is worthless.

For Group Discussion   Observe Figure 93, and disregard the graph of $y_2 = 136$. What remains is the graph of a linear function that gives the perimeter y of all rectangles satisfying the 16 by 9 aspect ratio as a *function* of its width x. Trace along the graph, back and forth, and describe what the display at the bottom is actually giving us. Why would nonpositive values of x be meaningless here? If such a television has a perimeter of 117 inches, approximately what would be its width and length?

  Television screens are usually advertised by their diagonal measure. The Panasonic model described in Example 2 is advertised as a "50-inch model." How can we further support our answer in Example 2? (*Hint:* Think Greek.)

EXAMPLE 3   *Solving a Mixture-of-Concentrations Problem*   How much pure alcohol should be added to 20 liters of a mixture that is 40% alcohol to increase the concentration to 50% alcohol?

SOLUTION   ANALYTIC   Since we are looking for the amount of pure alcohol, let x represent the number of liters of pure alcohol that must be added. The information can now be summarized in a "box diagram" as shown in Figure 94.

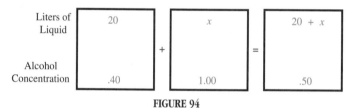

FIGURE 94

  In each box, we have the number of liters and the alcohol concentration. Using the formula $P = BR$ (Percentage = Base × Rate), we multiply the two items in each box to get the amount of pure alcohol in each case. The amount of pure alcohol on the left must equal the amount of pure alcohol on the right, so the equation to solve is

$$.40(20) \qquad + 1.00x \quad = \quad .50(20 + x).$$

Liters of pure alcohol   Liters of   Liters of pure
in starting mixture   pure   alcohol in final mixture
                          alcohol added

Now, solve this equation analytically.

$$40(20) + 100x = 50(20 + x) \qquad \text{Multiply by 100.}$$
$$800 + 100x = 1000 + 50x \qquad \text{Distributive property}$$
$$50x = 200 \qquad \text{Subtract } 50x \text{ and subtract 800.}$$
$$x = 4 \qquad \text{Divide by 50.}$$

  Therefore, 4 liters of pure alcohol must be added. We can check this solution analytically by direct substitution and/or graphically, using either method presented earlier.

  GRAPHICAL   Figure 95 shows the graph of $y_1 - y_2$, where $y_1 = .40(20) + 1.00x$ and $y_2 = .50(20 + x)$. Using the x-intercept method of graphical solution, our answer of 4 liters is supported.   ◆

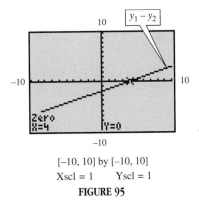

[−10, 10] by [−10, 10]
Xscl = 1   Yscl = 1
FIGURE 95

Caution   Just because we have correctly solved the *equation* that we have set up, it does not necessarily mean that we have solved the *problem*. What if our equation is set up incorrectly? This is why we should always check that our answer is reasonable in the context of the problem as stated.

DIRECT VARIATION

A common application involving linear functions deals with quantities that vary directly (or are in direct proportion). A formal definition of direct variation follows.

DIRECT VARIATION

A number y varies directly with x if there exists a nonzero number k such that

$$y = kx.$$

The number k is called the **constant of variation.**

Notice that the graph of $y = kx$ is simply a straight line with slope k, passing through the origin. See Figure 96.

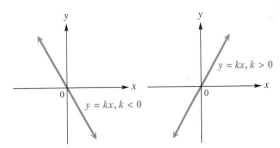

FIGURE 96

If we divide both sides of $y = kx$ by x, we get $\frac{y}{x} = k$, indicating that in a direct variation, the quotient (or proportion) of the two quantities is constant.

EXAMPLE 4   *Solving a Direct Variation Problem (Hooke's Law)*   Hooke's Law for an elastic spring states that the distance (y) a spring stretches varies directly with the force (x) applied. If a force of 15 pounds stretches a spring 8 inches, how much will a force of 35 pounds stretch the spring? Give the answer correct to the nearest unit.

SOLUTION ANALYTIC   Since there is direct variation here, we know that $y = kx$ for some number k. Using the fact that when $x = 15$, $y = 8$, we can determine the value of k.

$$y = kx$$
$$8 = k \cdot 15 \qquad \text{Let } x = 15, y = 8.$$
$$k = \frac{8}{15} \qquad \text{Divide by 15.}$$

Therefore, for this spring we have the linear function

$$y = \frac{8}{15}x$$

describing the relationship between the force (x) and the distance stretched (y). To answer the question of the problem, we let $x = 35$ in the equation.

$$y = \frac{8}{15}(35) \qquad \text{Let } x = 35.$$
$$y = \frac{56}{3} = 18.\overline{6},$$

or approximately 19 inches (to the nearest unit).

GRAPHICAL   Using a graphing calculator and locating the point where $x = 35$, we see $y = 18.\overline{6}$, supporting our solution. See Figure 97.   ◆

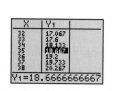

This table also supports the result of Example 4. Here Y_1 is defined as $\frac{8}{15} X$.

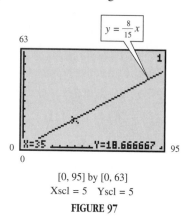

$$y = \frac{8}{15}x$$

[0, 95] by [0, 63]
Xscl = 5   Yscl = 5
FIGURE 97

BREAK-EVEN ANALYSIS

By expressing a company's cost of producing a product and the revenue from selling the product as linear functions, the company can determine at what point it will break even. In other words, we try to answer the question, "For what number of items sold will the revenue collected equal the cost of producing those items?"

EXAMPLE 5   *Determining the Break-Even Point in Business*   Peripheral Visions, Inc., produces studio-quality audiotapes of live concerts. The company places an ad in a trade newsletter. The cost of the ad is $100. Each tape costs $20 to produce, and the company charges $24 per tape.

(a)   Express the cost C as a function of x, the number of tapes produced.

(b) Express the revenue R as a function of x, the number of tapes sold.

(c) For what value of x does revenue equal cost?

(d) Graph $y_1 = 20x + 100$ and $y_2 = 24x$ in an appropriate window to support the answer in part (c).

(e) Use a table to support the answer in part (c).

SOLUTION

(a) The *fixed cost* is $100, and for each tape produced, the *variable cost* is $20. Therefore, the cost C can be expressed as a function of x, the number of tapes produced:

$$C(x) = 20x + 100 \quad (C \text{ in dollars}).$$

(b) Since each tape sells for $24, the revenue R is given by $R(x) = 24x$ (R in dollars).

(c) ANALYTIC The company will just break even (no profit and no loss) as long as revenue just equals cost, or $R(x) = C(x)$. This is true whenever

$$R(x) = C(x)$$
$$24x = 20x + 100 \qquad \text{Substitute for } R(x) \text{ and } C(x).$$
$$4x = 100$$
$$x = 25.$$

When 25 tapes are sold, the company will break even.

(d) GRAPHICAL A graphing calculator gives the graphs shown in Figure 98. Locating the point where the lines intersect, we find $x = 25$, confirming our solution. The y-value there, 600, indicates that when 25 tapes are sold, both the cost and the revenue are $600. Verify analytically that $C(25) = R(25) = 600$.

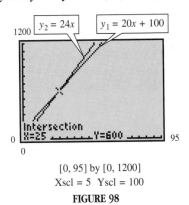

$y_2 = 24x$ $y_1 = 20x + 100$

Intersection
X=25   Y=600

$[0, 95]$ by $[0, 1200]$
Xscl = 5   Yscl = 100

FIGURE 98

(e) The table in Figure 99 shows that when the number of tapes is 25, both function values are 600, further supporting our answer in part (c). ◆

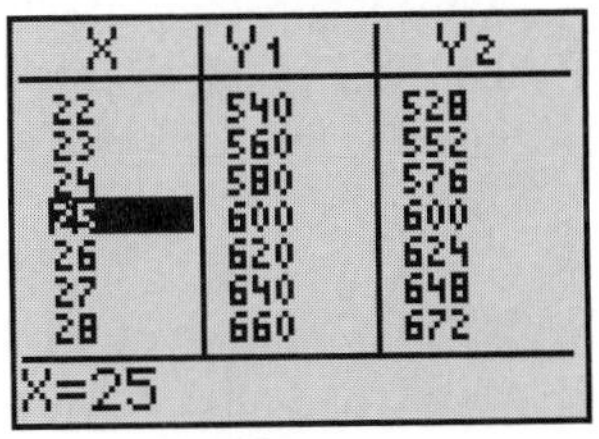

X	Y₁	Y₂
22	540	528
23	560	552
24	580	576
25	600	600
26	620	624
27	640	648
28	660	672

X=25

FIGURE 99

APPLICATIONS OF LINEAR INEQUALITIES

If we generalize the problem in Example 5, the graphs of a linear function $y_1 = R(x)$ representing revenue taken in and a linear function $y_2 = C(x)$ representing cost might look like what is seen in Figure 100.

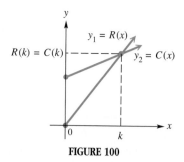

FIGURE 100

Notice that in each case, we have a function of x, where x represents the number of items in question. We choose a minimum domain value of 0, because, in practice, it makes no sense to deal with negative values of items. Now, recalling what we learned about inequalities and their graphical solutions in Section 1.6, we can make the following observations about the representation in the figure.

Observation 1: The break-even point is $(k, R(k))$, or equivalently, $(k, C(k))$. This means that when $C(x) = R(x)$, the company breaks even, selling k items and receiving $R(k)$ monetary units for these items.

Observation 2: The solution set of $R(x) < C(x)$ is $[0, k)$, indicating that when fewer than k items are produced and sold, revenue is less than cost and the company has not yet made a profit. Recall that $R(x)$ is less than $C(x)$ here, because the graph of $y_1 = R(x)$ is *below* that of $y_2 = C(x)$.

Observation 3: The solution set of $R(x) > C(x)$ is (k, ∞), indicating that when more than k items are produced and sold, revenue is greater than cost and the company is making a profit. Here, the graph of $y_1 = R(x)$ is *above* that of $y_2 = C(x)$.

EXAMPLE 6   *Determining Intervals of Loss and Profit in Business*   Refer to Figure 98 and answer the following questions.

(a)   For what numbers of tapes does the company not yet show a profit?

(b)   For what numbers of tapes does the company show a profit?

SOLUTION

(a)   The company does not show a profit when the number of tapes produced and sold is less than 25.

(b)   Since the graph of $y_2 = R(x)$ is above the graph of $y_1 = C(x)$ when x is greater than 25, the company is showing a profit if more than 25 tapes are produced and sold.   ◆

EXAMPLE 7   *Determining the Test Score Needed to Have a Certain Average*
K.B. Mello has scores of 84 and 92 on her first two tests in College Algebra with the Graphing Calculator. What score does she need on her third test in order to have an average of at least 90?

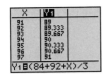

For values of X greater than or equal to 94, Y_1 is greater than or equal to 90. This supports the result of Example 7 in another way.

SOLUTION   ANALYTIC   To solve this problem analytically, we first create a linear function $A(x)$ that will give us her average in terms of her third test grade, x:

$$A(x) = \frac{84 + 92 + x}{3}.$$

(Notice that this can also be written as $A(x) = \frac{1}{3}x + \frac{176}{3}$, which is a linear function.)

Since she wants her average to be *at least* 90, we must solve $A(x) \geq 90$, or

$$\frac{84 + 92 + x}{3} \geq 90.$$

Using analytic methods described in Section 1.6, we find the solution set to be $[94, \infty)$. She must score 94 or greater to have an average of at least 90.

GRAPHICAL   To support this result graphically, we graph $y = (84 + 92 + x)/3$ and see that if we trace past the point at which $x = 94$, the y-values will be greater than 90. See Figure 101.   ◆

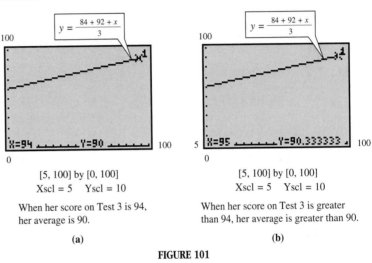

[5, 100] by [0, 100]	[5, 100] by [0, 100]
Xscl = 5 Yscl = 10	Xscl = 5 Yscl = 10
When her score on Test 3 is 94, her average is 90.	When her score on Test 3 is greater than 94, her average is greater than 90.
(a)	**(b)**

FIGURE 101

APPLICATIONS INVOLVING REAL DATA

The next example illustrates an application of linear equations using data obtained from A. Hines, T. Ghosh, S. Layalka, and R. Warder, *Indoor Air Quality & Control*, Prentice Hall, 1993.

EXAMPLE 8   *Examining the Effect of Formaldehyde on the Eyes*   Formaldehyde is a volatile organic compound that has come to be recognized as a highly toxic indoor air pollutant. Its source is found in building materials such as fiberboard, plywood, foam insulation, and carpeting. When concentrations of formaldehyde in the air exceed 33 μg/ft^3 (1 μg = 1 microgram = 10^{-6} gram), a strong odor and irritation to the eyes often occurs. One square foot of hardwood plywood paneling can emit 3365 μg of formaldehyde per day. A 4- by 8-foot sheet of this paneling is attached to an 8-foot wall in a room having dimensions of 10 by 10 feet.

(a)   How many cubic feet of air are there in the room?

(b)   Find the total number of micrograms of formaldehyde that are released into the air by the paneling each day.

(c)   If there is no ventilation in the room, write a linear equation that gives the amount of formaldehyde F in the room after x days.

(d)   How long will it take before a person's eyes become irritated in the room?

SOLUTION

(a)   The volume of the room is $8 \times 10 \times 10 = 800$ cubic feet.

(b)   The paneling releases 3365 μg for each square foot of area. The area of the sheet is 32 square feet, so it will release $32 \times 3365 = 107{,}680$ μg of formaldehyde into the air each day.

(c)   The paneling emits formaldehyde at a constant rate of 107,680 μg per day. Thus, $F = 107{,}680x$.

(d)   We must determine when concentration exceeds 33 μg/ft^3. Since the room has 800 cubic feet, this will occur when the total amount reaches $33 \times 800 = 26{,}400$ μg.

$$F = 107{,}680x = 26{,}400$$

$$x = \frac{26{,}400}{107{,}680}$$

$$x \approx .25$$

It will take approximately 1/4 day, or 6 hours.   ◆

An application that involves finding an equation from data points concludes this section.*

EXAMPLE 9   *Finding an Equation for Medicare Costs from Data Points*
Estimates for Medicare costs (in billions of dollars) are shown in the table below. The data are graphed in Figure 102. (*Source:* U.S. Office of Management and Budget.)

Year	Cost
1995	157
1996	178
1997	194
1998	211
1999	229
2000	247

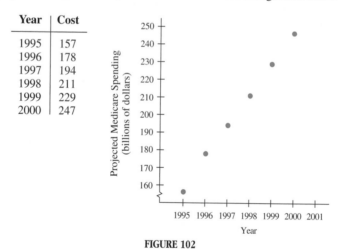

FIGURE 102

(a)   Find a linear function where $f(x)$ approximates the cost in year x.

(b)   Use $f(x)$ to predict the cost of Medicare in 2002.

SOLUTION

(a)   Start by choosing two data points that the line should go through. Suppose we choose the points (1995, 157) and (2000, 247). The slope of the line through these points is

$$\frac{247 - 157}{2000 - 1995} = 18.$$

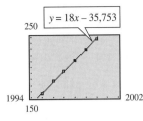

Now we use the point-slope form of the equation of a line with $(x_1, y_1) = (2000, 247)$ to get the equation.

$$y - 247 = 18(x - 2000)$$
$$y - 247 = 18x - 36{,}000$$
$$y = 18x - 35{,}753$$

Thus, $f(x) = 18x - 35{,}753$.

(b) $f(2002) = 18(2002) - 35{,}753 = 283$ billion dollars. ◆

The data in Example 9 can be analyzed using the statistics mode of a graphing calculator. The line
$y = 18x - 35{,}753$
is a "close fit" for the points as determined in the example.

1.8 EXERCISES Tape 2

Solve each formula for the specified variable.

1. $I = PRT$ for P (Simple interest)
2. $V = LWH$ for L (Volume of a box)
3. $P = 2L + 2W$ for W (Perimeter of a rectangle)
4. $P = a + b + c$ for c (Perimeter of a triangle)
5. $A = \frac{1}{2}h(b_1 + b_2)$ for h (Area of a trapezoid)
6. $S = 2LW + 2WH + 2HL$ for H (Surface area of a rectangular solid)
7. $S = 2\pi rh + 2\pi r^2$ for h (Surface area of a cylinder)
8. $V = \frac{1}{3}\pi r^2 h$ for h (Volume of a cone)
9. $F = \frac{9}{5}C + 32$ for C (Celsius to Fahrenheit)
10. $s = \frac{1}{2}gt^2$ for g (Distance traveled by a falling object)

Solve each of the following problems analytically, and support your solutions graphically.

11. *Dimensions of a Mailing Label* The length of a rectangular mailing label is 3 centimeters less than twice the width. The perimeter is 54 centimeters. Find the dimensions of the label. (In the figure, w represents the width and so $2w - 3$ represents the length.)

Side lengths are in centimeters.

12. *Dimensions of a Square* If the length of a side of a square is increased by 3 centimeters, the perimeter of the new square is 40 centimeters more than twice the length of the side of the original square. Find the dimensions of the original square.

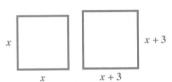

Original square · Side is increased by 3.

x and $x + 3$ are in centimeters.

13. *Dimensions of an Aquarium* The front face of a rectangular aquarium has a ratio of its length to its width of 5 to 3. The perimeter of the rectangle is 64 inches. Find the length and the width of the rectangle.

14. *Aspect Ratio of a Television Set* The aspect ratio of conventional television sets is 4:3. One such model Toshiba television has a rectangular viewing screen with a perimeter of 98 inches. What are the length and width of the screen? Since televisions are advertised by the diagonal measure of their screens, how would this set be advertised?

15. *World's Largest Tablecloth* The world's largest tablecloth has a perimeter of 28,803.2 inches. It is made of damask and was manufactured by Tonrose Limited of Manchester, England, in June 1988. Its length is 11,757.6 inches more than its width. What is its length in yards? (*Source: Guinness Book of Records 1995.*)

16. *Smallest Ticket* The smallest ticket ever produced was for admission to the Asian-Pacific Exposition Fukuoka '89 in Japan. It was rectangular, with length 4 mm more than its width. If the width had been increased by 1 mm and the length had been increased by 4 mm, the ticket would have had perimeter equal to 30 mm. What were the actual dimensions of the ticket? (*Source: Guinness Book of Records 1995.*)

Exercises 17 and 18 depend on the idea of the octane rating of gasoline, a measure of its antiknock qualities. In one measure of octane, a standard fuel is made with only two ingredients: heptane and isooctane. For this fuel, the octane rating is the percent of isooctane. An actual gasoline blend is then compared to a standard fuel. For example, a gasoline with an octane rating of 98 has the same antiknock properties as a standard fuel that is 98% isooctane.

17. *Octane Rating of Gasoline* How many gallons of 94-octane gasoline should be mixed with 400 gallons of 99-octane gasoline to obtain a mixture that is 97-octane?

18. *Octane Rating of Gasoline* How many gallons of 92-octane and 98-octane gasoline should be mixed together to provide 120 gallons of 96-octane gasoline?

Solve the following mixture problems.

19. *Alcohol Mixture* A pharmacist wishes to strengthen a mixture from 10% alcohol to 30% alcohol. How much pure alcohol should be added to 7 liters of the 10% mixture?

20. *Acid Mixture* A student needs 10% hydrochloric acid for a chemistry experiment. How much 5% acid should be mixed with 60 ml of 20% acid to get a 10% solution?

21. *Acid Mixture* How much pure acid should be added to 6 liters of 30% acid to increase the concentration to 50% acid?

22. *Saline Solution Mixture* How much water should be added to 8 ml of 6% saline solution to reduce the concentration to 4%?

23. *Acid Mixture* How much water should be added to 20 liters of an 18% acid solution to reduce the concentration to 15% acid?

24. *Antifreeze Mixture* An automobile radiator holds 8 liters of fluid. There is currently a mixture in the radiator that is 80% antifreeze and 20% water. How much of this mixture should be drained and replaced by pure antifreeze so that the resulting mixture is 90% antifreeze? (The total amount at the end must be the same as what was there at the beginning.)

Solve each of the following problems analytically, and support your solutions graphically.

25. *Pressure of a Liquid* The pressure exerted by a certain liquid at a given point is directly proportional to the depth of the point beneath the surface of the liquid. If the pressure at 30 feet is 15 pounds per square foot, what is the pressure exerted at 70 feet?

26. *Rate of Nerve Impulses* The rate at which impulses are transmitted along a nerve fiber is directly proportional to the diameter of the fiber. If the rate for a certain fiber is 40 meters per second when the diameter is 6 micrometers, what is the rate if the diameter is 8 micrometers?

27. *Height of a Shadow* The height of a vertical object is directly proportional to the length of its shadow assuming that the measure is made with the same angle of elevation of the sun. A certain tree casts a shadow 45 feet long. At the same time, the shadow cast by a vertical stick 2 feet high is 1.75 feet long. How tall is the tree?

28. *Height of a Shadow* (See Exercise 27.) A forest fire lookout tower casts a shadow 180 feet long at the same time that the shadow of a 15-foot tree is 9 feet long. What is the height of the lookout tower?

Biologists use direct variation to estimate the number of individuals of a species in a particular area. They first capture a sample of individuals from the area and mark each specimen with a harmless tag. Then later they return and capture another sample from the same area. They base their estimate on the theory that the proportion of tagged specimens in the new sample is the same as the proportion of tagged individuals in the entire area. Use this idea to work Exercises 29 and 30.

29. *Estimation of Fish in a Lake* Biologists tagged 250 fish in City Park Lake on October 12. On a later date they found 7 tagged fish in a sample of 350. Estimate the total number of fish in the lake to the nearest hundred.

30. *Estimation of Seal Pups in a Breeding Area* According to an actual survey in 1961, to estimate the number of seal pups in a certain breeding area in Alaska, 4963 pups were tagged in early August. In late August, a sample of 900 pups was examined and 218 of these were found to have been tagged earlier. Use this information to estimate, to the nearest hundred, the total number of seal pups in this breeding area. (*Source:* "Estimating the Size of Wildlife Populations" by S. Chatterjee in *Statistics by Example*, 1973, obtained from data in *Transactions of the American Fisheries Society*, July, 1968.)

In each of Exercises 31–34,

(a) Express the cost C as a function of x, where x represents the number of items as described.

(b) Express the revenue R as a function of x.

(c) Determine analytically the value of x for which revenue equals cost.

(d) Graph $y_1 = C(x)$ and $y_2 = R(x)$ in the same viewing window and interpret the graphs.

31. *Stuffing Envelopes* Rebecca Isaac-Fahey stuffs envelopes for extra income during her spare time. Her initial cost to obtain the necessary information for the job was $200.00. Each envelope costs $.02, and she gets paid $.04 per envelope stuffed. Let x represent the number of envelopes stuffed.

32. *Copier Service* Gerald Skidmore runs a copying service in his home. He paid $3500 for the copier and a lifetime service contract. Each sheet of paper he uses costs $.01, and he gets paid $.05 per copy he makes. Let *x* represent the number of copies he makes.

33. *Delivery Service* Tom Accardo operates a delivery service in a southern city. His start-up costs amounted to $2300. He estimates that it costs him (in terms of gasoline, wear and tear on his car, etc.) $3.00 per delivery. He charges $5.50 per delivery. Let *x* represent the number of deliveries he makes.

34. *Baking and Selling Cakes* Pat Stone bakes cakes and sells them at county fairs. Her initial cost for the Tomball fair in 1997 was $40.00. She figures that each cake costs $2.50 to make, and she charges $6.50 per cake. Let *x* represent the number of cakes sold. (Assume that there were no cakes left over.)

Work each problem involving grade averaging.

35. *Grade Averaging* Hien has grades of 84, 88, and 92 on his first three calculus tests. What grade on his fourth test will give him an average of 90?

36. *Grade Averaging* Jamal scored 78, 94, and 60 on his three trigonometry tests. If his final exam score is to be counted as two test grades in determining his course average, what grade must he make on his final exam to give him an average of 80?

37. *Grade Averaging* Gerry Fitch has grades of 88, 86, and 92 on her first three algebra tests. What score does she need on her fourth test to have an average of at least 90?

38. *Grade Averaging* (See Exercise 37.) What range of scores on her fourth test would assure Gerry of having an average between and inclusive of 80 and 90?

In Massachusetts, speeding fines are determined by the linear function

$$y = 10(x - 65) + 50, \quad x \geq 65,$$

where y is the cost in dollars of the fine if a person is caught driving x miles per hour. Use this information to work the problems in Exercises 39–42.

39. *Speeding Fines* José had to make an 8:00 A.M. final examination, but overslept after a big weekend in Boston. Radar clocked his speed at 76 mph. How much was his fine?

40. *Speeding Fines* While balancing his checkbook, Johnny ran across a canceled check that his wife Gwen had written to the Department of Motor Vehicles for a speeding fine. The check was written for $100. How fast was Gwen driving?

41. *Speeding Fines* Based on the formula above, at what speed do the troopers start giving tickets?

42. *Speeding Fines* For what range of speeds is the fine greater than $200? Solve analytically and support graphically.

In Exercises 43–46, assume that a linear relationship exists between the two quantities.

43. *Solar Heater Production* A company finds that it can produce 10 solar heaters for $7500, while producing 20 heaters costs $13,900.

(a) Express the cost, *y*, as a function of the number of heaters, *x*.

(b) Determine analytically the cost to produce 25 heaters.

(c) Support the result of part (b) graphically.

44. *Cricket Chirping* At 68° Fahrenheit, a certain species of cricket chirps 24 times per minute. At 40°F, the same cricket chirps 86 times per minute.*

(a) Express the number of chirps, *y*, as a function of the Fahrenheit temperature.

(b) If the temperature is 60°F, how many times will the cricket chirp per minute? Determine your answer analytically, and support it graphically.

(c) If you count the number of cricket chirps in one-half minute and hear 40 chirps, what is the temperature? Determine your answer analytically, and support it graphically.

45. *Appraised Value of a Home* In 1986, a house was purchased for $120,000. In 1996, it was appraised for $146,000.

(a) If *x* = 0 represents 1986 and *x* = 10 represents 1996, express the appraised value of the house, *y*, as a function of the number of years, *x,* after 1986.

(b) What will the house be worth in the year 2000? Determine your answer analytically, and support it graphically.

(c) What does the slope of the line represent (in your own words)?

46. *Depreciation of a Photocopier* A photocopier sold for $3000 in 1988 when it was purchased. Its value in 1996 had depreciated to $600.

(a) If *x* = 0 represents 1988 and *x* = 8 represents 1996, express the value of the machine, *y*, as a function of the number of years from 1988.

*Exercises 44, 53, and 58–62 are adapted from *A Sourcebook of Applications of School Mathematics* by Donald Bushaw et al. Copyright © 1980 by The Mathematical Association of America. Reprinted by permission.

(b) Graph the function from part (a) in a window [0, 10] by [0, 4000]. How would you interpret the *y*-intercept in terms of this particular problem situation?

(c) Use your calculator to determine the value of the machine in 1992, and verify this analytically.

Solve each problem. (See Example 8 for background information.)

47. *Ventilation in a Classroom*   Ventilation is an effective method for removing indoor air pollutants. According to the American Society of Heating, Refrigerating and Air-Conditioning Engineers, Inc. (ASHRAE), a non-smoking classroom should have a ventilation rate of 15 cubic feet per minute for each person in the classroom. (*Source:* ASHRAE, 1989.)

(a) Write a linear function that describes the total ventilation *V* (in cubic feet per hour) necessary for a classroom with *x* people.

(b) A common unit of ventilation is an air change per hour (ach). 1 ach is equivalent to exchanging all of the air in a room every hour. If *x* people are in a classroom having a volume of 15,000 cubic feet, determine how many air exchanges per hour are necessary to keep the room properly ventilated.

(c) Find the necessary number of ach *A* if the classroom has 40 people in it.

(d) In areas like bars and lounges that allow smoking, the ventilation rate should be increased to 50 cubic feet per minute per person. Compared to classrooms, ventilation should be increased by what factor in heavy-smoking areas?

48. *Snowmaking and Water Consumption*   Ski resorts require large amounts of water in order to make snow. Snowmass Ski Area in Colorado plans to pump between 1120 and 1900 gallons of water per minute at least 12 hours per day from Snowmass Creek between mid-October and late December. Environmentalists are concerned about the effects on the ecosystem. (*Source:* York Snow Incorporated.)

(a) Determine a linear function that will calculate the *minimum* amount of water *A* (in gallons) pumped after *x* days during mid-October to late December.

(b) Find the minimum amount of water pumped in 30 days.

(c) Suppose the water being pumped from Snowmass Creek was used to fill swimming pools. The average backyard swimming pool holds 20,000 gallons of water. Determine a linear function that will give the minimum number of pools *P* that could be filled after *x* days. How many pools could be filled each day?

(d) In how many days could a minimum of 1000 pools be filled?

49. *Cancer Risk from Pollutants*   The excess lifetime cancer risk *R* is a measure of the likelihood that an individual will develop cancer from a particular pollutant. For example, if *R* = .01, then a person has a 1% increased chance of developing cancer during a lifetime. This would translate into 1 case of cancer for every 100 people during an average lifetime. For nonsmokers exposed to environmental tobacco smoke (passive smokers), $R = 1.5 \times 10^{-3}$. (*Source:* Hines, A., T. Ghosh, S. Layalka, and R. Warder, *Indoor Air Quality & Control*, Prentice Hall, 1993.)

(a) If the average life expectancy is 72 years, what is the excess lifetime cancer risk per year for passive smokers?

(b) Write a linear function that will give the expected number of cancer cases *C* per year if there are *x* passive smokers.

(c) Estimate the number of cancer cases per 100,000 passive smokers.

(d) The excess lifetime risk of death from smoking is *R* = .44. Currently, 26% of the U.S. population smokes. If the U.S. population is 260 million, approximate the excess number of deaths caused by smoking each year.

50. *Cancer Risk from Formaldehyde*   (See Exercise 49.) The excess lifetime cancer risk *R* for formaldehyde can be calculated using the linear equation $R = kd$, where *k* is a constant and *d* is the daily dose in parts per million. The constant *k* for formaldehyde can be calculated using the formula $k = .132(B/W)$, where *B* is the total number of cubic meters of air a person breathes in one day and *W* is a person's weight in kilograms. (*Sources:* A. Hines, T. Ghosh, S. Layalka, and R. Warder, *Indoor Air Quality & Control*, Prentice Hall, 1993; I. Ritchie and R. Lehnen, "An Analysis of Formaldehyde Concentration in Mobile and Conventional Homes," *J. Env. Health* 47: 300–305.)

(a) Find *k* for a person who breathes in 20 cubic meters of air per day and weighs 75 kg.

(b) Mobile homes in Minnesota were found to have a mean daily dose *d* of .42 part per million. Calculate *R* using the value of *k* found in part (a).

(c) For every 5000 people, how many cases of cancer could be expected each year from these levels of formaldehyde? Assume an average life expectancy of 72 years.

The final group of exercises reviews material that was covered in detail in Section 1.7. The linear models are based on the least squares regression method discussed there.

Solve each problem analytically, and support your solutions graphically.

51. *Alcohol-Related Traffic Deaths*   The accompanying graph illustrates how the percent of alcohol-related traffic deaths in the United States over the period from 1982 to 1993 has declined. These data can be modeled by the linear function defined by $y = -1.18x + 57.03$, where $x = 0$ corresponds to 1982, $x = 1$ corresponds to 1983, and so on, and y is the percent of alcohol-related traffic deaths.

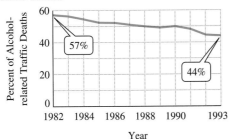

(*Source*: National Highway Traffic Safety Administration.)

(a) Use the model equation to find the year in which the percent was 50%. (*Hint:* Let $y = 50$ and solve.)

(b) Use the graph to answer the item in part (a). How closely do the answers correspond?

(c) If this model were used to calculate the percent in 1997, what would the percent be?

52. *Monthly Cable Television Rates*   The linear function defined by $y = 1.082x + 16.882$ provides the approximate average monthly rate for basic cable television subscribers between the years 1990 and 1993, where $x = 0$ corresponds to 1990, $x = 1$ corresponds to 1991, and so on, and y is in dollars. Use this model to answer the following questions.

Cable Rates

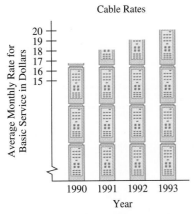

(*Source*: National Cable Television Association, Nations Bank, Paul Kagan Associates.)

(a) What was the approximate average monthly rate in 1991?

(b) What was the approximate average monthly rate in 1993?

(c) During 1994, the rates dropped substantially to $18.86. The model above is based on data from 1990 through 1993. If you were to use the model for 1994, what would the rate be?

(d) Why do you think there is such a discrepancy between the actual rate and the rate based on the model in part (c)? Discuss the pitfalls of using the model to predict rates for years following 1993.

53. *Pattern Sizes for Women's Clothing*   The Measurements Standard Committee of the Pattern Fashion Industry provides a table of body measurements (in inches) corresponding to misses' sizes. For misses' sizes 6 through 20, bust measurement y corresponds to misses' size x according to the model $y = .842x + 24.613$. If the size is 14, what is the corresponding bust size (rounded down to the nearest unit)?*

54. *Blood Alcohol Level*   If alcohol "burned up" by the body since the time of the first drink is disregarded, the number of drinks (1 drink = 12 oz. beer, 4 oz. wine, 1 oz. hard liquor), x, and the blood alcohol level, y, of a 240-pound person are related by the function $y = .0156x$. What would the blood alcohol level be if this person had 4 drinks? If .100 blood alcohol level is considered legally drunk, after what whole number of drinks would this person be legally drunk?

55. *Family Health Care Cost*   According to information provided by Families USA Foundation, the national average family health care cost (in dollars) between 1980 and 2000 (projected) can be approximated by the linear function $y = 382.75x + 1742$, where $x = 0$ corresponds to 1980 and $x = 20$ corresponds to 2000. Based on this model, what would be the expected national average health care cost in 1996?

56. *Family Health Care Cost*   In what year will (or did) the national average family health care cost first exceed $7000? (See Exercise 55.)

57. *Off-Track Betting*   The percentage y of off-track betting as a portion of all bets on horse racing between 1982 and 1992 can be estimated by the linear model $y = 1.8x + 15$, where $x = 0$ corresponds to 1982 and $x = 10$ corresponds to 1992. If this trend continues, what will the percentage of off-track betting as a portion of all bets on horse racing be in 1998? (*Source:* Christiansen/Cummings Associates, Inc.)

58. *Stopping Distance*   For speeds between 10 and 60 mph, the stopping distance y in ft for a person going x mph is estimated by the linear model $y = 4.5x - 46.7$. What would the stopping distance be at 40 mph?*

59. *Population of California*   From 1940 to 1960, the population y of California was approximated by the linear model $y = .4405x + 6.665$, where y is in millions and $x = 0$ corresponds to 1940 and $x = 20$ corresponds to 1960. If this trend had continued until 1965, what would have been the population then?*

60. *Tail Length of a Snake*   It has been reported that the total length x and the tail length y of females of the snake species *Lampropeltis polyzona* are nearly linearly related by the model $y = .134x - 1.18$, where x is the length of the snake in millimeters. If a snake of this species measures 1000 millimeters, what is its tail length to the nearest millimeter?*

61. *Expansion and Contraction of Gases*   In 1787, Jacques Charles noticed that gases expand when heated and contract when cooled. Suppose that a particular gas follows the model $y = \frac{5}{3}x + 455$, where x is the temperature in Celsius and y is the volume in cubic centimeters.*

(a) What is the volume when the temperature is 27° Celsius?

(b) What is the temperature when the volume is 605 cubic centimeters?

(c) Determine what temperature gives a volume of 0 cubic centimeters (that is, absolute zero, or the coldest possible temperature).

62. *Prices of Test Tubes*   According to information in the Edmund Scientific Catalog #761, the price y in dollars for a group of 72 test tubes of x milliliters is approximated by the linear function $y = .233x + 6.64$, for capacities between 4 milliliters and 36 milliliters. What would be the price for a group of 10-milliliter test tubes based on this model, rounded to the nearest 25 cents?*

CHAPTER 1   SUMMARY

Section	Important Concepts
1.1 Real Numbers, Logic, and Coordinate Systems	Natural numbers $\{1, 2, 3, 4, 5, \ldots\}$ Whole numbers $\{0, 1, 2, 3, 4, 5, \ldots\}$ Integers $\{\ldots, -4, -3, -2, -1, 0, 1, 2, 3, \ldots\}$ Rational numbers $\left\{\dfrac{p}{q} \,\middle|\, p, q \text{ are integers}, q \neq 0\right\}$ Real numbers $\{x \mid x \text{ corresponds to a point on a number line}\}$ (The real numbers include both the rational and the irrational numbers.) A graphing calculator returns a 1 for a true statement and a 0 for a false statement.
1.2 Introduction to Relations and Functions	**RELATION** A relation is a set of ordered pairs. **DOMAIN AND RANGE** If we denote the ordered pairs of a relation by (x, y), the set of all x-values is called the domain of the relation and the set of all y-values is called the range of the relation. **FUNCTION** A function is a relation in which each element in the domain corresponds to exactly one element in the range. **VERTICAL LINE TEST** If each vertical line intersects a graph in no more than one point, then the graph is the graph of a function.
1.3 Linear Functions	**DEFINITION OF LINEAR FUNCTION** A function f defined by $$f(x) = ax + b, \text{ where } a \text{ and } b \text{ are real numbers,}$$ is called a linear function.

SLOPE
The slope m of the line through the points (x_1, y_1) and (x_2, y_2) is

$$m = \frac{\Delta y}{\Delta x} = \frac{y_2 - y_1}{x_2 - x_1},$$

where $\Delta x \neq 0$.

GEOMETRIC ORIENTATION BASED ON SLOPE
For a line with slope m, if $m > 0$, the line rises from left to right. If $m < 0$, it falls from left to right. If $m = 0$, the line is horizontal.

VERTICAL LINE
A vertical line with x-intercept a has an equation of the form

$$x = a.$$

Its slope is undefined.

1.4 Equations of Lines and Geometric Considerations

POINT-SLOPE FORM
The line with slope m passing through the point (x_1, y_1) has an equation

$$y - y_1 = m(x - x_1).$$

This is called the point-slope form of the equation of a line.

SLOPE-INTERCEPT FORM
The slope-intercept form of the equation of a line with slope m and y-intercept b is

$$y = mx + b.$$

DISTANCE FORMULA
Suppose that $P(x_1, y_1)$ and $R(x_2, y_2)$ are two points in a coordinate plane. Then the distance between P and R, written $d(P, R)$, is given by the distance formula,

$$d(P, R) = \sqrt{(x_2 - x_1)^2 + (y_2 - y_1)^2}.$$

MIDPOINT FORMULA
The midpoint of the line segment with endpoints (x_1, y_1) and (x_2, y_2) is

$$\left(\frac{x_1 + x_2}{2}, \frac{y_1 + y_2}{2}\right).$$

PARALLEL AND PERPENDICULAR LINES
Two distinct nonvertical lines are parallel if and only if they have the same slope.
Two lines, neither of which is vertical, are perpendicular if and only if their slopes have a product of -1.

1.5 Solution of Linear Equations; Analytic Method and Graphical Support

INTERSECTION-OF-GRAPHS METHOD OF GRAPHICAL SOLUTION
To solve the equation

$$f(x) = g(x)$$

graphically, graph $y_1 = f(x)$ and $y_2 = g(x)$. The x-coordinate of any point of intersection of the two graphs is a solution of the equation.

x-INTERCEPT METHOD OF GRAPHICAL SOLUTION
To solve the equation

$$f(x) = g(x)$$

graphically, graph $y = f(x) - g(x) = F(x)$. Any x-intercept of the graph of $y = F(x)$ is a solution of the equation.

1.6 Solution of Linear Inequalities; Analytic Method and Graphical Support	**INTERSECTION-OF-GRAPHS METHOD OF SOLUTION OF LINEAR INEQUALITIES** Suppose that f and g are linear functions. The solution set of $f(x) > g(x)$ is the set of all real numbers x such that the graph of f is *above* the graph of g. The solution set of $f(x) < g(x)$ is the set of all real numbers x such that the graph of f is *below* the graph of g. **x-INTERCEPT METHOD OF SOLUTION OF LINEAR INEQUALITIES** The solution set of $F(x) > 0$ is the set of all real numbers x such that the graph of F is *above* the x-axis. The solution set of $F(x) < 0$ is the set of all real numbers x such that the graph of F is *below* the x-axis.
1.7 Linear Models	**ANALYSIS OF LINEAR DATA** If a collection of data points approximates a straight line, then we can find the equation of such a line. Choosing two data points, we apply the methods of Sections 1.3 and 1.4 to find this equation. The equation will vary, depending on which two points are chosen. The best-fitting line can be found using the linear regression feature of a graphing calculator.
1.8 Other Applications of Linear Functions	**POLYA'S FOUR-STEP PROCESS FOR PROBLEM SOLVING** **1.** Understand the problem. **2.** Devise a plan. **3.** Carry out the plan. **4.** Look back and check. **DIRECT VARIATION** A number y varies directly with x if there exists a nonzero number k such that $y = kx$. The number k is called the constant of variation.

CHAPTER 1   REVIEW EXERCISES

Let A represent the point with coordinates $(-1, 16)$ and let B represent the point with coordinates $(5, -8)$.

1. Find the exact distance between points A and B.

2. Find the coordinates of the midpoint of the line segment joining points A and B.

3. Find the slope of the line AB.

4. Find the equation of the line passing through points A and B. Write it in $y = mx + b$ form. (*Hint:* To check your answer, enter the equation into your calculator and use the capabilities of your calculator to verify that the line does indeed contain the points $(-1, 16)$ and $(5, -8)$.)

Consider the line with equation $3x + 4y = 144$ in Exercises 5–8.

5. What is the slope of this line?

6. What is the x-intercept of this line?

7. What is the y-intercept of this line?

8. Give a viewing window that will show a comprehensive graph. (There are many possible such windows.)

9. Suppose that f is a linear function such that $f(3) = 6$ and $f(-2) = 1$. Find $f(8)$.

10. Find the equation of the line perpendicular to the graph of $y = -4x + 3$, passing through the point $(-2, 4)$. Give it in $y = mx + b$ form. (*Hint:* To check your answer, use a square viewing window and graph the given equation, $y = -4x + 3$, and the equation you found. Then, verify that the equation you found does indeed contain $(-2, 4)$. The lines should also appear perpendicular.)

For each line shown,

(a) find the slope.

(b) find the slope-intercept form of the equation.

(c) find the midpoint of the *segment* joining the two points identified on the screen.

(d) find the distance between the two points identified on the screen.

11.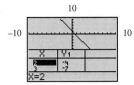

12.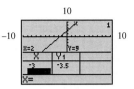

Choose the letter of the graph that would most closely resemble the graph of $f(x) = mx + b$, given the conditions on m and b.

13. $m < 0, b < 0$

14. $m > 0, b < 0$

15. $m < 0, b > 0$

16. $m > 0, b > 0$

17. $m = 0$

18. $b = 0$

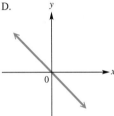

19. What are the domain and the range of the relation graphed on the screen? (*Hint:* Pay attention to scale.)

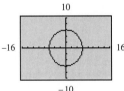

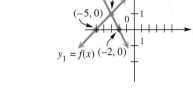

20. True or false? The graphs of $y_1 = 5.001x - 3$ and $y_2 = 5x + 6$ are shown on the accompanying screen. From this view, we may correctly conclude that these lines are parallel.

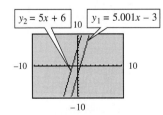

Refer to the graphs of the linear functions $y_1 = f(x)$ and $y_2 = g(x)$ in the figure to match the solution set in the columns on the right with the equation or inequality on the left. Choices may be used once, more than once, or not at all.

21. $f(x) = g(x)$ A. $(-\infty, -3]$ I. $\{-3\}$

22. $f(x) > g(x)$ B. $(-\infty, -3)$ J. $[-3, \infty)$

23. $f(x) < g(x)$ C. $\{3\}$ K. $(-3, \infty)$

24. $g(x) \geq f(x)$ D. $\{2\}$ L. $[-3]$

25. $y_2 - y_1 = 0$ E. $\{(3, 2)\}$ M. $(-\infty, -5)$

26. $f(x) < 0$ F. $\{-5\}$ N. $(-5, \infty)$

27. $g(x) > 0$ G. $\{-2\}$ O. $(-\infty, -2)$

28. $y_2 - y_1 < 0$ H. $\{0\}$ P. $(-2, \infty)$

Solve each equation using analytic methods.

29. $5[3 + 2(x - 6)] = 3x + 1$

30. $\dfrac{x}{4} - \dfrac{x + 4}{3} = -2$

31. Solve the inequality $-6 \leq \dfrac{4 - 3x}{7} < 2$ analytically.

32. Consider the linear function
$$f(x) = 5\pi x + (\sqrt{3})x - 6.24(x - 8.1) + (\sqrt[3]{9})x.$$

(a) Solve the equation $f(x) = 0$ using graphical methods. Give the solution to the nearest hundredth. Then, give an explanation of how you went about solving this equation graphically.

(b) Refer to the graph, and give the solution set of $f(x) < 0$.

(c) Refer to the graph, and give the solution set of $f(x) \geq 0$.

33. What is the solution set of $f(x) > 0$, based on the screen shown?

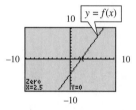

34. What is the solution set of $f(x) - g(x) = 0$, based on the screen shown?

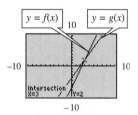

Production of Audiocassettes *A company produces studio-quality audiocassettes of live concerts. The company places an ad in a trade newsletter. The cost of the ad is $150. Each tape costs $30 to produce, and the company charges $37.50 per tape. Work Exercises 35–38.*

35. Express the company's cost C as a function of x, where x is the number of tapes produced and sold.

36. Assuming that the company sells x tapes, express the revenue as a function of x.

37. Determine analytically the value of x for which revenue equals cost.

38. The graph shows $y = C(x)$ and $y = R(x)$. Use the graph to discuss how it illustrates when the company is losing money, when it is breaking even, and when it is making a profit.

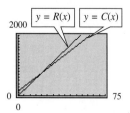

Solve the formula for the variable indicated.

39. $A = \dfrac{24f}{B(p + 1)}$ for f (approximate annual interest rate)

40. $A = \dfrac{24f}{B(p + 1)}$ for B (approximate annual interest rate)

The following problems involve linear models. Solve each problem.

41. *Temperature Levels above the Earth's Surface* The linear function $f(x) = -3.52x + 58.6$ gives an approximation for temperature in degrees Fahrenheit above the surface of the earth, where x is in thousands of feet and y is the temperature. (*Source:* Schwartz, Richard H., *Mathematics and Global Survival*, Ginn Press, 1991.)

(a) When the height is 5000 feet, what is the temperature? Solve analytically.

(b) When the temperature is $-15°F$, what is the height? Solve analytically.

(c) Explain how the answers in parts (a) and (b) can be supported graphically.

42. *Prevention of Indoor Pollutants* Kitchen gas ranges are a source of indoor pollutants such as carbon monoxide and nitrogen dioxide. One of the most effective ways of removing contaminants from the air while cooking is to use a *vented* range hood. If a range hood removes F liters of air per second, then the percent P of contaminants that are also removed from the surrounding air can be expressed by the linear function $P = 1.06F + 7.18$, where $10 \leq F \leq 75$. What flow rate must a range hood have to remove 50% of the contaminants from the air? (*Source:* Rezvan, R. L., "Effectiveness of Local Ventilation in Removing Simulated Pollutants from Point Sources," 65–75. In *Proceedings of the Third International Conference on Indoor Air Quality and Climate*, 1984.)

43. *Airline Passenger Growth at Various Airports* The table estimates the growth in airline passengers (in millions) at some of the fastest-growing airports in the United States between 1992 and 2005.

Airport	1992	2005
Harrisburg Intl.	.7	1.4
Dayton Intl.	1.1	2.4
Austin Robert Mueller	2.2	4.7
Milwaukee Gen. Mitchell Intl.	2.2	4.4
Sacramento Metropolitan	2.6	5.0
Fort Lauderdale–Hollywood	4.1	8.1
Washington Dulles Intl.	5.3	10.9
Greater Cincinnati Airport	5.8	12.3

(*Source:* FAA.)

(a) Determine a linear function $y = f(x)$ that approximates the data using the two points (.7, 1.4) and (5.3, 10.9).

(b) How does the slope of the graph of f relate to growth in airline passengers at these airports?

(c) Use the regression capability of a graphing calculator to find the least squares line that models the data.

(d) 4.9 million passengers used Raleigh-Durham International Airport in 1992. Use the equation from part (c) to approximate the number of passengers using this airport in 2005, and compare it with the Federal Aviation Administration's estimate of 10.3 million passengers.

44. *Minimum Hourly Wage*   The table shows the changes in the minimum hourly wage in the United States for selected years from 1938 to 1991.

Year	Wage
1938	$ .25
1950	.75
1956	1.00
1963	1.25
1974	2.00
1980	3.10
1981	3.35
1990	3.80
1991	4.25

(*Source:* U.S. Labor Department.)

(a) Let $x = 0$ represent 1938, $x = 1$ represent 1939, and so on. Use the data for 1938 and for 1991 to find a linear model for the changes in the minimum hourly wage.

(b) Use the statistical plot feature of a graphing calculator to plot the data. Comment on how well the data fit a linear model.

(c) Find the least squares regression line and the correlation coefficient for these data.

(d) The 1997 minimum wage was $5.15. Use the least squares regression line from part (c) to see how close an approximation it gives for 1997.

45. *Indianapolis 500 Pole Speeds*   The pole speeds for selected drivers in the Indianapolis 500 from 1980 to 1992 are given in the following table. Use the linear regression capability of a graphing calculator to predict the year in which the pole speed should reach 250 miles per hour.*

Year	Driver	Pole Speed
1980	Johnny Rutherford	192.256 mph
1981	Bobby Unser	200.546 mph
1982	Rick Mears	207.004 mph
1983	Teo Fabi	207.395 mph
1984	Tom Sneva	210.029 mph
1985	Pancho Carter	212.583 mph
1986	Rick Mears	216.828 mph
1987	Mario Andretti	215.390 mph
1988	Rick Mears	219.198 mph
1989	Rick Mears	223.885 mph
1990	Emerson Fittipaldi	225.301 mph
1991	Rick Mears	224.113 mph
1992	Roberto Guerrero	232.482 mph

46. *HIV Infection*   The time interval between a person's initial infection with HIV and that person's eventual development of AIDS symptoms is an important issue. The method of infection with HIV affects the time interval before AIDS develops. One study of HIV patients who were infected by intravenous drug use found that 17% of the patients had AIDS after 4 years and 33% had developed the disease after 7 years. The relationship between the time interval and the percentage of patients with AIDS can accurately be modeled using a linear function. (*Source:* Alcabes, P., A. Munoz, D. Vlahov, and G. Friedland, "Incubation Period of Human Immunodeficiency Virus," *Epidemiologic Review*, Vol. 15, No. 2, The Johns Hopkins University School of Hygiene and Public Health, 1993.)

(a) Write a linear function defined by $f(x) = mx + b$ that models this data, using the points (4, .17) and (7, .33).

(b) Find the slope of the graph of the linear function in part (a). What does the slope tell us about the percent of HIV patients who will develop AIDS?

47. *Speed of a Batted Baseball*   Suppose a baseball is thrown at 85 mph. The ball will travel 320 ft when hit by a bat swung at 50 mph and will travel 440 ft when hit by a bat swung at 80 mph. Let y be the number of feet traveled by the ball when hit by a bat swung at x mph. Find the equation of the line. (*Note:* This function is valid for $50 \leq x \leq 90$, where the bat is 35 inches long, weighs 32 oz, and is swung slightly upward to drive the ball at an angle of 35°.) How much farther will a ball travel for each one-mile-per-hour increase in the speed of the bat? (*Source:* Adair, Robert K., *The Physics of Baseball*, New York: HarperCollins Publishers, 1990.)

*The authors wish to thank Randall Leigh of the University of Southern Indiana for his input into this exercise.

48. *Growth of the Pro Billiards Tour*   The Pro Billiards Tour has grown substantially over the past few years. Based on data provided by the Tour, the total prize money offered from 1991 through 1994 can be modeled by the equation $y = 220,000x + 320,000$, where $x = 0$ corresponds to 1991 and y is in dollars. Based on this model, what would we expect the 1995 prize money to be?

Pro Billiards Tour Prize Money

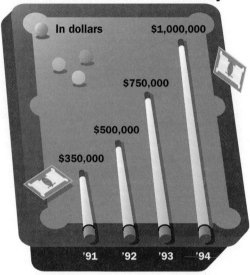

(*Source:* The Tour.)

Solve each application of linear equations.

49. *Dimensions of a Recycling Bin*   A recycling bin is in the shape of a rectangular box. Find the height of the box if its length is 18 ft, its width is 8 ft, and its surface area is 496 sq ft. (In the figure, let $h =$ height. Assume that the given surface area includes that of the top lid of the box.)

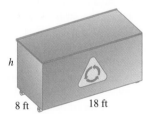

h is in feet.

50. *Temperature of Venus*   Venus is the hottest planet with a surface temperature of 864°F. What is this temperature in Celsius? (*Source: The Guinness Book of Records 1995.*)

51. *Running Speeds in Track Events*   In 1994, Leroy Burrell (USA) set a world record in the 100-meter dash with a time of 9.85 seconds. If this pace could be maintained for an entire 26-mile marathon, how would this time compare to the fastest time for a marathon of 2 hours, 6 minutes, and 50 seconds? (*Hint:* 1 meter ≈ 3.281 feet.) (*Source:* International Amateur Athletic Association.)

52. *Trust in e-mail and the Postal Service*   In a survey of 1014 households, ICR Survey Research Group asked the following question: "Which do you trust more to send a message: computer e-mail or the U.S. Postal Service?" The pie chart shows how the respondents answered. Determine the number of people responding in each of the four groups.

Perceived Reliability of U.S. Mail vs. e-mail

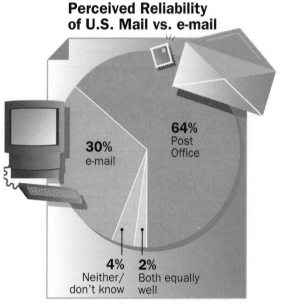

(*Source: USA Today,* 7/6/'95, ICR Survey Research Group, survey of 1014 households.)

53. *Pressure in a Liquid*   The pressure on a point in a liquid is directly proportional to the distance from the surface to the point. In a certain liquid, the pressure at a depth of 4 m is 60 kg per m². Find the pressure at a depth of 10 m.

54. *Take-Home Pay*   Meredith Many earns take-home pay of $592 a week. If her deductions for taxes, retirement, union dues, and medical plan amount to 26% of her wages, what is her weekly pay before deductions?

Effect of Ground-Level Ozone   *Tropospheric ozone (ground-level ozone) is a gas toxic to both plants and animals. Ozone causes respiratory problems and eye irritation in humans. Automobiles are a major source of this type of harmful ozone. It is often present when smog levels are significant. Ozone in outdoor air can enter buildings through ventilation systems. Guideline levels for indoor ozone are less than 50 parts per billion (ppb). Ozone can be removed from the air using filters. In a scientific study, a purafil air filter was used to reduce an initial ozone concentration of 140 ppb. The filter removed 43% of the ozone.* (*Source: Parmar and Grosjean, "Removal of Air Pollutants from Museum Display Cases," Getty Conservation Institute, Marina del Rey, CA, 1989.*)

55. Determine if this type of filter reduced the ozone concentration to acceptable levels.

56. What is the maximum initial concentration of ozone that this filter will reduce to an acceptable level?

57. *Alcohol Mixture* A chemist wishes to strengthen a mixture that is 10% alcohol to one that is 30% alcohol. How much pure alcohol should be added to 12 liters of the 10% mixture?

58. *Acid Mixture* A student needs 10% hydrochloric acid for a chemistry experiment. How much 5% acid should be mixed with 120 ml of 20% acid to get a 10% solution?

59. *Videotape Production* A company produces videotapes. The revenue from the sale of x units of tapes is $R(x) = 8x$. The cost to produce x units of tapes is $C(x) = 3x + 1500$. In what interval will the company at least break even?

60. *Intelligence Quotient* A person's intelligence quotient (IQ) is found by multiplying the mental age by 100 and dividing by the chronological age.

 (a) Jack is 7 years old. His IQ is 130. Find his mental age.

 (b) If a person is 16 years old with a mental age of 20, what is the person's IQ?

CHAPTER 1   TEST

1. For each function, determine

  (i) the domain.   (ii) the range.

  (iii) the x-intercept(s).   (iv) the y-intercept(s).

(a)

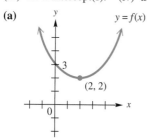

(b)

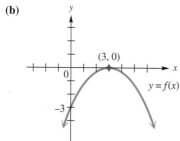

(c)

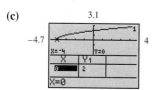

Assume that the graph extends up and to the right indefinitely.

2. Use the figure to solve each equation or inequality.

 (a) $f(x) = g(x)$
 (b) $f(x) < g(x)$
 (c) $f(x) \geq g(x)$
 (d) $y_2 - y_1 = 0$

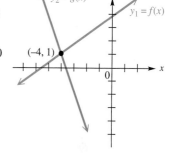

3. Use the screen to solve the equation or inequality. Here, the function is a linear function defined over the domain of real numbers.

 (a) $y_1 = 0$
 (b) $y_1 < 0$
 (c) $y_1 > 0$
 (d) $y_1 \leq 0$

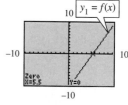

4. Consider the linear functions $f(x) = 3(x - 4) - 2(x - 5)$ and $g(x) = -2(x + 1) - 3$.

 (a) Solve $f(x) = g(x)$ analytically, showing all steps. Also show an analytic check.

 (b) Graph $y_1 = f(x)$ and $y_2 = g(x)$ in the window $[-10, 10]$ by $[-10, 10]$, and use your result in part (a) to find the solution set of $f(x) > g(x)$. Explain your answer.

 (c) Repeat part (b) for $f(x) < g(x)$.

5. Consider the linear function $f(x) = -\frac{1}{2}(8x + 4) + 3(x - 2)$.

(a) Solve the equation $f(x) = 0$ analytically.

(b) Solve the inequality $f(x) \leq 0$ analytically.

(c) Graph $y = f(x)$ in an appropriate viewing window and explain how the graph supports your answers in parts (a) and (b).

6. *Cable Television Rates* The graph depicts how the average monthly rates for cable television increased during the years 1980 through 1992.

Average Monthly Rates for Cable Television
in dollars

(*Sources:* Census Bureau, Paul Kagan Associates.)

(a) Use the midpoint formula to approximate the average monthly rate in 1986.

(b) Find the slope of the line, and explain its meaning in the context of this situation.

7. Find the equation of the line passing through the point $(-3, 5)$ and

(a) parallel to the line with equation $y = -2x + 4$.

(b) perpendicular to the line with equation $-2x + y = 0$.

8. *Surface Area* The formula for the surface area of a rectangular solid is
$$S = 2HW + 2LW + 2LH,$$
where S, H, W, and L represent surface area, height, width, and length, respectively. Solve this formula for W.

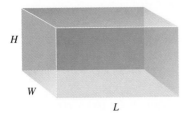

9. *Wind-Chill Factor* The table shows the wind-chill factor for various wind speeds when the Fahrenheit temperature is $40°$.

Wind Speed (mph)	Degrees
10	28
15	22
20	18
25	16
30	13
35	11

(a) Use the points $(10, 28)$ and $(35, 11)$ to find a linear model $f(x)$ that gives wind-chill factor in degrees as a function of the wind speed.

(b) Use the model obtained in part (a) to predict the wind-chill factor when the wind speed is 40 mph. How does it compare with the actual value of $10°F$?

(c) Use the regression capability of a calculator to find the least squares line for this data.

(d) Plot the points and graph the least squares line in the same window of a calculator. Also give the coefficient of correlation. Use this line to predict the wind-chill factor when the wind speed is 40 mph.

10. *Ozone Filtering of Ultraviolet Light* Stratospheric ozone occurs in the atmosphere between altitudes of 20 to 30 kilometers and is an important filter of ultraviolet light from the sun. Ozone is frequently measured in Dobson units, and these units vary directly with the thickness of the ozone layer. If 300 Dobson units correspond to an ozone layer of 3 millimeters, what was the thickness at the Antarctic ozone hole in 1991, which measured 10 Dobson units? (*Source:* Huffman, Robert E., *Atmospheric Ultraviolet Remote Sensing*, Academic Press, 1992.)

CHAPTER 1   PROJECT

 MODELING THE GROWTH OF THE WORLD'S TALLEST MAN

According to recent editions of *The Guinness Book of Records*, the tallest man in medical history was Robert Pershing Wadlow. Wadlow was born on February 22, 1918, in Alton, Illinois. Following a double hernia operation at age 2, he experienced a quite unusual growth rate. He died on July 15, 1940, in Manistee, Michigan, as a result of a septic blister on his right ankle caused by a poorly fitted ankle brace. Just prior to his death, his height had been measured at 8 feet, 11.1 inches.

Beginning at age 8, his height progressed as follows.

Age	Height
8	6 ft
9	6 ft, $2\frac{1}{4}$ in
10	6 ft, 5 in
11	6 ft, 7 in
12	6 ft, $10\frac{1}{2}$ in
13	7 ft, $1\frac{3}{4}$ in
14	7 ft, 5 in
15	7 ft, 8 in
16	7 ft, $10\frac{1}{4}$ in
17	8 ft, $\frac{1}{2}$ in
18	8 ft, $3\frac{1}{2}$ in
19	8 ft, $5\frac{1}{2}$ in
20	8 ft, $6\frac{3}{4}$ in
21	8 ft, $8\frac{1}{4}$ in
22	8 ft, $11\frac{1}{10}$ in

ACTIVITIES

1. Each student should prepare a scatterplot of the data above on his or her graphing calculator. Use the statistics function of the calculator to do this. See the owner's manual.

2. The data points lie almost in a straight line. What does this mean? What does this tell us about Wadlow's rate of growth?

One reason for fitting an equation to a set of data is that we can then use the equation to try to predict what will happen in the future and what might have happened in the past. A warning is required here: Using such a model to predict the past and the future requires that we assume the patterns exhibited in the data will (or did) remain the same. Sometimes this is not a valid assumption, and we must take our predictions "with a grain of salt."

Now, divide up into groups of two or three students each, and work through Activities 3 and 4.

3. Each group should randomly choose two data points from the scatterplot. Be sure that each group has chosen a different pair. Now, use these two data points to calculate an equation that models the data, and determine the slope-intercept form of the equation. Do this using traditional analytic methods. (See Sections 1.3

and 1.4.) Then, graph this line on the calculators over the scatterplot to check to see how well it fits. Each group should then write a short report, and answer the following:

(a) What two points were chosen?

(b) What is the slope-intercept form of the equation of the line containing these two points?

(c) What does m (the slope) tell us about Wadlow's growth rate?

(d) What does b (the y-intercept) imply about Wadlow's height?

(e) What assumption must we make when answering part (d)? Is this a reasonable assumption?

(f) Most people stop growing in their teens, but Wadlow continued to grow right up to his death at age 22. Let us assume that he continued to grow at the same rate. Use the equation to determine how tall he would have been at age 25 and at age 35.

4. As a class, discuss how the different results of the groups compare with the line of best fit, $y \approx .215x + 4.317$.

CHAPTER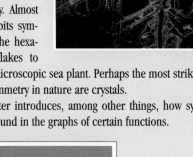

2

Analysis of Graphs of Functions

A figure has *rotational symmetry* around an axis *I* if it coincides with itself by all rotations about *I*. Because of their complete rotational symmetry, the circle in the plane and the sphere in space were considered by the early Greeks to be the most perfect geometric figures. Aristotle assumed a spherical shape for the celestial bodies because any other would detract from their heavenly perfection.

Symmetry has been an important characteristic of art from the earliest times. The art of M. C. Escher (1898–1972) is composed of symmetries and translations, and Leonardo da Vinci's sketches indicate a superior understanding of symmetry. Almost all nature exhibits symmetry—from the hexagons of snowflakes to the diatom, a microscopic sea plant. Perhaps the most striking examples of symmetry in nature are crystals.

This chapter introduces, among other things, how symmetry can be found in the graphs of certain functions.

A Diatom

(*Source: Mathematics*, Life Science Library, Time Inc., New York, 1963.)

A Cross-Section of Tourmaline

2.1   GRAPHS OF ELEMENTARY FUNCTIONS AND RELATIONS

Continuity; Increasing and Decreasing Functions ◆ The Identity Function ◆ The Squaring Function and Symmetry with Respect to the y-Axis ◆ The Cubing Function and Symmetry with Respect to the Origin ◆ The Square Root and Cube Root Functions ◆ The Absolute Value Function ◆ The Relation $x = y^2$ and Symmetry with Respect to the x-Axis ◆ Even and Odd Functions

CONTINUITY; INCREASING AND DECREASING FUNCTIONS

In Chapter 1, our work dealt mainly with linear functions. The graph of a linear function, a straight line, may be drawn by hand over any interval of its domain without picking the pencil up from the paper. In mathematics we say that a function with this property is **continuous** over any interval. The formal definition of continuity requires concepts from calculus, but we can give an informal definition at the college algebra level.

> **INFORMAL DEFINITION OF CONTINUITY**
>
> A function is continuous over an interval of its domain if its hand-drawn graph over that interval can be sketched without lifting the pencil from the paper.

If a function is not continuous at a point, then it may have a point of discontinuity (Figure 1(a)), or it may have a vertical *asymptote* (a vertical line which the graph does not intersect, as in Figure 1(b)). More will be said about asymptotes in Chapter 4.

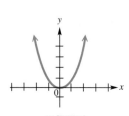

FIGURE 2

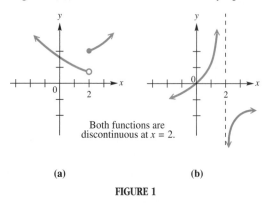

Both functions are discontinuous at $x = 2$.

(a)   (b)

FIGURE 1

Notice that both graphs in Figure 1 are graphs of functions, since they satisfy the conditions of the vertical line test.

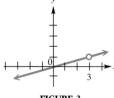

FIGURE 3

EXAMPLE 1   *Determining Intervals of Continuity*   The figures show graphs of functions and the descriptions indicate the intervals of the domain over which they are continuous.

SOLUTION

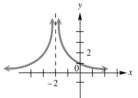

FIGURE 4

(a)   The function in Figure 2 is continuous over the entire domain of real numbers, $(-\infty, \infty)$.

(b)   The function in Figure 3 has a point of discontinuity at $x = 3$. It is continuous over the interval $(-\infty, 3)$ and the interval $(3, \infty)$.

(c)   The function in Figure 4 has a vertical asymptote at $x = -2$, as indicated by the dashed line. It is continuous over the interval $(-\infty, -2)$ and the interval $(-2, \infty)$.

◆

If a continuous function is not constant over an interval, then its graph will either rise from left to right or will fall from left to right. We use the words *increasing* and *decreasing* to describe this behavior. For example, a linear function with a positive slope is increasing over its entire domain, while one with a negative slope is decreasing. See Figure 5.

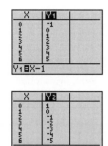

The graph of Y₁ has $m = 1 > 0$. As x increases, range values *increase*, as seen in the top table. The graph of Y₂ has $m = -1 < 0$. As x increases, range values *decrease*, as seen in the bottom table.

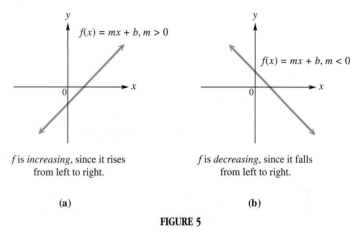

f is *increasing*, since it rises from left to right.

f is *decreasing*, since it falls from left to right.

(a)   **(b)**

FIGURE 5

Informally speaking, a function **increases** on an interval of its domain if its graph rises from left to right. It **decreases** on an interval if its graph falls from left to right. It is **constant** on an interval if its graph is horizontal on the interval.

INCREASING, DECREASING, AND CONSTANT FUNCTIONS

Suppose that a function f is defined over an interval I.

a. f increases on I if, whenever $x_1 < x_2$, $f(x_1) < f(x_2)$;
b. f decreases on I if, whenever $x_1 < x_2$, $f(x_1) > f(x_2)$;
c. f is constant on I if, for every x_1 and x_2, $f(x_1) = f(x_2)$.

Figure 6 illustrates these ideas.

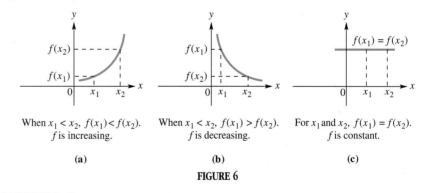

When $x_1 < x_2$, $f(x_1) < f(x_2)$.
f is increasing.

When $x_1 < x_2$, $f(x_1) > f(x_2)$.
f is decreasing.

For x_1 and x_2, $f(x_1) = f(x_2)$.
f is constant.

(a)   **(b)**   **(c)**

FIGURE 6

EXAMPLE 2 *Determining Intervals over Which a Function Is Increasing, Decreasing, or Constant* Figure 7 on the following page shows the graph of a function. Determine the intervals over which the function is increasing, decreasing, or constant.

SOLUTION In making our determination, we must always ask, "What is the y-value doing as x is getting larger?" For this graph, we see that on the interval $(-\infty, 1)$,

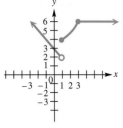

FIGURE 7

the y-values are *decreasing;* on the interval $[1, 3]$, the y-values are *increasing;* and on the interval $[3, \infty)$, the y-values are *constant* (all are 6). Therefore, the function is decreasing on $(-\infty, 1)$, increasing on $[1, 3]$, and constant on $[3, \infty)$.   ◆

Caution   A common error involves writing range values when determining intervals like those in Example 2. Remember that we are determining intervals of the domain, and thus are interested in x-values for our interval designations.

For Group Discussion

1. In the standard viewing window of your calculator, enter any linear function $y = mx + b$ with $m > 0$. Now, trace the graph from *left to right*. Watch the y-values as x gets larger. What is happening to y? How does this reinforce the concepts presented so far in this section?
2. Repeat this exercise, but with $m < 0$.
3. Repeat this exercise, but with $m = 0$.

The next part of this section is devoted to the introduction of several important basic functions and relations that will be investigated in detail as we progress through the chapters of this book.

THE IDENTITY FUNCTION

If we let $m = 1$ and $b = 0$ for the general form of the linear function $f(x) = mx + b$, we get the **identity function** $f(x) = x$. This function pairs every real number with itself.

IDENTITY FUNCTION

$f(x) = x$ (Figure 8)   Domain: $(-\infty, \infty)$ Range: $(-\infty, \infty)$

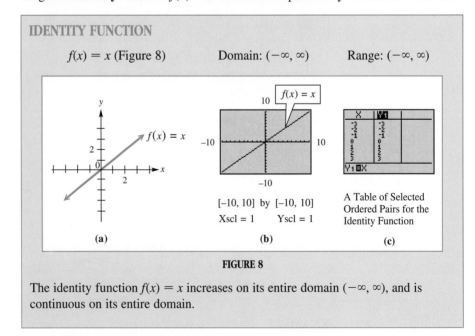

[−10, 10] by [−10, 10]
Xscl = 1   Yscl = 1

A Table of Selected Ordered Pairs for the Identity Function

(a)   (b)   (c)

FIGURE 8

The identity function $f(x) = x$ increases on its entire domain $(-\infty, \infty)$, and is continuous on its entire domain.

THE SQUARING FUNCTION AND SYMMETRY WITH RESPECT TO THE y-AXIS

We now look at the graph of the simplest degree 2 function, the **squaring function** $f(x) = x^2$. (The word *quadratic* refers to degree 2; we will investigate the general quadratic function in Chapter 3.) This function pairs every real number with its square. Its graph is called a **parabola**.

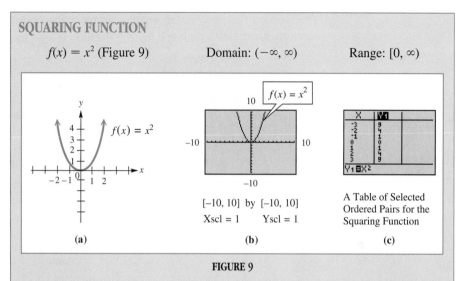

SQUARING FUNCTION

$f(x) = x^2$ (Figure 9)   Domain: $(-\infty, \infty)$ Range: $[0, \infty)$

[−10, 10] by [−10, 10]
Xscl = 1   Yscl = 1

A Table of Selected Ordered Pairs for the Squaring Function

(a)   (b)   (c)

FIGURE 9

The squaring function $f(x) = x^2$ decreases on the interval $(-\infty, 0]$ and increases on the interval $[0, \infty)$. It is continuous on its entire domain. The point at which the graph changes from decreasing to increasing (the point $(0, 0)$) is called the **vertex** of the parabola.

This table supports, but does not *prove*, that the graph of the squaring function is symmetric with respect to the y-axis. (These x-values were obtained from a TI-83 calculator by using the ASK option in the table setup.)

Notice that if we were able to "fold" the graph of $f(x) = x^2$ along the y-axis, the two halves would coincide exactly. In mathematics we refer to this property as symmetry, and we say that the graph of $f(x) = x^2$ is **symmetric with respect to the y-axis.** This may be generalized as follows.

SYMMETRY WITH RESPECT TO THE y-AXIS

If a function f is defined so that

$$f(x) = f(-x)$$

for all x in its domain, then the graph of f is symmetric with respect to the y-axis.

Some *particular* cases illustrating that the graph of $f(x) = x^2$ is symmetric with respect to the y-axis are as follows:

$$f(-4) = f(4) = 16$$
$$f(-3) = f(3) = 9$$
$$f(-2) = f(2) = 4$$
$$f(-1) = f(1) = 1$$
$$f(-0) = f(0) = 0.$$

This pattern holds for any real number x, since $f(-x) = (-x)^2 = x^2 = f(x)$.

THE CUBING FUNCTION AND SYMMETRY WITH RESPECT TO THE ORIGIN

The function $f(x) = x^3$ is the simplest degree 3 function, and it is called the **cubing** function. It pairs with each real number the third power, or cube, of the number.

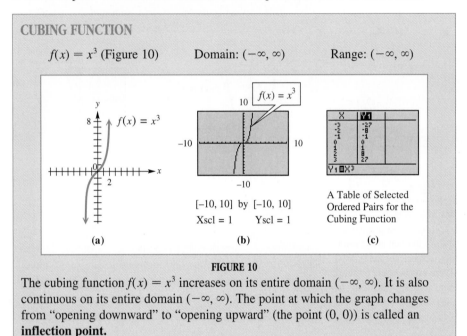

CUBING FUNCTION

$f(x) = x^3$ (Figure 10)   Domain: $(-\infty, \infty)$ Range: $(-\infty, \infty)$

[−10, 10] by [−10, 10]
Xscl = 1   Yscl = 1

(a)   **(b)**   **(c)**

A Table of Selected Ordered Pairs for the Cubing Function

FIGURE 10

The cubing function $f(x) = x^3$ increases on its entire domain $(-\infty, \infty)$. It is also continuous on its entire domain $(-\infty, \infty)$. The point at which the graph changes from "opening downward" to "opening upward" (the point $(0, 0)$) is called an **inflection point.**

Notice that if we were able to "fold" the graph of $f(x) = x^3$ along the y-axis and then along the x-axis, forming a "corner" at the origin, the two parts of the graph would coincide exactly. We say that the graph of $f(x) = x^3$ is **symmetric with respect to the origin.** This may be generalized as follows.

SYMMETRY WITH RESPECT TO THE ORIGIN

If a function f is defined so that

$$f(-x) = -f(x)$$

for all x in its domain, then the graph of f is symmetric with respect to the origin.

This table supports, but does not *prove*, that the graph of the cubing function is symmetric with respect to the origin.

Some *particular* cases illustrating that the graph of $f(x) = x^3$ is symmetric with respect to the origin are as follows:

$$f(-2) = -f(2) = -8$$
$$f(-1) = -f(1) = -1$$
$$f(-0) = -f(0) = -0 = 0.$$

This pattern holds for any real number x, since

$$f(-x) = (-x)^3 = (-1)^3 x^3 = -x^3 = -f(x).$$

EXAMPLE 3 *Determining Symmetry Analytically and Supporting it Graphically*

(a) Show analytically and support graphically the fact that $f(x) = x^4 - 3x^2 - 8$ has a graph that is symmetric with respect to the y-axis.

(b) Show analytically and support graphically the fact that $f(x) = x^3 - 4x$ has a graph that is symmetric with respect to the origin.

SOLUTION

(a) ANALYTIC We must show that $f(-x) = f(x)$ for any x.

$$f(-x) = (-x)^4 - 3(-x)^2 - 8$$
$$= (-1)^4 x^4 - 3(-1)^2 x^2 - 8$$
$$= x^4 - 3x^2 - 8$$
$$= f(x)$$

This *proves* that there is symmetry with respect to the y-axis.

GRAPHICAL The graph in Figure 11 supports this conclusion, since it appears to have this symmetry. (*Note:* Visual support is not a proof!) We will study graphs of this type in more detail in Chapter 3.

(b) ANALYTIC In this case, we must show that $f(-x) = -f(x)$ for any x.

$$f(-x) = (-x)^3 - 4(-x)$$
$$= (-1)^3 x^3 + 4x$$
$$= -x^3 + 4x \qquad *$$
$$= -(x^3 - 4x)$$
$$= -f(x)$$

In the line denoted *, note that the signs of the coefficients are all *opposites* of those in $f(x)$. We completed the argument by factoring out -1, showing that the final result is $-f(x)$.

GRAPHICAL The graph in Figure 12 supports our conclusion that the graph is symmetric with respect to the origin, since folding it along the y-axis and then along the x-axis would lead to coinciding curves. ◆

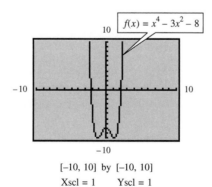

[−10, 10] by [−10, 10]
Xscl = 1   Yscl = 1

This graph is symmetric with respect to the y-axis.

FIGURE 11

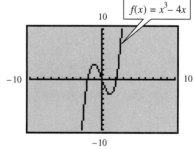

[−10, 10] by [−10, 10]
Xscl = 1   Yscl = 1

This graph is symmetric with respect to the origin.

FIGURE 12

THE SQUARE ROOT AND CUBE ROOT FUNCTIONS

We now investigate functions with expressions involving radicals. The first of these is the **square root function,** $f(x) = \sqrt{x}$. Notice that for the function value to be a real number, we must have $x \geq 0$. Thus, the domain is restricted to nonnegative numbers.

SQUARE ROOT FUNCTION

$f(x) = \sqrt{x}$ (Figure 13)   Domain: $[0, \infty)$ Range: $[0, \infty)$

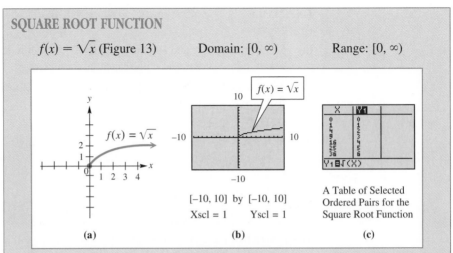

(a)   (b)   (c)

[−10, 10] by [−10, 10]
Xscl = 1   Yscl = 1

A Table of Selected
Ordered Pairs for the
Square Root Function

FIGURE 13

The square root function $f(x) = \sqrt{x}$ increases on $[0, \infty)$. It is also continuous on $[0, \infty)$. (The definition of rational exponents allows us to also enter $\sqrt{x}$ as $x^{1/2}$ on a calculator.)

The **cube root function,** $f(x) = \sqrt[3]{x}$, differs from the square root function in that *any* real number—positive, zero, or *negative*—has a real cube root, and thus the domain is $(-\infty, \infty)$. Also, when $x > 0$, $\sqrt[3]{x} > 0$, when $x = 0$, $\sqrt[3]{x} = 0$, and when $x < 0$, $\sqrt[3]{x} < 0$. As a result, the range is also $(-\infty, \infty)$.

CUBE ROOT FUNCTION

$f(x) = \sqrt[3]{x}$ (Figure 14)   Domain: $(-\infty, \infty)$ Range: $(-\infty, \infty)$

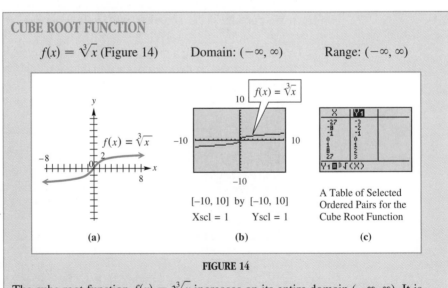

(a)   (b)   (c)

[−10, 10] by [−10, 10]
Xscl = 1   Yscl = 1

A Table of Selected
Ordered Pairs for the
Cube Root Function

FIGURE 14

The cube root function $f(x) = \sqrt[3]{x}$ increases on its entire domain $(-\infty, \infty)$. It is also continuous on $(-\infty, \infty)$. (The definition of rational exponents allows us to also enter $\sqrt[3]{x}$ as $x^{1/3}$ on a calculator.)

TECHNOLOGY NOTE

You should become familiar with the command on your particular calculator that allows you to graph the absolute value.

THE ABSOLUTE VALUE FUNCTION

On a number line, the absolute value of a real number x, denoted $|x|$, represents its undirected distance from the origin, 0. The **absolute value function,** which pairs every real number with its absolute value, is defined as follows.

DEFINITION OF ABSOLUTE VALUE $|x|$

$$f(x) = |x| = \begin{cases} x & \text{if } x \geq 0 \\ -x & \text{if } x < 0 \end{cases}$$

Notice that this function is defined in two parts. We use $|x| = x$ if x is positive or zero, and we use $|x| = -x$ if x is negative. Since x can be any real number, the domain of the absolute value function is $(-\infty, \infty)$, but since $|x|$ cannot be negative, the range is $[0, \infty)$.

ABSOLUTE VALUE FUNCTION

$f(x) = |x|$ (Figure 15)   Domain: $(-\infty, \infty)$ Range: $[0, \infty)$

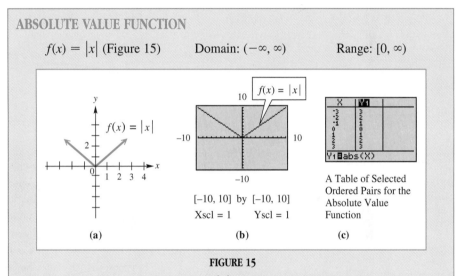

(a)   (b)   (c)

FIGURE 15

The absolute value function $f(x) = |x|$ decreases on the interval $(-\infty, 0]$ and increases on $[0, \infty)$. It is continuous on its entire domain.

For Group Discussion   Based on the discussion so far in this section, answer the following questions.

1. Which functions have graphs that are symmetric with respect to the y-axis?
2. Which functions have graphs that are symmetric with respect to the origin?
3. Which functions have graphs that show neither of these symmetries?
4. Why is it not possible for the graph of a function to be symmetric with respect to the x-axis?

THE RELATION $x = y^2$ AND SYMMETRY WITH RESPECT TO THE x-AXIS

Our discussion in this text since Section 1.2 has dealt almost exclusively with functions. Recall from Chapter 1 that a function is a relation that satisfies the condition that every domain value is paired with one and only one range value. However, there are cases where we are interested in graphing relations that are not functions, and one of the simplest of these is the relation defined by the equation $x = y^2$. Notice that the table of selected ordered pairs below indicates that this relation has two different y-values for each positive value of x.

Selected Ordered Pairs for $x = y^2$

x	y
0	0
1	± 1
4	± 2
9	± 3

Two different y-values for the same x-value

If we plot these points and join them with a smooth curve, we find that the graph of $x = y^2$ is a parabola opening to the right. See Figure 16.

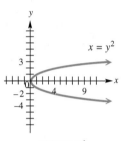

FIGURE 16

If a graphing calculator is set for the function mode, it is not possible to graph $x = y^2$ directly. (However, if it is set for the *parametric* mode, such a curve is possible with direct graphing.) To overcome this problem, we begin with $x = y^2$ and take the square root on each side, remembering to choose both the positive and negative square roots of x:

$$x = y^2 \qquad \text{Given equation}$$
$$y^2 = x \qquad \text{Transform so that } y \text{ is on the left.}$$
$$y = \pm\sqrt{x}. \qquad \text{Take square roots.}$$

Now, we have $x = y^2$ defined by two *functions*, $y_1 = \sqrt{x}$ and $y_2 = -\sqrt{x}$. Entering both of these into a calculator gives the graph shown in Figure 17.

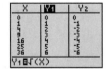

This screen shows selected ordered pairs for the relation $x = y^2$, where $Y_1 = \sqrt{x}$ and $Y_2 = -\sqrt{x}$.

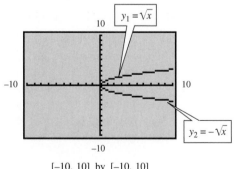

[–10, 10] by [–10, 10]
Xscl = 1   Yscl = 1

The graph of the relation $x = y^2$ is symmetric with respect to the x-axis.

FIGURE 17

It appears that if we were to fold the graph of $x = y^2$ along the x-axis, the two halves of the parabola would coincide. This is indeed the case, and this graph exhibits symmetry with respect to the x-axis.

SYMMETRY OF A GRAPH WITH RESPECT TO THE x-AXIS

If replacing y with $-y$ in an equation results in the same equation, then the graph is symmetric with respect to the x-axis.

To illustrate this, if we begin with $x = y^2$ and replace y with $-y$, we get

$$x = (-y)^2$$
$$x = (-1)^2 y^2$$
$$x = y^2. \qquad \text{The same equation with which we started}$$

A summary of the types of symmetry just discussed follows.

Type of Symmetry	Example	Basic Fact about Points on the Graph
y-axis symmetry	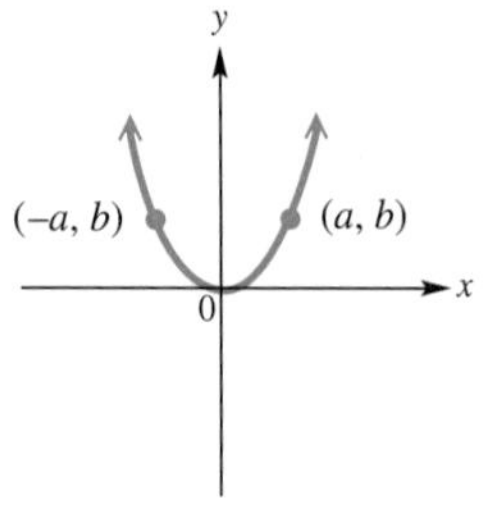	If (a, b) is on the graph, so is $(-a, b)$.
Origin symmetry	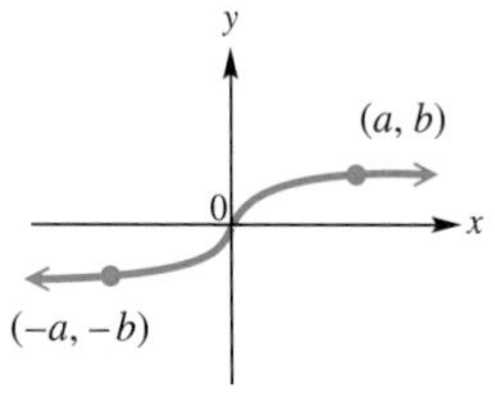	If (a, b) is on the graph, so is $(-a, -b)$.
x-axis symmetry (not possible for a function)	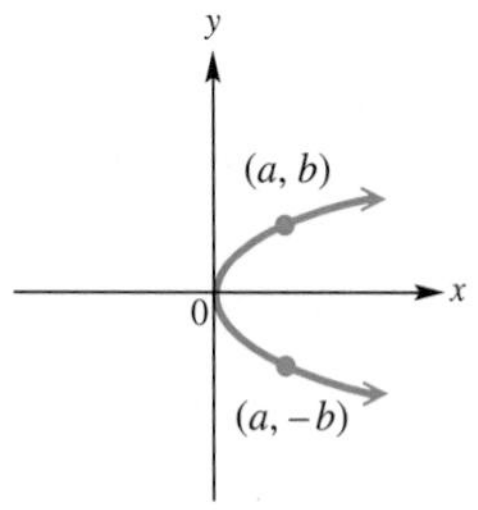	If (a, b) is on the graph, so is $(a, -b)$.

EVEN AND ODD FUNCTIONS

Closely associated with the concepts of symmetry with respect to the y-axis and symmetry with respect to the origin are the ideas of even and odd functions.

EVEN AND ODD FUNCTIONS

A function f is called an **even function** if $f(-x) = f(x)$ for all x in the domain of f. (Its graph is symmetric with respect to the y-axis.)

A function f is called an **odd function** if $f(-x) = -f(x)$ for all x in the domain of f. (Its graph is symmetric with respect to the origin.)

As an illustration, $f(x) = x^2$ is an even function because
$$f(-x) = (-x)^2 = x^2 = f(x).$$
The function $f(x) = x^3$ is an odd function because
$$f(-x) = (-x)^3 = -x^3 = -f(x).$$
A function may be neither even nor odd; for example, $f(x) = \sqrt{x}$ is neither even nor odd.

EXAMPLE 4 *Determining Analytically Whether a Function Is Even, Odd, or Neither*   Decide if the functions defined as follows are even, odd, or neither.

(a) $f(x) = 8x^4 - 3x^2$ **(b)** $f(x) = 6x^3 - 9x$ **(c)** $f(x) = 3x^2 + 5x$

SOLUTION

(a)   Replacing x with $-x$ gives
$$f(-x) = 8(-x)^4 - 3(-x)^2 = 8x^4 - 3x^2 = f(x).$$

Since $f(x) = f(-x)$ for each x in the domain of the function, f is an even function.

(b)   Here
$$f(-x) = 6(-x)^3 - 9(-x) = -6x^3 + 9x = -f(x).$$

This function is odd.

(c) $f(-x) = 3(-x)^2 + 5(-x)$
$$= 3x^2 - 5x$$

Since $f(-x) \neq f(x)$ and $f(-x) \neq -f(x)$, f is neither even nor odd.   ◆

For Group Discussion   The three functions discussed in Example 4 are graphed in Figure 18, but not necessarily in the same order as in the example. Without actually using your calculator, identify each function, remembering that an even function is symmetric with respect to the y-axis and an odd function is symmetric with respect to the origin.

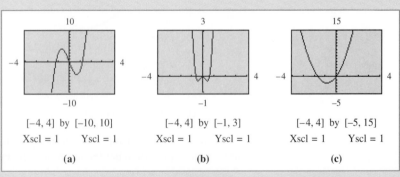

[-4, 4] by [-10, 10]   [-4, 4] by [-1, 3]   [-4, 4] by [-5, 15]
Xscl = 1   Yscl = 1   Xscl = 1   Yscl = 1   Xscl = 1   Yscl = 1
(a)   **(b)**   **(c)**

FIGURE 18

2.1   EXERCISES   **Tape 2**

Determine the intervals of the domain over which the given function is continuous.

1.
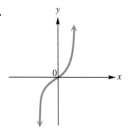

2.
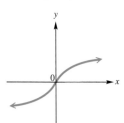

3.
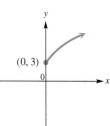

4.
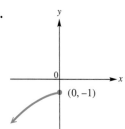

5.
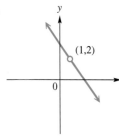

6.
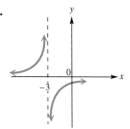

7. Graph the function $y = \frac{x^2 - 9}{x + 3}$ in the standard viewing window of your calculator. At first glance, does this graph seem to be continuous over the entire interval of the domain shown in the window? Now, try to locate the point for which $x = -3$. What happens? Why do you think this happens? (Functions of this kind, called *rational functions*, will be studied in detail in Chapter 4.)

8. Based on your work in Exercise 7, do you think that determination of continuity strictly by observation of a calculator-generated graph is foolproof?

*Determine the intervals of the domain over which the given function is (**a**) increasing, (**b**) decreasing, and (**c**) constant.*

9.
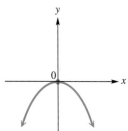

10.
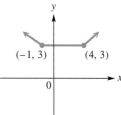

11.
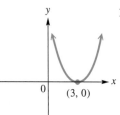

12.
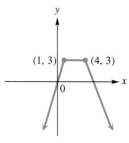

13.
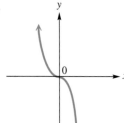

14.
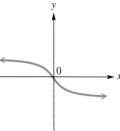

15.
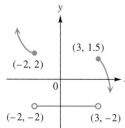

16.
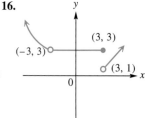

*In Exercises 17–22, you are given a function and an interval. Graph the function in the standard viewing window of your calculator, and trace from left to right along a representative portion of the specified interval. Then, fill in the blank of this sentence with either **increasing** or **decreasing**: OVER THE INTERVAL SPECIFIED, THIS FUNCTION IS _____.*

17. $f(x) = x^5; (-\infty, \infty)$

18. $f(x) = -x^3; (-\infty, \infty)$

19. $f(x) = x^4; (-\infty, 0]$

20. $f(x) = x^4; [0, \infty)$

21. $f(x) = -|x|; (-\infty, 0]$

22. $f(x) = -|x|; [0, \infty)$

Based on a visual observation, determine whether each graph is symmetric with respect to each of the following: (a) x-axis, (b) y-axis, (c) origin.

23. 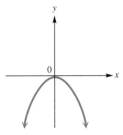 **24.** 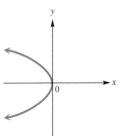 **25.** 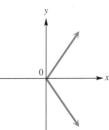 **26.**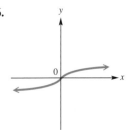

27. 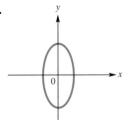 **28.** 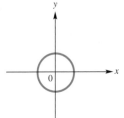 **29.** 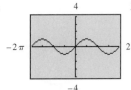 **30.**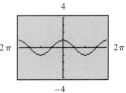

31. 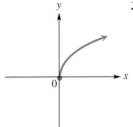 **32.**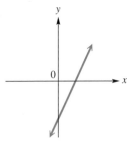

In Exercises 33–38, you are given a calculator-generated graph of a relation that exhibits a type of symmetry. Based on the given information at the bottom of each figure, determine the coordinates of another point that must also lie on the graph.

33. symmetric with respect to the y-axis

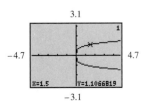

34. symmetric with respect to the y-axis

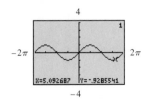

35. symmetric with respect to the x-axis

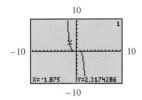

36. symmetric with respect to the x-axis

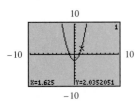

37. symmetric with respect to the origin

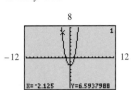

38. symmetric with respect to the origin

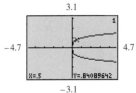

39. Complete the left half of the graph of $y = f(x)$ in the figure for each of the following conditions:

    (a) $f(-x) = f(x)$

    (b) $f(-x) = -f(x)$.

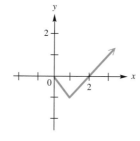

40. Complete the right half of the graph of $y = f(x)$ in the figure for each of the following conditions:

    (a) f is odd

    (b) f is even.

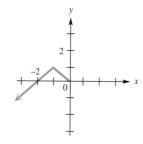

Based on the ordered pairs seen in each pair of tables, make a conjecture about whether the function defined in Y_1 *is even, odd, or neither even nor odd.*

41.

42.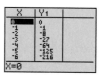

43.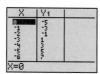

44.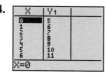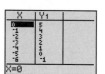

Prove analytically that each of the following functions is even.

45. $f(x) = x^4 - 7x^2 + 6$

46. $f(x) = -2x^6 - 8x^2$

Prove analytically that each of the following functions is odd.

47. $f(x) = 3x^3 - x$

48. $f(x) = -x^5 + 2x^3 - 3x$

Use the method of Example 3 to determine whether the given function is symmetric with respect to the y-axis, symmetric with respect to the origin, or neither of these. Then, graph the function on your calculator to support your conclusion, using the window specified.

49. $f(x) = -x^3 + 2x$;
    $[-10, 10]$ by $[-10, 10]$

50. $f(x) = x^5 - 2x^3$;
    $[-10, 10]$ by $[-10, 10]$

51. $f(x) = .5x^4 - 2x^2 + 1$;
    $[-10, 10]$ by $[-10, 10]$

52. $f(x) = .75x^2 + |x| + 1$;
    $[-10, 10]$ by $[-10, 10]$

53. $f(x) = x^3 - x + 3$;
    $[-10, 10]$ by $[-10, 10]$

54. $f(x) = x^4 - 5x + 2$;
    $[-10, 10]$ by $[-10, 10]$

Refer to the function described and determine whether it is even, odd, or neither even nor odd.

55. the function in Exercise 49

56. the function in Exercise 50

57. the function in Exercise 51

58. the function in Exercise 52

59. the function in Exercise 53

60. the function in Exercise 54

In Exercises 61–64, you are given a portion of the graph of a function near the origin, along with information regarding symmetry of the graph of the complete function. Sketch a simulated calculator graph, using the window coordinates given, to show what the graph would look like in that window.

61. Symmetric with respect to the y-axis
    Use the window $[-5, 5]$ by $[0, 5]$.

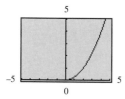

62. Symmetric with respect to the y-axis
    Use the window $[-5, 5]$ by $[0, 2]$.

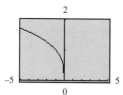

63. Symmetric with respect to the origin
Use the window $[-5, 5]$ by $[-2, 2]$.

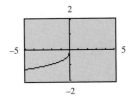

64. Symmetric with respect to the origin
Use the window $[-5, 5]$ by $[-2, 2]$.

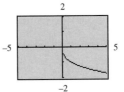

RELATING CONCEPTS (EXERCISES 65–68)

The line in the sketch and the line on the screen are said to be tangent to the curve at point P.

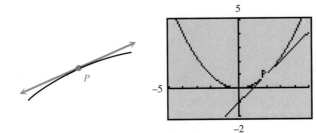

65. Sketch by hand the graph of $y = x^2$, and choose a point at which the function is *decreasing*. Draw a tangent line at that point. Is the slope of this line positive, negative, or zero? (Section 1.3)

66. Repeat Exercise 65, but choose a point at which the function is *increasing*. (Section 2.1)

67. Repeat Exercise 65, but choose the point at which the function changes from *decreasing* to *increasing*. (Section 2.1)

68. Based on your answers to Exercises 65–67, what conclusions can be drawn? These conclusions can be verified using methods discussed in calculus. (Section 2.1)

Draw sketches of the graphs of the basic algebraic functions introduced in this section, avoiding the temptation to look back at them in the text. Also, do not use your calculator. They are:

$$f(x) = x \quad f(x) = x^2 \quad f(x) = x^3$$
$$f(x) = \sqrt{x} \quad f(x) = \sqrt[3]{x} \quad f(x) = |x|.$$

Use your sketches to determine whether the following are true or false.

69. The range of $f(x) = x^2$ is the same as the range of $f(x) = |x|$.

70. The functions $f(x) = x^2$ and $f(x) = |x|$ increase on the same interval.

71. The functions $f(x) = \sqrt{x}$ and $f(x) = \sqrt[3]{x}$ have the same domain.

72. The function $f(x) = \sqrt[3]{x}$ decreases on its entire domain.

73. The function $f(x) = x$ has its domain equal to its range.

74. The function $f(x) = \sqrt{x}$ is continuous on the interval $(-\infty, 0)$.

75. None of the functions shown decreases on the interval $[0, \infty)$.

76. Both $f(x) = x$ and $f(x) = x^3$ have graphs that are symmetric with respect to the origin.

77. Both $f(x) = x^2$ and $f(x) = |x|$ have graphs that are symmetric with respect to the y-axis.

78. None of the graphs shown is symmetric with respect to the x-axis.

2.2 VERTICAL AND HORIZONTAL SHIFTS OF GRAPHS OF FUNCTIONS

Vertical Shifts ◆ Horizontal Shifts ◆ Combinations of Vertical and Horizontal Shifts ◆ Effects of Shifts on Domain and Range ◆ Horizontal Shifts Applied to Equations for Modeling

In this section, we will examine how the graphs of the elementary functions introduced in the previous section can be shifted vertically and horizontally in the plane. The basic ideas can then be generalized to apply to the graph of any function.

VERTICAL SHIFTS

For Group Discussion   In each group of functions below, we give four related functions. Graph the four functions in the first group (Group A), and then answer the questions below regarding those functions. Then repeat the process for Group B, Group C, and Group D. Use the standard viewing window in each case.

A	B	C	D
$y_1 = x^2$	$y_1 = x^3$	$y_1 = \sqrt{x}$	$y_1 = \sqrt[3]{x}$
$y_2 = x^2 + 3$	$y_2 = x^3 + 3$	$y_2 = \sqrt{x} + 3$	$y_2 = \sqrt[3]{x} + 3$
$y_3 = x^2 - 2$	$y_3 = x^3 - 2$	$y_3 = \sqrt{x} - 2$	$y_3 = \sqrt[3]{x} - 2$
$y_4 = x^2 + 5$	$y_4 = x^3 + 5$	$y_4 = \sqrt{x} + 5$	$y_4 = \sqrt[3]{x} + 5$

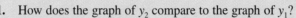

1. How does the graph of y_2 compare to the graph of y_1?
2. How does the graph of y_3 compare to the graph of y_1?
3. How does the graph of y_4 compare to the graph of y_1?
4. If $c > 0$, how do you think the graph of $y_1 + c$ would compare to the graph of y_1?
5. If $c > 0$, how do you think the graph of $y_1 - c$ would compare to the graph of y_1?

Choosing your own value of c, support your answers to Items 4 and 5 graphically. (Be sure that your choice is appropriate for the standard window.)

This screen shows how to minimize keystrokes in the activity in the "For Group Discussion" box. The entry for Y_1 can be altered as needed.

The objective of the preceding group discussion activity was to make conjectures about how the addition or subtraction of a constant c would affect the graph of a function $y = f(x)$. In each case, we obtained a vertical shift, or **translation,** of the graph of the basic function with which we started. Although our observations were based on the graphs of four different elementary functions, they can be generalized to any function.

VERTICAL SHIFTING OF THE GRAPH OF A FUNCTION

If $c > 0$, the graph of $y = f(x) + c$ is obtained by shifting the graph of $y = f(x)$ *upward* a distance of c units. The graph of $y = f(x) - c$ is obtained by shifting the graph of $y = f(x)$ *downward* a distance of c units.

In Figure 19, we give a graphical interpretation of the statement above.

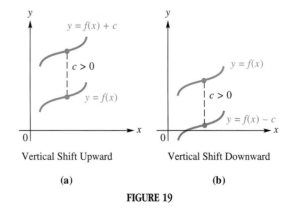

Vertical Shift Upward   Vertical Shift Downward

(a)   (b)

FIGURE 19

EXAMPLE 1   *Recognizing Vertical Shifts on Calculator-Generated Graphs*
Figure 20 shows the graphs of four functions. The graph labeled y_1 is that of the function $f(x) = |x|$. The viewing window is $[-10, 10]$ by $[-10, 10]$, with Xscl $= 1$ and Yscl $= 1$. Each of y_2, y_3, and y_4 are functions of the form $f(x) + c$ or $f(x) - c$, for $c > 0$. Give the rule for each of these functions.

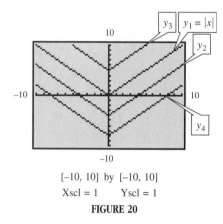

[-10, 10] by [-10, 10]
Xscl = 1   Yscl = 1
FIGURE 20

SOLUTION   Because the graph of y_2 lies 4 units below the graph of y_1, we have $y_2 = |x| - 4$. The graph of y_3 is a shift of the graph of y_1 a distance of 5 units upward, so the equation for y_3 is $y_3 = |x| + 5$. Finally, the graph of y_4 is a vertical shift of the graph of y_1 8 units downward, so its equation is $y_4 = |x| - 8$.   ◆

When using a graphing calculator to investigate shifts of graphs, it is important to use an appropriate window; otherwise, the graph may not appear. For example, the graph of $y = |x| + 12$ does not appear in the viewing window $[-10, 10]$ by $[-10, 10]$. Why is this so? There are many windows that would show the graph. (Name one.)

HORIZONTAL SHIFTS

For Group Discussion   This discussion parallels the one earlier in this section. Follow the same general directions.

A	B	C	D
$y_1 = x^2$	$y_1 = x^3$	$y_1 = \sqrt{x}$	$y_1 = \sqrt[3]{x}$
$y_2 = (x - 3)^2$	$y_2 = (x - 3)^3$	$y_2 = \sqrt{x - 3}$	$y_2 = \sqrt[3]{x - 3}$
$y_3 = (x - 5)^2$	$y_3 = (x - 5)^3$	$y_3 = \sqrt{x - 5}$	$y_3 = \sqrt[3]{x - 5}$
$y_4 = (x + 4)^2$	$y_4 = (x + 4)^3$	$y_4 = \sqrt{x + 4}$	$y_4 = \sqrt[3]{x + 4}$

1. How does the graph of y_2 compare to the graph of y_1?
2. How does the graph of y_3 compare to the graph of y_1?
3. How does the graph of y_4 compare to the graph of y_1?
4. If $c > 0$, how do you think the graph of $y_5 = f(x - c)$ would compare to the graph of $y_1 = f(x)$?
5. If $c > 0$, how do you think the graph of $y_5 = f(x + c)$ would compare to the graph of $y_1 = f(x)$?

Choosing your own value of c, support your answers to Items 4 and 5 graphically. Again, be sure that your choice is appropriate for the standard window.

The results of the preceding discussion should remind you of the results found earlier. There we saw how graphs of functions can be shifted vertically. Now, we see how they can be shifted *horizontally*. The observations can be generalized as follows.

HORIZONTAL SHIFTING OF THE GRAPH OF A FUNCTION

If $c > 0$, the graph of $y = f(x - c)$ is obtained by shifting the graph of $y = f(x)$ to the *right* a distance of c units. The graph of $y = f(x + c)$ is obtained by shifting the graph of $y = f(x)$ to the *left* a distance of c units.

Caution   Errors of interpretation frequently occur when horizontal shifts are involved. In order to determine the direction and magnitude of horizontal shifts, find the value of *x* that would cause the expression within the parentheses to equal 0. For example, the graph of $f(x) = (x - 5)^2$ would be shifted 5 units to the *right*, because +5 would cause $x - 5$ to equal 0. On the other hand, the graph of $f(x) = (x + 4)^2$ would be shifted 4 units to the *left*, because −4 would cause $x + 4$ to equal 0.

Figure 21 illustrates the effect of horizontal shifts of the graph of a function $y = f(x)$.

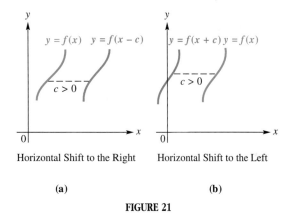

Horizontal Shift to the Right   Horizontal Shift to the Left

(a)   (b)

FIGURE 21

EXAMPLE 2   *Recognizing Horizontal Shifts on Calculator-Generated Graphs*
Figure 22 shows the graphs of four functions. As in Example 1, the function labeled y_1 is the function $f(x) = |x|$. The viewing window is $[-10, 10]$ by $[-10, 10]$, with Xscl = 1 and Yscl = 1. Each of y_2, y_3, and y_4 are functions of the form $f(x - c)$ or $f(x + c)$, where $c > 0$. Give the rule for each of these functions.

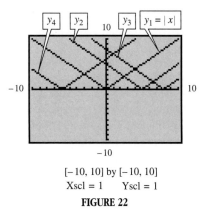

$[-10, 10]$ by $[-10, 10]$
Xscl = 1   Yscl = 1
FIGURE 22

SOLUTION   The graph of y_2 is the same as the graph of y_1, but it is shifted 5 units to the *right.* Therefore, we have $y_2 = |x - 5|$. Similarly, since y_3 is the graph of y_1 shifted 8 units to the *right,* $y_3 = |x - 8|$. The graph of y_4 is obtained by shifting that of y_1 6 units to the *left,* so its equation is $y_4 = |x + 6|$.   ◆

COMBINATIONS OF VERTICAL AND HORIZONTAL SHIFTS

Now that we have seen how graphs of functions can be shifted vertically and shifted horizontally, it is not difficult to extend these ideas to graphs that are obtained by applying *both* types of translations.

EXAMPLE 3   *Applying Both Vertical and Horizontal Shifts*   Describe how the graph of $y = |x + 15| - 20$ would be obtained by translating the graph of $y = |x|$. Determine an appropriate viewing window, and support the results by plotting both functions with a graphing calculator.

SOLUTION   The function defined by $y = |x + 15| - 20$ is translated 15 units to the *left* (because of the $|x + 15|$) and 20 units *downward* as compared to the graph of $y = |x|$. Because the point at which the graph changes from decreasing to increasing is now $(-15, -20)$, the standard viewing window is not appropriate. We must choose a window that contains the point $(-15, -20)$ in order to obtain a comprehensive graph. While many such windows are possible, one such window is shown in Figure 23. The display at the bottom of the screen indicates that the point $(-15, -20)$ lies on the graph.   ◆

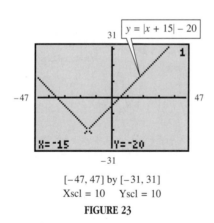

$[-47, 47]$ by $[-31, 31]$
Xscl = 10   Yscl = 10
FIGURE 23

EFFECTS OF SHIFTS ON DOMAIN AND RANGE

The domains and ranges of functions may or may not be affected by vertical and horizontal shifts. For example, if the domain of a function is $(-\infty, \infty)$, a horizontal shift will not affect the domain. Similarly, if the range is $(-\infty, \infty)$, a vertical shift will not affect the range. However, if the domain is not $(-\infty, \infty)$, a horizontal shift will affect the domain, and if the range is not $(-\infty, \infty)$, a vertical shift will affect the range. The next example illustrates this.

EXAMPLE 4   *Determining Domains and Ranges of Shifted Graphs*   Four functions are graphed in Figures 24–27. Give the domain and the range of each function.

SOLUTION   The graph of $y = (x - 12)^2 + 25$ is shown in Figure 24. It is a translation of the graph of $y = x^2$ 12 units to the right and 25 units upward. The original domain $(-\infty, \infty)$ is not affected. However, the range of this function is $[25, \infty)$, because of the vertical translation.

The graph in Figure 25, that of $y = (x + 10)^3 - 15$, was obtained by vertical and horizontal shifts of the graph of $y = x^3$, a function that has both domain and range equal to $(-\infty, \infty)$. Neither is affected here, and so the domain and range of $y = (x + 10)^3 - 15$ are also both $(-\infty, \infty)$.

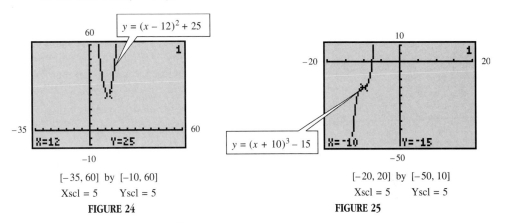

$[-35, 60]$ by $[-10, 60]$
Xscl = 5   Yscl = 5
FIGURE 24

$[-20, 20]$ by $[-50, 10]$
Xscl = 5   Yscl = 5
FIGURE 25

The function $y = \sqrt{x}$ has domain $[0, \infty)$. The function graphed in Figure 26, $y = \sqrt{x - 20} - 30$, was obtained by shifting the basic graph 20 units to the right, so the new domain is $[20, \infty)$. On the other hand, the original range, $[0, \infty)$, has been affected by the shift of the graph 30 units downward. The new range is $[-30, \infty)$.

The situation in Figure 27 is similar to that of Figure 25. A graph with domain and range both $(-\infty, \infty)$, that is, $y = \sqrt[3]{x}$, has been shifted 20 units to the left and 40 units upward. No matter what direction and magnitude these shifts might have been, the domain and the range are both unaffected. They both remain $(-\infty, \infty)$. ◆

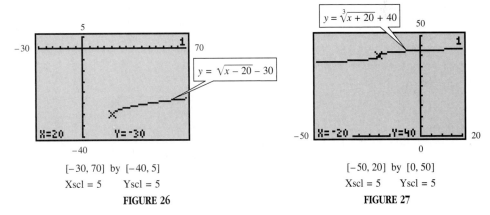

$[-30, 70]$ by $[-40, 5]$
Xscl = 5   Yscl = 5
FIGURE 26

$[-50, 20]$ by $[0, 50]$
Xscl = 5   Yscl = 5
FIGURE 27

The feature of graphing calculators that allows us to locate a point on a graph helps us support the results obtained in Example 4. You might wish to experiment with yours to confirm those results.

HORIZONTAL SHIFTS APPLIED TO EQUATIONS FOR MODELING

In Section 1.7, we examined data that showed the total amount (in millions of dollars) of government-guaranteed loans during the years 1986 through 1994. (See page 84.)

To make our work easier, we used $x = 0$ to represent the year 1986, $x = 1$ to represent the year 1987, and so on. We determined in Example 3 that the least squares regression line for the data is $y = 1.08833x + 8.7467$. The graph of this line is shown in Figure 86 on page 87. In order to use the values of the years 1986 through 1994 *directly* in a regression equation, we should shift the graph of the line 1986 units to the right. Based on the earlier discussion in this section, the equation of this new line would be $y = 1.08833(x - 1986) + 8.7467$.

With a graphing calculator, plot the set of points $\{(1986, 8.6), (1987, 9.8), (1988, 11.8), (1989, 12.5), (1990, 12.3), (1991, 13.5), (1992, 14.7), (1993, 16.5), (1994, 18.2)\}$ and graph $y = 1.08833(x - 1986) + 8.7467$ in the window [1986, 1996] by [1, 20]. Compare this to Figure 86 in Section 1.7 and notice that, except for the labels on the axes and the display of the equation, the two screens look exactly the same.

EXAMPLE 5   *Applying a Horizontal Shift to an Equation Model for Federal Debt*   The equation $y = 259.1x + 1843.4$ is the least squares regression line for the federal debt data first given in Exercise 14 of Section 1.7, where $x = 0$ corresponds to 1985, $x = 1$ to 1986, and so on, up to $x = 4$ for 1989. Write the equation that would allow the dates to be used directly as domain values.

SOLUTION   The expression $(x - 1985)$ becomes 0 if 1985 is substituted for x, becomes 1 if 1986 is substituted for x, and so on. Therefore, we replace x in the equation with $(x - 1985)$ to obtain $y = 259.1(x - 1985) + 1843.4$, the desired equation.   ◆

2.2   EXERCISES   Tape 2

Exercises 1–25 are grouped in "fives." For each group of five functions, match the correct graph A, B, C, D, or E to the function without using your calculator. You should use the concepts developed in this section to work these exercises based on visual observation. Then, after you have answered each group of five, use your calculator to check your answers. Every graph in these groups is plotted in the standard viewing window.

1. $y = x^2 - 3$

2. $y = (x - 3)^2$

3. $y = (x + 3)^2$

4. $y = (x - 3)^2 + 2$

5. $y = (x + 3)^2 + 2$

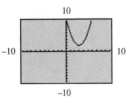

A

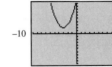

B

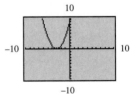

C

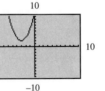

D

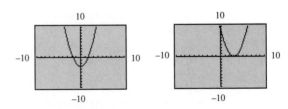

E

6. $y = |x| + 4$

7. $y = |x + 4|$

8. $y = |x - 4|$

9. $y = |x + 4| - 3$

10. $y = |x - 4| - 3$

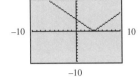

A

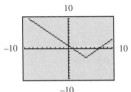

B

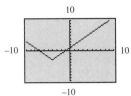

C

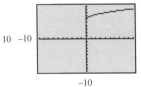

D

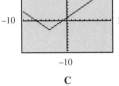

E

11. $y = \sqrt{x} + 6$

12. $y = \sqrt{x + 6}$

13. $y = \sqrt{x - 6}$

14. $y = \sqrt{x + 2} - 4$

15. $y = \sqrt{x - 2} - 4$

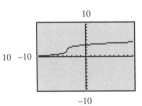

A

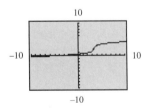

B

C

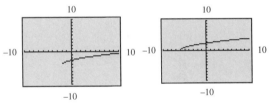

D

E

16. $y = \sqrt[3]{x} + 5$

17. $y = \sqrt[3]{x + 5}$

18. $y = \sqrt[3]{x - 4} + 2$

19. $y = \sqrt[3]{x + 4} + 2$

20. $y = \sqrt[3]{x - 4} - 2$

A

B

C

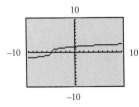

D

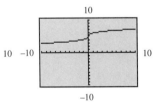

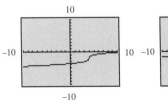

E

21. $y = x^3 + 3$

22. $y = (x - 3)^3$

23. $y = (x + 3)^3$

24. $y = (x + 2)^3 - 4$

25. $y = (x - 2)^3 - 4$

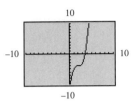

A

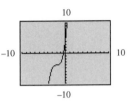

B

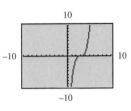

C

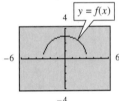

D

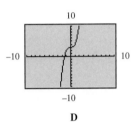

E

26. In which quadrant does the vertex of the graph of $y = (x - h)^2 + k$ lie, if $h < 0$ and $k < 0$?

*In Exercises 27–32, use the results of the specified earlier exercises and corresponding graphs to determine (**a**) the domain and (**b**) the range of the given function.*

27. $y = |x + 4| - 3$ *(Exercise 9)*   **28.** $y = |x - 4| - 3$ *(Exercise 10)*   **29.** $y = \sqrt{x - 2} - 4$ *(Exercise 15)*

30. $y = \sqrt{x + 2} - 4$ *(Exercise 14)*   **31.** $y = \sqrt[3]{x + 5}$ *(Exercise 17)*   **32.** $y = \sqrt[3]{x} + 5$ *(Exercise 16)*

The concepts introduced in this section can be applied to functions whose graphs may not be familiar to you. Given here is the graph of a function studied later in this text. We will call it $y = f(x)$. Each tick mark represents 1 unit.

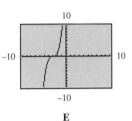

Now, match the function specified with the appropriate graph from the choices A, B, C, *or* D.

33. $y = f(x) + 1$

34. $y = f(x + 1)$

35. $y = f(x - 1)$

36. $y = f(x) - 1$

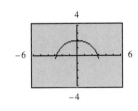

A

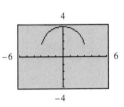

B

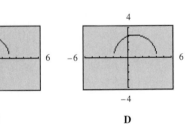

C

D

The function Y_2 *is defined as* $Y_1 + k$ *for some real number k. Based on the table shown, what is the value of k?*

37.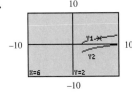

38.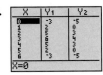

The function Y_2 *is defined as* $Y_1 + k$ *for some real number k. Based on the two views of the graphs of* Y_1 *and* Y_2 *and the displays at the bottoms of the screens, what is the value of k?*

39.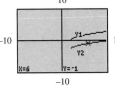

(6, 2) lies on the graph of Y_1.   (6, −1) lies on the graph of Y_2.
First View   Second View

40.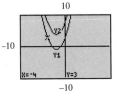

(−4, 3) lies on the graph of Y_1.   (−4, 8) lies on the graph of Y_2.
First View   Second View

Given the graph shown below, sketch by hand the graph of the function described, indicating how the three points labeled on the original graph have been translated.

41. $y = f(x) + 2$

42. $y = f(x) - 2$

43. $y = f(x + 2)$

44. $y = f(x - 2)$

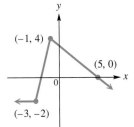

RELATING CONCEPTS (EXERCISES 45–52)

Recall from Chapter 1 that a unique line is determined by two different points on the line, and that the values of m and b then can be determined for the general form of the linear function $f(x) = mx + b$. *Work the exercises in order.*

45. Sketch by hand the line that passes through the points (1, −2) and (3, 2). (Section 1.3)

46. Use the slope formula to find the slope of this line. (Section 1.3)

47. Find the equation of this line and write it in the form $y_1 = mx + b$. (Section 1.4)

48. Keeping the same two *x*-values as indicated in Exercise 45, add 6 to each *y*-value. What are the coordinates of the two new points?

49. Find the slope of the line through the points determined in Exercise 48. (Section 1.3)

50. Find the equation of this new line and write it in the form $y_2 = mx + b$. (Section 1.4)

51. Graph both y_1 and y_2 in the standard viewing window of your calculator, and describe how the graph of y_2 can be obtained by vertically translating the graph of y_1. What is the value of the constant by which this vertical translation occurs? Where do you think this comes from? (Section 2.2)

52. Fill in the blanks with the correct responses, based on your work in Exercises 45–51. (Section 2.2)

If the points (x_1, y_1) and (x_2, y_2) lie on a line, then when we add the positive constant *c* to each *y*-value, we obtain the points $(x_1, y_1 + \underline{\hspace{1cm}})$ and $(x_2, y_2 + \underline{\hspace{1cm}})$. The slope of the new line is \underline{\hspace{2cm}} the slope of the original line. (the same as/different from)

The graph of the new line can be obtained by shifting the graph of the original line \underline{\hspace{1cm}} units in the \underline{\hspace{1cm}} direction.

Suppose that h and k are both positive numbers. Match the equation with the correct graph in Exercises 53–56.

53. $y = (x - h)^2 - k$

54. $y = (x + h)^2 - k$

55. $y = (x + h)^2 + k$

56. $y = (x - h)^2 + k$

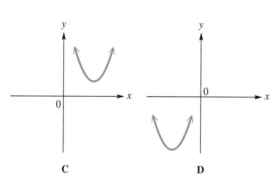

A   B

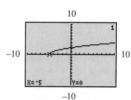

C   D

Each function in Exercises 57–60 is a translation of one of the basic functions $y = x^2$, $y = x^3$, $y = \sqrt{x}$, or $y = |x|$. Use the concepts of this section to find the equation that defines the function. Then, using the concepts of increasing and decreasing functions discussed in Section 2.1, determine the interval of the domain over which the function is (a) increasing and (b) decreasing.

57.

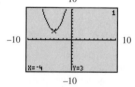

58.

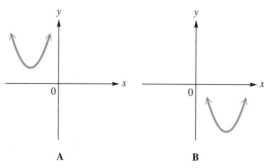

59.

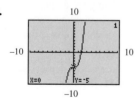

60.

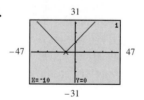

61. Suppose that the graph of $y = x^2$ is translated in such a way that its domain is $(-\infty, \infty)$ and its range is $[38, \infty)$. What are the possible values of h and k if the new function is of the form $y = (x - h)^2 + k$?

62. The graph shown is a translation of $y = |x|$. What are the values of h and k if the equation is of the form $y = |x - h| + k$?

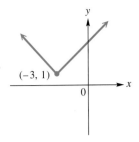

$(-3, 1)$

RELATING CONCEPTS (EXERCISES 63–66)

Use the x-intercept method of graphical solution of equations and inequalities to solve each equation or inequality, using the given graph of $y = f(x)$. See Sections 1.5 and 1.6.

63. (a) $f(x) = 0$
   (b) $f(x) > 0$
   (c) $f(x) < 0$

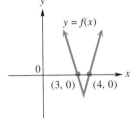

$y = f(x)$
$(3, 0)$ $(4, 0)$

64. (a) $f(x) = 0$
   (b) $f(x) > 0$
   (c) $f(x) < 0$

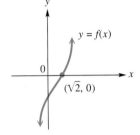

$y = f(x)$
$(\sqrt{2}, 0)$

65. (a) $f(x) = 0$
   (b) $f(x) \geq 0$
   (c) $f(x) \leq 0$

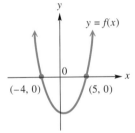

$y = f(x)$
$(-4, 0)$ $(5, 0)$

66. (a) $f(x) = 0$
   (b) $f(x) \geq 0$
   (c) $f(x) \leq 0$

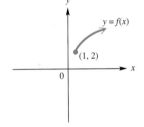

$y = f(x)$
$(1, 2)$

For Exercises 67–70, refer to Example 5.

67. *Cost of Private Education*   The linear equation $y = 588x + 3305$ provides an approximation for the annual cost y (in dollars) for tuition and fees at private four-year colleges and universities, where $x = 1$ represents 1981, $x = 2$ represents 1982, and so on. Write an equation that yields the same y-values when the exact year number is entered.

68. *Federal Debt*   The linear equation $y = 259x + 1843$ provides an approximation for the total federal debt y (in millions of dollars), where $x = 0$ represents 1985, $x = 1$ represents 1986, and so on. Write an equation that yields the same y-values when the exact year number is entered.

69. *International Mutual Funds Available*   The table in the next column shows the number of international mutual funds available during the years 1985–1989.

   (a)   Use a calculator to find the least squares regression line for this data, where $x = 0$ represents 1985, $x = 1$ represents 1986, and so on.

Year	Number of International Mutual Funds
1985	24
1986	35
1987	47
1988	59
1989	75

(*Source:* Investment Company Institute.)

   (b)   Based on your result from part (a), write an equation that yields the same y-values when the exact year number is entered.

   (c)   Use a calculator to find the least squares regression line for this data, where x represents the year number. Show algebraically that this line and the one you found in part (b) agree.

70. *U.S. Mutual Funds in Foreign Markets*   The table shows the amounts of U.S. mutual fund investments in foreign markets (in billions of dollars) during the years 1988–1992.

Year	U.S. Funds Invested in Foreign Markets (in billions of dollars)
1988	6.8
1989	9.9
1990	14.3
1991	19.1
1992	22.9

(*Source:* Investment Company Institute.)

(a) Use a calculator to find the least squares regression line for this data, where $x = 0$ corresponds to 1988, $x = 1$ corresponds to 1989, and so on.

(b) Based on your result from part (a), write an equation that yields the same y-values when the exact year number is entered.

(c) Use a calculator to find the least squares regression line for this data, where x represents the year number. Show algebraically that this line and the one you found in part (b) agree.

2.3 STRETCHING, SHRINKING, AND REFLECTING GRAPHS OF FUNCTIONS

Vertical Stretching ◆ Vertical Shrinking ◆ Reflecting across an Axis ◆ Combining Transformations of Graphs

We continue our discussion from the previous section on how the graphs of functions may be altered. We saw how adding or subtracting a constant can cause a vertical or horizontal shift. Now we will see how multiplication by a constant alters the graph of a function.

VERTICAL STRETCHING

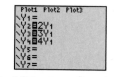

By defining Y_1 as directed in parts A, B, and C, and defining Y_2, Y_3, and Y_4 as shown here, you can minimize your keystrokes. (*Note:* These graphs will *not* appear unless Y_1 is defined.)

For Group Discussion   In each group of functions below, we give four related functions. Graph the four functions in the first group (Group A), and then answer the questions below regarding those functions. Then, repeat the process for Group B and Group C. Use the window specified for each group.

A	B	C
$[-5, 5]$ by $[-5, 20]$	$[-5, 15]$ by $[-5, 10]$	$[-20, 20]$ by $[-10, 10]$
$y_1 = x^2$	$y_1 = \sqrt{x}$	$y_1 = \sqrt[3]{x}$
$y_2 = 2x^2$	$y_2 = 2\sqrt{x}$	$y_2 = 2\sqrt[3]{x}$
$y_3 = 3x^2$	$y_3 = 3\sqrt{x}$	$y_3 = 3\sqrt[3]{x}$
$y_4 = 4x^2$	$y_4 = 4\sqrt{x}$	$y_4 = 4\sqrt[3]{x}$

1. How does the graph of y_2 compare to the graph of y_1?
2. How does the graph of y_3 compare to the graph of y_1?
3. How does the graph of y_4 compare to the graph of y_1?
4. If we choose $c > 4$, how do you think the graph of $y_5 = c \cdot y_1$ would compare to the graph of y_4?

Choosing your own value of c, support your answer to Item 4 graphically.

In each group of functions in the preceding activity, we started with an elementary function y_1 and observed how the graphs of functions of the form $y = c \cdot y_1$ compared with y_1 for positive values of c that began at 2 and became progressively larger. In each case, we obtained a *vertical stretch* of the graph of the basic function with which we started. These observations can be generalized to any function.

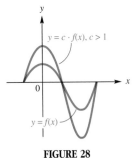

FIGURE 28

> ## VERTICAL STRETCHING OF THE GRAPH OF A FUNCTION
>
> If $c > 1$, the graph of $y = c \cdot f(x)$ is obtained by vertically stretching the graph of $y = f(x)$ by a factor of c. In general, the larger the value of c, the greater the stretch.

In Figure 28, we give a graphical interpretation of the statement above.

EXAMPLE 1 *Recognizing Vertical Stretches on Calculator-Generated Graphs*
Figure 29 shows the graphs of four functions. The function labeled y_1 is the function $f(x) = |x|$. The other three functions, y_2, y_3, and y_4, are defined as follows, but not necessarily in the given order: $2.4|x|, 3.2|x|, 4.3|x|$. Determine the correct rule for each graph.

SOLUTION The values of c here are 2.4, 3.2, and 4.3. The vertical heights of the points with the same x-coordinates on the three graphs will correspond to the magnitudes of these c values. Thus, the graph just above $y_1 = |x|$ will be that of $y = 2.4|x|$, the "highest" graph will be that of $y = 4.3|x|$, and the graph of $y = 3.2|x|$ will lie "between" the others. Therefore, based on our observation of the graphs in the figure, we have $y_2 = 4.3|x|$, $y_3 = 2.4|x|$, and $y_4 = 3.2|x|$.

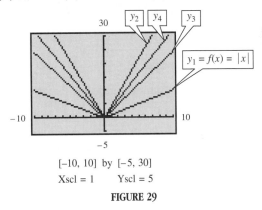

[−10, 10] by [−5, 30]
Xscl = 1   Yscl = 5

FIGURE 29

If we were to trace to any point on the graph of y_1 and then move our tracing cursor to the other graphs one by one, we would see that the y-values of the points would be multiplied by the appropriate values of c. You may wish to experiment with your calculator in this way.   ◆

VERTICAL SHRINKING

For Group Discussion   This discussion parallels the one given earlier in this section. Follow the same general directions. (*Note:* The fractions $\frac{3}{4}$, $\frac{1}{2}$, and $\frac{1}{4}$ may be entered as their decimal equivalents when plotting the graphs.)

A	B	C
$[-5, 5]$ by $[-5, 20]$	$[-5, 15]$ by $[-2, 5]$	$[-10, 10]$ by $[-2, 10]$
$y_1 = x^2$	$y_1 = \sqrt{x}$	$y_1 = |x|$
$y_2 = \frac{3}{4}x^2$	$y_2 = \frac{3}{4}\sqrt{x}$	$y_2 = \frac{3}{4}|x|$
$y_3 = \frac{1}{2}x^2$	$y_3 = \frac{1}{2}\sqrt{x}$	$y_3 = \frac{1}{2}|x|$
$y_4 = \frac{1}{4}x^2$	$y_4 = \frac{1}{4}\sqrt{x}$	$y_4 = \frac{1}{4}|x|$

1. How does the graph of y_2 compare to the graph of y_1?
2. How does the graph of y_3 compare to the graph of y_1?
3. How does the graph of y_4 compare to the graph of y_1?
4. If we choose $0 < c < \frac{1}{4}$, how do you think the graph of $y_5 = c \cdot y_1$ would compare to the graph of y_4? Provide support by choosing such a value of c.

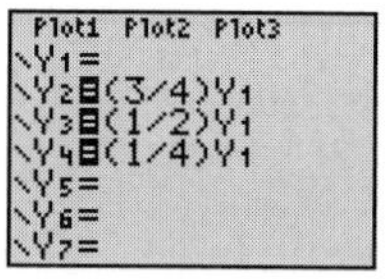

You can use a screen such as this to minimize your key-strokes in parts A, B, and C. Again, Y_1 must be defined in order to obtain the other graphs.

In this "For Group Discussion" activity, we began with an elementary function y_1 and observed the graphs of $y = c \cdot y_1$, where we started with $c = \frac{3}{4}$ and chose progressively smaller positive values of c. In each case, the graph of y_1 was *vertically shrunk*. These observations, like the ones for vertical stretching, can be generalized to any function.

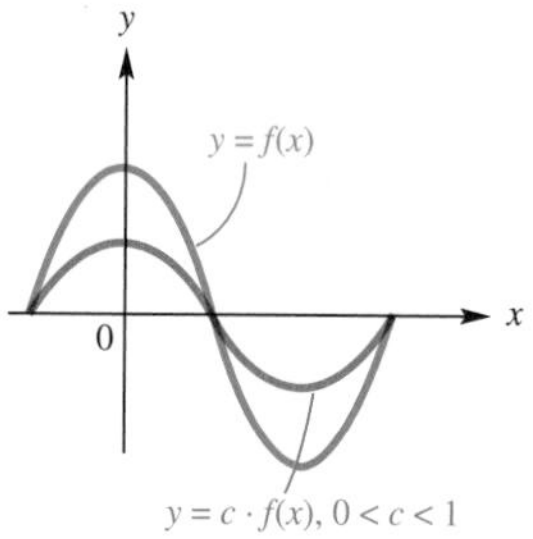

VERTICAL SHRINKING OF THE GRAPH OF A FUNCTION

If $0 < c < 1$, the graph of $y = c \cdot f(x)$ is obtained by vertically shrinking the graph of $y = f(x)$ by a factor of c. In general, the smaller the value of c, the greater the shrink.

FIGURE 30

Figure 30 shows a graphical interpretation of vertical shrinking.

EXAMPLE 2   *Recognizing Vertical Shrinks on Calculator-Generated Graphs*
Figure 31 shows the graphs of four functions. The function labeled y_1 is the function $f(x) = x^3$. The other three functions, y_2, y_3, and y_4, are defined as follows, but not necessarily in the given order: $.5x^3$, $.3x^3$, and $.1x^3$. Determine the correct rule for each graph.

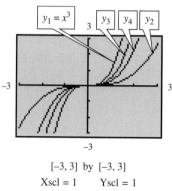

$[-3, 3]$ by $[-3, 3]$
Xscl = 1   Yscl = 1
FIGURE 31

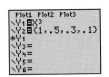

This method of defining Y_1 and Y_2—using a list of coefficients in Y_2—will allow you to duplicate Figure 31.

SOLUTION The smaller the positive value of c, where $0 < c < 1$, the more of a shrink toward the x-axis there will be. Since we have $c = .5, .3,$ and $.1$, the function rules must be as follows: $y_2 = .1x^3$, $y_3 = .5x^3$, and $y_4 = .3x^3$. ◆

REFLECTING ACROSS AN AXIS

In the previous section and so far in this one, we have seen how graphs can be transformed by shifting, stretching, and shrinking. We will now examine how graphs can be reflected across an axis.

For Group Discussion In each pair, we give two related functions. Graph $y_1 = f(x)$ and $y_2 = -f(x)$ in the standard viewing window, and then answer the questions below for the pair.

A	B	C	D
$y_1 = x^2$	$y_1 = |x|$	$y_1 = \sqrt{x}$	$y_1 = x^3$
$y_2 = -x^2$	$y_2 = -|x|$	$y_2 = -\sqrt{x}$	$y_2 = -x^3$

With respect to the x-axis,

1. how does the graph of y_2 compare with the graph of y_1?
2. how would the graph of $y = -\sqrt[3]{x}$ compare with the graph of $y = \sqrt[3]{x}$, based on your answer to Item 1? Confirm your answer by actual graphing.

Again, in each pair of functions, we give two related functions. Graph $y_1 = f(x)$ and $y_2 = f(-x)$ in the standard viewing window, and then answer the questions below for each pair.

E	F	G
$y_1 = \sqrt{x}$	$y_1 = \sqrt{x - 3}$	$y_1 = \sqrt[3]{x + 4}$
$y_2 = \sqrt{-x}$	$y_2 = \sqrt{-x - 3}$	$y_2 = \sqrt[3]{-x + 4}$

With respect to the y-axis,

3. how does the graph of y_2 compare with the graph of y_1?
4. how would the graph of $y = \sqrt[3]{-x}$ compare with the graph of $y = \sqrt[3]{x}$, based on your answer to Item 3? Confirm your answer by actual graphing.

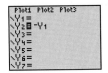

By defining Y_1 as directed in parts A, B, C, and D, and defining Y_2 as shown here, you can minimize your keystrokes.

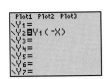

By defining Y_1 as directed in parts E, F, and G, and defining Y_2 as shown here (using function notation), you can minimize your keystrokes.

Based on the preceding group discussion, we can see how the graph of a function can be reflected across an axis. The results of that discussion are now formally summarized.

REFLECTING THE GRAPH OF A FUNCTION ACROSS AN AXIS

For a function $y = f(x)$,
(a) the graph of $y = -f(x)$ is a reflection of the graph of f across the x-axis.
(b) the graph of $y = f(-x)$ is a reflection of the graph of f across the y-axis.

Figure 32 shows how the reflections just described affect the graph of a function in general.

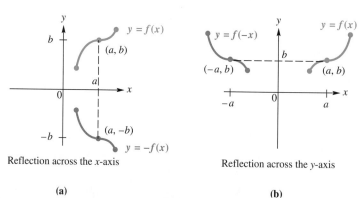

Reflection across the *x*-axis

(a)

Reflection across the *y*-axis

(b)

FIGURE 32

EXAMPLE 3   *Applying Reflections across Axes*   Figure 33 shows the graph of a function $y = f(x)$.

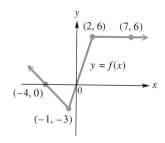

FIGURE 33

(a)   Sketch the graph of $y = -f(x)$.   **(b)**   Sketch the graph of $y = f(-x)$.

SOLUTION

(a)   We must reflect the graph across the *x*-axis. This means that if a point (a, b) lies on the graph of $y = f(x)$, then the point $(a, -b)$ must lie on the graph of $y = -f(x)$. Using the labeled points to assist us, we find the graph of $y = -f(x)$ in Figure 34(a).

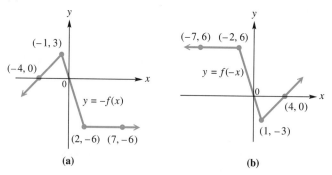

(a)

(b)

FIGURE 34

(b)   Here we must reflect the graph across the *y*-axis, meaning that if a point (a, b) lies on the graph of $y = f(x)$, then the point $(-a, b)$ must lie on the graph of $y = f(-x)$. Again, using the labeled points to guide us, we obtain the graph of $y = f(-x)$ as shown in Figure 34(b).   ◆

To illustrate how reflections appear on calculator-generated graphs, observe the two graphs shown in Figures 35 and 36. We use a higher-degree polynomial function (covered in detail in Chapter 3) here in Figure 35. The graph of y_2 is a reflection of the graph of y_1 across the x-axis. The display at the bottom of the screen shows that the point $(-2, 23)$ is on the graph of y_1. Therefore, if we were to use the capabilities of the calculator, we should locate the point $(-2, -23)$ on the graph of y_2. (*Note:* The function y_1 is defined as $y_1 = x^4 - 2x + 3$. You might wish to verify the above statement.)

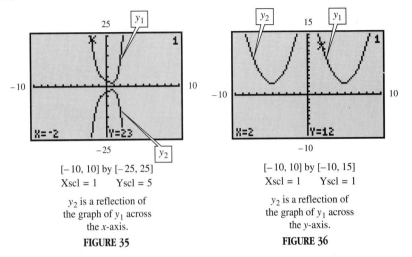

$[-10, 10]$ by $[-25, 25]$
Xscl = 1   Yscl = 5

y_2 is a reflection of
the graph of y_1 across
the x-axis.

FIGURE 35

$[-10, 10]$ by $[-10, 15]$
Xscl = 1   Yscl = 1

y_2 is a reflection of
the graph of y_1 across
the y-axis.

FIGURE 36

Figure 36 illustrates a reflection across the y-axis. The graph of y_2 is a reflection of the graph of y_1 across the y-axis. Notice that the point $(2, 12)$ lies on the graph of y_1. What point must lie on the graph of y_2? (To verify your answer, graph $y_1 = (x - 5)^2 + 3$. So we have $y_2 = (-x - 5)^2 + 3$. Now, see if you were correct.)

WHAT WENT WRONG ?   To observe how negative values of a affect the graph of $y = ax^2$, a student entered three functions into $Y_1, Y_2,$ and Y_3, as shown in the accompanying screen. The calculator graphed the first two as shown, but gave a syntax error when it attempted to graph the third.

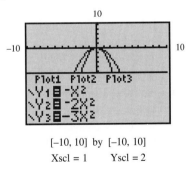

$[-10, 10]$ by $[-10, 10]$
Xscl = 1   Yscl = 2

1. What went wrong?
2. What must the student do in order to obtain the desired graph for $y = -3x^2$?

COMBINING TRANSFORMATIONS OF GRAPHS

The graphs of $y_1 = x^2$ and $y_2 = -2x^2$ are shown in the same viewing window in Figure 37.

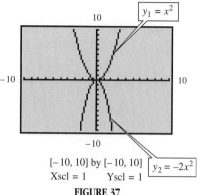

[−10, 10] by [−10, 10]
Xscl = 1   Yscl = 1

FIGURE 37

Notice that in terms of the types of transformations we have studied, the graph of y_2 is obtained by vertically stretching the graph of y_1 by a factor of 2 and then reflecting across the x-axis. Thus, we have a case of a combination of transformations. As you might expect, we can create an infinite number of functions by vertically stretching or shrinking, shifting left or right, and reflecting across an axis. The next example investigates examples of this type of function.

EXAMPLE 4   *Describing a Combination of Transformations of a Graph*

(a)   Describe how the graph of $y = -3(x - 4)^2 + 5$ can be obtained by transforming the graph of $y = x^2$. Illustrate with a graphing calculator.

(b)   Give the equation of the function that would be obtained by starting with the graph of $y = |x|$, shifting 3 units to the left, vertically shrinking the graph by a factor of $\frac{2}{3}$, reflecting across the x-axis, and shifting the graph 4 units down, in this order. Graph both functions using a graphing calculator.

SOLUTION

(a)   The fact that we have $(x - 4)^2$ in our function indicates that the graph of $y = x^2$ must be shifted 4 units to the *right*. Since the coefficient of $(x - 4)^2$ is -3 (a negative number with absolute value greater than 1), the graph is stretched vertically by a factor of 3 and then reflected across the x-axis. The constant $+5$ indicates that the graph is finally shifted up 5 units. Figure 38(a) shows the graph of both $y = x^2$ and $y = -3(x - 4)^2 + 5$.

(b)   Shifting 3 units to the left means that $|x|$ is transformed to $|x + 3|$. Vertically shrinking by a factor of $\frac{2}{3}$ means multiplying $|x + 3|$ by $\frac{2}{3}$, and reflecting across the x-axis changes $\frac{2}{3}$ to $-\frac{2}{3}$. Finally, shifting 4 units down means subtracting 4. Putting this all together leads to the following equation:

$$y = -\frac{2}{3}|x + 3| - 4.$$

The graphs of both $y = |x|$ and the new function are shown in Figure 38(b).

◆

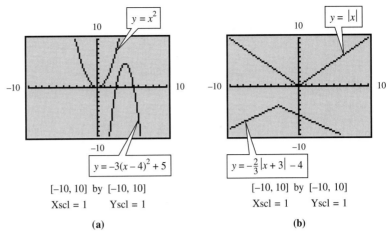

$y = x^2$

$y = -3(x - 4)^2 + 5$

[–10, 10] by [–10, 10]
Xscl = 1   Yscl = 1

(a)

$y = |x|$

$y = -\frac{2}{3}|x + 3| - 4$

[–10, 10] by [–10, 10]
Xscl = 1   Yscl = 1

(b)

FIGURE 38

Note The order in which the transformations are made is important. If the same transformations are made in a different order, a different equation can result.

2.3 EXERCISES

   Tape 3

Use the concepts of Section 2.2 and this section to draw a rough sketch by hand of the graphs of y_1, y_2, and y_3. Do not plot points. In each case, y_2 and y_3 can be graphed by one or more of these: a vertical and/or horizontal shift of the graph of y_1, a vertical stretch or shrink of the graph of y_1, or a reflection of the graph of y_1 across an axis. After you have made your sketches, check by graphing them in an appropriate viewing window of your calculator.

1. $y_1 = x$, $y_2 = x + 3$, $y_3 = x - 3$

2. $y_1 = x^2$, $y_2 = x^2 - 1$, $y_3 = x^2 + 1$

3. $y_1 = x^3$, $y_2 = x^3 + 4$, $y_3 = x^3 - 4$

4. $y_1 = \sqrt{x}$, $y_2 = \sqrt{x} + 6$, $y_3 = \sqrt{x} - 6$

5. $y_1 = |x|$, $y_2 = |x - 3|$, $y_3 = |x + 3|$

6. $y_1 = \sqrt[3]{x}$, $y_2 = \sqrt[3]{x - 4}$, $y_3 = \sqrt[3]{x + 4}$

7. $y_1 = |x|$, $y_2 = |x| - 3$, $y_3 = |x| + 3$

8. $y_1 = \sqrt[3]{x}$, $y_2 = \sqrt[3]{x} - 4$, $y_3 = \sqrt[3]{x} + 4$

9. $y_1 = \sqrt{x}$, $y_2 = \sqrt{x + 6}$, $y_3 = \sqrt{x - 6}$

10. $y_1 = x^3$, $y_2 = (x - 4)^3$, $y_3 = (x + 4)^3$

11. $y_1 = |x|$, $y_2 = 2|x|$, $y_3 = 2.5|x|$

12. $y_1 = |x|$, $y_2 = -2|x|$, $y_3 = -2.5|x|$

13. $y_1 = \sqrt[3]{x}$, $y_2 = -\sqrt[3]{x}$, $y_3 = -2\sqrt[3]{x}$

14. $y_1 = \sqrt[3]{x}$, $y_2 = 2\sqrt[3]{x}$, $y_3 = \frac{1}{2}\sqrt[3]{x}$

15. $y_1 = x^2$, $y_2 = (x - 2)^2 + 1$, $y_3 = -(x + 2)^2$

16. $y_1 = x^2$, $y_2 = -(x + 3)^2 - 2$, $y_3 = -(x - 4)^2$

17. $y_1 = |x|$, $y_2 = -2|x - 1| + 1$, $y_3 = -\frac{1}{2}|x| - 4$

18. $y_1 = |x|$, $y_2 = -|x + 1| - 4$, $y_3 = -|x - 1|$

In Exercises 19–24, fill in the blanks with the appropriate responses. (Remember that the vertical stretch or shrink factor is positive.)

19. The graph of $y = -4x^2$ can be obtained from the graph of $y = x^2$ by vertically stretching by a factor of _____ and reflecting across the _____-axis.

20. The graph of $y = -6\sqrt{x}$ can be obtained from the graph of $y = \sqrt{x}$ by vertically stretching by a factor of _____ and reflecting across the _____-axis.

21. The graph of $y = -\frac{1}{4}|x + 2| - 3$ can be obtained from the graph of $y = |x|$ by shifting horizontally _____ units to the _____, vertically shrinking by a factor of _____, reflecting across the _____-axis, and shifting vertically _____ units in the _____ direction.

22. The graph of $y = -\frac{2}{5}\left|-x\right| + 6$ can be obtained from the graph of $y = |x|$ by reflecting across the _____-axis, vertically shrinking by a factor of _____, reflecting a second time across the _____-axis, and shifting vertically _____ units in the _____ direction.

23. The graph of $y = 6\sqrt[3]{x-3}$ can be obtained from the graph of $y = \sqrt[3]{x}$ by shifting horizontally _____ units to the _____ and stretching vertically by a factor of _____.

24. The graph of $y = .5\sqrt[3]{x+2}$ can be obtained from the graph of $y = \sqrt[3]{x}$ by shifting horizontally _____ units to the _____ and shrinking vertically by a factor of _____.

Give the equation of the function whose graph is described.

25. The graph of $y = x^2$ is vertically shrunk by a factor of $\frac{1}{2}$, and the resulting graph is shifted 7 units downward.

26. The graph of $y = x^3$ is vertically stretched by a factor of 3. This graph is then reflected across the x-axis. Finally, the graph is shifted 8 units upward.

27. The graph of $y = \sqrt{x}$ is shifted 3 units to the right. This graph is then vertically stretched by a factor of 4.5. Finally, the graph is shifted 6 units downward.

28. The graph of $y = \sqrt[3]{x}$ is shifted 2 units to the left. This graph is then vertically stretched by a factor of 1.5. Finally, the graph is shifted 8 units upward.

Shown on the left is the graph of $Y_1 = (x - 2)^2 + 1$ *in the standard viewing window of a graphing calculator. Six other functions,* Y_2 *through* Y_7, *are graphed according to the rules shown in the screen on the right.*

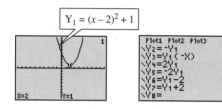

[–10, 10] by [–10, 10]

Match the function with its calculator-generated graph from choices A–F first without using a calculator, by applying the techniques of this section and the previous section. Then, confirm your answer by graphing the function on your calculator.

29. Y_2

30. Y_3

31. Y_4

32. Y_5

33. Y_6

34. Y_7

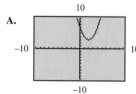

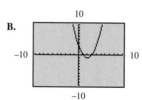

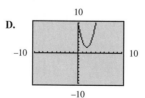

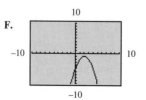

In Exercises 35 and 36, the graph of y = f(x) has been transformed to the graph of y = g(x). No shrinking or stretching is involved. Give the equation of y = g(x).

35.

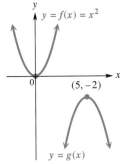

36.

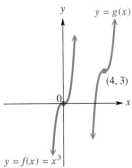

In each of Exercises 37–42, the figure shows the graph of a function y = f(x). Sketch by hand the graphs of the functions in parts (a), (b), and (c), and answer the question of part (d).

37. **(a)** $y = -f(x)$ **(b)** $y = f(-x)$ **(c)** $y = 2f(x)$

  (d) What is $f(0)$?

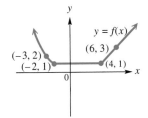

38. **(a)** $y = -f(x)$ **(b)** $y = f(-x)$ **(c)** $y = 3f(x)$

  (d) What is $f(4)$?

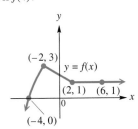

39. **(a)** $y = -f(x)$ **(b)** $y = f(-x)$

  (c) $y = f(x + 1)$

  (d) What are the x-intercepts of $y = f(x - 1)$?

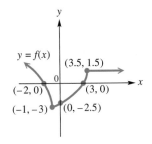

40. **(a)** $y = -f(x)$ **(b)** $y = f(-x)$

  (c) $y = -2f(x)$

  (d) On what interval of the domain is $f(x) < 0$?

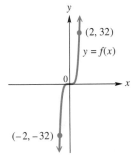

41. **(a)** $y = -f(x)$ **(b)** $y = f(-x)$ **(c)** $y = .5f(x)$

  (d) What symmetry does the graph of $y = f(x)$ exhibit?

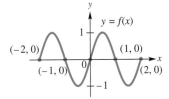

42. **(a)** $y = -f(x)$ **(b)** $y = f(-x)$ **(c)** $y = 3f(x)$

  (d) What symmetry does the graph of $y = f(x)$ exhibit?

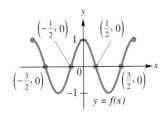

43. If r is an x-intercept of the graph of $y = f(x)$, what statement can be made about an x-intercept of the graph of each of the following? (*Hint:* Draw a picture.)

(a) $y = -f(x)$ (b) $y = f(-x)$

(c) $y = -f(-x)$

44. If b is the y-intercept of the graph of $y = f(x)$, what statement can be made about the y-intercept of the graph of each of the following? (*Hint:* Draw a picture.)

(a) $y = -f(x)$ (b) $y = f(-x)$ (c) $y = 5f(x)$

(d) $y = -3f(x)$

RELATING CONCEPTS (EXERCISES 45–48)

Each of the following functions will have a graph with an endpoint (a translation of the point $(0, 0)$*). Enter each into your calculator in an appropriate viewing window, and with your knowledge of the graph of* $y = \sqrt{x}$*, determine the domain and the range of the function. (*Hint:* Locate the endpoint.) See Section 1.2.*

45. $y = 10\sqrt{x - 20} + 5$ **46.** $y = -2\sqrt{x + 15} - 18$ **47.** $y = -.5\sqrt{x + 10} + 5$

48. Based on your observations in Exercise 45, what are the domain and the range of $f(x) = a\sqrt{x - h} + k$, if $a > 0$, $h > 0$, and $k > 0$?

*State the intervals over which the function is (**a**) increasing, (**b**) decreasing, and (**c**) constant.*

49. the function graphed in Figure 33

50. the function graphed in Figure 34(a)

51. the function graphed in Figure 34(b)

52. $y = -\dfrac{2}{3}|x + 3| - 4$ (see Figure 38(b))

Shown here are the graphs of $y_1 = \sqrt[3]{x}$ *and* $y_2 = 5\sqrt[3]{x}$*. The point whose coordinates are given at the bottom of the screen lies on the graph of* y_1*. Use this graph, and not your own calculator, to find the value of* y_2 *for the same value of x shown.*

53.

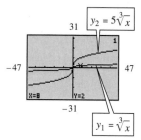

54.

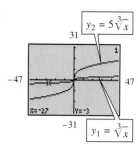

55.

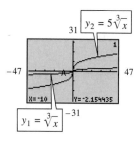

56.

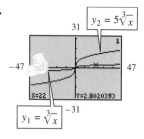

The following sketch shows an example of a function $y = f(x)$ *that increases on the interval* $[a, b]$ *of its domain.*

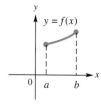

Use this graph as a visual aid, and apply the concepts of reflection introduced in this section to answer each of the following questions. (Make your own sketch if you wish.)

57. Does the function $y = -f(x)$ increase or decrease on the interval $[a, b]$?

58. Does the function $y = f(-x)$ increase or decrease on the interval $[-b, -a]$?

59. Does the function $y = -f(-x)$ increase or decrease on the interval $[-b, -a]$?

60. If $c > 0$, does the graph of $y = -c \cdot f(x)$ increase or decrease on the interval $[a, b]$?

61. If the graph of the function $y = f(x)$ is symmetric with respect to the y-axis, what can be said about the symmetry of **(a)** $y = f(-x)$ and **(b)** $y = -f(x)$?

62. If the graph of the function $y = f(x)$ is symmetric with respect to the origin, what can be said about the symmetry of **(a)** $y = f(-x)$ and **(b)** $y = -f(x)$?

RELATING CONCEPTS (EXERCISES 63–64)

Use either the x-intercept method of graphical solution or the intersection-of-graphs method of graphical solution to solve each equation or inequality. Express solutions or endpoints correct to the nearest hundredth. See Sections 1.5 and 1.6.

63. (a) $\sqrt[3]{\pi x + 3.6} = |\sqrt{2}x - 4.8|$

   (b) $\sqrt[3]{\pi x + 3.6} > |\sqrt{2}x - 4.8|$

   (c) $\sqrt[3]{\pi x + 3.6} < |\sqrt{2}x - 4.8|$

64. (a) $|2\pi x - 4| = 4(x - 1)^3 + 1.6$

   (b) $|2\pi x - 4| > 4(x - 1)^3 + 1.6$

   (c) $|2\pi x - 4| < 4(x - 1)^3 + 1.6$

2.4 THE ABSOLUTE VALUE FUNCTION: GRAPHS, EQUATIONS, INEQUALITIES, AND APPLICATIONS

The Graph of $y = |f(x)|$ ◆ Properties of Absolute Value ◆ Equations and Inequalities Involving Absolute Value ◆ An Application Involving Absolute Value

Of the elementary functions introduced so far in this chapter, the identity, squaring, and cubing functions are all examples of polynomial functions, a class of functions that will be more closely examined in Chapter 3. The square root and cube root functions are examples of root functions, and this class will be studied in Chapter 4 (along with another important class, rational functions). This leaves only the absolute value function, and in this section we will investigate this function in detail.

```
abs(-3)
               3
abs(0)
               0
abs(3)
               3
```

The command abs(x) is used by some graphing calculators to find absolute value.

THE GRAPH OF $y = |f(x)|$

The formal definition of the absolute value of a real number was given in Section 2.1. Geometrically, the absolute value of a real number is its undirected distance from 0 on the number line. As a result of this, the absolute value of a real number is never negative; it is always greater than or equal to 0. The absolute value function was defined in Section 2.1. Thus, the function

$$f(x) = |x| = \begin{cases} x & \text{if } x \geq 0 \\ -x & \text{if } x < 0 \end{cases}$$

is just an extension of the definition of absolute value. The expression within the absolute value bars, x, is the defining expression for the identity function $y = x$.

Now, let us extend the concept further and consider the definition of a function defined by the *absolute value of a function f*:

$$|f(x)| = \begin{cases} f(x) & \text{if } f(x) \geq 0 \\ -f(x) & \text{if } f(x) < 0. \end{cases}$$

In order to graph a function of the form $y = |f(x)|$, the definition indicates that the graph is the same as that of $y = f(x)$ for values of $f(x)$ (that is, range values) that are nonnegative. The second part of the definition indicates that for range values that are negative, the graph of $y = f(x)$ is reflected across the *x*-axis.

The domain of $y = |f(x)|$ is the same as the domain of *f*, while the range of $y = |f(x)|$ will be a subset of $[0, \infty)$.

EXAMPLE 1   *Comparing the Calculator-Generated Graphs of $y = f(x)$ and $y = |f(x)|$*   Figure 39 shows the graph of $y = (x - 4)^2 - 3$, which is the graph of $y = x^2$ shifted 4 units to the right and 3 units downward. Figure 40 shows the graph of $y = |(x - 4)^2 - 3|$. Notice that all points with negative *y*-values in the first graph have been reflected across the *x*-axis, while all points with nonnegative *y*-values are the same for both graphs. Give the domain and the range of each function.

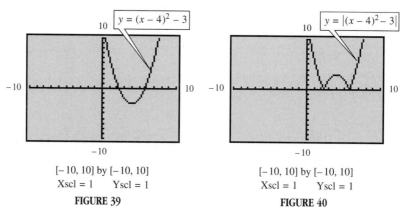

[−10, 10] by [−10, 10]
Xscl = 1   Yscl = 1

FIGURE 39

[−10, 10] by [−10, 10]
Xscl = 1   Yscl = 1

FIGURE 40

SOLUTION   The domain of both functions is $(-\infty, \infty)$. The range of $y = (x - 4)^2 - 3$ is $[-3, \infty)$, while the range of $y = |(x - 4)^2 - 3|$ is $[0, \infty)$.   ◆

EXAMPLE 2   *Sketching the Graph of $y = |f(x)|$ Given the Graph of $y = f(x)$*   Figure 41 shows the graph of a function $y = f(x)$. Use the figure to sketch the graph of $y = |f(x)|$. Give the domain and the range of each.

SOLUTION   As stated earlier, the graph will remain the same for points whose *y*-values are nonnegative, while it will be reflected across the *x*-axis for all other points. Figure 42 shows the graph of $y = |f(x)|$. The domain of both functions is $(-\infty, \infty)$. The range of $y = f(x)$ is $[-3, \infty)$, while the range of $y = |f(x)|$ is $[0, \infty)$.   ◆

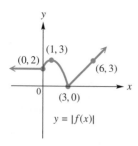

FIGURE 41

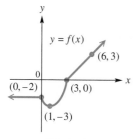

FIGURE 42

EXAMPLE 3   *Answering a Conceptual Question Concerning the Graph of $y = |f(x)|$*   Why is the graph of $y = (x + 2)^2 + 1$ the same as the graph of $y = |(x + 2)^2 + 1|$?

SOLUTION   The function $y = (x + 2)^2 + 1$ has as its graph the graph of $y = x^2$ shifted 2 units to the left, and 1 unit upward, causing its range to be $[1, \infty)$. Since the range consists of only positive numbers, the graph of $y = |(x + 2)^2 + 1|$ will not be affected, since there is no reflection across the *x*-axis. Therefore, the two graphs are identical. (Verify this on your calculator.)   ◆

PROPERTIES OF ABSOLUTE VALUE

There are several properties of absolute value that will be useful in our later work.

PROPERTIES OF ABSOLUTE VALUE

For all real numbers a and b:

1. $|ab| = |a| \cdot |b|$
   The absolute value of a product is equal to the product of the absolute values.

2. $\left|\dfrac{a}{b}\right| = \dfrac{|a|}{|b|}$ $(b \neq 0)$
   The absolute value of a quotient is equal to the quotient of the absolute values.

3. $|a| = |-a|$
   The absolute value of a number is equal to the absolute value of its additive inverse.

4. $|a| + |b| \geq |a + b|$ (the triangle inequality)
   The sum of the absolute values of two numbers is greater than or equal to the absolute value of their sum.

Among other applications, these properties can be used to explain the behavior of graphs of functions involving absolute value. For example, consider the function $y = |2x + 11|$ and observe the following sequence of transformations.

$$y = |2x + 11| \qquad \text{Given}$$

$$y = \left|2\left(x + \frac{11}{2}\right)\right| \qquad \text{Factor out a 2.}$$

$$y = |2| \cdot \left|x + \frac{11}{2}\right| \qquad \text{Property 1}$$

$$y = 2\left|x + \frac{11}{2}\right| \qquad |2| = 2$$

Using the concepts of the previous two sections, we conclude that the graph of this function can be found by starting with the graph of $y = |x|$, shifting $\frac{11}{2}$ units to the left, and then vertically stretching by a factor of 2. The graphs of $y_1 = |x|$ and $y_2 = |2x + 11|$ in Figure 43 give support to this statement.

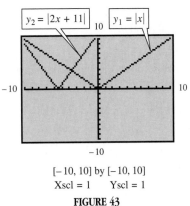

$[-10, 10]$ by $[-10, 10]$
Xscl = 1   Yscl = 1
FIGURE 43

We will often be interested in examining absolute value functions of the form

$$f(x) = |ax + b|,$$

where the expression inside the absolute value bars is linear. For purposes of discussion, we now give a definition of a comprehensive graph of $f(x) = |ax + b|$.

COMPREHENSIVE GRAPH OF $f(x) = |ax + b|$

A comprehensive graph of $f(x) = |ax + b|$ will include all intercepts and the lowest point on the "V-shaped" graph.

EQUATIONS AND INEQUALITIES INVOLVING ABSOLUTE VALUE

There are numerous types of equations (and related inequalities) involving absolute value. We will investigate those types that involve absolute values of linear functions, and constants. As first explained in Chapter 1, we will solve them analytically, and then support our solutions graphically.

The summary that follows indicates the general method for solving some of the simpler absolute value equations and inequalities analytically.

SOLVING ABSOLUTE VALUE EQUATIONS AND INEQUALITIES

Let k be a positive number.

1. To solve $|ax + b| = k$, solve the compound equation

$$ax + b = k \quad \text{or} \quad ax + b = -k.$$

2. To solve $|ax + b| > k$, solve the compound inequality

$$ax + b > k \quad \text{or} \quad ax + b < -k.$$

3. To solve $|ax + b| < k$, solve the three-part inequality

$$-k < ax + b < k.$$

Inequalities involving $\leq$ or $\geq$ are solved similarly, using the equality part of the symbol as well.

EXAMPLE 4   *Solving $|ax + b| = k$, $k > 0$ Analytically*   Solve $|2x + 1| = 7$ analytically.

SOLUTION   ANALYTIC   For $|2x + 1|$ to equal 7, $2x + 1$ must be 7 units from 0 on the number line. This can happen only when $2x + 1 = 7$ or $2x + 1 = -7$. Solve this compound equation as follows.

$$
\begin{array}{rcl}
2x + 1 = 7 & \text{or} & 2x + 1 = -7 \\
2x = 6 & \text{or} & 2x = -8 \\
x = 3 & \text{or} & x = -4
\end{array}
$$

The solution set is $\{-4, 3\}$.   ◆

GRAPHICAL   Figure 44 shows the graphs of $y_1 = |2x + 1|$ and $y_2 = 7$. A calculator will show that they intersect at points for which $x = -4$ and $x = 3$. This gives graphical support to our solution in Example 4. Using the intersection-of-graphs method for solving *inequalities*, it would seem to suggest that the solution set of

$$|2x + 1| < 7$$

is the open interval $(-4, 3)$, since the graph of y_1 lies *below* that of y_2 there. On the other hand, the solution set of

$$|2x + 1| > 7$$

would be $(-\infty, -4) \cup (3, \infty)$, since for these intervals, the graph of y_1 lies *above* that of y_2. Notice that these observations agree with the way $|ax + b| < k$ and $|ax + b| > k$ were defined. In the following example, we provide analytic justification for these observations.

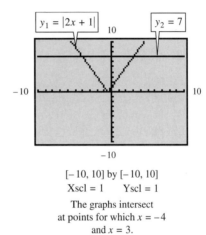

$y_1 = |2x + 1|$ $y_2 = 7$

$[-10, 10]$ by $[-10, 10]$
Xscl = 1   Yscl = 1

The graphs intersect
at points for which $x = -4$
and $x = 3$.

FIGURE 44

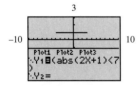

This screen uses the "test" capability of a TI-83 graphing calculator to simulate a number line graph of the solution set of $|2x + 1| < 7$. Notice that the three discrete dots to the left of Y_1 indicate that the calculator will graph in dot mode. See Example 5(a).

EXAMPLE 5   *Solving* $|ax + b| < k$ *and* $|ax + b| > k, k > 0$ *Analytically*

(a)   Solve $|2x + 1| < 7$ analytically.   **(b)**   Solve $|2x + 1| > 7$ analytically.

SOLUTION

(a)   ANALYTIC   Here the expression $2x + 1$ must represent a number that is less than 7 units from 0 on the number line. Another way of thinking of this is to realize that $2x + 1$ must be between -7 and 7. This is written as the three-part inequality

$$-7 < 2x + 1 < 7.$$

We solved such inequalities in Section 1.6 by working with all three parts at the same time.

$$-7 < 2x + 1 < 7$$
$$-8 < 2x < 6 \qquad \text{Subtract 1 from each part.}$$
$$-4 < x < 3 \qquad \text{Divide each part by 2.}$$

The solution set is the open interval $(-4, 3)$, as we observed earlier.

(b)   ANALYTIC   This absolute value inequality must be rewritten as

$$2x + 1 > 7 \qquad \text{or} \qquad 2x + 1 < -7,$$

because $2x + 1$ must represent a number that is *more* than 7 units from 0 on either side of the number line. Now, solve the compound inequality.

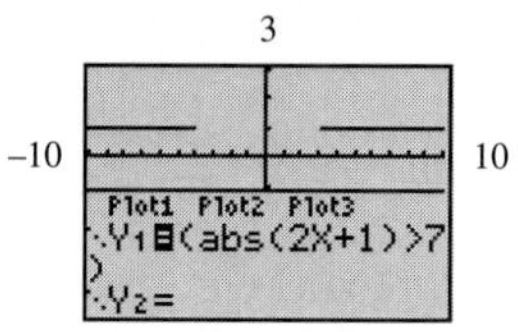

This screen uses the "test" capability of a graphing calculator to simulate a number line graph of the solution set of $|2x + 1| > 7$. Again, the calculator will graph in dot mode. See Example 5(b).

$$\begin{aligned} 2x + 1 > 7 \quad &\text{or} \quad 2x + 1 < -7 \\ 2x > 6 \quad &\text{or} \quad 2x < -8 \\ x > 3 \quad &\text{or} \quad x < -4 \end{aligned}$$

The solution set, $(-\infty, -4) \cup (3, \infty)$, again confirms our earlier observation. ◆

An absolute value equation of the form $|ax + b| = k$, where $k < 0$, will have no solution, since the absolute value of a real number cannot be negative. The related inequalities, $|ax + b| > k$ and $|ax + b| < k$, will have solution sets $(-\infty, \infty)$ and $\emptyset$, respectively. (What is the solution set of $|ax + b| = 0$?) The following example provides graphical support for a particular case.

EXAMPLE 6 *Solving Absolute Value Equations and Inequalities with No Solutions or Infinitely Many Solutions Graphically* Solve the following graphically:

$$|3x + 10| = -5, \qquad |3x + 10| > -5, \qquad |3x + 10| < -5.$$

Use the x-intercept method of solution.

SOLUTION Let $y_1 = |3x + 10|$ and $y_2 = -5$ and graph $y_1 - y_2$ in the standard viewing window. See Figure 45.

The equation $|3x + 10| = -5$ is equivalent to $|3x + 10| + 5 = 0$. The left side of this equation is our function $y_1 - y_2$. Since the graph has no x-intercepts, the solution set of this equation is $\emptyset$.

The inequality $|3x + 10| > -5$ is equivalent to $|3x + 10| + 5 > 0$. Since the graph of $y_1 - y_2$ lies completely *above* the x-axis, the solution set of this inequality is $(-\infty, \infty)$. On the other hand, the solution set of the inequality $|3x + 10| < -5$ is $\emptyset$, since there are no x-values for which the graph lies *below* the x-axis. ◆

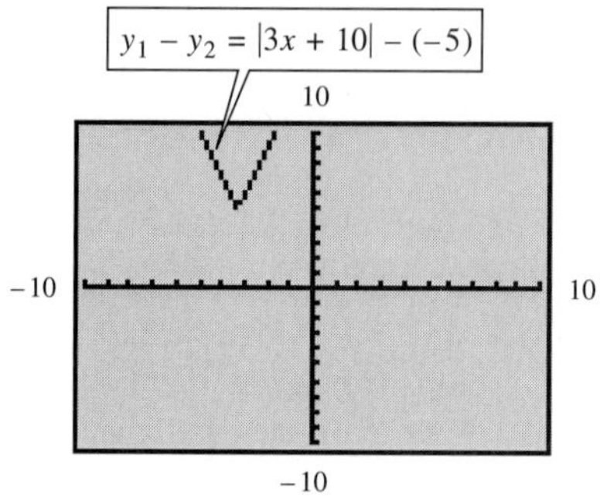

[−10, 10] by [−10, 10]
Xscl = 1 Yscl = 1

FIGURE 45

If two quantities have the same absolute value, they must either be equal to each other or be negatives of each other. This fact allows us to solve absolute value equations of the form

$$|ax + b| = |cx + d|.$$

Once we have solved this type of equation analytically and supported our result graphically, we can then solve the related inequalities $|ax + b| < |cx + d|$ and $|ax + b| > |cx + d|$.

SOLVING $|ax + b| = |cx + d|$

To solve the equation $|ax + b| = |cx + d|$ analytically, solve the compound equation

$$ax + b = cx + d \qquad \text{or} \qquad ax + b = -(cx + d).$$

EXAMPLE 7   *Solving an Equation and Related Inequalities Involving Two Absolute Values*

(a)   Solve the equation $|x + 6| = |2x - 3|$ analytically, and support the solution graphically.

(b)   Use Figure 46 and the results of part (a) to solve the inequalities

$$|x + 6| < |2x - 3| \qquad \text{and} \qquad |x + 6| > |2x - 3|.$$

SOLUTION

(a)   ANALYTIC   This equation is satisfied if

$$x + 6 = 2x - 3 \qquad \text{or} \qquad x + 6 = -(2x - 3).$$

Solve each equation.

$$
\begin{array}{ll}
x + 6 = 2x - 3 & x + 6 = -2x + 3 \\
9 = x & 3x = -3 \\
& x = -1
\end{array}
$$

The solution set of the equation is $\{-1, 9\}$.

GRAPHICAL   By graphing $y_1 = |x + 6|$ and $y_2 = |2x - 3|$ in the window $[-20, 20]$ by $[-20, 20]$, we see that the points of intersection of y_1 and y_2 appear to have x-coordinates -1 and 9. Use your calculator to verify this statement. (Why is the standard window not appropriate for this problem?) See Figure 46.

TECHNOLOGY NOTE

You might wish to support the results of Example 7 graphically by using the x-intercept method of solution. To do this, enter Y_1 as $|x + 6|$, Y_2 as $|2x - 3|$, and Y_3 as $Y_1 - Y_2$. Then, graph only Y_3. You will notice that your graph differs from the one seen in Figure 46, but the support should be evident. (You may need to adjust the window to obtain a better view.)

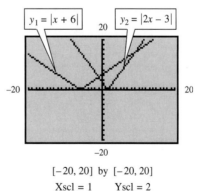

$[-20, 20]$ by $[-20, 20]$
Xscl = 1   Yscl = 2

FIGURE 46

(b)   The graph of $y_1 = |x + 6|$ lies *below* the graph of $y_2 = |2x - 3|$ on the intervals $(-\infty, -1)$ and $(9, \infty)$. Therefore, the solution set of $|x + 6| < |2x - 3|$ is

$$(-\infty, -1) \cup (9, \infty).$$

The solution set of $|x + 6| > |2x - 3|$ consists of all x-values for which the graph of y_1 is *above* that of y_2. This happens between -1 and 9, so the solution set of $|x + 6| > |2x - 3|$ is the interval $(-1, 9)$.   ◆

Note   In Examples 5, 6, and 7, all of the inequalities that were solved used the symbols $<$ or $>$. If such an inequality involves $\leq$ or $\geq$, the method of solution is the same, but the endpoints are included. The graphical support will not tell us this; we must pay attention to the type of inequality symbol in order to determine whether to use parentheses or brackets when writing the solution set.

EXAMPLE 8   *Solving Equations and Inequalities Involving a Sum of Absolute Values*

(a)   Solve graphically $|x + 5| + |x - 3| = 16$ by the intersection-of-graphs method and verify analytically.

(b)   Use Figure 47 and the result of part (a) to solve the inequalities

$$|x + 5| + |x - 3| \leq 16 \quad \text{and} \quad |x + 5| + |x - 3| \geq 16.$$

SOLUTION

(a)   Let $y_1 = |x + 5| + |x - 3|$ and $y_2 = 16$. Graphing them in an appropriate window gives us the graphs shown in Figure 47. Using a calculator to locate the points of intersection of the graphs, we find the x-coordinates of the points are -9 and 7. To verify these solutions analytically, we substitute them into the equation.

$$\text{Let } x = -9: \qquad |(-9) + 5| + |(-9) - 3| = |-4| + |-12|$$
$$= 4 + 12 = 16. \ \checkmark$$
$$\text{Let } x = 7: \qquad |7 + 5| + |7 - 3| = |12| + |4|$$
$$= 12 + 4 = 16. \ \checkmark$$

Therefore, the solution set of the equation is $\{-9, 7\}$.

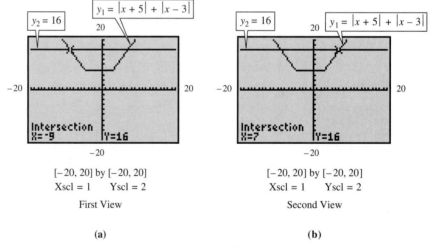

FIGURE 47

(b)   To solve the first inequality, we look for the interval(s) on which the graph of the sum of the absolute values (y_1, the V-shaped graph with the "flat bottom" in Figure 47) is *below* or *coincides* with the graph of y_2 (the horizontal line). We see that this occurs on the closed interval $[-9, 7]$. The solution set of the second inequality consists of the intervals on which the graph of y_1 is *above* or *coincides* with the graph of y_2. This occurs on $(-\infty, -9] \cup [7, \infty)$.

Summarizing the two parts of this example, we have the following:

The solution set of $|x + 5| + |x - 3| = 16$ is $\{-9, 7\}$.

The solution set of $|x + 5| + |x - 3| \leq 16$ is the closed interval $[-9, 7]$.

The solution set of $|x + 5| + |x - 3| \geq 16$ is $(-\infty, -9] \cup [7, \infty)$. ◆

AN APPLICATION INVOLVING ABSOLUTE VALUE

This section concludes with an application of absolute value that deals with carbon dioxide emissions during breathing.

EXAMPLE 9   *Solving an Application Involving* CO_2 *Emission Using Absolute Value*   When humans breathe, carbon dioxide is emitted. In one study, the emission rates of carbon dioxide by college students were measured during both lectures and exams. The average individual rate R_L (in grams per hour) during a lecture class satisfied the inequality $|R_L - 26.75| \leq 1.42$, whereas during an exam the rate R_E satisfied the inequality $|R_E - 38.75| \leq 2.17$. (*Source:* Wang, T. C., *ASHRAE* Trans., 81 (Part 1), 32 (1975).)

(a)   Find the range of values for R_L and R_E.

(b)   The class had 225 students. If T_L and T_E represent the total amounts of carbon dioxide in grams emitted during a one-hour lecture and exam, respectively, write inequalities that describe the ranges for T_L and T_E.

SOLUTION

(a)   First, we find the range for R_L.

$$|R_L - 26.75| \leq 1.42$$

$$-1.42 \leq R_L - 26.75 \leq 1.42 \qquad \text{Write as a compound inequality.}$$

$$25.33 \leq R_L \leq 28.17 \qquad \text{Add 26.75 to each expression.}$$

During a lecture class, the individual rate for emission was between 25.33 and 28.17 grams per hour (inclusive).

Next, we find the range for R_E.

$$|R_E - 38.75| \leq 2.17$$

$$-2.17 \leq R_E - 38.75 \leq 2.17$$

$$36.58 \leq R_E \leq 40.92$$

During an exam, the individual rate for emission was between 36.58 and 40.92 grams per hour (inclusive).

(b)   Since there are 225 students, and R_L and R_E are individual rates, the total amounts of carbon dioxide emitted would be $T_L = 225R_L$ and $T_E = 225R_E$. Thus,

$$225(25.33) \leq T_L \leq 225(28.17)$$

$$5699.25 \leq T_L \leq 6338.25 \qquad \text{Range for } T_L$$

and

$$225(36.58) \leq T_E \leq 225(40.92)$$

$$8230.5 \leq T_E \leq 9207. \qquad \text{Range for } T_E \quad ◆$$

2.4 EXORCISES   Tape 3

In Exercises 1–12, you are given graphs of functions $y = f(x)$. *Sketch by hand the graph of*
$y = |f(x)|$.

1.

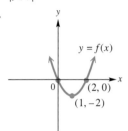

2.

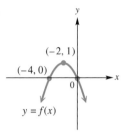

3.

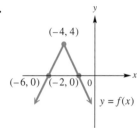

4.

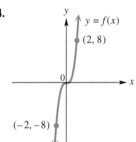

5.

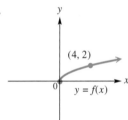

6.

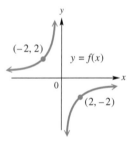

7.

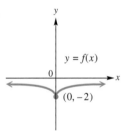

8.

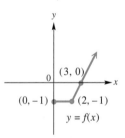

9.

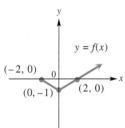

10.

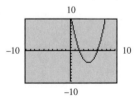

11.

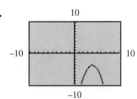

12.

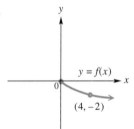

13. Explain in your own words the process you used to find the graphs of $y = |f(x)|$ in Exercises 1–12.

In Exercises 14–17, the graph of a function $y = f(x)$ *is shown. Match the graph of* $y = f(x)$
with the graph of $y = |f(x)|$ *from choices A–D on the following page.*

14.

15.

16.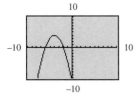

17.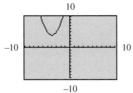

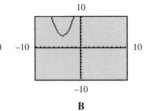

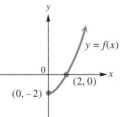

A

B

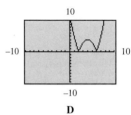

C

D

18. Shown here is the graph of a function $y = f(x)$. Sketch by hand, in order, the graph of each of the following. Use the concept of reflecting introduced in Section 2.3, and the concept of graphing $y = |f(x)|$ introduced in this section.

(a) $y = f(-x)$

(b) $y = -f(-x)$

(c) $y = |-f(-x)|$

19. Repeat Exercise 18 for the graph of $y = f(x)$ shown here.

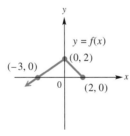

20. If the range of $y = f(x)$ is $[-2, \infty)$, what is the range of $y = |f(x)|$?

21. If the range of $y = f(x)$ is $(-\infty, -2]$, what is the range of $y = |f(x)|$?

In Exercises 22 and 23, one of the graphs is that of $y = f(x)$ and the other is that of $y = |f(x)|$. State which is which.

22. (a)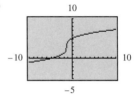

(b)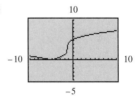

23. (a)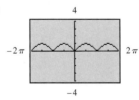

(b)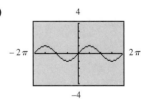

Verify properties 1, 2, 3, and 4 of absolute value for the given values of a and b.

24. $a = 7, b = 14$ **25.** $a = -10, b = -20$ **26.** $a = -26, b = 13$

27.

(a)   (b)

(a) In the figure on the left above, how does the sum of the lengths p and q compare to length r? (*less than, greater than,* or *equal to*)

(b) In the figure on the right above, how does the sum of the lengths p and q compare to length r? (*less than, greater than,* or *equal to*)

(c) Complete the following well-known fact: "The shortest distance between two points is _____."

(d) In the triangle inequality (absolute value property 4), when does the order operation $>$ hold? When does equality hold?

Solve each group of equations and inequalities analytically. Support your solutions graphically.

28. (a) $|x + 4| = 9$ **29.** (a) $|x - 3| = 5$ **30.** (a) $|2x + 7| = 3$
    (b) $|x + 4| > 9$ (b) $|x - 3| > 5$ (b) $|2x + 7| \geq 3$
    (c) $|x + 4| < 9$ (c) $|x - 3| < 5$ (c) $|2x + 7| \leq 3$

31. (a) $|3x - 9| = 6$ **32.** (a) $|2x + 1| + 3 = 5$ **33.** (a) $|4x + 7| = 0$
    (b) $|3x - 9| \geq 6$ (b) $|2x + 1| + 3 \leq 5$ (b) $|4x + 7| > 0$
    (c) $|3x - 9| \leq 6$ (c) $|2x + 1| + 3 \geq 5$ (c) $|4x + 7| < 0$

34. (a) $|7x - 5| = 0$ **35.** (a) $|\pi x + 8| = -4$ **36.** (a) $|\sqrt{2}x - 3.6| = -1$
    (b) $|7x - 5| \geq 0$ (b) $|\pi x + 8| < -4$ (b) $|\sqrt{2}x - 3.6| \leq -1$
    (c) $|7x - 5| \leq 0$ (c) $|\pi x + 8| > -4$ (c) $|\sqrt{2}x - 3.6| \geq -1$

Use the graph, along with the indicated points, to give the solution set for each equation or inequality.

37. (a) $y_1 = y_2$
    (b) $y_1 < y_2$
    (c) $y_1 > y_2$

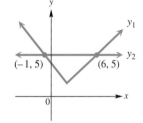

38. (a) $y_1 = y_2$
    (b) $y_1 < y_2$
    (c) $y_1 > y_2$

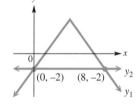

39. (a) $y_1 = y_2$
    (b) $y_1 \leq y_2$
    (c) $y_1 \geq y_2$

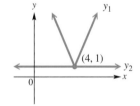

40. (a) $y_1 = y_2$
    (b) $y_1 \leq y_2$
    (c) $y_1 \geq y_2$

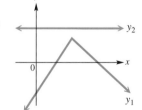

Solve the equation in part (a) by analytic methods. Then, using your graphing calculator, solve the related inequalities in parts (b) and (c).

41. (a) $|3x + 1| = |2x - 7|$
    (b) $|3x + 1| > |2x - 7|$
    (c) $|3x + 1| < |2x - 7|$

42. (a) $|7x + 12| = |x - 4|$
    (b) $|7x + 12| < |x - 4|$
    (c) $|7x + 12| > |x - 4|$

43. (a) $|.5x - 2| = |x - .5|$
    (b) $|.5x - 2| \le |x - .5|$
    (c) $|.5x - 2| \ge |x - .5|$

44. (a) $|2x - 6| = |2x + 5|$
    (b) $|2x - 6| < |2x + 5|$
    (c) $|2x - 6| > |2x + 5|$

45. (a) $|3x - 1| = |3x + 4|$
    (b) $|3x - 1| > |3x + 4|$
    (c) $|3x - 1| < |3x + 4|$

The equations in Exercises 46–49 have integer solutions. Solve the equations and related inequalities graphically.

46. (a) $|x + 1| + |x - 6| = 11$
    (b) $|x + 1| + |x - 6| < 11$
    (c) $|x + 1| + |x - 6| > 11$

47. (a) $|2x + 2| + |x + 1| = 9$
    (b) $|2x + 2| + |x + 1| > 9$
    (c) $|2x + 2| + |x + 1| < 9$

48. (a) $|x| + |x - 4| = 8$
    (b) $|x| + |x - 4| \le 8$
    (c) $|x| + |x - 4| \ge 8$

49. (a) $|.5x + 2| + |.25x + 4| = 9$
    (b) $|.5x + 2| + |.25x + 4| \ge 9$
    (c) $|.5x + 2| + |.25x + 4| \le 9$

RELATING CONCEPTS (EXERCISES 50–55)

The figure shows the graphs of two functions,
$f(x) = |.5x + 6|$ *and* $g(x) = 3x - 14$.

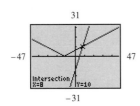

[−47, 47] by [−31, 31]
Xscl = 10   Yscl = 10

Answer Exercises 50–55 in order, without actually using your graphing calculator.

50. Which graph is that of $y = f(x)$? How do you know? (Section 2.4)

51. Which graph is that of $g(x) = 3x - 14$? How do you know? (Section 1.4)

52. Solve $f(x) = g(x)$ based on the display at the bottom of the screen. (Section 1.5)

53. Solve $f(x) > g(x)$ based on the graphs and the display. (Section 1.6)

54. Solve $f(x) < g(x)$ based on the graphs and the display. (Section 1.6)

55. What is the solution set of $|.5x + 6| - (3x - 14) = 0$? (Section 1.5)

Solve graphically the following equations and inequalities. Express solutions or endpoints of intervals rounded to the nearest hundredth if necessary.

56. $|2x + 7| = 6x - 1$

57. $-|3x - 12| \ge -x - 1$

58. $|x - 4| > .5x - 6$

59. $2x + 8 > -|3x + 4|$

60. $|3x + 4| < -3x - 14$

61. $|x - \sqrt{13}| + \sqrt{6} \le -x - \sqrt{10}$

In quality control and other applications, as well as in more advanced mathematics, we often wish to keep the difference between two quantities within some predetermined amount. For example, suppose $y = 2x + 1$, and we want y to be within .01 unit of 4. This can be written using absolute value as $|y - 4| < .01$. To find the values of x that will satisfy this condition on y, we use properties of absolute value as follows.

$$|y - 4| < .01$$
$$|2x + 1 - 4| < .01 \qquad \text{Substitute } 2x + 1 \text{ for } y.$$
$$|2x - 3| < .01$$
$$-.01 < 2x - 3 < .01$$
$$2.99 < 2x < 3.01 \qquad \text{Add 3 to each part.}$$
$$1.495 < x < 1.505 \qquad \text{Divide each part by 2.}$$

By reversing these steps, we can show that keeping x between 1.495 and 1.505 will ensure that the difference between y and 4 is less than .01.

Find the open interval in which x must lie in order for the given condition to hold.

62. $y = 2x + 1$ and the difference between y and 1 is less than .1.

63. $y = 3x - 6$ and the difference between y and 2 is less than .01.

64. $y = 4x - 8$ and the difference between y and 3 is less than .001.

65. $y = 5x + 12$ and the difference between y and 4 is less than .0001.

Wind-Chill Factor   *The wind-chill factor is a measure of the cooling effect that the wind has on a person's skin. It calculates the equivalent cooling temperature if there were no wind. The table gives the wind-chill factor for various wind speeds and temperatures.*

Wind/°F	40°	30°	20°	10°	0°	−10°	−20°	−30°	−40°	−50°
5 mph	37	27	16	6	−5	−15	−26	−36	−47	−57
10 mph	28	16	4	−9	−21	−33	−46	−58	−70	−83
15 mph	22	9	−5	−18	−36	−45	−58	−72	−85	−99
20 mph	18	4	−10	−25	−39	−53	−67	−82	−96	−110
25 mph	16	0	−15	−29	−44	−59	−74	−88	−104	−118
30 mph	13	−2	−18	−33	−48	−63	−79	−94	−109	−125
35 mph	11	−4	−20	−35	−49	−67	−82	−98	−113	−129
40 mph	10	−6	−21	−37	−53	−69	−85	−100	−116	−132

(*Source:* Miller, A., and J. Thompson, *Elements of Meteorology*, Second Edition, Charles E. Merrill Publishing Co., 1975.)

Suppose that we wish to determine the difference between two of these entries, and are interested only in the magnitude, or absolute value, of this difference. Then, we subtract the two entries and find the absolute value. For example, the difference in wind-chill factors for wind at 20 miles per hour with a 20°F temperature and wind at 30 miles per hour with a 40°F temperature is $\left| -10° - 13° \right| = 23°F$, or equivalently, $\left| 13° - (-10°) \right| = 23°F$.

Find the absolute value of the difference of the two indicated wind-chill factors.

66. wind at 15 miles per hour with a 30°F temperature and wind at 10 miles per hour with a −10°F temperature

67. wind at 20 miles per hour with a −20°F temperature and wind at 5 miles per hour with a 30°F temperature

68. wind at 30 miles per hour with a −30°F temperature and wind at 15 miles per hour with a −20°F temperature

69. wind at 40 miles per hour with a 40°F temperature and wind at 25 miles per hour with a −30°F temperature

70. *State and Local Government Finances*   The accompanying graphs depict the amounts of revenue and expenditures for state and local governments in the United States during the years 1990, 1991, and 1992, in billions of dollars. Determine the absolute value of the difference between the revenue and the expenditure for each year, and tell whether the governments were "in the red" or "in the black." (Note: These descriptions go back to the days when bookkeepers used red ink to indicate losses and black ink to indicate gains.)

(a) 1990   **(b)** 1991   **(c)** 1992

State and Local Government
Revenue (in billions of dollars)

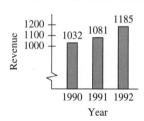

(*Source:* U.S. Bureau of the Census, *Government Finances*, Series GF, No. 5, annual.)

State and Local Government
Expenditures (in billions of dollars)

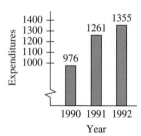

(*Source:* U.S. Bureau of the Census, *Government Finances*, Series GF, No. 5, annual.)

71. *Blood Pressure* Systolic blood pressure is the maximum pressure produced by each heartbeat. Both low blood pressure and high blood pressure are cause for medical concern. Therefore, health care professionals are interested in a patient's "pressure difference from normal," or P_d. If 120 is considered a normal systolic pressure, $P_d = |P - 120|$, where P is the patient's recorded systolic pressure. For example, a patient with a systolic pressure, P, of 113 would have a pressure difference from normal of $P_d = |P - 120| = |113 - 120| = |-7| = 7$.

(a) Calculate the P_d value for a woman whose actual systolic pressure is 116 and whose normal value should be 125.

(b) If a patient's P_d value is 17 and the normal pressure for his sex and age should be 130, what are the two possible values for his systolic blood pressure?

72. *Temperatures on Mars* The temperatures on the surface of Mars in degrees Celsius approximately satisfy the inequality

$$|C + 84| \le 56.$$

What range of temperatures corresponds to this inequality?

73. *Weights of Babies* Dr. Tydings has found that, over the years, 95% of the babies he has delivered weighed y pounds, where $|y - 8.0| \le 1.5$. What range of weights corresponds to this inequality?

74. *Conversion of Methanol to Gasoline* The industrial process that is used to convert methanol to gasoline is carried out at a temperature range of 680°F to 780°F. Using F as the variable, write an absolute value inequality that corresponds to this range.

75. *Kite Flying* When a model kite was flown in crosswinds in tests to determine its limits of power extraction, it attained speeds of 98 to 148 feet per second in winds of 16 to 26 feet per second. Using x as the variable in each case, write absolute value inequalities that correspond to these ranges.

2.5 PIECEWISE DEFINED FUNCTIONS

Graphing Functions Defined Piecewise ◆ The Greatest Integer Function ◆ Applications of Piecewise Defined Functions

GRAPHING FUNCTIONS DEFINED PIECEWISE

The absolute value function, defined in Section 2.1, is a simple example of a function defined by different rules over different subsets of its domain. This is called a piecewise defined function. Recall that the domain of $f(x) = |x|$ is $(-\infty, \infty)$. For the interval $[0, \infty)$ of the domain, the rule that we use is $f(x) = x$. On the other hand, for the interval $(-\infty, 0)$, we use the rule $f(x) = -x$. Thus, the graph of $f(x) = |x|$ is composed of two "pieces." One piece comes from the graph of $y = x$, and the other from $y = -x$.

EXAMPLE 1 *Finding Function Values for a Piecewise Defined Function*
Consider the function f defined piecewise:

$$f(x) = \begin{cases} x + 2 & \text{if } x \le 0 \\ \dfrac{1}{2}x^2 & \text{if } x > 0. \end{cases}$$

Find the following function values.

(a) $f(-3)$ (b) $f(0)$ (c) $f(3)$

SOLUTION

(a) Since $-3 \le 0$ (specifically, $-3 < 0$), we use the rule $f(x) = x + 2$. Thus, $f(-3) = -3 + 2 = -1$. Note that this means the graph of f will contain the point $(-3, -1)$.

(b) Since $0 \le 0$ (because $0 = 0$), we again use the rule $f(x) = x + 2$. So we have $f(0) = 0 + 2 = 2$. The point $(0, 2)$ will lie on the graph of f, meaning that the y-intercept of the graph will be 2.

(c)   The number 3 comes from the interval $(0, \infty)$, and the second part of the rule f indicates that we must now use the rule $f(x) = \frac{1}{2}x^2$. Therefore, $f(3) = \frac{1}{2}(3)^2 = \frac{1}{2}(9) = 4.5$. (What point on the graph of f does this result lead to?)   ◆

If we look at the function f defined in Example 1, we can visualize that for negative values of x or 0, the graph of f will be a portion of a line, because $y = x + 2$ defines a linear function. However, if x is positive, the graph of f will consist of a portion of a parabola. Specifically, it will be the "right half" of the graph of the squaring function $y = x^2$, with a shrink factor of $\frac{1}{2}$.

EXAMPLE 2   *Graphing a Function Defined Piecewise*   Using your knowledge of the graphs of linear functions and the squaring function, sketch the graph of the function f in Example 1 by hand.

SOLUTION   Graph the ray $y = x + 2$, choosing x so that $x \leq 0$, with a solid endpoint at $(0, 2)$. The ray has slope 1 and y-intercept 2. Then, graph $y = \frac{1}{2}x^2$ for $x > 0$. This graph will be half of a parabola with an open endpoint at $(0, 0)$. See Figure 48.   ◆

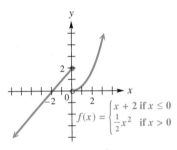

FIGURE 48

Graphing calculators can be used to graph functions defined piecewise. The procedure involves using the test menu. Recall from Chapter 1 that a true statement returns a 1, while a false statement returns a 0. We will rely on the fact that an expression divided by 1 is equal to itself, while an expression divided by 0 is undefined.

The piecewise defined function in Example 1 is defined in the screen in Figure 49(a). We use Y_1 to define the ray $y = x + 2$ (where $x \leq 0$), and we use Y_2 to define the portion of the parabola $y = \frac{1}{2}x^2$ (where $x > 0$). Using the window $[-6, 4]$ by $[-4, 5]$ to match Figure 48, we obtain the calculator-generated graph in Figure 49(b).

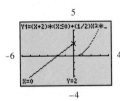

The function defined and graphed in Figure 49 can also be defined as shown in the top screen. See the "Note" following Example 3 for more on this method.

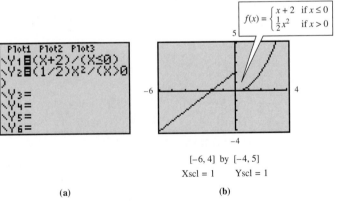

$$f(x) = \begin{cases} x + 2 & \text{if } x \leq 0 \\ \frac{1}{2}x^2 & \text{if } x > 0 \end{cases}$$

$[-6, 4]$ by $[-4, 5]$
Xscl = 1   Yscl = 1

(a)   **(b)**

FIGURE 49

Functions defined piecewise often exhibit discontinuities. The function graphed in Example 2 is an example of such a function. However, depending on the manner in which the function is defined, a piecewise defined function may be continuous on its entire domain, as seen in Example 3.

EXAMPLE 3   *Graphing a Function Defined Piecewise*   Graph the function, first by hand, and then with a calculator.

$$f(x) = \begin{cases} x + 1 & \text{if } x \le 2 \\ -2x + 7 & \text{if } x > 2 \end{cases}$$

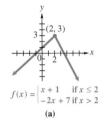

$f(x) = \begin{cases} x+1 & \text{if } x \le 2 \\ -2x+7 & \text{if } x > 2 \end{cases}$

(a)

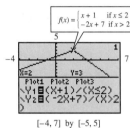

$f(x) = \begin{cases} x+1 & \text{if } x \le 2 \\ -2x+7 & \text{if } x > 2 \end{cases}$

[-4, 7] by [-5, 5]
Xscl = 1   Yscl = 1

(b)

FIGURE 50

SOLUTION   ANALYTIC   We must graph each part of the domain separately. If $x \le 2$, this portion of the graph has an endpoint at $x = 2$. Find the y-value by substituting 2 for x in $y = x + 1$ to get $y = 3$. Another point is needed to graph this part of the graph. Choose an x-value less than 2. Choosing $x = -1$ gives $y = -1 + 1 = 0$. Draw the graph through $(2, 3)$ and $(-1, 0)$ as a ray with an endpoint at $(2, 3)$. Graph the ray for $x > 2$ similarly. This ray will have an open endpoint when $x = 2$ and $y = -2(2) + 7 = 3$. Choosing $x = 4$ gives $y = -2(4) + 7 = -1$. The ray through $(2, 3)$ and $(4, -1)$ completes the graph. In this example, the two rays meet at $(2, 3)$, although this is not always the case, as seen in Example 2. The graph is shown in Figure 50(a).

GRAPHICAL   A calculator-generated graph is shown in Figure 50(b). Notice how the function is defined using the test capability of the calculator.   ◆

Note   The piecewise defined function in Example 3 can also be graphed with a graphing calculator by defining it as follows:

$$Y_1 = (x + 1)(x \le 2) + (-2x + 7)(x > 2).$$

For x-values less than or equal to 2, the second factor in the first expression will be 1, while the second factor in the second expression will be 0. Thus, only the first expression will have an effect on the graph. Similarly, for x-values greater than 2, only the second expression will have an effect.

THE GREATEST INTEGER FUNCTION

An important example of a function defined piecewise is the **greatest integer function**. The notation $[\![x]\!]$ represents the greatest integer less than or equal to x. The definition of $[\![x]\!]$ follows.

$$f(x) = [\![x]\!] = \begin{cases} x \text{ if } x \text{ is an integer} \\ \text{the greatest integer less than } x \text{ if } x \text{ is not an integer} \end{cases}$$

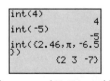

The command "int" is used by many graphing calculators for the greatest integer function. The results of Example 4 are supported in this screen.

EXAMPLE 4   *Evaluating $[\![x]\!]$ for Different Values of x*   Evaluate $[\![x]\!]$ for each of the following values of x.

(a) 4   **(b)** -5 **(c)** 2.46   **(d)** π **(e)** $-6\frac{1}{2}$

SOLUTION

(a)   Since 4 is an integer, $[\![4]\!] = 4$.

(b) $[\![-5]\!] = -5$, since -5 itself is an integer.

(c) $[\![2.46]\!] = 2$, since 2.46 is not an integer, and 2 is the greatest integer less than 2.46.

(d) $[\![\pi]\!] = 3$, since $\pi \approx 3.14$.

(e) $\left[\!\left[-6\frac{1}{2}\right]\!\right] = -7$.   ◆

The important facts concerning the greatest integer function and its graph are summarized in the following.

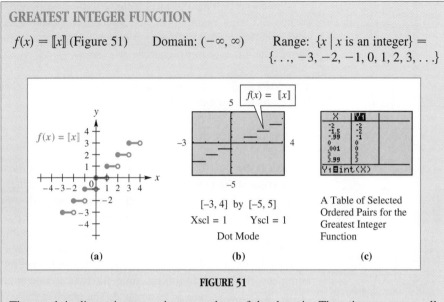

GREATEST INTEGER FUNCTION

$f(x) = [\![x]\!]$ (Figure 51)   Domain: $(-\infty, \infty)$ Range: $\{x \mid x \text{ is an integer}\} = \{\ldots, -3, -2, -1, 0, 1, 2, 3, \ldots\}$

(a)

(b)

[-3, 4] by [-5, 5]
Xscl = 1   Yscl = 1
Dot Mode

(c)

A Table of Selected Ordered Pairs for the Greatest Integer Function

FIGURE 51

The graph is discontinuous at integer values of the domain. The *x*-intercepts are all real numbers in the interval [0, 1), and the *y*-intercept is 0.

The greatest integer function is often called a **step function**. Do you see why?

Graphing functions with discontinuities, such as the greatest integer function, often poses problems for graphing calculators. For instance, Figure 51(b) shows an accurate graph of $f(x) = [\![x]\!]$. However, if we graph the greatest integer function with a calculator, and the calculator is in the *connected* graphing mode, the calculator attempts to literally connect the portions of the graph at integer values. This is why calculators can also be directed to graph in the *dot* graphing mode. In Figure 52(a), we show an accurate graph of $f(x) = [\![x]\!]$ in the standard window, with the calculator in the dot mode. Figure 52(b) shows a distorted graph of this function, since it was graphed in the connected mode.

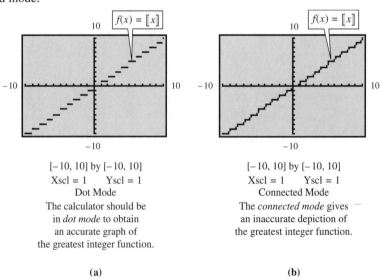

[-10, 10] by [-10, 10]
Xscl = 1   Yscl = 1
Dot Mode
The calculator should be in *dot mode* to obtain an accurate graph of the greatest integer function.

(a)

[-10, 10] by [-10, 10]
Xscl = 1   Yscl = 1
Connected Mode
The *connected mode* gives an inaccurate depiction of the greatest integer function.

(b)

FIGURE 52

Notice from the graph in Figure 52(a) that the inclusion or exclusion of the endpoint for each segment is not readily determined from the calculator-generated graph. However, analysis of the hand-drawn graph in Figure 51(a) does show whether endpoints are included or excluded. Once again, we state the following important conclusion.

A CAUTION AGAINST RELYING TOO HEAVILY ON TECHNOLOGY

Analysis of a calculator-generated graph is often not sufficient to draw correct conclusions. While this technology is incredibly powerful and pedagogically useful, we cannot rely on it alone in our study of the graphs of functions. We must understand the basic concepts of functional analysis as well.

EXAMPLE 5   *Graphing a Step Function*   Graph the function $y = \left[\!\left[\frac{1}{2}x + 1 \right]\!\right]$

(a) by hand and

(b) with a calculator in dot mode. Give the domain and the range.

SOLUTION

(a)   ANALYTIC   Try some values of x in the equation to see how the values of y behave. Some sample ordered pairs are given here.

x	0	$\frac{1}{2}$	1	2	3	4	-1	-2	-3
y	1	1	1	2	2	3	0	0	-1

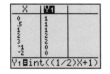

This table shows some selected ordered pairs for the function in Example 5.

These ordered pairs suggest that if x is in the interval $[0, 2)$, then $y = 1$. For x in $[2, 4)$, $y = 2$, and so on. The graph is shown in Figure 53. Again, the domain is $(-\infty, \infty)$. The range is $\{\ldots, -1, 0, 1, 2, \ldots\}$.

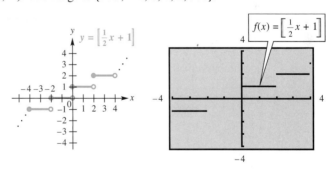

The dots indicate that the graph continues indefinitely in the same pattern.

FIGURE 53

$[-4, 4]$ by $[-4, 4]$
Xscl = 1   Yscl = 1
Dot Mode

FIGURE 54

(b)   GRAPHICAL   Figure 54 shows the graph of $y = \left[\!\left[\frac{1}{2}x + 1 \right]\!\right]$ in a window of $[-4, 4]$ by $[-4, 4]$. Use the trace feature to convince yourself that the range of the function is $\{\ldots, -2, -1, 0, 1, 2, \ldots\}$.   ◆

APPLICATIONS OF PIECEWISE DEFINED FUNCTIONS

A certain type of pricing procedure can be defined by a function that incorporates the greatest integer function concept.

EXAMPLE 6   *Applying the Greatest Integer Function to Parking Rates*
Downtown Parking charges a $5 base fee for parking through 1 hour, and $1 for each
additional hour or fraction thereof. The maximum fee for 24 hours is $15. Sketch a
graph of the function that describes this pricing scheme.

SOLUTION   For any amount of time during and up to the first hour, the rate is $5.
Thus, some sample ordered pairs for the function in the interval (0, 1] would be

$$(.25, 5), (.5, 5), (.75, 5), \text{ and } (1, 5).$$

After the first hour is completed, the price immediately jumps (or steps up) to $6, and
remains $6 until the time equals 2 hours. It then jumps to $7 during the third hour, and
so on. During the 11th hour, it will have jumped to $15, and will remain at $15 for the
rest of the 24-hour period. Figure 55 shows the graph of this function, for the interval
(0, 24]. The range of the function is {5, 6, 7, . . . , 15}.   ◆

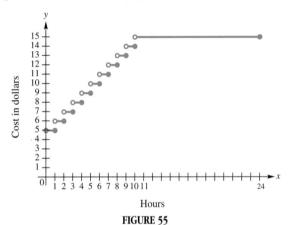

Hours

FIGURE 55

EXAMPLE 7   *Applying a Piecewise Defined Function to Shoe Sizes*   Professional basketball player Shaquille O'Neal is 7-foot 1-inch tall and weighs 300 pounds.
The table lists his age and shoe size.

Age	Size
20	19
21	20
22	21
23	22

(a)   Determine a linear function defined by $f(x)$ that models the data, where x is
Shaquille O'Neal's age and $f(x)$ computes his shoe size. Interpret the slope of the
graph of f.

(b)   Could $f(x)$ be used to predict O'Neal's shoe size at any age?

(c)   Suppose his feet continue to grow at the present rate and then stop growing at age
24. Use a calculator to graph a function defined piecewise that describes his shoe
size between the ages of 20 and 30.

SOLUTION

(a)   The shoe size is 1 less than his age, so we let $f(x) = x - 1$. The slope of 1 indicates that his shoe size is increasing at a rate of 1 size per year.

(b)   No, because most people's feet eventually stop growing at some age.

(c) On the interval [20, 24], $y = x - 1$, and on (24, 30], $y = 23$. Using function notation, the piecewise function is

$$g(x) = \begin{cases} x - 1 & \text{if } 20 \le x \le 24 \\ 23 & \text{if } 24 < x \le 30. \end{cases}$$

The calculator-generated graph of this function is shown in Figure 56. ◆

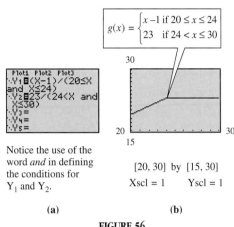

$$g(x) = \begin{cases} x - 1 & \text{if } 20 \le x \le 24 \\ 23 & \text{if } 24 < x \le 30 \end{cases}$$

Notice the use of the word *and* in defining the conditions for Y_1 and Y_2.

[20, 30] by [15, 30]

Xscl = 1   Yscl = 1

(a)   **(b)**

FIGURE 56

2.5 EXERCISES   Tape 3

*For each of the following functions defined piecewise, find (**a**) $f(-5)$, (**b**) $f(-1)$, (**c**) $f(0)$, (**d**) $f(3)$, and (**e**) $f(5)$.*

1. $f(x) = \begin{cases} 2x & \text{if } x \le -1 \\ x - 1 & \text{if } x > -1 \end{cases}$
　　　　　　　　　2. $f(x) = \begin{cases} x - 2 & \text{if } x < 3 \\ 4 - x & \text{if } x \ge 3 \end{cases}$

3. $f(x) = \begin{cases} 3x + 5 & \text{if } x \le 0 \\ 4 - 2x & \text{if } 0 < x < 2 \\ x & \text{if } x \ge 2 \end{cases}$
　　　4. $f(x) = \begin{cases} 4x + 1 & \text{if } x < 2 \\ 3x & \text{if } 2 \le x \le 5 \\ 3 - 2x & \text{if } x > 5 \end{cases}$

Sketch by hand the graph of each of the following piecewise defined functions.

5. $f(x) = \begin{cases} x - 1 & \text{if } x \le 3 \\ 2 & \text{if } x > 3 \end{cases}$
　6. $f(x) = \begin{cases} 6 - x & \text{if } x \le 3 \\ 3x - 6 & \text{if } x > 3 \end{cases}$
　7. $f(x) = \begin{cases} 4 - x & \text{if } x < 2 \\ 1 + 2x & \text{if } x \ge 2 \end{cases}$

8. $f(x) = \begin{cases} 2x + 1 & \text{if } x \ge 0 \\ x & \text{if } x < 0 \end{cases}$
　9. $f(x) = \begin{cases} 2 + x & \text{if } x < -4 \\ -x & \text{if } -4 \le x \le 5 \\ 3x & \text{if } x > 5 \end{cases}$
　10. $f(x) = \begin{cases} -2x & \text{if } x < -3 \\ 3x - 1 & \text{if } -3 \le x \le 2 \\ -4x & \text{if } x > 2 \end{cases}$

11. $f(x) = \begin{cases} -\dfrac{1}{2}x^2 + 2 & \text{if } x \le 2 \\ \dfrac{1}{2}x & \text{if } x > 2 \end{cases}$
　12. $f(x) = \begin{cases} \sqrt[3]{x} & \text{if } x < 0 \\ \sqrt{x} + 4 & \text{if } x \ge 0 \end{cases}$
　13. $f(x) = \begin{cases} x^3 + 5 & \text{if } x \le 0 \\ -x^2 & \text{if } x > 0 \end{cases}$

14. $f(x) = \begin{cases} -|x| + 4 & \text{if } x \ne 6 \\ 3 & \text{if } x = 6 \end{cases}$

Use a graphing calculator to graph the piecewise defined function in the specified exercise, using the window indicated.

15. Exercise 5, window $[-4, 6]$ by $[-2, 4]$
16. Exercise 6, window $[-2, 8]$ by $[-2, 10]$

17. Exercise 7, window $[-4, 6]$ by $[-2, 8]$
18. Exercise 8, window $[-5, 4]$ by $[-3, 8]$

19. Exercise 9, window $[-12, 12]$ by $[-6, 20]$

20. Exercise 10, window $[-6, 6]$ by $[-10, 8]$

21. Exercise 11, window $[-5, 6]$ by $[-2, 4]$

22. Exercise 12, window $[-5, 6]$ by $[-3, 6]$

23. Exercise 13, window $[-3, 4]$ by $[-3, 6]$

RELATING CONCEPTS (EXERCISES 24–26)

The screen shown here will simulate a piecewise defined function on a graphing calculator. Answer the questions asked in Exercises 24–26.

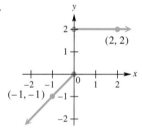

24. For what values of x will the graph be a portion of a straight line?

25. For what values of x will the graph be a portion of a parabola?

26. For what values of x will the graph be a portion of a square root function graph?

 27. Why is the following not truly a piecewise defined function?

$$f(x) = \begin{cases} x + 7 & \text{if } x \le 4 \\ x^2 & \text{if } x \ge 4 \end{cases}$$

28. Consider the "function" defined in Exercise 27. Suppose you were asked to find $f(4)$. How would you respond?

Give a rule of a function f defined piecewise for the graph shown. Give the domain and the range.

29.

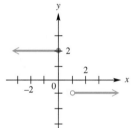

30.

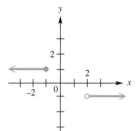

31.

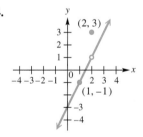

32.

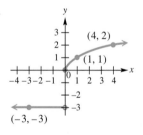

33.

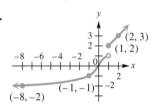

34.

35. Graph the function $y = [\![x]\!]$ on your calculator, using the window $[-100, 100]$ by $[-100, 100]$, with Xscl = 10 and Yscl = 10. Observe the graph, and discuss why an incorrect conclusion might be drawn concerning the behavior of this function if one does not know the mathematical definition of $[\![x]\!]$.

36. Graph the function $y = 3[\![x]\!]$ on your calculator, using the window $[0, 10]$ by $[0, 40]$, with Xscl = 1 and Yscl = 3. Use the dot mode of your calculator. Now, trace from $x = 0$ to $x = 40$. Describe what is happening. What does this exercise reinforce about the special name given to functions of this type?

Use a graphing calculator in dot mode with the specified window to create an accurate graph of each of the following functions, defined using the greatest integer function.

37. $y = [\![x]\!] - 1.5$
window $[-5, 5]$ by $[-3, 3]$

38. $y = [\![-x]\!]$
window $[-5, 5]$ by $[-3, 3]$

39. $y = -[\![x]\!]$
window $[-5, 5]$ by $[-3, 3]$

40. $y = [\![x + 2]\!]$
window $[-5, 5]$ by $[-3, 3]$

RELATING CONCEPTS (EXERCISES 41–44)

Use the terminology of Sections 2.2 and 2.3 to describe how the graph of the given function can be obtained from the graph of $y = [\![x]\!]$.

41. $y = [\![x]\!] - 1.5$ **42.** $y = [\![-x]\!]$ **43.** $y = -[\![x]\!]$ **44.** $y = [\![x + 2]\!]$

Work each of the following problems.

45. *Minimum Wage* The table lists the federal minimum hourly wage from 1978 to 1995. Sketch a graph of the data as a piecewise defined function. (*Source:* U.S. Department of Labor.)

Year(s)	Wage
1978	$2.65
1979	$2.90
1980	$3.10
1981–90	$3.35
1991–95	$4.50

46. *Rabies Cases* The following table lists the approximate number of animal rabies cases in the United States from 1988 to 1992. (*Source: USA Today* research.)

Year	Cases
1988	4800
1989	4900
1990	5000
1991	6700
1992	8400

(a) Describe the change in the data from one year to the next.

(b) Determine a piecewise defined function $f(x)$ that approximates the data. Let $x = 0$ correspond to the year 1988.

47. *Express Mail Charges* An express mail company charges $10 for a package weighing 1 pound or less. Each additional pound or part of a pound costs $3 more. Find the cost to send a package weighing 2 pounds; 2.5 pounds; 5.8 pounds. Graph the function on the interval (0, 7]. What is the range of the function for this domain?

48. *Montreal Taxi Rates* Montreal taxi rates in a recent year were $1.80 for the first $\frac{1}{9}$ mi and $.20 for each additional $\frac{1}{9}$ mi or fraction of $\frac{1}{9}$. Let $C(x)$ be the cost for a taxi ride of $\frac{x}{9}$ mi. Find (a) $C(1)$, (b) $C(2.3)$, and (c) $C(8)$. (d) Graph $y = C(x)$ for $0 < x \le \frac{4}{9}$. (e) Give the domain and range of C, assuming a ride of unlimited length.

49. *Mail Charges* A mail-order firm charges $.30 to mail a package weighing 1 ounce or less, and then $.27 for each additional ounce or fraction of an ounce. Let $M(x)$ be the cost of mailing a package weighing x oz. Find (a) $M(.75)$, (b) $M(1.6)$, and (c) $M(4)$. (d) Graph

$y = M(x)$ for $0 < x \le 4$. (e) Give the domain and range of M, assuming an unlimited package weight.

50. *Car Rental* A car rental costs $37 for 1 day, which includes 50 free miles. Each additional 25 mile or portion costs $10. Graph the function for $0 < x \le 150$, where x represents the number of miles driven.

51. *Lift Truck Rental* For a lift truck rental of no more than 3 days, the charge is $300. An additional charge of $75 is made for each day or portion of a day after 3. Graph the function for $0 < x \le 7$, where x represents the number of days the truck is rented.

52. *Floor Polisher Rental* Let f be a function that gives the cost to rent a floor polisher for x days. The cost is a flat $3 for renting the polisher plus $4 per day or fraction of a day for using the polisher. Graph the function for $0 < x \le 4$, where x represents the number of days rented.

53. *Insulin Level* When a diabetic takes long-acting insulin, the insulin reaches its peak effect on the blood sugar level in about 3 hr. This effect remains fairly constant for 5 hr, then declines, and is very low until the next injection. In a typical patient, the level of insulin might be given by the following function.

$$f(t) = \begin{cases} 40t + 100 & \text{if } 0 \le t \le 3 \\ 220 & \text{if } 3 < t \le 8 \\ -80t + 860 & \text{if } 8 < t \le 10 \\ 60 & \text{if } 10 < t \le 24 \end{cases}$$

Here $f(t)$ is the blood sugar level, in appropriate units, at time t measured in hours from the time of the injection. Chuck takes his insulin at 6 A.M. Find the blood sugar level at each of the following times.

(a) 7 A.M.   (b) 9 A.M.   (c) 10 A.M.   (d) Noon

(e) 2 P.M.   (f) 5 P.M.   (g) Midnight

(h) Graph $y = f(t)$ for $0 \le t \le 18$.

54. *Snow Depth* The snow depth in Michigan's Isle Royale National Park varies throughout the winter. In a typical winter, the snow depth in inches is approximated by the following function.

$$f(x) = \begin{cases} 6.5x & \text{if } 0 \le x \le 4 \\ -5.5x + 48 & \text{if } 4 < x \le 6 \\ -30x + 195 & \text{if } 6 < x \le 6.5 \end{cases}$$

Here x represents the time in months, with $x = 0$ representing the beginning of October, $x = 1$ representing the beginning of November, and so on.

(a) Graph the function for $0 \le x \le 6.5$.

(b) In what month is the snow deepest? What is the deepest snow depth?

(c) In what months does the snow begin and end?

First-Class Postage Charges   First-class postage with the U.S. Postal Service in 1997 cost
\$.32 for all weights through 1 ounce and then \$.23 for each ounce or fraction of an ounce
thereafter. The function

$$Y_1 = .32 + .23 \, \text{int}(x - .001)$$

gives a fairly good model for these rates. Using a table, as shown in the screen here, we can get
an idea of how much a letter will cost depending upon its weight. Here we use $\Delta\text{Tbl} = .1$.

*Use a graphing calculator with a table feature to determine the cost of a letter weighing each
of the following amounts.*

55. .8 ounce   **56.** 1.7 ounces   **57.** 2.3 ounces   **58.** 5.9 ounces

2.6 FURTHER TOPICS IN THE STUDY OF FUNCTIONS

Operations on Functions ◆ The Difference Quotient ◆ Composition of Functions ◆ Applications of Operations and Composition

OPERATIONS ON FUNCTIONS

Just as we add, subtract, multiply, and divide real numbers, we can also perform these operations on functions. Given two functions f and g, their *sum*, written $f + g$, is defined as

$$(f + g)(x) = f(x) + g(x),$$

for all x such that both $f(x)$ and $g(x)$ exist. Similar definitions can be given for the difference, $f - g$, product fg, and quotient f/g, of functions; however, the quotient,

$$\left(\frac{f}{g}\right)(x) = \frac{f(x)}{g(x)},$$

is defined only for those values of x where both $f(x)$ and $g(x)$ exist, with the additional
condition $g(x) \ne 0$. The various operations on functions are defined as follows.

OPERATIONS ON FUNCTIONS

If f and g are functions, then for all values of x for which both $f(x)$ and $g(x)$ exist,
the **sum** of f and g is defined by

$$(f + g)(x) = f(x) + g(x),$$

the **difference** of f and g is defined by

$$(f - g)(x) = f(x) - g(x),$$

the **product** of f and g is defined by

$$(fg)(x) = f(x) \cdot g(x),$$

and the **quotient** of f and g is defined by

$$\left(\frac{f}{g}\right)(x) = \frac{f(x)}{g(x)}, \quad \text{where } g(x) \ne 0.$$

The domains of $f + g$, $f - g$, fg, and f/g are summarized below. (Recall that the intersection of two sets is the set of all elements belonging to *both* sets.)

DOMAINS OF $f + g$, $f - g$, fg, f/g

For functions f and g, the domains of $f + g$, $f - g$, and fg include all real numbers in the intersection of the domains of f and g, while the domain of f/g includes those real numbers in the intersection of the domains of f and g for which $g(x) \neq 0$.

As an example of a sum of functions, suppose that we let $f(x) = x^2 + 1$ and $g(x) = 3x + 5$. Then, $f + g$ is found as follows:

$$(f + g)(x) = f(x) + g(x)$$
$$= (x^2 + 1) + (3x + 5)$$
$$= x^2 + 3x + 6.$$

Since the domains of both f and g are $(-\infty, \infty)$, the intersection of their domains is also $(-\infty, \infty)$, and this is the domain of $f + g$. The graph of $f + g$ can be found on a graphing calculator by entering it directly, or by letting $y_1 = x^2 + 1$, $y_2 = 3x + 5$, and $y_3 = y_1 + y_2$. Then, instruct the calculator to graph y_3. Figure 57(a) shows the graphs of y_1 and y_2, and Figure 57(b) shows the graph of y_3. Notice that the graph of y_3 is a parabola. (In the next chapter, we will determine analytically the transformations on $y = x^2$ that will give us y_3.)

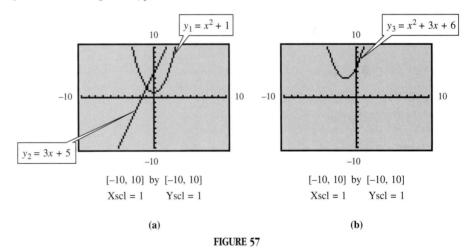

[−10, 10] by [−10, 10]	[−10, 10] by [−10, 10]
Xscl = 1 Yscl = 1	Xscl = 1 Yscl = 1
(a)	**(b)**

FIGURE 57

EXAMPLE 1   *Using the Operations on Functions*   Let $f(x) = x^2 + 1$ and $g(x) = 3x + 5$. Find each of the following.

(a) $(f + g)(1)$ **(b)** $(f - g)(-3)$ **(c)** $(fg)(5)$ **(d)** $\left(\dfrac{f}{g}\right)(0)$

SOLUTION

(a) Since $f(1) = 2$ and $g(1) = 8$, use the definition above to get

$$(f + g)(1) = f(1) + g(1) = 2 + 8 = 10.$$

This result indicates that the point $(1, 10)$ lies on the graph of $f + g$. This can be verified graphically by using the calculator to locate the point. It can also be verified

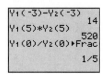

These screens support some of the results of Example 1. Note the use of function notation.

by finding $(f + g)(1)$, using the rule for $(f + g)(x)$ found earlier: $(f + g)(x) = x^2 + 3x + 6$, so $(f + g)(1) = 1^2 + 3(1) + 6 = 10$.

(b) $(f - g)(-3) = f(-3) - g(-3) = 10 - (-4) = 14$

(c) $(fg)(5) = f(5) \cdot g(5) = 26 \cdot 20 = 520$

(d) $\left(\dfrac{f}{g}\right)(0) = \dfrac{f(0)}{g(0)} = \dfrac{1}{5}$ ◆

EXAMPLE 2   *Using the Operations on Functions*   Let $f(x) = 8x - 9$ and $g(x) = \sqrt{2x - 1}$. Find each of the following.

(a) $(f + g)(x)$ **(b)** $(f - g)(x)$ **(c)** $(fg)(x)$ **(d)** $\left(\dfrac{f}{g}\right)(x)$

(e)   the domains of f, g, $f + g$, $f - g$, fg, and f/g

SOLUTION

(a) $(f + g)(x) = f(x) + g(x) = 8x - 9 + \sqrt{2x - 1}$

(b) $(f - g)(x) = f(x) - g(x) = 8x - 9 - \sqrt{2x - 1}$

(c) $(fg)(x) = f(x) \cdot g(x) = (8x - 9)\sqrt{2x - 1}$

(d) $\left(\dfrac{f}{g}\right)(x) = \dfrac{f(x)}{g(x)} = \dfrac{8x - 9}{\sqrt{2x - 1}}$

(e)   The domain of f is the set of all real numbers, while the domain of $g(x) = \sqrt{2x - 1}$ includes just those real numbers that make $2x - 1 \geq 0$; the domain of g is the interval $\left[\frac{1}{2}, \infty\right)$. The domain of $f + g$, $f - g$, and fg is thus $\left[\frac{1}{2}, \infty\right)$. With f/g, the denominator cannot be zero, so the value $\frac{1}{2}$ is excluded from the domain. The domain of f/g is $\left(\frac{1}{2}, \infty\right)$.   ◆

THE DIFFERENCE QUOTIENT

Suppose that the point P lies on the graph of $y = f(x)$, and suppose that h is a positive number. If we let $(x, f(x))$ denote the coordinates of P and $(x + h, f(x + h))$ denote the coordinates of Q, then the line joining P and Q has slope

$$m = \frac{f(x + h) - f(x)}{(x + h) - x} = \frac{f(x + h) - f(x)}{h}.$$

This expression, called the **difference quotient,** is important in the study of calculus.

Figure 58 shows the graph of the line PQ (called a secant line). If h is allowed to approach 0, the slope of this secant line approaches the slope of the line tangent to the curve at P. Important applications of this idea are developed in calculus, where the concepts of *limit* and *derivative* are investigated.

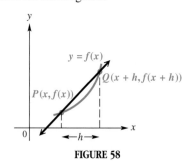

FIGURE 58

EXAMPLE 3 *Finding the Difference Quotient* Let $f(x) = 2x^2 - 3x$. Find the difference quotient and simplify the expression.

SOLUTION To find $f(x + h)$, replace x in $f(x)$ with $x + h$ to get

$$f(x + h) = 2(x + h)^2 - 3(x + h).$$

Then,

$$\frac{f(x + h) - f(x)}{h}$$

$$= \frac{2(x + h)^2 - 3(x + h) - (2x^2 - 3x)}{h}$$

$$= \frac{2(x^2 + 2xh + h^2) - 3x - 3h - 2x^2 + 3x}{h} \qquad \text{Square } x + h; \text{ use the distributive property.}$$

$$= \frac{2x^2 + 4xh + 2h^2 - 3x - 3h - 2x^2 + 3x}{h}$$

$$= \frac{4xh + 2h^2 - 3h}{h} \qquad \text{Combine terms.}$$

$$= \frac{h(4x + 2h - 3)}{h} \qquad \text{Factor out } h.$$

$$= 4x + 2h - 3. \qquad \text{Divide.} \qquad \blacklozenge$$

Caution Notice that $f(x + h)$ is not the same as $f(x) + f(h)$. For $f(x) = 2x^2 - 3x$, as shown in Example 3,

$$f(x + h) = 2(x + h)^2 - 3(x + h) = 2x^2 + 4xh + 2h^2 - 3x - 3h$$

but

$$f(x) + f(h) = (2x^2 - 3x) + (2h^2 - 3h) = 2x^2 - 3x + 2h^2 - 3h.$$

These expressions differ by $4xh$.

COMPOSITION OF FUNCTIONS

The diagram in Figure 59 shows a function f that assigns to each element x of set X some element y of set Y. Suppose also that a function g takes each element of set Y and assigns a value z of set Z. Using both f and g, then, an element x in X is assigned to an element z in Z. The result of this process is a new function, h, that takes an element x in X and assigns an element z in Z. This function h is called the *composition* of functions g and f, written $g \circ f$, and is defined as follows.

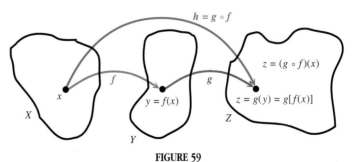

FIGURE 59

> ## COMPOSITION OF FUNCTIONS
>
> If f and g are functions, then the **composite function,** or **composition,** of g and f is
>
> $$(g \circ f)(x) = g[f(x)]$$
>
> for all x in the domain of f such that $f(x)$ is in the domain of g.

EXAMPLE 4 *Finding and Simplifying Composite Functions* Suppose that $f(x) = 2x^2 + 1$ and $g(x) = x - 1$.

(a) Find $f \circ g$ and simplify the expression.

(b) Find $g \circ f$ and simplify the expression.

SOLUTION

(a)
$$\begin{aligned} (f \circ g)(x) &= f[g(x)] = f(x - 1) \\ &= 2(x - 1)^2 + 1 \qquad * \\ &= 2(x^2 - 2x + 1) + 1 \\ &= 2x^2 - 4x + 2 + 1 \\ &= 2x^2 - 4x + 3 \end{aligned}$$

By examining the form of $f \circ g$ in the line marked *, we can see that its graph is that of the function $y = x^2$, shifted 1 unit to the right, stretched vertically by a factor of 2, and shifted 1 unit upward.

(b)
$$\begin{aligned} (g \circ f)(x) &= g[f(x)] = g(2x^2 + 1) \\ &= (2x^2 + 1) - 1 \\ &= 2x^2 \end{aligned}$$

The simplified form shows us that the graph of $g \circ f$ is that of the function $y = x^2$, stretched vertically by a factor of 2.   ◆

Comparing the results in the two parts in Example 4 shows that, *in general, $f \circ g$ is not equal to $g \circ f$.*

In calculus, it is sometimes useful to treat a function as a composition of two functions. The next example shows how this can be done.

EXAMPLE 5 *Finding Functions That Form a Given Composite* Suppose $h(x) = \sqrt{2x + 3}$. Find functions f and g, so that $(f \circ g)(x) = h(x)$.

SOLUTION Since there is a quantity, $2x + 3$, under a radical, one possibility is to choose $f(x) = \sqrt{x}$ and $g(x) = 2x + 3$. Then, $(f \circ g)(x) = \sqrt{2x + 3}$, as required. Other combinations are possible. For example, we could choose $f(x) = \sqrt{x + 3}$ and $g(x) = 2x$.   ◆

APPLICATIONS OF OPERATIONS AND COMPOSITION

There are numerous applications in business, economics, physics, and other fields that can be viewed from the perspective of combining functions. For example, in manufacturing, the cost of producing a product usually consists of two parts. One part is a *fixed cost* for designing the product, setting up a factory, training workers, and so on. Usually, the fixed cost is constant for a particular product and does not change as more items are made. The other part of the cost is a *variable cost* per item for labor, materials, packaging, shipping, and so on. The variable cost is often the same per item, so that the total amount of variable cost increases as more items are produced. A *linear cost function* has the form $C(x) = mx + b$, where m represents the variable cost per item and b represents the fixed cost. The revenue from selling a product depends on the price per item and the number of items sold, as given by the *revenue function*, $R(x) = px$, where p is the price per item and $R(x)$ is the revenue from the sale of x items. The profit is described by the *profit function*, given by $P(x) = R(x) - C(x)$.

EXAMPLE 6 *Finding and Analyzing Cost, Revenue, and Profit* Suppose that a businessman invests $1500 as his fixed cost in a new venture that produces and sells a device that makes programming a VCR easier. Each such device costs $100 to manufacture.

(a) Write a cost function for the product, if x represents the number of devices produced. Assume that the function is linear.

(b) Find the revenue function if each device in part (a) sells for $125.

(c) Give the profit function for the item in part (a).

(d) How many items must be produced and sold before the company makes a profit?

(e) Support the result of part (d) graphically.

SOLUTION

(a) Since the cost function is linear, it will have the form $C(x) = mx + b$, with $m = 100$ and $b = 1500$. That is,

$$C(x) = 100x + 1500.$$

(b) The revenue function is

$$R(x) = px = 125x. \quad \text{Let } p = 125.$$

(c) The profit function is given by

$$P(x) = R(x) - C(x)$$
$$= 125x - (100x + 1500)$$
$$= 125x - 100x - 1500$$
$$= 25x - 1500.$$

(d) ANALYTIC To make a profit, $P(x)$ must be positive. Set $P(x) = 25x - 1500 > 0$ and solve for x.

$$25x - 1500 > 0$$
$$25x > 1500 \quad \text{Add 1500 to each side.}$$
$$x > 60 \quad \text{Divide by 25.}$$

At least 61 items must be sold for the company to make a profit.

(e) GRAPHICAL On a calculator, let $y_1 = 100x + 1500$ be the cost function $C(x)$ and let $y_2 = 125x$ be the revenue function $R(x)$. Then we can graph $y_3 = y_2 - y_1$, with y_3 representing the profit function (that is, $R(x) - C(x)$). We must decide on the smallest whole number value of x for which y_3 is greater than 0. Using an appropriate window, the calculator will allow us to find the point at which the graph intersects the x-axis; we see in Figure 60 on the following page that the x-intercept is 60. Thus, the company must sell at least 61 devices to earn a profit.

(Notice that in Figure 60, we have used the connected mode to emphasize that the profit function has a straight-line graph. However, we should understand that only whole number values of x are in the domain of this function, since it would not make sense to consider fractional parts of devices produced.) ◆

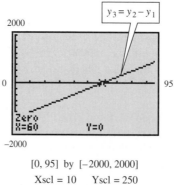

[0, 95] by [−2000, 2000]
Xscl = 10   Yscl = 250
FIGURE 60

In Section 1.8, we saw how a quantity y is in direct proportion, or varies directly, with another quantity x if there exists a constant k such that $y = kx$. Sometimes, y varies directly as a power of x.

DIRECT VARIATION AS A POWER

Let n be a positive real number. Then, y **varies directly as the nth power** of x, or y is **directly proportional to the nth power** of x, if a nonzero real number k exists such that

$$y = kx^n.$$

While the definition above allows n to be any positive real number, we will now only consider positive integer powers. Furthermore, if $n = 2$, we have the case where one quantity varies as the square of another. The familiar formula $A = \pi r^2$ for the area of a circle is an example of this type of variation; that is, the area of a circle varies directly as the square of its radius. The constant of proportionality here is π.

The next example shows how composition of functions can be applied to a physical phenomenon.

EXAMPLE 7 *Applying Composition of Functions to an Oil Well Leak* Suppose an oil well off the California coast is leaking, with the leak spreading oil in a circular layer over the surface. At any time t, in minutes, after the beginning of the leak, the radius of the circular oil slick is $r(t) = 5t$ feet. Express the area A of the leak as a function of the time that it has been spreading, and determine graphically how much of the surface is covered 20 minutes after the leak begins.

SOLUTION   Since $A(r) = \pi r^2$ gives the area of a circle of radius r, the area can be expressed as a function of time by substituting $5t$ for r in $A(r) = \pi r^2$ to get

$$A(r) = \pi r^2$$
$$A[r(t)] = \pi(5t)^2 = 25\pi t^2.$$

The function $A[r(t)]$ is a composite function of the functions A and r.

To determine *graphically* how much of the surface is covered 20 minutes after the leak begins, we graph $y_1 = 25\pi x^2$. (Note that we use y rather than A, and x rather than t. The variables used are immaterial, and it is more appropriate for graphing calculator interpretation to use x and y.) Figure 61 indicates that if y_1 is graphed in an appropriate window, when $x = 20$, $y \approx 31{,}415.927$. Thus, the leak will cover approximately 31,400 square feet 20 minutes after it begins.   ◆

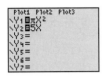

Notice that using function notation twice allows us to support the result in Example 7.

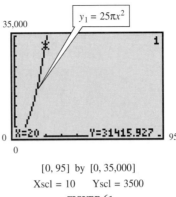

[0, 95] by [0, 35,000]
Xscl = 10   Yscl = 3500

FIGURE 61

EXAMPLE 8 *Applying a Difference of Functions to Surface Area of a Ball*
The formula for the surface area S of a sphere is $S = 4\pi r^2$, where r is the radius of the sphere.

(a) Construct a model $S(r)$ that describes the amount of surface area gained if the radius r inches of a ball is increased by 2 inches.

(b) The function found in part (a) is graphed in Figure 62. What classification of function does this seem to be?

(c) Use the graph to determine the amount of extra material needed to manufacture a ball of radius 22 inches as compared to a ball of radius 20 inches.

(d) Confirm analytically the result obtained graphically in part (c).

SOLUTION

(a) In words, we have

Surface area gained = Larger surface area − smaller surface area.

Symbolically, this translates as a difference of functions:

$$S(r) = 4\pi(r + 2)^2 - 4\pi r^2.$$

(b) Figure 62 appears to be a linear function with $x = r$. (Verify this.)

(c) GRAPHICAL   We observe that for the graph of S, when $x = 20$, $y \approx 1055.5751$. See Figure 63. Thus, it would take about 1056 extra square inches of material.

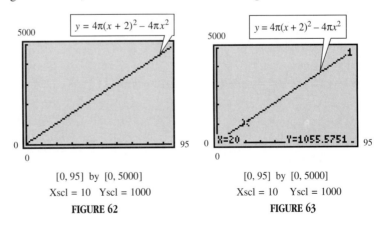

[0, 95] by [0, 5000]
Xscl = 10   Yscl = 1000

FIGURE 62

[0, 95] by [0, 5000]
Xscl = 10   Yscl = 1000

FIGURE 63

    (d)   ANALYTIC   Let $r = 20$ in the model found in part (a).

$$S(r) = 4\pi(r + 2)^2 - 4\pi r^2$$
$$S(20) = 4\pi(20 + 2)^2 - 4\pi(20)^2$$
$$= 1936\pi - 1600\pi$$
$$= 336\pi$$
$$\approx 1056$$

This result confirms our earlier conclusion.   ◆

2.6   EXERCISES   Tape 3

*For each pair of functions in Exercises 1–8, (**a**) find $f + g$, $f - g$, and fg, (**b**) give the domains of the functions in part (a), (**c**) find f/g and give its domain, (**d**) find $f \circ g$ and give its domain, and (**e**) find $g \circ f$ and give its domain.*

1. $f(x) = 4x - 1$, $g(x) = 6x + 3$ **2.** $f(x) = 9 - 2x$, $g(x) = -5x + 2$

3. $f(x) = |x + 3|$, $g(x) = 2x$ **4.** $f(x) = |2x - 4|$, $g(x) = x + 1$

5. $f(x) = \sqrt[3]{x + 4}$, $g(x) = x^3 + 5$ **6.** $f(x) = \sqrt[3]{6 - 3x}$, $g(x) = 2x^3 + 1$

7. $f(x) = \sqrt{x^2 + 3}$, $g(x) = x + 1$ **8.** $f(x) = \sqrt{2 + 4x^2}$, $g(x) = x$

Let $f(x) = 4x^2 - 2x$ and $g(x) = 8x + 1$. Find each of the following.

9. $(f \circ g)(x)$ **10.** $(g \circ f)(x)$ **11.** $(f \circ g)(3)$ **12.** $(g \circ f)(-2)$ **13.** $(f + g)(3)$

14. $(f + g)(-5)$ **15.** $(fg)(4)$ **16.** $(fg)(-3)$ **17.** $\left(\dfrac{f}{g}\right)(-1)$ **18.** $\left(\dfrac{f}{g}\right)(4)$

19. $(f \circ g)(2)$ **20.** $(f \circ g)(-5)$ **21.** $(g \circ f)(2)$ **22.** $(g \circ f)(-5)$

The graphs of functions f and g are shown. Use these graphs to find the value.

23. $f(1) + g(1)$ **24.** $f(4) - g(3)$

25. $f(-2) \cdot g(4)$ **26.** $\dfrac{f(4)}{g(2)}$

27. $(f \circ g)(2)$ **28.** $(g \circ f)(2)$

29. $(g \circ f)(-4)$ **30.** $(f \circ g)(-2)$

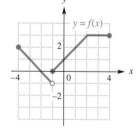 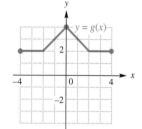

The graphs of two functions f and g are shown in the figure here.

31. Find $(f \circ g)(2)$.

32. Find $(g \circ f)(3)$.

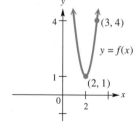 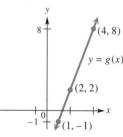

The tables below give some selected ordered pairs for functions f and g.

x	3	4	6
$f(x)$	1	3	9

x	2	7	1	9
$g(x)$	3	6	9	12

Find each of the following.

33. $(f \circ g)(2)$ **34.** $(f \circ g)(7)$ **35.** $(g \circ f)(3)$ **36.** $(g \circ f)(6)$ **37.** $(f \circ f)(4)$ **38.** $(g \circ g)(1)$

39. Why can you not determine $(f \circ g)(1)$ given the information in the tables for Exercises 33–38?

40. Extend the concept of composition of functions to evaluate $(g \circ (f \circ g))(7)$, using the tables for Exercises 33–38.

The graphing calculator screen on the left shows three functions, Y_1, Y_2, and Y_3. The last of these, Y_3, is defined as $Y_1 \circ Y_2$, indicated by the notation $Y_3 = Y_1(Y_2)$. The table on the right shows selected values of X, along with the calculated values of Y_3. Verify by direct calculation that the result is true for the given value of X.

41. $X = 0$

42. $X = 1$

43. $X = 2$

44. $X = 3$

For certain pairs of functions f and g, $(f \circ g)(x) = x$ and $(g \circ f)(x) = x$. Show that this is true for the pairs in Exercises 45–48.

45. $f(x) = 4x + 2$, $g(x) = \frac{1}{4}(x - 2)$ **46.** $f(x) = -3x$, $g(x) = -\frac{1}{3}x$

47. $f(x) = \sqrt[3]{5x + 4}$, $g(x) = \frac{1}{5}x^3 - \frac{4}{5}$ **48.** $f(x) = \sqrt[3]{x + 1}$, $g(x) = x^3 - 1$

RELATING CONCEPTS (EXERCISES 49–50)

Functions such as the pairs in Exercises 45–48 are called inverse functions, *because upon composition in both directions, the result is the identity function. (Inverse functions will be discussed in detail in Section 4.5.)*

49. In a square viewing window, graph $y_1 = \sqrt[3]{x - 6}$ and $y_2 = x^3 + 6$, an example of a pair of inverse functions. Now, graph $y_3 = x$. Describe how the graph of y_2 can be obtained from the graph of y_1, using the graph of $y_3 = x$ as a basis for your description. (*Hint:* Review the terminology of Section 2.3.)

50. Repeat Exercise 49 for $y_1 = 5x - 3$ and $y_2 = \frac{1}{5}(x + 3)$.

Determine the difference quotient

$$\frac{f(x + h) - f(x)}{h} \quad (h \neq 0)$$

for each function f in Exercises 51–56. Simplify completely.

51. $f(x) = 4x + 3$ **52.** $f(x) = 5x - 6$ **53.** $f(x) = -6x^2 - x + 4$

54. $f(x) = \frac{1}{2}x^2 + 4x$ **55.** $f(x) = x^3$ **56.** $f(x) = -2x^3$

In each of Exercises 57–62, a function h is defined. Find functions f and g such that $(f \circ g)(x) = h(x)$. (There are many possible ways to do this.)

57. $h(x) = (6x - 2)^2$ **58.** $h(x) = (11x^2 + 12x)^2$ **59.** $h(x) = \sqrt{x^2 - 1}$

60. $h(x) = (2x - 3)^3$ **61.** $h(x) = \sqrt{6x + 12}$ **62.** $h(x) = \sqrt[3]{2x + 3} - 4$

Cost/Revenue/Profit Analysis   For each of the following, if *x* represents the number of items produced, **(a)** write a cost function, **(b)** find a revenue function if each item sells for the price given, **(c)** give the profit function, **(d)** determine analytically how many items must be produced before a profit is shown (assume whole numbers of items), and **(e)** support the result of part (d) graphically.

63. The fixed cost is \$500, the cost to produce an item is \$10, and the selling price of the item is \$35.

64. The fixed cost is \$180, the cost to produce an item is \$11, and the selling price of the item is \$20.

65. The fixed cost is \$2700, the cost to produce an item is \$100, and the selling price of the item is \$280.

66. The fixed cost is \$1000, the cost to produce an item is \$200, and the selling price of the item is \$240.

Solve each application of the arithmetic and composition of functions.

67. *Volume of a Sphere*   **(a)** The formula for the volume of a sphere is $V = \frac{4}{3}\pi r^3$, where *r* represents the radius of the sphere. Construct a model representing the amount of volume gained when a sphere of radius *r* inches is increased by 3 inches.

 (b) Graph the model found in part (a), using *y* for *V* and *x* for *r*, in the window [0, 10] by [0, 1500]. What classification of function does this appear to be?

 (c) Use your calculator to find graphically the amount of volume gained when a sphere of 4-inch radius is increased to a 7-inch radius.

 (d) Verify your conjecture in part (b) analytically.

68. *Surface Area of a Sphere*   Rework Example 8, parts (a) and (b), but consider what happens when the radius is doubled (rather than increased by 2 inches).

69. *Dimensions of a Rectangle*   Suppose that the length of a rectangle is twice its width. Let *x* represent the width of the rectangle.

 (a) Write a formula for the perimeter *P* of the rectangle in terms of *x* alone. Then, use *P(x)* notation to describe it as a function. What classification of function is this?

 (b) Graph the function *P* found in part (a) in the window [0, 10] by [0, 100]. Locate the point for which *x* = 4 and explain what *x* represents and what *y* represents.

 (c) Locate on the graph of *P* the point with *x*-value 4. Now, sketch a rectangle satisfying the conditions

described earlier, and evaluate its perimeter if its width is this *x*-value. Use the standard perimeter formula. How does the result compare with the *y*-value shown on your screen?

 (d) Locate on the graph of *P* a point with an integer *y*-value. Describe in words the meaning of the *x*- and *y*-coordinates here.

70. *Perimeter of a Square*   The perimeter *x* of a square with side of length *s* is given by the formula

$$x = 4s.$$

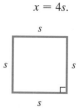

 (a) Solve for *s* in terms of *x*.

 (b) If *y* represents the area of this square, write *y* as a function of the perimeter *x*.

 (c) Use the composite function of part (b) to analytically find the area of a square with perimeter 6.

 (d) Support the result of part (c) graphically, and explain the result.

71. *Area of an Equilateral Triangle*   The area of an equilateral triangle with sides of length *x* is given by the function $A(x) = \frac{\sqrt{3}}{4}x^2$.

 (a) Find *A(2x)*, the function representing the area of an equilateral triangle with sides of length twice the original length.

 (b) Find analytically the area of an equilateral triangle with side length 16. Use the formula for *A(x)* given above.

 (c) Support graphically your result of part (b).

72. *Textbook Author Royalties*   A textbook author invests his royalties in two accounts for 1 year.

 (a) The first account pays 4% simple interest. If he invests *x* dollars in this account, write an expression for y_1 in terms of *x*, where y_1 represents the amount of interest earned.

 (b) He invests in a second account \$500 more than he invested in the first account. This second account pays 2.5% simple interest. Write an expression for y_2, where y_2 represents the amount of interest earned.

(c) What does $y_1 + y_2$ represent?

(d) Graph $y_1 + y_2$ in the window [0, 1000] by [0, 200]. Use the graph to find the amount of interest he will receive if he invests $250 in the first account.

(e) Support the result of part (d) analytically.

73. *Emission of Pollutants* When a thermal inversion layer is over a city, pollutants cannot rise vertically but are trapped below the layer and must disperse horizontally. Assume that a factory smokestack begins emitting a pollutant at 8 A.M. Assume that the pollutant disperses horizontally over a circular area. If t represents the time, in hours, since the factory began emitting pollutants ($t = 0$ represents 8 A.M.), assume that the radius of the circle of pollution is $r(t) = 2t$ mi. Let $A(r) = \pi r^2$ represent the area of a circle of radius r.

(a) Find $(A \circ r)(t)$.

(b) Interpret $(A \circ r)(t)$.

(c) What is the area of the circular region covered by the layer at noon?

(d) Support your result graphically.

74. *Relationship of Measurement Units* The function $f(x) = 12x$ computes the number of inches in x feet and the function $g(x) = 5280x$ computes the number of feet in x miles. What does $(f \circ g)(x)$ compute?

CHAPTER 2   SUMMARY

Section	Important Concepts
2.1 Graphs of Elementary Functions and Relations	**INFORMAL DEFINITION OF CONTINUITY** A function is continuous over an interval of its domain if its hand-drawn graph over that interval can be sketched without lifting the pencil from the paper. **INCREASING, DECREASING, AND CONSTANT FUNCTIONS** 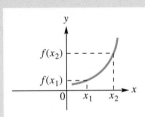 When $x_1 < x_2$, $f(x_1) < f(x_2)$. *f* is increasing. When $x_1 < x_2$, $f(x_1) > f(x_2)$. *f* is decreasing. 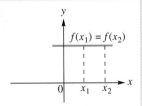 For x_1 and x_2, $f(x_1) = f(x_2)$. *f* is constant. **THREE TYPES OF SYMMETRY** 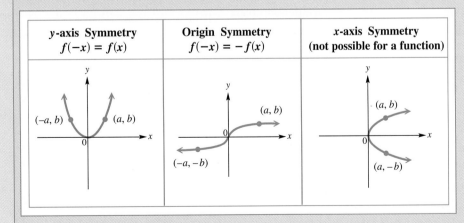

y-axis Symmetry $f(-x) = f(x)$	Origin Symmetry $f(-x) = -f(x)$	*x*-axis Symmetry (not possible for a function)

EVEN AND ODD FUNCTIONS

A function f is called an **even function** if $f(-x) = f(x)$ for all x in the domain of f. (Its graph is symmetric with respect to the y-axis.)

A function f is called an **odd function** if $f(-x) = -f(x)$ for all x in the domain of f. (Its graph is symmetric with respect to the origin.)

BASIC FUNCTIONS AND RELATIONS

For the Graph of:	See Page:
The Identity Function ($f(x) = x$)	124
The Squaring Function ($f(x) = x^2$)	125
The Cubing Function ($f(x) = x^3$)	126
The Square Root Function ($f(x) = \sqrt{x}$)	128
The Cube Root Function ($f(x) = \sqrt[3]{x}$)	128
The Absolute Value Function ($f(x) = |x|$)	129
The Relation $x = y^2$	130

2.2 Vertical and Horizontal Shifts of Graphs of Functions

VERTICAL SHIFTING OF THE GRAPH OF A FUNCTION

If $c > 0$, the graph of $y = f(x) + c$ is obtained by shifting the graph of $y = f(x)$ *upward* a distance of c units. The graph of $y = f(x) - c$ is obtained by shifting the graph of $y = f(x)$ *downward* a distance of c units.

HORIZONTAL SHIFTING OF THE GRAPH OF A FUNCTION

If $c > 0$, the graph of $y = f(x - c)$ is obtained by shifting the graph of $y = f(x)$ to the *right* a distance of c units. The graph of $y = f(x + c)$ is obtained by shifting the graph of $y = f(x)$ to the *left* a distance of c units.

2.3 Stretching, Shrinking, and Reflecting Graphs of Functions

VERTICAL STRETCHING OF THE GRAPH OF A FUNCTION

If $c > 1$, the graph of $y = c \cdot f(x)$ is obtained by vertically stretching the graph of $y = f(x)$ by a factor of c. In general, the larger the value of c, the greater the stretch.

VERTICAL SHRINKING OF THE GRAPH OF A FUNCTION

If $0 < c < 1$, the graph of $y = c \cdot f(x)$ is obtained by vertically shrinking the graph of $y = f(x)$ by a factor of c. In general, the smaller the value of c, the greater the shrink.

REFLECTING THE GRAPH OF A FUNCTION ACROSS AN AXIS

For a function $y = f(x)$,

(a) the graph of $y = -f(x)$ is a reflection of the graph of f across the x-axis.

(b) the graph of $y = f(-x)$ is a reflection of the graph of f across the y-axis.

2.4 The Absolute Value Function: Graphs, Equations, Inequalities, and Applications

PROPERTIES OF ABSOLUTE VALUE

For all real numbers a and b,

1. $|ab| = |a| \cdot |b|$.

2. $\left|\dfrac{a}{b}\right| = \dfrac{|a|}{|b|}$ $(b \neq 0)$.

3. $|a| = |-a|$.

4. $|a| + |b| \geq |a + b|$ (the triangle inequality).

THE GRAPH OF $y = |f(x)|$

The graph of $y = |f(x)|$ is obtained from the graph of $y = f(x)$ by reflecting the portion of the graph below the x-axis across the x-axis, and leaving the graph unchanged for the portion on or above the x-axis.

SOLUTION OF ABSOLUTE VALUE EQUATIONS AND INEQUALITIES

To solve $|ax + b| = c, c > 0$, solve the compound statement

$$ax + b = c \quad \text{or} \quad ax + b = -c.$$

To solve $|ax + b| < c, c > 0$, solve the compound statement

$$-c < ax + b < c.$$

To solve $|ax + b| > c, c > 0$, solve the compound statement

$$ax + b > c \quad \text{or} \quad ax + b < -c.$$

2.5 Piecewise Defined Functions

PIECEWISE DEFINED FUNCTION

A function defined piecewise is defined by different rules over different subsets of its domain.

THE GREATEST INTEGER FUNCTION

$$f(x) = [\![x]\!] = \begin{cases} x \text{ if } x \text{ is an integer} \\ \text{the greatest integer less than } x \text{ if } x \text{ is not an integer} \end{cases}$$

See the graph of $f(x) = [\![x]\!]$ on page 176.

2.6 Further Topics in the Study of Functions

OPERATIONS ON FUNCTIONS

If f and g are functions, then for all values for which both $f(x)$ and $g(x)$ exist, these four operations are defined:

$$\text{Sum} \quad (f + g)(x) = f(x) + g(x)$$

$$\text{Difference} \quad (f - g)(x) = f(x) - g(x)$$

$$\text{Product} \quad (fg)(x) = f(x) \cdot g(x)$$

$$\text{Quotient} \quad \left(\frac{f}{g}\right)(x) = \frac{f(x)}{g(x)}, \text{ where } g(x) \neq 0.$$

For functions f and g, the domains of $f + g$, $f - g$, and fg include all real numbers in the intersection of the domains of f and g, while the domain of f/g includes those real numbers in the intersection of the domains of f and g for which $g(x) \neq 0$.

COMPOSITION OF FUNCTIONS

If f and g are functions, then the composite function, or composition, of g and f is

$$(g \circ f)(x) = g[f(x)]$$

for all x in the domain of f such that $f(x)$ is in the domain of g.

DIRECT VARIATION AS A POWER

Let n be a positive real number. Then, y varies directly as the nth power of x, or y is directly proportional to the nth power of x, if a nonzero real number k exists such that

$$y = kx^n.$$

CHAPTER 2   REVIEW EXERCISES

Match the equation with the graph that most closely resembles its graph.

1. $y = \sqrt{x} + 2$

2. $y = \sqrt{x + 2}$

3. $y = 2\sqrt{x}$

4. $y = -2\sqrt{x}$

5. $y = \sqrt[3]{x} - 2$

6. $y = \sqrt[3]{x - 2}$

7. $y = 2\sqrt[3]{x}$

8. $y = -2\sqrt[3]{x}$

A.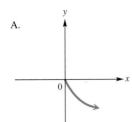

B.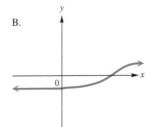

C.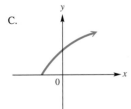

D.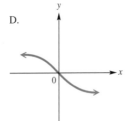

E.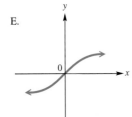

F.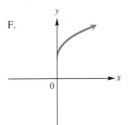

G.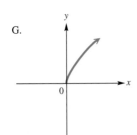

H.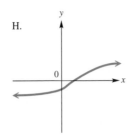

Give the interval that describes the following.

9. domain of $f(x) = \sqrt{x}$ **10.** range of $f(x) = |x|$

11. range of $f(x) = \sqrt[3]{x}$ **12.** domain of $f(x) = x^2$

13. the largest interval over which $f(x) = \sqrt[3]{x}$ is increasing

14. the largest interval over which $f(x) = |x|$ is increasing

15. domain of $x = y^2$

16. range of $x = y^2$

17. The screen shows the graph of $x = y^2 - 4$. Give the two functions Y_1 and Y_2 that must be used to graph this relation if the calculator is in function mode.

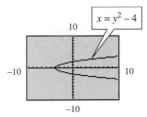

18. *Job Market in Sacramento* The figure shows the number of jobs gained or lost in the Sacramento area in a recent period from September to May. Assume that the y-value represents the total change in the number of jobs for the month.

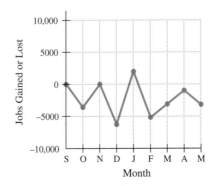

(a) Is this the graph of a function?

(b) In what month were the most jobs lost? The most gained?

(c) What was the largest number of lost jobs? The most gained?

(d) Do these data show an upward or downward trend? If so, which is it?

19. Consider the function whose graph is shown here.

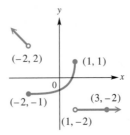

Give the interval(s) over which the function

(a) is continuous.   (b) increases.

(c) decreases.   (d) is constant.

(e) What is the domain of the function?

(f) What is the range of the function?

20. Consider the function $f(x) = -(x + 3)^2 - 5$. Give the interval(s) over which the function

(a) is continuous.   (b) increases.

(c) decreases.   (d) is constant.

(e) What is the domain of the function?

(f) What is the range of the function?

Determine whether the given relation has x-axis symmetry, y-axis symmetry, origin symmetry, or none of these symmetries. (More than one choice is possible.) Also, if the relation is a function, determine whether it is an even function, an odd function, or neither.

21.

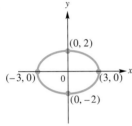

22. $F(x) = x^3 - 6$ **23.** $y = |x| + 4$

24. $f(x) = \sqrt{x - 5}$ **25.** $y^2 = x - 5$

26. $f(x) = 3x^4 + 2x^2 + 1$

27. Use the terminology of Sections 2.2 and 2.3 to describe how the graph of $y = -3(x + 4)^2 - 8$ can be obtained from the graph of $y = x^2$.

28. Find the rule for the function whose graph is obtained by reflecting the graph of $y = \sqrt{x}$ across the y-axis, then reflecting across the x-axis, shrinking vertically by a factor of $\frac{2}{3}$, and, finally, translating 4 units upward.

The graph of $y = f(x)$ is given here. Sketch by hand the graph of each function listed, and indicate three points on the graph.

29. $y = -f(x + 1) - 2$

30. $y = |f(x)| + 1$

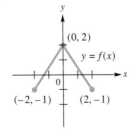

Solve the equation or inequality analytically.

31. $|2x + 5| = 7$

32. $|2x + 5| \leq 7$

33. $|2x + 5| \geq 7$

34. The graphs of $y_1 = |2x + 5|$ and $y_2 = 7$ are shown, along with the two points of intersection of the graphs. Write a paragraph explaining how these screens support the answers in Exercises 31–33.

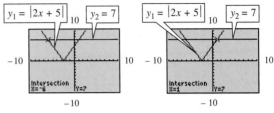

First View   Second View

35. (a) Solve the equation $|x + 1| + |x - 3| = 8$ graphically. Then, give an analytic check by substituting the values in the solution set directly into the left-hand side of the equation.

(b) Use the graph found in part (a), along with the solution set, to give the solution set for each inequality.

(i) $|x + 1| + |x - 3| < 8$

(ii) $|x + 1| + |x - 3| > 8$

Give the solution set of the equation or inequality, based on the graphs of $y = f(x)$ and $y = g(x)$.

36. $f(x) = g(x)$

37. $f(x) < g(x)$

38. $f(x) \geq g(x)$

39. $f(x) \geq 1$

40. $f(x) \leq 0$

41. $g(x) < 0$

42. $f(x) \leq 1$

Sketch the graph of each function by hand.

43. $f(x) = \begin{cases} -4x + 2 & \text{if } x \leq 1 \\ 3x - 5 & \text{if } x > 1 \end{cases}$

44. $f(x) = \begin{cases} 3x + 1 & \text{if } x < 2 \\ -x + 4 & \text{if } x \geq 2 \end{cases}$

45. $f(x) = \begin{cases} |x| & \text{if } x < 3 \\ 6 - x & \text{if } x \geq 3 \end{cases}$

46. Graph the function in Exercise 44, using a graphing calculator with the window $[-10, 10]$ by $[-10, 10]$.

Use a graphing calculator to graph each function defined by the greatest integer function.

47. $f(x) = [\![x - 3]\!]$ window: $[-5, 5]$ by $[-5, 5]$

48. $f(x) = [\![x]\!] + 1$ window: $[-5, 5]$ by $[-5, 5]$

Cesarean Births   The percentages of babies delivered by Cesarean birth are listed in the table.

Year	Percentage
1970	5%
1975	11%
1980	17%
1985	23%
1990	23%

(*Source:* Teutsch S., R. Churchill, *Principles and Practice of Public Health Surveillance*, Oxford University Press, New York, 1994.)

49. Determine a piecewise linear function defined by $f(x)$ that models these data, where $x = 0$ corresponds to 1970.

50. Use f to estimate the percentage of Cesarean deliveries in 1973 and 1982.

Let $f(x) = 3x^2 - 4$ and $g(x) = x^2 - 3x - 4$. Find each of the following.

51. $(f + g)(x)$

52. $(fg)(x)$

53. $(f - g)(4)$

54. $(f + g)(-4)$

55. $(f + g)(2k)$

56. $\left(\dfrac{f}{g}\right)(3)$

57. $\left(\dfrac{f}{g}\right)(-1)$

58. Give the domain of $(fg)(x)$.

59. Give the domain of $\left(\dfrac{f}{g}\right)(x)$.

60. Composition is an operation that is unique to functions. Is composition of functions commutative? That is, does $f \circ g = g \circ f$ for all functions f and g? Explain.

For the given function, find and simplify $\dfrac{f(x + h) - f(x)}{h}$.

61. $f(x) = 2x + 9$

62. $f(x) = x^2 - 5x + 3$

Find functions f and g such that $(f \circ g)(x) = h(x)$.

63. $h(x) = (x^3 - 3x)^2$

64. $h(x) = \dfrac{1}{x - 5}$

Solve each problem.

65. *Mammal Population*   The population P of a certain mammal depends on the number x (in hundreds) of a smaller mammal that serves as its primary food supply. The number x (in hundreds) of the smaller mammal depends upon the amount (in appropriate units) of its food supply, a type of plant. Suppose $P(x) = 2x^2 + 1$ and $x = f(a) = 3a + 2$. Find $(P \circ f)(a)$, the relationship between the population P of the larger mammal and the amount a of plants available to serve as food for the smaller mammal.

66. *Volume of a Sphere*   The formula for the volume of a sphere is $V(r) = \dfrac{4}{3}\pi r^3$, where r represents the radius of the sphere. Construct a model representing the amount of volume gained when a sphere of radius r inches is increased by 6 inches.

67. *Dimensions of a Rectangle*   Suppose the length of a rectangle is 2 inches greater than its width. Let x represent the width of the rectangle in inches.

Write a simplified formula for the perimeter of the rectangle in terms of x alone, using $P(x)$ notation to describe it as a function. What type of function is this?

68. *Dimensions of a Cylinder*   Cylindrical cans make the most efficient use of materials when their height is the same as the diameter of their top.

 (a) Express the volume V of such a can as a function of the diameter d of its top.

 (b) Express the surface area S of such a can as a function of the diameter d of its top. (*Hint:* The curved side is made from a rectangle whose length is the circumference of the top of the can.)

CHAPTER 2   TEST

1. Match the set described in Column I with the correct interval notation from Column II. Choices in Column II may be used once, more than once, or not at all.

I	II
(a) domain of $f(x) = \sqrt{x} + 3$	**A.** $[-3, \infty)$
(b) range of $f(x) = \sqrt{x} - 3$	**B.** $[3, \infty)$
(c) domain of $f(x) = x^2 - 3$	**C.** $(-\infty, \infty)$
(d) range of $f(x) = x^2 + 3$	**D.** $[0, \infty)$
(e) domain of $f(x) = \sqrt[3]{x} - 3$	**E.** $(-\infty, 3)$
(f) range of $f(x) = \sqrt[3]{x} + 3$	**F.** $(-\infty, 3]$
(g) domain of $f(x) = |x| - 3$	**G.** $(3, \infty)$
(h) range of $f(x) = |x + 3|$	**H.** $(-\infty, 0]$
(i) domain of $x = y^2$	
(j) range of $x = y^2$	

2. The graph of $y = f(x)$ is shown here.

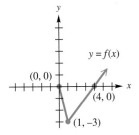

Sketch the graph of each of the following. Use ordered pairs to indicate three points on the graph.

(a) $y = f(x) + 2$ **(b)** $y = f(x + 2)$ **(c)** $y = -f(x)$ **(d)** $y = f(-x)$ **(e)** $y = 2f(x)$ **(f)** $y = |f(x)|$

3. Observe the coordinates displayed at the bottom of the screen showing only the right half of the graph of $y = f(x)$. Answer each of the following based on your observation.

   (a) If the graph is symmetric with respect to the y-axis, what are the coordinates of another point on the graph?

   (b) If the graph is symmetric with respect to the origin, what are the coordinates of another point on the graph?

   (c) Suppose the graph is symmetric with respect to the y-axis. Sketch a typical viewing window with dimensions $[-4, 4]$ by $[0, 8]$. Then, draw the graph you would expect to see in this window.

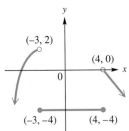

$[0, 4]$ by $[0, 8]$
Xscl = 1   Yscl = 1

4. **(a)** Write a description of how the graph of $y = 4\sqrt[3]{x + 2} - 5$ can be obtained by a translation of the graph of $y = \sqrt[3]{x}$.

   (b) Sketch by hand the graph of $y = -\frac{1}{2}|x - 3| + 2$. Give the domain and the range.

5. Consider the graph of the function shown here.

   (a) Give the interval over which the function is increasing.

   (b) Give the interval over which the function is decreasing.

   (c) Give the interval over which the function is constant.

   (d) Give the intervals over which the function is continuous.

   (e) What is the domain of this function?

   (f) What is the range of this function?

6. Graph $y_1 = |4x + 8|$ and $y_2 = 4$ in the standard viewing window of a graphing calculator. Solve each of the following, showing all steps. Then, state how the graphs support your solution in each case.

(a) $|4x + 8| = 4$ (b) $|4x + 8| < 4$ (c) $|4x + 8| > 4$

7. Given $f(x) = 2x^2 - 3x + 2$ and $g(x) = -2x + 1$, find each of the following. Simplify the expressions when possible.

(a) $(f - g)(x)$ (b) $\dfrac{f}{g}(x)$ (c) the domain of $\dfrac{f}{g}$ (d) $(f \circ g)(x)$ (e) $\dfrac{f(x + h) - f(x)}{h}$ $(h \neq 0)$

8. (a) Graph by hand the piecewise defined function

$$f(x) = \begin{cases} -x^2 + 3 & \text{if } x \leq 1 \\ \sqrt{x} + 2 & \text{if } x > 1. \end{cases}$$

(b) State the two functions defined by y_1 and y_2 that will successfully graph on a graphing calculator the piecewise defined function of part (a). Verify your answer with a graph.

9. *Long-Distance Call Charges*   A certain long-distance carrier provides service between Podunk and Nowheresville. If x represents the number of minutes for the call, where $x > 0$, then the function f defined by

$$f(x) = .40[\![x]\!] + .75$$

gives the total cost of the call in dollars.

(a) Using dot mode and window [0, 10] by [0, 6], graph this function on a graphing calculator.

(b) Use the graph to find the cost of a call that is 5.5 minutes long.

(c) Use the table feature of the calculator to construct a table of prices for .5 to 3.5 minutes, with ΔTbl = .5.

10. *Cost/Revenue/Profit Analysis*   Christine O'Brien starts up a small business, hoping to cash in on the Beanie Baby craze that is sweeping the country. Her initial cost is $3300. Each Beanie Baby costs $4.50 to manufacture.

(a) Write a cost function C, where x represents the number of Beanie Babies manufactured.

(b) Find the revenue function R, if each Beanie Baby in part (a) sells for $10.50.

(c) Give the profit function P.

(d) How many items must be produced and sold before Christine earns a profit?

(e) Support the result of part (d) graphically.

CHAPTER 2   PROJECT

MODELING A TRIP FROM ONE POINT TO ANOTHER, USING A PIECEWISE DEFINED FUNCTION

A piecewise defined function can be used to model a trip that a person makes from one point to another. In this discussion, we will assume that on each leg of the trip, the person travels at a constant rate. To illustrate, consider the following situation.

On a typical morning, it takes Sara exactly 25 minutes (at a constant rate) to pedal the 3.5 miles from her home to the Daybreak Cafe. She stays there for 15 minutes, long enough for juice, a bagel, and a grapefruit. Her pedaling is a little slower after she has eaten, so it takes her another 18 minutes (at a constant rate) to pedal the final 1.8 miles to her job.

We will create a table with a graphing calculator that will display Sara's distance from home at 1-minute intervals. To do this, we must model her movement with a piecewise defined function. Begin by entering the given data into the list editor of the calculator, and graph a scatterplot of this data. See Figures A, B, and C.

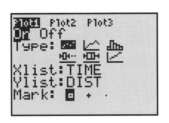

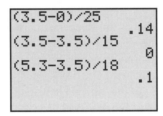

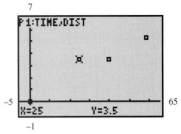

The trace feature shows that when her time is 25 minutes, she is 3.5 miles from home.

FIGURE A   **FIGURE B**   **FIGURE C**

The adjacent points in Figure C will be joined with line segments to determine an appropriate model. Figure D shows how the three slopes are determined, and Figure E shows how the *y*-intercept of the third linear piece is calculated.

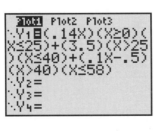

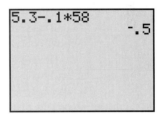

Determining the slopes of the segments

Determining the *y*-intercept of the third segment

FIGURE D   **FIGURE E**

The piecewise function that models Sara's trip is given by

$$S(x) = \begin{cases} .14x & 0 \le x \le 25 \\ 3.5 & 25 < x \le 40 \\ .1x - .5 & 40 < x \le 58. \end{cases}$$

This can be entered into the calculator as

$$Y_1 = (.14x)(x \ge 0)(x \le 25) + 3.5(x > 25)(x \le 40) + (.1x - .5)(x > 40)(x \le 58).$$

The inequalities in the statement above are evaluated by the calculator as 1 if true and 0 if false. So, for example, if an *x*-value is in the interval (25, 40], then the two inequalities following 3.5 would yield a product of 1, but the other two pairs would yield a product of 0. Thus, only the "middle" expression would influence the graph of Y_1. Figure F shows the function as defined above, and Figure G shows its graph (in dot mode) on top of the scatterplot.

FIGURE F   **FIGURE G**

The three sample screens in Figure H show a table of values of the function. The *x*-value gives the number of minutes, and the Y_1-value gives Sara's distance from home after *x* minutes.

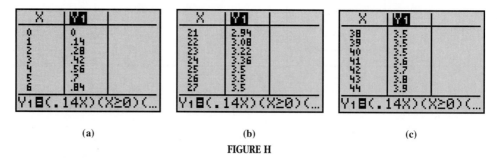

(a)   (b)   (c)

FIGURE H

ACTIVITY

Divide the class into groups of three or four students each. One student in each group should be willing to provide his or her automobile for this activity and act as driver. A second student will act as timekeeper and provide a watch or stopwatch, and a third student will observe the odometer to record mileage.

The group will drive from one point to another, and stop for a brief period of time at least twice along the way. The timekeeper should make accurate timings of how long it takes between legs of the trip, and how long each stop lasts. The odometer reader will record the distances traveled on each leg.

The group should then prepare a written report on the trip that includes specific information about where the trip started and ended and where the stops were made. It should be in paragraph form and contain the piecewise function that models the trip, and include sketches or printouts of the calculator plots, graphs, and tables. As in the previous example, we will assume that all rates are constant (although, as you might expect, this is not a completely valid assumption). The report should explain how the graph and table can be used to determine how far the group was from the starting point at a particular time.

CHAPTER 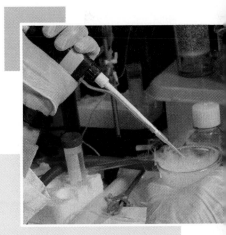3

Polynomial Functions

*I*n 1981, the first case of AIDS (acquired immune deficiency syndrome) was reported in the United States. According to the Centers for Disease Control and Prevention, over 360,000 individuals have been diagnosed with AIDS, and of them over 220,000 have died. AIDS is one of the most devastating diseases of our time. The World Health Organization estimates that over 17 million people have been infected with HIV (human immunodeficiency virus), and by the year 2000, this number will increase to 30–40 million.

The emergence of new diseases and drug-resistant strains of old ones have eliminated modern medicine's hope of eradicating infectious diseases. AIDS, toxic shock, Lyme disease, and Legionnaires' disease were unknown 30 years ago. In order to understand how diseases spread, mathematicians and scientists analyze data that have been reported to health officials. This information can be used to create a mathematical model. Mathematical models help officials forecast future needs for health care and determine risk factors for different populations of people.

How can this data be used to predict the number of new AIDS cases in the following years? First, mathematicians and scientists must create a model. A model not only explains present data but also makes predictions about future phenomena. Polynomial functions are common functions used to model data. By using these functions and their graphs, we can predict future trends.

Year	AIDS Cases
1982	1563
1983	4647
1984	10,845
1985	22,620
1986	41,662
1987	70,222
1988	105,489
1989	147,170
1990	193,245
1991	248,023
1992	315,329
1993	361,509

3.1 COMPLEX NUMBERS

Defining Complex Numbers ◆ Operations with Complex Numbers

DEFINING COMPLEX NUMBERS

Our work in the two previous chapters has involved only real numbers—that is, numbers that are positive, negative, or zero. In order to fully develop some of the concepts in this chapter, we must now investigate some of the ideas of the complex number system.

Observe the graph of $y = x^2 + 1$ in Figure 1. Notice that the graph does not intersect the *x*-axis and, therefore, there are no real solutions to the equation $x^2 + 1 = 0$. This equation is equivalent to $x^2 = -1$, and we know from experience that no real number has a square of -1 (or *any* negative number, for that matter). To handle this situation, mathematicians have developed an expanded number system that includes the set of real numbers as a subset, called the complex number system. Its basic unit is *i*, which is defined to be a square root of -1. Thus, $i^2 = -1$.

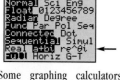

Some graphing calculators such as the TI-83 are capable of complex number operations, as indicated by $a + bi$ here.

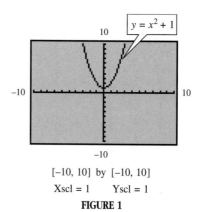

$y = x^2 + 1$

$[-10, 10]$ by $[-10, 10]$
Xscl = 1   Yscl = 1

FIGURE 1

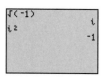

The screen supports the definition of *i*. $\sqrt{-1} = i$ and $i^2 = -1$.

DEFINITION OF *i*

$$i = \sqrt{-1} \quad \text{or} \quad i^2 = -1$$

Numbers of the form $a + bi$, where *a* and *b* are real numbers, are called **complex numbers.** Each real number is a complex number, since a real number *a* may be thought of as the complex number $a + 0i$. A complex number of the form $a + bi$, where *b* is nonzero, is called an **imaginary number.** Both the set of real numbers and the set of imaginary numbers are subsets of the set of complex numbers. (See Figure 2, which is an extension of Figure 5 in Section 1.1.) A complex number that is written in the form $a + bi$ or $a + ib$ is in **standard form.** (The form $a + ib$ is used to simplify certain symbols such as $i\sqrt{5}$, since $\sqrt{5}\,i$ could be too easily mistaken for $\sqrt{5i}$.) The real number *a* is called the **real part** of $a + bi$, and the real number *b* is called the **imaginary part.** For example, in $-7 + 2i$, -7 is the real part, and 2 is the imaginary part.

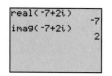

The real part of $-7 + 2i$ is -7, and the imaginary part of $-7 + 2i$ is 2.

---

*In some texts, the term *bi* is defined to be the imaginary part.

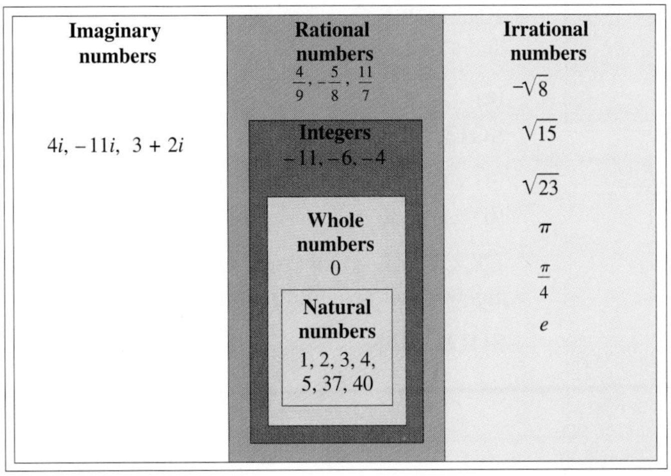

Imaginary numbers	Rational numbers $\frac{4}{9}, -\frac{5}{8}, \frac{11}{7}$	Irrational numbers $-\sqrt{8}$
		$\sqrt{15}$
	Integers $-11, -6, -4$	$\sqrt{23}$
$4i, -11i, \ 3+2i$		
	Whole numbers 0	π
		$\frac{\pi}{4}$
	Natural numbers $1, 2, 3, 4,$ $5, 37, 40$	e

Complex numbers (Real numbers are shaded.)

FIGURE 2

EXAMPLE 1 *Identifying Kinds of Complex Numbers* Consider the following numbers:

$$-8, 3i, \sqrt{7}, \pi, i\sqrt{14}, 5+i, -11i.$$

Classify these numbers as one or more of the following: complex number, real number, or imaginary number.

SOLUTION $-8, \sqrt{7}$, and π are both real numbers and complex numbers, while $3i$, $-11i, i\sqrt{14}$, and $5+i$ are both imaginary numbers and complex numbers. ◆

EXAMPLE 2 *Writing Complex Numbers in Standard Form* Write the standard form of each of the following complex numbers: $6i, -9, 0, -i+2, 8+i\sqrt{3}$.

SOLUTION The standard forms are listed in the table that follows.

Number	Standard Form
$6i$	$6i$
-9	-9
0	0
$-i+2$	$2-i$
$8+i\sqrt{3}$	$8+i\sqrt{3}$

◆

In later sections of this chapter, we will solve equations that will have solutions that lead to expressions involving terms of the form $\sqrt{-a}$, where $a > 0$. This kind of term may be rewritten as a product of a real number and i, using the following definition.

DEFINITION OF $\sqrt{-a}$

If $a > 0$, then

$$\sqrt{-a} = i\sqrt{a}.$$

Some of the latest models of graphing calculators are capable of handling complex numbers and operations with them. The margin screens in this section support many of the analytic results that we will give in the examples.

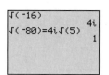

This screen supports the results in Example 3.

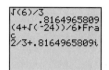

This screen supports the result in Example 4.

EXAMPLE 3   *Writing $\sqrt{-a}$ as $i\sqrt{a}$*   Write each expression as the product of i and a real number.

(a) $\sqrt{-16}$ **(b)** $\sqrt{-80}$

SOLUTION

(a) $\sqrt{-16} = \sqrt{-1 \cdot 16} = i\sqrt{16} = 4i$

(b) $\sqrt{-80} = i\sqrt{80} = i\sqrt{16 \cdot 5} = 4i\sqrt{5}$ ◆

EXAMPLE 4   *Writing a Complex Number in Standard Form*   Use the rules of algebra and the definition of $\sqrt{-a}$ above to show that $\frac{4 + \sqrt{-24}}{6}$ is equivalent to $\frac{2}{3} + i\frac{\sqrt{6}}{3}$.

SOLUTION

$$\frac{4 + \sqrt{-24}}{6} = \frac{4 + i\sqrt{24}}{6} \qquad \sqrt{-a} = i\sqrt{a}$$

$$= \frac{4 + i\sqrt{4 \cdot 6}}{6} \qquad \text{Factor 24 as } 4 \cdot 6.$$

$$= \frac{4 + 2i\sqrt{6}}{6} \qquad \sqrt{ab} = \sqrt{a} \cdot \sqrt{b}; \sqrt{4} = 2$$

$$= \frac{2(2 + i\sqrt{6})}{2 \cdot 3} \qquad \text{Factor out a 2 in both numerator and denominator.}$$

$$= \frac{2 + i\sqrt{6}}{3} \qquad \text{Divide out the common factor 2.}$$

$$= \frac{2}{3} + i\frac{\sqrt{6}}{3} \qquad \text{Write the complex number in standard form.} \qquad ◆$$

The procedure shown in Example 4 will be used extensively in simplifying solutions of quadratic equations, covered in the following sections.

OPERATIONS WITH COMPLEX NUMBERS

Complex numbers may be added, subtracted, multiplied, and divided using the properties of real numbers, as shown by the following definitions and examples.

The *sum* of two complex numbers $a + bi$ and $c + di$ is defined as follows.

DEFINITION OF ADDITION OF COMPLEX NUMBERS

$$(a + bi) + (c + di) = (a + c) + (b + d)i$$

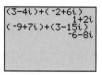

This screen supports the results in Example 5.

EXAMPLE 5   *Adding Complex Numbers*   Find each sum.

(a) $(3 - 4i) + (-2 + 6i)$ **(b)** $(-9 + 7i) + (3 - 15i)$

SOLUTION

(a) $(3 - 4i) + (-2 + 6i) = [3 + (-2)] + [-4 + 6]i = 1 + 2i$

(b) $(-9 + 7i) + (3 - 15i) = -6 - 8i$ ◆

Since $(a + bi) + (0 + 0i) = a + bi$ for all complex numbers $a + bi$, the number $0 + 0i$ is called the *additive identity* for complex numbers. The sum of $a + bi$ and $-a - bi$ is $0 + 0i$ or 0, so the number $-a - bi$ is called the *negative* or *additive inverse* of $a + bi$.

Using this definition of additive inverse, *subtraction* of complex numbers $a + bi$ and $c + di$ is defined as

$$(a + bi) - (c + di) = (a + bi) + (-c - di)$$
$$= (a - c) + (b - d)i.$$

DEFINITION OF SUBTRACTION OF COMPLEX NUMBERS

$$(a + bi) - (c + di) = (a - c) + (b - d)i$$

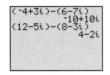

This screen supports the results in Example 6.

EXAMPLE 6 *Subtracting Complex Numbers* Find each difference.

(a) $(-4 + 3i) - (6 - 7i)$ **(b)** $(12 - 5i) - (8 - 3i)$

SOLUTION

(a) $(-4 + 3i) - (6 - 7i) = (-4 - 6) + [3 - (-7)]i$
$$= -10 + 10i$$

(b) $(12 - 5i) - (8 - 3i) = (12 - 8) + (-5 + 3)i$
$$= 4 - 2i \qquad \blacklozenge$$

To summarize, we add complex numbers by adding their real parts and adding their imaginary parts. We subtract complex numbers by subtracting their real parts and subtracting their imaginary parts.

The *product* of two complex numbers can be found by multiplying as if the numbers were binomials and using the fact that $i^2 = -1$, as follows.

$$(a + bi)(c + di) = ac + adi + bic + bidi$$
$$= ac + adi + bci + bdi^2$$
$$= ac + (ad + bc)i + bd(-1)$$
$$(a + bi)(c + di) = (ac - bd) + (ad + bc)i$$

Based on this result, the product of the complex numbers $a + bi$ and $c + di$ is defined in the following way.

DEFINITION OF MULTIPLICATION OF COMPLEX NUMBERS

$$(a + bi)(c + di) = (ac - bd) + (ad + bc)i$$

In practice, we seldom use this definition to multiply complex numbers. Instead, we usually use the customary method of multiplying binomials (known as FOIL, standing for First, Outside, Inside, Last), then replace i^2 with -1, and combine the real and imaginary parts.

EXAMPLE 7 *Multiplying Complex Numbers* Find each of the following products.

(a) $(2 - 3i)(3 + 4i)$ **(b)** $(5 - 4i)(7 - 2i)$

(c) $(4 + 3i)^2$ **(d)** $(6 + 5i)(6 - 5i)$

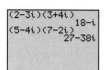

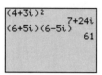

These screens support the results in Example 7.

SOLUTION

(a) $(2 - 3i)(3 + 4i) = 2(3) + 2(4i) - 3i(3) - 3i(4i)$

$\qquad\qquad\qquad\quad = 6 + 8i - 9i - 12i^2$

$\qquad\qquad\qquad\quad = 6 - i - 12(-1) \qquad i^2 = -1$

$\qquad\qquad\qquad\quad = 18 - i$

(b) $(5 - 4i)(7 - 2i) = 5(7) + 5(-2i) - 4i(7) - 4i(-2i)$

$\qquad\qquad\qquad\quad = 35 - 10i - 28i + 8i^2$

$\qquad\qquad\qquad\quad = 35 - 38i + 8(-1)$

$\qquad\qquad\qquad\quad = 27 - 38i$

(c) $(4 + 3i)^2 = 4^2 + 2(4)(3i) + (3i)^2$ Square of a binomial

$\qquad\qquad\quad = 16 + 24i + (-9)$

$\qquad\qquad\quad = 7 + 24i$

(d) $(6 + 5i)(6 - 5i) = 6^2 - 25i^2$ Product of the sum and difference of two terms $i^2 = -1$

$\qquad\qquad\qquad\quad = 36 - 25(-1)$

$\qquad\qquad\qquad\quad = 36 + 25$

$\qquad\qquad\qquad\quad = 61$ Standard form   ◆

Example 7(d) illustrates an important property of complex numbers. Notice that the factors $6 + 5i$ and $6 - 5i$ have the same real parts but opposite imaginary parts. Pairs of complex numbers satisfying these conditions are called **complex conjugates,** or simply conjugates. An important property of conjugates is that their product is always a real number, determined by the sum of the squares of their real and imaginary parts.

> **PRODUCT OF COMPLEX CONJUGATES**
>
> $$(a + bi)(a - bi) = a^2 + b^2$$

This screen supports several results in Example 8.

EXAMPLE 8   *Examining Conjugates and Their Products*   For each of the following, determine the conjugate and then find the product of the number and its conjugate.

$$3 - i \qquad 2 + 7i \qquad -6i \qquad 4$$

SOLUTION   The table that follows shows each number, its conjugate, and the product of the number and its conjugate.

Number	Conjugate	Product
$3 - i$	$3 + i$	$(3 - i)(3 + i) = 9 + 1 = 10$
$2 + 7i$	$2 - 7i$	$(2 + 7i)(2 - 7i) = 53$
$-6i$	$6i$	$(-6i)(6i) = 36$
4	4	$4 \cdot 4 = 16$

◆

The conjugate of the divisor is used to find the *quotient* of two complex numbers. The quotient is found by multiplying both the numerator and the denominator by the conjugate of the denominator. The result should be written in standard form.

This screen supports the result in Example 9. Notice that $1/2i$ is actually $\frac{1}{2}i$. Once again, we see how we must be careful when interpreting calculator displays.

EXAMPLE 9   *Dividing Complex Numbers*   Write the quotient $\frac{3 + 2i}{5 - i}$ in standard form.

SOLUTION   Multiply numerator and denominator by the conjugate of $5 - i$.

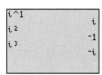

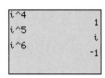

These screens show how powers of *i* follow a cycle.

$$\frac{3 + 2i}{5 - i} = \frac{(3 + 2i)(5 + i)}{(5 - i)(5 + i)}$$

$$= \frac{15 + 3i + 10i + 2i^2}{25 - i^2} \quad \text{Multiply.}$$

$$= \frac{13 + 13i}{26} \quad i^2 = -1$$

$$= \frac{13}{26} + \frac{13i}{26} \quad \frac{a + bi}{c} = \frac{a}{c} + \frac{bi}{c}$$

$$= \frac{1}{2} + \frac{1}{2}i \quad \text{Lowest terms}$$

To check this answer, show that

$$(5 - i)\left(\frac{1}{2} + \frac{1}{2}i\right) = 3 + 2i. \quad \blacklozenge$$

By definition, $i^1 = i$ and $i^2 = -1$. Now, observe the following pattern.

$$i^1 = i$$
$$i^2 = -1$$
$$i^3 = i^2 \cdot i = -1 \cdot i = -i$$
$$i^4 = i^3 \cdot i = -i \cdot i = -i^2 = -(-1) = 1$$

The results in Example 10 are supported in this screen. In the first case, the real part 0 is approximated as -3×10^{-13}, and in the second case, the imaginary part is approximated as -4×10^{-13}.

Here we see another example of the limitations of technology, because the results may be misinterpreted by the user if the mathematical concepts are not understood.

Because $i^4 = 1$, any larger power of *i* may be found by writing the power as a product of two powers of *i*, one exponent being a multiple of 4, and then simplifying.

EXAMPLE 10   *Simplifying Powers of i*   Simplify each of the following powers of *i*.

(a) i^{13} **(b)** i^{56}

SOLUTION

(a) $i^{13} = i^{12} \cdot i = (i^4)^3 \cdot i = 1^3 \cdot i = i$

(b) $i^{56} = (i^4)^{14} = 1^{14} = 1 \quad \blacklozenge$

3.1   EXERCISES   ◈   Tape 3

Note: For the exercises in this section, use your graphing calculator with complex number capability to confirm your answers when possible.

For each complex number, (a) state the real part, (b) state the imaginary part, and (c) state whether the number is real or imaginary.

1. $-9i$ **2.** 6   **3.** π **4.** $-\sqrt{7}$

5. $i\sqrt{6}$ **6.** $-3i$ **7.** $2 + 5i$ **8.** $-7 - 6i$

Write each of the following without negative radicands.

9. $\sqrt{-100}$ **10.** $\sqrt{-169}$ **11.** $-\sqrt{-400}$ **12.** $-\sqrt{-225}$

13. $-\sqrt{-39}$ **14.** $-\sqrt{-95}$ **15.** $5 + \sqrt{-4}$ **16.** $-7 + \sqrt{-100}$

Predict the answer the graphing calculator will give for the indicated entry.

17. `real(-2.87-3.2i)`   **18.** `real(-5.14+2.1i)`   **19.** `imag(√(5)+6.4i)`

20. `imag(-3√(12)+i)`   **21.** `conj(16-24i)`   **22.** `conj(-18i)`

23. Explain why a real number must be a complex number, but a complex number need not be a real number.

24. If the complex number $a + bi$ is real, then what can be said about the value of b?

Use the rules of algebra and the definition of $\sqrt{-a}$ for $a > 0$ to show that the first expression is equivalent to the second expression.

25. $\dfrac{4 + \sqrt{-60}}{8}$; $\dfrac{1}{2} + \dfrac{1}{4}i\sqrt{15}$ **26.** $\dfrac{-2 - \sqrt{-88}}{6}$; $-\dfrac{1}{3} - \dfrac{1}{3}i\sqrt{22}$ **27.** $\dfrac{-12 - \sqrt{-18}}{6}$; $-2 - \dfrac{1}{2}i\sqrt{2}$

28. $\dfrac{-13 + \sqrt{-338}}{13}$; $-1 + i\sqrt{2}$ **29.** $\dfrac{-10 + \sqrt{-100}}{10}$; $-1 + i$ **30.** $\dfrac{5 - \sqrt{-25}}{5}$; $1 - i$

Add or subtract as indicated. Write each result in standard form.

31. $(3 + 2i) + (4 - 3i)$ **32.** $(4 - i) + (2 + 5i)$ **33.** $(-2 + 3i) - (-4 + 3i)$

34. $(-3 + 5i) - (-4 + 3i)$ **35.** $(2 - 5i) - (3 + 4i) - (-2 + i)$ **36.** $(-4 - i) - (2 + 3i) + (-4 + 5i)$

Multiply as indicated. Write each result in standard form.

37. $(2 + 4i)(-1 + 3i)$ **38.** $(1 + 3i)(2 - 5i)$ **39.** $(-3 + 2i)^2$

40. $(2 + i)^2$ **41.** $(2 + 3i)(2 - 3i)$ **42.** $(6 - 4i)(6 + 4i)$

43. $(\sqrt{6} + i)(\sqrt{6} - i)$ **44.** $(\sqrt{2} - 4i)(\sqrt{2} + 4i)$ **45.** $i(3 - 4i)(3 + 4i)$

46. $i(2 + 7i)(2 - 7i)$

Predict the answer the graphing calculator will give for the indicated entry.

47. `(6-i)(1+i)²`   **48.** `(3-4i)²(1+3i)`

49. `7i(5-5i)(1+3i)`   **50.** `-4i(2-5i)²`

RELATING CONCEPTS (EXERCISES 51–54)

Recall that a solution, or root, of an equation is a number that, when substituted for the variable, gives a true statement. In earlier chapters, we have only considered real number solutions of equations. In this chapter, we will see that equations may also have solutions that are not real numbers. For the equation

$$x^3 - x^2 - 7x + 15 = 0,$$

show that each of the following is a solution by substituting it for x.

51. the real number -3 (Section 1.5)   **52.** the complex number $2 - i$ **53.** the complex number $2 + i$

54. What relationship do the solutions in Exercises 52 and 53 have?

Divide as indicated. Write the quotient in standard form.

55. $\dfrac{-19 - 9i}{4 + i}$

56. $\dfrac{-12 - 5i}{3 - 2i}$

57. $\dfrac{1 - 3i}{1 + i}$

58. $\dfrac{-3 + 4i}{2 - i}$

59. $\dfrac{-6 + 8i}{4 + 3i}$

60. $\dfrac{-14 - 14i}{2 - 2i}$

61. $\dfrac{2 - i}{2 + i}$

62. $\dfrac{4 - 3i}{4 + 3i}$

63. $\dfrac{3}{-i}$

64. Predict the answer the graphing calculator will give for the indicated entry.

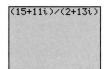

65. Explain why the method of dividing complex numbers (that is, multiplying both the numerator and the denominator by the conjugate of the denominator) works. That is, what property justifies this process?

66. Suppose that your friend, Anne Kelly, tells you that she has discovered a method of simplifying a positive power of i. "Just divide the exponent by 4," she says, "and then look at the remainder. Then, refer to the short table of powers of i in this section. The large power of i is equal to i to the power indicated by the remainder. And if the remainder is 0, the result is $i^0 = 1$." Explain why Anne's method works.

Simplify each of the following powers of i to i, 1, $-i$, or -1.

67. i^{15} **68.** i^{42} **69.** i^{61} **70.** i^{28}

71. i^{102} **72.** i^{19} **73.** i^{32} **74.** i^{69}

3.2 QUADRATIC FUNCTIONS AND THEIR GRAPHS

Basic Terminology ◆ Graphs of Quadratic Functions ◆ The Zero-Product Property ◆ Extreme Values, End Behavior, and Concavity

BASIC TERMINOLOGY

In Chapter 1, we introduced linear functions. Recall that a linear function is defined by an equation of the form $y = f(x) = ax + b$. It is the simplest example of a larger group of functions known as *polynomial functions*.

> ### DEFINITION OF POLYNOMIAL FUNCTION
>
> **A polynomial function of degree n in the variable x is a function defined by**
> $$P(x) = a_n x^n + a_{n-1} x^{n-1} + \cdots + a_1 x + a_0,$$
> where each a_i is a real number, $a_n \neq 0$, and n is a whole number.

While the letter used to name a function is immaterial, we will often use P (as above) to name a polynomial function. Polynomial functions of degree 1, 2, and 3 occur so often that we give them special names, as shown in the chart that follows.

Example	Degree	Special Name
$P(x) = \quad 4x - 7$	1	Linear
$P(x) = \quad 2x^2 + 4x - 16$	2	Quadratic
$P(x) = -3x^3 + 5x$	3	Cubic

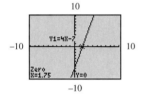

The TI-83 calculator allows you to locate a zero of a function on the graph of the function.

The coefficient a_n for a polynomial function of degree n is called the **leading coefficient,** and a_0 is called the **constant.** In the three examples shown above, the leading coefficients are, respectively, 4, 2, and -3. The constants are -7, -16, and 0. Notice (as in the example of the cubic polynomial function) that a power of the variable may not be present. There is no x^2-term in the cubic polynomial, but we may consider it to actually be there with coefficient 0.

In our study of polynomial functions, we will often be interested in finding the values of x that satisfy $P(x) = 0$. As mentioned earlier in Chapter 1, such a value is called a zero of the function. Because of the importance of this concept, we repeat its definition here.

> **DEFINITION OF ZERO OF A FUNCTION**
>
> For any function f, the number c is a **zero** of f if $f(c) = 0$.

For example, the linear function $P(x) = 4x - 7$ has 1.75 as its only zero, since $P(1.75) = 4(1.75) - 7 = 7 - 7 = 0$. We learned how to find zeros of linear functions in Chapter 1, and with respect to the graph, we can see that *a real zero of a function is an x-intercept of its graph.* We will devote a lot of effort to determining zeros of polynomial functions of higher degree in this chapter.

[–10, 10] by [–10, 10]

Xscl = 1   Yscl = 1

The only zero of Y1 = $P(x)$ = $4x - 7$ is 1.75.

GRAPHS OF QUADRATIC FUNCTIONS

In Chapter 2, we saw that the graph of $y = x^2$ is a parabola. The function $P(x) = x^2$ is the simplest example of a quadratic function.

> **DEFINITION OF QUADRATIC FUNCTION**
>
> A function of the form
> $$P(x) = ax^2 + bx + c, a \neq 0,$$
> is called a **quadratic function.**

When discussing graphs of linear functions, we refer to a comprehensive graph as a graph in a viewing window that shows all intercepts. For a quadratic function, whose graph is a parabola, we define a comprehensive graph as follows.

> **COMPREHENSIVE GRAPH OF A QUADRATIC FUNCTION**
>
> A comprehensive graph of a quadratic function will show all intercepts and the vertex of the parabola.

Let us consider $P(x) = 2x^2 + 4x - 16$. A comprehensive graph is shown in a viewing window of $[-10, 10]$ by $[-20, 10]$ in Figure 3. When compared to the graph of $y = x^2$, also shown in the figure, it appears that the graph of P can be obtained by a vertical stretch with a factor greater than 1, a shift to the left, and a shift downward. It would be possible to determine the magnitudes if the function were written in the form $P(x) = a(x - h)^2 + k$.

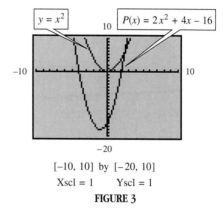

[−10, 10] by [−20, 10]
Xscl = 1   Yscl = 1
FIGURE 3

It is possible to transform an equation of the form $P(x) = ax^2 + bx + c$ into this desired form by the method of *completing the square*. The steps in this procedure are summarized below.

COMPLETING THE SQUARE

To transform the equation $P(x) = ax^2 + bx + c$ into the form $P(x) = a(x - h)^2 + k$:

1. Divide both sides of the equation by a so that the coefficient of x^2 is 1.
2. Add to both sides the square of half the coefficient of x; that is, $\left(\frac{b}{2a}\right)^2$.
3. Factor the right-hand side as the square of a binomial and combine terms on the left.
4. Isolate the term involving $P(x)$ on the left.
5. Multiply both sides by a.

We will now apply this procedure to $P(x) = 2x^2 + 4x - 16$.

$$P(x) = 2x^2 + 4x - 16 \qquad \text{Given function}$$

$$\frac{P(x)}{2} = x^2 + 2x - 8 \qquad \text{Divide by 2 to make the coefficient of } x^2 \text{ equal to 1.}$$

$$\frac{P(x)}{2} + 8 = x^2 + 2x \qquad \text{Add 8 to both sides.}$$

$$\frac{P(x)}{2} + 8 + 1 = x^2 + 2x + 1 \qquad \text{Add } \left[\frac{1}{2}(2)\right]^2 = 1 \text{ to both sides to complete the square on the right.}$$

$$\frac{P(x)}{2} + 9 = (x + 1)^2 \qquad \text{Combine terms on the left and factor on the right.}$$

$$\frac{P(x)}{2} = (x + 1)^2 - 9 \qquad \text{Add } -9 \text{ to both sides.}$$

$$P(x) = 2(x + 1)^2 - 18 \qquad \text{Multiply both sides by 2.}$$

Using the concepts of Chapter 2, we can now make the following statements: The graph of $P(x) = 2x^2 + 4x - 16$ can be obtained from the graph of $y = x^2$ by a vertical stretch factor of 2, shifting 1 unit to the left, and 18 units down. The vertex of the parabola has coordinates $(-1, -18)$, its domain is $(-\infty, \infty)$, and its range is $[-18, \infty)$. It decreases on the interval $(-\infty, -1]$ and increases on the interval $[-1, \infty)$.

EXAMPLE 1   *Analyzing the Graph of a Quadratic Function*   The graph of $P(x) = -x^2 - 6x - 8 = -(x + 3)^2 + 1$ is shown in Figure 4, using both a traditional graph and a calculator-generated graph in a viewing window of $[-8, 2]$ by $[-10, 2]$. Discuss the features of the graph.

Notice the symmetry in the table. The y-values repeat "above" and "below" the entry for the vertex $(-3, 1)$. Compare with the discussion in Example 1.

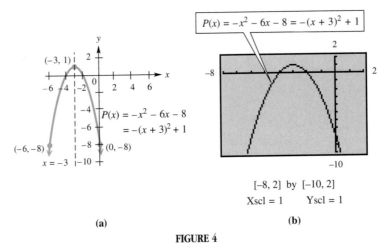

$$P(x) = -x^2 - 6x - 8 = -(x + 3)^2 + 1$$

(a)

$[-8, 2]$ by $[-10, 2]$
Xscl = 1   Yscl = 1

(b)

FIGURE 4

SOLUTION   The parabola is the graph of $y = x^2$, translated 3 units to the left and 1 unit upward. It opens downward because of the negative sign preceding $(x + 3)$. (This is a reflection across the x-axis.) The line $x = -3$ is its **axis of symmetry,** since if it were folded along this line, the two halves would coincide. The vertex, $(-3, 1)$, is the highest point on the graph. The domain is $(-\infty, \infty)$ and the range is $(-\infty, 1]$. The function increases on the interval $(-\infty, -3]$ and decreases on $[-3, \infty)$. Since $P(0) = -8$, the y-intercept is -8, and since $P(-4) = P(-2) = 0$, the x-intercepts are -4 and -2.   ◆

Our discussion up to this point leads to the following generalizations about the graph of a quadratic function in the form $P(x) = a(x - h)^2 + k$.

GRAPH OF $P(x) = a(x - h)^2 + k$

The graph of $P(x) = a(x - h)^2 + k, a \neq 0,$

a.   is a parabola with vertex (h, k), and the vertical line $x = h$ as axis of symmetry;

b.   opens upward if $a > 0$ and downward if $a < 0$;

c.   is broader than $y = x^2$ if $0 < |a| < 1$ and narrower than $y = x^2$ if $|a| > 1$.

Determining the coordinates of the vertex of the graph of a quadratic function can be done by using the method described earlier for the function $P(x) = 2x^2 + 4x - 16$. Rather than go through the procedure for each individual function, we may generalize it for the standard form of the quadratic function, $P(x) = ax^2 + bx + c$.

$$P(x) = ax^2 + bx + c \quad (a \neq 0)$$
Standard form

$$y = ax^2 + bx + c \quad (a \neq 0)$$
Replace $P(x)$ with y to simplify notation.

$$\frac{y}{a} = x^2 + \frac{b}{a}x + \frac{c}{a}$$
Divide by a.

$$\frac{y}{a} - \frac{c}{a} = x^2 + \frac{b}{a}x$$
Subtract $\frac{c}{a}$.

$$\frac{y}{a} - \frac{c}{a} + \frac{b^2}{4a^2} = x^2 + \frac{b}{a}x + \frac{b^2}{4a^2}$$
Add $\frac{b^2}{4a^2}$.

$$\frac{y}{a} + \frac{b^2 - 4ac}{4a^2} = \left(x + \frac{b}{2a}\right)^2$$
Combine terms on left and factor on right.

$$\frac{y}{a} = \left(x + \frac{b}{2a}\right)^2 - \frac{b^2 - 4ac}{4a^2}$$
Get y-term alone on the left.

$$y = a\left(x + \frac{b}{2a}\right)^2 + \frac{4ac - b^2}{4a}$$
Multiply by a.

$$P(x) = a\left[x - \left(-\frac{b}{2a}\right)\right]^2 + \frac{4ac - b^2}{4a}$$
Write in the form $P(x) = a(x - h)^2 + k$.

h k

TECHNOLOGY NOTE

Most current models of graphing calculators are capable of determining the coordinates of the "highest point" or "lowest point" in a designated interval of a graph. See if your calculator is capable of this; these are usually designated with commands like "maximum" and "minimum." With this capability, you can find the coordinates of the vertex of a parabola graphically, to support the analytic discussion in this section.

The final equation shows that the vertex (h, k) can be expressed in terms of a, b, and c. It is not necessary to memorize the expression for k, since it is equal to $P\left(-\frac{b}{2a}\right)$.

VERTEX FORMULA

The vertex of the graph of $P(x) = ax^2 + bx + c$ $(a \neq 0)$ is the point

$$\left(-\frac{b}{2a},\, P\left(-\frac{b}{2a}\right)\right).$$

EXAMPLE 2   *Using the Vertex Formula*   Use the vertex formula to find the coordinates of the vertex of the graph of $P(x) = -.65x^2 + \sqrt{2}x + 4$.

(a)   Give the exact vertex values of x and y.

(b)   Give the approximate vertex values of x and y to the nearest hundredth.

(c)   Support your answer in part (b) by graphing the function in an appropriate window and using the capabilities of your calculator to find the vertex.

SOLUTION

(a)   ANALYTIC   For this function, $a = -.65$ and $b = \sqrt{2}$, so, applying the vertex formula,

$$x = -\frac{b}{2a} = -\frac{\sqrt{2}}{2(-.65)} = \frac{\sqrt{2}}{2(.65)}$$

and $$y = P\left(-\frac{b}{2a}\right) = -.65\left(\frac{\sqrt{2}}{2(.65)}\right)^2 + \sqrt{2}\left(\frac{\sqrt{2}}{2(.65)}\right) + 4.$$

These are the *exact* (but not simplified) values of x and y.

(b)   Using the arithmetic, squaring, and square root functions of a calculator, we find that to the nearest hundredth, $x \approx 1.09$ and $y \approx 4.77$.

(c)   GRAPHICAL   Graphing the function in a window of $[-2, 4]$ by $[-2, 5]$ and locating the highest point on the graph, we see a display of $x = 1.0878556$ and $y = 4.7692308$ in Figure 5. These values support our answer in part (b).   ◆

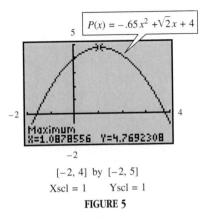

$P(x) = -.65x^2 + \sqrt{2}x + 4$

$[-2, 4]$ by $[-2, 5]$
Xscl = 1   Yscl = 1

FIGURE 5

THE ZERO-PRODUCT PROPERTY

An important property that will allow us to solve for the zeros of certain polynomial functions (and thus find the x-intercepts of their graphs) is the zero-product property.

ZERO-PRODUCT PROPERTY

If a and b are complex numbers and $ab = 0$, then $a = 0$ or $b = 0$ or both.

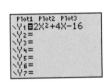

The zero-product property says that if the product of two complex numbers is 0, then at least one of the factors must equal 0.

Consider the following problems:

1. Find the zeros of the quadratic function $P(x) = 2x^2 + 4x - 16$.
2. Find the x-intercepts of the graph of $P(x) = 2x^2 + 4x - 16$.
3. Find the solution set of the equation $2x^2 + 4x - 16 = 0$.

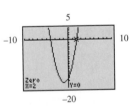

All three problems may be solved the same way: we must find the numbers that make the expression $2x^2 + 4x - 16$ equal to 0. The following example shows how this can be done using the zero-product property.

EXAMPLE 3   *Using the Zero-Product Property*   Solve the quadratic equation $2x^2 + 4x - 16 = 0$.

SOLUTION   ANALYTIC   To make our work easier, we may divide both sides by 2. Then, factor, and set each factor equal to 0. Solve the resulting linear equations.

$$2x^2 + 4x - 16 = 0$$
$$x^2 + 2x - 8 = 0$$
$$(x + 4)(x - 2) = 0$$
$$x + 4 = 0 \quad \text{or} \quad x - 2 = 0$$
$$x = -4 \qquad\qquad x = 2$$

$[-10, 10]$ by $[-20, 5]$
Xscl = 1   Yscl = 1

These screens show how a graphing calculator supports the results in Example 3.

The solution set of the equation is $\{-4, 2\}$.

GRAPHICAL   The margin screens at the side show how we can provide graphical support for these results.   ◆

EXAMPLE 4   *Using the Zero-Product Property*   Find all zeros of the quadratic function $P(x) = x^2 - 6x + 9$. Support your answer graphically.

SOLUTION ANALYTIC We must solve $x^2 - 6x + 9 = 0$.

$$x^2 - 6x + 9 = 0$$
$$(x - 3)^2 = 0 \qquad \text{Factor.}$$
$$x - 3 = 0 \quad \text{or} \quad x - 3 = 0 \qquad \text{Use the zero-product property.}$$
$$x = 3 \qquad\qquad x = 3$$

There is only one *distinct* zero, 3. It is sometimes called a double zero, or double solution (root) of the equation.

GRAPHICAL   Graphing the function in the standard viewing window allows us to support our result, as seen in Figure 6. The vertex has coordinates (3, 0).   ◆

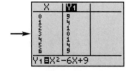

This table indicates that when X = 3, Y$_1$ = 0, supporting the result in Example 4.

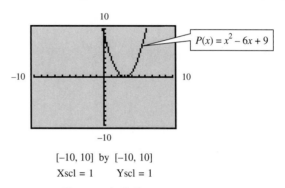

[−10, 10] by [−10, 10]
Xscl = 1   Yscl = 1

The vertex is (3, 0),
so 3 is a double solution (root).

FIGURE 6

Caution   A purely graphical approach may not prove to be as useful with a graph like the one in Figure 6, since the vertex may be slightly above or slightly below the x-axis. (See the discussion of "hidden behavior" in Section 3.5.) **This is why we need to understand the algebraic concepts presented in this section—only when we know the mathematics can we use the technology to its utmost.**

For Group Discussion   Figure 7 shows the possible numbers of x-intercepts of the graph of a quadratic function that opens upward.

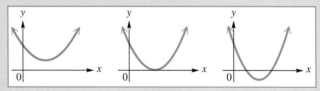

FIGURE 7

Figure 8 shows the possible numbers of x-intercepts of the graph of a quadratic function that opens downward.

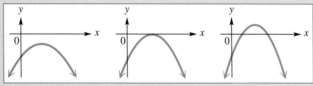

FIGURE 8

Use these figures to discuss the following.
1.   What is the maximum number of real solutions of a quadratic equation?
2.   What is the minimum number of real solutions of a quadratic equation?
3.   If a quadratic function has only one real zero, what do we know about the vertex of its graph?

Note The zero-product property is quite limited in its practical applications. If a quadratic polynomial cannot easily be factored, then it is of little use. For this reason, we will develop a more powerful method of finding zeros of quadratic functions in the next section. It is called the quadratic formula.

EXTREME VALUES, END BEHAVIOR, AND CONCAVITY

The vertex of the graph of $P(x) = ax^2 + bx + c$ is the lowest point on the graph of the function if $a > 0$, and is the highest point if $a < 0$. Such points are called **extreme points** (also **extrema,** singular: **extremum**). As we extend our study of polynomial functions, we will examine extrema on a more general basis.

EXTREME POINT AND EXTREME VALUE OF A QUADRATIC FUNCTION

For the quadratic function $P(x) = ax^2 + bx + c$,

a. if $a > 0$, the vertex (h, k) is called the *minimum point* of the graph. The minimum value of the function is $P(h) = k$.

b. if $a < 0$, the vertex (h, k) is called the *maximum point* of the graph. The *maximum value* of the function is $P(h) = k$.

Figure 9 illustrates these ideas.

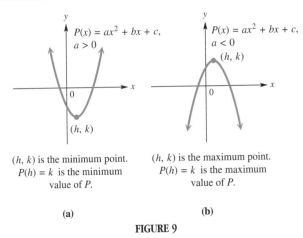

(a) (h, k) is the minimum point. $P(h) = k$ is the minimum value of P.

(b) (h, k) is the maximum point. $P(h) = k$ is the maximum value of P.

FIGURE 9

Note Modern graphing calculators have the capabilities of locating extrema to great accuracy.

EXAMPLE 5 *Identifying Extreme Points and Extreme Values* Give the coordinates of the extreme point of the graph of each function, and the corresponding maximum or minimum value of the function.

(a) $P(x) = 2x^2 + 4x - 16$ (Figure 3)

(b) $P(x) = -x^2 - 6x - 8$ (Example 1, Figure 4)

(c) $P(x) = -.65x^2 + \sqrt{2}x + 4$ (Example 2, Figure 5)

SOLUTION

(a) In the discussion preceding Example 1, we found that the vertex of the graph of this function is $(-1, -18)$. It opens upward since $a > 0$ (as seen in Figure 3), so the vertex $(-1, -18)$ is the minimum point and -18 is the minimum value of the function.

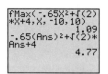

This screen supports the result in Example 5(c). Here the calculator has been directed to round to two decimal places.

(b) The vertex $(-3, 1)$ is the maximum point and $P(-3) = 1$ is the maximum value of the function.

(c) Based on our work in Example 2, the vertex has approximate coordinates $(1.09, 4.77)$. It is the highest point on the graph, so it is a maximum point, and the maximum value of the function is approximately 4.77. (*Note:* The exact maximum value is the y-value indicated in part (a) of Example 2.) ◆

We know that if the value of a is positive for the quadratic function $P(x) = ax^2 + bx + c$, the graph opens upward, and if a is negative, the graph opens downward. The sign of a determines the *end behavior* of the graph. If $a > 0$, as x approaches $-\infty$ or ∞ (written $x \to -\infty$ or $x \to \infty$), the value of $P(x)$ approaches $+\infty$ (written $P(x) \to \infty$). The other situations similar to this are summarized in the following box.

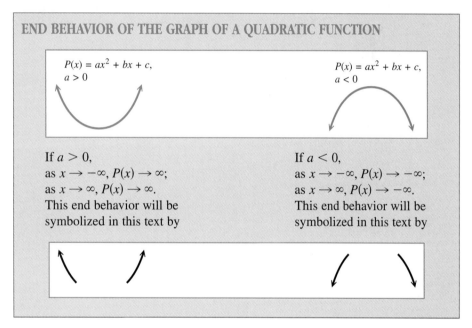

END BEHAVIOR OF THE GRAPH OF A QUADRATIC FUNCTION

$P(x) = ax^2 + bx + c,$
$a > 0$

$P(x) = ax^2 + bx + c,$
$a < 0$

If $a > 0$,
as $x \to -\infty$, $P(x) \to \infty$;
as $x \to \infty$, $P(x) \to \infty$.
This end behavior will be symbolized in this text by

If $a < 0$,
as $x \to -\infty$, $P(x) \to -\infty$;
as $x \to \infty$, $P(x) \to -\infty$.
This end behavior will be symbolized in this text by

We conclude this section with a brief discussion of *concavity*. Using the quadratic function graph as an illustration, we see that if $a > 0$, the graph is at all times opening upward. If water were to be poured from above, the graph would, in a sense, "hold water." We say that this graph is *concave up* for all values in its domain. On the other hand, if $a < 0$, the graph opens downward at all times, and it would similarly "dispel water" if it were poured from above. In this case, the graph is *concave down* for all values in its domain. See Figure 10.

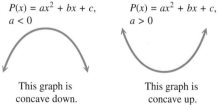

$P(x) = ax^2 + bx + c,$
$a < 0$

$P(x) = ax^2 + bx + c,$
$a > 0$

This graph is
concave down.

This graph is
concave up.

FIGURE 10

Note   A *formal* discussion of concavity requires concepts beyond the scope of this text. It is studied more rigorously in calculus.

3.2   EXERCISES   Tape 4

In Exercises 1–8, you are given an equation and the graph of a quadratic function. Without using your calculator, do each of the following: (a) Give the domain and the range. (b) Give the coordinates of the vertex. (c) Give the equation of the axis of symmetry. (d) Give the interval over which the function is increasing. (e) Give the interval over which the function is decreasing. (f) State whether the vertex is a maximum or minimum point, and give the corresponding maximum or minimum value of the function. (g) Tell whether the graph is concave up or concave down.

1. $P(x) = (x - 2)^2$

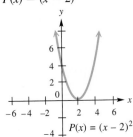

2. $P(x) = (x + 4)^2$

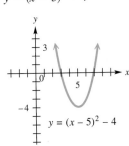

3. $y = (x + 3)^2 - 4$

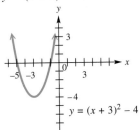

4. $y = (x - 5)^2 - 4$

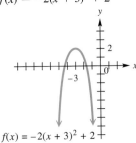

5. $f(x) = -2(x + 3)^2 + 2$

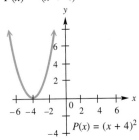

6. $f(x) = -3(x - 2)^2 + 1$

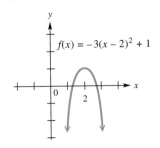

7. $P(x) = -.5(x + 1)^2 - 3$

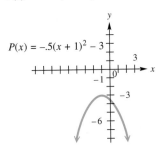

8. $P(x) = \frac{2}{3}(x - 2)^2 - 1$

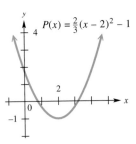

In Exercises 9–16, an equation of a quadratic function is given. Do each of the following. (a) Find the coordinates of the vertex of the graph analytically, using the method of completing the square or the vertex formula given in this section. (b) Find the x-intercepts of the graph using the zero-product property. (c) Find the y-intercept. (Hint: Evaluate the function for x = 0.) (d) Support your answers to parts (a)–(c), using a calculator-generated graph in an appropriate viewing window that will show a comprehensive graph of the function.

9. $P(x) = 2x^2 - 2x - 24$ **10.** $P(x) = 3x^2 + 3x - 6$ **11.** $y = x^2 - 2x - 15$ **12.** $y = -x^2 - 3x + 10$

13. $f(x) = -2x^2 + 6x$ **14.** $f(x) = 4x^2 - 4x$ **15.** $P(x) = 4x^2 - 22x - 12$ **16.** $P(x) = 6x^2 - 16x - 6$

The graphs of the functions in Exercises 17–20 are shown in Figures A–D. Match each function with its graph, using the concepts of this section without actually entering it into your calcula-

tor. Then, after you have completed the exercises, check your answers with your calculator. Use the standard viewing window.

17. $y = (x - 4)^2 - 3$

18. $y = -(x - 4)^2 + 3$

19. $y = (x + 4)^2 - 3$

20. $y = -(x + 4)^2 + 3$

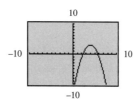

A

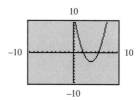

B

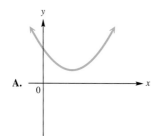

C

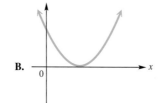

D

In Exercises 21–26, a quadratic function with decimal and/or irrational coefficients is given. Graph the function in a viewing window that will allow you to use your calculator to approximate **(a)** *the coordinates of the vertex, and* **(b)** *the x-intercepts. Give values to the nearest hundredth.*

21. $P(x) = -.32x^2 + \sqrt{3}x + 2.86$ **22.** $P(x) = -\sqrt{2}x^2 + .45x + 1.39$ **23.** $y = 1.34x^2 - 3x + \sqrt{5}$

24. $y = 2.53x^2 - 2x + \sqrt{19}$ **25.** $f(x) = \sqrt{10}x^2 + 3.26x - 4.16$ **26.** $f(x) = \sqrt[3]{20}x^2 + 6.48x - \sqrt{2}$

Refer to the graphs in Figures A–F to answer Exercises 27–34. There may be one correct choice, more than one correct choice, or no correct choices.

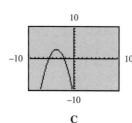

A.

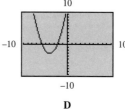

B.

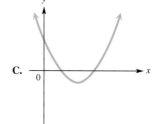

C.

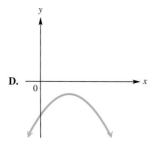

D.

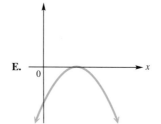

E.

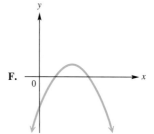

F.

27. Which functions have two real zeros?

28. Which functions have exactly one real zero?

29. Which functions have no real zeros?

30. If each function is of the form $P(x) = ax^2 + bx + c$, for which functions is $a < 0$? For which is $a > 0$?

31. If each function is of the form $P(x) = ax^2 + bx + c$, for which functions is $c > 0$? For which is $c < 0$? For which is $c = 0$?

32. Which graphs are concave up? Which graphs are concave down?

33. If each function is of the form $P(x) = a(x - h)^2 + k$, which function would satisfy the conditions that a, h, and k are all positive?

34. Which function could possibly have the equation $f(x) = -2(x - 4)^2$?

Draw an end behavior diagram (that is, ⌣ *or* ⌢ *) for each of the following quadratic functions. Do not rely on your calculator to do this.*

35. $P(x) = 19(x + 12)^2 - 48$ **36.** $P(x) = 27(x - 3)^2 + 84$ **37.** $y = -200x^2 - 480x + 1993$

38. $y = -300x^2 + 1280x - 1936$ **39.** $f(x) = -\sqrt{483}\,(x + \sqrt{2})^2 - \sqrt{13}$ **40.** $f(x) = -\sqrt{276}\,(x - 1.7)^2 + .483$

41. $P(x) = 129\pi x^2 - \dfrac{\pi}{2}x + 12$ **42.** $P(x) = 486\pi x^2 - 13\pi x + \pi$

RELATING CONCEPTS (EXERCISES 43–44)

Recall that if we are given the graph of a function $y = f(x)$, the solutions of $f(x) = 0$ are the x-intercepts, the solutions of $f(x) < 0$ are those x-values for which the graph is below the x-axis, and the solutions of $f(x) > 0$ are those x-values for which the graph is above the x-axis. These concepts were first presented in Chapter 1. Use them to answer the following.

43. The figure shows the graph of $f(x) = 2x^2 + 5x - 3$. The x-intercepts, which may be determined by the zero-product property, are -3 and $\frac{1}{2}$.

    (a) Give the solution set of $2x^2 + 5x - 3 = 0$. (Section 1.5)

    (b) Give the solution set of $2x^2 + 5x - 3 < 0$. (Section 1.6)

    (c) Give the solution set of $2x^2 + 5x - 3 > 0$. (Section 1.6)

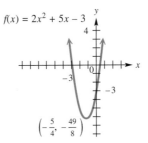

44. For the function $f(x) = x^2 - x - 1$, use your calculator to determine the x-intercepts to the nearest hundredth, and then answer the following.

    (a) Give the solution set of $f(x) = 0$. (Section 1.5)

    (b) Give the solution set of $f(x) \geq 0$. (Section 1.6)

    (c) Give the solution set of $f(x) \leq 0$. (Section 1.6)

*In each table, Y_1 is defined as a quadratic function. Use the table to do each of the following. **(a)** Determine the coordinates of the vertex of the graph. **(b)** Determine whether the vertex is a minimum point or a maximum point. **(c)** Find the minimum or maximum value of the function, and tell which one it is. **(d)** Determine the range of the function. **(e)** Give an end behavior diagram.*

45.

X	Y₁
1	15
2	0
3	-9
4	-12
5	-9
6	0
7	15

X=1

46.

X	Y₁
-6	27
-5	17
-4	11
-3	9
-2	11
-1	17
0	27

X=-6

47.

X	Y₁
0	-7
.5	-2
1	1
1.5	2
2	1
2.5	-2
3	-7

X=0

48.

X	Y₁
-4	-9.75
-3.5	-6
-3	-3.75
-2.5	-3
-2	-3.75
-1.5	-6
-1	-9.75

X=-4

Find the equation of the quadratic function satisfying the given conditions. (Hint: Find a, h, and k that satisfy $P(x) = a(x - h)^2 + k$.) Use your calculator to support your results. Express your answer in the form $P(x) = ax^2 + bx + c$.

49. vertex: $(-1, -4)$; through: $(5, 104)$

50. vertex: $(-2, -3)$; through: $(0, -19)$

51. vertex: $(8, 3)$; through: $(10, 5)$

52. vertex: $(-6, -12)$; through: $(6, 24)$

53. vertex: $(-4, -2)$; through: $(2, -26)$

54. vertex: $(5, 6)$; through: $(1, -6)$

In Exercises 55 and 56, find a polynomial function P whose graph matches the one in the figure. Then, use a graphing calculator to graph the function and verify your result.

55.
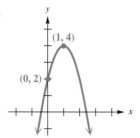

56.
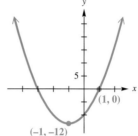

3.3 SOLUTION OF QUADRATIC EQUATIONS AND INEQUALITIES

Solving $x^2 = k$ ◆ The Quadratic Formula and the Discriminant ◆ Solving Quadratic Equations ◆ Solving Quadratic Inequalities ◆ Literal Equations Involving Quadratics

We know that if $a \neq 0$, the solution of $ax + b = 0$ is $-b/a$. In this section, we will develop methods of solving general quadratic equations, and extend them to solving quadratic inequalities.

DEFINITION OF QUADRATIC EQUATION IN ONE VARIABLE

An equation that can be written in the form
$$ax^2 + bx + c = 0,$$
where a, b, and c are real numbers with $a \neq 0$, is a **quadratic equation in standard form.**

SOLVING $x^2 = k$

We are often interested in solving quadratic equations of the form $x^2 = k$, where k is a real number. This type of equation can be solved by factoring, using the following sequence of equivalent equations.

$$x^2 = k$$
$$x^2 - k = 0$$
$$(x - \sqrt{k})(x + \sqrt{k}) = 0$$
$$x - \sqrt{k} = 0 \quad \text{or} \quad x + \sqrt{k} = 0$$
$$x = \sqrt{k} \quad \text{or} \quad x = -\sqrt{k}$$

We have proved the following statement, which we will call the square root property for solving quadratic equations.

SQUARE ROOT PROPERTY FOR SOLVING QUADRATIC EQUATIONS

The solution set of $x^2 = k$ is

a. $\{\pm\sqrt{k}\}$ if $k > 0$ **b.** $\{0\}$ if $k = 0$ **c.** $\{\pm i\sqrt{|k|}\}$ if $k < 0$.

As shown in Figure 11, the graph of $y_1 = x^2$ intersects the graph of $y_2 = k$ twice if $k > 0$, once if $k = 0$, and not at all if $k < 0$.

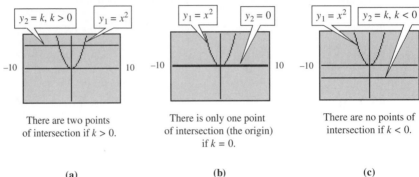

There are two points of intersection if $k > 0$.

There is only one point of intersection (the origin) if $k = 0$.

There are no points of intersection if $k < 0$.

(a)   (b)   (c)

FIGURE 11

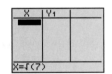

By defining Y_1 as $x^2 - 7$ and inputting $x = \sqrt{7}$, we should get $Y_1 = 0$. (See the screen below.)

Both $x = \sqrt{7}$ and $x = -\sqrt{7}$ lead to $Y_1 = 0$ for $Y_1 = x^2 - 7$.

EXAMPLE 1   *Using the Square Root Property*   Solve each of the following quadratic equations.

(a) $x^2 = 7$ **(b)** $x^2 = -5$

SOLUTION

(a)   ANALYTIC   Since $7 > 0$, there will be two real solutions.

$$x^2 = 7$$
$$x = \pm\sqrt{7}$$

GRAPHICAL   This result may be supported graphically by using the intersection-of-graphs method. If we graph $y_1 = x^2$ and $y_2 = 7$ in a standard viewing window, and then locate the points of intersection, we will find that the x-coordinates are approximately -2.65 and 2.65, which are approximations for $\pm\sqrt{7}$. See Figure 12. The solution set is $\{\pm\sqrt{7}\}$.

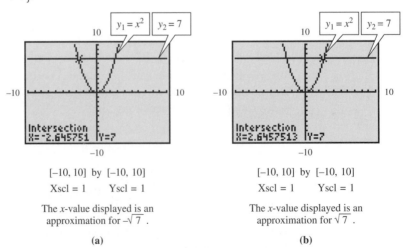

[–10, 10] by [–10, 10]
Xscl = 1   Yscl = 1

The x-value displayed is an approximation for $-\sqrt{7}$.

(a)

[–10, 10] by [–10, 10]
Xscl = 1   Yscl = 1

The x-value displayed is an approximation for $\sqrt{7}$.

(b)

FIGURE 12

(b) ANALYTIC There is no real number whose square is -5. However, this equation has two complex imaginary solutions.

$$x^2 = -5, \quad \text{so} \quad x = \pm\sqrt{-5} = \pm i\sqrt{5}$$

The solution set is $\left\{ \pm i\sqrt{5} \right\}$.

GRAPHICAL Notice that the graphs of $y_1 = x^2$ and $y_2 = -5$ do not intersect. This indicates that there are no *real* solutions. See Figure 13.   ◆

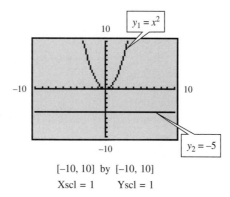

$$[-10, 10] \text{ by } [-10, 10]$$
$$\text{Xscl} = 1 \qquad \text{Yscl} = 1$$

There are no points of intersection,
and thus no *real* solutions.

FIGURE 13

TECHNOLOGY NOTE

Programs for the various makes and models of graphing calculators are available from the manufacturers, users' groups, the Website for this text, etc.

THE QUADRATIC FORMULA AND THE DISCRIMINANT

We saw in Section 3.2 that certain quadratic equations can be solved by factoring and then using the zero-product property. However, this method is quite limited—what if the polynomial cannot be factored? For this reason, we develop the quadratic formula.

We begin with the standard form of the general quadratic equation,

$$ax^2 + bx + c = 0 \quad (a \neq 0),$$

and will solve for x in terms of the constants $a, b,$ and c. To do this, we employ the method of completing the square. For now, assume $a > 0$ and divide both sides by a to obtain

$$x^2 + \frac{b}{a}x + \frac{c}{a} = 0.$$

Add $-\frac{c}{a}$ to both sides.

$$x^2 + \frac{b}{a}x = -\frac{c}{a}$$

Now, take half of $\frac{b}{a}$, and square the result:

$$\frac{1}{2} \cdot \frac{b}{a} = \frac{b}{2a} \quad \text{and} \quad \left(\frac{b}{2a}\right)^2 = \frac{b^2}{4a^2}.$$

Add the square to both sides, producing

$$x^2 + \frac{b}{a}x + \frac{b^2}{4a^2} = \frac{b^2}{4a^2} - \frac{c}{a}.$$

The expression on the left side of the equals sign can be written as the square of a binomial, while the expression on the right can be simplified.

$$\left(x + \frac{b}{2a}\right)^2 = \frac{b^2 - 4ac}{4a^2}$$

By the square root property, this last statement leads to

$$x + \frac{b}{2a} = \sqrt{\frac{b^2 - 4ac}{4a^2}} \quad \text{or} \quad x + \frac{b}{2a} = -\sqrt{\frac{b^2 - 4ac}{4a^2}}.$$

Since $4a^2 = (2a)^2$, or $4a^2 = (-2a)^2$,

$$x + \frac{b}{2a} = \frac{\sqrt{b^2 - 4ac}}{2a} \quad \text{or} \quad x + \frac{b}{2a} = \frac{-\sqrt{b^2 - 4ac}}{2a}.$$

Adding $-\frac{b}{2a}$ to both sides of each result gives

$$x = \frac{-b + \sqrt{b^2 - 4ac}}{2a} \quad \text{or} \quad x = \frac{-b - \sqrt{b^2 - 4ac}}{2a}.$$

It can be shown that these two results are also valid if $a < 0$. A compact form of these two equations, called the *quadratic formula*, follows.

THE QUADRATIC FORMULA

The solutions of the quadratic equation $ax^2 + bx + c = 0$, where $a \neq 0$, are

$$x = \frac{-b \pm \sqrt{b^2 - 4ac}}{2a}.$$

Caution   Notice that the fraction bar in the quadratic formula extends under the $-b$ term in the numerator.

The expression under the radical in the quadratic formula, $b^2 - 4ac$, is called the **discriminant.** The value of the discriminant determines whether the quadratic equation has two real solutions, one real solution, or no real solutions. In the latter case, there will be two imaginary solutions. The following chart summarizes how the discriminant affects the number and nature of the solutions.

EFFECT OF THE DISCRIMINANT

If a, b, and c are real numbers, $a \neq 0$, then the complex solutions of $ax^2 + bx + c = 0$ are described as follows, based on the value of the discriminant, $b^2 - 4ac$.

Value of $b^2 - 4ac$	Number of Solutions	Nature of Solutions
Positive	Two	Complex, real
Zero	One (a double solution)	Complex, real
Negative	Two	Complex, imaginary

Furthermore, if a, b, and c are *integers*, $a \neq 0$, the real solutions are *rational* if $b^2 - 4ac$ is the square of an integer.

Note   The final sentence in the preceding box suggests that the quadratic equation may be solved by factoring if $b^2 - 4ac$ is a "perfect square."

SOLVING QUADRATIC EQUATIONS

EXAMPLE 2   *Using the Quadratic Formula*   Solve the equation

$$x(x - 2) = 2x - 2$$

using the quadratic formula, and support your solutions graphically using the intersection-of-graphs method and the x-intercept method.

SOLUTION   ANALYTIC   Before we can apply the quadratic formula, we must rewrite the equation in the form $ax^2 + bx + c = 0$.

$$x(x - 2) = 2x - 2 \qquad \text{Given equation}$$
$$x^2 - 2x = 2x - 2 \qquad \text{Distributive property}$$
$$x^2 - 4x + 2 = 0 \qquad \text{Subtract } 2x \text{ and add 2 on both sides.}$$

Here $a = 1$, $b = -4$, and $c = 2$. Substitute these values into the quadratic formula to get

$$x = \frac{-b \pm \sqrt{b^2 - 4ac}}{2a}$$

$$= \frac{-(-4) \pm \sqrt{(-4)^2 - 4(1)2}}{2(1)} \qquad a = 1,\ b = -4,\ c = 2$$

$$= \frac{4 \pm \sqrt{16 - 8}}{2} = \frac{4 \pm \sqrt{8}}{2}$$

$$= \frac{4 \pm 2\sqrt{2}}{2} = 2 \pm \sqrt{2}.$$

The solution set is $\{2 + \sqrt{2}, 2 - \sqrt{2}\}$, abbreviated as $\{2 \pm \sqrt{2}\}$.

GRAPHICAL   We can support our solution graphically, using the intersection-of-graphs method, by considering the graphs of $y_1 = x(x - 2)$ and $y_2 = 2x - 2$. (Note that the original form of our equation is $y_1 = y_2$.) By using the capabilities of the calculator, we can find that the x-coordinates of the points of intersection are approximations for $2 - \sqrt{2}$ and $2 + \sqrt{2}$. See Figures 14(a) and (b).

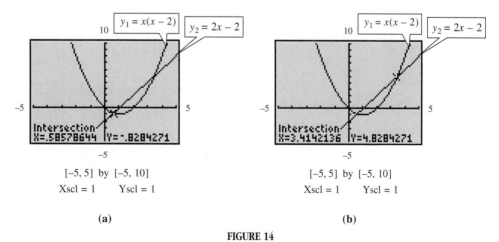

[−5, 5] by [−5, 10]	[−5, 5] by [−5, 10]
Xscl = 1 Yscl = 1	Xscl = 1 Yscl = 1
(a)	**(b)**

FIGURE 14

To support the solutions using the x-intercept method, we graph $y_1 = x(x - 2) - (2x - 2)$ and use the calculator to locate the zeros. Notice that the approximations in Figures 15(a) and (b) on the following page correspond to the x-coordinates of the points of intersection in Figures 14(a) and (b).   ◆

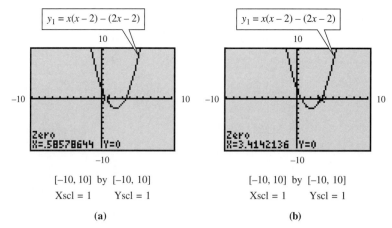

FIGURE 15

Note   For the equation in Example 2, $2 - \sqrt{2}$ and $2 + \sqrt{2}$ are the *exact* solutions; the decimal values shown in the screens are *approximations* of the exact solutions.

EXAMPLE 3   *Using the Quadratic Formula*   Solve $2x^2 - x + 4 = 0$ by the quadratic formula, and support your solutions graphically by the *x*-intercept method of solution.

SOLUTION   ANALYTIC   Here we have $a = 2$, $b = -1$, and $c = 4$. By the quadratic formula,

$$x = \frac{-(-1) \pm \sqrt{(-1)^2 - 4(2)(4)}}{2(2)}$$

$$x = \frac{1 \pm \sqrt{1 - 32}}{4} = \frac{1 \pm \sqrt{-31}}{4}.$$

Because the discriminant is negative, we know that there are no real solutions. Writing the solutions in $a + bi$ form, we get

$$x = \frac{1 \pm i\sqrt{31}}{4} = \frac{1}{4} \pm i\frac{\sqrt{31}}{4}.$$

The solution set is $\left\{\frac{1}{4} \pm i\frac{\sqrt{31}}{4}\right\}$.

GRAPHICAL   If we graph $y = 2x^2 - x + 4$, we see that there are no *x*-intercepts, supporting our result that the only solutions are imaginary. See Figure 16.   ◆

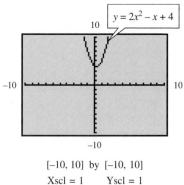

There are no *x*-intercepts and thus no *real* solutions.

FIGURE 16

For Group Discussion Solve $x^2 = 4x - 4$ analytically as a class. Then, graph $y_1 = x^2$ and $y_2 = 4x - 4$, and use the intersection-of-graphs method to support your result.

1. Do you encounter a problem if you use the standard viewing window?
2. Why do you think that analytic methods of solution are essential for understanding graphical methods?

SOLVING QUADRATIC INEQUALITIES

If P is a quadratic function, the solution sets of $P(x) = 0$, $P(x) < 0$, and $P(x) > 0$ can be found graphically based on the following summary.

POSSIBLE ORIENTATIONS FOR QUADRATIC FUNCTION GRAPHS IN THE PLANE

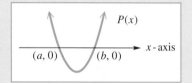

Solution Set of	Is
$P(x) = 0$	$\{a, b\}$
$P(x) < 0$	the interval (a, b)
$P(x) > 0$	$(-\infty, a) \cup (b, \infty)$

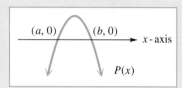

Solution Set of	Is
$P(x) = 0$	$\{a, b\}$
$P(x) < 0$	$(-\infty, a) \cup (b, \infty)$
$P(x) > 0$	the interval (a, b)

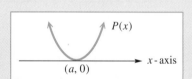

Solution Set of	Is
$P(x) = 0$	$\{a\}$
$P(x) < 0$	$\emptyset$
$P(x) > 0$	$(-\infty, a) \cup (a, \infty)$

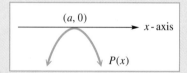

Solution Set of	Is
$P(x) = 0$	$\{a\}$
$P(x) < 0$	$(-\infty, a) \cup (a, \infty)$
$P(x) > 0$	$\emptyset$

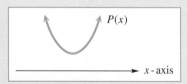

$P(x) = 0$ has no *real* solutions, but two complex imaginary solutions. Real solution set of $P(x) < 0$ is $\emptyset$. Real solution set of $P(x) > 0$ is $(-\infty, \infty)$.

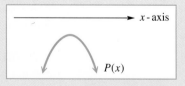

$P(x) = 0$ has no *real* solutions, but two complex imaginary solutions. Real solution set of $P(x) < 0$ is $(-\infty, \infty)$. Real solution set of $P(x) > 0$ is $\emptyset$.

Suppose that the graph of a quadratic polynomial intersects the *x*-axis in two points. Then, the two solutions of the polynomial *equation* divide the real number line (*x*-axis) into three intervals. Within each interval, the polynomial is either always positive or always negative. This idea is used in explaining how a quadratic inequality may be solved using analytic methods, employing a *sign graph*.

EXAMPLE 4   *Solving a Quadratic Inequality Analytically and Graphically*
Solve the quadratic inequality $x^2 - x - 12 < 0$ analytically using a sign graph, and then support the answer graphically using a calculator-generated graph.

SOLUTION   ANALYTIC   Here we have $a = 1$, $b = -1$, and $c = -12$. The discriminant of the corresponding quadratic *equation* is $b^2 - 4ac = 49$. Since 49 is a perfect square, we can solve the quadratic equation $x^2 - x - 12 = 0$ by factoring.

$$x^2 - x - 12 = 0$$
$$(x + 3)(x - 4) = 0 \qquad \text{Factor.}$$
$$x = -3 \quad \text{or} \quad x = 4 \qquad \text{Use the zero-product property.}$$

These two points, -3 and 4, divide a number line into the three regions shown in Figure 17. If a point in region *B*, for example, makes the polynomial $x^2 - x - 12$ negative, then all points in region *B* will make that polynomial negative.

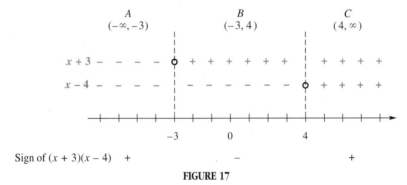

FIGURE 17

To find the regions that make $x^2 - x - 12$ negative (<0), draw a number line that shows where factors are positive or negative, as in Figure 17. First, decide on the sign of the factor $x + 3$ in each of the three regions; then, do the same thing for the factor $x - 4$. The results are shown in Figure 17.

Now, consider the sign of the product of the two factors in each region. As Figure 17 shows, both factors are negative in the interval $(-\infty, -3)$; therefore, their product is positive in that interval. For the interval $(-3, 4)$, one factor is positive and the other is negative, giving a negative product. In the last interval, $(4, \infty)$, both factors are positive, so their product is positive. The polynomial $x^2 - x - 12$ is negative (what the original inequality calls for) when the product of its factors is negative, that is, for the interval $(-3, 4)$. Therefore, the solution set is the open interval $(-3, 4)$.

GRAPHICAL   The graph of $y = x^2 - x - 12$ is shown in Figure 18. Notice that the graph lies *below* the *x*-axis between -3 and 4, supporting our analytic result.   ◆

Note   Graphical solution methods for solving inequalities will not be sufficient for determining whether endpoints should be included or excluded from the solution set. Therefore, we must make our decision based on the symbol in the given inequality. The symbol $<$ or the symbol $>$ indicates that the endpoints are excluded, while $\leq$ or $\geq$ indicates that the endpoints are included.

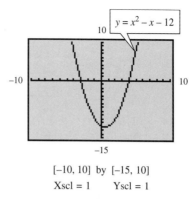

[−10, 10] by [−15, 10]

Xscl = 1   Yscl = 1

The *x*-intercepts are −3 and 4.
The graph lies *below* the *x*-axis in
the open interval (−3, 4).

FIGURE 18

The steps used in solving a quadratic inequality analytically are summarized below.

SOLVING A QUADRATIC INEQUALITY ANALYTICALLY

1. Solve the corresponding quadratic equation.
2. Identify the intervals determined by the solutions of the equation.
3. Use a sign graph to determine which intervals are in the solution set.
4. Decide whether or not the endpoints are included.

EXAMPLE 5 *Solving a Quadratic Inequality Analytically and Graphically*
Solve the quadratic inequality

$$2x^2 \geq -5x + 12$$

analytically. Then, support your answer graphically using the *x*-intercept method.

SOLUTION ANALYTIC   We begin by writing the quadratic inequality in standard form:
$2x^2 + 5x - 12 \geq 0$. The corresponding quadratic equation can be solved by factoring.

$$2x^2 + 5x - 12 = 0$$
$$(2x - 3)(x + 4) = 0$$
$$x = \frac{3}{2} = 1.5 \quad \text{or} \quad x = -4$$

These two points divide the number line into the three regions shown in the sign graph
in Figure 19 on the following page. Since both factors are negative in the first interval,
their product, $2x^2 + 5x - 12$, is positive there. In the second interval, the factors have
opposite signs, and, therefore, their product is negative. Both factors are positive in the
third interval, and their product also is positive there. Thus, the polynomial
$2x^2 + 5x - 12$ is positive or zero in the interval $(-\infty, -4]$ and also in the interval
$[1.5, \infty)$. Since both of the intervals belong to the solution set, the result can be written
as the *union** of the two intervals, $(-\infty, -4] \cup [1.5, \infty)$.

---

*The **union** of sets *A* and *B*, written $A \cup B$, is defined as $A \cup B = \{x \mid x$ is an element of *A* or *x* is an element of *B*}.

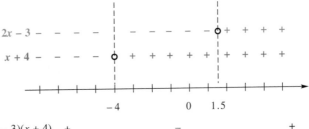

Sign of $(2x - 3)(x + 4)$ +   —   +

FIGURE 19

GRAPHICAL   Using the x-intercept method to support our result involves writing the original inequality in the form $y_1 - y_2 \geq 0$, and then finding the domain values for which the graph of $y_1 - y_2$ *lies above* or *on* the x-axis. It appears that the intervals $(-\infty, -4]$ and $[1.5, \infty)$ give these values, supporting our earlier result. See Figure 20.   ◆

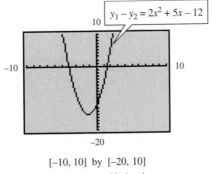

Choosing values of x less than or equal to -4 *or* greater than or equal to 1.5 will always give $Y_1 \geq Y_2$. (See Example 5.)

$[-10, 10]$ by $[-20, 10]$
Xscl = 1   Yscl = 1

The x-intercepts are -4 and 1.5.
The solution set of $y_1 - y_2 \geq 0$ is
$(-\infty, -4] \cup [1.5, \infty)$.

FIGURE 20

Note   While we solved the quadratic equations in Examples 4 and 5 by factoring, we could also have used the quadratic formula. The formula will work for *any* quadratic equation, whether or not factoring is applicable.

For Group Discussion

1.   The function $P(x) = x^2 - x - 20$ has integer zeros. Graph the function in a window that will show both x-intercepts, and, without performing any analytic work, solve each of the following:
     a. $x^2 - x - 20 = 0$ **b.** $x^2 - x - 20 < 0$ **c.** $x^2 - x - 20 > 0$.
2.   The function $P(x) = x^2 + x + 20$ has no real zeros. Graph the function in a window that will show the vertex and the y-intercept. (Why won't the standard window work?) Then, without performing any analytic work, solve each of the following:
     a. $x^2 + x + 20 = 0$ **b.** $x^2 + x + 20 < 0$ **c.** $x^2 + x + 20 > 0$.
     (Give only real solutions.)

From the preceding group discussion, we can see that solving a quadratic equation will lead quite easily to the solution sets of the corresponding inequalities.

EXAMPLE 6 *Solving Quadratic Equations and Inequalities Graphically*
Solve the quadratic equation

$$2.57x^2 - 1.56x - \sqrt{7.04} = 0$$

graphically, giving solutions to the nearest hundredth. Then, use the graph to solve

$$2.57x^2 - 1.56x - \sqrt{7.04} > 0 \quad \text{and} \quad 2.57x^2 - 1.56x - \sqrt{7.04} < 0.$$

SOLUTION A viewing window of $[-3, 3]$ by $[-3, 5]$ gives a comprehensive graph of $P(x) = 2.57x^2 - 1.56x - \sqrt{7.04}$. See Figure 21. By using a root identification procedure, we find that the solutions, to the nearest hundredth, are $-.76$ and 1.36. The graph is below the x-axis between these x-intercepts, and above the x-axis to the left of $x = -.76$ and to the right of $x = 1.36$. Putting all this information together leads to the following conclusions:

The solution set of $2.57x^2 - 1.56x - \sqrt{7.04} = 0$ is $\{-.76, 1.36\}$.

The solution set of $2.57x^2 - 1.56x - \sqrt{7.04} < 0$ is the open interval $(-.76, 1.36)$.

The solution set of $2.57x^2 - 1.56x - \sqrt{7.04} > 0$ is $(-\infty, -.76) \cup (1.36, \infty)$.

Remember that the numbers given in the solutions above are approximations. Application of the quadratic formula would have given exact (but messy) answers. ◆

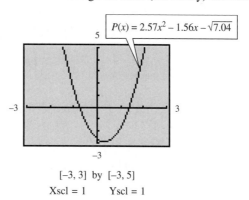

$[-3, 3]$ by $[-3, 5]$
Xscl = 1   Yscl = 1

The x-intercepts are approximately
$-.76$ and 1.36.
FIGURE 21

EXAMPLE 7 *Solving Quadratic Inequalities for Exact Solutions*   Use the graph in Figure 22 on the following page and the result of Example 2 to solve the following inequalities for intervals with *exact values* at endpoints.

(a) $x^2 - 4x + 2 \leq 0$ **(b)** $x^2 - 4x + 2 \geq 0$

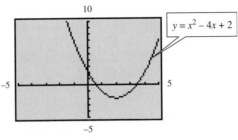

$$[-5, 5] \text{ by } [-5, 10]$$
$$\text{Xscl} = 1 \qquad \text{Yscl} = 1$$

The *exact* x-intercepts are $2 - \sqrt{2}$
and $2 + \sqrt{2}$.

FIGURE 22

SOLUTION

(a) We found by the quadratic formula that the exact solutions of $x^2 - 4x + 2 = 0$ are $2 - \sqrt{2}$ and $2 + \sqrt{2}$. Since the graph of $y = x^2 - 4x + 2$ lies below or intersects the x-axis between or at these values (as shown in Figure 22), the solution set for this inequality is the closed interval $[2 - \sqrt{2}, 2 + \sqrt{2}]$.

(b) Using the same concept as in Example 6, we conclude that the solution set is $(-\infty, 2 - \sqrt{2}] \cup [2 + \sqrt{2}, \infty)$. ◆

LITERAL EQUATIONS INVOLVING QUADRATICS

Sometimes it is necessary to solve a literal equation for a variable that is squared. The next example shows how this is done.

EXAMPLE 8   *Solving for a Quadratic Variable*   Solve each equation for the indicated variable.

(a) $A = \dfrac{\pi d^2}{4}$ for d **(b)** $rt^2 - st = k \ (r \neq 0)$ for t

SOLUTION

(a) Start by multiplying both sides by 4 to get

$$4A = \pi d^2.$$

Now, divide by π.

$$d^2 = \frac{4A}{\pi}$$

Use the square root property and rationalize the denominator on the right.

$$d = \pm\sqrt{\frac{4A}{\pi}} = \frac{\pm 2\sqrt{A}}{\sqrt{\pi}} = \frac{\pm 2\sqrt{A\pi}}{\pi}$$

(b) Because this equation has a term with t as well as t^2, we use the quadratic formula. Subtract k from both sides to get

$$rt^2 - st - k = 0.$$

Now, use the quadratic formula to find t, with $a = r$, $b = -s$, and $c = -k$.

$$t = \frac{-b \pm \sqrt{b^2 - 4ac}}{2a}$$

$$t = \frac{-(-s) \pm \sqrt{(-s)^2 - 4(r)(-k)}}{2(r)}$$

$$t = \frac{s \pm \sqrt{s^2 + 4rk}}{2r} \qquad \blacklozenge$$

3.3 EXERCISES   Tape 4

1. Which one of the following equations is set up for direct use of the zero-product property? Solve it.

    (a) $3x^2 - 17x - 6 = 0$ **(b)** $(2x + 5)^2 = 7$ **(c)** $x^2 + x = 12$ **(d)** $(3x + 1)(x - 7) = 0$

2. Which one of the following equations is set up for direct use of the square root property? Solve it.

    (a) $3x^2 - 17x - 6 = 0$ **(b)** $(2x + 5)^2 = 7$ **(c)** $x^2 + x = 12$ **(d)** $(3x + 1)(x - 7) = 0$

3. Only one of the following equations does not require step 1 of the method of completing the square described in the previous section. Which one is it? Solve it.

    (a) $3x^2 - 17x - 6 = 0$ **(b)** $(2x + 5)^2 = 7$ **(c)** $x^2 + x = 12$ **(d)** $(3x + 1)(x - 7) = 0$

4. Only one of the following equations is set up so that the values of a, b, and c can be determined immediately. Which one is it? Solve it.

    (a) $3x^2 - 17x - 6 = 0$ **(b)** $(2x + 5)^2 = 7$ **(c)** $x^2 + x = 12$ **(d)** $(3x + 1)(x - 7) = 0$

Find all solutions, both real and imaginary, of the following quadratic equations. Use the square root property. For the equations with real solutions, support your answers graphically.

5. $x^2 = 16$ **6.** $x^2 = 144$ **7.** $3x^2 = 27$ **8.** $2x^2 = 48$

9. $x^2 = -16$ **10.** $x^2 = -100$ **11.** $x^2 = -18$ **12.** $x^2 = -32$

Solve each of the following equations by the quadratic formula. Find all solutions, both real and imaginary. If the equation is not in the form $P(x) = 0$, you will need to write it this way in order to identify a, b, and c. For the equations with real solutions, support your answers graphically by using the x-intercept method, graphing $y = P(x)$, and then locating those intercepts.

13. $x^2 - 2x - 4 = 0$ **14.** $x^2 + 8x + 13 = 0$ **15.** $2x^2 + 2x = -1$ **16.** $9x^2 - 12x = -8$

17. $x(x - 1) = 1$ **18.** $x(x - 3) = 2$ **19.** $x^2 - 5x = x - 7$ **20.** $11x^2 - 3x + 2 = 4x + 1$

21. $4x^2 - 12x = -11$ **22.** $x^2 = 2x - 5$

Write each equation so that 0 is on the right side, and then evaluate the discriminant. Use the discriminant to determine the number of real solutions the equation has. If the equation has real solutions, tell whether they are rational or irrational. Do not actually solve the equation.

23. $x^2 + 8x + 16 = 0$ **24.** $8x^2 = 14x - 3$ **25.** $4x^2 = 6x + 3$

26. $2x^2 - 4x + 1 = 0$ **27.** $9x^2 + 11x + 4 = 0$ **28.** $3x^2 = 4x - 5$

The figures below show several possible graphs of $f(x) = ax^2 + bx + c$. *For the restrictions on a, b, and c given in Exercises 29–34, select the corresponding graph from A–F.*

29. $a < 0, b^2 - 4ac = 0$

30. $a > 0, b^2 - 4ac < 0$

31. $a < 0, b^2 - 4ac < 0$

32. $a < 0, b^2 - 4ac > 0$

33. $a > 0, b^2 - 4ac > 0$

34. $a > 0, b^2 - 4ac = 0$

A.

B.

C.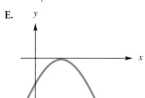

D.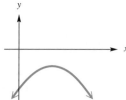

E.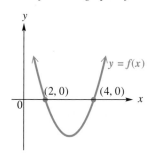

F.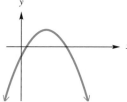

Exercises 35–44 refer to the graphs of the quadratic functions f, g, and h shown here.

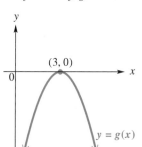

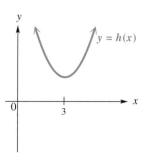

35. What is the solution set of $f(x) = 0$?

36. What is the solution set of $f(x) < 0$?

37. What is the solution set of $f(x) > 0$?

38. What is the solution set of $g(x) = 0$?

39. What is the solution set of $g(x) < 0$?

40. What is the solution set of $g(x) > 0$?

41. Solve $h(x) > 0$.

42. Solve $h(x) < 0$.

43. How many real solutions does $h(x) = 0$ have? How many complex solutions does it have?

44. What is the value of the discriminant of $g(x)$?

RELATING CONCEPTS (EXERCISES 45–50)

Recall the concepts presented in Section 3.2 involving vertex, axis of symmetry, and intercepts, and answer the following questions about the graphs of f, g, and h. (See the graphs accompanying Exercises 35–44.)

45. What is the *x*-coordinate of the vertex of the graph of $y = f(x)$?

46. What is the equation of the axis of symmetry of the graph of $y = g(x)$?

47. Does the graph of $y = g(x)$ have a y-intercept? If so, is it positive or is it negative?

48. Is the minimum value of h positive or negative?

49. Which function satisfies this description? It is concave up and the solution set of $y < 0$ is not empty.

50. Which function is decreasing on the interval $[3, \infty)$?

The built-in equation solver of a graphing calculator can be used to solve quadratic equations. For example, if we wish to solve $2x^2 - 11x - 40 = 0$, we obtain the solutions -2.5 and 8 as shown in the screens below.

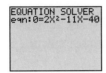 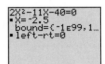 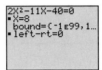

The same results obtained above are shown in the following screen, using an alternative equation-solving routine.

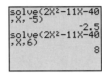

Some models require you to make a "guess" to determine the solution. In the two examples shown above, we used -5 and 6 as guesses. Solve the following quadratic equations using an equation solver. If your model requires guesses, use those given.

51. $2x^2 - 13x - 7 = 0$ (guesses: -5 and 5)

52. $4x^2 - 11x = 3$ (guesses: 0 and 5)

53. $\sqrt{6}x^2 - 2x - 1.4 = 0$ (guesses: -1 and 2)

54. $\sqrt{10}x(x - 1) = 8.6$ (guesses: -2 and 2)

Solve each inequality analytically using a sign graph. Then, support your answer graphically, using a calculator-generated graph. Give exact values for endpoints.

55. (a) $x^2 + 4x + 3 \geq 0$

    (b) $x^2 + 4x + 3 < 0$

56. (a) $x^2 + 6x + 8 < 0$

    (b) $x^2 + 6x + 8 \geq 0$

57. (a) $2x^2 - 9x + 4 > 0$

    (b) $2x^2 - 9x + 4 \leq 0$

58. (a) $3x^2 + 13x + 10 \leq 0$

    (b) $3x^2 + 13x + 10 > 0$

59. (a) $-x^2 + 2x + 1 \geq 0$

    (b) $-x^2 + 2x + 1 < 0$

60. (a) $-x^2 - 5x + 2 > 0$

    (b) $-x^2 - 5x + 2 \leq 0$

RELATING CONCEPTS (EXERCISES 61–66)

We saw in Chapter 1 that in order to solve an inequality analytically, if we multiply both sides by a negative number we must reverse the direction of the inequality sign. Work through Exercises 61–66 in order. They illustrate this concept for quadratic inequalities.

61. Graph $y_1 = x^2 + 2x - 8$ in the standard viewing window. This function has two integer-valued x-intercepts. What are they?

62. Based on the graph, what is the solution set of $x^2 + 2x - 8 < 0$?

63. Now, graph $y_2 = -y_1 = -x^2 - 2x + 8$ on the same screen. Using the terminology of Chapter 2, how is the graph of y_2 obtained by transforming the graph of y_1?

64. Based on the graph of y_2, what is the solution set of $-x^2 - 2x + 8 > 0$?

65. How do the two solution sets of the inequalities in Exercises 62 and 64 compare?

66. Write a short paragraph explaining how Exercises 61–65 illustrate the property involving multiplying an inequality by a negative number.

In each screen, the graph of a quadratic function Y_1 *and a table are given. Use the information to give the solution set of the equation or inequality.*

67. (a) $Y_1 = 0$
 (b) $Y_1 < 0$
 (c) $Y_1 > 0$

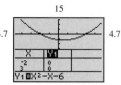

68. (a) $Y_1 = 0$
 (b) $Y_1 < 0$
 (c) $Y_1 > 0$

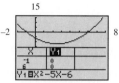

69. (a) $Y_1 = 0$
 (b) $Y_1 \leq 0$
 (c) $Y_1 \geq 0$

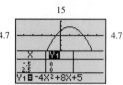

70. (a) $Y_1 = 0$
 (b) $Y_1 \leq 0$
 (c) $Y_1 \geq 0$

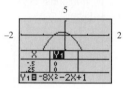

In Exercises 71–74, **(a)** *solve the equation* $P(x) = 0$ *graphically, giving solutions to the nearest hundredth. Then, give the solution set of* **(b)** $P(x) > 0$ *and* **(c)** $P(x) < 0$, *based on your graph.*

71. $P(x) = 3.15x^2 + .65x - 3.24$

72. $P(x) = 2.78x^2 + .47x - 6.13$

73. $P(x) = -\pi x^2 + 9.8x - \sqrt{7}$

74. $P(x) = -\sqrt{5}x^2 + 5.4x - \sqrt{3}$

Solve the equation for the indicated variable. Assume that no denominators are zero.

75. $s = \dfrac{1}{2} gt^2$ for t

76. $A = \pi r^2$ for r

77. $F = \dfrac{kMv^4}{r}$ for v

78. $s = s_0 + gt^2 + k$ for t

79. $P = \dfrac{E^2 R}{(r + R)^2}$ for R

80. $S = 2\pi rh + 2\pi r^2$ for r

3.4   APPLICATIONS OF QUADRATIC FUNCTIONS AND MODELS

Applications of Quadratic Functions   ◆   Quadratic Models

In Section 1.8, we saw how the concepts of linear functions, equations, and inequalities could be applied to solving certain types of problems. We will now see how certain problems can be solved using quadratic functions, and solving the corresponding equations and inequalities. Furthermore, determining the coordinates of the vertex of the graph of a quadratic function will often enable us to solve problems requiring the minimum or maximum value of the function. Such problems are called **optimization problems**.

APPLICATIONS OF QUADRATIC FUNCTIONS

EXAMPLE 1   *Solving a Problem Involving Area of a Rectangular Region*   A farmer wishes to enclose a rectangular region. He has 120 feet of fencing, and plans to use one side of his barn as a part of the enclosure. See Figure 23. Let x represent the length of one of the parallel sides of the fencing, and respond to each of the following.

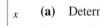

FIGURE 23

(a)   Determine a function A that represents the area of the region in terms of x.

(b)   For this particular problem, what are the restrictions on x?

(c)   Find a viewing window that will show both x-intercepts and the vertex of the graph of this quadratic function.

(d)   Figure 24 shows the cursor at $(18, 1512)$. Interpret this information.

(e)   What is the maximum area the farmer can enclose? Determine the answer analytically, and support it graphically.

SOLUTION

(a) The lengths of the sides of the region bordered by the fencing are x, x, and $120 - 2x$, as shown in the figure. Since area = width × length, the function is

$$A(x) = x(120 - 2x) \qquad \text{or} \qquad A(x) = -2x^2 + 120x.$$

(b) Since x represents a length, we must have $x > 0$. Furthermore, the side of length $120 - 2x$ must also be positive. Therefore, $120 - 2x > 0$, or $x < 60$. Putting these two restrictions together, we have $0 < x < 60$.

(c) While many viewing windows will satisfy these requirements, one such window is shown in Figure 24.

(d) If the parallel sides of fencing each measure 18 feet, the area of the enclosure is 1512 square feet. This can be written $A(18) = 1512$, and checked as follows: If the width is 18 feet, the length is $120 - 2(18) = 84$ feet, and $18 \times 84 = 1512$.

(e) ANALYTIC We must find the maximum value of the function. This occurs at the vertex. For $A(x) = -2x^2 + 120x$, we have $a = -2$ and $b = 120$. Using the vertex formula from Section 3.2, $x = -\frac{b}{2a} = -\frac{120}{2(-2)} = 30$. Evaluating $A(30)$ gives $-2(30)^2 + 120(30) = 1800$. Therefore, the farmer can enclose a maximum of 1800 square feet when the parallel sides of fencing measure 30 feet.

GRAPHICAL To support this answer, use a calculator to locate the vertex of the parabola and observe the x- and y-values there. Figure 25 gives support for our answer. ◆

TECHNOLOGY NOTE

In Figures 24 and 25, the display at the bottom obscures the view of the intercepts. This problem is simple to overcome: just lower the minimum values of x and y enough so that when the display appears, the x- and y-axes are visible.

This table supports the result found analytically and graphically in Example 1(e).

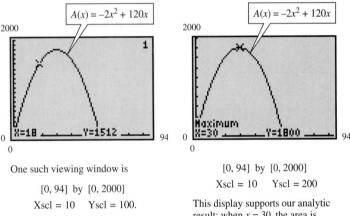

One such viewing window is

[0, 94] by [0, 2000]

Xscl = 10   Yscl = 100.

FIGURE 24

[0, 94] by [0, 2000]

Xscl = 10   Yscl = 200

This display supports our analytic result: when $x = 30$, the area is maximized at 1800 square feet.

FIGURE 25

Caution As seen in Example 1(e), it is important to be careful when interpreting the meanings of the coordinates of the vertex in optimization problems. The first coordinate, x, gives the *domain* value for which the *function value* is a maximum or minimum. It is always necessary to read the problem carefully to determine whether you are asked to find the value of the independent variable x, the function value y, or both.

EXAMPLE 2 *Solving a Problem Involving the Volume of a Box* A piece of machinery is capable of producing rectangular sheets of metal satisfying the condition that the length is three times the width. Furthermore, equal size squares measuring 5 inches on a side can be cut from the corners so that the resulting piece of metal can be shaped into an open box by folding up the flaps. See Figure 26(a) on the next page.

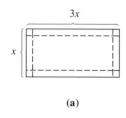

(a)

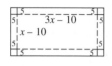

(b)

FIGURE 26

(a) Determine a function V that expresses the volume of the box in terms of the width x of the original sheet of metal.

(b) What restrictions must be placed on x in this particular problem?

(c) If specifications call for the volume of such a box to be 1435 cubic inches, what should the dimensions of the original piece of metal be? Solve analytically and support graphically.

(d) What dimensions of the original piece of metal will assure us of a volume greater than 2000 but less than 3000 cubic inches? Solve graphically.

SOLUTION

(a) If x represents the width, then $3x$ represents the length. Figure 26(b) indicates that the width of the bottom of the box is $x - 10$, the length of the bottom of the box is $3x - 10$, and the height is 5 inches (the length of the side of each cut-out square). Since volume = length × width × height, the function is

$$V(x) = (3x - 10)(x - 10)(5) = 15x^2 - 200x + 500.$$

(b) Since the dimensions of the box must represent positive numbers, we must have $3x - 10 > 0$ and $x - 10 > 0$, or

$$x > \frac{10}{3} \quad \text{and} \quad x > 10.$$

Both conditions are satisfied when $x > 10$. Therefore, the theoretical domain of x in this problem is $(10, \infty)$.

(c) ANALYTIC We must find x such that $V(x) = 1435$.

$$1435 = 15x^2 - 200x + 500 \qquad \text{Set } V(x) = 1435.$$
$$0 = 15x^2 - 200x - 935 \qquad \text{Subtract 1435.}$$
$$0 = (15x + 55)(x - 17) \qquad \text{Factor.}$$
$$15x + 55 = 0 \quad \text{or} \quad x - 17 = 0 \qquad \text{Use the zero-product property.}$$

$$x = -\frac{11}{3} \quad \text{or} \quad x = 17 \qquad \text{Solve.}$$

Of these two solutions, only 17 satisfies the condition that $x > 10$. Therefore, the dimensions of the original piece of metal should be 17 inches by $3(17) = 51$ inches.

Since $(51 - 10) \cdot (17 - 10) \cdot 5 = 1435$, our answer is correct. (Notice that we could have solved the quadratic equation by the quadratic formula as well.)

GRAPHICAL One way to support this result graphically is to graph $y_1 = 15x^2 - 200x + 500$ in a window with minimum x-value 10 and $y_2 = 1435$ in the same window. The point at which they intersect should have an x-value of 17. See Figure 27 on the following page.

Note If we were to graph y_1 and y_2 in a window containing the x-value $-\frac{11}{3}$, the parabola and the line would intersect again. (Where have we seen the analytic justification for this?) A window of $[-6, 20]$ by $[-200, 1600]$ will allow us to see this other point of intersection.

(d) Using the graphs of the functions $y_1 = 15x^2 - 200x + 500$, $y_2 = 2000$, and $y_3 = 3000$, we find that the points of intersection of the graphs are *approximately* $(18.7, 2000)$ and $(21.2, 3000)$ for $x > 10$. See Figure 28. Therefore, the width of the rectangle should be between 18.7 and 21.2 inches, with the corresponding length three times these values (that is, between $3(18.7) \approx 56.1$ and $3(21.2) \approx 63.6$ inches). ◆

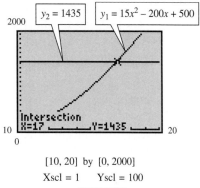

[10, 20] by [0, 2000]

Xscl = 1   Yscl = 100

FIGURE 27

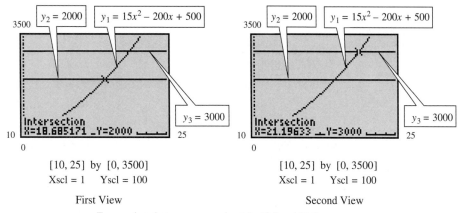

[10, 25] by [0, 3500]   [10, 25] by [0, 3500]
Xscl = 1   Yscl = 100   Xscl = 1   Yscl = 100

First View   Second View

For *x*-values between approximately 18.7 and 21.2,
the *y*-values will be between 2000 and 3000.

FIGURE 28

An important application of quadratic functions deals with the height of a propelled object as a function of the time elapsed after it is propelled.

FORMULA FOR THE HEIGHT OF A PROPELLED OBJECT

If air resistance is neglected, the height s (in feet) of an object propelled directly upward from an initial height s_0 feet with initial velocity v_0 feet per second is described by the function

$$s(t) = -16t^2 + v_0 t + s_0,$$

where t is the number of seconds after the object is propelled.

In this formula, the coefficient of t^2 (that is, -16) is a constant based on the gravitational force of the earth. This constant varies on other surfaces, such as the moon and other planets. Here we have an example of a quadratic function in which height is a function of time. We use t to represent the independent variable; however, when graphing this type of function on our calculator, we will use x, since graphing calculators are so equipped. In reality, there is no difference between the functions $s(t) = -16t^2 + v_0 t + s_0$ and $s(x) = -16x^2 + v_0 x + s_0$.

EXAMPLE 3   *Solving a Problem Involving Projectile Motion*   A ball is thrown directly upward from an initial height of 100 feet with an initial velocity of 80 feet per second.

(a)   What is the function that describes the height of the ball in terms of the time *t* elapsed?

(b)   Graph this function so that the *y*-intercept, the positive *x*-intercept, and the vertex are visible.

(c)   The cursor in Figure 29 shows that the point (4.8, 115.36) lies on the graph of the function. What does this mean for this particular problem?

(d)   After how many seconds does the projectile reach its maximum height? What is this maximum height? Solve analytically and support graphically.

(e)   For what interval of time is the height of the ball greater than 160 feet? Determine the answer graphically.

(f)   After how many seconds will the ball fall to the ground? Determine the answer analytically and support graphically.

SOLUTION

(a)   We use the projectile height formula with $v_0 = 80$ and $s_0 = 100$:

$$s(t) = -16t^2 + 80t + 100.$$

(b)   There are many suitable choices for such a window. One such choice is $[0, 10]$ by $[0, 300]$, as shown in Figure 29. It shows the graph of $y = -16x^2 + 80x + 100$. (Here, $x = t$.)

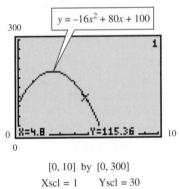

[0, 10] by [0, 300]
Xscl = 1   Yscl = 30
FIGURE 29

(c)   When $x = 4.8$, $y = 115.36$. Therefore, when 4.8 seconds have elapsed, the projectile is at a height of 115.36 feet.

(d)   ANALYTIC   To answer this question, we must find the coordinates of the vertex of the parabola. Using the vertex formula from Section 3.2, we find

$$x = -\frac{b}{2a} = -\frac{80}{2(-16)} = 2.5$$

and

$$y = -16(2.5)^2 + 80(2.5) + 100 = 200.$$

Therefore, after 2.5 seconds the ball reaches its maximum height of 200 feet.

GRAPHICAL   To support this graphically, we use the capabilities of the calculator to find that the vertex coordinates are indeed (2.5, 200). See Figure 30.

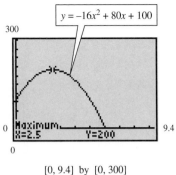

[0, 9.4] by [0, 300]

Xscl = 1   Yscl = 30

When the time is 2.5 seconds, the
height is at a maximum, 200 feet.

FIGURE 30

Caution It is easy to misinterpret the graph shown in Figure 30. This graph does not define the *path*
followed by the ball; rather, it defines height as a function of time.

(e) With $y_1 = -16x^2 + 80x + 100$ and $y_2 = 160$ graphed, we locate the two points
of intersection. See Figure 31. We find that the *x*-coordinates there are approximately .92 and 4.08. Therefore, between .92 and 4.08 seconds, the ball is more
than 160 feet above the ground.

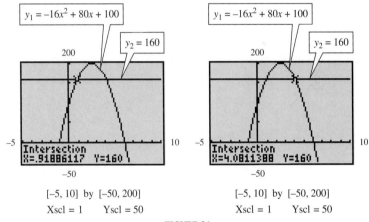

[−5, 10] by [−50, 200]   [−5, 10] by [−50, 200]

Xscl = 1   Yscl = 50   Xscl = 1   Yscl = 50

FIGURE 31

(f) ANALYTIC When the ball falls to the ground, its height will be 0 feet, so we must
solve the quadratic equation $0 = -16x^2 + 80x + 100$. Using the quadratic formula, we find

$$x = \frac{-80 \pm \sqrt{80^2 - 4(-16)(100)}}{2(-16)}$$

$$x \approx -1.04 \text{ or } 6.04.$$

We must choose the positive solution. Therefore, the ball will return to the ground
about 6.04 seconds after it was projected.

   GRAPHICAL To support this result, we use the capabilities of the calculator to
find the positive *x*-intercept of the graph; it is indeed about 6.04. See Figure 32 on
the next page.

(Notice that if the negative *x*-intercept were shown, it would be about −1.04, the rejected solution above.)   ◆

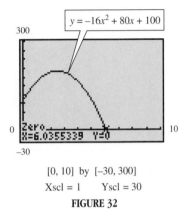

[0, 10] by [−30, 300]

Xscl = 1   Yscl = 30

FIGURE 32

QUADRATIC MODELS

In Section 1.7, we saw how a linear model can help us predict and interpret data in some cases. By extending the statistical concept of regression to polynomials, we can sometimes fit data to a quadratic function to provide a model for the data.

EXAMPLE 4   *Using a Quadratic Function to Model Data Involving AIDS Cases*   The table below lists the total (cumulative) AIDS cases diagnosed in the United States up to 1993. For example, a total of 22,620 cases were diagnosed between 1981 and 1985.

TECHNOLOGY NOTE

Some graphing calculators will fit a window to data points. Consult your owner's manual for details.

Year	AIDS Cases
1982	1563
1983	4647
1984	10,845
1985	22,620
1986	41,662
1987	70,222
1988	105,489
1989	147,170
1990	193,245
1991	248,023
1992	315,329
1993	361,509

(*Source*: U.S. Dept. of Health and Human Services, Centers for Disease Control and Prevention, *HIV/AIDS Surveillance*, March 1994.)

(a)   Plot the data, letting $x = 0$ correspond to the year 1980.

(b)   Find a quadratic function $f(x) = a(x - h)^2 + k$ that models the data by using (2, 1563) as the vertex and a second point such as (13, 361,509) to determine a.

(c)   Plot the data together with f in the same window. How well does f model the number of AIDS cases?

(d) Use the quadratic regression feature of a graphing calculator to determine the quadratic function *g* that provides the best fit for the data.

(e) Use the functions *f* and *g* to predict the cumulative number of cases for the year 1994.

SOLUTION

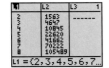

This is a portion of the data found in Example 4, as entered into lists L1 and L2 for the purpose of statistical analysis.

(a) Using the statistical feature of a graphing calculator and letting the *x*-list L1 be {2, 3, 4, . . . , 12, 13} and the *y*-list L2 be {1563, 4647, 10,845, . . . , 315,329, 361,509}, we get the statistical plot seen in Figure 33.

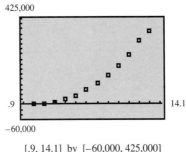

[.9, 14.1] by [−60,000, 425,000]
Xscl = 1   Yscl = 25,000
FIGURE 33

(b) We substitute 2 for *h* and 1563 for *k* in the given form of the quadratic function *f*, leading to

$$f(x) = a(x - 2)^2 + 1563.$$

To solve for *a*, let $x = 13$ and $f(x) = f(13) = 361{,}509$.

$$361{,}509 = a(13 - 2)^2 + 1563$$

$$361{,}509 - 1563 = 121a$$

$$\frac{359{,}946}{121} = a$$

$$a \approx 2974.76$$

The desired function is $f(x) = 2974.76(x - 2)^2 + 1563$. (Note that choosing other second points would produce other models.)

(c) Figure 34 shows the graph of *f* plotted over the data points. There is a relatively good fit.

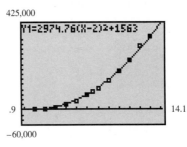

[.9, 14.1] by [−60,000, 425,000]
Xscl = 1   Yscl = 25,000
FIGURE 34

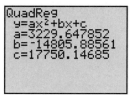

FIGURE 35

(d) Figure 35 shows the result of finding the quadratic model that best fits the data points. The model is approximately $g(x) = 3229.65x^2 - 14,805.89x + 17,750.15$. Notice that the calculator gives the quadratic expression in $ax^2 + bx + c$ form.

(e) Using function f from part (b), with $x = 14$ for the year 1994, we get $f(14) = 2974.76(14 - 2)^2 + 1563 \approx 429,928$. Function g from part (d) gives $g(14) = 3229.65(14)^2 - 14,805.89(14) + 17,750.15 \approx 443,479.$ ◆

For Group Discussion

After reading Example 4, answer the following.

1. Have each class member choose a point other than (13, 361,509) and determine another function h in a manner similar to the one in which f was found in part (b). Then, graph $y = h(x)$ over the data points and discuss the accuracy of the various models found.

2. Figure 36 shows the best fitting quadratic model $g(x) = 3229.65x^2 - 14,805.89x + 17,750.15$ and the data points. Is it *noticeably* a better fit than the function f found in part (b) and graphed in Figure 34?

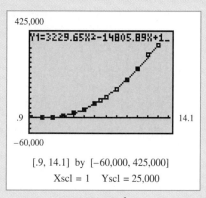

[.9, 14.1] by [−60,000, 425,000]
Xscl = 1   Yscl = 25,000

FIGURE 36

3.4   EXERCISES   Tape 4

1. *Area of a Parking Lot*   For the rectangular parking area of the shopping center shown, which one of the following equations says that the area is 40,000 square yards?

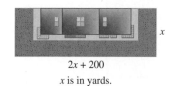

2x + 200

x is in yards.

(a) $x(2x + 200) = 40,000$

(b) $2x + 2(2x + 200) = 40,000$

(c) $x + (2x + 200) = 40,000$

(d) none of the above

2. *Area of a Picture*   The mat around the picture shown measures x inches across. Which one of the following equations says that the area of the picture itself is 600 square inches?

34 in

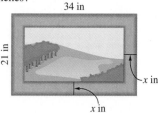

21 in

x in

x in

(a) $2(34 - 2x) + 2(21 - 2x) = 600$

(b) $(34 - 2x)(21 - 2x) = 600$

(c) $(34 - x)(21 - x) = 600$

(d) $x(34)(21) = 600$

3. *Sum of Two Numbers* Suppose that x represents one of two *positive* numbers whose sum is 30.

  (a) Represent the other of the two numbers in terms of x.

  (b) What are the restrictions on x?

  (c) Describe a function P that represents the product of these two numbers.

  (d) Determine analytically and support graphically the two such numbers whose product is a maximum. What is this maximum product?

4. *Sum of Two Numbers* Suppose that x represents one of two *positive* numbers whose sum is 45.

  (a) Represent the other of the two numbers in terms of x.

  (b) What are the restrictions on x?

  (c) Describe a function P that represents the product of these two numbers.

  (d) For what two such numbers is the product equal to 504? Determine analytically.

  (e) Determine analytically and support graphically the two such numbers whose product is a maximum.

5. *Area of a Parking Lot* American River College has plans to construct a rectangular parking lot on land bordered on one side by a highway. There are 640 feet of fencing available to fence the other three sides. Let x represent the length of each of the two parallel sides of fencing.

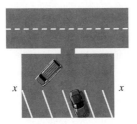

  (a) Express the length of the remaining side to be fenced in terms of x.

  (b) What are the restrictions on x?

  (c) Describe a function A that represents the area of the parking lot in terms of x.

  (d) Graph the function of part (c) in a viewing window of [0, 320] by [0, 55,000]. Determine graphically the values of x that will give an area between 30,000 and 40,000 square feet.

  (e) What dimensions will give a maximum area, and what will this area be? Determine analytically and support graphically.

6. *Area of a Rectangular Region* A farmer wishes to enclose a rectangular region bordering a river with fencing, as shown in the diagram. Suppose that x represents the length of each of the three parallel pieces of fencing. She has 600 feet of fencing available.

  (a) What is the length of the remaining piece of fencing in terms of x?

  (b) Describe a function A that represents the total area of the enclosed region. Give any restrictions on x.

  (c) What dimensions for the total enclosed region would give an area of 22,500 square feet? Determine the answer analytically.

  (d) What is the maximum area that can be enclosed? Use a graph to answer this question.

7. *Volume of a Box* A piece of cardboard is twice as long as it is wide. It is to be made into a box with an open top by cutting 2-inch squares from each corner and folding up the sides. Let x represent the width of the original piece of cardboard.

  (a) Represent the length of the original piece of cardboard in terms of x.

  (b) What will be the dimensions of the bottom rectangular base of the box? Give restrictions on x.

  (c) Describe a function V that represents the volume of the box in terms of x.

  (d) For what bottom dimensions will the volume be 320 cubic inches? Determine the answer analytically and support it graphically.

  (e) Determine graphically (to the nearest tenth of an inch) the values of x if such a box is to have a volume between 400 and 500 cubic inches.

8. *Volume of a Box*   A piece of sheet metal is 2.5 times as long as it is wide. It is to be made into a box with an open top by cutting 3-inch squares from each corner and folding up the sides. Let x represent the width of the original piece of sheet metal.

(a) Represent the length of the original piece of sheet metal in terms of x.

(b) What are the restrictions on x?

(c) Describe a function V that represents the volume of the box in terms of x.

(d) For what values of x (that is, original widths) will the volume of the box be between 600 and 800 cubic inches? Determine the answer using graphical methods, and give values to the nearest tenth of an inch.

In Exercises 9–12, use the formula for the height of a propelled object, given in this section.

9. *Height of a Propelled Rock*   A rock is thrown upward from ground level with an initial velocity of 90 feet per second. Let t represent the amount of time elapsed after it is thrown.

(a) Explain why t cannot be a negative number in this problem situation.

(b) Explain why $s_0 = 0$ in this problem situation.

(c) Give the function s that describes the height of the rock as a function of t.

(d) How high will the rock be 1.5 seconds after it is thrown?

(e) What is the maximum height attained by the rock? After how many seconds will this happen? Determine the answer analytically and support graphically.

(f) After how many seconds will the rock hit the ground? Determine analytically and support graphically.

10. *Height of a Toy Rocket*   A toy rocket is launched from the top of a building 50 feet tall at an initial velocity of 200 feet per second. Let t represent the amount of time elapsed after the launch.

(a) Express the height s as a function of the time t.

(b) Determine both analytically and graphically the time at which it reaches its highest point. How high will it be at that time?

(c) For what time interval will the rocket be more than 300 feet above ground level? Determine the answer using graphical methods, and give times to the nearest tenth of a second.

(d) After how many seconds will it hit the ground? Determine the answer both analytically and graphically.

11. *Height of a Propelled Ball*   **(a)** Determine graphically whether a ball thrown upward from ground level with an initial velocity of 150 feet per second will reach a height of 355 feet. If it will, determine the time(s) at which this happens. If it will not, explain why, using a graphical interpretation.

(b) Repeat part (a) for a ball thrown from a height of 30 feet with an initial velocity of 250 feet per second.

12. *Height of a Propelled Ball on the Moon*   An astronaut on the moon throws a baseball upward. The astronaut is 6 feet, 6 inches tall, and the initial velocity of the ball is 30 feet per second. The height of the ball is approximated by the function $s(t) = -2.7t^2 + 30t + 6.5$, where t is the number of seconds after the ball was thrown.

(a) After how many seconds is the ball 12 feet above the moon's surface?

(b) How many seconds will it take for the ball to return to the surface?

(c) The ball will never reach a height of 100 feet. How can this be determined analytically?

In Exercises 13 and 14, assume that x and y represent positive numbers.

13. *Minimum Value of an Expression*   Find the minimum value of the expression $x^2 + y^2$, if the sum of x and y is 16.

14. *Maximum Value of an Expression*   Find the maximum value of the expression xy^2, if $x + y^2 = 6$.

15. *Apartment Rental*   The manager of an 80-unit apartment complex knows from experience that at a rent of $400 per month, all units will be rented. However, for each increase of $20 in rent, he can expect one unit to be vacated. Let x represent the number of $20 increases over $400.

(a) Express, in terms of x, the number of apartments that will be rented if x increases of $20 are made. (For example, if 3 such increases are made, the number of apartments rented will be $80 - 3 = 77$.)

(b) Express the rent per apartment if x increases of $20 are made. (For example, if he increases rent $60 = 3 \times \$20$, the rent per apartment is $400 + 3(20) = 460$ dollars.)

(c) Describe a revenue function R in terms of x that will give the revenue generated as a function of the number of $20 increases.

(d) For what number of increases will the revenue be $37,500?

(e) What rent should he charge in order to achieve the maximum revenue?

16. *Seminar Fee*   When *Money Means Power* charges $600 for a seminar on management techniques, it attracts 1000 people. For each decrease of $20 in the charge, an additional 100 people will attend the seminar. Let x represent the number of $20 decreases in the charge.

(a) Describe a revenue function R that will give revenue generated as a function of the number of $20 decreases.

(b) Find the value of x that maximizes the revenue. What should the company change to maximize the revenue?

(c) What is the maximum revenue the company can generate?

17. *Path of a Frog's Leap*   A frog leaps from a stump 3 feet high and lands 4 feet from the base of the stump. We can consider the initial position of the frog to be at $(0, 3)$ and its landing position to be at $(4, 0)$. See the figure.

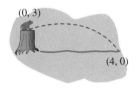

(0, 3)

(4, 0)

It is determined that the height of the frog as a function of its distance x from the base of the stump is given by the function $h(x) = -.5x^2 + 1.25x + 3$, where h is in feet.

(a) How high was the frog when its horizontal distance from the base of the stump was 2 feet?

(b) At what two times after it jumped from the base of the stump was the frog 3.25 feet above the ground?

(c) At what distance from the base of the stump did the frog reach its highest point?

(d) What was the maximum height reached by the frog?

18. *Path of a Frog's Leap*   Refer to Exercise 17. Suppose that the initial position of the frog is $(0, 4)$ and its landing position is $(6, 0)$. The height of the frog is given by $h(x) = -\frac{1}{3}x^2 + \frac{4}{3}x + 4$.

(a) After how many seconds did it reach its maximum height?

(b) What was this maximum height?

19. *Carbon Monoxide Exposure*   Carbon monoxide (CO) is a dangerous combustion product. It combines with the hemoglobin of the blood to form carboxyhemoglobin (COHb), which reduces transport of oxygen to tissues. A person's health is affected by both the concentration of carbon monoxide in the air and the exposure time. Smokers routinely have a 4% to 6% COHb level in their blood, which can cause symptoms such as blood flow alterations, visual impairment, and poorer vigilance ability. The quadratic function $T(x) = .00787x^2 - 1.528x + 75.89$ approximates the exposure time in hours necessary to reach this 4% to 6% level, where $50 \le x \le 100$ is the amount of carbon monoxide present in the air in parts per million (ppm). (*Source: Indoor Air Quality Environmental Information Handbook: Combustion Sources*, Report No. DOE/EV/10450-1, U.S. Department of Energy, 1985.)

(a) A kerosene heater or a room full of smokers is capable of producing 50 ppm of carbon monoxide. How long would it take for a nonsmoking person to start feeling the above symptoms?

(b) Find the carbon monoxide concentration necessary for a person to reach the 4% to 6% COHb level in 3 hours.

20. *Carbon Monoxide Exposure*   High concentrations of carbon monoxide can cause coma and possible death. The time required for a person to reach a COHb level capable of causing a coma can be approximated by the quadratic function $T(x) = .0002x^2 - .316x + 127.9$, where T is the exposure time in hours necessary to reach this level and $500 \le x \le 800$ is the amount of carbon monoxide in parts per million (ppm). (*Source: Indoor Air Quality Environmental Information Handbook: Combustion Sources*, Report No. DOE/EV/10450-1, U.S. Department of Energy, 1985.)

(a) What is the exposure time when $x = 600$ ppm?

(b) Estimate the concentration of CO necessary to produce a coma in 4 hours.

21. *Postsecondary Degrees in the United States*   In the United States, postsecondary degrees below bachelor's degrees earned during the years 1985 to 1990 can be approximated by the model $y = 5334.59x^2 - 23,024.00x + 617,519.11$, where $x = 0$ corresponds to 1985. Use the graph of this function in the window $[0, 6]$ by $[500,000, 800,000]$ to determine the year during this interval that the number of these degrees earned reached a minimum. To the nearest 10,000, what was this minimum number?

22. *Visitors to Federal Recreation Areas*   Based on information provided by the National Park Service, the number y of visitor hours in millions to Federal Recreation Areas between 1985 and 1992 can be approximated by the model $y = -21.99x^2 + 353.44x +$

6507.5, where $x = 0$ corresponds to 1985. Based on this model, in what year was the number of visitor hours about 7500 million?

23. *AIDS Cases in the United States*   The following table lists the total (cumulative) number of known deaths caused by AIDS in the United States up to 1993. (*Source:* U.S. Dept. of Health and Human Services, Centers for Disease Control and Prevention, *HIV/AIDS Surveillance*, March 1994.)

Year	Deaths
1982	620
1983	2122
1984	5600
1985	12,529
1986	24,550
1987	40,820
1988	61,723
1989	89,172
1990	119,821
1991	154,567
1992	191,508
1993	220,592

(a) Plot the data. Let $x = 0$ correspond to the year 1980.

(b) Find a quadratic function $f(x) = a(x - h)^2 + k$ that models the data. Use (2, 620) as the vertex and (13, 220,592) as the other point to determine a.

(c) Plot the data together with f in the same window. How well does f model the number of deaths caused by AIDS?

(d) Use the quadratic regression feature of a graphing calculator to determine the quadratic function g that provides the best fit for the data.

(e) Use the functions f and g to predict the total number of known deaths in the year 1994.

24. *Americans Over 100 Years of Age*   The table lists the number of Americans (in thousands) who are expected to be over 100 years old for selected years. (*Source:* U.S. Census Bureau.)

Year	Number
1994	50
1996	56
1998	65
2000	75
2002	94
2004	110

(a) Plot the data. Let $x = 4$ correspond to the year 1994, $x = 6$ correspond to 1996, and so on.

(b) Find a quadratic function $f(x) = a(x - h)^2 + k$ that models the data. Use (4, 50) as the vertex and (14, 110) as the other point to determine a.

(c) Plot the data together with f in the same window. How well does f model the number of Americans (in thousands) who are expected to be over 100 years old?

(d) Use the quadratic regression feature of a graphing calculator to determine the quadratic function g that provides the best fit for the data.

(e) Use the functions f and g to predict the number of Americans, in thousands, who will be over 100 years old in the year 2006.

25. *International Mutual Funds*   The number of international mutual funds has grown since 1984. The table shows the year and the number of such funds.

Year	Number of Funds
1984	13
1985	24
1986	35
1987	47
1988	59
1989	75
1990	95
1991	128
1992	138
1993	196
1994	288
1995	367

(*Source:* Investment Company Institute.)

(a) Plot the data. Let $x = 0$ represent 1984.

(b) Find a quadratic function of the form $f(x) = a(x - h)^2 + k$ that models the data. Use (0, 13) as the vertex and (11, 367) as the other data point to determine a.

(c) Use the quadratic regression capability of a graphing calculator to find the quadratic function g that best models the data.

(d) Graph both f and g over the data points.

(e) Which of the two models best predicts the actual 1996 number of international mutual funds, 485?

26. *Personal Computers in the United States*   The total numbers of multimedia personal computers in U.S. homes for the years 1992–1996 are given in the table. (Numbers are in millions.)

Year	Number of PCs (in millions)
1992	.47
1993	1.98
1994	3.63
1995	5.58
1996	7.98

(*Source:* Dataquest, Inc.)

(a) Plot the data. Let $x = 0$ represent 1992.

(b) Find a quadratic function of the form $f(x) = a(x - h)^2 + k$ that models the data. Use $(0, .47)$ as the vertex and $(4, 7.98)$ as the other data point to determine a.

(c) Use the quadratic regression capability of a graphing calculator to find the quadratic function g that best models the data.

(d) Graph both f and g over the data points.

(e) Which of the two models best predicts the actual 1997 number of 10.82 million?

3.5   HIGHER DEGREE POLYNOMIAL FUNCTIONS AND THEIR GRAPHS

Introduction ◆ Extrema ◆ Cubic Functions and Other Functions of Odd Degree ◆ Number of *x*-Intercepts (Real Zeros) ◆ Hidden Behavior ◆ Quartic Functions and Other Functions of Even Degree ◆ Comprehensive Graphs

INTRODUCTION

The linear functions discussed in Chapter 1 and the quadratic functions discussed earlier in this chapter are the simplest examples of a larger class of functions known as polynomial functions. The definition given in Section 3.2 is repeated here.

> **DEFINITION OF POLYNOMIAL FUNCTION**
>
> A **polynomial function of degree *n* in the variable *x*** is a function defined by
> $$P(x) = a_n x^n + a_{n-1} x^{n-1} + \cdots + a_1 x + a_0,$$
> where each a_i is a real number, $a_n \neq 0$, and n is a whole number.

We will see later in this section that the behavior of the graph of a polynomial function is due largely to the value of the coefficient a_n and the *parity* (that is, "evenness" or "oddness") of the exponent on the term of highest degree. For this reason, we will refer to a_n as the *leading coefficient*, and $a_n x^n$ as the *dominating term*. The term a_0 is the constant term of the polynomial function, and since $P(0) = a_0$, it is the *y*-intercept of the graph.

As we study the graphs of polynomial functions, we will use the following general properties (which are proved in higher courses):

1.   A polynomial function (unless otherwise specified) has domain $(-\infty, \infty)$.
2.   The graph of a polynomial function is a smooth, continuous curve with no sharp turns.

EXAMPLE 1   *Observing Some Sample Graphs of Polynomial Functions*   Figure 37 shows graphs of several typical polynomial functions generated on a graphing calculator. Each was graphed in the connected mode, in the specified window.   ◆

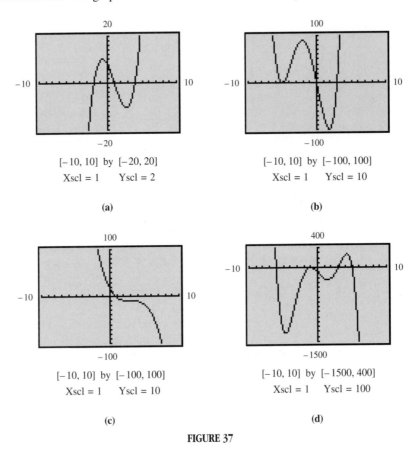

(a)
[−10, 10] by [−20, 20]
Xscl = 1   Yscl = 2

(b)
[−10, 10] by [−100, 100]
Xscl = 1   Yscl = 10

(c)
[−10, 10] by [−100, 100]
Xscl = 1   Yscl = 10

(d)
[−10, 10] by [−1500, 400]
Xscl = 1   Yscl = 100

FIGURE 37

EXTREMA

TECHNOLOGY NOTE

The feature described in the Technology Note in Section 3.2 is not limited to the vertex of a parabola. Local maxima and minima for polynomial functions of higher degree can also be obtained, provided an appropriate interval is designated.

Notice in Figure 37 that several of the graphs have **turning points** where the function changes from increasing to decreasing or vice versa. We first saw this when we studied quadratic functions, noticing that the vertex could be a maximum or minimum point on the graph. In general, the highest point at a "peak" is known as a **local maximum point**, and the lowest point at a "valley" is known as a **local minimum point**. Function values at such points are called **local maxima** (plural of maximum) and **local minima** (plural of minimum). Collectively, they are called **extrema** (plural of **extremum**). Figure 38 and the accompanying chart illustrate these ideas for typical graphs.

Refer again to Figure 38(a). Notice that the point P_2 is the absolute highest point on the graph, and the range of the function is $(-\infty, y_2]$. We call P_2 the **absolute maximum point** on the graph, and y_2 the **absolute maximum value** of the function. Because the y-values approach $-\infty$, this function has no absolute minimum value. On the other hand, because the graph in Figure 38(b) is that of a function with range $(-\infty, \infty)$, it has neither an absolute maximum nor an absolute minimum.

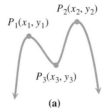
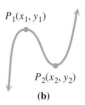

FIGURE 38

Extreme Point	Specifics	Extreme Point	Specifics
P_1	P_1 is a local maximum point. The function has a local maximum value of y_1 at $x = x_1$.	P_1	P_1 is a local maximum point. The function has a local maximum value of y_1 at $x = x_1$.
P_2	P_2 is a local maximum point. The function has a local maximum value of y_2 at $x = x_2$.	P_2	P_2 is a local minimum point. The function has a local minimum value y_2 at $x = x_2$.
P_3	P_3 is a local minimum point. The function has a local minimum value of y_3 at $x = x_3$.		

EXAMPLE 2 *Identifying Local and Absolute Extrema*   Consider the graphs in Figure 39.

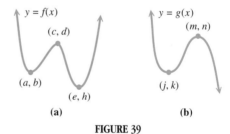

FIGURE 39

(a) Name and classify the local extrema of f.

(b) Name and classify the local extrema of g.

(c) Discuss absolute extrema for f and g.

SOLUTION

(a) The points (a, b) and (e, h) are local minimum points. The point (c, d) is a local maximum.

(b) The point (j, k) is a local minimum and the point (m, n) is a local maximum.

(c) The absolute minimum value of function f is the number h, since the range of f is $[h, \infty)$. It has no absolute maximum value. Function g has no absolute extrema, since its range is $(-\infty, \infty)$. ◆

We now state an important property of the graph of any polynomial function.

MAXIMUM NUMBER OF LOCAL EXTREMA

The maximum number of local extrema of the graph of a polynomial function of degree n is $n - 1$.

The property above can be applied to the polynomial functions we have previously studied. The degree of a linear function is 1, and since its graph is a straight line, it has no local extrema. A quadratic function is of degree 2, and since it has only 1 extreme point (its vertex), the property is satisfied. Notice that the property states that the *maximum* number of local extrema is $n - 1$ for a polynomial function of degree n. The graph may have fewer than $n - 1$ local extrema.

For Group Discussion

1. Recall the graph of the function $f(x) = x^3$ studied in Chapter 2. How many local extrema does it have? How does your answer support the property stated above?

2. Consider the polynomial function graph in Figure 40.

FIGURE 40

   a. What is the least possible degree of this function?
   b. Explain why this function cannot be of degree 4.

3. Repeat Item 2 for the polynomial function graph in Figure 41.

FIGURE 41

CUBIC FUNCTIONS AND OTHER FUNCTIONS OF ODD DEGREE

A polynomial function of the form $P(x) = ax^3 + bx^2 + cx + d$, $a \neq 0$, is a third degree, or **cubic function**. We studied the simplest cubic function, $f(x) = x^3$, in Chapter 2. If we graph a cubic function in an appropriate window, the graph will resemble, in general, one of the shapes shown in Figure 42.

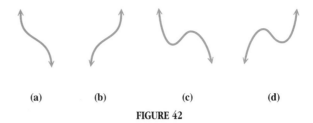

(a)   (b)   (c)   (d)

FIGURE 42

For Group Discussion

1. Without graphing, tell which of the shapes in Figure 42 depicts the general form of the graph of $f(x) = x^3$.
2. Using the concepts of reflection of graphs from Chapter 2, tell which of the shapes depicted in Figure 42 most closely resembles the graph of $g(x) = -x^3$.
3. Graph each of the following functions in the window specified, and determine which one of the shapes in Figure 42 the graph most closely resembles.
    a. $P(x) = x^3 + 5x^2 + 5x - 2$, $[-10, 10]$ by $[-10, 10]$
    b. $f(x) = -x^3 + 5x - 1$, $[-10, 10]$ by $[-10, 10]$
    c. $g(x) = x^3 + 3x^2 - 3x + 1$, $[-10, 10]$ by $[-10, 10]$
    d. $h(x) = -2x^3 - 6x^2 - 6x - 1$, $[-2, 1]$ by $[-1, 2]$

In Section 3.2, we studied end behavior of quadratic function graphs. The end behavior of the graph of a polynomial function is determined by the sign of the leading coefficient and the parity of the degree. A cubic function is an example of an odd degree polynomial function, and we can make the following observations about the end behavior of an odd degree polynomial function.

END BEHAVIOR OF ODD DEGREE POLYNOMIAL FUNCTIONS

Suppose that ax^n is the dominating term of a polynomial function P of *odd degree*.

1. If $a > 0$, then as $x \to \infty$, $P(x) \to \infty$, and as $x \to -\infty$, $P(x) \to -\infty$. Therefore, the end behavior of the graph is of the type shown in Figure 43(a), and is symbolized ⤴.

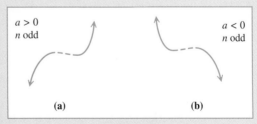

(a)   (b)

FIGURE 43

2. If $a < 0$, then as $x \to \infty$, $P(x) \to -\infty$, and as $x \to -\infty$, $P(x) \to \infty$. Therefore, the end behavior of the graph is of the type shown in Figure 43(b), and is symbolized ⤵.

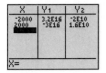

If we define Y_1 to be $f(x) = -x^5 + 2x - 1$ and Y_2 to be $g(x) = 2x^3 - x^2 + x - 1$, we see that for a negative x-value with large absolute value (-2000), $Y_1 > 0$. For a large positive x-value (2000), $Y_1 < 0$. This confirms the graphical interpretation in Example 3; the graph of $Y_1 = f(x)$ is the graph in Figure 44(b). A similar argument confirms that the graph of $Y_2 = g(x)$ is the one in Figure 44(a).

EXAMPLE 3   *Determining End Behavior Given the Defining Polynomial*
One of the graphs shown in Figure 44 is that of

$$f(x) = -x^5 + 2x - 1,$$

and the other is that of

$$g(x) = 2x^3 - x^2 + x - 1.$$

Use end behavior to determine which is which.

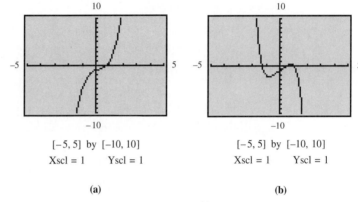

[−5, 5] by [−10, 10]
Xscl = 1   Yscl = 1

(a)

[−5, 5] by [−10, 10]
Xscl = 1   Yscl = 1

(b)

FIGURE 44

SOLUTION   The function $f(x) = -x^5 + 2x - 1$ is of odd degree, and the dominating term, $-x^5$, has coefficient -1 (a negative number). Therefore, it has end behavior ↖↘, and its graph is in part (b) of Figure 44. The function $g(x) = 2x^3 - x^2 + x - 1$ also is of odd degree, but the leading coefficient, 2, is positive. Therefore, it has end behavior ↙↗, and its graph is in part (a) of Figure 44.   ◆

NUMBER OF x-INTERCEPTS (REAL ZEROS)

Figure 45 shows how a cubic polynomial function may have one, two, or three x-intercepts.

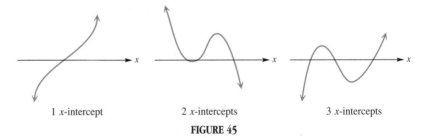

1 x-intercept   2 x-intercepts   3 x-intercepts

FIGURE 45

We saw earlier how a linear function may have no more than one x-intercept, and a quadratic function may have no more than two x-intercepts. Now we see that a cubic polynomial function may have no more than three x-intercepts. These observations suggest an important property of polynomial functions.

> ### NUMBER OF *x*-INTERCEPTS (REAL ZEROS) OF A POLYNOMIAL FUNCTION
>
> A polynomial function of degree *n* will have a maximum of *n* *x*-intercepts (real zeros).

EXAMPLE 4   *Determining x-Intercepts Graphically*   Find the *x*-intercepts (that is, the real zeros) of the polynomial function

$$P(x) = x^3 + 5x^2 + 5x - 2$$

graphically.

SOLUTION   If we use the standard window, we get the graph shown in Figure 46. Notice that it has three *x*-intercepts. By using graphing calculator capabilities, we find that the *x*-intercepts (real zeros) are -2, approximately -3.30, and approximately .30.

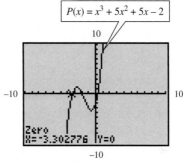

$$P(x) = x^3 + 5x^2 + 5x - 2$$

[–10, 10] by [–10, 10]

Xscl = 1   Yscl = 1

The other *x*-intercepts are –2 and
approximately .30.

FIGURE 46

It can be shown, using the methods of Section 3.6, that the two approximations have *exact* values

$$\frac{-3 - \sqrt{13}}{2} \quad \text{and} \quad \frac{-3 + \sqrt{13}}{2}. \qquad \blacklozenge$$

EXAMPLE 5   *Analyzing a Polynomial Function*   For the fifth degree polynomial function

$$P(x) = x^5 + 2x^4 - x^3 + x^2 - x - 4,$$

do each of the following.

(a)   Determine its domain.

(b)   Determine its range.

(c)   Use its graph to find approximations of its local extrema.

(d)   Use its graph to find its approximate and/or exact *x*-intercepts.

SOLUTION

(a)   Because it is a polynomial function, its domain is $(-\infty, \infty)$.

(b)   Because it is of odd degree, its range is $(-\infty, \infty)$.

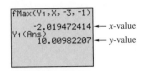

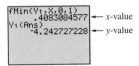

The results found graphically in Example 5(c) can also be found as shown in these two screens. Here, Y_1 is defined as $x^5 + 2x^4 - x^3 + x^2 - x - 4$, the maximum is in the interval $[-3, -1]$, and the minimum is in the interval $[0, 1]$.

(c)   The graph of P is shown in Figure 47, using a window of $[-5, 5]$ by $[-20, 50]$. It appears that there are only two extreme points. (Methods of calculus can verify that these are the only two.) Using the capabilities of a calculator, we find that the local maximum has approximate coordinates $(-2.02, 10.01)$, and the local minimum has approximate coordinates $(.41, -4.24)$.

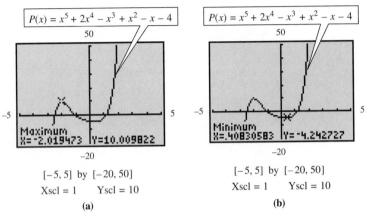

$[-5, 5]$ by $[-20, 50]$
Xscl = 1   Yscl = 10
(a)

$[-5, 5]$ by $[-20, 50]$
Xscl = 1   Yscl = 10
(b)

FIGURE 47

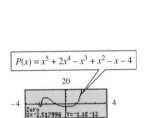

$[-4, 4]$ by $[-20, 20]$
Xscl = 1   Yscl = 5

The display and the table show the three zeros of P. See Example 5(d).

(d)   Once again, using the capabilities of a calculator, we find that the x-intercepts are -1 (exact), 1.14 (approximate), and -2.52 (approximate). The first of these can be verified analytically quite easily by evaluating $P(-1)$:

$$P(-1) = (-1)^5 + 2(-1)^4 - (-1)^3 + (-1)^2 - (-1) - 4$$
$$= -1 + 2(1) - (-1) + 1 + 1 - 4$$
$$= -1 + 2 + 1 + 1 + 1 - 4$$
$$= 0.$$

This again points out the fact that an x-intercept of the graph of a function is a real zero of the function.

Notice that this function has only three x-intercepts, and thus three real zeros. This supports the earlier statement that a polynomial function of degree n will have *at most* n x-intercepts. It may have fewer, as in this case.   ◆

HIDDEN BEHAVIOR

It is possible for the graph of a polynomial function to have a maximum or a minimum point that is not apparent in a particular window. This is an example of **hidden behavior**.

EXAMPLE 6   *Examining Hidden Behavior*   Figure 48 shows the graph of $P(x) = x^3 - 2x^2 + x - 2$ in the standard viewing window. Make a conjecture concerning possible hidden behavior, and verify the conjecture.

SOLUTION   Based on the view in Figure 48, it seems that there may be behavior in the domain interval $[0, 2]$ that is not apparent in the given window. (The horizontal line segment in this window leads us to suspect this.) By changing the window to $[-2.5, 2.5]$ by $[-4.5, .5]$, we see that there are two extrema there. The local maximum point, as seen in Figure 49, has approximate coordinates $(.33, -1.85)$. There is also a local minimum point at $(1, -2)$.   ◆

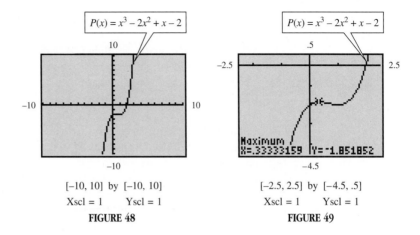

$$P(x) = x^3 - 2x^2 + x - 2$$

[−10, 10] by [−10, 10]

Xscl = 1   Yscl = 1

FIGURE 48

$$P(x) = x^3 - 2x^2 + x - 2$$

Maximum
X=.33333159   Y=-1.851852

[−2.5, 2.5] by [−4.5, .5]

Xscl = 1   Yscl = 1

FIGURE 49

QUARTIC FUNCTIONS AND OTHER FUNCTIONS OF EVEN DEGREE

A polynomial function of the form $P(x) = ax^4 + bx^3 + cx^2 + dx + e$, $a \neq 0$, is a fourth degree, or **quartic function**. The simplest quartic function, $P(x) = x^4$, is graphed in Figure 50. Notice that it resembles the graph of the squaring function; however, it is not actually a parabola (based on a formal definition of parabola, presented later in this text).

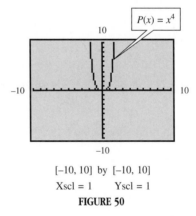

$$P(x) = x^4$$

[−10, 10] by [−10, 10]

Xscl = 1   Yscl = 1

FIGURE 50

If we graph a quartic function in an appropriate window, the graph will resemble, in general, one of the shapes shown in Figure 51. The dashed portions in C and D indicate that there may be irregular, but smooth, behavior.

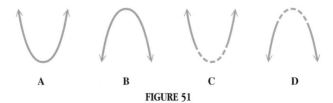

A   B   C   D

FIGURE 51

For Group Discussion

1. Which one of the shapes in Figure 51 depicts the general form of the graphs of $f(x) = x^2$ and $g(x) = x^4$?

2. Using the concepts of reflection of graphs from Chapter 2, tell which of the shapes depicted in Figure 51 most closely resembles the graph of $h(x) = -x^4$.

3. Graph each of the following functions in the window specified, and determine which one of the shapes in Figure 51 the graph most closely resembles.

    a. $y = 3x^4 + x - 2$, $[-10, 10]$ by $[-10, 10]$

    b. $y = -2x^4 - x^3 + x - 3$, $[-10, 10]$ by $[-10, 10]$

    c. $y = x^4 - 5x^2 + 4$, $[-10, 10]$ by $[-10, 10]$

    d. $y = -x^4 + 12x^3$, $[-15, 15]$ by $[-5000, 2500]$

A quartic function is an example of an even degree polynomial function (as was a quadratic function). We can see from Figure 51 that the end behavior of quartic functions corresponds to the type of end behavior exhibited by quadratic functions. This type of end behavior is typical of even degree polynomial functions, and we now make the following observations.

END BEHAVIOR OF EVEN DEGREE POLYNOMIAL FUNCTIONS

Suppose that ax^n is the dominating term of a polynomial function P of *even degree*.

1. If $a > 0$, then as $|x| \to \infty$, $P(x) \to \infty$. Therefore, the end behavior of the graph is of the type shown in Figure 52, and is symbolized $\cup$.

2. If $a < 0$, then as $|x| \to \infty$, $P(x) \to -\infty$. Therefore, the end behavior of the graph is of the type shown in Figure 53, and is symbolized $\cap$.

FIGURE 52

FIGURE 53

EXAMPLE 7 *Determining End Behavior Given the Defining Polynomial*
The graphs shown in Figure 54 are of the functions

$$f(x) = x^4 - x^2 + 5x - 4 \qquad g(x) = -x^6 + x^2 - 3x - 4$$

$$h(x) = 3x^3 - x^2 + 2x - 4 \quad \text{and} \quad k(x) = -x^7 + x - 4.$$

Based on dominating term analysis, determine which graph is which.

SOLUTION Because f is of even degree with positive leading coefficient, its graph is in C. Because g is of even degree with negative leading coefficient, its graph is in A. Using the ideas presented earlier (see Example 3), the graph of h is in B and the graph of k is in D.   ◆

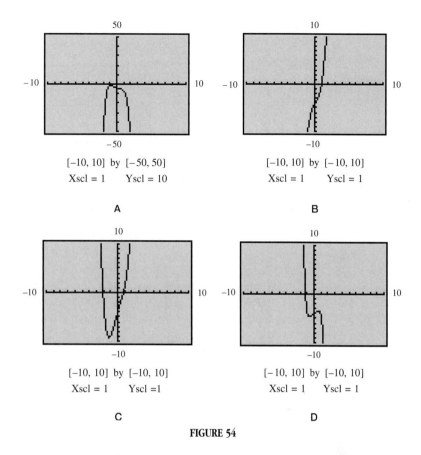

[−10, 10] by [−50, 50]
Xscl = 1   Yscl = 10

A

[−10, 10] by [−10, 10]
Xscl = 1   Yscl = 1

B

[−10, 10] by [−10, 10]
Xscl = 1   Yscl =1

C

[−10, 10] by [−10, 10]
Xscl = 1   Yscl = 1

D

FIGURE 54

EXAMPLE 8   *Analyzing a Polynomial Function*   For the fourth degree polynomial function

$$P(x) = x^4 + 2x^3 - 15x^2 - 12x + 36,$$

do each of the following.

(a) Determine its domain.

(b) Use its graph to find its local extrema. Does it have an absolute minimum? What is the range of the function?

(c) Use its graph to find its *x*-intercepts.

SOLUTION

(a) Because it is a polynomial function, its domain is $(-\infty, \infty)$.

(b) After experimenting with various windows, we see that a window of $[-6, 6]$ by $[-80, 50]$ provides a view of the extreme points, as well as all intercepts. See Figure 55 on the next page. Using the capabilities of a calculator, we find that the two local minimum points have approximate coordinates $(-3.43, -41.61)$ and $(2.31, -18.63)$, and the local maximum has approximate coordinates $(-.38, 38.31)$. Because the end behavior is ⌣, and the point $(-3.43, -41.61)$ is the lowest point on the graph, the absolute minimum value of the function is approximately -41.61, and, therefore, the range is approximately $[-41.61, \infty)$.

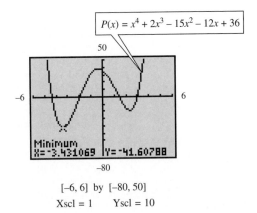

$$[-6, 6] \text{ by } [-80, 50]$$
$$Xscl = 1 \qquad Yscl = 10$$

The two other extreme points are $(-.38, 38.31)$ and $(2.31, -18.63)$.

FIGURE 55

(c) This fourth degree function has the maximum number of x-intercepts possible (four). Using the calculator capabilities, we find that two exact values for the x-intercepts are -2 and 3, while, to the nearest hundredth, the other two are -4.37 and 1.37. (Using concepts to follow in this chapter, we can show that these latter two are approximations for the *exact* values

$$\frac{-3 - \sqrt{33}}{2} \qquad \text{and} \qquad \frac{-3 + \sqrt{33}}{2}.) \qquad \blacklozenge$$

COMPREHENSIVE GRAPHS

The most important features of the graph of a polynomial function are its intercepts, its extrema, and its end behavior. For this reason, we give the following criteria for the comprehensive graph of a polynomial function.

COMPREHENSIVE GRAPH CRITERIA FOR A POLYNOMIAL FUNCTION

A comprehensive graph of a polynomial function will exhibit the following features.
a. all x-intercepts (if any)
b. the y-intercept
c. all extreme points (if any)
d. enough of the graph to exhibit the correct end behavior

EXAMPLE 9   *Determining an Appropriate Window for a Comprehensive Graph*   The window $[-1.25, 1.25]$ by $[-400, 50]$ is used in Figure 56 to give a view of the graph of

$$P(x) = x^6 - 36x^4 + 288x^2 - 256.$$

Is this a comprehensive graph?

SOLUTION   Since the function is of even degree and the dominating term has positive coefficient, the end behavior seems to be correct. The y-intercept, -256, is shown, and two x-intercepts are shown. One local minimum is shown. This function may, however, have up to six x-intercepts, since it is of degree 6. By experimenting with other viewing windows, we see that a window of $[-8, 8]$ by $[-1000, 600]$ shows a total of *five* local extrema, and four *x*-intercepts that were not apparent in the earlier figure. See Figure 57. Since there can be no more than five local extrema, this second view (and *not* the first view in Figure 56) gives us a comprehensive graph. ◆

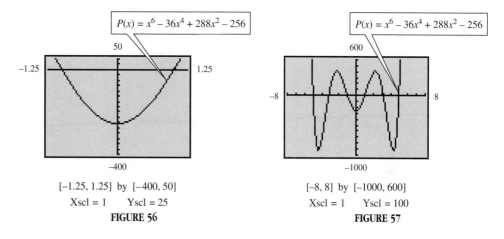

$P(x) = x^6 - 36x^4 + 288x^2 - 256$

$[-1.25, 1.25]$ by $[-400, 50]$
Xscl = 1   Yscl = 25
FIGURE 56

$P(x) = x^6 - 36x^4 + 288x^2 - 256$

$[-8, 8]$ by $[-1000, 600]$
Xscl = 1   Yscl = 100
FIGURE 57

For Group Discussion How does Example 9 illustrate the warning "Don't always believe what you see"? How does it illustrate the shortcomings of learning technology without learning mathematical concepts?

WHAT WENT WRONG ? After studying the concepts of this section, a student graphed $y = .045x^4 - 2x^2 + 2$ in the decimal window of a popular model of graphing calculator, $[-4.7, 4.7]$ by $[-3.1, 3.1]$. Because the polynomial has even degree and positive leading coefficient, she expected to find end behavior ⌣. However, this graph indicates ⌢ as end behavior.

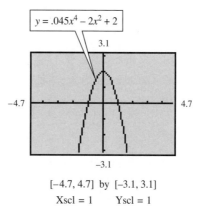

$y = .045x^4 - 2x^2 + 2$

$[-4.7, 4.7]$ by $[-3.1, 3.1]$
Xscl = 1   Yscl = 1

1. What went wrong?

2. Is there a way to graph this function so that the correct end behavior is apparent?

3.5   EXERCISES   ◆   🖥   《📼》   **Tape 4**

Figure 37 is repeated below. The graphs are of $y = x^3 - 3x^2 - 6x + 8$, $y = x^4 + 7x^3 - 5x^2 - 75x$, $y = -x^3 + 9x^2 - 27x + 17$, *and* $y = -x^5 + 36x^3 - 22x^2 - 147x - 90$, *but not necessarily in that order. Assuming that each is a comprehensive graph as specified in this section, answer each of the questions in Exercises 1–10.*

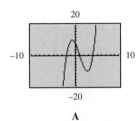

A

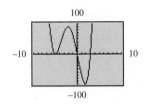

B

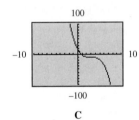

C

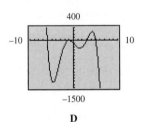

D

1. Which one of the graphs is that of $y = x^3 - 3x^2 - 6x + 8$?

2. Which one of the graphs is that of $y = x^4 + 7x^3 - 5x^2 - 75x$?

3. How many real zeros does the graph in C have?

4. Which one of C and D is the graph of $y = -x^3 + 9x^2 - 27x + 17$? (*Hint:* Look at the y-intercept.)

5. Which of the graphs cannot be that of a cubic polynomial function?

6. How many positive real zeros does the function graphed in D have?

7. How many negative real zeros does the function graphed in A have?

8. Is the absolute minimum value of the function graphed in B a positive number or a negative number?

9. Which one of the graphs is that of a function whose range is *not* $(-\infty, \infty)$?

10. One of the following is an approximation for the local maximum point of the graph in Figure A. Which one is it?

    (a) $(.73, 10.39)$ **(b)** $(-.73, 10.39)$ **(c)** $(-.73, -10.39)$ **(d)** $(.73, -10.39)$

Use an end behavior diagram (⌣, ⌢, ↘, *or* ↗ *) to describe the end behavior of the given function. Then, verify your answer by graphing the function on your calculator.*

11. $P(x) = \sqrt{5}x^3 + 2x^2 - 3x + 4$ 12. $P(x) = -\sqrt{7}x^3 - 4x^2 + 2x - 1$ 13. $P(x) = -\pi x^5 + 3x^2 - 1$

14. $P(x) = \pi x^7 - x^5 + x - 1$ 15. $P(x) = 2.74x^4 - 3x^2 + x - 2$ 16. $P(x) = \sqrt{6}x^6 - x^5 + 2x - 2$

17. $P(x) = -\pi x^6 + x^5 - x^4 - x + 3$ 18. $P(x) = -2.84x^4 - 3.2x^3 + x^2 - x + 3$

The functions in Exercises 19–22 are graphed in Figures A–D. Use end behavior and analysis of dominating term to match each equation with the correct graph.

19. $f(x) = 2x^3 + x^2 - x + 3$

20. $g(x) = -2x^3 - x + 3$

21. $h(x) = -2x^4 + x^3 - 2x^2 + x + 3$

22. $k(x) = 2x^4 - x^3 - 2x^2 + 3x + 3$

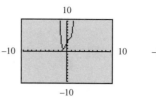

A

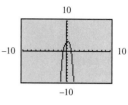

B

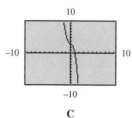

C

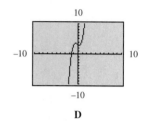

D

23. Using a window of $[-1, 1]$ by $[-1, 1]$, graph the odd degree polynomial functions

$$y = x, \quad y = x^3, \quad \text{and} \quad y = x^5.$$

Describe the behavior of these functions relative to each other. Predict the behavior of the graph of $y = x^7$ in the same window, and then graph it to support your prediction.

24. Repeat the activity of Exercise 23 for the even degree polynomial functions

$$y = x^2, \quad y = x^4, \quad \text{and} \quad y = x^6.$$

Predict the behavior of the graph of $y = x^8$ in the same window, and then graph it to support your prediction.

25. The graphs of $f(x) = x^n$ for $n = 3, 5, 7, \ldots$ all resemble each other. Describe how they are the same. As n gets larger, what happens to the graph?

26. Repeat Exercise 25 for $f(x) = x^n$, where $n = 2, 4, 6, \ldots$.

For the functions defined in Exercises 27–36, find a comprehensive graph and do each of the following. Use your graphing calculator to its maximum capability.

(a) Determine the domain.

(b) Determine all local minimum points, and tell if any is an absolute minimum point. (Approximate coordinates to the nearest hundredth.)

(c) Determine all local maximum points, and tell if any is an absolute maximum point. (Approximate coordinates to the nearest hundredth.)

(d) Determine the range. (If an approximation is necessary, give it to the nearest hundredth.)

(e) Determine all intercepts. For each function, there is at least one x-intercept that is an integer. For those that are not integers, give an approximation to the nearest hundredth. You should determine the y-intercept analytically.

(f) Give the interval(s) over which the function is increasing.

(g) Give the interval(s) over which the function is decreasing.

27. $y = x^3 - 4x^2 + x + 6$

28. $y = x^3 + x^2 - 22x - 40$

29. $y = -2x^3 - 14x^2 + 2x + 84$

30. $y = -3x^3 + 6x^2 + 39x - 60$

31. $y = x^5 + 4x^4 - 3x^3 - 17x^2 + 6x + 9$

32. $y = -2x^5 + 7x^4 + x^3 - 20x^2 + 4x + 16$

33. $y = 2x^4 + 3x^3 - 17x^2 - 6x - 72$

34. $y = 3x^4 - 33x^2 + 54$

35. $y = -x^6 + 24x^4 - 144x^2 + 256$

36. $y = -3x^6 + 2x^5 + 9x^4 - 8x^3 + 11x^2 + 4$

RELATING CONCEPTS (EXERCISES 37–40)

In Section 3.2 we informally introduced the concept of concavity of a graph. (A formal discussion of concavity requires concepts from calculus.) A polynomial function graph may change concavity several times throughout its domain, or its concavity may remain the same. Use any method you wish to answer the following questions about concavity.

37. For a function of the form $P(x) = x^n$, where n is even, describe the concavity of the graph.

38. For a function of the form $P(x) = x^n$, where n is odd, discuss the concavity of the graph.

39. Graph the following function in the window $[-4, 4]$ by $[-20, 20]$, and discuss the concavity of the graph: $y = x^3 + x^2 + x$.

40. Explain why an even degree polynomial function with negative leading coefficient must be concave down for some interval of its domain.

Determine a window that will provide a comprehensive graph (as defined in this section) of each of the following polynomial functions. (In each case, there are many possible such windows.)

41. $y = 4x^5 - x^3 + x^2 + 3x - 16$

42. $y = 3x^5 - x^4 + 12x^2 - 25$

43. $y = 2.9x^3 - 37x^2 + 28x - 143$

44. $y = -5.9x^3 + 16x^2 - 120$

45. $y = \pi x^4 - 13x^2 + 84$

46. $y = 2\pi x^4 - 12x^2 + 100$

Each of the following functions is graphed in a window such that hidden behavior is not evident. Experiment with various windows to locate the extrema of the function.

47. $y = \dfrac{1}{3}x^3 - \dfrac{5}{2}x^2 + 6x - 1$

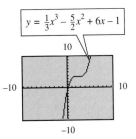

48. $y = -\dfrac{1}{3}x^3 - \dfrac{9}{2}x^2 - 20x - \dfrac{59}{3}$

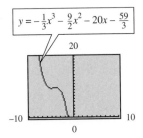

49. $y = -x^3 - 11x^2 - 40x - 50$

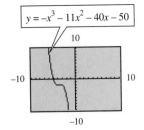

50. $y = 2x^3 - 3.3x^2 + 1.8x + 3$

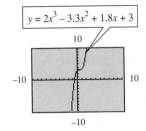

In Exercises 51–54, it is not apparent from the standard viewing window whether the graph of the quadratic function intersects the x-axis twice, once, or not at all. This is another example of hidden behavior. Experiment with various windows to determine the number of x-intercepts. If there are x-intercepts, give their values to the nearest hundredth.

51. $y = x^2 - 4.25x + 4.515$

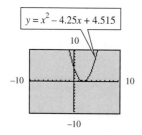

52. $y = x^2 + 6.95x + 12.07$

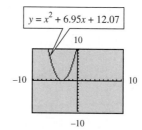

53. $y = -x^2 + 6.5x - 10.60$

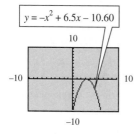

54. $y = -2x^2 + .2x - .15$

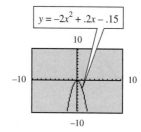

RELATING CONCEPTS (EXERCISES 55–58)

The concepts of stretching, translating, and reflecting graphs presented in Sections 2.2 and 2.3 can be applied to polynomial functions of the form $P(x) = x^n$. For example, the graph of $y = -2(x + 4)^4 - 6$ can be obtained from the graph of $y = x^4$ by shifting 4 units to the left, stretching vertically by a factor of 2, reflecting across the x-axis, and shifting down 6 units. Thus, the graph should resemble the graph at the right.

If we expand the expression $-2(x + 4)^4 - 6$ algebraically, we get

$$-2x^4 - 32x^3 - 192x^2 - 512x - 518.$$

Thus, the graph of $y = -2(x + 4)^4 - 6$ is the same as that of

$$y = -2x^4 - 32x^3 - 192x^2 - 512x - 518.$$

In Exercises 55–58, two forms of the same polynomial function are given. Sketch by hand the general shape of the graph of the function, using the concepts of Chapter 2, and describe the transformations. Then, support your answer by graphing it on your calculator in a suitable window.

55. $y = 2(x + 3)^4 - 7$
    $y = 2x^4 + 24x^3 + 108x^2 + 216x + 155$

56. $y = -3(x + 1)^4 + 12$
    $y = -3x^4 - 12x^3 - 18x^2 - 12x + 9$

57. $y = -3(x - 1)^3 + 12$
    $y = -3x^3 + 9x^2 - 9x + 15$

58. $y = .5(x - 1)^5 + 13$
    $y = .5x^5 - 2.5x^4 + 5x^3 - 5x^2 + 2.5x + 12.5$

Without graphing, answer true or false to each of the following. Then, support your answer, if you wish, by graphing.

59. The function defined by $f(x) = x^3 + 2x^2 - 4x + 3$ has four real zeros.

60. The function defined by $f(x) = x^3 + 3x^2 + 3x + 1$ must have at least one real zero.

61. If a polynomial function of even degree has a negative leading coefficient and a positive y-intercept, it must have at least two real zeros.

62. The function defined by $f(x) = 3x^4 + 5$ has no real zeros.

63. The function defined by $f(x) = -3x^4 + 5$ has two real zeros.

64. The graph of the cubic function defined by $f(x) = x^3 - 3x^2 + 3x - 1 = (x - 1)^3$ has exactly one x-intercept.

65. A fifth degree polynomial function cannot have a single real zero.

66. An even degree polynomial function must have at least one real zero.

RELATING CONCEPTS (EXERCISES 67–70)

67. If the graph of a polynomial function P satisfies the property $P(x) = P(-x)$ for all real numbers x, then the graph of P is symmetric with respect to the y-axis, and P is called an even function. (These ideas were first presented in Sections 2.1 and 2.3.) Consider the polynomial function defined by

$$P(x) = x^4 - 8x^2 - 9.$$

  (a) Show *analytically* that the graph of P is symmetric with respect to the y-axis.

  (b) Graph the function in the window $[-10, 0]$ by $[-30, 10]$, and use your calculator to find the single *negative* x-intercept of the graph.

  (c) Verify your result from part (b) analytically.

  (d) Now, using the results from parts (a)–(c), predict the value of the *positive* x-intercept of the graph, and then graph in the window $[-10, 10]$ by $[-30, 10]$ to support your result.

68. In Section 2.3, we learned that graphs of functions may be shifted horizontally. For example, the graph of $y = P(x + 4)$ may be obtained from the graph of $y = P(x)$ by shifting the graph of P 4 units to the left.

  (a) Graph the polynomial function defined by $P(x) = x^3 - 3x^2 + x - 5$ in the standard viewing window.

  (b) Predict what the graph of $P(x - 4) = (x - 4)^3 - 3(x - 4)^2 + (x - 4) - 5$ will look like when compared to the graph of $y = P(x)$.

  (c) Support your prediction from part (b) by graphing $y_2 = P(x - 4)$ in the same window as $y_1 = P(x)$.

69. Use the capabilities of your calculator to find the intervals over which the function defined by $y = 2x^3 - 5x^2 - 3x + 2$

  (a) is increasing.   **(b)** is decreasing.

Express endpoints of intervals as approximations to the nearest hundredth.

70. (a) Graph the functions defined by $f(x) = \frac{2}{3}x^3 + \frac{1}{2}x^2 - 21x + 4$ and $g(x) = 2x^2 + x - 21$ in the window $[-8, 8]$ by $[-50, 75]$.

  (b) Use the capabilities of your calculator to show that the solutions of $g(x) = 0$ are the x-coordinates of the local extrema of f.

  (c) Confirm that the intervals over which f is increasing correspond to the intervals over which g is non-negative.

  (d) Confirm that the interval over which f is decreasing corresponds to the interval over which g is non-positive.

(The ideas of this exercise relate to concepts from the study of calculus. The function g is called the *derivative* of f.)

3.6 TOPICS IN THE THEORY OF POLYNOMIAL FUNCTIONS (I)

The Intermediate Value Theorem and the Root Location Theorem ◆ Division of Polynomials and Synthetic Division ◆ The Remainder and Factor Theorems

In this section and the next, we will examine some important topics in the theory of polynomial functions. These topics will complement the graphical work done in Section 3.5, and will also help prepare us for the work to follow in Section 3.8 (equations, inequalities, and applications of polynomial functions).

THE INTERMEDIATE VALUE THEOREM AND THE ROOT LOCATION THEOREM

Two theorems presented here apply to the zeros of every polynomial function with real coefficients. The first theorem uses the fact that graphs of polynomial functions are continuous curves, with no gaps or sudden jumps. The proof requires advanced methods.

> **THE INTERMEDIATE VALUE THEOREM FOR POLYNOMIAL FUNCTIONS**
>
> If $P(x)$ defines a polynomial function, and if $P(a) \neq P(b)$, then P takes on every value between $P(a)$ and $P(b)$.

A result of this theorem is the root location theorem, which is used to determine the locations of x-intercepts of graphs of polynomial functions.

> **ROOT LOCATION THEOREM FOR POLYNOMIAL FUNCTIONS**
>
> Suppose that a polynomial function P is defined in such a way that for real numbers a and b, $P(a)$ and $P(b)$ differ in sign. Then there exists at least one real number c between a and b such that $P(c) = 0$.

To see how the root location theorem is applied, note that in Figure 58, $P(a)$ and $P(b)$ are opposite in sign, so 0 is between $P(a)$ and $P(b)$. Then, by the root location theorem, there must be a number c between a and b such that $P(c) = 0$.

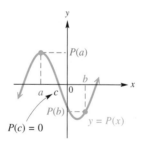

FIGURE 58

EXAMPLE 1   *Applying the Root Location Theorem*

(a) Show *analytically* that the polynomial function defined by

$$P(x) = x^3 - 2x^2 - x + 1$$

has a real zero between 2 and 3.

(b) Support the result of part (a) graphically.

SOLUTION

(a)   ANALYTIC   Begin by evaluating $P(2)$ and $P(3)$.

$$P(2) = 2^3 - 2(2)^2 - 2 + 1 = -1$$
$$P(3) = 3^3 - 2(3)^2 - 3 + 1 = 7$$

Since $P(2) = -1$ and $P(3) = 7$ differ in sign, the root location theorem assures us that there is a real zero between 2 and 3.

(b)   GRAPHICAL   If we graph $P(x) = x^3 - 2x^2 - x + 1$ in the standard viewing window, we see that there is an x-intercept between 2 and 3, confirming our result that P has a zero between 2 and 3. See Figure 59. (P has two other real zeros, but this does not affect our discussion here.) Using the root-finding capabilities of the calculator, we can determine that, to the nearest hundredth, this zero is 2.25.   ◆

A zero of the function discussed in Example 1 lies between 2.246 and 2.247, since there is a sign change for the function values.

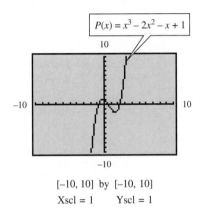

[−10, 10] by [−10, 10]
Xscl = 1   Yscl = 1

There is an x-intercept between 2 and 3.

FIGURE 59

Caution   Be careful how you interpret the root location theorem. If $P(a)$ and $P(b)$ are *not* opposite in sign, it does not necessarily mean that there is no zero between a and b. For example, in Figure 60, $P(a)$ and $P(b)$ are both negative, but −3 and −1, which are between a and b, are zeros of P.

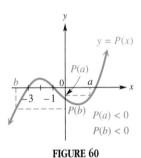

FIGURE 60

DIVISION OF POLYNOMIALS AND SYNTHETIC DIVISION

As we shall see later in this section, it is important to be able to determine whether a binomial of the form $x - k$ is a factor of a polynomial. Just as we can use the old-fashioned "long division method" to determine whether one whole number is a factor of another, we can also use it to determine whether one polynomial is a factor of another. Example 2 illustrates the method of long division of a polynomial by a binomial.

EXAMPLE 2   *Dividing a Polynomial by a Binomial*   Divide the polynomial $3x^3 - 2x + 5$ by $x - 3$. Determine the quotient and the remainder.

SOLUTION   In performing this division, we make sure that the powers of the variable in the dividend ($3x^3 - 2x + 5$) are descending, which they are. We also insert the term $0x^2$ to act as a placeholder. We then follow the same procedure as for long division of whole numbers.

Missing term

$$x - 3 \overline{)3x^3 + 0x^2 - 2x + 5}$$

Start with $\frac{3x^3}{x} = 3x^2$.

$$\begin{array}{r} 3x^2 \\ x - 3 \overline{)3x^3 + 0x^2 - 2x + 5} \\ 3x^3 - 9x^2 \end{array}$$ $\leftarrow \frac{3x^3}{x} = 3x^2$
$\leftarrow 3x^2(x-3)$

Subtract by changing the signs on $3x^3 - 9x^2$ and adding.

$$\begin{array}{r} 3x^2 \\ x - 3 \overline{)3x^3 + 0x^2 - 2x + 5} \\ 3x^3 - 9x^2 \\ \hline 9x^2 \end{array}$$ $\leftarrow$ Subtract.

Bring down the next term.

$$\begin{array}{r} 3x^2 \\ x - 3 \overline{)3x^3 + 0x^2 - 2x + 5} \\ 3x^3 - 9x^2 \\ \hline 9x^2 - 2x \end{array}$$ $\leftarrow$ Bring down $-2x$.

In the next step, $\frac{9x^2}{x} = 9x$.

$$\begin{array}{r} 3x^2 + 9x \\ x - 3 \overline{)3x^3 + 0x^2 - 2x + 5} \\ 3x^3 - 9x^2 \\ \hline 9x^2 - 2x \\ 9x^2 - 27x \\ \hline 25x + 5 \end{array}$$ $\leftarrow \frac{9x^2}{x} = 9x$
$\leftarrow 9x(x-3)$
$\leftarrow$ Subtract and bring down 5.

Finally, $\frac{25x}{x} = 25$.

$$\begin{array}{r} 3x^2 + 9x + 25 \\ x - 3 \overline{)3x^3 + 0x^2 - 2x + 5} \\ 3x^3 - 9x^2 \\ \hline 9x^2 - 2x \\ 9x^2 - 27x \\ \hline 25x + 5 \\ 25x - 75 \\ \hline 80 \end{array}$$ $\leftarrow \frac{25x}{x} = 25$
$\leftarrow 25(x-3)$
$\leftarrow$ Subtract.

The quotient is $3x^2 + 9x + 25$ with a remainder of 80. ◆

The division shown in Example 2 allows us to make several observations about polynomial division in general. Notice that we divided a cubic polynomial (degree 3) by a linear polynomial (degree 1) and obtained a quadratic polynomial quotient (degree 2). Notice also that $3 - 1 = 2$, so we observe that the degree of the quotient polynomial is found by subtracting the degree of the divisor from the degree of the dividend. Also, since the remainder is a nonzero constant (80), we can write it as the numerator of a fraction with denominator $x - 3$ to express the fractional part of the quotient.

Dividend $\rightarrow \dfrac{3x^3 - 2x + 5}{x - 3}$ $\leftarrow$ Divisor $= 3x^2 + 9x + 25 + \dfrac{80}{x - 3}$ $\leftarrow$ Remainder $\leftarrow$ Divisor

Quotient polynomial

Fractional part of the quotient

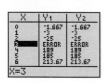

A table supports the result of Example 2. Why is there an error message for $x = 3$?

In general, the following rules apply to the division of a polynomial by a binomial.

DIVISION OF A POLYNOMIAL BY $x - k$

1. If the degree n polynomial $P(x)$ is divided by $x - k$, the quotient polynomial, $Q(x)$, has degree $n - 1$.

2. The remainder R is a constant (and may be 0). The complete quotient for $\frac{P(x)}{x - k}$ may be written as follows:

$$\frac{P(x)}{x - k} = Q(x) + \frac{R}{x - k}.$$

The procedure of long division of a polynomial by a binomial of the form $x - k$ can be condensed using **synthetic division**. To see how synthetic division works in the case of the division performed in Example 2, observe the following.

$$
\begin{array}{r}
3x^2 + 9x + 25 \\
x - 3\overline{)3x^3 + 0x^2 - 2x + 5} \\
\underline{3x^3 - 9x^2} \\
9x^2 - 2x \\
\underline{9x^2 - 27x} \\
25x + 5 \\
\underline{25x - 75} \\
80
\end{array}
\qquad
\begin{array}{r}
3 \quad\; 9 \quad\; 25 \\
1 - 3\overline{)3 \quad\;\; 0 \quad -2 \quad\; 5} \\
(3) \; -9 \\
9 \quad -2 \\
(9) \; -27 \\
25 \quad\; 5 \\
(25) \; -75 \\
80
\end{array}
$$

On the right, exactly the same division is shown written without the variables. All the numbers in parentheses on the right are repetitions of the numbers directly above them, so they may be omitted, as shown on the left below.

$$
\begin{array}{r}
3 \quad\; 9 \quad\;\; 25 \\
1 - 3\overline{)3 \quad\;\; 0 \quad -2 \quad\; 5} \\
-9 \\
9 \quad (-2) \\
-27 \\
25 \quad (5) \\
-75 \\
80
\end{array}
\qquad
\begin{array}{r}
3 \quad\; 9 \quad\;\; 25 \\
1 - 3\overline{)3 \quad\;\; 0 \quad -2 \quad\; 5} \\
-9 \\
9 \\
-27 \\
25 \\
-75 \\
80
\end{array}
$$

The numbers in parentheses on the left are again repetitions of the numbers directly above them; they too may be omitted, as shown on the right above.

Now the problem can be condensed. If the 3 in the dividend is brought down to the beginning of the bottom row, the top row can be omitted, since it duplicates the bottom row.

$$
\begin{array}{r}
1 - 3\overline{)3 \quad\;\; 0 \quad -2 \quad\;\; 5} \\
-9 \quad -27 \quad -75 \\
\hline
3 \quad\;\; 9 \quad\;\; 25 \quad\;\; 80
\end{array}
$$

Finally, the 1 at the upper left can be omitted. Also, to simplify the arithmetic, subtraction in the second row is replaced by addition. We compensate for this by changing the -3 at upper left to its additive inverse 3. The result of doing all this is now shown.

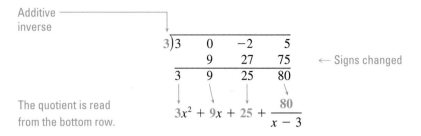

Additive inverse

← Signs changed

The quotient is read from the bottom row.

$$3x^2 + 9x + 25 + \frac{80}{x - 3}$$

EXAMPLE 3   *Using Synthetic Division*   Use synthetic division to divide $5x^3 - 6x^2 - 28x + 8$ by $x + 2$.

SOLUTION   To use synthetic division, the divisor must be of the form $x - k$. Writing $x + 2$ as $x - (-2)$ shows that the value of k here is -2. We begin by writing

$$-2)\overline{5 \quad -6 \quad -28 \quad 8}.$$

Next, bring down the 5.

$$\begin{array}{r} -2)\overline{5 \quad -6 \quad -28 \quad 8} \\ \hline 5 \end{array}$$

Now, multiply -2 by 5 to get -10, and add it to -6 from the first row. The result is -16.

$$\begin{array}{r} -2)\overline{5 \quad -6 \quad -28 \quad 8} \\ \quad\quad -10 \\ \hline 5 \quad -16 \end{array}$$

Next, multiply -2 by -16 to get 32. Add this to -28 from the first row.

$$\begin{array}{r} -2)\overline{5 \quad -6 \quad -28 \quad 8} \\ \quad\quad -10 \quad 32 \\ \hline 5 \quad -16 \quad 4 \end{array}$$

Finally, $(-2)(4) = -8$. Add this result to 8 to obtain 0.

$$\begin{array}{r} -2)\overline{5 \quad -6 \quad -28 \quad 8} \\ \quad\quad -10 \quad 32 \quad -8 \\ \hline 5 \quad -16 \quad 4 \quad 0 \end{array}$$

The coefficients of the quotient polynomial and the remainder are read directly from the bottom row. Since the degree of the quotient will always be one less than the degree of the polynomial to be divided, and here the remainder is 0,

$$\frac{5x^3 - 6x^2 - 28x + 8}{x + 2} = 5x^2 - 16x + 4.$$

Notice that the divisor $x + 2$ is a *factor* of $5x^3 - 6x^2 - 28x + 8$, because $5x^3 - 6x^2 - 28x + 8 = (x + 2)(5x^2 - 16x + 4)$.   ◆

THE REMAINDER AND FACTOR THEOREMS

In Example 2, we divided $3x^3 - 2x + 5$ by $x - 3$ and obtained a remainder of 80. If we evaluate the function $P(x) = 3x^3 - 2x + 5$ at $x = 3$, we get

$$P(3) = 3(3)^3 - 2(3) + 5 = 81 - 6 + 5 = 80.$$

Notice that the remainder is equal to $P(3)$. Also, in Example 3 we divided $5x^3 - 6x^2 - 28x + 8$ by $x - (-2)$ and obtained a remainder of 0. If we evaluate the function $P(x) = 5x^3 - 6x^2 - 28x + 8$ at $x = -2$, we get

$$P(-2) = 5(-2)^3 - 6(-2)^2 - 28(-2) + 8$$
$$= -40 - 24 + 56 + 8$$
$$= 0.$$

Notice that here also the remainder is equal to $P(-2)$. These two examples illustrate an important theorem in the study of polynomial functions, the remainder theorem, and an important corollary to the theorem, the factor theorem. (A *corollary* is a theorem that follows directly from another theorem.)

THE REMAINDER THEOREM

If a polynomial $P(x)$ is divided by $x - k$, the remainder is equal to $P(k)$.

EXAMPLE 4   *Using the Remainder Theorem and Supporting the Result Graphically*

(a)   Use the remainder theorem and synthetic division to find $P(-2)$ if

$$P(x) = -x^4 + 3x^2 - 4x - 5.$$

(b)   Support the result of part (a) graphically.

SOLUTION

(a)   ANALYTIC   We use synthetic division to find the remainder when $P(x)$ is divided by $x - (-2)$.

$$
\begin{array}{r|rrrrr}
-2） & -1 & 0 & 3 & -4 & -5 \\
     &   & 2 & -4 & 2 & 4 \\
\hline
     & -1 & 2 & -1 & -2 & -1 \quad \leftarrow \text{Remainder}
\end{array}
$$

Since the remainder is -1, by the remainder theorem we have $P(-2) = -1$.

(b)   GRAPHICAL   If we graph $P(x) = -x^4 + 3x^2 - 4x - 5$, we should see that the point $(-2, -1)$ lies on the graph. Figure 61 supports this fact.   ◆

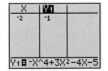

This screen shows how the result of Example 4 can be supported with a table.

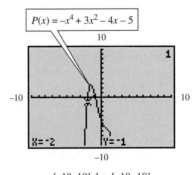

[−10, 10] by [−10, 10]
Xscl = 1   Yscl = 1
$P(-2) = -1$

FIGURE 61

EXAMPLE 5   *Deciding Whether a Number Is a Zero of a Polynomial Function*
Decide whether the given number is a zero of the function defined by the given polynomial.

(a)   2; $P(x) = x^3 - 4x^2 + 9x - 10$ **(b)** -2; $P(x) = 3x^3 - 2x^2 + 4x$

SOLUTION

(a)   Use synthetic division.

$$
\begin{array}{r|rrrr}
2 & 1 & -4 & 9 & -10 \\
  &   & 2 & -4 & 10 \\
\hline
  & 1 & -2 & 5 & 0
\end{array}
$$

Since the remainder is 0, $P(2) = 0$, and 2 is a zero of the polynomial function $P(x) = x^3 - 4x^2 + 9x - 10$.

(b)   Remember to use a coefficient of 0 for the missing constant term in the synthetic division.

$$
\begin{array}{r|rrrr}
-2 & 3 & -2 & 4 & 0 \\
   &   & -6 & 16 & -40 \\
\hline
   & 3 & -8 & 20 & -40
\end{array}
$$

The remainder is not zero, so -2 is not a zero of P, where $P(x) = 3x^3 - 2x^2 + 4x$. In fact, $P(-2) = -40$. From this, we know that the point $(-2, -40)$ lies on the graph of P.   ◆

   In Example 5(a), we showed that 2 is a zero of the polynomial function $P(x) = x^3 - 4x^2 + 9x - 10$. The first three numbers in the bottom row of the synthetic division process used there indicate the coefficients of the quotient polynomial, and thus

$$
\frac{P(x)}{x - 2} = x^2 - 2x + 5.
$$

Multiplying both sides of this equation by $x - 2$ gives

$$
P(x) = (x - 2)(x^2 - 2x + 5),
$$

indicating that $x - 2$ is a *factor* of $P(x)$.

   By the remainder theorem, if $P(k) = 0$, then the remainder when $P(x)$ is divided by $x - k$ is zero. This means that $x - k$ is a factor of $P(x)$. Conversely, if $x - k$ is a factor of $P(x)$, then $P(k)$ must equal 0. This is summarized in the following theorem, a corollary of the remainder theorem.

THE FACTOR THEOREM

The polynomial $x - k$ is a factor of the polynomial $P(x)$ if and only if $P(k) = 0$.

EXAMPLE 6   *Using the Factor Theorem*   Determine whether the second polynomial listed is a factor of the first.

(a) $4x^3 + 24x^2 + 48x + 32$; $x + 2$ **(b)** $2x^4 + 3x^2 - 5x + 7$; $x - 1$

SOLUTION

(a)   We use synthetic division with $k = -2$, since $x + 2 = x - (-2)$.

$$
\begin{array}{r|rrrr}
-2 & 4 & 24 & 48 & 32 \\
   &   & -8 & -32 & -32 \\
\hline
   & 4 & 16 & 16 & 0
\end{array}
\quad \leftarrow \text{Remainder is 0.}
$$

Since the remainder is 0, we know that $x + 2$ is a factor of $4x^3 + 24x^2 + 48x + 32$. A factored form (but not necessarily *completely* factored form) of the latter is $(x + 2)(4x^2 + 16x + 16)$.

(b)   By the factor theorem, $x - 1$ will be a factor of $P(x)$ only if $P(1) = 0$. Use synthetic division and the remainder theorem to decide.

$$
\begin{array}{r|rrrrr}
1 & 2 & 0 & 3 & -5 & 7 \\
  &   & 2 & 2 & 5 & 0 \\
\hline
  & 2 & 2 & 5 & 0 & 7
\end{array}
$$

Since the remainder is 7, $P(1) = 7$, not 0, so $x - 1$ is not a factor of $P(x)$.   ◆

Note   An easy way to determine $P(1)$ for a polynomial function P is simply to add the coefficients of $P(x)$. This method works since every power of 1 is equal to 1. For example, using $P(x) = 2x^4 + 3x^2 - 5x + 7$ as shown in Example 6(b), we have $P(1) = 2 + 3 - 5 + 7 = 7$, confirming our result found by synthetic division earlier.

The next example illustrates the close relationship among the ideas of x-intercepts of the graph of a polynomial function, real zeros of the function, and solutions of the corresponding polynomial equation.

EXAMPLE 7   *Examining Relationships among x-Intercepts, Zeros, and Solutions*   Consider the polynomial function defined by $P(x) = 2x^3 + 5x^2 - x - 6$.

(a)   Show by synthetic division that -2, $-\frac{3}{2}$, and 1 are zeros of P, and write $P(x)$ in factored form with all factors linear.

(b)   Graph P in a suitable viewing window and locate the x-intercepts.

(c)   Solve the polynomial equation $2x^3 + 5x^2 - x - 6 = 0$ analytically.

SOLUTION

(a)
$$
\begin{array}{r|rrrr}
-2 & 2 & 5 & -1 & -6 \\
   &   & -4 & -2 & 6 \\
\hline
   & 2 & 1 & -3 & 0
\end{array}
\quad \leftarrow P(-2) = 0
$$

Since $P(-2) = 0$, $x + 2$ is a factor, and thus $P(x) = (x + 2)(2x^2 + x - 3)$. Rather than show that $-\frac{3}{2}$ and 1 are zeros of $P(x)$, we need only show that they are zeros of $2x^2 + x - 3$. This can be shown by elementary factoring methods, or by synthetic division, as follows.

$$
\begin{array}{r|rrr}
-\dfrac{3}{2} & 2 & 1 & -3 \\
   &   & -3 & 3 \\
\hline
   & 2 & -2 & 0
\end{array}
\quad \leftarrow P\!\left(-\tfrac{3}{2}\right) = 0 \qquad -\tfrac{3}{2} \text{ is a zero of } 2x^2 + x - 3.
$$

$$
\begin{array}{r|rr}
1 & 2 & -2 \\
  &   & 2 \\
\hline
  & 2 & 0
\end{array}
\quad \leftarrow P(1) = 0 \qquad 1 \text{ is a zero of } 2x - 2.
$$

↑ This 2 is the constant factor.

The completely factored form of $P(x)$ is $2(x + 2)\left(x + \frac{3}{2}\right)(x - 1)$, or

$$P(x) = (x + 2)(2x + 3)(x - 1).$$

(b) GRAPHICAL   Figure 62 shows the graph of this function. The calculator will determine the x-intercepts: -2, $-\frac{3}{2}$, and 1.

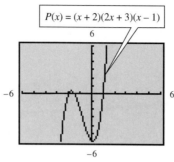

$$P(x) = (x + 2)(2x + 3)(x - 1)$$

[−6, 6] by [−6, 6]
Xscl = 1   Yscl = 1
The x-intercepts are -2, $-\frac{3}{2}$, and 1.

FIGURE 62

(c) ANALYTIC   The x-intercepts of the graph of P are the real solutions of the equation $P(x) = 0$. Therefore, the solution set of the equation is $\left\{-2, -\frac{3}{2}, 1\right\}$. To justify this analytically, we can extend the zero-product property (Section 3.2) to more than two factors and solve this equation. Since $P(x) = (x + 2)(2x + 3)(x - 1)$, we set each factor equal to 0 and solve.

$$x + 2 = 0 \quad\text{or}\quad 2x + 3 = 0 \quad\text{or}\quad x - 1 = 0$$

$$x = -2 \qquad\qquad x = -\frac{3}{2} \qquad\qquad x = 1$$

We see once again that the solution set is $\left\{-2, -\frac{3}{2}, 1\right\}$.   ◆

3.6   EXERCISES   **Tape 4**

Use the root location theorem to show that each function has a real zero between the two numbers given. Then, use your calculator to approximate the zero to the nearest hundredth.

1. $P(x) = 3x^2 - 2x - 6$;   1 and 2

2. $P(x) = x^3 + x^2 - 5x - 5$;   2 and 3

3. $P(x) = 2x^3 - 8x^2 + x + 16$;   2 and 2.5

4. $P(x) = 3x^3 + 7x^2 - 4$; $\frac{1}{2}$ and 1

5. $P(x) = 2x^4 - 4x^2 + 3x - 6$;   2 and 1.5

6. $P(x) = x^4 - 4x^3 - x + 1$;   1 and .3

7. Suppose that a polynomial function P is defined in such a way that $P(2) = -4$ and $P(2.5) = 2$. What conclusion does the root location theorem allow you to make?

8. Suppose that a polynomial function P is defined in such a way that $P(3) = -4$ and $P(4) = -10$. Can we be certain that there is no zero between 3 and 4? Explain, using a graph.

Use synthetic division to find the quotient polynomial Q(x) and the remainder R when P(x) is divided by the binomial following it.

9. $P(x) = x^3 + 2x^2 - 17x - 10;$ $x + 5$

10. $P(x) = x^4 + 4x^3 + 2x^2 + 9x + 4;$ $x + 4$

11. $P(x) = 3x^3 - 11x^2 - 20x + 3;$ $x - 5$

12. $P(x) = x^4 - 3x^3 - 5x^2 + 2x - 16;$ $x - 3$

13. $P(x) = x^4 - 3x^3 - 4x^2 + 12x;$ $x - 2$

14. $P(x) = x^5 - 1;$ $x - 1$

Use synthetic division to find P(k) for the given value of k and the given function P.

15. $k = 3;$ $P(x) = x^2 - 4x + 5$

16. $k = -2;$ $P(x) = x^2 + 5x + 6$

17. $k = -2;$ $P(x) = 5x^3 + 2x^2 - x + 5$

18. $k = 2;$ $P(x) = 2x^3 - 3x^2 - 5x + 4$

19. $k = 2;$ $P(x) = x^2 - 5x + 1$

20. $k = 3;$ $P(x) = x^2 - x + 3$

Use synthetic division to determine whether the given number is a zero of the polynomial.

21. $2;$ $P(x) = x^2 + 2x - 8$

22. $-1;$ $P(x) = x^2 + 4x - 5$

23. $4;$ $P(x) = 2x^3 - 6x^2 - 9x + 6$

24. $-4;$ $P(x) = 9x^3 + 39x^2 + 12x$

25. $-.5;$ $P(x) = 4x^3 + 12x^2 + 7x + 1$

26. $-.25;$ $P(x) = 8x^3 + 6x^2 - 3x - 1$

RELATING CONCEPTS (EXERCISES 27–32)

The close relationships among x-intercepts of a graph of a function, real zeros of the function, and real solutions of the associated equation should, by now, be apparent to you. Using the concepts presented so far in this text, consider the graph of the polynomial function $P(x) = x^3 - 2x^2 - 11x + 12$ shown below, and respond to Exercises 27–32.

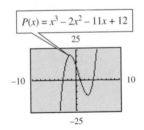

The x-intercepts are –3, 1, and 4.

27. What are the linear factors of $P(x)$? (Section 3.6)

28. What are the solutions of the equation $P(x) = 0$? (Section 3.5)

29. What are the zeros of the function P? (Section 3.5)

30. If $P(x)$ is divided by $x - 2$, what is the remainder? What is $P(2)$? (Section 3.6)

31. Give the solution set of $P(x) > 0$, using interval notation. (Section 3.5)

32. Give the solution set of $P(x) < 0$, using interval notation. (Section 3.5)

For each of the following, one zero is given. Find all others analytically.

33. $P(x) = x^3 - 2x + 1;$ 1

34. $P(x) = 2x^3 + 8x^2 - 11x - 5;$ -5

35. $P(x) = 3x^3 + 5x^2 - 3x - 2;$ -2

36. $P(x) = x^3 - 7x^2 + 13x - 3;$ 3

Factor P(x) into linear factors given that k is a zero of P.

37. $P(x) = 2x^3 - 3x^2 - 17x + 30;$ $k = 2$

38. $P(x) = 2x^3 - 3x^2 - 5x + 6;$ $k = 1$

39. $P(x) = 6x^3 + 25x^2 + 3x - 4;$ $k = -4$

40. $P(x) = 8x^3 + 50x^2 + 47x - 15;$ $k = -5$

There exists a useful theorem that helps us determine the number of positive and the number of negative real zeros of a polynomial function. It is known as Descartes' Rule of Signs.

DESCARTES' RULE OF SIGNS

Let $P(x)$ define a polynomial function with real coefficients and a nonzero constant term, with terms in descending powers of x.

a. The number of positive real zeros of P either equals the number of variations in sign occurring in the coefficients of $P(x)$, or is less than the number of variations by a positive even integer.
b. The number of negative real zeros of P either equals the number of variations in sign occurring in the coefficients of $P(-x)$, or is less than the number of variations by a positive even integer.

In the theorem, *variation in sign* is a change from positive to negative or negative to positive in successive terms of the polynomial. Missing terms (those with 0 coefficients) are counted as no change in sign and can be ignored. For example, we consider the polynomial function $P(x) = x^4 - 6x^3 + 8x^2 + 2x - 1$.

$P(x)$ has 3 variations in sign:

$$+x^4 - 6x^3 + 8x^2 + 2x - 1.$$
$$\quad\; 1 \quad 2 \qquad\quad 3$$

Thus, by Descartes' Rule of Signs, P has either 3 or $3 - 2 = 1$ positive real zeros. Since

$$P(-x) = (-x)^4 - 6(-x)^3 + 8(-x)^2 + 2(-x) - 1$$
$$= x^4 + 6x^3 + 8x^2 - 2x - 1$$

has only one variation in sign, P has only one negative real zero. If you graph the function in the window $[-5, 5]$ by $[-10, 10]$, you can interpret the theorem in terms of x-intercepts. Verify that there are 3 positive x-intercepts and 1 negative x-intercept.

Use Descartes' Rule of Signs to determine the possible number of positive real zeros and the possible number of negative real zeros for the following functions. Then, use a graph to determine the actual numbers of positive and negative real zeros.

41. $P(x) = 2x^3 - 4x^2 + 2x + 7$

42. $P(x) = x^3 + 2x^2 + x - 10$

43. $P(x) = 5x^4 + 3x^2 + 2x - 9$

44. $P(x) = 3x^4 + 2x^3 - 8x^2 - 10x - 1$

45. $P(x) = x^5 + 3x^4 - x^3 + 2x + 3$

46. $P(x) = 2x^5 - x^4 + x^3 - x^2 + x + 5$

3.7 TOPICS IN THE THEORY OF POLYNOMIAL FUNCTIONS (II)

Complex Zeros and the Fundamental Theorem of Algebra ◆ Multiplicity of Zeros ◆ The Rational Zeros Theorem

COMPLEX ZEROS AND THE FUNDAMENTAL THEOREM OF ALGEBRA

In Example 3 of Section 3.3, we found that the imaginary solutions of $2x^2 - x + 4 = 0$ are $\frac{1}{4} + i\frac{\sqrt{31}}{4}$ and $\frac{1}{4} - i\frac{\sqrt{31}}{4}$. Notice that these two solutions are complex conjugates. This is not a coincidence; it can be shown that if $a + bi$ is a zero of a polynomial function with *real* coefficients, then its complex conjugate $a - bi$ is also a zero. This is given as the next theorem. Its proof is left for the exercises.

CONJUGATE ZEROS THEOREM

If $P(x)$ is a polynomial having only real coefficients, and if $a + bi$ is a zero of P, then the conjugate $a - bi$ is also a zero of P.

EXAMPLE 1   *Defining a Polynomial Function Satisfying Given Conditions*

(a) Find a cubic polynomial in standard form with real coefficients having zeros 3 and $2 + i$.

(b) Find a polynomial function P satisfying the conditions of part (a), with the additional requirement $P(-2) = 4$. Support the result graphically.

SOLUTION

(a) By the conjugate zeros theorem, $2 - i$ must also be a zero of the function. Since the polynomial will be cubic, it will have three linear factors, and by the factor theorem they must be $x - 3$, $x - (2 + i)$, and $x - (2 - i)$. Therefore, one such cubic polynomial $P(x)$ can be defined as follows:

$$P(x) = (x - 3)[x - (2 + i)][x - (2 - i)]$$
$$= (x - 3)(x - 2 - i)(x - 2 + i)$$
$$= x^3 - 7x^2 + 17x - 15.$$

Multiplying this polynomial by any real nonzero constant will also yield a function satisfying the given conditions, so a more general form of $P(x)$ is $a(x^3 - 7x^2 + 17x - 15)$.

(b)   ANALYTIC   Let a represent a real nonzero constant. We must have

$$P(x) = a(x^3 - 7x^2 + 17x - 15)$$

defined in such a way that $P(-2) = 4$. To find a, let $x = -2$, and set the result equal to 4. Then solve.

$$a[(-2)^3 - 7(-2)^2 + 17(-2) - 15] = 4$$
$$a(-8 - 28 - 34 - 15) = 4$$
$$-85a = 4$$
$$a = -\frac{4}{85}$$

Therefore, the desired function is

$$P(x) = -\frac{4}{85}(x^3 - 7x^2 + 17x - 15)$$
$$= -\frac{4}{85}x^3 + \frac{28}{85}x^2 - \frac{4}{5}x + \frac{12}{17}.$$

GRAPHICAL   We can support this result graphically by graphing $P(x) = -\frac{4}{85}x^3 + \frac{28}{85}x^2 - \frac{4}{5}x + \frac{12}{17}$, and showing that the point $(-2, 4)$ lies on the graph. See Figure 63.   ◆

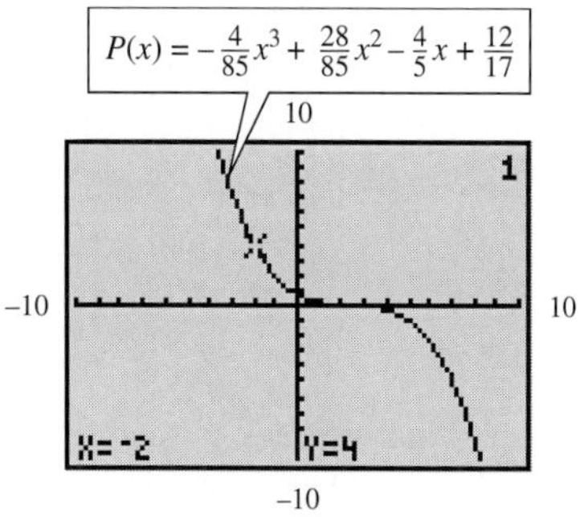

$$P(x) = -\frac{4}{85}x^3 + \frac{28}{85}x^2 - \frac{4}{5}x + \frac{12}{17}$$

[−10, 10] by [−10, 10]
Xscl = 1   Yscl = 1
$P(-2) = 4$

FIGURE 63

The next theorem says that every polynomial of degree 1 or more has a zero, so that every such polynomial can be factored. This theorem was first proved by the mathematician Karl F. Gauss in his doctoral thesis in 1799, when he was 22 years old. Although many proofs of this result have been given, all of them involve mathematics beyond the algebra in this book, so no proof is included here.

THE FUNDAMENTAL THEOREM OF ALGEBRA

Every function defined by a polynomial of degree 1 or more has at least one complex zero.

From the fundamental theorem, if $P(x)$ is of degree 1 or more, then there is some number k such that $P(k) = 0$. By the factor theorem, then, $P(x) = (x - k) \cdot Q(x)$ for some polynomial $Q(x)$. The fundamental theorem and the factor theorem can be used to factor $Q(x)$ in the same way. Assuming that $P(x)$ has degree n, repeating this process n times gives

$$P(x) = a(x - k_1)(x - k_2) \cdots (x - k_n),$$

where a is the leading coefficient of $P(x)$. Each of these factors leads to a zero of $P(x)$, so $P(x)$ has the n zeros $k_1, k_2, k_3, \ldots, k_n$. This result can be used to prove the next theorem. The proof is left for the exercises.

ZEROS OF A POLYNOMIAL FUNCTION (I)

A function defined by a polynomial of degree n has at most n distinct complex zeros.

Notice that the statement above says that a polynomial function has *at most n* complex (real or imaginary) zeros. Some zeros may be repeated, and this will be addressed later in this section.

EXAMPLE 2 *Finding All Zeros of a Polynomial Function* Find all complex zeros of the function $P(x) = x^4 - 7x^3 + 18x^2 - 22x + 12$, given that $1 - i$ is a zero.

SOLUTION   This quartic function will have at most four complex zeros. Since $1 - i$ is a zero and the coefficients are real numbers, by the conjugate zeros theorem $1 + i$ is also a zero. The remaining zeros are found by first dividing the original polynomial by $x - (1 - i)$.

$$1 - i \overline{)\begin{array}{ccccc} 1 & -7 & 18 & -22 & 12 \\ & 1 - i & -7 + 5i & 16 - 6i & -12 \\ \hline 1 & -6 - i & 11 + 5i & -6 - 6i & 0 \end{array}}$$

Rather than go back to the original polynomial, divide the quotient from the first division by $x - (1 + i)$ as follows.

$$1 + i \overline{)\begin{array}{cccc} 1 & -6 - i & 11 + 5i & -6 - 6i \\ & 1 + i & -5 - 5i & 6 + 6i \\ \hline 1 & -5 & 6 & 0 \end{array}}$$

Find the zeros of the function defined by the quadratic polynomial $x^2 - 5x + 6$ by solving the equation $x^2 - 5x + 6 = 0$. By the quadratic formula or by factoring, we determine the other zeros to be 2 and 3. Thus, this function has exactly four complex zeros: $1 - i$, $1 + i$, 2, and 3.   ◆

MULTIPLICITY OF ZEROS

Consider the polynomial function

$$P(x) = x^6 + x^5 - 5x^4 - x^3 + 8x^2 - 4x$$
$$= x(x + 2)^2(x - 1)^3.$$

Each factor will lead to a zero of the function. The factor x leads to a *single* zero, 0, the factor $(x + 2)^2$ leads to a zero of -2 appearing *twice*, and the factor $(x - 1)^3$ leads to a zero of 1 appearing *three* times. The number of times a zero appears is referred to as the **multiplicity of the zero**. We can now state the rule for the number of zeros of a polynomial function more precisely.

> ### ZEROS OF A POLYNOMIAL FUNCTION (II)
> A function defined by a polynomial of degree n has exactly n complex zeros if zeros of multiplicity m are counted m times.

EXAMPLE 3   *Defining a Polynomial Function Satisfying Given Conditions*
Find a polynomial function with real coefficients of lowest possible degree having a zero 2 of multiplicity 3, a zero 0 of multiplicity 2, and a zero i of single multiplicity.

SOLUTION   This polynomial function must also have a zero $-i$ of single multiplicity. (Why?) Its lowest possible degree is 7. (Why?) By the factor theorem, one possible such polynomial function in factored form is

$$P(x) = x^2(x - 2)^3(x - i)(x + i).$$

Multiplying the factors on the right leads to

$$P(x) = x^7 - 6x^6 + 13x^5 - 14x^4 + 12x^3 - 8x^2.$$

This is one of infinitely many such functions. Multiplying $P(x)$ by a nonzero constant will yield another function satisfying these conditions.

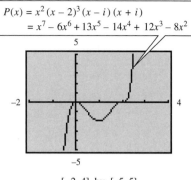

$$P(x) = x^2 (x - 2)^3 (x - i) (x + i)$$
$$= x^7 - 6x^6 + 13x^5 - 14x^4 + 12x^3 - 8x^2$$

$[-2, 4]$ by $[-5, 5]$
Xscl = 1   Yscl = 1

The graph is tangent to the x-axis at $x = 0$,
and crosses the x-axis at $x = 2$.

FIGURE 64

Graphing this function shows that it has only two distinct x-intercepts, corresponding to the real zeros 0 and 2. See Figure 64.   ◆

For Group Discussion   Graph each of the following functions one at a time, and then respond to the items below.

$P(x) = (x + 3)(x - 2)^2$ $P(x) = (x + 3)^2(x - 2)^3$ $P(x) = x^2(x - 1)(x + 2)^2$
Use the window   Use the window   Use the window
$[-10, 10]$ by $[-30, 30]$. $[-4, 4]$ by $[-125, 50]$. $[-4, 4]$ by $[-5, 5]$.

1. Describe the behavior of each graph at each x-intercept that corresponds to a zero of odd multiplicity.
2. Describe the behavior of each graph at each x-intercept that corresponds to a zero of even multiplicity.

The observations made in the preceding group discussion activity should indicate that the behavior of the graph of a polynomial function near an x-intercept depends on the parity of multiplicity of the zero that leads to the x-intercept. If the zero is of odd multiplicity, the graph will cross the x-axis at the corresponding x-intercept. If the zero is of even multiplicity, the graph will be tangent to the x-axis at the corresponding x-intercept (that is, it will touch but not cross the x-axis). See Figure 65.

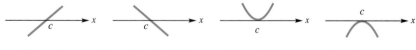

The graph crosses the x-axis at $(c, 0)$ if   The graph is tangent to the x-axis at $(c, 0)$ if
c is a zero of odd multiplicity. c is a zero of even multiplicity.

(a)   **(b)**

FIGURE 65

By observing the dominating term and noting the parity of multiplicities of zeros of a polynomial function in factored form, we can sketch a rough graph of a polynomial function by hand. This is shown in the next example.

EXAMPLE 4   *Sketching a Polynomial Function Graph by Hand*   Consider the polynomial function defined by

$$P(x) = -2x^5 - 18x^4 - 38x^3 + 42x^2 + 112x - 96,$$

with factored form

$$P(x) = -2(x + 4)^2(x + 3)(x - 1)^2.$$

Sketch the graph of P by hand, and then support the result with a calculator-generated graph.

SOLUTION   Because the dominating term is $-2x^5$, the end behavior of the graph will be ↖↘ . Because -4 and 1 are both x-intercepts determined by zeros of even multiplicity, the graph will be tangent to the x-axis at these x-intercepts. Because -3 is a zero of multiplicity one, the graph will cross the x-axis at $x = -3$. The y-intercept is easily determined to be -96. Combining all of this information leads to the following rough sketch of the graph. See Figure 66(a).

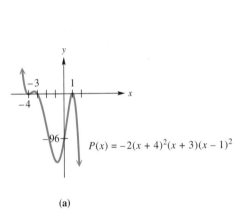

$$P(x) = -2(x + 4)^2(x + 3)(x - 1)^2$$

(a)

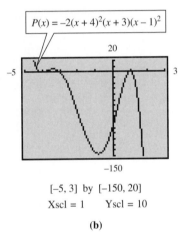

$[-5, 3]$ by $[-150, 20]$

Xscl = 1   Yscl = 10

(b)

FIGURE 66

Using a window of $[-5, 3]$ by $[-150, 20]$, we get the graph shown in Figure 66(b), confirming our rough sketch. (Notice that the hand-drawn method does not necessarily give a good indication of local extrema.)   ◆

THE RATIONAL ZEROS THEOREM

The final theorem discussed in this section gives us a method to determine all possible candidates for rational zeros of a polynomial function with integer coefficients.

THE RATIONAL ZEROS THEOREM

Let $P(x) = a_n x^n + a_{n-1}x^{n-1} + \cdots + a_1 x + a_0$, where $a_n \neq 0$, define a polynomial function with integer coefficients. If p/q is a rational number written in lowest terms, and if p/q is a zero of P, then p is a factor of the constant term a_0, and q is a factor of the leading coefficient a_n.

Proof $P(p/q) = 0$ since p/q is a zero of $P(x)$, so

$$a_n(p/q)^n + a_{n-1}(p/q)^{n-1} + \cdots + a_1(p/q) + a_0 = 0.$$

This also can be written as

$$a_n(p^n/q^n) + a_{n-1}(p^{n-1}/q^{n-1}) + \cdots + a_1(p/q) + a_0 = 0.$$

Multiply both sides of this last result by q^n and add $-a_0q^n$ to both sides.

$$a_np^n + a_{n-1}p^{n-1}q + \cdots + a_1pq^{n-1} = -a_0q^n$$

Factoring out p gives

$$p(a_np^{n-1} + a_{n-1}p^{n-2}q + \cdots + a_1q^{n-1}) = -a_0q^n.$$

This result shows that $-a_0q^n$ equals the product of the two factors, p and $(a_np^{n-1} + \cdots + a_1q^{n-1})$. For this reason, p must be a factor of $-a_0q^n$. Since it was assumed that p/q is written in lowest terms, p and q have no common factor other than 1, so p is not a factor of q^n. Thus, p must be a factor of a_0. In a similar way it can be shown that q is a factor of a_n.

EXAMPLE 5   *Using the Rational Zeros Theorem*   For the polynomial function $P(x) = 6x^4 + 7x^3 - 12x^2 - 3x + 2$, do each of the following.

(a)   List all possible rational zeros.

(b)   Use a graph to eliminate some of the possible zeros listed in part (a).

(c)   Find all rational zeros and factor $P(x)$.

SOLUTION

(a)   For a rational number p/q to be a zero, p must be a factor of $a_0 = 2$ and q must be a factor of $a_4 = 6$. Thus, p can be ±1 or ±2, and q can be ±1, ±2, ±3, or ±6. The possible rational zeros, p/q, are

$$\pm1, \qquad \pm2, \qquad \pm1/2, \qquad \pm1/3, \qquad \pm1/6, \qquad \pm2/3.$$

(b)   From Figure 67 we see that the zeros are no less than -2 and no greater than 1. Therefore, we can eliminate 2. Furthermore, it is obvious that -1 is not a zero, since the graph does not intersect the x-axis at $(-1, 0)$.

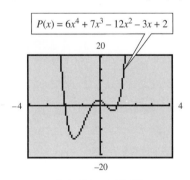

$$P(x) = 6x^4 + 7x^3 - 12x^2 - 3x + 2$$

$$[-4, 4] \text{ by } [-20, 20]$$
$$\text{Xscl} = 1 \qquad \text{Yscl} = 5$$

FIGURE 67

(c)   We will use the remainder theorem to show that 1 and -2 are zeros.

$$
\begin{array}{r|rrrrr}
1) & 6 & 7 & -12 & -3 & 2 \\
   &   & 6 & 13 & 1 & -2 \\
\hline
   & 6 & 13 & 1 & -2 & 0
\end{array}
$$

[−4, 4] by [−20, 20]

Xscl = 1   Yscl = 5

This graph/table shows that the four zeros of the function of Example 5 are -2, $-\frac{1}{2}$, $\frac{1}{3}$, and 1.

The 0 remainder shows that 1 is a zero. Now, use the quotient polynomial $6x^3 + 13x^2 + x - 2$ and synthetic division to find that -2 is also a zero.

$$
\begin{array}{r|rrrr}
-2 & 6 & 13 & 1 & -2 \\
   &   & -12 & -2 & 2 \\
\hline
   & 6 & 1 & -1 & 0
\end{array}
$$

The new quotient polynomial is $6x^2 + x - 1$. Use the quadratic formula or factor to solve the equation $6x^2 + x - 1 = 0$. The remaining two zeros are $1/3$ and $-1/2$.

Factor the polynomial $P(x)$ in the following way. Since the four zeros of $P(x) = 6x^4 + 7x^3 - 12x^2 - 3x + 2$ are 1, -2, $1/3$, and $-1/2$, the factors are $x - 1$, $x + 2$, $x - 1/3$, and $x + 1/2$, and

$$P(x) = a(x - 1)(x + 2)\left(x - \frac{1}{3}\right)\left(x + \frac{1}{2}\right).$$

Since the leading coefficient of $P(x)$ is 6, let $a = 6$. Then,

$$P(x) = 6(x - 1)(x + 2)\left(x - \frac{1}{3}\right)\left(x + \frac{1}{2}\right)$$

$$= (x - 1)(x + 2)(3)\left(x - \frac{1}{3}\right)(2)\left(x + \frac{1}{2}\right)$$

$$= (x - 1)(x + 2)(3x - 1)(2x + 1).   \blacklozenge$$

Caution   The rational zeros theorem has limited usefulness, since it gives only possible rational zeros; it does not tell us whether these rational numbers are actual zeros. We must rely on other methods to determine whether or not they are indeed zeros. Furthermore, the function must have integer coefficients. To begin to apply the rational zeros theorem to a polynomial with fractional coefficients, multiply through by the least common denominator of all the fractions. For example, any rational zeros of

$$P(x) = x^4 - \frac{1}{6}x^3 + \frac{2}{3}x^2 - \frac{1}{6}x - \frac{1}{3}$$

will also be rational zeros of

$$Q(x) = 6x^4 - x^3 + 4x^2 - x - 2.$$

The function Q was obtained by multiplying the terms of P by 6.

3.7   EXERCISES   Tape 5

Find a cubic polynomial in standard form with real coefficients, having the given zeros.

1. 4 and $2 + i$

2. -3 and $6 + 2i$

3. 5 and i

4. -9 and $-i$

5. 0 and $3 + i$

6. 0 and $4 - 3i$

For each of the following, find a function P defined by a polynomial of degree 3 with real coefficients that satisfies the given conditions.

7. Zeros of -3, -1, and 4; $P(2) = 5$

8. Zeros of 1, -1, and 0; $P(2) = -3$

9. Zeros of -2, 1, and 0; $P(-1) = -1$

10. Zeros of 2, 5, and -3; $P(1) = -4$

11. Zeros of 4 and $1 + i$; $P(2) = 4$

12. Zeros of -7 and $2 - i$; $P(1) = 9$

For each of the following, one or more zeros are given. Find all remaining zeros.

13. $P(x) = x^3 - x^2 - 4x - 6$;   3 is a zero

14. $P(x) = x^3 - 5x^2 + 17x - 13$;   1 is a zero

15. $P(x) = x^4 + 2x^3 - 10x^2 - 18x + 9$; -3 and 3 are zeros

16. $P(x) = 2x^4 - x^3 - 27x^2 + 16x - 80$; -4 and 4 are zeros

17. $P(x) = x^4 - x^3 + 10x^2 - 9x + 9$; $3i$ is a zero

18. $P(x) = 2x^4 - 2x^3 + 55x^2 - 50x + 125$; $-5i$ is a zero

For each of the following, find a polynomial function P of lowest possible degree, having real coefficients, with the given zeros.

19. 5 and -4

20. 6 and -2

21. -3, 2, and i

22. 5 (multiplicity 2) and $-2i$

23. -3 (multiplicity 2) and $2 + i$

24. 2 (multiplicity 2) and $1 + 2i$

25. Show that -2 is a zero of multiplicity 2 of P, where $P(x) = x^4 + 2x^3 - 7x^2 - 20x - 12$, and find all other complex zeros. Then, write $P(x)$ in factored form.

26. Show that -1 is a zero of multiplicity 3 of P, where $P(x) = x^5 + 9x^4 + 33x^3 + 55x^2 + 42x + 12$, and find all other complex zeros. Then, write $P(x)$ in factored form.

27. What are the possible numbers of real zeros (counting multiplicities) for a polynomial function with real coefficients of degree 5?

28. Explain why a function defined by a polynomial of degree 4 with real coefficients has either zero, two, or four real zeros (counting multiplicities).

29. Explain why it is not possible for a function defined by a polynomial of degree 3 with real coefficients to have zeros of 1, 2, and $1 + i$.

30. Suppose that k, a, b, and c are real numbers, $a \neq 0$, and a polynomial function $P(x)$ may be expressed in factored form as $(x - k)(ax^2 + bx + c)$.

  (a) What is the degree of P?

  (b) What are the possible numbers of *real* zeros of P?

  (c) What are the possible numbers of *imaginary* zeros of P?

  (d) Use the discriminant to explain how to determine the number and kind of zeros P has.

In each of the following, a polynomial function is given in both standard form and factored form. Use dominating term and multiplicity of zeros to draw by hand a rough sketch of the graph of the function. Then, support your answer by using a calculator-generated graph.

31. $P(x) = 2x^3 - 5x^2 - x + 6$
$= (x + 1)(2x - 3)(x - 2)$

32. $P(x) = x^3 + x^2 - 8x - 12$
$= (x + 2)^2(x - 3)$

33. $P(x) = x^4 - 18x^2 + 81$
$= (x - 3)^2(x + 3)^2$

34. $P(x) = x^4 - 8x^2 + 16$
$= (x + 2)^2(x - 2)^2$

35. $P(x) = 2x^4 + x^3 - 6x^2 - 7x - 2$
$= (2x + 1)(x - 2)(x + 1)^2$

36. $P(x) = 3x^4 - 7x^3 - 6x^2 + 12x + 8$
$= (3x + 2)(x + 1)(x - 2)^2$

For each of the following polynomial functions, (a) list all possible rational zeros, (b) use a graph to eliminate some of the possible zeros listed in part (a), (c) find all rational zeros, and (d) factor P(x).

37. $P(x) = x^3 - 2x^2 - 13x - 10$

38. $P(x) = x^3 + 5x^2 + 2x - 8$

39. $P(x) = x^3 + 6x^2 - x - 30$

40. $P(x) = x^3 - x^2 - 10x - 8$

41. $P(x) = 6x^3 + 17x^2 - 31x - 12$

42. $P(x) = 15x^3 + 61x^2 + 2x - 8$

43. $P(x) = 12x^3 + 20x^2 - x - 6$

44. $P(x) = 12x^3 + 40x^2 + 41x + 12$

Find all rational zeros of the polynomial function.

45. $P(x) = x^3 + \frac{1}{2}x^2 - \frac{11}{2}x - 5$

46. $P(x) = \frac{10}{7}x^4 - x^3 - 7x^2 + 5x - \frac{5}{7}$

47. $P(x) = \frac{1}{6}x^4 - \frac{11}{12}x^3 + \frac{7}{6}x^2 - \frac{11}{12}x + 1$

48. $P(x) = x^4 - \frac{1}{6}x^3 + \frac{2}{3}x^2 - \frac{1}{6}x - \frac{1}{3}$

49. For any polynomial $P(x)$ and any complex number k, there exists a unique polynomial $Q(x)$ and number R such that

$$P(x) = (x - k) \cdot Q(x) + R.$$

This statement is known as the division algorithm. In order to prove the remainder theorem, let $x = k$ in this statement. Write a proof of the remainder theorem.

50. Suppose that c and d represent complex numbers: $c = a + bi$ and $d = m + ni$. Let $\overline{c}$ and $\overline{d}$ represent the complex conjugates of c and d, respectively. Prove each of the following statements. (These properties will be used in Exercise 51 to prove the conjugate zeros theorem.)

(a) $\overline{c + d} = \overline{c} + \overline{d}$

(b) $\overline{cd} = \overline{c} \cdot \overline{d}$

(c) $\overline{x} = x$ for any real number x

(d) $\overline{c^n} = \left(\overline{c}\right)^n$, n is a positive integer

51. Complete the proof of the conjugate zeros theorem, outlined below. Assume that

$$P(x) = a_n x^n + a_{n-1} x^{n-1} + \cdots + a_1 x + a_0,$$

where all coefficients are real numbers.

(a) Suppose the complex number z is a zero of P; find $P(z)$.

(b) Take the conjugate of both sides of the result from part (a).

(c) Use generalizations of the properties given in Exercise 50 on the result of part (b) to show that $a_n\left(\overline{z}\right)^n + a_{n-1}\left(\overline{z}\right)^{n-1} + \cdots + a_1\left(\overline{z}\right) + a_0 = 0$.

(d) Why does the result in part (c) mean that $\overline{z}$ is a zero of P?

52. The function defined by $P(x) = x^2 - x + (i + 1)$ has i as a zero, but does not have its conjugate, $-i$, as a zero. Explain why this does not violate the conjugate zeros theorem.

3.8 SOLUTION OF POLYNOMIAL EQUATIONS AND INEQUALITIES AND THEIR APPLICATIONS

Polynomial Equations and Inequalities ◆ Complex *n*th Roots ◆ Applications of Polynomial Functions ◆ Polynomial Models

While methods of solving quadratic equations were known to ancient civilizations, for hundreds of years mathematicians wrestled with finding methods of solving higher degree equations (by analytic methods, of course). It was not until the sixteenth century that progress in this area was made, and the European mathematicians Scipione del Ferro, Nicolo Fontana (a.k.a. Tartaglia), Girolamo Cardano, and François Viete were able to derive formulas allowing the solution of cubic equations. Work progressed and methods of solving quartics followed. While these methods were quite complicated, they showed that, in theory, third and fourth degree polynomial equations could be solved analytically. It was not until 1824 that the Norwegian mathematician Niels Henrik Abel proved that it is *impossible* to find a formula that will yield solutions to the general quintic (fifth degree) equation. A similar result holds for polynomial functions of degree greater than five.

We can use elementary methods to solve *some* higher degree polynomial equations analytically, as we will show in this section. The technology of graphing calculators also allows us to support our analytic work, and allows us to find accurate approximations of solutions of such equations that cannot be solved easily by elementary methods or at all by analytic methods.

POLYNOMIAL EQUATIONS AND INEQUALITIES

EXAMPLE 1   *Solving a Polynomial Equation and Associated Inequalities*

(a)   Solve the polynomial equation

$$x^3 + 3x^2 - 4x - 12 = 0$$

by using the zero-product property.

(b)   Support the result of part (a) graphically.

(c)   Use the graph from part (b) to find the solution set of

$$x^3 + 3x^2 - 4x - 12 > 0.$$

(d)   Use the graph from part (b) to find the solution set of

$$x^3 + 3x^2 - 4x - 12 \leq 0.$$

SOLUTION

(a)   ANALYTIC   Since the right-hand side of the equation is 0, we begin by factoring the left side. Then, set each factor equal to 0 and solve each equation.

$x^3 + 3x^2 - 4x - 12 = 0$	Given equation
$(x^3 + 3x^2) + (-4x - 12) = 0$	Group terms with common factors.
$x^2(x + 3) - 4(x + 3) = 0$	Factor out common factors in each group.
$(x + 3)(x^2 - 4) = 0$	Factor out $x + 3$.
$(x + 3)(x - 2)(x + 2) = 0$	Factor the difference of two squares.
$x + 3 = 0$ or $x - 2 = 0$ or $x + 2 = 0$	Use the zero-product property. Solve.
$x = -3$ $x = 2$ $x = -2$	

The solution set is $\{-3, -2, 2\}$.

(b)   GRAPHICAL   We graph $y = x^3 + 3x^2 - 4x - 12$ in the window $[-10, 10]$ by $[-15, 10]$ and see that the x-intercepts are -3, -2, and 2, supporting our analytic solution in part (a). See Figure 68.

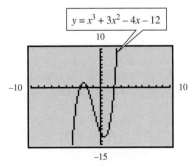

$[-10, 10]$ by $[-15, 10]$
Xscl = 1   Yscl = 1
The x-intercepts are –3, –2, and 2.

FIGURE 68

(c)   Recall that the solution set of $f(x) > 0$ consists of all numbers in the domain of f for which the graph lies *above* the x-axis. We see in Figure 68 that this occurs in the intervals $(-3, -2)$ and $(2, \infty)$. Therefore, the solution set of this inequality is $(-3, -2) \cup (2, \infty)$.

(d)   Here we must locate the intervals where the graph lies *below or intersects* the *x*-axis. From the figure, we find the solution set is

$$(-\infty, -3] \cup [-2, 2].$$

Notice that endpoints are included here due to the nonstrict inequality $\leq$.   ◆

For Group Discussion

1.   Consider the equation

$$x^3 + 3x^2 = 4x + 12.$$

Why are the solutions of this equation the same as those in part (a) of Example 1?

2.   Figure 69 shows the graph of

$$y_1 = x^3 + 3x^2 \qquad \text{and} \qquad y_2 = 4x + 12.$$

How do we interpret the solutions of the equation in Item 1 above using this graph?

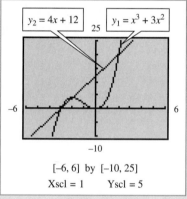

$[-6, 6]$ by $[-10, 25]$
Xscl = 1   Yscl = 5

FIGURE 69

3.   Consider the inequality

$$x^3 + 3x^2 > 4x + 12.$$

Why are the solutions of this inequality the same as those in part (c) of Example 1? How does the graph in Figure 69 support our result there?

4.   Consider the inequality

$$x^3 + 3x^2 \leq 4x + 12.$$

Why are the solutions of this inequality the same as those in part (d) of Example 1? How does the graph in Figure 69 support our result there?

EXAMPLE 2   *Solving an Equation Quadratic in Form*

(a)   Solve the polynomial equation

$$x^4 - 6x^2 - 40 = 0$$

analytically. Find all complex solutions.

(b) Use a graph to support the real solutions of the equation in part (a).

(c) Use the graph from part (b) to solve the inequalities

$$x^4 - 6x^2 - 40 \geq 0 \quad \text{and} \quad x^4 - 6x^2 - 40 < 0.$$

Give endpoints of intervals in both exact and approximate form.

SOLUTION

(a) ANALYTIC   This equation is said to be **quadratic in form**. Notice that if we let $t = x^2$, then the equation becomes quadratic in the variable t. Then we can solve for t, using methods of solving quadratic equations, and finally go back and solve for x.

$x^4 - 6x^2 - 40 = 0$	Given equation
$(x^2)^2 - 6x^2 - 40 = 0$	
$t^2 - 6t - 40 = 0$	Let $t = x^2$.
$(t - 10)(t + 4) = 0$	Factor.
$t = 10 \quad \text{or} \quad t = -4$	Use the zero-product property.
$x^2 = 10 \quad \text{or} \quad x^2 = -4$	Go back and solve for x.
$x = \pm\sqrt{10} \quad \text{or} \quad x = \pm 2i$	Use the square root property for solving quadratic equations.

The solution set is $\left\{ -\sqrt{10}, \sqrt{10}, -2i, 2i \right\}$.

(b) GRAPHICAL   We graph $y = x^4 - 6x^2 - 40$ in the window $[-4, 4]$ by $[-80, 50]$. See Figure 70. By using the appropriate feature of a calculator, we can find that the x-intercepts are approximately -3.16 and 3.16, which are approximations of $-\sqrt{10}$ and $\sqrt{10}$. Notice that the graph will not provide support for the imaginary solutions.

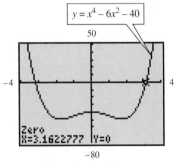

$[-4, 4]$ by $[-80, 50]$
Xscl = 1   Yscl = 10

The x-intercept displayed corresponds to $\sqrt{10}$. The other x-intercept is $-\sqrt{10} \approx -3.1622777$.

FIGURE 70

(c) GRAPHICAL   Since the graph lies above or intersects the x-axis for real numbers less than or equal to $-\sqrt{10}$ and for real numbers greater than or equal to $\sqrt{10}$, the solution set for $x^4 - 6x^2 - 40 \geq 0$ is

$$\left(-\infty, -\sqrt{10} \right] \cup \left[\sqrt{10}, \infty \right) \quad \leftarrow \text{Exact form}$$

$$\text{or} \qquad (-\infty, -3.16] \cup [3.16, \infty). \quad \leftarrow \text{Approximate form}$$

By similar reasoning, the solution set of $x^4 - 6x^2 - 40 < 0$ is

$$\left(-\sqrt{10}, \sqrt{10}\right) \quad \leftarrow \text{Exact form}$$

or $(-3.16, 3.16).$ ← Approximate form

(Note that the imaginary solutions do not affect the solution sets of the inequalities.)   ◆

EXAMPLE 3   *Solving a Polynomial Equation and Associated Inequalities*
The graph of $P(x) = x^3 + 3x^2 - 11x + 2$ is shown in Figure 71.

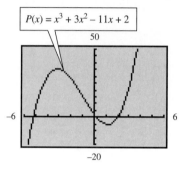

$[-6, 6]$ by $[-20, 50]$
Xscl = 1   Yscl = 5
FIGURE 71

(a)   Explain why 2 is a real solution of the equation $x^3 + 3x^2 - 11x + 2 = 0$, and then find all solutions of this equation.

(b)   Support the result of part (a) graphically.

(c)   Give the exact solution set of $x^3 + 3x^2 - 11x + 2 \le 0$.

SOLUTION

(a)   ANALYTIC   We can use a calculator to confirm that 2 is an x-intercept of the graph of P, or we can show it by synthetic division.

$$\begin{array}{r|rrrr} 2) & 1 & 3 & -11 & 2 \\ & & 2 & 10 & -2 \\ \hline & 1 & 5 & -1 & 0 \end{array} \quad \leftarrow P(2) = 0 \text{ by the}$$
$$\underbrace{}_{\substack{\text{Coefficients of the} \\ \text{quotient polynomial}}} \qquad \text{remainder theorem.}$$

By the factor theorem, $x - 2$ is a factor of $P(x)$, and

$$P(x) = (x - 2)(x^2 + 5x - 1).$$

To find the other zeros of P, we must solve

$$x^2 + 5x - 1 = 0.$$

We use the quadratic formula, with $a = 1, b = 5$, and $c = -1$.

$$x = \frac{-5 \pm \sqrt{5^2 - 4(1)(-1)}}{2(1)}$$

$$= \frac{-5 \pm \sqrt{29}}{2}$$

The complete solution set is $\left\{\frac{-5 - \sqrt{29}}{2}, \frac{-5 + \sqrt{29}}{2}, 2\right\}$.

(b)   GRAPHICAL   Using the capabilities of a calculator, we find that the x-intercepts are 2, approximately -5.19, and .19. Since these latter two approximations correspond to the approximations for $\left(-5 - \sqrt{29}\right)/2$ and $\left(-5 + \sqrt{29}\right)/2$, our analytic work is supported.

(c)   By finding the intervals for which the graph of P lies below or intersects the x-axis, we conclude that the solution set is $\left(-\infty, \frac{-5 - \sqrt{29}}{2}\right] \cup \left[\frac{-5 + \sqrt{29}}{2}, 2\right]$.   ◆

EXAMPLE 4   *Using Purely Graphical Methods to Solve a Polynomial Equation and Associated Inequalities*   Let $P(x) = 2.45x^3 - 3.14x^2 - 6.99x + 2.58$. Use the graph of P to find the solution sets of $P(x) = 0$, $P(x) > 0$, and $P(x) < 0$. Express solutions of the equation and endpoints of the intervals for the inequalities to the nearest hundredth.

SOLUTION   The graph of P is shown in Figure 72. Using the capabilities of a calculator, we find that the approximate x-intercepts are -1.37, .33, and 2.32.

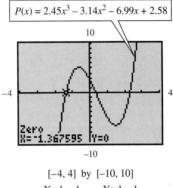

$$P(x) = 2.45x^3 - 3.14x^2 - 6.99x + 2.58$$

[−4, 4] by [−10, 10]

Xscl = 1   Yscl = 1

The other two x-intercepts are approximately .33 and 2.32.

FIGURE 72

Therefore, the solution set of the equation $P(x) = 0$ is

$$\{-1.37, .33, 2.32\}.$$

Based on the graph, the solution set of $P(x) > 0$ is

$$(-1.37, .33) \cup (2.32, \infty),$$

while that of $P(x) < 0$ is

$$(-\infty, -1.37) \cup (.33, 2.32). \quad ◆$$

Note   The graphical method of solving $P(x) = 0$ in Example 4 would not have yielded imaginary solutions had there been any. Only real solutions are obtained using this method.

COMPLEX nth ROOTS

If n is a positive integer and k is a nonzero complex number, then a solution of $x^n = k$ is called an **nth root of k.** For example, since -1 and 1 are solutions of $x^2 = 1$, they are called second or square roots of 1. Similarly, $-2i$ and $2i$ are called square roots of -4, since $(\pm 2i)^2 = -4$.

The real number 2 is a sixth root of 64, since $2^6 = 64$. However, it can be shown that 64 has five more complex sixth roots. While a complete discussion of the following theorem requires concepts from trigonometry, we will state it and use it to solve particular problems involving nth roots.

COMPLEX *n*TH ROOTS THEOREM

If n is a positive integer and k is a nonzero complex number, then the equation $x^n = k$ has *exactly n* complex roots.

EXAMPLE 5   *Finding the nth Roots of a Number*   Find all six complex sixth roots of 64.

SOLUTION   We must find all six complex roots of $x^6 = 64$.

$$x^6 = 64 \qquad \text{Equation to solve}$$

$$x^6 - 64 = 0 \qquad \text{Subtract 64.}$$

$$(x^3 - 8)(x^3 + 8) = 0 \qquad \text{Factor the difference of two squares.}$$

$$(x - 2)(x^2 + 2x + 4)(x + 2)(x^2 - 2x + 4) = 0 \qquad \text{Factor the difference of cubes and the sum of cubes.}$$

Now we apply the zero-product theorem to obtain the real roots 2 and -2. Setting the quadratic factors equal to zero and applying the quadratic formula twice gives us the remaining four complex roots (all imaginary).

$$x^2 + 2x + 4 = 0 \qquad \text{or} \quad x^2 - 2x + 4 = 0$$

$$x = \frac{-2 \pm \sqrt{2^2 - 4(1)(4)}}{2(1)} \qquad\qquad x = \frac{2 \pm \sqrt{(-2)^2 - 4(1)(4)}}{2(1)}$$

$$= \frac{-2 \pm \sqrt{-12}}{2} \qquad\qquad\qquad = \frac{2 \pm \sqrt{-12}}{2}$$

$$= \frac{-2 \pm 2i\sqrt{3}}{2} \qquad\qquad\qquad = \frac{2 \pm 2i\sqrt{3}}{2}$$

$$= -1 \pm i\sqrt{3} \qquad\qquad\qquad = 1 \pm i\sqrt{3}$$

Therefore, the six complex sixth roots of 64 are

$$2, \quad -2, \quad -1 + i\sqrt{3}, \quad -1 - i\sqrt{3}, \quad 1 + i\sqrt{3}, \quad \text{and} \quad 1 - i\sqrt{3}.$$

The graph of $y = x^6 - 64$ confirms the two real sixth roots of 64. See Figure 73. The x-intercepts are -2 and 2.   ◆

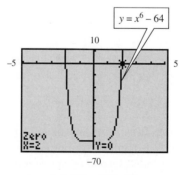

$$y = x^6 - 64$$

[−5, 5] by [−70, 10]
Xscl = 1   Yscl = 10

The other zero is −2. The two real sixth roots of 64 are −2 and 2.

FIGURE 73

APPLICATIONS OF POLYNOMIAL FUNCTIONS

In Section 3.4, we saw how one class of polynomial functions, quadratics, can be used to solve certain types of applied problems. We will now see how higher degree polynomial functions can be used similarly.

EXAMPLE 6   *Using a Polynomial Function to Describe the Volume of a Box*
A box with an open top is to be constructed from a rectangular 12-inch by 20-inch piece of cardboard by cutting equal size squares from each corner and folding up the sides.

(a)   If x represents the length of the side of each such square, determine a function V that describes the volume of the box in terms of x.

(b)   Graph V in the window [0, 6] by [0, 300], and locate a point on the graph. Interpret the displayed values of x and y.

(c)   Determine the value of x for which the volume of the box is maximized. What is this volume?

(d)   For what values of x is the volume equal to 200 cubic units? Greater than 200 cubic units? Less than 200 cubic units?

SOLUTION

(a)   As shown in Figure 74, the dimensions of the box to be formed will be

$$\text{length} = 20 - 2x$$
$$\text{width} = 12 - 2x$$
$$\text{height} = x. \qquad \text{All in inches}$$

Furthermore, x must be positive, and both $20 - 2x$ and $12 - 2x$ must be positive, implying that $0 < x < 6$. Since the volume of the box can be found by multiplying length times width times height, the desired function is

$$V(x) = (20 - 2x)(12 - 2x)x \qquad 0 < x < 6$$
$$= 4x^3 - 64x^2 + 240x.$$

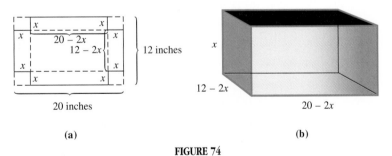

(a)   (b)

FIGURE 74

(b)   Figure 75 on the next page shows the graph of V with the cursor at the arbitrarily chosen point (3.6, 221.184). This means that when the side of each cut-out square measures 3.6 inches, the volume of the resulting box is 221.184 cubic inches.

(c)   We use the capabilities of the calculator to find the local maximum point on the graph of V. To the nearest hundredth, the coordinates of this point are (2.43, 262.68). See Figure 76 on the next page. Therefore, when $x \approx 2.43$ is the length of the side of each square, the volume of the box is at its maximum, approximately 262.68 cubic inches.

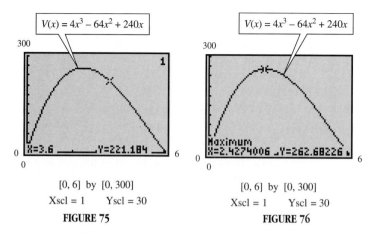

[0, 6] by [0, 300]
Xscl = 1   Yscl = 30

FIGURE 75

[0, 6] by [0, 300]
Xscl = 1   Yscl = 30

FIGURE 76

(d)   Graphing $y_1 = V(x)$ and $y_2 = 200$ gives the graphs shown in Figure 77. The points of intersection of the line and the cubic curve are approximately $(1.17, 200)$ and $(3.90, 200)$. By the intersection-of-graphs method of solving equations and inequalities, the volume is equal to 200 cubic units for $x \approx 1.17$ or 3.90, is greater than 200 cubic units for $1.17 < x < 3.90$, and is less than 200 cubic units for $0 < x < 1.17$ or $3.90 < x < 6$.   ◆

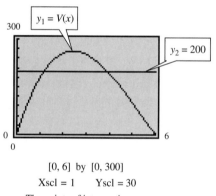

[0, 6] by [0, 300]
Xscl = 1   Yscl = 30
The points of intersection are
approximately (1.17, 200) and (3.90, 200).

FIGURE 77

For Group Discussion

Refer to Example 6 to answer the following.

1.   While we can enter $y_1 = (20 - 2x)(12 - 2x)x$ or $y_1 = 4x^3 - 64x^2 + 240x$ to obtain the desired graph, is it really necessary for us to multiply out the factors when using a graphing technique for solving? What would be a good reason for *not* actually performing the multiplication?

2.   Figure 78, on the following page, shows a comprehensive graph of V in the window $[-5, 20]$ by $[-300, 300]$. Explain why a comprehensive graph was not necessary in solving the problem.

3.   Graph $y_3 = y_1 - 200$ in the window $[0, 6]$ by $[-50, 100]$, and explain how part (d) could be solved using the graph of y_3.

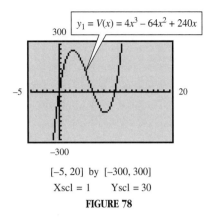

$$[-5, 20] \text{ by } [-300, 300]$$
$$\text{Xscl} = 1 \qquad \text{Yscl} = 30$$
FIGURE 78

POLYNOMIAL MODELS

We saw in Chapter 1 that data can sometimes be fit to linear models, while earlier in this chapter we saw how quadratic functions can be used to model data. Higher degree polynomial functions may be used as well, and modern graphing calculators provide in their statistical functions the capability for determining many types of models.

EXAMPLE 7   *Examining Polynomial Models for Debit Card Use*   The graph in Figure 79 shows the number of transactions made, in millions, by users of off-line (the kind you sign for) debit cards issued by MasterCard and Visa during the years 1991 through 1995. (*Source*: Debit Card News.)

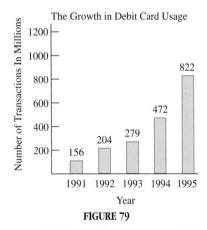

FIGURE 79

(a)   Using $x = 1$ to represent 1991, $x = 2$ to represent 1992, and so on, use the polynomial-fitting capability of a calculator to determine a quadratic function that best fits these data. Plot the data and the graph.

(b)   Repeat part (a) for a cubic function.

(c)   Repeat part (a) for a quartic function.

(d)   Use the three functions in (a)–(c) to predict the number of transactions for 1996. Compare the results to the actual number, 1213 million.

SOLUTION

(a)   The best-fitting quadratic function is approximately $y = 51.57x^2 - 149.43x + 267.6$. The data points and the graph are seen in Figure 80(a). (*Note*: The graph and those to follow in parts (b) and (c) were generated by the equations determined with the calculator, which contain more decimal digits in the coefficients than given here.)

(b) The best-fitting cubic function is approximately $y = 10.83x^3 - 45.93x^2 + 106.24x + 85.60$. See Figure 80(b).

(c) The best-fitting quartic function is approximately $y = -2.17x^4 + 36.83x^3 - 153.33x^2 + 282.67x - 8.00$. See Figure 80(c).

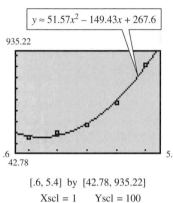

$y \approx 51.57x^2 - 149.43x + 267.6$

935.22

.6   5.4
42.78

[.6, 5.4] by [42.78, 935.22]
Xscl = 1   Yscl = 100

(a)

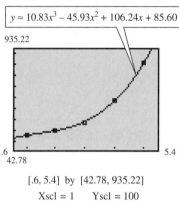

$y \approx 10.83x^3 - 45.93x^2 + 106.24x + 85.60$

935.22

.6   5.4
42.78

[.6, 5.4] by [42.78, 935.22]
Xscl = 1   Yscl = 100

(b)

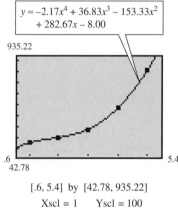

$y \approx -2.17x^4 + 36.83x^3 - 153.33x^2 + 282.67x - 8.00$

935.22

.6   5.4
42.78

[.6, 5.4] by [42.78, 935.22]
Xscl = 1   Yscl = 100

(c)

FIGURE 80

TECHNOLOGY NOTE

The values of Xmin, Xmax, Ymin, and Ymax in Figure 80 were determined by the automatic statistical zoom function of the calculator.

(d) Using the actual equations determined by the calculator and evaluating each of the functions in parts (a)–(c) for the year 1996 ($x = 6$), we obtain the following:

part (a), quadratic function: 1227.6 million

part (b), cubic function: 1409.6 million

part (c), quartic function: 1316 million.

The quadratic function in part (a) gives the closest approximation, 1227.6 million, for the actual 1996 number of 1213 million. ◆

3.8 EXERCISES   **Tape 5**

Solve each equation analytically for all complex solutions, giving exact *forms in your solution set. Then, graph the left-hand side of the equation as y_1 in the suggested viewing window, and support the real solutions, using the capabilities of the calculator.*

1. $4x^4 - 25x^2 + 36 = 0$;
[−5, 5] by [−5, 100]

2. $4x^4 - 29x^2 + 25 = 0$;
[−5, 5] by [−50, 100]

3. $x^4 - 15x^2 - 16 = 0$;
[−5, 5] by [−100, 100]

4. $9x^4 + 35x^2 - 4 = 0$;
[−3, 3] by [−10, 100]

5. $x^3 - x^2 - 64x + 64 = 0$;
[−10, 10] by [−300, 300]

6. $x^3 + 6x^2 - 100x - 600 = 0$;
[−15, 15] by [−1000, 300]

7. $-2x^3 - x^2 + 3x = 0$;
[−4, 4] by [−10, 10]

8. $-5x^3 + 13x^2 + 6x = 0$;
[−4, 4] by [−2, 30]

9. $x^3 + x^2 - 7x - 7 = 0$;
[−10, 10] by [−20, 20]

10. $x^3 + 3x^2 - 19x - 57 = 0$;
[−10, 10] by [−100, 50]

11. $3x^3 + x^2 - 6x = 0$;
[−4, 4] by [−10, 10]

12. $-4x^3 - x^2 + 4x = 0$;
[−4, 4] by [−10, 10]

13. $3x^3 + 3x^2 + 3x = 0$;
[−5, 5] by [−5, 5]

14. $2x^3 + 2x^2 + 12x = 0$;
[−10, 10] by [−20, 20]

15. $x^4 + 17x^2 + 16 = 0$;
[−4, 4] by [−10, 40]

16. $36x^4 + 85x^2 + 9 = 0$;
$[-4, 4]$ by $[-10, 40]$

17. $x^6 + 19x^3 - 216 = 0$;
$[-4, 4]$ by $[-350, 200]$

18. $8x^6 + 7x^3 - 1 = 0$;
$[-4, 4]$ by $[-5, 100]$

19. $3x^4 - 12x^2 + 1 = 0$;
$[-10, 10]$ by $[-15, 10]$

20. $4x^4 - 13x^2 + 2 = 0$;
$[-10, 10]$ by $[-10, 10]$

RELATING CONCEPTS (EXERCISES 21–24)

The graph of $y = x^4 - 28x^2 + 75$ *is shown in the window* $[-6, 0]$ *by* $[-150, 100]$ *in the figure.*

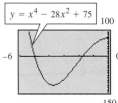

$y = x^4 - 28x^2 + 75$ 100

-6 0

-150

21. Based on the discussion of comprehensive graphs of polynomial functions in Section 3.5, is this a comprehensive graph of the function? Why or why not?

22. If this function is graphed over the domain of all real numbers, what symmetry is exhibited? (Recall the concepts of symmetry introduced in Section 2.1.)

23. The x-intercepts for the window shown are -5 and $-\sqrt{3}$. What is the *complete* solution set of $x^4 - 28x^2 + 75 = 0$? (Section 2.1)

24. Does the equation $x^4 - 28x^2 + 75 = 0$ have any imaginary solutions? Explain your answer. (Sections 3.6, 3.7)

In Exercises 25–30, a polynomial $P(x)$ is given in both descending powers of the variable and factored form. Graph the polynomial by hand, as shown in Section 3.7, or in the window suggested, and solve the equation and the inequalities given.

25. $P(x) = x^3 - 3x^2 - 6x + 8$
$= (x - 4)(x - 1)(x + 2)$
Window: $[-10, 10]$ by $[-15, 15]$

  (a) $P(x) = 0$

  (b) $P(x) < 0$

  (c) $P(x) > 0$

26. $P(x) = x^3 + 4x^2 - 11x - 30$
$= (x - 3)(x + 2)(x + 5)$
Window: $[-10, 10]$ by $[-40, 40]$

  (a) $P(x) = 0$

  (b) $P(x) < 0$

  (c) $P(x) > 0$

27. $P(x) = 2x^4 - 9x^3 - 5x^2 + 57x - 45$
$= (x - 3)^2(2x + 5)(x - 1)$
Window: $[-5, 5]$ by $[-120, 50]$

  (a) $P(x) = 0$

  (b) $P(x) < 0$

  (c) $P(x) > 0$

28. $P(x) = 4x^4 + 27x^3 - 42x^2 - 445x - 300$
$= (x + 5)^2(4x + 3)(x - 4)$
Window: $[-10, 10]$ by $[-1200, 400]$

  (a) $P(x) = 0$

  (b) $P(x) < 0$

  (c) $P(x) > 0$

29. $P(x) = -x^4 - 4x^3 + 3x^2 + 18x$
$= x(2 - x)(x + 3)^2$
Window: $[-5, 5]$ by $[-30, 30]$

  (a) $P(x) = 0$

  (b) $P(x) \geq 0$

  (c) $P(x) \leq 0$

30. $P(x) = -x^4 + 2x^3 + 8x^2$
$= x^2(4 - x)(x + 2)$
Window: $[-6, 6]$ by $[-10, 50]$

  (a) $P(x) = 0$

  (b) $P(x) \geq 0$

  (c) $P(x) \leq 0$

Prior to the 1970s, courses in the theory of equations were taught at the undergraduate level and formed part of a typical mathematics major's curriculum. Such courses dealt with various algebraic techniques for solving cubic and quartic equations, as well as other topics dealing with analysis of polynomial functions. In Exercises 31–34, you are given equations with one of the real roots in its exact value, as determined in the classic text Theory of Equations *by J. V. Uspensky (New York: McGraw-Hill, 1948). Graph the function on the left-hand side of the equation as* y_1 *in the suggested window, and then use the capabilities of your calculator to locate the x-intercept whose approximation corresponds to that of the exact root given. Give the approximation of the root to as many places as your calculator will provide.*

31. $x^3 - 6x - 6 = 0$;   window: $[-6, 6]$ by $[-20, 10]$

Exact root: $\sqrt[3]{2} + \sqrt[3]{4}$

32. $x^3 - 12x - 34 = 0$;   window: $[-6, 6]$ by $[-100, 20]$

Exact root: $\sqrt[3]{2} + 2\sqrt[3]{4}$

33. $x^3 + 9x - 2 = 0$;   window: $[-5, 5]$ by $[-20, 20]$

Exact root: $\sqrt[3]{\sqrt{28} + 1} - \sqrt[3]{\sqrt{28} - 1}$

34. $x^4 + 5x^3 + x^2 - 13x + 6 = 0$;   window: $[-6, 6]$ by $[-5, 50]$

Exact root: $\dfrac{-3 + \sqrt{17}}{2}$

35. One of the most interesting stories in the history of mathematics involves the dispute between Nicolo Fontana (Tartaglia) and Girolamo Cardano, two sixteenth-century Italian mathematicians. The source of the dispute was the origin of a formula for solving for one root of a cubic equation of the form $x^3 + mx = n$. The formula is

$$x = \sqrt[3]{\frac{n}{2} + \sqrt{\left(\frac{n}{2}\right)^2 + \left(\frac{m}{3}\right)^3}} - \sqrt[3]{-\left(\frac{n}{2}\right) + \sqrt{\left(\frac{n}{2}\right)^2 + \left(\frac{m}{3}\right)^3}}.$$

(a) Solve the equation $x^3 + 9x = 26$ for its single real root using this formula.

(b) Support your answer in part (a) by finding the x-intercept of the graph of $y = x^3 + 9x - 26$.

(c) Find the two imaginary roots by synthetically dividing $x^3 + 9x - 26$ by $x - k$, where k is the real root, and then solving the equation $Q(x) = 0$, where $Q(x)$ is the quadratic quotient polynomial.

36. A method for solving fourth degree polynomial equations is described in Rees and Sparks, *College Algebra*, 5th Edition, New York: McGraw-Hill, 1967. As an example, the equation

$$x^4 + 2x^3 - x^2 + x + \frac{1}{4} = 0$$

is solved and shown to have the four solutions

$$\underbrace{\frac{-1 + \sqrt{3} \pm \sqrt{2}\sqrt{1 - \sqrt{3}}}{2}}_{\text{Imaginary}}, \quad \underbrace{\frac{-1 - \sqrt{3} \pm \sqrt{2}\sqrt{1 + \sqrt{3}}}{2}}_{\text{Real}}.$$

Use a calculator to support the two real solutions graphically.

Use graphical methods (either intersection-of-graphs method or x-intercept method) or the equation-solving feature of a graphing calculator to find all real solutions of the following equations. Express solutions rounded to the nearest hundredth.

37. $.86x^3 - 5.24x^2 + 3.55x + 7.84 = 0$

38. $-2.47x^3 - 6.58x^2 - 3.33x + .14 = 0$

39. $-\sqrt{7}x^3 + \sqrt{5}x^2 + \sqrt{17} = 0$

40. $\sqrt{10}x^3 - \sqrt{11}x - \sqrt{8} = 0$

41. $2.45x^4 - 3.22x^3 = -.47x^2 + 6.54x + 3$

42. $\sqrt{17}x^4 - \sqrt{22}x^2 = -1$

Find all n complex solutions of each of the following equations of the form $x^n = k$.

43. $x^2 = -1$ **44.** $x^2 = -4$ **45.** $x^3 = -1$ **46.** $x^3 = -8$ **47.** $x^3 = 27$

48. $x^3 = 64$ **49.** $x^4 = 16$ **50.** $x^4 = 81$ **51.** $x^6 = 1$

52. Consider the equation $x^8 = 1$. This equation has eight distinct solutions, each of which is an eighth root of 1. (Roots of 1 are often called *roots of unity*.)

(**a**) Graph $y = x^8 - 1$ to determine the two real eighth roots of unity.

(**b**) Show analytically that i is also an eighth root of unity.

(**c**) Based on the conjugate zeros theorem and the result of part (b), what must be another eighth root of unity? Verify that this is a root analytically.

(**d**) Using concepts from trigonometry, it can be shown that

$$\frac{\sqrt{2}}{2} + i\frac{\sqrt{2}}{2} \quad \text{and} \quad -\frac{\sqrt{2}}{2} + i\frac{\sqrt{2}}{2}$$

are imaginary eighth roots of unity. Based on the conjugate zeros theorem, what two other imaginary numbers must also be eighth roots of unity?

(**e**) List all eight eighth roots of unity.

Solve each of the following problems. Use a graphical method to find numerical answers, and give approximations to the nearest hundredth.

53. *Volume of a Box*   A rectangular piece of cardboard measuring 12 inches by 18 inches is to be made into a box with an open top by cutting equal size squares from each corner and folding up the sides. Let x represent the length of a side of each such square.

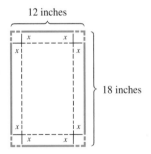

(**a**) Give the restrictions on x.

(**b**) Describe a function V that gives the volume of the box as a function of x.

(**c**) For what value of x will the volume be a maximum? What is this maximum volume?

(**d**) For what values of x will the volume be greater than 80 cubic inches?

54. *Construction of a Rain Gutter*   A piece of rectangular sheet metal is 20 inches wide. It is to be made into a

rain gutter by turning up the edges to form parallel sides. Let x represent the length of each of the parallel sides.

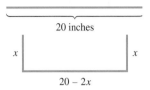

(**a**) Give the restrictions on x.

(**b**) Describe a function A that gives the area of a cross-section of the gutter.

(**c**) For what value of x will A be a maximum (and thus maximize the amount of water that the gutter will hold)? What is this maximum area?

(**d**) For what values of x will the area of a cross-section be less than 40 square inches?

55. *Buoyancy of a Spherical Object*   It has been determined that a spherical object of radius 4 inches with specific gravity .25 will sink in water to a depth of x inches, where x is a positive root of the equation $x^3 - 12x^2 + 64 = 0$. To what depth will this object sink given that $x < 10$?

56. *Area of a Rectangle*   Find the value of x in the figure that will maximize the area of rectangle $ABCD$.

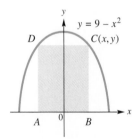

57. *Sides of a Right Triangle*   A certain right triangle has an area of 84 square inches. One leg of the triangle measures 1 inch less than the hypotenuse. Let x represent the length of the hypotenuse.

(**a**) Express the length of the leg mentioned above in terms of x.

(**b**) Express the length of the other leg in terms of x.

(**c**) Write an equation based on the information determined thus far. Square both sides, and then write the equation with one side as a polynomial with integer coefficients, in descending powers, and the other side equal to 0.

(d) Solve the equation in part (c) graphically. Find the lengths of the three sides of the triangle.

58. *Butane Gas Storage*   A storage tank for butane gas is to be built in the shape of a right circular cylinder of altitude 12 feet, with a half sphere attached to each end. If x represents the radius of each half sphere, what radius should be used to cause the volume of the tank to be 144π cubic feet?

Volume of a Box   A standard piece of notebook paper measuring 8.5 inches by 11 inches is to be made into a box with an open top by cutting equal size squares from each corner and folding up the sides. Let x represent the length of a side of each such square.

59. Use the table feature of your graphing calculator to find the maximum volume of the box.

60. Use the table feature to determine when the volume of the box will be greater than 40 cubic inches.

Solve each problem involving modeling with polynomial functions.

61. *Community College Attendance*   Use the points (0, 118,000), (2, 122,000), (3, 63,000), and (4, 70,000) to determine a cubic (third degree) polynomial model for these data, where $x = 0$ corresponds to 1990 and $x = 4$ corresponds to 1994.

Students Attending Community Colleges after Earning a B.A. or B.S. Degree

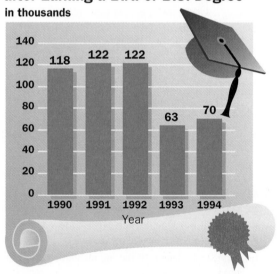

(*Source:* Chancellor's Office, California Community Colleges.)

(a) Graph the function in the window [0, 4] by [0, 180,000].

(b) Based on the cubic model, what was the statewide total of students with bachelor's degrees in 1991?

(c) How does your answer in part (b) compare with the bar graph?

62. *Community College Enrollment*   Use the points (0, 1.24), (2, 1.18), (7, 1.5), (9, 1.5), and (11, 1.36) to determine a quartic (fourth degree) model for these data, where $x = 0$ corresponds to 1983 and $x = 11$ corresponds to 1994.

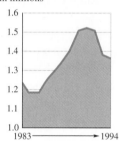

Statewide Fall Enrollment at Community Colleges
in millions

(*Source:* Chancellor's Office, California Community Colleges.)

(a) Graph the function in the window [0, 11] by [1.0, 1.6].

(b) Based on the quartic model, what was the statewide fall enrollment, in millions, in 1988?

(c) How does your answer in part (b) compare with the line graph?

63. *Breast Cancer Cases*   From 1930 to 1990, the rate of breast cancer was nearly constant at 30 cases per 100,000 females, whereas the rate of lung cancer in females over the same period increased. The number of lung cancer cases per 100,000 females in the year t (where $t = 0$ corresponds to 1930) can be modeled, using the function defined by $f(t) = 2.8 \times 10^{-4}t^3 - .011t^2 + .23t + .93$. (*Source:* Valanis, B., *Epidemiology in Nursing and Health Care*, Norwalk, Conn.: Appleton & Lange, 1992.)

(a) Use a graphing calculator to graph the rates of breast and lung cancer for $0 \le t \le 60$. Use the window [0, 60] by [0, 40].

(b) Determine the year when rates for lung cancer first exceeded those for breast cancer.

(c) Discuss reasons for the rapid increase of lung cancer in females.

64. *Concentration of Toxin*   A survey team measures the concentration (in parts per million) of a particular toxin in a local river. On a normal day, the concentration of the toxin at time x (in hours) after the factory upstream dumps its waste is given by $g(x) = -.006x^4 + .14x^3 - .05x^2 + .02x$, where $0 \le x \le 24$.

(a) Graph $y = g(x)$ in the window [0, 24] by [0, 200].

(b) Estimate the time at which the concentration is greatest.

(c) A concentration greater than 100 parts per million is considered pollution. Using the graph from part (a), estimate the period during which the river is polluted.

65. *Use of Debit Cards* The bar graph shows the number of transactions made, in millions, by users of on-line (the kind you need a PIN code to use) debit cards issued by MasterCard and Visa during the years 1991 through 1995. (*Source:* Debit Card News.)

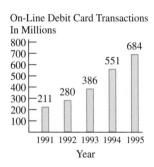

On-Line Debit Card Transactions In Millions

(a) Using $x = 1$ to represent 1991, $x = 2$ to represent 1992, and so on, use the polynomial-fitting capability of a calculator to determine a quadratic function that best fits these data. Plot the data and the graph.

(b) Repeat part (a) for a cubic function.

(c) Repeat part (a) for a quartic function.

(d) Use the three functions in (a)–(c) to predict the number of transactions for 1996. Compare the results to the actual number of 905 million.

66. *U.S. School Enrollment* School enrollment in the United States, in millions, for the years 1986–1993 is shown in the bar graph. (*Source:* National Education Association.)

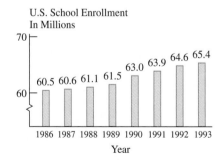

U.S. School Enrollment In Millions

(a) Using $x = 1$ to represent 1986, $x = 2$ to represent 1987, and so on, use the polynomial-fitting capabil-

ity of a calculator to determine a quadratic function that best fits these data. Plot the data and the graph.

(b) Repeat part (a) for a cubic function.

(c) Repeat part (a) for a quartic function.

(d) Use the three functions in (a)–(c) to predict the school enrollment for 1994. Compare the results to the actual number, 69.3 million.

67. *Swing of a Pendulum* A simple pendulum will swing back and forth in regular time intervals (periods). Grandfather clocks use pendulums to keep accurate time. The relationship between the length of a pendulum L and the time T for one complete oscillation can be expressed by the equation $L = kT^n$, where k is a constant and n is a positive integer to be determined. The following data were taken for different lengths of pendulums.

L (ft)	T (sec)
1.0	1.11
1.5	1.36
2.0	1.57
2.5	1.76
3.0	1.92
3.5	2.08
4.0	2.22

(a) As L increases, what happens to T?

(b) Discuss how n and k could be found.

(c) Use the data to approximate k, and determine the best value for n.

(d) Using the values of k and n from part (c), predict T for a pendulum having a length of 5 feet.

(e) If the length L of a pendulum doubles, what happens to the period T?

68. *Deer Population* During the early part of the twentieth century, the deer population of the Kaibab Plateau in Arizona experienced a rapid increase because hunters had reduced the number of natural predators and because the deer were protected from hunters. The increase in population depleted the food resources and eventually caused the population to decline. For the period from 1905 to 1930, the deer population was approximated by $D(x) = -.125x^5 + 3.125x^4 + 4000$, where x is time in years from 1905.

(a) Graph $y = D(x)$ in the window [0, 50] by [0, 120,000].

(b) From the graph, over what period of time (from 1905 to 1930) was the deer population at its maximum?

CHAPTER 3   SUMMARY

Section	Important Concepts
3.1 Complex Numbers	**DEFINITION OF i** $$i = \sqrt{-1} \quad \text{or} \quad i^2 = -1$$ **DEFINITION OF COMPLEX NUMBER** A number in the form $a + bi$, where a and b are real numbers and i is the imaginary unit, is called a complex number. **DEFINITION OF $\sqrt{-a}$** If $a > 0$, then $$\sqrt{-a} = i\sqrt{a}.$$ **DEFINITION OF COMPLEX CONJUGATE** The complex conjugate of $a + bi$ is $a - bi$. **OPERATIONS WITH COMPLEX NUMBERS** Addition: $(a + bi) + (c + di) = (a + c) + (b + d)i$ Subtraction: $(a + bi) - (c + di) = (a - c) + (b - d)i$ Multiplication: $(a + bi)(c + di) = (ac - bd) + (ad + bc)i$ Rule for Division: To divide complex numbers, multiply both the numerator and the denominator by the conjugate of the denominator.

3.2 Quadratic Functions and Their Graphs

GRAPH OF $P(x) = a(x - h)^2 + k$

The graph of $P(x) = a(x - h)^2 + k, a \neq 0$,

a.   is a parabola with vertex (h, k), and the vertical line $x = h$ as axis of symmetry;

b.   opens upward if $a > 0$ and downward if $a < 0$;

c.   is broader than $y = x^2$ if $0 < |a| < 1$ and narrower than $y = x^2$ if $|a| > 1$.

VERTEX FORMULA

The vertex of the graph of $P(x) = ax^2 + bx + c$ $(a \neq 0)$ is the point $\left(-\frac{b}{2a}, P\left(-\frac{b}{2a}\right)\right)$.

ZERO-PRODUCT PROPERTY

If a and b are complex numbers, and $ab = 0$, then $a = 0$ or $b = 0$ or both.

EXTREME POINT AND EXTREME VALUE OF A QUADRATIC FUNCTION

For the quadratic function $P(x) = ax^2 + bx + c$,

a.   if $a > 0$, the vertex (h, k) is called the *minimum point* of the graph. The *minimum value* of the function is $P(h) = k$.

b.   if $a < 0$, the vertex (h, k) is called the *maximum point* of the graph. The *maximum value* of the function is $P(h) = k$.

END BEHAVIOR AND CONCAVITY OF THE GRAPH OF A QUADRATIC FUNCTION

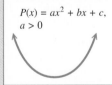 $P(x) = ax^2 + bx + c,$ $a > 0$

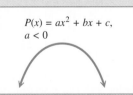 $P(x) = ax^2 + bx + c,$ $a < 0$

If $a > 0$,
as $x \to -\infty, P(x) \to \infty$;
as $x \to \infty, P(x) \to \infty$.
This graph is concave up.

If $a < 0$,
as $x \to -\infty, P(x) \to -\infty$;
as $x \to \infty, P(x) \to -\infty$.
This graph is concave down.

3.3 Solution of
Quadratic Equations
and Inequalities

SQUARE ROOT PROPERTY FOR SOLVING QUADRATIC EQUATIONS
The solution set of $x^2 = k$ is
a. $\left\{ \pm \sqrt{k} \right\}$ if $k > 0$
b. $\{0\}$ if $k = 0$
c. $\left\{ \pm i \sqrt{|k|} \right\}$ if $k < 0$.

QUADRATIC FORMULA
The solutions of the quadratic equation $ax^2 + bx + c = 0$, where $a \neq 0$, are

$$x = \frac{-b \pm \sqrt{b^2 - 4ac}}{2a}.$$

EFFECT OF THE DISCRIMINANT
If a, b, and c are real numbers, $a \neq 0$, then the complex solutions of $ax^2 + bx + c = 0$ are
described as follows, based on the value of the discriminant, $b^2 - 4ac$.

Value of $b^2 - 4ac$	Number of Solutions	Nature of Solutions
Positive	Two	Complex, real
Zero	One (a double solution)	Complex, real
Negative	Two	Complex, imaginary

SOLVING A QUADRATIC INEQUALITY ANALYTICALLY

1. Solve the corresponding quadratic equation.
2. Identify the intervals determined by the solutions of the equation.
3. Use a sign graph to determine which intervals are in the solution set.
4. Decide whether or not the endpoints are included.

3.4 Applications of
Quadratic Functions and
Models

FORMULA FOR THE HEIGHT OF A PROPELLED OBJECT
If air resistance is neglected, the height s (in feet) of an object propelled directly upward from
an initial height s_0 feet with initial velocity v_0 feet per second is described by the function

$$s(t) = -16t^2 + v_0 t + s_0,$$

where t is the number of seconds after the object is propelled.

It is possible that a quadratic function can be used to model data. A model can be determined by choosing an appropriate data point (h, k) as vertex, and using another point to find the
value of a in $f(x) = a(x - h)^2 + k$.

3.5 Higher Degree
Polynomial Functions
and Their Graphs

EXTREMA OF POLYNOMIAL FUNCTIONS

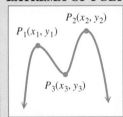

P_1 is a local maximum point.
P_2 is a local maximum point.
P_3 is a local minimum point.

MAXIMUM NUMBER OF LOCAL EXTREMA
The maximum number of local extrema of the graph of a polynomial function of degree n is
$n - 1$.

END BEHAVIOR OF ODD DEGREE POLYNOMIAL FUNCTIONS

NUMBER OF x-INTERCEPTS (REAL ZEROS) OF A POLYNOMIAL FUNCTION

A polynomial function of degree n will have a maximum of n x-intercepts (real zeros).

END BEHAVIOR OF EVEN DEGREE POLYNOMIAL FUNCTIONS

3.6 Topics in the Theory of Polynomial Functions (I)

THE INTERMEDIATE VALUE THEOREM FOR POLYNOMIAL FUNCTIONS

If $P(x)$ defines a polynomial function, and if $P(a) \neq P(b)$, then P takes on every value between $P(a)$ and $P(b)$.

ROOT LOCATION THEOREM FOR POLYNOMIAL FUNCTIONS

Suppose that a polynomial function P is defined in such a way that for real numbers a and b, $P(a)$ and $P(b)$ differ in sign. Then there exists at least one real number c between a and b such that $P(c) = 0$.

DIVISION OF A POLYNOMIAL BY $x - k$

1. If the degree n polynomial $P(x)$ is divided by $x - k$, the quotient polynomial, $Q(x)$, has degree $n - 1$.
2. The remainder R is a constant (and may be 0). The complete quotient for $\frac{P(x)}{x-k}$ may be written as follows:

$$\frac{P(x)}{x - k} = Q(x) + \frac{R}{x - k}.$$

THE REMAINDER THEOREM

If a polynomial $P(x)$ is divided by $x - k$, the remainder is equal to $P(k)$.

THE FACTOR THEOREM

The polynomial $x - k$ is a factor of the polynomial $P(x)$ if and only if $P(k) = 0$.

3.7 Topics in the Theory of Polynomial Functions (II)

CONJUGATE ZEROS THEOREM

If $P(x)$ is a polynomial having only real coefficients, and if $a + bi$ is a zero of P, then the conjugate $a - bi$ is also a zero of P.

THE FUNDAMENTAL THEOREM OF ALGEBRA

Every function defined by a polynomial of degree 1 or more has at least one complex zero.

ZEROS OF A POLYNOMIAL FUNCTION

1. A function defined by a polynomial of degree n has at most n distinct complex zeros.
2. A function defined by a polynomial of degree n has exactly n complex zeros if zeros of multiplicity m are counted m times.

BEHAVIOR OF THE GRAPH OF A POLYNOMIAL FUNCTION NEAR THE x-AXIS

| The graph crosses the x-axis at $(c, 0)$ if c is a zero of odd multiplicity. | The graph is tangent to the x-axis at $(c, 0)$ if c is a zero of even multiplicity. |

THE RATIONAL ZEROS THEOREM

Let $P(x) = a_n x^n + a_{n-1} x^{n-1} + \cdots + a_1 x + a_0$, where $a_n \neq 0$, define a polynomial function with integer coefficients. If p/q is a rational number written in lowest terms, and if p/q is a zero of P, then p is a factor of the constant term a_0, and q is a factor of the leading coefficient a_n.

3.8 Solution of Polynomial Equations and Inequalities and Their Applications

COMPLEX nTH ROOTS THEOREM

If n is a positive integer and k is a nonzero complex number, then the equation $x^n = k$ has *exactly n* complex roots.

Graphing calculators can be used to find the best-fitting cubic and quartic models for a set of data points.

CHAPTER 3   REVIEW EXERCISES

Let $w = 17 - i$ and let $z = 1 - 3i$. Write each complex number in $a + bi$ form.

1. $w + z$ **2.** $w - z$ **3.** wz **4.** w^2 **5.** $\dfrac{1}{z}$ **6.** $\dfrac{w}{z}$

Consider the function defined by $P(x) = 2x^2 - 6x - 8$ for Exercises 7–17.

7. What is the domain of P?

8. Determine analytically the coordinates of the vertex of the graph.

9. Use an end behavior diagram to describe the end behavior of the graph of P.

10. Determine analytically the x-intercepts, if any, of the graph of P.

11. Determine analytically the y-intercept of the graph of P.

12. What is the range of P?

13. Over what interval is the function increasing? Over what interval is it decreasing?

14. Give the solution set of each of the following.

(a) $2x^2 - 6x - 8 = 0$ (b) $2x^2 - 6x - 8 > 0$ (c) $2x^2 - 6x - 8 \leq 0$

15. The graph of *P* is shown here. Explain how the graph supports your solution sets in Exercise 14.

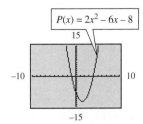

$P(x) = 2x^2 - 6x - 8$

16. What is the equation of the axis of symmetry of the graph in Exercise 15?

17. Discuss the concavity of the graph in Exercise 15.

Consider the function defined by $P(x) = -2.64x^2 + 5.47x + 3.54$ for Exercises 18–22.

18. Use the discriminant to explain how you can determine the number of *x*-intercepts the graph of *P* will have even before graphing it on your calculator.

19. Graph the function in the standard window of your calculator, and use the root-finding capabilities to solve the equation $P(x) = 0$. Express solutions as approximations to the nearest hundredth.

20. Use your answer to Exercise 19 and the graph of *P* to solve **(a)** $P(x) > 0$ and **(b)** $P(x) < 0$.

21. Use the capabilities of your calculator to find the coordinates of the vertex of the graph. Express coordinates to the nearest hundredth.

22. Verify *analytically* that your answer in Exercise 21 is correct.

23. The figure shows the graph of a quadratic function $y = f(x)$.

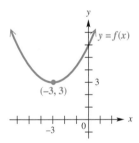

(a) What is the minimum value of $f(x)$?

(b) For what value of *x* is $f(x)$ a minimum?

(c) How many real solutions are there to the equation $f(x) = 2$?

(d) How many real solutions are there to the equation $f(x) = 4$?

Height of a Projectile A projectile is fired vertically upward, and its height $s(t)$ in feet after *t* seconds is given by the function

$$s(t) = -16t^2 + 800t + 600.$$

Graph this function in the window $[0, 60]$ *by* $[0, 11,000]$, *and use either analytic or graphical methods to answer Exercises 24–28.*

24. From what height was the projectile fired?

25. After how many seconds will it reach its maximum height?

26. What is the maximum height it will reach?

27. Between what two times (in seconds, to the nearest tenth) will it be more than 5000 feet above the ground?

28. How long will the projectile be in the air? Give your answer to the nearest tenth of a second.

Solve each problem involving modeling with a quadratic function.

29. *Volume of a Box* A piece of cardboard is 3 times as long as it is wide. Equal size squares measuring 4 inches on each side are to be cut from the corners of the piece of cardboard, and the flaps formed will be folded up to form a box with an open top.

(a) Determine a function *V* that will describe the volume of the box as a function of *x*, where *x* is the original width in inches.

(b) What should be the original dimensions of the piece of cardboard if the box is to have a volume of 2496 cubic inches? Solve this problem analytically.

(c) Support the answer in part (b) graphically.

30. *Concentration of Atmospheric CO_2* The International Panel on Climate Change (IPCC) in 1990 published that if current trends of burning fossil fuel and deforestation continue, then future amounts of atmospheric carbon dioxide in parts per million (ppm) will increase, as shown in the table.

Year	Carbon Dioxide
1990	353
2000	375
2075	590
2175	1090
2275	2000

(a) Plot the data. Let $x = 0$ correspond to 1990.

(b) Find a function $f(x) = a(x - h)^2 + k$ that models the data. Use (0, 353) as the vertex and (285, 2000) as another point to determine *a*.

(c) Use the regression capability of a graphing calculator to find the best-fitting quadratic function *g* for this data.

Use synthetic division to find the quotient $Q(x)$ and the remainder R.

31. $\dfrac{x^3 + x^2 - 11x - 10}{x - 3}$

32. $\dfrac{3x^3 + 8x^2 + 5x + 10}{x + 2}$

Use synthetic division to find P(2).

33. $P(x) = -x^3 + 5x^2 - 7x + 1$

34. $P(x) = 2x^3 - 3x^2 + 7x - 12$

35. $P(x) = 5x^4 - 12x^2 + 2x - 8$

36. $P(x) = x^5 + 4x^2 - 2x - 4$

37. If $P(x)$ is defined by a polynomial with real coefficients, and $7 + 2i$ is a zero of the function, what other complex number must also be a zero?

Find a polynomial function with real coefficients and of lowest degree having the given zeros.

38. $-1, 4, 7$

39. $8, 2, 3$

40. $\sqrt{3}, -\sqrt{3}, 2, 3$

41. $-2 + \sqrt{5}, -2 - \sqrt{5}, -2, 1$

42. Is -1 a zero of $P(x) = 2x^4 + x^3 - 4x^2 + 3x + 1$?

43. Is $x + 1$ a factor of $P(x) = x^3 + 2x^2 + 3x + 2$?

44. Find a polynomial function P with real coefficients of degree 4 with 3, 1, and $-1 - 3i$ as zeros, and $P(2) = -36$.

45. Find a polynomial function P of degree 3 with -2, 1, and 4 as zeros, and $P(2) = 16$.

46. Give an example of a fourth degree polynomial function having exactly two distinct real zeros, and then graph it, using a calculator.

47. Give an example of a cubic polynomial function having exactly one real zero, and then graph it, using a calculator.

48. Find all zeros of $P(x) = x^4 - 3x^3 - 8x^2 + 22x - 24$, given that $1 - i$ is a zero.

49. Find all zeros of $P(x) = 2x^4 - x^3 + 7x^2 - 4x - 4$, given that 1 and $2i$ are zeros.

50. Find all rational zeros of $P(x) = 3x^5 - 4x^4 - 26x^3 - 21x^2 - 14x + 8$. Do this by first listing the possible rational zeros based on the rational zeros theorem.

Consider the function defined by $P(x) = x^3 - 2x^2 - 4x + 3$ for Exercises 51–55.

51. Graph the function in the window $[-10, 10]$ by $[-10, 10]$, giving a comprehensive graph. Based on the graph, how many real solutions does the equation $x^3 - 2x^2 - 4x + 3 = 0$ have? Then, use the root-finding capabilities of your calculator to find the real root that is an integer.

52. Use your answer in Exercise 51 along with synthetic division to factor $x^3 - 2x^2 - 4x + 3$ so that one factor is linear and the other factor is quadratic.

53. Find the exact values of any remaining zeros of P analytically.

54. Use the root-finding capabilities of your calculator to support your answer in Exercise 53.

55. Give the solution set of each inequality, using *exact* values:

(a) $x^3 - 2x^2 - 4x + 3 > 0$

(b) $x^3 - 2x^2 - 4x + 3 \leq 0$

56. Use an analytic method to find all solutions of the equation $x^3 + 2x^2 + 5x = 0$. Then, without graphing, give the exact values of all x-intercepts of the graph. Using your knowledge of end behavior of the graph of $P(x) = x^3 + 2x^2 + 5x$, give the solution set of $P(x) > 0$ and of $P(x) < 0$.

57. The graph of $P(x) = x^4 - 5x^3 + x^2 + 21x - 18$ is shown here. Suppose that you know that all zeros of P are integers, and each linear factor is of degree 1 or 2. Give the factored form of $P(x)$. (*Hint:* Once you have determined your answer, graph the factored form to see if it matches the one shown here.)

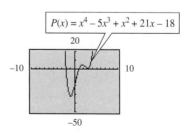

The x-intercepts are -2, 1, and 3.

Comprehensive graphs of polynomial functions f and g are shown here. They have only real coefficients. Answer Exercises 58–66 based on the graphs.

58. Is the degree of g even or odd?

59. Is the degree of f even or odd?

60. Is the leading coefficient of f positive or negative?

61. How many real solutions does $g(x) = 0$ have?

62. Express the solution set of $f(x) < 0$ in interval form.

63. What is the solution set of $f(x) > g(x)$?

64. What is the solution set of $f(x) - g(x) = 0$?

65. If $r + pi$ is an imaginary solution of $g(x) = 0$, what must be another imaginary solution?

66. Suppose that f is of degree 3. Explain why f cannot have imaginary zeros.

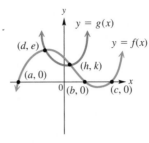

67. The table shows several ordered pairs that lie on the graph of Y_1. Use one of the ordered pairs to find a factor of Y_1. Then, find the exact values of all zeros of Y_1, and write it as a product of three linear factors.

X	Y1
-6	-134
-5	-51
-4	0
-3	25
-2	30
-1	21
0	4

$Y_1 \boxminus X^3 - X^2 - 19X + 4$

68. Use the root location theorem to determine which two x-values from the table have a zero between them. Explain your answer.

X	Y1
.6	.784
.7	.157
.8	-.512
.9	-1.229
1	-2
1.1	-2.831
1.2	-3.728

$Y_1 \boxminus -X^3 - 5X + 4$

Answer true or false to each statement in Exercises 69–74.

69. The function defined by $f(x) = 3x^7 - 8x^6 + 9x^5 + 12x^4 - 18x^3 + 26x^2 - x + 500$ has 8 x-intercepts.

70. The function f in Exercise 69 may have up to 6 local extrema.

71. The function f in Exercise 69 has a positive y-intercept.

72. Based on end behavior of the function f in Exercise 69 and your answer in Exercise 71, the graph must have at least one negative x-intercept.

73. If a polynomial function of even degree has a positive leading coefficient and a negative y-intercept, it must have at least two real zeros.

74. Because $-\dfrac{1}{2} + i\dfrac{\sqrt{3}}{2}$ is a complex zero of $f(x) = x^2 + x + 1$, another zero must be $\dfrac{1}{2} + i\dfrac{\sqrt{3}}{2}$.

Graph the function defined by $P(x) = -2x^5 + 15x^4 - 21x^3 - 32x^2 + 60x$ in the window $[-8, 8]$ by $[-100, 200]$ to obtain a comprehensive graph. Then, use your calculator and the concepts of this chapter to answer Exercises 75–79.

75. How many local maxima does this function have?

76. One local minimum lies on the x-axis and has an integer as its x-value. What are the coordinates of this point?

77. The greatest *x*-intercept is 5. Therefore, $x - 5$ is a factor of $P(x)$. Use synthetic division to find the quotient polynomial $Q(x)$ obtained when $P(x)$ is divided by $x - 5$.

78. What is the range of P?

79. The graph of P has a local minimum with a negative *x*-value. Use your calculator to find its coordinates. Express them to the nearest hundredth.

80. Solve the equation $3x^3 + 2x^2 - 21x - 14 = 0$ analytically for all complex solutions, giving exact values in your solution set. Then, graph the left-hand side of the equation in the viewing window $[-6, 6]$ by $[-35, 35]$, and support the real solutions, using the capabilities of the calculator.

81. Use the results of Exercise 80 to give the solution set of each inequality, expressing all values in exact form.

 (a) $3x^3 + 2x^2 - 21x - 14 \geq 0$

 (b) $3x^3 + 2x^2 - 21x - 14 < 0$

82. Consider the polynomial function $P(x) = -x^4 + 3x^3 + 3x^2 - 7x - 6$. The factored form of the polynomial is $(-x + 2)(x - 3)(x + 1)^2$. Graph the polynomial by hand as explained in Section 3.7, and solve the equation or inequality.

 (a) $P(x) = 0$ **(b)** $P(x) > 0$ **(c)** $P(x) < 0$

Solve each application of polynomial functions.

83. *Dimensions of a Cube* After a 2-inch slice is cut off the top of a cube, the resulting solid has a volume of 32 cubic inches. Find the dimensions of the original cube.

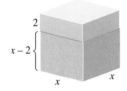

84. *Dimensions of a Box* The width of a rectangular box is 3 times its height, and its length is 11 inches more than its height. Find the dimensions of the box if its volume is 720 cubic inches.

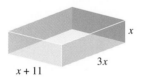

85. *Military Personnel on Active Duty* The number of military personnel on active duty in the United States during the period 1985 to 1990 can be determined by the cubic model $y = -7.66x^3 + 52.71x^2 - 93.43x + 2151$, where $x = 0$ corresponds to 1985, and y is in thousands. Based on this model, how many military personnel were on active duty in 1990? (*Source:* U.S. Department of Defense.)

86. *Catholic Elementary Schools* The number of Catholic elementary schools in the United States for selected years between 1970 and 1990 is shown in the table.

Year	Number
1970	9362
1975	8340
1980	8043
1985	7806
1990	7291

(*Source:* National Catholic Education Association.)

(a) Let $x = 0$ represent 1970, $x = 5$ represent 1975, and so on. Plot the data.

(b) Use the quadratic regression capability of a calculator to determine the best-fitting quadratic model for these data.

(c) Repeat part (b) for a cubic model.

(d) Repeat part (b) for a quartic model. Then, graph all three equations over the data points.

(e) The actual number of Catholic elementary schools in 1992 was 7174. Which one of the three models gives the most accurate prediction for this figure?

CHAPTER 3   TEST

1. Perform each operation with complex numbers. Give answers in $a + bi$ form.

  (a) $(8 - 7i) - (-12 + 2i)$

  (b) $\dfrac{11 + 10i}{2 + 3i}$

  (c) Simplify i^{65}.

  (d)
  $$2i(3-i)^2$$

2. For the quadratic function $f(x) = -2x^2 - 4x + 6$, do each of the following.

  (a) Find the vertex, using an analytic method.

  (b) Give a comprehensive graph and use a calculator to support your result in part (a).

  (c) Find the zeros of f and support your result, using a graph for one zero and a table for the other.

  (d) Find the y-intercept analytically.

  (e) State the domain and range of f.

  (f) Give the interval over which the function is increasing and the interval over which it is decreasing.

3. (a) Solve the quadratic equation $3x^2 + 3x - 2 = 0$ analytically. Give solutions in exact form.

  (b) Graph $f(x) = 3x^2 + 3x - 2$ with a calculator. Use your results in part (a) along with this graph to give the solution set of each inequality. Express endpoints of intervals in exact form.

   (i) $f(x) < 0$ (ii) $f(x) \geq 0$

4. *Height of a Propelled Object*   If air resistance is neglected, an object projected straight upward with an initial velocity of 40 meters per second from a height of 50 meters will be at a height s meters after t seconds, where $s(t) = -4.9t^2 + 40t + 50$.

  (a) Graph the function s in the window [0, 10] by [0, 200].

  (b) Use the graph to find the time at which the object reaches its highest point.

  (c) Use the graph to find the maximum height reached by the object.

  (d) Write the equation that allows you to find the time at which the object reaches the ground. Then, solve the equation, using the method of your choice. Give your answer to the nearest hundredth of a second.

5. *Live Births in the United States*   The table gives the number of live births in the United States (in thousands) for the years 1981 through 1990.

Year	Number of Births
1981	3629
1982	3681
1983	3639
1984	3669
1985	3761
1986	3757
1987	3809
1988	3910
1989	4041
1990	4158

(*Source:* U.S. Center for Health Statistics.)

  (a) Plot the data, letting $x = 1$ represent 1981, $x = 2$ represent 1982, and so on.

  (b) Find a function of the form $f(x) = a(x - h)^2 + k$ that models this data. Let (1, 3629) represent the vertex, and use (10, 4158) to determine the value of a.

  (c) Use the statistical capability of a graphing calculator to find the best-fitting quadratic function for this data. Graph both this function and the function determined in part (b) over the data points.

6. (a) Given that $f(x) = x^6 - 5x^5 + 3x^4 + x^3 + 40x^2 - 24x - 72$ has 3 as a zero of multiplicity two, 2 as a single zero, and -1 as a single zero, find all other zeros of f.

  (b) Use the information from part (a) to sketch the graph of f by hand. Give an end behavior diagram.

7. For the function $f(x) = 4x^4 - 21x^2 - 25$, do each of the following.

  (a) Find all zeros analytically.

  (b) Find a comprehensive graph of f, and support the real zeros found in part (a).

  (c) Discuss the symmetry of the graph of f.

  (d) Use the graph and the results of part (a) to find the solution set of each inequality.

   (i) $f(x) \geq 0$ (ii) $f(x) < 0$

8. (a) Use only a graphical method to find the real solutions of the following.

$$x^5 - 4x^4 + 2x^3 - 4x^2 + 6x - 1 = 0$$

(b) Based on the degree and your answer in part (a), how many nonreal complex solutions does the equation have?

9. *Volume of a Box*   A rectangular sheet of metal is to be formed into a box with an open top by cutting equal size squares from each corner and folding up the sides. The sheet measures 16 inches by 24 inches. Let *x* represent the length of the side of each square.

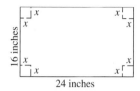

16 inches

24 inches

(a) Express the volume *V* as a function of *x*. State any restrictions on *x*.

(b) Use a graph to determine the length of the side of each square that will maximize the volume.

(c) What is the maximum volume?

10. *Live Births in the United States*   Refer to the data in Item 5. Again, letting $x = 1$ represent 1981, $x = 2$ represent 1982, and so on, do each of the following.

(a) Use a graphing calculator to find the best-fitting cubic function for the data.

(b) Repeat part (a) for a quartic function.

(c) Graph the functions of parts (a) and (b) over the data points.

(d) Which one of these functions gives the best prediction of the 1991 figure of 4111 thousand births?

CHAPTER 3   PROJECT

CREATING YOUR PERSONAL SOCIAL SECURITY POLYNOMIAL

Your Social Security number is unique to you, and with it you can construct your own personal Social Security polynomial. Let's agree that the polynomial function will be defined as follows, where a_i represents the *i*th digit in your Social Security number:

$$SSN(x) = (x - a_1)(x + a_2)(x - a_3)(x + a_4)(x - a_5)(x + a_6)(x - a_7)(x + a_8)(x - a_9).$$

For example, if the Social Security number is 539-58-0954, the polynomial is

$$SSN(x) = (x - 5)(x + 3)(x - 9)(x + 5)(x - 8)(x + 0)(x - 9)(x + 5)(x - 4).$$

A comprehensive graph of this function is shown in Figure A. In Figure B, we show a screen obtained by zooming in on the positive zeros, as the comprehensive graph does not show the local behavior well in this region.

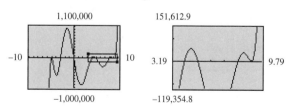

FIGURE A   **FIGURE B**

ACTIVITIES

(These activities can be performed by individual students, or by students in groups, where one group member's Social Security number is used.)

   1. Construct your personal Social Security polynomial *SSN(x)*.

2. What is the degree of this polynomial? What is the dominating term?

3. Construct an end behavior diagram.

4. List all zeros of multiplicity one.

5. List all zeros of multiplicity two.

6. List all zeros of multiplicity three or higher.

7. Using your graphing calculator, find a comprehensive graph of $SSN(x)$. It should include all local extreme points, all intercepts, and illustrate end behavior. Sketch the graph on paper, or print it out, using the appropriate computer software. Regardless of which type of graph you submit, give the values of Xmin, Xmax, Xscl, Ymin, Ymax, and Yscl. (*Hint:* Ymin and Ymax will be quite large in comparison to values you might be accustomed to using.) You may need to include "enlargements" of certain parts of the graph, because they may not show enough detail in one window.

8. List the coordinates of each local maximum.

9. List the coordinates of each local minimum.

10. What are the domain and the range of $SSN(x)$?

11. List the intervals of the domain over which $SSN(x)$ is increasing.

12. List the intervals of the domain over which $SSN(x)$ is decreasing.

13. What must be true of a Social Security number that leads to a polynomial whose graph passes through the origin?

Rational, Root, and Inverse Functions

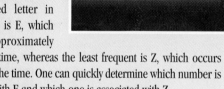
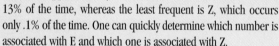

*A*lthough cryptography has been used by the military for centuries, only recently has the public found a need for it. Because of the large amount of electronic communication today, businesses, banks, and private individuals use cryptography to keep information secret. The process of disguising plain text into a cipher is called encryption. When a message is received, the cipher must be decoded into the original plain text using decryption.

Codes that encrypt a letter of the alphabet in the same way each time are usually easy to break. Suppose a simple code changes letters into numbers using the following technique: A = 01, B = 02, C = 03, . . . , Z = 26. After a few messages, a person may be able to decipher the code because letters in the English language do not occur with equal likelihood. The most frequently used letter in the alphabet is E, which occurs approximately 13% of the time, whereas the least frequent is Z, which occurs only .1% of the time. One can quickly determine which number is associated with E and which one is associated with Z.

In the final section of this chapter, we show a simple way to encode and decode messages using inverse functions.

4.1 GRAPHS OF RATIONAL FUNCTIONS

Introduction to Rational Functions ◆ Graphing Calculator Considerations for Rational Functions ◆ Graphs of Simple Rational Functions ◆ Determination of Asymptotes ◆ Graphs of More Complicated Rational Functions

INTRODUCTION TO RATIONAL FUNCTIONS

Our study of the various types of functions now leads us to a new type: the rational function.

Sources: Schneier, B., *Applied Cryptography: Protocols, Algorithms, and Source Code in C,* John Wiley & Sons, Inc., 1994; Sinkov, A., *Elementary Cryptanalysis: A Mathematical Approach,* Random House, 1968.

> **DEFINITION OF RATIONAL FUNCTION**
>
> A function f of the form $\frac{p}{q}$ defined by
>
> $$f(x) = \frac{p(x)}{q(x)},$$
>
> where $p(x)$ and $q(x)$ are polynomials, is called a **rational function.**

Since any values of x such that $q(x) = 0$ are excluded from the domain of a rational function, this type of function often has a graph that has one or more breaks in it.

Some examples of rational functions are listed below. Their graphs will be studied later in this section.

$$f(x) = \frac{1}{x} \qquad f(x) = \frac{x + 1}{2x^2 + 5x - 3} \qquad f(x) = \frac{3x^2 - 3x - 6}{x^2 + 8x + 16}$$

The simplest rational function with a variable denominator is the reciprocal function, defined by

$$f(x) = \frac{1}{x}.$$

The domain of this function is the set of all real numbers except 0. The number 0 cannot be used as a value of x, but for analysis it is helpful to find the values of $f(x)$ for some values of x close to 0. We can use the table feature of a graphing calculator to do this. See Figure 1.

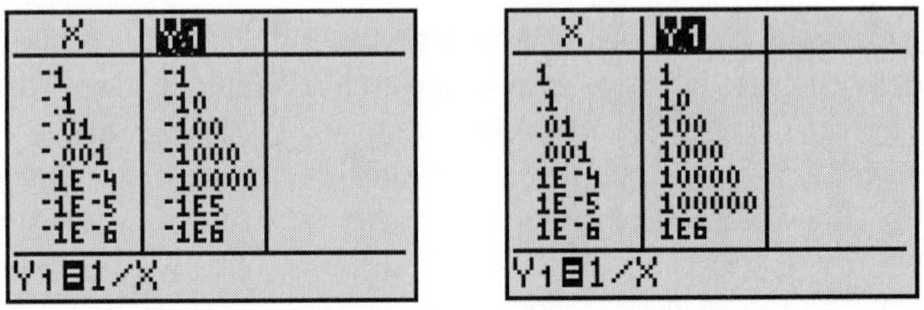

X	Y1
-1	-1
-.1	-10
-.01	-100
-.001	-1000
-1E-4	-10000
-1E-5	-1E5
-1E-6	-1E6

Y1=1/X

X	Y1
1	1
.1	10
.01	100
.001	1000
1E-4	10000
1E-5	100000
1E-6	1E6

Y1=1/X

As x approaches 0 from the left, $Y_1 = \frac{1}{x}$ approaches $-\infty$.

As x approaches 0 from the right, $Y_1 = \frac{1}{x}$ approaches ∞.

(a)

(b)

FIGURE 1

The table suggests that $|f(x)|$ gets larger and larger as x gets closer and closer to 0, which is written in symbols as

$$|f(x)| \to \infty \text{ as } x \to 0.$$

(The symbol $x \to 0$ means that x approaches 0, without necessarily ever being equal to 0.) Since x cannot equal 0, the graph of $f(x) = \frac{1}{x}$ will never intersect the vertical line $x = 0$. This line is called a *vertical asymptote.*

On the other hand, as $|x|$ gets larger and larger, the values of $f(x) = \frac{1}{x}$ get closer and closer to 0, as shown in the tables of Figure 2.

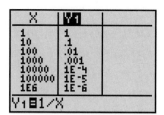

As x approaches ∞, $Y_1 = \frac{1}{x}$
approaches 0 through positive values.

(a)

As x approaches $-\infty$, $Y_1 = \frac{1}{x}$
approaches 0 through negative values.

(b)

FIGURE 2

Letting $|x|$ get larger and larger without bound (written $|x| \rightarrow \infty$) causes the graph of $f(x) = \frac{1}{x}$ to move closer and closer to the horizontal line $y = 0$. This line is called a *horizontal asymptote*.

The graphs and important features of $f(x) = \frac{1}{x}$ and $f(x) = \frac{1}{x^2}$ are now summarized.

THE RATIONAL FUNCTION $f(x) = \dfrac{1}{x}$ **(THE RECIPROCAL FUNCTION)**

$f(x) = \frac{1}{x}$ (Figure 3)   Domain: $(-\infty, 0) \cup (0, \infty)$ Range: $(-\infty, 0) \cup (0, \infty)$

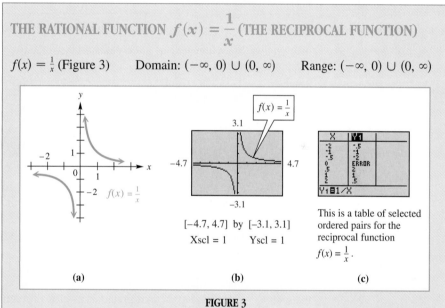

$f(x) = \frac{1}{x}$

$[-4.7, 4.7]$ by $[-3.1, 3.1]$
Xscl = 1   Yscl = 1

This is a table of selected ordered pairs for the reciprocal function $f(x) = \frac{1}{x}$.

(a)   **(b)**   **(c)**

FIGURE 3

The rational function $f(x) = \frac{1}{x}$ decreases on the interval $(-\infty, 0)$ and also decreases on the interval $(0, \infty)$. It is discontinuous at $x = 0$. The y-axis is a vertical asymptote, and the x-axis is a horizontal asymptote. It is an odd function, and its graph is symmetric with respect to the origin.

For any positive integer n, if $x \neq 0$, then $x^{-n} = 1/x^n$. Therefore, we may graph $f(x) = 1/x^n$ with a graphing calculator by using this definition of a negative exponent. For example, $f(x) = 1/x$ can be graphed by using the $\boxed{x^{-1}}$ key.

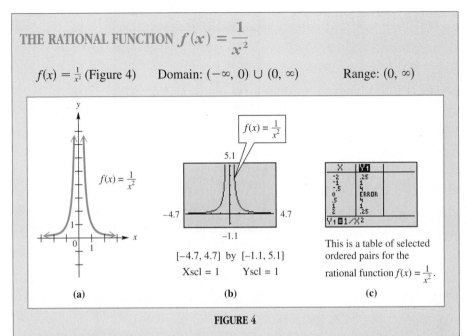

THE RATIONAL FUNCTION $f(x) = \dfrac{1}{x^2}$

$f(x) = \frac{1}{x^2}$ (Figure 4)   Domain: $(-\infty, 0) \cup (0, \infty)$ Range: $(0, \infty)$

$f(x) = \frac{1}{x^2}$

$[-4.7, 4.7]$ by $[-1.1, 5.1]$

Xscl = 1   Yscl = 1

This is a table of selected ordered pairs for the rational function $f(x) = \frac{1}{x^2}$.

(a)   (b)   (c)

FIGURE 4

The rational function $f(x) = \frac{1}{x^2}$ increases on the interval $(-\infty, 0)$ and decreases on the interval $(0, \infty)$. It is discontinuous at $x = 0$. The y-axis is a vertical asymptote, and the x-axis is a horizontal asymptote. It is an even function, and its graph is symmetric with respect to the y-axis.

GRAPHING CALCULATOR CONSIDERATIONS FOR RATIONAL FUNCTIONS

In Section 2.5, we studied the graph of the greatest integer function. Recall that because of the many discontinuities of the graph, in order to obtain a more realistic picture of the graph we used the dot mode of our calculator (rather than the connected mode). Because rational functions often have values for which the denominator is zero, there will be discontinuities in them as well. As a result, the dot mode of the calculator will often give a more realistic picture. If the calculator is in the connected mode, it may attempt to connect adjacent lighted pixels, giving the appearance of a vertical line on the screen. While this may be interpreted as a vertical asymptote, the student should be aware that this line is not a part of the graph.

To illustrate, consider the function $y = \frac{1}{x+3}$. This graph is obtained by shifting the graph of $f(x) = \frac{1}{x}$ three units to the left, since it is the same as $f(x + 3)$. If this graph is generated by a calculator in connected mode in the window $[-6, 3]$ by $[-3, 3]$, we get the display shown in Figure 5. Notice the appearance of the vertical line at $x = -3$. This line cannot be part of the graph because the function is not defined for $x = -3$. (Why is this so?) Furthermore, we know that it cannot be part of the graph because the graph would then fail the vertical line test (Section 1.2).

There are two ways to obtain a more realistic picture for this graph. The first involves using the dot mode, as seen in Figure 6(a). We must realize, however, that the function is continuous on the intervals $(-\infty, -3)$ and $(-3, \infty)$. The second method involves experimenting with various windows so that the calculator does not show the "vertical line." See Figure 6(b).

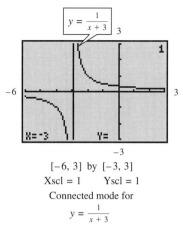

$Y_1 = \frac{1}{x+3}$ is undefined at $x = -3$, as indicated by the error message in the table.

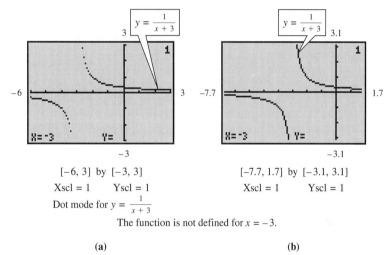

[−6, 3] by [−3, 3]
Xscl = 1   Yscl = 1
Connected mode for
$y = \frac{1}{x+3}$

FIGURE 5

[−6, 3] by [−3, 3]
Xscl = 1   Yscl = 1
Dot mode for $y = \frac{1}{x+3}$

[−7.7, 1.7] by [−3.1, 3.1]
Xscl = 1   Yscl = 1

The function is not defined for $x = -3$.

(a)   **(b)**

FIGURE 6

GRAPHS OF SIMPLE RATIONAL FUNCTIONS

The graph of $y = \frac{1}{x}$ can be shifted, translated, and reflected in the same way that we observed such transformations on other elementary functions in Chapter 2.

EXAMPLE 1   *Analyzing the Graph of a Simple Rational Function*   Discuss how the graph of $y = -\frac{2}{x}$ may be obtained from the graph of $f(x) = \frac{1}{x}$. Then, graph the function both by hand and with a calculator.

SOLUTION   The expression $-\frac{2}{x}$ can be written as $-2\left(\frac{1}{x}\right)$, indicating that the graph may be obtained by stretching vertically by a factor of 2, and reflecting across the *y*-axis (or *x*-axis). The *x*- and *y*-axes remain the horizontal and vertical asymptotes. The domain and the range are still both $(-\infty, 0) \cup (0, \infty)$. Figure 7 on the following page shows both a traditional graph and a calculator-generated graph.   ◆

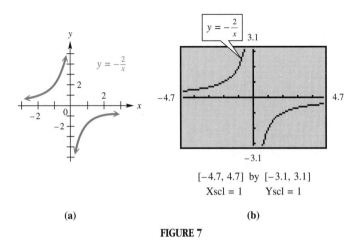

$$[-4.7, 4.7] \text{ by } [-3.1, 3.1]$$
$$\text{Xscl} = 1 \qquad \text{Yscl} = 1$$

(a)   (b)

FIGURE 7

EXAMPLE 2   *Analyzing the Graph of a Simple Rational Function*   Discuss how the graph of $y = \frac{2}{x+1}$ may be obtained from the graph of $f(x) = \frac{1}{x}$. Then, graph the function both by hand and with a calculator.

SOLUTION   The expression $\frac{2}{x+1}$ can be written as $2\left(\frac{1}{x+1}\right)$, indicating that the graph may be obtained by shifting the graph of $y = \frac{1}{x}$ one unit to the left, and stretching vertically by a factor of 2. The graphs are shown in Figure 8. Notice that the horizontal shift affects the domain; the domain of this new function is $(-\infty, -1) \cup (-1, \infty)$. The line $x = -1$ is the vertical asymptote. The range is still $(-\infty, 0) \cup (0, \infty)$.   ◆

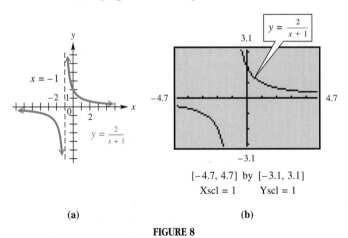

$$[-4.7, 4.7] \text{ by } [-3.1, 3.1]$$
$$\text{Xscl} = 1 \qquad \text{Yscl} = 1$$

(a)   (b)

FIGURE 8

EXAMPLE 3   *Analyzing the Graph of a Simple Rational Function*   Discuss how the graph of $f(x) = \frac{7}{x-3} + 2$ may be obtained from the graph of $y = \frac{1}{x}$. Then, graph the function both by hand and with a calculator.

SOLUTION   For the function f, we have

$$f(x) = 7 \cdot \frac{1}{x-3} + 2,$$

indicating a horizontal shift of 3 units to the right, a vertical stretch by a factor of 7, and a vertical shift of 2 units up. The graph is shown in Figure 9. Notice that the vertical asymptote has the equation $x = 3$, the horizontal asymptote has the equation $y = 2$, the domain is $(-\infty, 3) \cup (3, \infty)$, and the range is $(-\infty, 2) \cup (2, \infty)$.   ◆

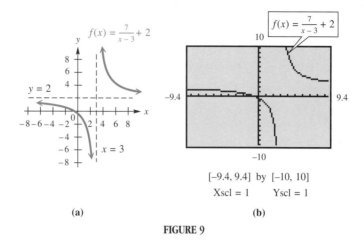

[-9.4, 9.4] by [-10, 10]
Xscl = 1   Yscl = 1

(a)   (b)

FIGURE 9

DETERMINATION OF ASYMPTOTES

In our discussion thus far, we have seen how rational functions approach certain values as $|x| \to \infty$ and as $|x| \to a$ if the function is undefined at $x = a$. We now give formal definitions for vertical and horizontal asymptotes.

DEFINITIONS OF VERTICAL AND HORIZONTAL ASYMPTOTES

For the rational function f with $f(x) = \frac{p(x)}{q(x)}$, written in lowest terms, if $|f(x)| \to \infty$ as $x \to a$, then the line $x = a$ is a **vertical asymptote;** and if $f(x) \to a$ as $|x| \to \infty$, then the line $y = a$ is a **horizontal asymptote.**

Locating asymptotes is an important part of analysis of the graphs of rational functions. Vertical asymptotes are found by determining the values of x which make the denominator equal to 0 but do not make the numerator equal to 0. Horizontal asymptotes (and, in some cases, *oblique* asymptotes) are found by considering what happens to $f(x)$ as $|x| \to \infty$. The next example shows how to find asymptotes.

EXAMPLE 4 *Finding Asymptotes of Graphs of Rational Functions* For each rational function f, find all asymptotes.

(a) $f(x) = \dfrac{x + 1}{2x^2 + 5x - 3}$ **(b)** $f(x) = \dfrac{2x + 1}{x - 3}$ **(c)** $f(x) = \dfrac{x^2 + 1}{x - 2}$

SOLUTION

(a) To find the vertical asymptotes, set the denominator equal to zero and solve.

$$2x^2 + 5x - 3 = 0$$

$$(2x - 1)(x + 3) = 0 \qquad \text{Factor.}$$

$$2x - 1 = 0 \quad \text{or} \quad x + 3 = 0 \qquad \text{Zero-product property}$$

$$x = \frac{1}{2} \quad \text{or} \qquad x = -3$$

The equations of the vertical asymptotes are $x = \frac{1}{2}$ and $x = -3$.

To find the equation of the horizontal asymptote, we divide each term by the variable factor of greatest degree in the expression for $f(x)$. In this case, we divide each term by x^2.

$$f(x) = \frac{\dfrac{x}{x^2} + \dfrac{1}{x^2}}{\dfrac{2x^2}{x^2} + \dfrac{5x}{x^2} - \dfrac{3}{x^2}} = \frac{\dfrac{1}{x} + \dfrac{1}{x^2}}{2 + \dfrac{5}{x} - \dfrac{3}{x^2}}$$

As $|x|$ gets larger and larger, the quotients $\frac{1}{x}, \frac{1}{x^2}, \frac{5}{x}$, and $\frac{3}{x^2}$ all approach 0, and the value of $f(x)$ approaches

$$\frac{0 + 0}{2 + 0 - 0} = \frac{0}{2} = 0.$$

The line $y = 0$ (that is, the x-axis) is therefore the horizontal asymptote.

(b) Set the denominator equal to zero to find that the vertical asymptote has the equation $x = 3$. To find the horizontal asymptote, divide each term in the rational expression by x.

$$f(x) = \frac{2x + 1}{x - 3} = \frac{\dfrac{2x}{x} + \dfrac{1}{x}}{\dfrac{x}{x} - \dfrac{3}{x}} = \frac{2 + \dfrac{1}{x}}{1 - \dfrac{3}{x}}$$

As $|x|$ gets larger and larger, both $\frac{1}{x}$ and $\frac{3}{x}$ approach 0, and $f(x)$ approaches

$$\frac{2 + 0}{1 - 0} = \frac{2}{1} = 2,$$

so the line $y = 2$ is the horizontal asymptote.

(c) Setting the denominator equal to zero shows that the vertical asymptote has the equation $x = 2$. If we divide by the variable factor of greatest degree of x as before (x^2 in this case), we see that there is no horizontal asymptote because

$$f(x) = \frac{\dfrac{x^2}{x^2} + \dfrac{1}{x^2}}{\dfrac{x}{x^2} - \dfrac{2}{x^2}} = \frac{1 + \dfrac{1}{x^2}}{\dfrac{1}{x} - \dfrac{2}{x^2}}$$

does not approach any real number as $|x| \to \infty$, since $\frac{1}{0}$ is undefined. This will happen whenever the degree of the numerator is greater than the degree of the denominator. In such cases, divide the denominator into the numerator to write the expression in another form.

$$\begin{array}{r} x + 2 \\ x - 2 \overline{\smash{)}x^2 + 0x + 1} \\ \underline{x^2 - 2x } \\ 2x + 1 \\ \underline{2x - 4} \\ 5 \end{array}$$

(Notice that we could have used synthetic division here.) The function can now be written as

$$f(x) = \frac{x^2 + 1}{x - 2} = x + 2 + \frac{5}{x - 2}.$$

For very large values of $|x|$, $\frac{5}{x-2}$ is close to 0, and the graph approaches the line $y = x + 2$. This line is an **oblique asymptote** (neither vertical nor horizontal) for the graph of the function.

In general, if the degree of the numerator is exactly one more than the degree of the denominator, a rational function may have an oblique asymptote. The equation of this asymptote is found by dividing the numerator by the denominator and disregarding the remainder. ◆

The results of Example 4 can be summarized as follows.

DETERMINING ASYMPTOTES

In order to find asymptotes of a rational function defined by a rational expression *in lowest terms,* use the following procedures.

1. **Vertical Asymptotes**
    Find any vertical asymptotes by setting the denominator equal to zero and solving for x. If a is a zero of the denominator but not the numerator, then the line $x = a$ is a vertical asymptote.
2. **Other Asymptotes**
    Determine any other asymptotes. We consider three possibilities:
    a. If the numerator has lower degree than the denominator, there is a horizontal asymptote, $y = 0$ (the x-axis).
    b. If the numerator and denominator have the same degree, and the function is of the form
    $$f(x) = \frac{a_n x^n + \cdots + a_0}{b_n x^n + \cdots + b_0}, \quad \text{where } b_n \neq 0,$$
    dividing by x^n in the numerator and denominator produces the horizontal asymptote
    $$y = \frac{a_n}{b_n}.$$
    c. If the numerator is of degree exactly one more than the denominator, there may be an oblique asymptote. To find it, divide the numerator by the denominator and disregard any remainder. Set the rest of the quotient equal to y to get the equation of the asymptote.*

Note The graph of a rational function may have more than one vertical asymptote, or it may have none at all. The graph cannot intersect any vertical asymptote. There can be only one other (non-vertical) asymptote, and the graph *may* intersect that asymptote. This will be seen in Example 7. The method of graphing a rational function that is not in lowest terms will be covered in Example 9.

GRAPHS OF MORE COMPLICATED RATIONAL FUNCTIONS

As rational functions become more complicated, obtaining accurate graphing calculator-generated graphs becomes more difficult. We suggest that the student become proficient at sketching the graphs by hand before attempting to graph them with a calculator, because a calculator can lead to a distorted view if an inappropriate window is used. We now outline the procedure for graphing rational functions by hand.

---

*More involved rational functions, such as $f(x) = \frac{8x^3 - 1}{x}$, are not covered in this book.

GRAPHING RATIONAL FUNCTIONS

Let $f(x) = \frac{p(x)}{q(x)}$ define a function where the rational expression is written in lowest terms. To sketch its graph, follow the steps below.

1. Find any vertical asymptotes.
2. Find any horizontal or oblique asymptotes.
3. Find the y-intercept by evaluating $f(0)$.
4. Find the x-intercepts, if any, by solving $f(x) = 0$. (These will be the zeros of the numerator, p.)
5. Determine whether the graph will intersect its nonvertical asymptote by solving $f(x) = k$, where k is the y-value of the horizontal asymptote, or $f(x) = mx + b$, where $y = mx + b$ is the equation of the oblique asymptote.
6. Plot a few selected points, as necessary. Choose an x-value in each interval of the domain as determined by the vertical asymptotes and x-intercepts.
7. Complete the sketch.

TECHNOLOGY NOTE

Determining the locations of the intercepts and asymptotes analytically will help in obtaining a realistic comprehensive graph of a rational function. Choosing Xmin and Xmax so that the calculator will attempt to compute a value for which the rational function is undefined will eliminate the misleading vertical line that occasionally appears when the calculator is in connected mode.

Here are the criteria for a comprehensive graph of a rational function.

COMPREHENSIVE GRAPH CRITERIA FOR A RATIONAL FUNCTION

A comprehensive graph of a rational function will exhibit the following features:

1. all intercepts, both x- and y-;
2. location of all asymptotes: vertical, horizontal, and/or oblique;
3. the point at which the graph intersects its nonvertical asymptote (if there is any such point);
4. enough of the graph to exhibit the correct end behavior (i.e., behavior as the graph approaches its nonvertical asymptote).

EXAMPLE 5 *Graphing a Rational Function Defined by an Expression with Degree of Numerator Less than Degree of Denominator* Graph $f(x) = \dfrac{x + 1}{2x^2 + 5x - 3}$.

SOLUTION

Step 1 As shown in Example 4(a), the vertical asymptotes have equations $x = \frac{1}{2}$ and $x = -3$.

Step 2 Again, as shown in Example 4(a), the horizontal asymptote is the x-axis.

Step 3 Since $f(0) = \frac{0 + 1}{2(0)^2 + 5(0) - 3} = -\frac{1}{3}$, the y-intercept is $-\frac{1}{3}$.

Step 4 The x-intercept is found by solving $f(x) = 0$.

$$\frac{x + 1}{2x^2 + 5x - 3} = 0$$
$$x + 1 = 0 \qquad \text{If a rational expression is equal to 0, then its}$$
$$\text{numerator must equal 0.}$$
$$x = -1$$

The x-intercept is -1.

Step 5 To determine whether the graph intersects its horizontal asymptote, solve

$$f(x) = 0. \leftarrow y\text{-value of horizontal asymptote}$$

Since the horizontal asymptote is the x-axis, the solution of this equation was found in Step 4. The graph intersects its horizontal asymptote at $(-1, 0)$.

Step 6 Plot a point in each of the intervals determined by the x-intercepts and vertical asymptotes, $(-\infty, -3)$, $(-3, -1)$, $\left(-1, \frac{1}{2}\right)$, and $\left(\frac{1}{2}, \infty\right)$ to get an idea of how the graph behaves in each region.

Step 7 Complete the sketch. Keep in mind that the graph approaches its asymptotes as the points on the graph become farther away from the origin. The traditional graph is shown in Figure 10(a), and the calculator-generated graph is shown in Figure 10(b). ◆

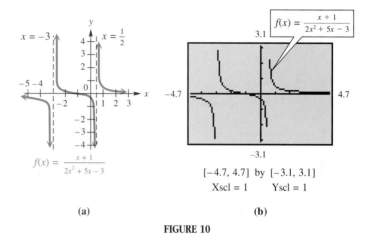

$$f(x) = \frac{x+1}{2x^2+5x-3}$$

(a)   (b)

FIGURE 10

The function f discussed in Example 3 may be written in the form $\frac{p(x)}{q(x)}$ by using methods of adding rational expressions introduced in elementary algebra courses.

$$f(x) = \frac{7}{x-3} + 2$$

$$= \frac{7}{x-3} + \frac{2(x-3)}{x-3} \qquad \text{Find a common denominator.}$$

$$= \frac{7 + 2(x-3)}{x-3}$$

$$= \frac{7 + 2x - 6}{x-3}$$

$$= \frac{2x+1}{x-3}$$

Its graph is, of course, the same as that shown in Figure 9. We will use the step-by-step method to graph this function by hand in Example 6. (In the remaining examples, we will not specifically number the steps.)

EXAMPLE 6 *Graphing a Rational Function Defined by an Expression with Degree of Numerator Equal to Degree of Denominator* Graph $f(x) = \dfrac{2x+1}{x-3}$.

SOLUTION As shown in Example 4(b), the equation of the vertical asymptote is $x = 3$, and the equation of the horizontal asymptote is $y = 2$. Since $f(0) = -\frac{1}{3}$, the y-intercept is $-\frac{1}{3}$. The solution of $f(x) = 0$ is $-\frac{1}{2}$, so the only x-intercept is $-\frac{1}{2}$. The graph does not intersect its horizontal asymptote, since $f(x) = 2$ has no solution. (Verify this.) The points $(-4, 1)$ and $\left(6, \frac{13}{3}\right)$ are on the graph and can be used to complete the hand-drawn sketch, as shown in Figure 11(a). A calculator-generated graph is shown in Figure 11(b). Notice that these graphs are the same as those in Figure 9. ◆

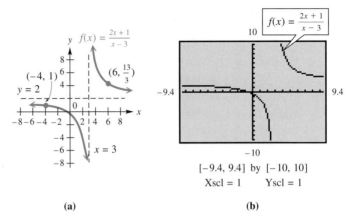

(a)   (b)

FIGURE 11

EXAMPLE 7   *Graphing a Rational Function Whose Graph Intersects Its Horizontal Asymptote*   Graph $f(x) = \dfrac{3x^2 - 3x - 6}{x^2 + 8x + 16}$.

SOLUTION   To find the vertical asymptote(s), solve $x^2 + 8x + 16 = 0$.

$$x^2 + 8x + 16 = 0$$
$$(x + 4)^2 = 0$$
$$x = -4$$

The only vertical asymptote has the equation $x = -4$. As explained in the preceding guidelines, by dividing all terms by x^2, the equation of the horizontal asymptote can be shown to be

$$y = \frac{3}{1} \qquad \begin{array}{l} \leftarrow \text{Leading coefficient of numerator} \\ \leftarrow \text{Leading coefficient of denominator} \end{array}$$

or $y = 3$. The y-intercept is $f(0) = -\frac{3}{8}$. To find the x-intercept(s), if any, we solve $f(x) = 0$.

$$f(x) = \frac{3x^2 - 3x - 6}{x^2 + 8x + 16} = 0$$
$$3x^2 - 3x - 6 = 0$$
$$x^2 - x - 2 = 0 \qquad \text{Divide by 3.}$$
$$(x - 2)(x + 1) = 0$$
$$x = 2 \quad \text{or} \quad x = -1$$

The x-intercepts are -1 and 2. By setting $f(x) = 3$ and solving, we can locate the point where the graph intersects the horizontal asymptote.

$$f(x) = \frac{3x^2 - 3x - 6}{x^2 + 8x + 16}$$
$$3 = \frac{3x^2 - 3x - 6}{x^2 + 8x + 16}$$
$$3x^2 - 3x - 6 = 3x^2 + 24x + 48 \qquad \text{Multiply by } x^2 + 8x + 16.$$
$$-3x - 6 = 24x + 48 \qquad \text{Subtract } 3x^2.$$
$$-27x = 54$$
$$x = -2$$

The graph intersects its horizontal asymptote at $(-2, 3)$.

Some other points that lie on the graph are $(-10, 9)$, $(-3, 30)$ and $\left(5, \frac{2}{3}\right)$. These can be used to complete the hand-drawn graph, as shown in Figure 12(a). The calculator-generated graph in Figure 12(b) confirms our result.   ◆

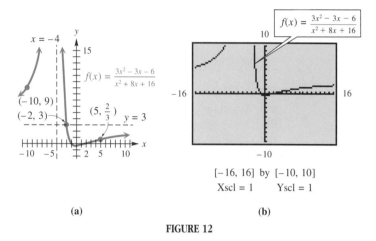

(a)   (b)

FIGURE 12

Notice the behavior of the graph of the function in Example 7 near the line $x = -4$. As x approaches -4 from either side, we have $f(x)$ approaching ∞. On the other hand, if we examine the behavior of the graph in Figure 11 near the line $x = 3$, we have $f(x) \to -\infty$ as x approaches 3 from the left, while $f(x) \to \infty$ as x approaches 3 from the right. The behavior of the graph of a rational function near a vertical asymptote $x = a$ will partially depend on the exponent on $x - a$ in the denominator.

BEHAVIOR OF GRAPHS OF RATIONAL FUNCTIONS NEAR VERTICAL ASYMPTOTES

If n is the largest positive integer such that $(x - a)^n$ is a factor of the denominator of $f(x)$, the graph will behave in the manner illustrated.

if n is even:

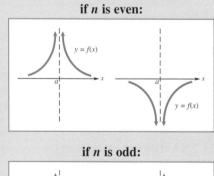

if n is odd:

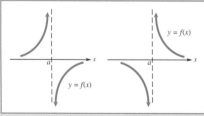

In both cases, we assume that $f(x)$ is in lowest terms.

The next example discusses a rational function defined by an expression having the degree of its numerator greater than the degree of its denominator.

EXAMPLE 8   *Graphing a Rational Function Defined by an Expression with the Degree of the Numerator Greater than the Degree of the Denominator*
Graph $f(x) = \dfrac{x^2 + 1}{x - 2}$. Then use a table to illustrate the behavior near the oblique asymptote.

SOLUTION   As shown in Example 4(c), the vertical asymptote has the equation $x = 2$, and the graph has an oblique asymptote with the equation $y = x + 2$. Refer to the preceding box to determine the behavior near the asymptote $x = 2$. The y-intercept is $-\frac{1}{2}$, and the graph has no x-intercepts, since the numerator, $x^2 + 1$, has no real zeros. It can be shown that the graph does not intersect its oblique asymptote. Using the intercepts, asymptotes, the points $\left(4, \frac{17}{2}\right)$ and $\left(-1, -\frac{2}{3}\right)$, and the general behavior of the graph near its asymptotes, we obtain the graph shown in Figure 13(a). Compare with the calculator-generated graph shown in Figure 13(b).

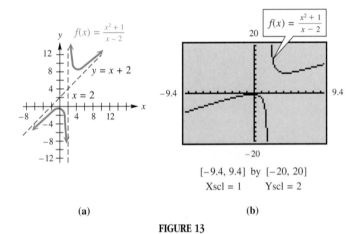

(a)

(b)

$$[-9.4, 9.4] \text{ by } [-20, 20]$$
$$\text{Xscl} = 1 \qquad \text{Yscl} = 2$$

FIGURE 13

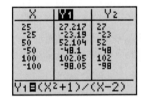

For large values of $|x|$, $Y_1 \approx Y_2$.
(Here, $Y_2 = x + 2$.)

FIGURE 14

To illustrate how the values of f and $y = x + 2$ are approximately equal for large values of $|x|$, the table in Figure 14 has f defined as Y_1 and $Y_2 = x + 2$. Note that the function values are approximately equal for the values of x shown.   ◆

As mentioned earlier, a rational function must be defined by an expression in lowest terms before we can use the methods discussed thus far in this section to hand-sketch the graph. The final example shows a typical rational function defined by an expression that is not in lowest terms.

EXAMPLE 9   *Graphing a Rational Function Defined by an Expression that Is Not in Lowest Terms*   Graph $f(x) = \dfrac{x^2 - 4}{x - 2}$.

SOLUTION   Notice that the domain of this function cannot contain 2. The rational expression $\frac{x^2 - 4}{x - 2}$ can be reduced to lowest terms by factoring the numerator, and dividing both the numerator and denominator by $x - 2$.

$$f(x) = \frac{x^2 - 4}{x - 2} = \frac{(x + 2)(x - 2)}{x - 2} = x + 2 \quad (x \neq 2)$$

TECHNOLOGY NOTE

Experiment with your calculator to see if you can obtain a graph like the one in Figure 15(b), where the point of discontinuity is obvious. Then, experiment with different windows to see if you can find one where the point of discontinuity is not visible.

Therefore, the graph of this function will be the same as the graph of $y = x + 2$ (a straight line), with the exception of the point with x-value 2. A "hole" appears in the graph at $(2, 4)$. See Figure 15(a).

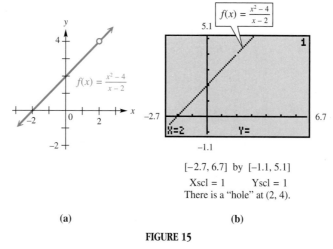

[−2.7, 6.7] by [−1.1, 5.1]
Xscl = 1   Yscl = 1
There is a "hole" at (2, 4).

(a)   (b)

FIGURE 15

This table supports the fact that $Y_1 = Y_2$ (where Y_2 is defined as $x + 2$) for all values of x except 2, where Y_1 is undefined. Compare with the graphical approach in Example 9.

If the window of a graphing calculator is set in a manner so that an x-value for the location of the cursor is 2, then we can see from the display that the calculator cannot determine a value for y, as in Figure 15(b). A close examination of the screen will show an unlighted pixel at $x = 2$. However, such points of discontinuity will often *not* be evident from calculator-generated graphs—once again showing us a reason for studying the concepts hand-in-hand with the technology. ◆

4.1   EXERCISES   Tape 5

Use the graphs of the rational functions in A through D to answer the questions in Exercises 1–8. Give all possible answers, as there may be more than one correct choice.

1. Which choices have domain $(-\infty, 3) \cup (3, \infty)$?

2. Which choices have range $(-\infty, 3) \cup (3, \infty)$?

3. Which choices have range $(-\infty, 0) \cup (0, \infty)$?

4. Which choices have range $(0, \infty)$?

5. If f represents the function, only one choice has a single solution to the equation $f(x) = 3$. Which one is it?

6. What is the range of the function in D?

7. Which choices have the x-axis as a horizontal asymptote?

8. Which choices are symmetric with respect to a vertical line?

A.
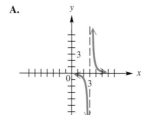

B.
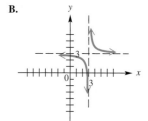

C.
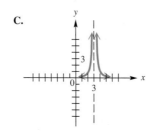

D.
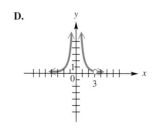

Use the terminology of Sections 2.2 and 2.3 involving shifting, stretching, shrinking, and reflecting to explain how the graph of f can be obtained from the graph of $y = \frac{1}{x}$ or $y = \frac{1}{x^2}$. Then, draw a sketch of the graph of f by hand.

9. $f(x) = \dfrac{2}{x}$ **10.** $f(x) = -\dfrac{3}{x}$ **11.** $f(x) = \dfrac{1}{x + 2}$ **12.** $f(x) = \dfrac{1}{x - 3}$ **13.** $f(x) = \dfrac{1}{x} + 1$

14. $f(x) = \dfrac{1}{x} - 2$ **15.** $f(x) = -\dfrac{2}{x^2}$ **16.** $f(x) = \dfrac{1}{x^2} - 3$ **17.** $f(x) = \dfrac{1}{(x - 3)^2}$ **18.** $f(x) = \dfrac{-2}{(x - 3)^2}$

The figures below show the four ways that the graph of a rational function can approach the vertical line x = 2 as an asymptote. Identify the graph of each of the rational functions in Exercises 19–22.

19. $f(x) = \dfrac{1}{(x - 2)^2}$

20. $f(x) = \dfrac{1}{x - 2}$

21. $f(x) = \dfrac{-1}{x - 2}$

22. $f(x) = \dfrac{-1}{(x - 2)^2}$

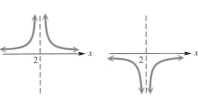

A.   **B.**

C.   **D.**

Give the equations of any vertical, horizontal, or oblique asymptotes for the graphs of the rational functions in Exercises 23–30. For those with horizontal asymptotes, graph the function on your calculator and support your answer by tracing and letting $x \to \infty$ and $x \to -\infty$. You should be able to see how the y-value approaches a constant.

23. $f(x) = \dfrac{3}{x - 5}$ **24.** $f(x) = \dfrac{-6}{x + 9}$ **25.** $f(x) = \dfrac{4 - 3x}{2x + 1}$ **26.** $f(x) = \dfrac{2x + 6}{x - 4}$

27. $f(x) = \dfrac{x^2 - 1}{x + 3}$ **28.** $f(x) = \dfrac{x^2 + 4}{x - 1}$ **29.** $f(x) = \dfrac{x^2 - 2x - 3}{2x^2 - x - 10}$ **30.** $f(x) = \dfrac{3x^2 - 6x - 24}{5x^2 - 26x + 5}$

31. Which one of the following has a graph that does not have a vertical asymptote?

   (a) $f(x) = \dfrac{1}{x^2 + 2}$ **(b)** $f(x) = \dfrac{1}{x^2 - 2}$ **(c)** $f(x) = \dfrac{3}{x^2}$ **(d)** $f(x) = \dfrac{2x + 1}{x - 8}$

32. Which one of the following has a graph that does not have a horizontal asymptote?

   (a) $f(x) = \dfrac{2x - 7}{x + 3}$ **(b)** $f(x) = \dfrac{3x}{x^2 - 9}$ **(c)** $f(x) = \dfrac{x^2 - 9}{x + 3}$ **(d)** $f(x) = \dfrac{x + 5}{(x + 2)(x - 3)}$

RELATING CONCEPTS (EXERCISES 33–36)

Consider the rational function

$$f(x) = \frac{x + 3}{x^2 + x + 4},$$

which is defined by a rational expression in lowest terms, whose denominator is a quadratic polynomial.

33. Explain the procedure you would use to find any vertical asymptotes of the graph of f. (Section 4.1)

34. Under what conditions does a quadratic equation have no real solutions? (Section 3.3)

35. Apply the procedure of Exercise 33 to this function *f*. What are the complex solutions of the equation you solved? What are the real solutions? Does *f* have any vertical asymptotes?

36. With your calculator in connected mode and using a window of $[-10, 10]$ by $[-1, 3]$, graph the function. Why is the connected mode acceptable here to get a realistic view of the graph?

Graph each of the following rational functions, using a graphing calculator. You may wish to first make a rough sketch by hand, using the guidelines of this section, so that it will be easier to find an appropriate window. Your graph should be a comprehensive one.

37. $f(x) = \dfrac{x + 1}{x - 4}$

38. $f(x) = \dfrac{x - 5}{x + 3}$

39. $f(x) = \dfrac{3x}{(x + 1)(x - 2)}$

40. $f(x) = \dfrac{2x + 1}{(x + 2)(x + 4)}$

41. $f(x) = \dfrac{5x}{x^2 - 1}$

42. $f(x) = \dfrac{x}{4 - x^2}$

43. $f(x) = \dfrac{(x - 3)(x + 1)}{(x - 1)^2}$

44. $f(x) = \dfrac{x(x - 2)}{(x + 3)^2}$

45. $f(x) = \dfrac{x}{x^2 - 9}$

46. $f(x) = \dfrac{-5}{2x + 4}$

47. $f(x) = \dfrac{1}{x^2 + 1}$

48. $f(x) = \dfrac{(x - 5)(x - 2)}{x^2 + 9}$

49. $f(x) = \dfrac{x^2 + 1}{x + 3}$

50. $f(x) = \dfrac{2x^2 + 3}{x - 4}$

51. $f(x) = \dfrac{x^2 + 2x}{2x - 1}$

52. $f(x) = \dfrac{x^2 - x}{x + 2}$

53. $f(x) = \dfrac{x^2 - 9}{x + 3}$

54. $f(x) = \dfrac{x^2 - 16}{x + 4}$

In each table, Y_1 is defined by a rational function of the form $\dfrac{x - p}{x - q}$. Use the table to find the values of p and q.

55.

56.

57.

58.

Find an equation for the rational function graph.

59.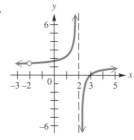

60.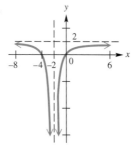

61.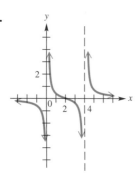

62.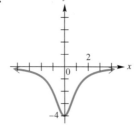

RELATING CONCEPTS (EXERCISES 63–66)

*Recall from Section 2.3 that if we are given the graph of $y = f(x)$, we can obtain the graph of $y = -f(x)$ by reflecting across the x-axis, and we can obtain the graph of $y = f(-x)$ by reflecting across the y-axis. In Exercises 63–66, you are given the graph of a rational function $y = f(x)$. Draw a sketch by hand of the graph of **(a)** $y = -f(x)$ and **(b)** $y = f(-x)$.*

63.   **64.**

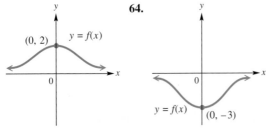

65.   **66.**

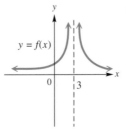 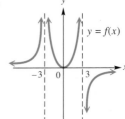

Each of the following rational functions has an oblique asymptote. Determine the equation of this asymptote. Then, use a graphing calculator to graph both the function and the asymptote in the window indicated. (These windows are appropriate for TI-83 calculators.)

67. $f(x) = \dfrac{2x^2 + 3}{4 - x}$; window: $[-18.8, 18.8]$ by $[-50, 25]$

68. $f(x) = \dfrac{x^2 + 9}{x + 3}$; window: $[-9.4, 9.4]$ by $[-25, 25]$

69. $f(x) = \dfrac{x - x^2}{x + 2}$; window: $[-9.4, 9.4]$ by $[-15, 25]$

70. $f(x) = \dfrac{x^2 + 2x}{1 - 2x}$; window: $[-4.7, 4.7]$ by $[-5, 5]$

71. Consider the rational function $f(x) = \dfrac{x^3 - 4x^2 + x + 6}{x^2 + x - 2}$. Divide the numerator by the denominator and use the method of Example 4(c) to determine the equation of the oblique asymptote. Then, determine the coordinates of the point where the graph of f intersects its oblique asymptote. Use a calculator to support your answer.

72. Use long division of polynomials to show that for the function

$$f(x) = \dfrac{x^4 - 5x^2 + 4}{x^2 + x - 12},$$

if we divide the numerator by the denominator, the quotient polynomial is $x^2 - x + 8$, and the remainder is $-20x + 100$. Then, graph both f and g, where $g(x) = x^2 - x + 8$ in the window $[-50, 50]$ by $[0, 1000]$. Comment on the appearance of the two graphs. Explain how the graph of f approaches that of g as $|x|$ gets very large.

Each of the rational functions in Exercises 73 and 74 has a "hole" in its graph.

  (a) *Reduce the fraction to lowest terms, and call this new function g(x).*

  (b) *Use a table with $Y_1 = f(x)$ and $Y_2 = g(x)$ to support the fact that for all values except those for which f is undefined, $f(x) = g(x)$.*

  (c) *What are the coordinates of the "hole"? Support your answer with a graph.*

73. $f(x) = \dfrac{x^2 - 9}{x + 3}$

74. $f(x) = \dfrac{x^2 - 36}{6 - x}$

75. Let $f(x) = p(x)/q(x)$ define a rational function where the expression is reduced to lowest terms. Suppose the degree of $p(x)$ is m, and the degree of $q(x)$ is n. Write an explanation of how you would determine the nonvertical asymptote in each of the following situations.

  (a) $m < n$

  (b) $m = n$

  (c) $m = n + 1$

76. Suppose a friend tells you that the graph of

$$f(x) = \dfrac{x^2 - 25}{x + 5}$$

has a vertical asymptote with equation $x = -5$. Is this correct? If not, describe the behavior of the graph at $x = -5$.

77. Use the table feature of a graphing calculator to confirm that the horizontal asymptote of $f(x) = \dfrac{6x^2 + 3}{2x^2}$ has equation $y = 3$. (*Hint:* Let Tblmin $= 0$, and ΔTbl $= 10$. Then, scroll through the table to see what happens to $f(x)$ as $x \to \infty$ and $x \to -\infty$.)

78. Use the table feature of a graphing calculator to confirm that the horizontal asymptote of $f(x) = \dfrac{2x^3 + 3}{-5x^3 - x^2 + 8}$ has equation $y = -.4$.

4.2 RATIONAL EQUATIONS, INEQUALITIES, AND APPLICATIONS

Solving Rational Equations and Inequalities ◆ Applications of Rational Functions ◆ Inverse Variation ◆ Combined and Joint Variation

SOLVING RATIONAL EQUATIONS AND INEQUALITIES

In this section, we will examine methods of solving equations, inequalities, and applications involving rational functions. A rational equation (or inequality) is an equation (or inequality) that involves at least one term having a variable expression in a denominator, or at least one term having a variable expression raised to a negative integer power. Some examples of such equations and inequalities are

$$\frac{2x-1}{3x+4}=5, \qquad \frac{2x-1}{3x+4}\le 5, \qquad \frac{x}{x-2}+\frac{1}{x+2}=\frac{8}{x^2-4},$$

and

$$7x^{-4}-8x^{-2}+1=0.$$

In solving rational equations and inequalities, it is important to remember that an expression with a variable denominator or with a negative exponent may be undefined for certain values of the variable. Each time we begin solving such an equation or inequality, we will identify such values (which are actually values at which the associated rational function will have a vertical asymptote or "hole"). The general procedure for solving rational equations is to multiply both sides of the equation by the least common denominator of all the terms in the equation, as shown in the first example.

EXAMPLE 1 *Solving a Rational Equation and Supporting the Solution Graphically* Solve the equation

$$\frac{2x-1}{3x+4}=5$$

analytically, and support the solution graphically.

SOLUTION ANALYTIC We begin by noticing that the rational expression is undefined for $x=-\frac{4}{3}$. Multiplying both sides of the equation by $3x+4$ gives

$$2x-1=5(3x+4).$$

Now, solve this equation.

$$2x-1=15x+20$$
$$-13x=21$$
$$x=-\frac{21}{13}$$

TECHNOLOGY NOTE
Some graphing calculators have the capability to convert decimals representing rational numbers to quotients of integers. For example, the solution of the equation in Example 1 has its decimal displayed in Figure 16. Consult your owner's manual to see if your model is capable of converting this decimal numeral to its fraction form, $-21/13$.

The solution set is $\left\{-\frac{21}{13}\right\}$. An analytic check will verify this.

GRAPHICAL To support this solution, we notice that the original equation is equivalent to $\frac{2x-1}{3x+4}-5=0$. If we let the left side of this equation be represented by $f(x)$, then the x-intercept of f is the solution of the equation $f(x)=0$. The graph in Figure 16 on the next page suggests that $-\frac{21}{13}\approx -1.62$ is indeed the solution. By using the capabilities of the calculator, we can confirm that the decimal value of the x-intercept is approximately -1.62, which is the correct decimal approximation for $-\frac{21}{13}$ to the nearest hundredth. ◆

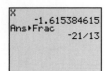

The zero found on the screen in Figure 16 is shown here to equal −21/13.

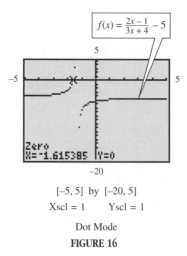

$$f(x) = \frac{2x-1}{3x+4} - 5$$

[−5, 5] by [−20, 5]

Xscl = 1   Yscl = 1

Dot Mode

FIGURE 16

In Example 2, we will solve the rational inequality $\frac{2x-1}{3x+4} \le 5$ analytically. It would be incorrect to multiply both sides of the inequality by $3x + 4$, since it may represent a negative number, and this would require reversing the inequality symbol. We will use the sign graph method, first introduced in Section 3.3 to solve quadratic inequalities analytically.

EXAMPLE 2 *Solving a Rational Inequality Analytically and Supporting Graphically*   Use a sign graph to solve

$$\frac{2x-1}{3x+4} \le 5,$$

and use the graph in Figure 16 to support the answer.

SOLUTION   ANALYTIC   We begin by subtracting 5 on both sides and combining the terms on the left into a single fraction.

$$\frac{2x-1}{3x+4} \le 5$$

$$\frac{2x-1}{3x+4} - 5 \le 0 \qquad \text{Subtract 5.}$$

$$\frac{2x-1-5(3x+4)}{3x+4} \le 0 \qquad \text{Common denominator is } 3x+4.$$

$$\frac{-13x-21}{3x+4} \le 0 \qquad \text{Combine terms.}$$

To draw a sign graph, first solve the equations

$$-13x - 21 = 0 \quad \text{and} \quad 3x + 4 = 0,$$

getting the solutions

$$x = -\frac{21}{13} \quad \text{and} \quad x = -\frac{4}{3}.$$

Use the values $-\frac{21}{13}$ and $-\frac{4}{3}$ to divide the number line into three intervals. Now, complete a sign graph and find the intervals where the quotient is negative. See Figure 17.

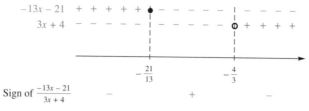

FIGURE 17

From the sign graph, values of x in the two intervals $\left(-\infty, -\frac{21}{13}\right)$ and $\left(-\frac{4}{3}, \infty\right)$ make the quotient negative, as required. Because the inequality is not strict, the endpoint $-\frac{21}{13}$ satisfies it. However, the endpoint $-\frac{4}{3}$ causes the denominator to equal 0, so it is not included in the solution set. Therefore, the solution set should be written $\left(-\infty, -\frac{21}{13}\right] \cup \left(-\frac{4}{3}, \infty\right)$.

GRAPHICAL   The original inequality is equivalent to $\frac{2x-1}{3x+4} - 5 \leq 0$. If we observe the graph in Figure 16, we see that it lies *below* the x-axis for x-values less than $-\frac{21}{13}$, and for x-values greater than the x-value of the vertical asymptote, which is $-\frac{4}{3}$. Therefore, the graph supports our analytic solution.   ◆

Caution   As suggested by Example 2, be very careful with the endpoints of the intervals in the solution of rational inequalities.

The next example shows the importance of determining the values for which the rational expressions in an equation are undefined before solving the equation analytically.

EXAMPLE 3   *Solving a Rational Equation Analytically and Supporting Graphically*   Solve the rational equation

$$\frac{x}{x-2} + \frac{1}{x+2} = \frac{8}{x^2-4}$$

analytically, and support the solution graphically.

SOLUTION   ANALYTIC   First, note that for this equation, $x \neq \pm 2$.

$$\frac{x}{x-2} + \frac{1}{x+2} = \frac{8}{x^2-4} \qquad \text{Given equation}$$

$$x(x+2) + 1(x-2) = 8 \qquad \text{Multiply by } (x-2)(x+2).$$

$$x^2 + 2x + x - 2 = 8$$

$$x^2 + 3x - 10 = 0 \qquad \text{Put in standard form.}$$

$$(x+5)(x-2) = 0 \qquad \text{Factor.}$$

$$x+5 = 0 \quad \text{or} \quad x-2 = 0 \qquad \text{Use the zero-product property.}$$

$$x = -5 \qquad\qquad x = 2$$

The numbers -5 and 2 are the *possible* solutions of the equation. Recall that 2 is not in the domain of the original equation and, therefore, must be rejected. (Such a value is called *extraneous.*) The solution set is $\{-5\}$. (The number 2 is a solution of the equation obtained after the multiplication step. But multiplying by $x-2$ when $x = 2$ is multiplying by 0, which does not lead to an equivalent equation.)

GRAPHICAL   To support this solution, we graph $f(x) = \frac{x}{x-2} + \frac{1}{x+2} - \frac{8}{x^2-4}$ and notice that the x-intercept is -5, as expected. The line $x = -2$ is a vertical asymptote, and for $x = 2$, the graph has a "hole." See Figure 18 on the next page.   ◆

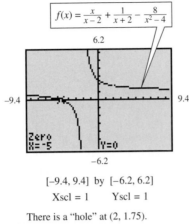

$$f(x) = \frac{x}{x-2} + \frac{1}{x+2} - \frac{8}{x^2-4}$$

[−9.4, 9.4] by [−6.2, 6.2]
Xscl = 1   Yscl = 1

There is a "hole" at (2, 1.75).

FIGURE 18

Example 3 illustrates the general procedure for solving rational equations analytically.

SOLVING EQUATIONS INVOLVING RATIONAL FUNCTIONS ANALYTICALLY

1. Determine all values for which the rational functions will be undefined.
2. Multiply both sides of the equation by the least common denominator of all rational expressions in the equation.
3. Solve the resulting equation.
4. Reject any values that were determined in Step 1.

For Group Discussion   In solving the rational equation in Example 3, our first step was to multiply both sides by $(x-2)(x+2)$. This eventually led to the quadratic equation $x^2 + 3x - 10 = 0$. The graph of $y = x^2 + 3x - 10$ is shown in Figure 19.

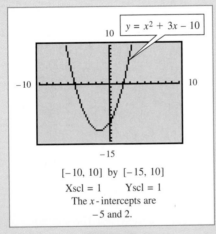

$$y = x^2 + 3x - 10$$

[−10, 10] by [−15, 10]
Xscl = 1   Yscl = 1
The x-intercepts are
−5 and 2.

FIGURE 19

1. Discuss how the solution of the equation in Example 3 and the extraneous solution found there relate to the solutions of the equation $x^2 + 3x - 10 = 0$. How does the extraneous solution show up on the graph of the parabola?
2. Why should you always pay attention to the domain for a rational equation?

EXAMPLE 4 *Solving a Rational Equation and Associated Inequalities* Consider the rational function $f(x) = 7x^{-4} - 8x^{-2} + 1$.

(a) Solve the equation $f(x) = 0$ analytically, and support the solutions graphically.

(b) Use a graph to find the solution set of $f(x) \le 0$.

(c) Use a graph to find the solution set of $f(x) \ge 0$.

SOLUTION

(a) ANALYTIC The expression for $f(x)$ may be written equivalently as

$$\frac{7}{x^4} - \frac{8}{x^2} + 1.$$

We begin by noticing that for this expression, $x \ne 0$. Now we solve the equation.

$$\frac{7}{x^4} - \frac{8}{x^2} + 1 = 0$$

$$7 - 8x^2 + x^4 = 0 \qquad \text{Multiply by } x^4.$$

$$x^4 - 8x^2 + 7 = 0$$

$$(x^2 - 7)(x^2 - 1) = 0$$

$$(x + \sqrt{7})(x - \sqrt{7})(x + 1)(x - 1) = 0$$

$$x + \sqrt{7} = 0 \quad \text{ or } \quad x - \sqrt{7} = 0 \quad \text{ or } \quad x + 1 = 0 \quad \text{ or } \quad x - 1 = 0$$

$$x = -\sqrt{7} \qquad\qquad x = \sqrt{7} \qquad\qquad x = -1 \qquad\qquad x = 1$$

The solution set is $\left\{ \pm\sqrt{7},\ \pm 1 \right\}$.

GRAPHICAL The graph in Figure 20 suggests the accuracy of our solutions, as the four x-intercepts correspond to the four solutions. This can be verified by using the calculator capabilities to find solutions, noting that $\sqrt{7} \approx 2.65$.

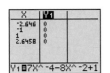

This table further supports the solutions of the equation in Example 4(a).

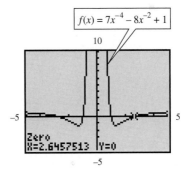

$$[-5, 5] \text{ by } [-5, 10]$$
$$\text{Xscl} = 1 \qquad \text{Yscl} = 1$$

A decimal approximation for $\sqrt{7}$ is displayed. The other zeros of the function are $-\sqrt{7}$, -1, and 1.

FIGURE 20

(b) The graph of f lies *below* or *on* the x-axis for values of x between $-\sqrt{7}$ and -1 (inclusive of both values) or between 1 and $\sqrt{7}$ (inclusive of both values). Therefore, the solution set is $\left[-\sqrt{7}, -1 \right] \cup \left[1, \sqrt{7} \right]$.

(c) The graph of f lies *above* or *on* the x-axis for the intervals $\left(-\infty, -\sqrt{7} \right]$, $[-1, 0)$, $(0, 1]$, $\left[\sqrt{7}, \infty \right)$. Therefore, the solution set is the union of these intervals: $\left(-\infty, -\sqrt{7} \right] \cup [-1, 0) \cup (0, 1] \cup \left[\sqrt{7}, \infty \right)$. Notice that 0 is not included because it is not in the domain. ◆

APPLICATIONS OF RATIONAL FUNCTIONS

We now look at two examples of applications of rational functions. The first is related to traffic control.

EXAMPLE 5   *Applying Rational Functions to Traffic Intensity*   Vehicles arrive randomly at a parking ramp at an average rate of 2.6 vehicles per minute. The parking attendant can admit 3.2 vehicles per minute. However, since arrivals are random, lines form at various times. (*Source:* Mannering, F., and W. Kilareski, *Principles of Highway Engineering and Traffic Control,* John Wiley & Sons, 1990.)

(a)   The traffic intensity x is defined as the ratio of the average arrival rate to the average admittance rate. Determine x for this parking ramp.

(b)   The average number of vehicles waiting in line to enter the ramp is given by the rational function $f(x) = \frac{x^2}{2(1 - x)}$, where $0 \leq x < 1$ is the traffic intensity. Compute $f(x)$ for this parking ramp.

(c)   Graph $y = f(x)$. What happens to the number of vehicles waiting as the traffic intensity approaches 1?

SOLUTION

(a)   The average arrival rate is 2.6 vehicles, and the average admittance rate is 3.2 vehicles, so

$$x = \frac{2.6}{3.2} = .8125.$$

(b)   In part (a), we found that $x = .8125$. Thus,

$$f(x) = \frac{.8125^2}{2(1 - .8125)} \approx 1.76 \text{ vehicles.}$$

(c)   From the graph shown in Figure 21, we see that as x approaches 1, $y = f(x)$ gets very large. This is not surprising; it is what we would expect.   ◆

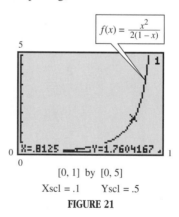

[0, 1] by [0, 5]

Xscl = .1   Yscl = .5

FIGURE 21

Sometimes the model for solving an optimization problem is a rational function. This is shown in the next example.

EXAMPLE 6   *Solving an Optimization Problem Involving Aluminum Can Manufacture*   A manufacturer wants to construct cylindrical aluminum cans with a volume of 2000 cubic centimeters (2 liters). What radius and what height of the can will minimize the amount of aluminum used? What will this amount be?

FIGURE 22

SOLUTION   The two unknowns in this problem are the radius and the height of the can. We will label the radius x and the height h, as shown in Figure 22. Minimizing the amount of aluminum used requires minimizing the surface area of the can, which we will designate S. From the list of geometric formulas on the inside covers, we find that the surface area S is given by the formula

$$S = 2\pi xh + 2\pi x^2. \qquad \text{(where } x \text{ is the radius and } h \text{ is the height)}$$

The formula involves two variables on the right, so we will solve for one in terms of the other in order to obtain a function of a single variable. Since the volume of the can is to be 2000 cubic centimeters, and the formula for volume is $V = \pi x^2 h$ (where x is the radius and h is the height), we have

$$V = \pi x^2 h$$
$$2000 = \pi x^2 h \qquad \text{Let } V = 2000.$$
$$h = \frac{2000}{\pi x^2}. \qquad \text{Solve for } h.$$

Now we can write the surface area S as a function of x alone:

$$S(x) = 2\pi x \left(\frac{2000}{\pi x^2} \right) + 2\pi x^2$$
$$= \frac{4000}{x} + 2\pi x^2 \qquad\qquad\qquad *$$
$$= \frac{4000 + 2\pi x^3}{x}. \qquad \text{Combine terms.*}$$

Since x represents the radius, it must be a positive number. We graph the function, using the window $[0, 20]$ by $[-500, 5000]$, and find that the local minimum is at approximately $(6.83, 878.76)$. See Figure 23. Therefore, the radius should be 6.83 cm to the nearest hundredth of a centimeter, and the height should be approximately $\frac{2000}{\pi (6.83)^2}$, or 13.65 centimeters. These dimensions lead to a minimum amount of 878.76 cubic centimeters of aluminum used.   ◆

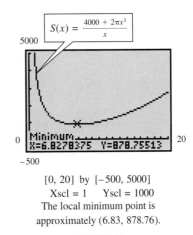

$[0, 20]$ by $[-500, 5000]$
Xscl = 1   Yscl = 1000
The local minimum point is
approximately $(6.83, 878.76)$.

FIGURE 23

---

*These steps are not necessary to obtain the appropriate graph.

INVERSE VARIATION

In Section 1.8, we saw how direct variation is an example of an application of linear functions. Direct variation as a *power* was discussed in Section 2.6. Another type of variation, inverse variation, is an example of an application of rational functions.

INVERSE VARIATION

Let n be a positive real number. Then y **varies inversely** as the nth power of x, or y is **inversely proportional** to the nth power of x, if a nonzero real number k exists such that

$$y = \frac{k}{x^n}.$$

If $n = 1$, then $y = \frac{k}{x}$, and y **varies inversely** as x.

EXAMPLE 7 *Solving an Inverse Variation Problem In Manufacturing*   In a certain manufacturing process, the cost of producing a single item varies inversely as the square of the number of items produced. If 100 items are produced, each costs $2. Find the cost per item if 400 items are produced. Support the result with a graph.

SOLUTION   ANALYTIC   Let x represent the number of items produced and y the cost per item, and write

$$y = \frac{k}{x^2}$$

for some nonzero constant k. Since $y = 2$ when $x = 100$,

$$2 = \frac{k}{100^2} \quad \text{or} \quad k = 20{,}000.$$

Thus, the relationship between x and y is given by

$$y = \frac{20{,}000}{x^2}.$$

When 400 items are produced, the cost per item is

$$y = \frac{20{,}000}{400^2} = .125, \text{ or } 12.5\text{¢}.$$

GRAPHICAL   The graph of $y = \frac{20{,}000}{x^2}$ in Figure 24 supports this result. See the display of coordinates.   ◆

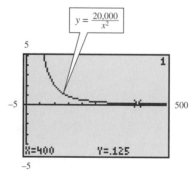

$[-5, 500]$ by $[-5, 5]$

Xscl = 50   Yscl = 1

FIGURE 24

COMBINED AND JOINT VARIATION

One variable may depend on more than one other variable. Such variation is called *combined variation*. More specifically, when a variable depends on the *product* of two or more other variables, it is referred to as *joint variation*.

JOINT VARIATION

Let m and n be real numbers. Then y **varies jointly** as the nth power of x and the mth power of z, if a nonzero real number k exists such that

$$y = kx^n z^m.$$

EXAMPLE 8 *Solving a Joint Variation Problem to Find the Area of a Triangle*
The area of a triangle varies jointly as the lengths of the base and the height. A triangle with a base of 10 feet and a height of 4 feet has an area of 20 square feet. Find the area of a triangle with a base of 3 centimeters and a height of 8 centimeters.

SOLUTION Let A represent the area, b the base, and h the height of the triangle. Then, $A = kbh$ for some number k. Since A is 20 when b is 10 and h is 4,

$$20 = k(10)(4)$$

$$\frac{1}{2} = k.$$

Then,

$$A = \frac{1}{2}bh,$$

which is the familiar formula for the area of a triangle. When $b = 3$ centimeters and $h = 8$ centimeters,

$$A = \frac{1}{2}(3)(8) = 12 \text{ square centimeters.} \qquad \blacklozenge$$

EXAMPLE 9 *Solving a Combined Variation Problem In Photography* Variation can be seen extensively in the field of photography. The formula $L = \frac{25F^2}{st}$ represents a combined variation. The luminance, L, varies directly as the square of the F-stop, F. It also varies inversely as the product of the film ASA number, s, and the shutter speed, t. The constant of variation is 25.

Suppose we want to use 200 ASA film and a shutter speed of $\frac{1}{250}$ when 500 footcandles of light are available. What would be an appropriate F-stop?

SOLUTION We begin with the given formula $L = \frac{25F^2}{st}$ and substitute the given values for the variables: $L = 500$, $s = 200$, and $t = \frac{1}{250}$. Then, solve for F.

$$L = \frac{25F^2}{st}$$

$$500 = \frac{25F^2}{200\left(\dfrac{1}{250}\right)}$$

$$400 = 25F^2$$

$$16 = F^2$$

$$4 = F \qquad (F > 0)$$

An F-stop of 4 would be appropriate. $\qquad \blacklozenge$

4.2   EXERCISES   Tape 5

*In Exercises 1–8, the graph of a rational function $y = f(x)$ is given. Use the graph to give the solution set of (**a**) $f(x) = 0$, (**b**) $f(x) < 0$, and (**c**) $f(x) > 0$. Use set braces for (a) and interval notation for (b) and (c).*

1.

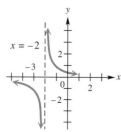

2.

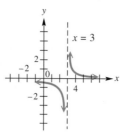

3.

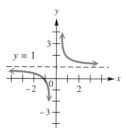

4.

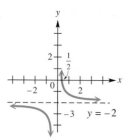

5.

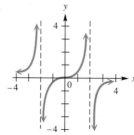

6.

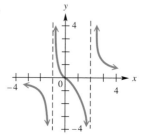

7.

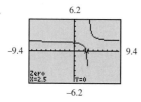

The line $x = 3$ is a vertical asymptote.

8.

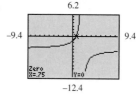

The line $x = 2$ is a vertical asymptote.

Use the methods of Examples 1 and 2 to solve the rational equation and rational inequalities given in each of Exercises 9–20. Then, support your answer using the x-intercept method with a calculator-generated graph in the suggested window.

9. (a) $\dfrac{x-3}{x+5} = 0$

(b) $\dfrac{x-3}{x+5} \leq 0$

(c) $\dfrac{x-3}{x+5} \geq 0$

Window: $[-10, 10]$ by $[-5, 8]$

10. (a) $\dfrac{x+1}{x-4} = 0$

(b) $\dfrac{x+1}{x-4} \geq 0$

(c) $\dfrac{x+1}{x-4} \leq 0$

Window: $[-10, 10]$ by $[-5, 10]$

11. (a) $\dfrac{x-1}{x+2} = 1$

(b) $\dfrac{x-1}{x+2} > 1$

(c) $\dfrac{x-1}{x+2} < 1$

Window: $[-10, 10]$ by $[-5, 10]$

12. (a) $\dfrac{x-6}{x+2} = -1$

(b) $\dfrac{x-6}{x+2} < -1$

(c) $\dfrac{x-6}{x+2} > -1$

Window: $[-10, 10]$ by $[-10, 10]$

13. (a) $\dfrac{1}{x-1} = \dfrac{5}{4}$

(b) $\dfrac{1}{x-1} < \dfrac{5}{4}$

(c) $\dfrac{1}{x-1} > \dfrac{5}{4}$

Window: $[-5, 5]$ by $[-5, 5]$

14. (a) $\dfrac{6}{5-3x} = 2$

(b) $\dfrac{6}{5-3x} \leq 2$

(c) $\dfrac{6}{5-3x} \geq 2$

Window: $[-5, 5]$ by $[-5, 5]$

15. (a) $\dfrac{4}{x-2} = \dfrac{3}{x-1}$

(b) $\dfrac{4}{x-2} \le \dfrac{3}{x-1}$

(c) $\dfrac{4}{x-2} \ge \dfrac{3}{x-1}$

Window: $[-3, 3]$ by $[-20, 20]$

16. (a) $\dfrac{4}{x+1} = \dfrac{2}{x+3}$

(b) $\dfrac{4}{x+1} < \dfrac{2}{x+3}$

(c) $\dfrac{4}{x+1} > \dfrac{2}{x+3}$

Window: $[-8, 5]$ by $[-10, 10]$

17. (a) $\dfrac{1}{(x-2)^2} = 0$

(b) $\dfrac{1}{(x-2)^2} < 0$

(c) $\dfrac{1}{(x-2)^2} > 0$

Window: $[-5, 10]$ by $[-5, 10]$

18. (a) $\dfrac{-2}{(x+3)^2} = 0$

(b) $\dfrac{-2}{(x+3)^2} > 0$

(c) $\dfrac{-2}{(x+3)^2} < 0$

Window: $[-10, 5]$ by $[-10, 5]$

19. (a) $\dfrac{5}{x+1} = \dfrac{12}{x+1}$

(b) $\dfrac{5}{x+1} > \dfrac{12}{x+1}$

(c) $\dfrac{5}{x+1} < \dfrac{12}{x+1}$

Window: $[-10, 10]$ by $[-10, 10]$

20. (a) $\dfrac{7}{x+2} = \dfrac{1}{x+2}$

(b) $\dfrac{7}{x+2} \ge \dfrac{1}{x+2}$

(c) $\dfrac{7}{x+2} \le \dfrac{1}{x+2}$

Window: $[-10, 10]$ by $[-10, 10]$

21. The graph of $y = f(x)$, where $f(x) = \frac{4}{x-2} - \frac{3}{x-1} = \frac{x+2}{x^2-3x+2}$, is shown in the window $[-4.7, 4.7]$ by $[-3.1, 12.4]$ in the accompanying figure. The solution set of $f(x) \le 0$ was required in Exercise 15(b) (in a different form, but equivalent to this one). From the graph, it appears that no part of the curve lies below the x-axis, which would lead to an empty solution set. Yet, the solution set of $f(x) \le 0$ is $(-\infty, -2] \cup (1, 2)$. Use this observation to explain why relying on graphical analysis alone is not sufficient for solving equations and inequalities.

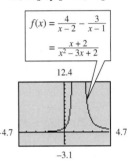

$f(x) = \dfrac{4}{x-2} - \dfrac{3}{x-1}$

$= \dfrac{x+2}{x^2-3x+2}$

The lines $x = 1$ and $x = 2$ are vertical asymptotes.

22. Consider the rational function $f(x) = \frac{1}{x^2+1}$. Without graphing, answer the following series of items in order.

(a) For all real numbers x, $x^2 + 1$ is _____ 0.
(equal to, greater than, less than)

(b) Because $1 > 0$, for all real numbers x, $f(x)$ is _____ 0.
(equal to, greater than, less than)

(c) The solution set of $f(x) = 0$ is _____, of $f(x) < 0$ is _____, and of $f(x) > 0$ is _____.

(d) Support your answers in part (c) with a graph and an explanation.

Use the methods explained in Examples 3 and 4 to find all complex solutions for each equation. Support your real solutions with a graph, using an appropriate window.

23. $1 - \dfrac{13}{x} + \dfrac{36}{x^2} = 0$

24. $1 - \dfrac{3}{x} - \dfrac{10}{x^2} = 0$

25. $1 + \dfrac{3}{x} = \dfrac{5}{x^2}$

26. $4 + \dfrac{7}{x} = -\dfrac{1}{x^2}$

27. $\dfrac{x}{2-x} + \dfrac{2}{x} - 5 = 0$

28. $\dfrac{2x}{x-3} + \dfrac{4}{x} - 6 = 0$

29. $x^{-4} - 3x^{-2} - 4 = 0$

30. $x^{-4} - 5x^{-2} + 4 = 0$

31. $\dfrac{1}{x+2} + \dfrac{3}{x+7} = \dfrac{5}{x^2+9x+14}$

32. $\dfrac{1}{x+3} + \dfrac{4}{x+5} = \dfrac{2}{x^2+8x+15}$

33. $\dfrac{x}{x-3} + \dfrac{4}{x+3} = \dfrac{18}{x^2-9}$

34. $\dfrac{2x}{x-3} + \dfrac{4}{x+3} = \dfrac{24}{9-x^2}$

35. $9x^{-1} + 4(6x-3)^{-1} = 2(6x-3)^{-1}$ **36.** $x(x-2)^{-1} + x(x+2)^{-1} = 8(x^2-4)^{-1}$

RELATING CONCEPTS (EXERCISES 37–42)

In Example 2, we illustrated how a sign graph may be used to solve a rational inequality analytically. There is another method of solving rational inequalities analytically. For example, suppose that we wish to solve

$$\frac{x-2}{x+3} \le 2.$$

Since -3 is not in the domain, it cannot possibly be part of the solution. We should not multiply both sides by $x + 3$, because it may be negative, which would require changing the direction of the inequality sign. However, $(x + 3)^2$ is positive. (Why?) Therefore, we multiply both sides by $(x + 3)^2$.

$$\frac{x-2}{x+3} \le 2 \qquad\qquad \text{Given inequality}$$

$$(x + 3)^2\left(\frac{x-2}{x+3}\right) \le 2(x + 3)^2 \qquad\qquad \text{Multiply by } (x + 3)^2, \text{ which is positive.}$$

$$(x + 3)(x - 2) \le 2(x^2 + 6x + 9)$$

$$x^2 + x - 6 \le 2x^2 + 12x + 18$$

$$0 \le x^2 + 11x + 24$$

$$x^2 + 11x + 24 \ge 0 \qquad\qquad \text{Rewrite with 0 on the right side.}$$

The polynomial $x^2 + 11x + 24$ defines a quadratic function whose graph is shown in the figure. Its x-intercepts are the solutions of $x^2 + 11x + 24 = 0$. From the graph we see that the solution set of $x^2 + 11x + 24 \ge 0$ is $(-\infty, -8] \cup [-3, \infty)$. Because -3 must be excluded from the solution set of the original inequality, the solution set of $\frac{x-2}{x+3} \le 2$ is $(-\infty, -8] \cup (-3, \infty)$.

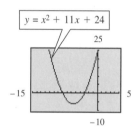

The x-intercepts are
-8 and -3.

Use the method described above to solve the following rational inequalities.

37. $\dfrac{x+5}{x-3} \ge 1$

38. $\dfrac{2-x}{x+6} \ge -1$

39. $\dfrac{(x-3)^2}{x+1} \ge x+1$

40. $\dfrac{(x-6)^2}{x+5} \ge x+5$

41. $\dfrac{x+3}{(x+1)^2} < \dfrac{1}{x+1}$

42. $\dfrac{x-4}{(x+2)^2} > \dfrac{1}{x+2}$

Use a purely graphical method to solve the equation in part (a), expressing solutions to the nearest hundredth. Then, use the graph to solve the associated inequalities in parts (b) and (c), expressing endpoints to the nearest hundredth.

43. (a) $\dfrac{\sqrt{2x+5}}{x^3 - \sqrt{3}} = 0$

 (b) $\dfrac{\sqrt{2x+5}}{x^3 - \sqrt{3}} > 0$

 (c) $\dfrac{\sqrt{2x+5}}{x^3 - \sqrt{3}} < 0$

44. (a) $\dfrac{\sqrt[3]{7x^3} - 1}{x^2 + 2} = 0$

 (b) $\dfrac{\sqrt[3]{7x^3} - 1}{x^2 + 2} > 0$

 (c) $\dfrac{\sqrt[3]{7x^3} - 1}{x^2 + 2} < 0$

RELATING CONCEPTS (EXERCISES 45–48)

Consider the rational equation below, and work Exercises 45–48 in order.

$$1 - \frac{2}{x} - \frac{2}{x^2} + \frac{3}{x^3} = 0$$

45. Graph $f(x) = 1 - \frac{2}{x} - \frac{2}{x^2} + \frac{3}{x^3}$, and determine its x-intercepts graphically. (*Hint:* One is rational and two are irrational.) Give approximations of the irrational solutions correct to the nearest hundredth.

46. What is the domain of f?

47. Use the rational solution determined in Exercise 45 with synthetic division to obtain a quadratic equation that will give the other two solutions in their exact forms. (*Hint:* Multiply both sides by x^3 first.) (See Sections 3.3 and 3.6.)

48. Verify that the approximations found in Exercise 45 are indeed approximations of the two irrational solutions whose exact values were found in Exercise 47.

49. A student attempted to solve the inequality

$$\frac{2x - 1}{x + 2} \le 0$$

by multiplying both sides by $x + 2$ to get

$$2x - 1 \le 0$$

$$x \le \frac{1}{2}.$$

He wrote the solution set as $\left(-\infty, \frac{1}{2}\right]$. Is his solution correct? Explain.

50. The inequality in Exercise 49 may be solved using this alternative analytic method:

Case 1: For $x > -2$, $x + 2$ is positive. Multiply both sides by $x + 2$, keeping the inequality symbol pointing in the same direction. Take the intersection of the interval where $x > -2$, $(-2, \infty)$, and the interval obtained in the solution process.

Case 2: For $x < -2$, $x + 2$ is negative. Multiply both sides by $x + 2$ and reverse the direction of the inequality symbol. Take the intersection of the interval where $x < -2$, $(-\infty, -2)$, and the interval obtained in the solution process.

The *union* of the intervals obtained in the two cases is the solution set of the original inequality. Solve the inequality in Exercise 49 using this method.

Solve each problem involving rational functions.

51. *Traffic Intensity*   Refer to Example 5. Let the average number of vehicles arriving at the gate of an amusement park per minute be equal to k, and let the average number of vehicles admitted by the park attendants be equal

to r. Then, the average waiting time T (in minutes) for each vehicle arriving at the park is given by the rational function $T(r) = \dfrac{2r - k}{2r^2 - 2kr}$, where $r > k$. (*Source:* Mannering, F., and W. Kilareski, *Principles of Highway Engineering and Traffic Control*, John Wiley & Sons, Inc., 1990.)

(a) It is known from experience that on Saturday afternoon k is equal to 25 vehicles per minute. Use graphing to estimate the admittance rate r that is necessary to keep the average waiting time T for each vehicle to 30 seconds.

(b) If one park attendant can serve 5.3 vehicles per minute, how many park attendants will be needed to keep the average wait to 30 seconds?

52. *Braking Distance*   The rational function defined by

$$d(x) = \frac{(8.71 \times 10^3)x^2 - (6.94 \times 10^4)x + (4.70 \times 10^5)}{(1.08)x^2 - (3.24 \times 10^2)x + (8.22 \times 10^4)}$$

can be used to accurately model the braking distance for automobiles traveling at x miles per hour, where $20 \le x \le 70$. (*Source:* Mannering, F., and W. Kilareski, *Principles of Highway Engineering and Traffic Control*, John Wiley & Sons, Inc., 1990.)

(a) Use graphing to estimate x when $d(x) = 300$.

(b) Complete the table for each value of x.

x	20	25	30	35	40	45	50	55	60	65	70
$d(x)$											

(c) If a car doubles its speed, does the stopping distance double or more than double? Explain.

(d) Suppose the stopping distance doubled whenever the speed doubled. What type of relationship would exist between the stopping distance and the speed?

53. *Diseases and Smoking*   The table contains incidence ratios by age for deaths due to coronary heart disease (CHD) and lung cancer (LC) when comparing smokers (21–39 cigarettes per day) to nonsmokers.

Age	CHD	LC
55–64	1.9	10
65–74	1.7	9

The incidence ratio of 10 means that smokers are 10 times more likely than nonsmokers to die of lung cancer between the ages of 55 and 64. If the incidence ratio is x, then the percent P (in decimal form) of deaths caused by smoking can be calculated using the rational function $P(x) = \dfrac{x - 1}{x}$. (*Source:* Walker, A.,

Observation and Inference: An Introduction to the Methods of Epidemiology, Epidemiology Resources Inc., Newton Lower Falls, MA, 1991.)

(a) As x increases, what value does $P(x)$ approach?

(b) Why do you suppose the incidence ratios are slightly smaller for ages 65–74 than for ages 55–64?

54. *Deaths Due to AIDS*   Refer to Example 4 and Exercise 23 in Section 3.4.

(a) Make a table listing the ratio of total deaths caused by AIDS to total cases of AIDS in the United States for each year from 1982 to 1993. (For example, in 1982 there were 620 deaths and 1563 cases, so the ratio is $\dfrac{620}{1563} \approx .397$.)

(b) As time progresses, what happens to the values of the ratio?

(c) Using the polynomial functions f and g that were found in the sources cited above, define the rational function h, where $h(x) = \dfrac{g(x)}{f(x)}$. Graph $h(x)$ on the interval $[2, 20]$. Compare $h(x)$ to the values for the ratio found in your table.

(d) Use $h(x)$ to write an equation that approximates the relationship between the functions $f(x)$ and $g(x)$ as x increases.

(e) The ratio of AIDS deaths to AIDS cases can be used to estimate the total number of AIDS deaths. According to the World Health Organization, in 1994 there had been 4 million AIDS cases diagnosed worldwide since the disease began. Predict the total number of deaths caused by AIDS.

55. Computers often use rational functions to approximate other types of functions. Use graphing to match each function in (a)–(d) with its rational approximation in (i)–(iv) on the interval $[1, 15]$.

(a) $f_1(x) = \sqrt{x}$

(b) $f_2(x) = \sqrt{4x + 1}$

(c) $f_3(x) = \sqrt[3]{x}$

(d) $f_4(x) = \dfrac{1 - \sqrt{x}}{1 + \sqrt{x}}$

(i) $r_1(x) = \dfrac{2 - 2x^2}{3x^2 + 10x + 3}$

(ii) $r_2(x) = \dfrac{15x^2 + 75x + 33}{x^2 + 23x + 31}$

(iii) $r_3(x) = \dfrac{10x^2 + 80x + 32}{x^2 + 40x + 80}$

(iv) $r_4(x) = \dfrac{7x^3 + 42x^2 + 30x + 2}{2x^3 + 30x^2 + 42x + 7}$

56. *Average Cost per Unit*   If the cost of producing x items of a particular commodity is given by y, where $y = 200{,}000 + .50x$ (in dollars), then the *average cost per unit* for the commodity is given by the function

$$C(x) = \frac{y}{x} = \frac{200{,}000 + .50x}{x}, \quad x > 0.$$

Use the graph of $C(x)$ to find the average cost when 25,000 units are produced.

57. *Ordering and Storage Expenses*   A company finds that if it places orders for stereo receivers x times per year, the expense of ordering and storage is given by the function

$$E(x) = 40x + \frac{1000}{x}, \quad x > 0,$$

dollars. Find the number of times that the company should order per year so that its ordering and storage expenses are minimized. (Use the graph of $y = E(x)$.)

58. *Volume of a Cylindrical Can*   A metal cylindrical can with an *open top* and *closed bottom* is to have a volume of 4 cubic feet. Find the dimensions that require the least amount of material. (Compare this problem to Example 6.)

Fill in the blanks with the correct responses.

59. $b = \dfrac{24}{h}$ is the formula for the base of a parallelogram with area 24. The base of this parallelogram varies _____ as its _____. The constant of variation is _____.

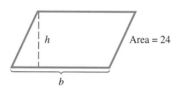

60. $h = \dfrac{300/\pi}{r^2}$ is the formula for the height of a right circular cone with volume 100. The height of this cone varies _____ as the _____ of the _____ of its base. The constant of variation is _____.

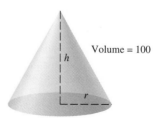

Solve the following problems involving various types of variation.

61. Suppose r varies directly as the square of m, and inversely as s. If $r = 12$ when $m = 6$ and $s = 4$, find r when $m = 4$ and $s = 10$.

62. Suppose p varies directly as the square of z, and inversely as r. If $p = \frac{32}{5}$ when $z = 4$ and $r = 10$, find p when $z = 2$ and $r = 16$.

63. Let a be proportional to m and n^2, and inversely proportional to y^3. If $a = 9$ when $m = 4$, $n = 9$, and $y = 3$, find a if $m = 6$, $n = 2$, and $y = 5$.

64. If y varies directly as x, and inversely as m^2 and r^2, and $y = \frac{5}{3}$ when $x = 1$, $m = 2$, and $r = 3$, find y if $x = 3$, $m = 1$, and $r = 8$.

65. For $k > 0$, if y varies directly as x, when x increases, y _____, and when x decreases, y _____.

66. For $k > 0$, if y varies inversely as x, when x increases, y _____, and when x decreases, y _____.

Solve each problem involving inverse, joint, or combined variation.

67. *Current Flow* In electric current flow, it is found that the resistance (measured in units called ohms) offered by a fixed length of wire of a given material varies inversely as the square of the diameter of the wire. If a wire .01 inch in diameter has a resistance of .4 ohm, what is the resistance of a wire of the same length and material with a diameter of .03 inch?

68. *Illumination* The illumination produced by a light source varies inversely as the square of the distance from the source. The illumination of a light source at 5 m is 70 candela. What is the illumination 12 m from the source?

69. *Vibrations of a Guitar String* The number of vibrations per second (the pitch) of a steel guitar string varies directly as the square root of the tension and inversely as the length of the string. If the number of vibrations per second is 5 when the tension is 225 kilograms and the length is .60 meter, find the number of vibrations per second when the tension is 196 kilograms and the length is .65 meter.

70. *Period of a Pendulum* The period of a pendulum varies directly as the square root of the length of the pendulum and inversely as the square root of the acceleration due to gravity. Find the period when the length is 121 cm and the acceleration due to gravity is 980 cm per sec squared, if the period is 6π sec when the length is 289 cm and the acceleration due to gravity is 980 cm per sec squared.

71. *Ripening of Fruit* Under certain conditions, the length of time that it takes for fruit to ripen during the growing season varies inversely as the average maximum temperature during the season. If it takes 25 days for fruit to ripen with an average maximum temperature of 80°, find the number of days it would take at 75°.

72. *Long-Distance Phone Calls* The number of long-distance phone calls between two cities in a certain time period varies directly as the populations p_1 and p_2 of the cities, and inversely as the distance between them. If 10,000 calls are made between two cities 500 mi apart, having populations of 50,000 and 125,000, find the number of calls between two cities 800 mi apart, having populations of 20,000 and 80,000.

73. *Car Skidding* The force needed to keep a car from skidding on a curve varies inversely as the radius of the curve and jointly as the weight of the car and the square of the speed. It takes 3000 lb of force to keep a 2000-lb car from skidding on a curve of radius 500 ft at 30 mph. What force is needed to keep the same car from skidding on a curve of radius 800 ft at 60 mph?

74. *Sports Arena Construction* The roof of a new sports arena rests on round concrete pillars. The maximum load a cylindrical column of circular cross-section can hold varies directly as the fourth power of the diameter and inversely as the square of the height. The arena has 9 m tall columns that are 1 m in diameter and will support a load of 8 metric tons. How many metric tons will be supported by a column 12 m high and $\frac{2}{3}$ m in diameter?

Photography   Refer to Example 9 to work Exercises 75–77.

75. Suppose we want to use 200 ASA film and a shutter speed of .004 when 500 footcandles of light (L) are available. What would be an appropriate F-stop?

76. Determine the luminance needed when a photographer is using 400 ASA film, a shutter speed of $\frac{1}{60}$ of a second, and an F-stop of 5.6.

77. If 125 footcandles of light are available, and an F-stop of 2 is used with 200 ASA film, what shutter speed should be used?

78. *Aluminum Can Construction* A packaging company is studying various shapes for cylindrical aluminum cans. The plan is to build a can with a volume of 1000 cubic centimeters. While minimizing the amount of material used ultimately will be a consideration, for now they are only interested in aesthetics and want to determine the height the can needs to be depending on a given radius. Use the table feature of your graphing calculator to find the heights of cans when given a radius of 2, 4, 6, 8, 10, and 12 centimeters. (Let x be the radius, and solve the volume formula for height. Then, y_1 will be the height of the can.)

In each table, $Y_1 = \frac{k}{x}$ *for some value of k. Find the value of k.*

79.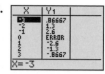

80.

81.

82.

4.3 GRAPHS OF ROOT FUNCTIONS

Radicals and Rational Exponents ◆ Functions Defined with Rational Exponents ◆ Graphs of $f(x) = \sqrt[n]{ax + b}$ ◆ Special Functions Defined by Roots

RADICALS AND RATIONAL EXPONENTS*

Up to this point, we have used radicals in a very informal manner. We observed the graphs of $y = \sqrt{x}$ and $y = \sqrt[3]{x}$ in Chapter 2; now we will investigate radical notation and rational numbers used as exponents in a more formal way.

DEFINITION OF $\sqrt[n]{a}$

Suppose that n is a positive integer greater than 1. Let a be any real number.

(i)   If $a > 0$, $\sqrt[n]{a}$ is the positive real number b such that $b^n = a$.
(ii) If $a < 0$ and n is odd, $\sqrt[n]{a}$ is the negative real number b such that $b^n = a$.
(iii) If $a < 0$ and n is even, then there is no real number $\sqrt[n]{a}$.
(iv) If $a = 0$, then $\sqrt[n]{a} = 0$.

In the radical expression $\sqrt[n]{a}$, a is called the **radicand**, and n is called the **root index**.

EXAMPLE 1   *Using Radical Notation*   The display in Figure 25 illustrates how a calculator will compute roots that are rational numbers.

$$\sqrt{25} = 5, \qquad \sqrt[3]{-27} = -3, \qquad \sqrt[5]{-32} = -2$$

The display in Figure 26 illustrates how a calculator will give rational approximations of roots that are irrational numbers.

$$\sqrt[3]{-7} \approx -1.912931183 \quad \sqrt[5]{36.7} \approx 2.055574459 \quad \sqrt{21} \approx 4.582575695 \qquad ◆$$

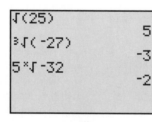

FIGURE 25   FIGURE 26

A familiar rule for exponents states that $(a^m)^n = a^{mn}$. If this rule is to be extended to rational numbers, then it makes sense to define $a^{1/n}$ as $\sqrt[n]{a}$, since $(a^{1/n})^n = a^1 = a$, while $\left(\sqrt[n]{a}\right)^n = a^1 = a$ as well. We now define $a^{1/n}$.

---

*At this point, you may wish to review the material on radicals and rational exponents in Chapter R.

DEFINITION OF $a^{1/n}$

   (i)   If n is an *even* positive integer and if $a > 0$, then $a^{1/n}$ is the *positive* real number whose nth power is a. That is, $a^{1/n} = \sqrt[n]{a}$.

   (ii) If n is an *odd* positive integer and if a is any real number, then $a^{1/n}$ is the single real number whose nth power is a. That is, $a^{1/n} = \sqrt[n]{a}$.

EXAMPLE 2   *Using $a^{1/n}$ Notation*   The display in Figure 27 illustrates how a calculator will compute numbers raised to the power $\frac{1}{n}$, leading to rational number results. Compare with Figure 25.

$$25^{1/2} = 5, \qquad (-27)^{1/3} = -3, \qquad (-32)^{1/5} = -2$$

The display in Figure 28 illustrates how a calculator will compute numbers raised to the power $\frac{1}{n}$, leading to rational approximations of irrational results. Compare with Figure 26.

$$(-7)^{1/3} \approx -1.912931183 \qquad 36.7^{1/5} \approx 2.055574459 \qquad 21^{1/2} \approx 4.582575695 \qquad \blacklozenge$$

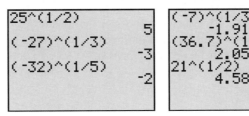

| **FIGURE 27** | **FIGURE 28** |

The definition of more general rational exponents is consistent with the familiar rule of exponents. For the power rule to hold, $(a^{1/n})^m$ must equal $a^{m/n}$.

DEFINITION OF $a^{m/n}$

If $\frac{m}{n}$ is a rational number, where n is a positive integer greater than 1 and a is a real number such that $\sqrt[n]{a}$ is also real, then

$$a^{m/n} = \left(\sqrt[n]{a}\right)^m = \sqrt[n]{a^m}.$$

EXAMPLE 3   *Using $a^{m/n}$ Notation*   Use the definition of $a^{m/n}$ to evaluate each of the expressions. Support the results with a calculator.

   (a) $125^{2/3}$ **(b)** $32^{7/5}$ **(c)** $(-27)^{2/3}$

   (d) $16^{-3/4}$ **(e)** $(-4)^{5/2}$ **(f)** $-4^{5/2}$

SOLUTION

ANALYTIC

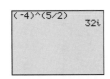

A calculator in complex number mode *will* return an imaginary value for $(-4)^{5/2}$. Compare to the result in Figure 29(c), where the calculator was in real number mode.

   (a) $125^{2/3} = \left(\sqrt[3]{125}\right)^2 = 5^2 = 25$, or $\sqrt[3]{125^2} = \sqrt[3]{15,625} = 25$

   (b) $32^{7/5} = \left(\sqrt[5]{32}\right)^7 = 2^7 = 128$

   (c) $(-27)^{2/3} = \left(\sqrt[3]{-27}\right)^2 = (-3)^2 = 9$

   (d) $16^{-3/4} = \dfrac{1}{16^{3/4}} = \dfrac{1}{\left(\sqrt[4]{16}\right)^3} = \dfrac{1}{2^3} = \dfrac{1}{8}$ or .125

   (e) $(-4)^{5/2}$ is not real because $\sqrt{-4}$ is not real.

   (f) $-4^{5/2} = -\left(\sqrt{4}\right)^5 = -(2^5) = -32$

GRAPHICAL   See Figure 29 on the next page for calculator support. $\blacklozenge$

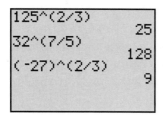

This screen supports the results in Example 3 (a), (b), and (c).

(a)

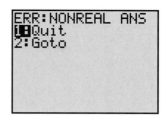

This screen supports the results in Example 3 (d) and (f).

(b)

This screen supports the result in Example 3 (e).

(c)

FIGURE 29

TECHNOLOGY NOTE

Some early models of graphing calculators will not allow you to calculate the expression in part (c) of Example 3, $(-27)^{2/3}$, by inputting the symbols as they appear. Use this fact to explain why it is essential for mathematical concepts to be studied hand-in-hand with technology.

Caution   Notice the difference between the expressions in parts (e) and (f) of Example 3. In part (e), the base is −4, while in part (f), the base is 4. Be very careful when dealing with expressions of these types.

FUNCTIONS DEFINED WITH RATIONAL EXPONENTS

Let us for the moment consider the function $f(x) = x^{3/2}$. Because $x^{3/2} = (\sqrt{x})^3$, the domain of this function must be $[0, \infty)$. It seems reasonable to expect that the graph of f will lie "between" the graphs of $y = x^1 = x$ and $y = x^2$, since $1 < \frac{3}{2} < 2$. Figure 30 confirms this expectation.

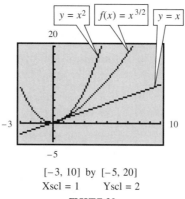

$[-3, 10]$ by $[-5, 20]$
Xscl = 1   Yscl = 2

FIGURE 30

Now, let us examine several different ways to obtain a rational approximation of the number $14^{3/2}$. See Figure 31.

$$14^{3/2} \approx 52.38320341 \qquad 14^{1.5} \approx 52.38320341 \qquad \left(\text{Because } 1.5 = \tfrac{3}{2}\right)$$

$$\left(\sqrt{14}\right)^3 \approx 52.38320341 \qquad (14^{.5})^3 \approx 52.38320341$$

$$\sqrt{14^3} \approx 52.38320341 \qquad (14^3)^{.5} \approx 52.38320341$$

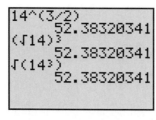

(a)

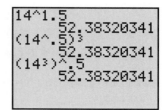

(b)

FIGURE 31

TECHNOLOGY NOTE

The number of digits shown in the display can be set by the user of a graphing calculator. For example, the irrational number seen in Figure 31 has a total of 10 digits displayed, 8 of which follow the decimal point.

Read your owner's manual to see how your model allows you to determine the number of digits that will be displayed.

As seen in Figure 31, there are several options as to how this approximation can be found. Now, look at Figure 32, which shows the graph of $f(x) = x^{3/2}$. The display shows graphical support of what we examined numerically in Figure 31.

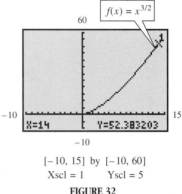

$f(x) = x^{3/2}$

X=14   Y=52.383203

$[-10, 15]$ by $[-10, 60]$
Xscl = 1   Yscl = 5

FIGURE 32

EXAMPLE 4 *Applying Rational Exponents to Storm Duration* Meteorologists can approximate the duration of a storm by using the formula $T = .07D^{3/2}$, where D is the diameter of the storm in miles and T is the time in hours. Suppose that radar shows a storm in the Gulf of Mexico to have a diameter of 3.8 miles. Approximately how long will the storm last? Support the answer with a graph.

SOLUTION   ANALYTIC   Let $D = 3.8$, and use a calculator to find T.

$$T = .07(3.8)^{3/2}   \quad \text{Let } D = 3.8.$$

$$T = .52   \quad \text{(To the nearest hundredth)}$$

We see that the storm can be expected to last approximately .52 hour, or a little more than 30 minutes.

GRAPHICAL   This result can be supported by a graph of $y = .07x^{3/2}$. When $x = 3.8$, $y \approx .52$. See Figure 33.   ◆

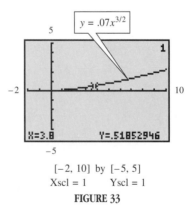

$y = .07x^{3/2}$

X=3.8   Y=.51852946

$[-2, 10]$ by $[-5, 5]$
Xscl = 1   Yscl = 1

FIGURE 33

GRAPHS OF $f(x) = \sqrt[n]{ax + b}$

Figure 13 in Section 2.1 shows the graph of $f(x) = \sqrt{x}$. When n is even, the graph of $f(x) = \sqrt[n]{x}$ resembles the graph of the square root function; as n gets larger, the graph lies closer to the x-axis as $x \to \infty$. Figure 34 on the next page shows the graphs of $y = \sqrt{x}$, $y = \sqrt[4]{x}$, and $y = \sqrt[6]{x}$ as examples of the graph of $f(x) = \sqrt[n]{x}$, where n is even, along with other pertinent information.

ROOT FUNCTION, *n* EVEN

$f(x) = \sqrt[n]{x}$ (Figure 34 for $n = 2, 4, 6$)   Domain: $[0, \infty)$ Range: $[0, \infty)$

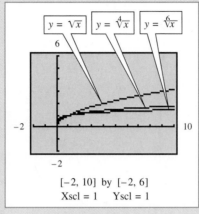

FIGURE 34

For *n* even, the root function $f(x) = \sqrt[n]{x}$ increases on $[0, \infty)$. It is also continuous on $[0, \infty)$. The definition of $x^{1/n}$ allows us to use a rational exponent as well as a radical when graphing a root function.

A discussion similar to the one above pertains to the graph of $f(x) = \sqrt[n]{x}$, where *n* is odd. We saw the graph of $f(x) = \sqrt[n]{x}$ in Figure 14 of Section 2.1. When *n* is odd, the graph of $f(x) = \sqrt[n]{x}$ resembles the graph of the cube root function; as *n* gets larger, the graph lies closer to the *x*-axis as $|x| \to \infty$. Figure 35 shows the graphs of $y = \sqrt[3]{x}$, $y = \sqrt[5]{x}$, and $y = \sqrt[7]{x}$ as examples of the graph of $f(x) = \sqrt[n]{x}$, where *n* is odd, along with other pertinent information.

ROOT FUNCTION, *n* ODD

$f(x) = \sqrt[n]{x}$ (Figure 35 for $n = 3, 5, 7$)   Domain: $(-\infty, \infty)$ Range: $(-\infty, \infty)$

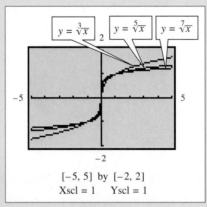

FIGURE 35

For *n* odd, the root function $f(x) = \sqrt[n]{x}$ increases on the domain $(-\infty, \infty)$. It is also continuous on $(-\infty, \infty)$. The definition of $x^{1/n}$ allows us to use a rational exponent as well as a radical when graphing a root function.

In order to determine the domain of a function of the form $f(x) = \sqrt[n]{ax + b}$, we must note the parity of n. If n is even, $ax + b$ must be greater than or equal to 0; if n is odd, $ax + b$ can be any real number.

EXAMPLE 5 *Finding Domains of Root Functions* Find the domain of each function.

(a) $f(x) = \sqrt{4x + 12}$ **(b)** $g(x) = \sqrt[3]{-8x + 8}$

SOLUTION

(a) For the function to be defined, $4x + 12$ must be greater than or equal to 0, since this is an even root ($n = 2$).

$$4x + 12 \geq 0$$
$$4x \geq -12$$
$$x \geq -3$$

The domain of f is $[-3, \infty)$.

(b) Because $n = 3$ and 3 is an odd number, the domain of g is $(-\infty, \infty)$. ◆

EXAMPLE 6 *Transforming Graphs of Root Functions*

(a) Use the terminology of Sections 2.2 and 2.3 to explain how the graph of $y = \sqrt{4x + 12}$ can be obtained from the graph of $y = \sqrt{x}$. Then, graph it in an appropriate window.

(b) Repeat part (a) for the graph of $y = \sqrt[3]{-8x + 8}$, as compared to the graph of $y = \sqrt[3]{x}$.

SOLUTION

(a) We begin by writing the expression in an equivalent form.

$$y = \sqrt{4x + 12} \qquad \text{Given form}$$
$$y = \sqrt{4(x + 3)} \qquad \text{Factor.}$$
$$y = \sqrt{4}\sqrt{x + 3} \qquad \sqrt{ab} = \sqrt{a} \cdot \sqrt{b}$$
$$y = 2\sqrt{x + 3} \qquad \sqrt{4} = 2$$

The graph of this function can be obtained from the graph of $y = \sqrt{x}$ by shifting horizontally 3 units to the left and stretching vertically by a factor of 2. Figure 36 shows both graphs for comparison. Notice that the domain is $[-3, \infty)$, as determined in Example 5(a). The range is $[0, \infty)$.

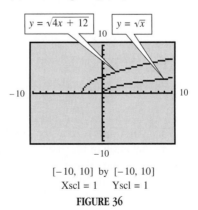

$$y = \sqrt{4x + 12} \qquad y = \sqrt{x}$$

$[-10, 10]$ by $[-10, 10]$
Xscl = 1   Yscl = 1

FIGURE 36

(b) $y = \sqrt[3]{-8x + 8}$ Given form

$\quad y = \sqrt[3]{-8(x - 1)}$ Factor.

$\quad y = \sqrt[3]{-8} \cdot \sqrt[3]{x - 1}$ $\sqrt[3]{ab} = \sqrt[3]{a} \cdot \sqrt[3]{b}$

$\quad y = -2\sqrt[3]{x - 1}$ $\sqrt[3]{-8} = -2$

The graph can be obtained by shifting the graph of $y = \sqrt[3]{x}$ one unit to the right, stretching vertically by a factor of 2, and reflecting across the x-axis (because of the negative sign in -2). Figure 37 shows both graphs. The domain and range are both $(-\infty, \infty)$.   ◆

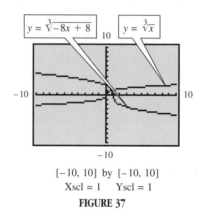

$y = \sqrt[3]{-8x + 8}$ $y = \sqrt[3]{x}$

[−10, 10] by [−10, 10]
Xscl = 1   Yscl = 1

FIGURE 37

(In Exercise 74, we will examine the function in Example 6(b) in a different manner.)

SPECIAL FUNCTIONS DEFINED BY ROOTS

Suppose that we consider the graph of all points (x, y) that lie a fixed distance r $(r > 0)$ from the origin. Then, by the distance formula, $\sqrt{(x - 0)^2 + (y - 0)^2} = r$. Simplifying and squaring both sides of this equation leads to $x^2 + y^2 = r^2$. The graph of this set of points is a circle with center at $(0, 0)$ and radius r. See Figure 38. While we will study circles in more detail later in the text, this brief introduction gives us an opportunity to investigate a special type of function defined by a root.

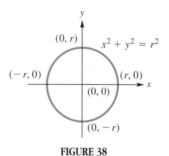

FIGURE 38

Because a circle is not the graph of a function, a graphing calculator in function mode is not appropriate for graphing a circle. However, if we imagine the circle as being the union of the graphs of two functions, one the top semicircle and the other the bottom semicircle, then we can graph both functions in the same square viewing window to obtain the desired graph. To accomplish this, we solve for y in the equation $x^2 + y^2 = r^2$.

$$x^2 + y^2 = r^2$$
$$y^2 = r^2 - x^2 \qquad \text{Subtract } x^2.$$
$$y = \pm\sqrt{r^2 - x^2} \qquad \text{Use the square root property of solving}$$
equations from Section 3.3.

The final line can be interpreted as two equations, both of which define functions:

$$y_1 = \sqrt{r^2 - x^2} \qquad \text{The semicircle above the } x\text{-axis}$$

and

$$y_2 = -\sqrt{r^2 - x^2}. \qquad \text{The semicircle below the } x\text{-axis}$$

Notice that $y_2 = -y_1$, indicating that the graph of the "bottom" semicircle is simply a reflection of the graph of the "top" semicircle across the x-axis.

EXAMPLE 7   *Graphing a Circle Using a Calculator in the Function Mode*
Use a calculator in function mode to graph the circle $x^2 + y^2 = 36$.

SOLUTION   Based on the argument above, this graph can be obtained by graphing both

$$y_1 = \sqrt{36 - x^2} \quad \text{and} \quad y_2 = -y_1 = -\sqrt{36 - x^2}$$

in the same window. In order to obtain a graph that is visually in correct proportions, we use a square window on the calculator. Figure 39 shows the graph of these two root functions, forming a circle.   ◆

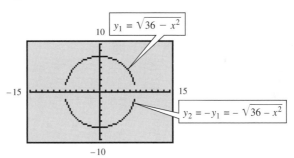

[−15, 15] by [−10, 10]
A square window is
necessary to obtain the
correct perspective.

FIGURE 39

TECHNOLOGY NOTE

The circle shown in Figure 39 is formed by the union of two functions. Because of the manner in which the calculator plots points in the function mode, the two semicircles do not completely "connect." This will often happen, and you should realize that, *mathematically, this is a complete circle.* This is an excellent example illustrating how we must understand the concepts in order to interpret what we see on the screen.

For Group Discussion   The two functions that form the circle in Example 7 both have the same domain.

1. Discuss how a sign graph can be used to solve the inequality $36 - x^2 \geq 0$. How does this inequality pertain to the graphs found in Example 7?
2. Figure 40 on the next page shows the graph of $y = 36 - x^2$. Use the graph to find the solution set of $36 - x^2 \geq 0$. Discuss how this solution set pertains to the graphs found in Example 7.

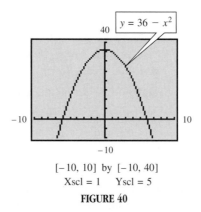

$$[-10, 10] \text{ by } [-10, 40]$$
$$\text{Xscl} = 1 \qquad \text{Yscl} = 5$$

FIGURE 40

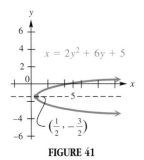

FIGURE 41

From the discussion in Sections 3.2 and 3.3, we know that the graph of $y = ax^2 + bx + c$ $(a \neq 0)$ is a parabola with a vertical axis of symmetry. Such an equation defines a function. If we reverse the roles of x and y, however, we obtain an equation whose graph is also a parabola, but having a horizontal axis of symmetry. (These, too, will be examined more closely later in the text.) Figure 41 shows the graph of $x = 2y^2 + 6y + 5$. Notice that this is not the graph of a function. However, if we consider it to be the union of the graphs of two functions, one the top half-parabola and the other the bottom, then it can be graphed by a calculator in the function mode. The two equations can be found by the method of completing the square. The final example shows how this can be done.

EXAMPLE 8   *Graphing a Horizontal Parabola Using a Calculator in Function Mode*   Graph $x = 2y^2 + 6y + 5$ in the window $[-2, 8]$ by $[-8, 2]$.

SOLUTION   We must solve for y by completing the square.

$$x = 2y^2 + 6y + 5$$

$$\frac{x}{2} = y^2 + 3y + \frac{5}{2} \qquad \text{Divide by 2.}$$

$$\frac{x}{2} - \frac{5}{2} = y^2 + 3y \qquad \text{Subtract } \tfrac{5}{2}.$$

$$\frac{x}{2} - \frac{5}{2} + \frac{9}{4} = y^2 + 3y + \frac{9}{4} \qquad \text{Add } \left[\tfrac{1}{2}(3)\right]^2 = \tfrac{9}{4}.$$

$$\frac{x}{2} - \frac{1}{4} = \left(y + \frac{3}{2}\right)^2 \qquad \begin{array}{l}\text{Combine terms on the left and factor}\\ \text{on the right.}\end{array}$$

$$\left(y + \frac{3}{2}\right)^2 = \frac{x}{2} - \frac{1}{4}$$

$$y + \frac{3}{2} = \pm\sqrt{\frac{x}{2} - \frac{1}{4}} \qquad \text{Use the square root property.}$$

$$y = -\frac{3}{2} \pm \sqrt{\frac{x}{2} - \frac{1}{4}} \qquad \text{Subtract } \tfrac{3}{2}.$$

Two functions are now defined. It is easier to use decimal notation to input the equations into a calculator. Therefore, let

$$y_1 = -1.5 + \sqrt{.5x - .25} \quad \text{and} \quad y_2 = -1.5 - \sqrt{.5x - .25}.$$

The graphs of these two functions together form the parabola with horizontal axis of symmetry $y = -1.5$, both of which are shown in Figure 42.   ◆

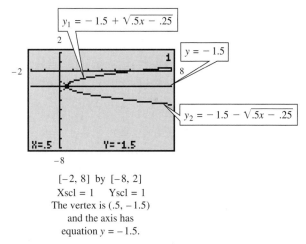

[−2, 8] by [−8, 2]
Xscl = 1   Yscl = 1
The vertex is (.5, −1.5)
and the axis has
equation y = −1.5.

FIGURE 42

For Group Discussion   The two functions that form the parabola in Example 8 both have the same domain.

1. Solve the inequality $.5x - .25 \geq 0$, using analytic methods. How does the solution set of this inequality pertain to the graphs found in Example 8?
2. Figure 43 shows the graph of $y = .5x - .25$. Use this graph to find the solution set of $.5x - .25 \geq 0$. Discuss how this solution set pertains to the graphs found in Example 8.

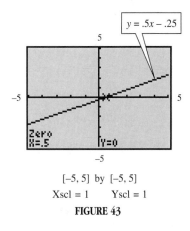

[−5, 5] by [−5, 5]
Xscl = 1   Yscl = 1
FIGURE 43

4.3 EXERCISES   Tape 5

Evaluate each of the following without *the use of a calculator, using the definitions of $\sqrt[n]{a}$ and $a^{m/n}$. Then, check your result with your calculator.*

1. $\sqrt{169}$
2. $-\sqrt[3]{64}$
3. $\sqrt[5]{-32}$
4. $\sqrt[4]{16}$
5. $81^{3/2}$
6. $27^{4/3}$
7. $125^{-2/3}$
8. $\left(\sqrt[3]{-27}\right)^2$
9. $(-1000)^{2/3}$
10. $(-125)^{-4/3}$

11. The screen shows how a table of square roots can be generated. Use such a table generated by your own calculator to find an approximation of each of the following.

(a) $\sqrt{39}$

(b) $\sqrt{143.8}$

(c) $\sqrt{9071}$

In this table, $Y_1 = \sqrt{x}$.
Thus, $\sqrt{2} \approx 1.41421356237$.

12. The screen shows how a table of cube roots can be generated. Use such a table generated by your own calculator to find an approximation of each of the following.

(a) $\sqrt[3]{39}$

(b) $\sqrt[3]{-143.8}$

(c) $\sqrt[3]{9071}$

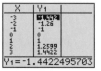

In this table, $Y_1 = \sqrt[3]{x}$.
Thus, $\sqrt[3]{-3} \approx -1.4422495703$.

Use the appropriate root function of your calculator to find each of the following roots. Tell whether the display represents the exact value or a rational approximation of an irrational number. Give as many digits as your display shows.

13. $\sqrt[6]{9}$

14. $\sqrt[4]{12}$

15. $\sqrt[3]{18.609625}$

16. $\sqrt[5]{286.29151}$

17. $\sqrt[3]{-17}$

18. $\sqrt[5]{-8}$

19. $\sqrt[6]{\pi^2}$

20. $\sqrt[6]{\pi^{-1}}$

Use the exponentiation function of your calculator to find each of the following powers. Tell whether the display represents the exact value or a rational approximation of an irrational number. Give as many digits as your display shows.

21. $5^{.1}$

22. $12^{.37}$

23. $\left(\dfrac{5}{6}\right)^{-1.3}$

24. $\left(\dfrac{4}{7}\right)^{-.6}$

25. π^{-3}

26. $(2\pi)^{4/3}$

27. $17^{1/17}$

28. $17^{-1/17}$

29. Consider the number $16^{-3/4}$.

    (a) Simplify this expression without the use of a calculator. Give the answer in both decimal and $\frac{a}{b}$ forms.

    (b) Write two different radical expressions that are equivalent to it, and use your calculator to evaluate them to show that the result is the same as the decimal form you found in part (a).

    (c) If your calculator has the capability to convert decimal numbers to fractions, use it to verify your results in part (a).

30. If a number is raised to a negative rational power, should we consider the negative sign to be in the numerator or in the denominator of the exponent to simplify the expression without the use of a calculator? Explain.

31. If we wish to compute a radical expression whose root index is a power of 2, we may simply use the $\boxed{\sqrt{\ \ }}$ key repeatedly to do so. For example,

$$\sqrt[4]{16} = \sqrt{\sqrt{16}} = 2 \quad \text{and} \quad \sqrt[8]{6561} = \sqrt{\sqrt{\sqrt{6561}}} = 3.$$

Explain why this is so, and calculate $\sqrt[16]{65{,}536}$ using this method. Then, support your result by calculating $65{,}536^{1/16}$.

32. Consider the expression $5^{.47}$.

    (a) Use the exponentiation capability of your calculator to find an approximation. Give as many digits as your calculator displays.

    (b) Use the fact that $.47 = \frac{47}{100}$ to write the expression as a radical, and then use the root-finding capability of your calculator to find an approximation that agrees with the one found in part (a).

In Exercises 33–38, you are given the graph of a function defined by $y = x^n$ for some value of n, along with a display indicating a point on the graph. Use your calculator to support the display by using the exponentiation capability of the calculator and then by using the root-finding capability.

33. $y = x^{1/3}$

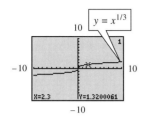

34. $y = x^{1/4}$

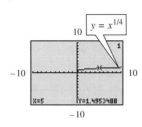

35. $y = x^{-1/2}$

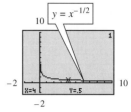

36. $y = x^{-1/3}$

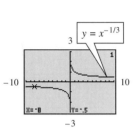

37. $y = x^{3/5}$

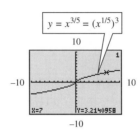

38. $y = x^{4/7}$

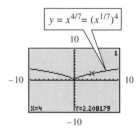

RELATING CONCEPTS (EXERCISES 39–42)

Duplicate each of the following screens on your calculator.

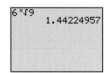

In this table, $Y_1 = \sqrt[6]{x}$.

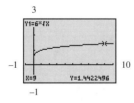

These screens show multiple ways of finding an approximation for $\sqrt[6]{9}$. Work Exercises 39–42 as directed, using your calculator.

39. Use a radical expression to approximate $\sqrt[6]{81}$ to as many decimal places as the calculator will give.

40. Use a rational exponent to repeat Exercise 39.

41. Use a table to repeat Exercise 39.

42. Use the graph of $y = \sqrt[6]{x}$ to repeat Exercise 39.

43. Does the graphing calculator screen shown indicate that the number π is exactly equal to $\sqrt[4]{\dfrac{2143}{22}}$? Explain.

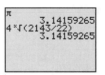

44. Would a calculator indicate that the display on the screen is true or false, assuming that the calculator is capable of returning an accurate answer? Support your answer, if you wish, by using your own calculator.

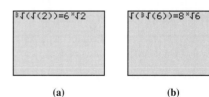

(a)   **(b)**

Determine analytically the domain of each function defined as follows.

45. $f(x) = \sqrt{5 + 4x}$

46. $f(x) = \sqrt{9x + 18}$

47. $f(x) = -\sqrt{6 - x}$

48. $f(x) = -\sqrt{2 - .5x}$

49. $f(x) = \sqrt[3]{8x - 24}$

50. $f(x) = \sqrt[5]{x + 32}$

51. $f(x) = \sqrt{49 - x^2}$

52. $f(x) = \sqrt{81 - x^2}$

53. $f(x) = \sqrt{x^3 - x}$

54. Explain why determining the domain of functions of the form $f(x) = \sqrt[n]{ax + b}$ requires two different considerations, depending upon the parity of n.

RELATING CONCEPTS (EXERCISES 55–60)

55. Use a graph to determine the interval(s) over which the function defined by $f(x) = x^2 - 6x - 7$ is nonnegative. (Section 3.3)

56. Use your calculator to graph $g(x) = \sqrt{x^2 - 6x - 7}$ in the window $[-6, 12]$ by $[-10, 10]$. What is the domain of g?

57. Graph f from Exercise 55 and g from Exercise 56 together in the window $[-6, 12]$ by $[-20, 10]$. Explain how the sign of $f(x)$ relates to the domain of g.

58. Without graphing, determine the domain of $h(x) = \sqrt{x^2 + 10}$. Why is this a relatively "easy" problem? Support your answer with a graph.

59. Discuss the symmetry of the graph of h from Exercise 58. (Section 2.2)

60. Use the graph of h in Exercise 58 to solve each of the following. (Section 1.6)

    (a) $h(x) = 0$ **(b)** $h(x) > 0$ **(c)** $h(x) < 0$

---

*Graph each function in the window specified. The domain of each function was determined analytically in Exercises 45–52. Use the graph to **(a)** find the range, **(b)** give the interval over which the function is increasing, **(c)** give the interval over which the function is decreasing, and **(d)** solve graphically the equation $f(x) = 0$.*

61. $f(x) = \sqrt{5 + 4x}$

62. $f(x) = \sqrt{9x + 18}$

63. $f(x) = -\sqrt{6 - x}$

    Window: $[-2, 10]$ by $[-10, 10]$ Window: $[-10, 10]$ by $[-10, 10]$ Window: $[-10, 10]$ by $[-10, 10]$

64. $f(x) = -\sqrt{2 - .5x}$

65. $f(x) = \sqrt[3]{8x - 24}$

66. $f(x) = \sqrt[5]{x + 32}$

    Window: $[-10, 10]$ by $[-10, 10]$ Window: $[-10, 10]$ by $[-10, 10]$ Window: $[-40, 5]$ by $[-10, 10]$

67. $f(x) = \sqrt{49 - x^2}$

68. $f(x) = \sqrt{81 - x^2}$

    Window: Standard square   Window: Standard square

Follow the procedure of Example 6 to explain how the graph of the given function can be obtained from the graph of the appropriate root function $\left(y = \sqrt{x} \text{ or } y = \sqrt[3]{x} \right)$.

69. $y = \sqrt{9x + 27}$

70. $y = \sqrt{16x + 16}$

71. $y = \sqrt{7x + 28} + 4$

72. $y = \sqrt{32 - 4x} - 3$

73. $y = \sqrt[3]{27x + 54} - 5$

74. In Example 6(b), we began by factoring -8 from the radicand. Repeat the problem, but start by factoring 8 from the radicand. Then, explain how the graph of $y = \sqrt[3]{-8x + 8}$ can alternatively be obtained from the graph of $y = \sqrt[3]{x}$.

In Exercises 75–82, describe the graph of the equation as one of the following: circle or parabola with a horizontal axis of symmetry. Then, analytically determine two functions, designated by y_1 and y_2, such that their union will give the graph of the given equation. Finally, graph the equation in the viewing window given. (These are appropriate for a TI-83 calculator.)

75. $x^2 + y^2 = 100$;

76. $x^2 + y^2 = 81$;

77. $(x - 2)^2 + y^2 = 9$;

78. $(x + 3)^2 + y^2 = 16$;

    $[-15, 15]$ by $[-10, 10]$ $[-15, 15]$ by $[-10, 10]$ $[-15, 15]$ by $[-10, 10]$ $[-15, 15]$ by $[-10, 10]$

79. $x = y^2 + 6y + 9$;

80. $x = y^2 - 8y + 16$;

81. $x = 2y^2 + 8y + 1$;

82. $x = -3y^2 - 6y + 2$;

    $[-10, 10]$ by $[-10, 10]$ $[-10, 10]$ by $[-10, 10]$ $[-10, 10]$ by $[-10, 10]$ $[-10, 10]$ by $[-10, 10]$

4.4   ROOT EQUATIONS, INEQUALITIES, AND APPLICATIONS

Solving Equations Involving Roots ◆ Solving Inequalities Involving Roots ◆ Applications of Root Functions

SOLVING EQUATIONS INVOLVING ROOTS

We will now look at the analytic procedure used to solve equations involving roots, such as

$$\sqrt{5 - 5x} + x = 1, \quad (11 - x)^{1/2} - x = 1, \quad \text{and} \quad \sqrt{2x + 3} - \sqrt{x + 1} = 1.$$

The procedure we will use is based on the following property.

PROPERTY FOR SOLVING EQUATIONS INVOLVING ROOTS

If P and Q are algebraic expressions, then every solution of the equation $P = Q$ is also a solution of the equation $(P)^n = (Q)^n$, for any positive integer n.

Caution   Be very careful when using this result. It does *not* say that the equations $P = Q$ and $(P)^n = (Q)^n$ are equivalent; it says only that each solution of the original equation $P = Q$ is also a solution of the new equation $(P)^n = (Q)^n$.

When using this property to solve equations, we must be aware that the new equation may have *more* solutions than the original equation. For example, the solution set of the equation $x = -2$ is $\{-2\}$. If we square both sides of the equation $x = -2$, we get the new equation $x^2 = 4$, which has solution set $\{-2, 2\}$. Since the solution sets are not equal, the equations are not equivalent. Because of this, when an equation contains radicals or rational exponents, it is *essential* to check all proposed solutions in the original equation.

The analytic procedure for solving equations involving roots is outlined below.

SOLVING EQUATIONS INVOLVING ROOT FUNCTIONS ANALYTICALLY

1. Isolate a term involving a root on one side of the equation.
2. Raise both sides of the equation to a power that will eliminate the radical or rational exponent.
3. Solve the resulting equation. (If a root is still present after Step 2, repeat Steps 1 and 2.)
4. Check all proposed solutions in the original equation.

EXAMPLE 1   *Solving an Equation Involving Roots*   Solve the equation

$$\sqrt{5 - 5x} + x = 1$$

analytically, and support the result graphically.

SOLUTION   ANALYTIC

$\sqrt{5 - 5x} + x = 1$	Given equation
$\sqrt{5 - 5x} = 1 - x$	Isolate the radical term.
$\left(\sqrt{5 - 5x}\right)^2 = (1 - x)^2$	Square both sides.
$5 - 5x = 1 - 2x + x^2$	
$0 = x^2 + 3x - 4$	Put in standard form of a quadratic equation.
$0 = (x + 4)(x - 1)$	Factor.
$x + 4 = 0 \quad \text{or} \quad x - 1 = 0$	Use the zero-product property.
$x = -4 \qquad\qquad x = 1$	

The proposed solutions are -4 and 1. They must be checked in the *original* equation.

$$\text{Let } x = -4.$$
$$\sqrt{5 - 5(-4)} + (-4) = 1 \qquad ?$$
$$\sqrt{25} + (-4) = 1 \qquad ?$$
$$5 + (-4) = 1 \qquad ?$$
$$1 = 1 \qquad \text{True}$$

$$\text{Let } x = 1.$$
$$\sqrt{5 - 5(1)} + 1 = 1 \qquad ?$$
$$\sqrt{0} + 1 = 1 \qquad ?$$
$$0 + 1 = 1 \qquad ?$$
$$1 = 1 \qquad \text{True}$$

Both proposed solutions are indeed solutions, and the solution set is $\{-4, 1\}$.

GRAPHICAL   To support this result graphically, we use the x-intercept method, letting $y_1 = \sqrt{5 - 5x} + x$ and $y_2 = 1$. Graphing $y_1 - y_2$ produces the curve shown in Figures 44(a) and (b). Notice that the two x-intercepts are -4 and 1, as determined analytically.   ◆

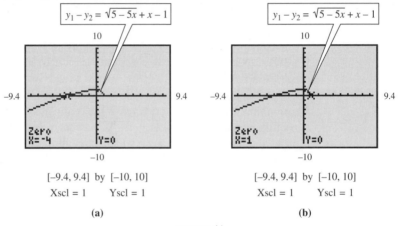

[-9.4, 9.4]   by   [-10, 10]   [-9.4, 9.4]   by   [-10, 10]

Xscl = 1   Yscl = 1   Xscl = 1   Yscl = 1

(a)   **(b)**

FIGURE 44

In Example 1, both proposed solutions proved to be actual solutions. However, this is not always the case, as seen in Example 2.

EXAMPLE 2   *Solving an Equation Involving Roots*   Solve the equation

$$(11 - x)^{1/2} - x = 1$$

analytically, and support the result graphically.

SOLUTION   ANALYTIC   We use the same procedure as in Example 1.

$$(11 - x)^{1/2} - x = 1 \qquad \text{Given equation}$$

$$(11 - x)^{1/2} = 1 + x \qquad \text{Isolate the term involving the rational exponent.}$$

$$11 - x = 1 + 2x + x^2 \qquad \text{Square both sides.}$$

$$0 = x^2 + 3x - 10$$

$$0 = (x + 5)(x - 2)$$

$$x = -5 \quad \text{or} \quad x = 2$$

The proposed solutions are -5 and 2. These must be checked in the original equation.

Let $x = -5$.			Let $x = 2$.	
$(11 - (-5))^{1/2} - (-5) = 1$	?		$(11 - 2)^{1/2} - 2 = 1$	?
$16^{1/2} + 5 = 1$	?		$9^{1/2} - 2 = 1$	?
$4 + 5 = 1$	?		$3 - 2 = 1$	?
$9 = 1$	False		$1 = 1$	True

The procedure of squaring both sides of the equation led to the *extraneous* root -5, as indicated by the false statement $9 = 1$. Therefore, the only solution of the equation is 2. The solution set is $\{2\}$.

GRAPHICAL   While there are several ways to support our answer graphically, an interesting method involves using the second step in the previous solution. The equation in the second step has the same solution set as the original equation, because the squaring of both sides has not yet been done. If we graph $y_1 = (11 - x)^{1/2}$ and $y_2 = 1 + x$, and observe the x-coordinate of the only point of intersection of the graphs, we see that it is 2, supporting our analytic solution. See Figure 45.   ◆

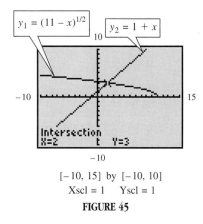

$[-10, 15]$ by $[-10, 10]$
Xscl = 1   Yscl = 1
FIGURE 45

EXAMPLE 3   *Solving an Equation Involving Roots (Cube Root Radicals)*
Solve the equation

$$\sqrt[3]{x^2 + 3x} = \sqrt[3]{5}$$

analytically, and support the result graphically.

SOLUTION   ANALYTIC

$$\sqrt[3]{x^2 + 3x} = \sqrt[3]{5} \qquad \text{Given equation}$$
$$x^2 + 3x = 5 \qquad \text{Cube both sides.}$$
$$x^2 + 3x - 5 = 0 \qquad \text{Put in standard form.}$$

We must now use the quadratic formula, with $a = 1$, $b = 3$, and $c = -5$.

$$x = \frac{-3 \pm \sqrt{3^2 - 4(1)(-5)}}{2(1)} = \frac{-3 \pm \sqrt{29}}{2}$$

GRAPHICAL   At this point, an analytic check would be rather messy, so we will use our calculator. The points of intersection of the curve $y_1 = \sqrt[3]{x^2 + 3x}$ and the line $y_2 = \sqrt[3]{5}$ should have these two x-values:

$$\frac{-3 + \sqrt{29}}{2} \approx 1.19 \quad \text{and} \quad \frac{-3 - \sqrt{29}}{2} \approx -4.19.$$

Figures 46(a), (b), and (c) support this conclusion. Therefore, the solution set is $\left\{ \frac{-3 \pm \sqrt{29}}{2} \right\}$. ◆

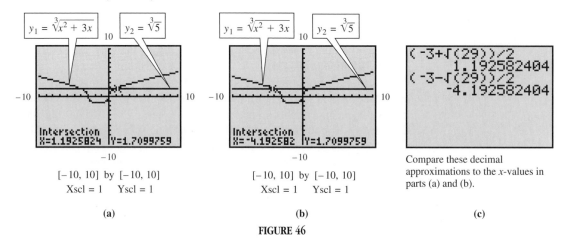

$[-10, 10]$ by $[-10, 10]$
Xscl = 1   Yscl = 1

(a)

$[-10, 10]$ by $[-10, 10]$
Xscl = 1   Yscl = 1

(b)

Compare these decimal approximations to the x-values in parts (a) and (b).

(c)

FIGURE 46

EXAMPLE 4   *Solving an Equation Involving Roots (Squaring Twice)*   Solve the equation

$$\sqrt{2x + 3} - \sqrt{x + 1} = 1$$

analytically, and support the result graphically.

SOLUTION   ANALYTIC   We begin by isolating a radical. When a choice must be made as to which radical to isolate, it is usually easier to isolate the more complicated radical first. We will isolate $\sqrt{2x + 3}$, and then proceed.

$$\sqrt{2x + 3} = 1 + \sqrt{x + 1}$$
$$\left(\sqrt{2x + 3}\right)^2 = \left(1 + \sqrt{x + 1}\right)^2 \qquad \text{Square both sides.}$$
$$2x + 3 = 1 + 2\sqrt{x + 1} + x + 1$$
$$x + 1 = 2\sqrt{x + 1} \qquad \text{Isolate the radical.}$$

One side of the equation still contains a radical; to eliminate it, square both sides again.

$$x^2 + 2x + 1 = 4(x + 1)$$
$$x^2 - 2x - 3 = 0$$
$$(x - 3)(x + 1) = 0$$
$$x = 3 \quad \text{or} \quad x = -1$$

A check shows that both proposed solutions 3 and -1 are solutions of the original equation, giving $\{3, -1\}$ as the solution set.

GRAPHICAL   The two solutions may be supported graphically by graphing $y_1 - y_2$, where $y_1 = \sqrt{2x + 3} - \sqrt{x + 1}$ and $y_2 = 1$. The x-intercepts indicated in Figures 47(a) and (b) are -1 and 3, as determined analytically. ◆

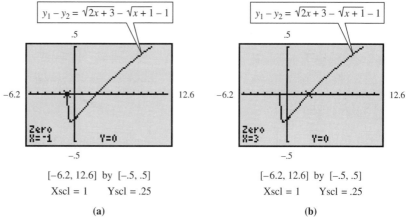

$$y_1 - y_2 = \sqrt{2x+3} - \sqrt{x+1} - 1$$

$$[-6.2, 12.6]\ \text{by}\ [-.5, .5]$$
$$\text{Xscl} = 1 \qquad \text{Yscl} = .25$$

(a)

$$y_1 - y_2 = \sqrt{2x+3} - \sqrt{x+1} - 1$$

$$[-6.2, 12.6]\ \text{by}\ [-.5, .5]$$
$$\text{Xscl} = 1 \qquad \text{Yscl} = .25$$

(b)

FIGURE 47

SOLVING INEQUALITIES INVOLVING ROOTS

We have seen in previous sections how the solution sets of inequalities can be determined once the solution set of the associated equations is found. The solution set of $f(x) > 0$ consists of all x-values for which the graph of f is above the x-axis, while that of $f(x) < 0$ consists of all x-values for which the graph is below the x-axis. We will use this approach in the following examples, referring to Examples 1–3 and their associated figures. We will always pay close attention to the domain of the function involved.

EXAMPLE 5   *Solving an Inequality Involving Roots (Radical Form)*   Use Figure 44 to find the solution set of

$$\sqrt{5 - 5x} + x \geq 1.$$

SOLUTION   The given inequality is equivalent to

$$\sqrt{5 - 5x} + x - 1 \geq 0.$$

Notice that the domain of $y = \sqrt{5 - 5x} + x - 1$ is $(-\infty, 1]$ because it is for only these real numbers that $5 - 5x$ is nonnegative. Figure 44 shows that the graph of this function lies above or on the x-axis in the interval $[-4, 1]$; therefore, this interval is the solution set of the inequality.   ◆

EXAMPLE 6   *Solving an Inequality Involving Roots (Rational Exponent)*   Solve each inequality.

(a) $(11 - x)^{1/2} - x > 1$ **(b)** $(11 - x)^{1/2} - x < 1$

SOLUTION   In Example 2, we found that the solution set of the equation $(11 - x)^{1/2} - x = 1$ is $\{2\}$. The domain of the function $y = (11 - x)^{1/2} - x - 1$ is $(-\infty, 11]$, and its x-intercept is 2, as seen in Figure 48 on the next page.

The solution set of the inequality in (a) is the same as that of $y > 0$, which is $(-\infty, 2)$. The solution set of the inequality in (b) is the same as that of $y < 0$, which is $(2, 11]$.   ◆

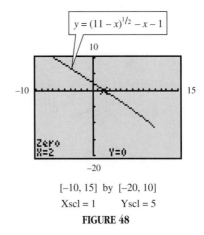

$$y = (11 - x)^{1/2} - x - 1$$

[−10, 15] by [−20, 10]
Xscl = 1   Yscl = 5
FIGURE 48

EXAMPLE 7   *Solving an Inequality Involving Roots (Cube Root Radicals)*
Solve the inequality

$$\sqrt[3]{x^2 + 3x} \le \sqrt[3]{5}.$$

SOLUTION   The associated equation was solved in Example 3, where we found its solutions to be $\frac{-3 + \sqrt{29}}{2}$ and $\frac{-3 - \sqrt{29}}{2}$. We can use the x-intercept method to solve this inequality. As seen in Figure 49, the graph of $y = \sqrt[3]{x^2 + 3x} - \sqrt[3]{5}$ lies below or on the x-axis in the interval between the two x-intercepts, including the endpoints. Therefore, the solution set of this inequality is $\left[\frac{-3 - \sqrt{29}}{2}, \frac{-3 + \sqrt{29}}{2}\right]$.   ◆

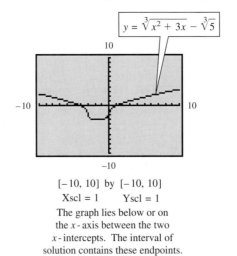

$$y = \sqrt[3]{x^2 + 3x} - \sqrt[3]{5}$$

[−10, 10] by [−10, 10]
Xscl = 1   Yscl = 1
The graph lies below or on
the x-axis between the two
x-intercepts. The interval of
solution contains these endpoints.

FIGURE 49

APPLICATIONS OF ROOT FUNCTIONS

Certain types of problems can be solved using functions involving roots, as shown in the next example.

EXAMPLE 8   *Solving a Cable Installation Problem Involving a Root Function*
A company wishes to run a utility cable from point *A* on the shore (as shown in Fig-

ure 50) to an installation at point B on the island. The island is 6 miles from the shore. It costs \$400 per mile to run the cable on land and \$500 per mile underwater. Assume that the cable starts at A and runs along the shoreline, then angles and runs underwater to the island. Let x represent the distance from C at which the underwater portion of the cable run begins, and the distance between A and C be 9 miles.

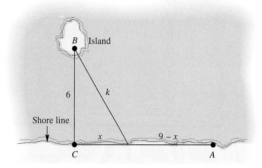

FIGURE 50

(a) What are the possible values of x in this problem?

(b) Express the cost of laying the cable as a function of x.

(c) Find the total cost if 3 miles of cable are on land. Determine analytically, and support graphically.

(d) Find the point at which the line should begin to angle in order to minimize the total cost. What is this total cost?

SOLUTION

(a) The value of x must be a real number greater than 0 and less than 9, meaning that x must be in the interval $(0, 9)$. (If we assume that the cable may be laid so that it is all underwater, then 9 would be included, giving the interval $(0, 9]$.)

(b) The total cost is determined by adding the cost of the cable on land to the cost of the cable underwater. If we let k represent the length of the cable underwater, by the Pythagorean theorem,

$$k^2 = 6^2 + x^2$$
$$k^2 = 36 + x^2$$
$$k = \sqrt{36 + x^2}. \qquad (k > 0)$$

The cost of the cable on land is $400(9 - x)$ dollars, while the cost of the cable underwater is $500k$ or $500\sqrt{36 + x^2}$ dollars. Therefore, the total cost is given by the function $C(x) = 400(9 - x) + 500\sqrt{36 + x^2}$, where $C(x)$ is in dollars.

(c) ANALYTIC According to Figure 50, if 3 miles of cable are on land, then $3 = 9 - x$, giving $x = 6$. We must evaluate $C(6)$.

$$C(6) = 400(9 - 6) + 500\sqrt{36 + 6^2}$$
$$\approx 5442.64 \text{ dollars}$$

GRAPHICAL The graph of the function in Figure 51 on the following page supports this result, since the point $(6, \ 5442.64)$ lies on the graph.

(d) The absolute minimum value of the function on the interval $(0, 9)$ is found when $x = 8$, meaning that $9 - 8 = 1$ mile should be along land and $\sqrt{36 + 1^2} \approx 6.08$

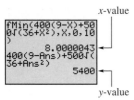

x-value

y-value

This screen provides another method of support for the result of Example 8.

miles should be underwater. $C(8) = 5400$ (dollars) is the minimum total cost. These results may be found using the capabilities of a graphing calculator. See Figure 52. ◆

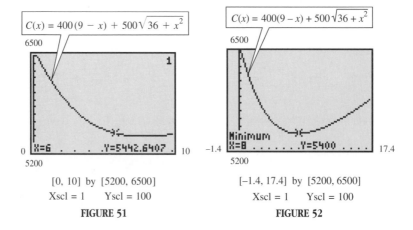

[0, 10] by [5200, 6500]
Xscl = 1   Yscl = 100
FIGURE 51

[−1.4, 17.4] by [5200, 6500]
Xscl = 1   Yscl = 100
FIGURE 52

WHAT WENT WRONG ?   A student was directed to solve the equation $x + 4 = \sqrt{x + 6}$. He squared both sides to get $x^2 + 8x + 16 = x + 6$, and then subtracted x and 6 from both sides to obtain $x^2 + 7x + 10 = 0$. Factoring led to $(x + 5)(x + 2) = 0$, leading to the possible solutions -5 and -2. He found that, of these, only -2 is an actual solution. To support his solution, he graphed $y_1 = x + 4$ and $y_2 = \sqrt{x + 6}$. Using the intersection-of-graphs method, he obtained the screen on the left, supporting his solution.

Feeling great about his success so far, he decided to use the *x*-intercept method to show that -2 is a solution. He got the screen shown at the right, and was flabbergasted. Everything had been going great, but now he knew there was a problem.

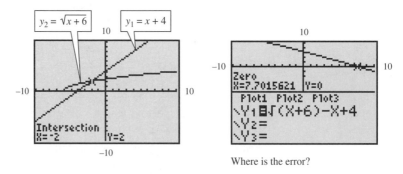

Where is the error?

1.   What went wrong?

2.   How can he correct his work so that the screen on the right shows a zero of -2?

4.4   EXERCISES   **Tape 6**

Use the calculator-generated graph to find the solution set of the given equation and inequalities.

1. (a) $\sqrt{x+5} = x - 1$
 (b) $\sqrt{x+5} \le x - 1$
 (c) $\sqrt{x+5} \ge x - 1$

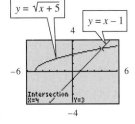

2. (a) $-\sqrt{x+12} = x$
 (b) $-\sqrt{x+12} \le x$
 (c) $-\sqrt{x+12} \ge x$

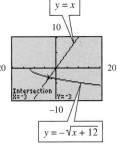

3. (a) $\sqrt[3]{2-2x} = \sqrt[3]{2x+14}$
 (b) $\sqrt[3]{2-2x} > \sqrt[3]{2x+14}$
 (c) $\sqrt[3]{2-2x} < \sqrt[3]{2x+14}$

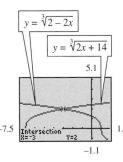

4. (a) $\sqrt[3]{3x} = \sqrt[3]{7-4x}$
 (b) $\sqrt[3]{3x} > \sqrt[3]{7-4x}$
 (c) $\sqrt[3]{3x} < \sqrt[3]{7-4x}$

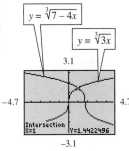

Use an analytic method to solve the equation in part (a). Then, use a graph, along with your result from part (a), to solve the inequalities in parts (b) and (c).

5. (a) $\sqrt{3x+7} - 3x = 5$
 (b) $\sqrt{3x+7} - 3x > 5$
 (c) $\sqrt{3x+7} - 3x < 5$

6. (a) $\sqrt{4x+13} - 2x = -1$
 (b) $\sqrt{4x+13} - 2x > -1$
 (c) $\sqrt{4x+13} - 2x < -1$

7. (a) $\sqrt{3x+4} = 8 - x$
 (b) $\sqrt{3x+4} > 8 - x$
 (c) $\sqrt{3x+4} < 8 - x$

8. (a) $\sqrt{1+5x} = 2x - 2$
 (b) $\sqrt{1+5x} > 2x - 2$
 (c) $\sqrt{1+5x} < 2x - 2$

9. (a) $(x+5)^{1/2} - 2 = (x-1)^{1/2}$
 (b) $(x+5)^{1/2} - 2 \ge (x-1)^{1/2}$
 (c) $(x+5)^{1/2} - 2 \le (x-1)^{1/2}$

10. (a) $(2x-5)^{1/2} - 2 = (x-2)^{1/2}$
 (b) $(2x-5)^{1/2} - 2 \ge (x-2)^{1/2}$
 (c) $(2x-5)^{1/2} - 2 \le (x-2)^{1/2}$

11. (a) $\sqrt{3x+4} - \sqrt{2x-4} = 2$
 (b) $\sqrt{3x+4} - \sqrt{2x-4} > 2$
 (c) $\sqrt{3x+4} - \sqrt{2x-4} < 2$

12. (a) $\sqrt{5x-6} = 2 + \sqrt{3x-6}$
 (b) $\sqrt{5x-6} > 2 + \sqrt{3x-6}$
 (c) $\sqrt{5x-6} < 2 + \sqrt{3x-6}$

13. (a) $\sqrt[3]{(2x-1)^2} - \sqrt[3]{x} = 0$
 (b) $\sqrt[3]{(2x-1)^2} - \sqrt[3]{x} > 0$
 (c) $\sqrt[3]{(2x-1)^2} - \sqrt[3]{x} < 0$

14. (a) $\sqrt[3]{x^2-2x} - \sqrt[3]{x} = 0$
 (b) $\sqrt[3]{x^2-2x} - \sqrt[3]{x} > 0$
 (c) $\sqrt[3]{x^2-2x} - \sqrt[3]{x} < 0$

15. (a) $\sqrt[4]{x-15} = 2$
 (b) $\sqrt[4]{x-15} > 2$
 (c) $\sqrt[4]{x-15} < 2$

16. (a) $\sqrt[4]{3x+1} = 1$
 (b) $\sqrt[4]{3x+1} > 1$
 (c) $\sqrt[4]{3x+1} < 1$

17. (a) $(x^2+2x)^{1/4} = \sqrt[4]{3}$
 (b) $(x^2+2x)^{1/4} > \sqrt[4]{3}$
 (c) $(x^2+2x)^{1/4} < \sqrt[4]{3}$

18. (a) $(x^2+6x)^{1/4} = 2$
 (b) $(x^2+6x)^{1/4} > 2$
 (c) $(x^2+6x)^{1/4} < 2$

19. (a) $(2x-1)^{2/3} = x^{1/3}$
 (b) $(2x-1)^{2/3} > x^{1/3}$
 (c) $(2x-1)^{2/3} < x^{1/3}$

20. (a) $(x-3)^{2/5} = (4x)^{1/5}$
 (b) $(x-3)^{2/5} > (4x)^{1/5}$
 (c) $(x-3)^{2/5} < (4x)^{1/5}$

21. (a) $\sqrt{3-3x} = 3 + \sqrt{3x+2}$
 (b) $\sqrt{3-3x} > 3 + \sqrt{3x+2}$
 (c) $\sqrt{3-3x} < 3 + \sqrt{3x+2}$

22. (a) $\sqrt{2\sqrt{7x+2}} = \sqrt{3x+2}$
 (b) $\sqrt{2\sqrt{7x+2}} > \sqrt{3x+2}$
 (c) $\sqrt{2\sqrt{7x+2}} < \sqrt{3x+2}$

In Exercises 23–28, a pair of functions is given. The function defined by y_1 is a root function, while the function defined by y_2 is a polynomial function. Use your knowledge of the general appearance of each to draw by hand a sketch of both functions on the same set of axes. Then, use the sketch to determine the number of points of intersection of the two graphs. Finally, use an analytic method to support your answer, and find the solution set of $y_1 = y_2$.

23. $y_1 = \sqrt{x}$
$y_2 = -x + 3$

24. $y_1 = \sqrt{x}$
$y_2 = x - 7$

25. $y_1 = \sqrt{x}$
$y_2 = 2x + 5$

26. $y_1 = \sqrt{x}$
$y_2 = 3x$

27. $y_1 = \sqrt[3]{x}$
$y_2 = x^2$

28. $y_1 = \sqrt[3]{x}$
$y_2 = 2x$

29. Use a hand-drawn graph to explain why $\sqrt{x} = -x - 5$ has no real solutions. (*Hint:* Sketch $y_1 = \sqrt{x}$ and $y_2 = -x - 5$ on the same axes. What do you notice about the number of points of intersection?)

30. Explain why the equation $\sqrt[3]{x} = ax + b$ must have at least one real solution for any values of a and b. (*Hint:* What is the range of $y_1 = \sqrt[3]{x}$? What kind of graph does $y_2 = ax + b$ have?)

RELATING CONCEPTS (EXERCISES 31–42)

Exercises 31–42 incorporate many concepts from Chapter 3 with the method of solving equations involving roots. They should be worked in order. Consider the equation

$$\sqrt[3]{4x - 4} = \sqrt{x + 1}.$$

31. Rewrite the equation, using rational exponents.

32. What is the least common denominator of the rational exponents found in Exercise 31?

33. Raise both sides of the equation in Exercise 31 to the power indicated by your answer in Exercise 32.

34. Show that the equation in Exercise 33 is equivalent to $x^3 - 13x^2 + 35x - 15 = 0$.

35. Graph the cubic function defined by the polynomial on the left side of the equation in Exercise 34 in the window $[-5, 10]$ by $[-100, 100]$. How many real roots does the equation have? (Section 3.5)

36. Use synthetic division to show that 3 is a zero of $P(x) = x^3 - 13x^2 + 35x - 15$. (Section 3.6)

37. Use the result of Exercise 36 to factor $P(x)$ so that one factor is linear and the other is quadratic. (Section 3.6)

38. Set the quadratic factor of $P(x)$ from Exercise 37 equal to 0, and solve the equation, using the quadratic formula. (Section 3.3)

39. What are the three proposed solutions of the original equation $\sqrt[3]{4x - 4} = \sqrt{x + 1}$?

40. Let $y_1 = \sqrt[3]{4x - 4}$ and $y_2 = \sqrt{x + 1}$. Graph $y_1 - y_2$ in the viewing window $[-2, 20]$ by $[-.5, .5]$ to determine the number of real solutions of the original equation. (Section 1.5)

41. Use both an analytic method and your calculator to find values of the real roots of the original equation. (Section 1.5)

42. Write a paragraph explaining how the solutions of the equation in Exercise 34 relate to the solutions of the original equation. Discuss any extraneous solutions that may be involved.

The problems in Exercises 43–48 involve formulas having radicals or rational exponents. Solve each problem.

43. *Illumination*   The illumination I in footcandles produced by a light source is related to the distance d in feet from the light source by the equation

$$d = \left(\frac{k}{I}\right)^{1/2},$$

where k is a constant. If $k = 400$, how far from the source will the illumination be 14 footcandles? Round to the nearest hundredth of a foot.

44. *Velocity of a Meteorite*   The velocity v of a meteorite approaching the earth is given by $v = kd^{-1/2}$, measured in kilometers per second, where d is its distance from the center of the earth and k is a constant. If $k = 350$, what is the velocity of a meteorite that is 6000 kilometers away from the center of the earth? Round to the nearest tenth.

45. *Visibility from an Airplane*   A formula for calculating the distance d one can see from an airplane to the horizon on a clear day is $d = 1.22x^{1/2}$, where x is the altitude of the plane in feet and d is given in miles. How far can one see to the horizon in a plane flying at the following altitudes? Give answers to the nearest mile.

(a) 20,000 feet   (b) 30,000 feet

46. *Period of a Pendulum*   The period of a pendulum in seconds depends on its length L in feet, and is given by

$$P = 2\pi\sqrt{\frac{L}{32}}.$$

If the length of a pendulum is 5 feet, what is its period? Round to the nearest tenth.

47. *Plant Species and Land Area*   A biologist has shown that the number of different plant species S on a Galápagos Island is related to the area of the island, A, by $S = 28.6A^{1/3}$. How many plant species would exist on such an island with the following areas?

(a) 100 square miles   **(b)** 1500 square miles

48. *Speed of a Car in an Accident*   To estimate the speed s at which a car was traveling at the time of an accident, police sometimes use the following procedure. A police officer drives the car involved in the accident under conditions similar to those during which the accident took place, and then skids to a stop. If the car is driven at 30 miles per hour, the speed at the time of the accident is given by

$$s = 30\sqrt{\frac{a}{p}},$$

where a is the length of the skid marks and p is the length of the marks in the police test. Find s if $a = 900$ feet and $p = 97$ feet.

RELATING CONCEPTS (EXERCISES 49–52)

Wind-Chill Factor   The chart describes the wind-chill factor for various wind speeds and temperatures.

Wind/°F	40°	30°	20°	10°	0°	−10°	−20°	−30°	−40°	−50°
5 mph	37	27	16	6	−5	−15	−26	−36	−47	−57
10 mph	28	16	4	−9	−21	−33	−46	−58	−70	−83
15 mph	22	9	−5	−18	−36	−45	−58	−72	−85	−99
20 mph	18	4	−10	−25	−39	−53	−67	−82	−96	−110
25 mph	16	0	−15	−29	−44	−59	−74	−88	−104	−118
30 mph	13	−2	−18	−33	−48	−63	−79	−94	−109	−125
35 mph	11	−4	−20	−35	−49	−67	−82	−98	−113	−129
40 mph	10	−6	−21	−37	−53	−69	−85	−100	−116	−132

Work Exercises 49–52 in order, so that you can see how an equation involving radicals can model these data.

49. Consider the expression $T - \left(\frac{v}{4} + 7\sqrt{v}\right)\left(1 - \frac{T}{90}\right)$ as a model, where T represents the temperature and v represents the wind velocity. Evaluate it for **(a)** $T = -10$ and $v = 30$, and **(b)** $T = -40$ and $v = 5$.

50. Consider the expression $91.4 - (91.4 - T) \times \left(.478 + .301\sqrt{v} - .02v\right)$ as a model, where once again T represents the temperature and v represents the wind velocity. Repeat parts (a) and (b) of Exercise 49.

51. Use the chart to find the wind-chill factors for the information in parts (a) and (b) of Exercise 49.

52. Based on your results in Exercises 49–51, make a conjecture about which formula models the wind-chill factor better.

Solve each of the following applications of root functions.

53. *Wire between Two Poles*   Two vertical poles of lengths 12 feet and 16 feet are situated on level ground, 20 feet apart, as shown in the figure in the next column. A piece of wire is to be strung from the top of the 12-foot pole to the top of the 16-foot pole, attached to a stake in the ground at a point P on a line formed by the vertical poles. Let x represent the distance from P to D, the base of the 12-foot pole.

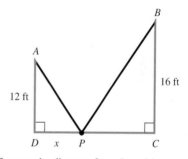

(a) Express the distance from P to C in terms of x.

(b) What are the restrictions on the value of x in this problem?

(c) Use the Pythagorean theorem to express the lengths AP and BP in terms of x.

(d) Form a function f that expresses the total length of the wire used.

(e) Graph f in the window $[0, 20]$ by $[0, 50]$. Use a function of your calculator to find $f(4)$, and interpret your result.

(f) Find the value of x that will minimize the amount of wire used.

(g) Write a short paragraph summarizing what this problem has examined, and the results you have obtained.

54. *Wire between Two Poles*   Repeat Exercise 53 if the heights of the poles are 9 feet and 12 feet, and the distance between the poles is 16 feet. Let *P* be *x* feet from the 9-foot pole. In part (e), use the window [0, 16] by [0, 50].

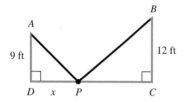

55. *Hunter Returns to His Cabin*   A hunter is at a point on a river bank. He wants to get to his cabin, located 3 miles north and 8 miles west. (See the figure.) He can travel 5 mph on the river but only 2 mph on this very rocky land. How far upriver should he go in order to reach the cabin in a minimum amount of time? (*Hint:* Distance = Rate × Time.)

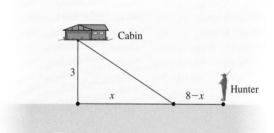

56. *Homing Pigeon Flight*   Homing pigeons avoid flying over large bodies of water, preferring to fly around them instead. (One possible explanation is the fact that extra energy is required to fly over water because air pressure drops over water in the daytime.) Assume that a pigeon released from a boat 1 mile from the shore of a lake (point *B* in the figure) flies first to point *P* on the shore and then along the straight edge of the lake to reach its home at *L*. If *L* is 2 miles from point *A*, the point on the shore closest to the boat, and if a pigeon needs $\frac{4}{3}$ as much energy to fly over water as over land, find the location of point *P*.

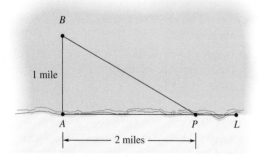

57. *Cruise Ship Travel*   At noon, the cruise ship *Queen Anne* is 60 miles due south of the cruise ship *King Bill* and is sailing north at a rate of 30 mph. If the *King Bill* is sailing west at a rate of 20 mph, find the time at which the distance *d* between the ships is a minimum. What is this distance?

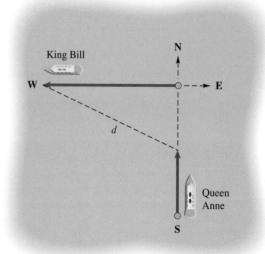

58. *Fisherman Returns to Camp*   Brent Simon is in his bass boat, the *Mido,* 3 miles from the nearest point on the shore. He wishes to reach his camp at La Branche, 6 miles farther down the shoreline. (See the figure.) If Brent's motor is disabled, and he can row his boat at a rate of 4 mph and walk at a rate of 5 mph, find the least amount of time that he will need to reach the camp.

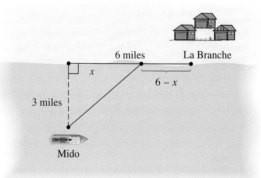

59. Explain why the table shows error messages for all values of *x* less than 15.

X	Y1	
12	ERROR	
13	ERROR	
14	ERROR	
15	0	
16	1	
17	1.4142	
18	1.7321	

Y₁ = √(X−15)

60. Explain how this table allows you to solve the equation $\sqrt{x} = \sqrt[3]{x}$. Here, $Y_1 = \sqrt{x}$ and $Y_2 = \sqrt[3]{x}$.

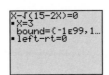

The screen here shows how a graphing calculator with an equation solver can find a solution (3) of the equation $x - \sqrt{15 - 2x} = 0.$

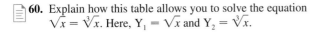

Use a graphing calculator to find the single real solution of the equation. If the equation-solving feature of the calculator requires a guess, use the guess suggested.

61. $2\sqrt{x} - \sqrt{3x + 4} = 0$ (Guess: 0)

62. $2\sqrt{x} - \sqrt{5x - 16} = 0$ (Guess: 10)

63. $\sqrt{x^2 - \sqrt{5}x + 6} = x + \pi$ (Guess: 0)

64. $\sqrt{x^2 - \pi x + \pi} = x + \sqrt{2}$ (Guess: 0)

4.5 INVERSE FUNCTIONS

Basic Concepts ◆ One-to-One Functions ◆ Inverse Functions and Their Graphs ◆ An Application of Inverse Functions

BASIC CONCEPTS

In ordinary arithmetic, the real numbers 0 and 1 serve as *identity* elements for addition and multiplication, respectively.

> **IDENTITY PROPERTIES**
>
> For all real numbers a, $a + 0 = 0 + a = a$.
> For all real numbers a, $a \cdot 1 = 1 \cdot a = a$.

For the operation of composition of functions (Section 2.6), the function $f(x) = x$ serves as the identity function.

Every real number a has an *additive inverse*, a number which when added to a produces a sum of 0, the identity element for addition. Also, every nonzero real number a has a *multiplicative inverse*, or *reciprocal*, a number which when multiplied by a produces a product of 1, the identity element for multiplication.

> **INVERSE PROPERTIES**
>
> For every real number a, there exists a unique real number $-a$ such that
> $a + (-a) = (-a) + a = 0.$
> For every nonzero real number a, there exists a unique real number a^{-1} such that
> $a \cdot a^{-1} = a^{-1} \cdot a = 1.$

It would seem that the operation of composition of functions would lend itself to a natural extension of the concept of inverses. This is indeed the case, and in this section we will investigate inverse functions and the conditions under which they exist.

As a simple example, consider the functions

$$f(x) = 3x \quad \text{and} \quad g(x) = \frac{1}{3}x.$$

Let us choose a value for x, say $x = 6$. If we evaluate f at 6, we get $f(6) = 3(6) = 18$. Now, if we take this result and evaluate g, we get $g(18) = \frac{1}{3}(18) = 6$, which is the value with which we started. Recall from Section 2.6 that $(g \circ f)(x) = g(f(x))$, so

$$(g \circ f)(6) = g(f(6)) = g(18) = 6.$$

Similarly, we can show that $(f \circ g)(6) = 6$. This type of result is typical of *inverse functions*. For these functions, we can show that for all x,

$$(f \circ g)(x) = (g \circ f)(x) = x.$$

Notice that the final result is the identity function.

ONE-TO-ONE FUNCTIONS

In order for a function to have an inverse, it must be *one-to-one*. For the function $y = 5x - 8$, any two different values of x produce two different values of y. On the other hand, for the function $y = x^2$, two different values of x can lead to the *same* value of y; for example, both $x = 4$ and $x = -4$ give $y = 4^2 = (-4)^2 = 16$. A function such as $y = 5x - 8$, where different elements from the domain always lead to different elements from the range, is called a *one-to-one function*.

DEFINITION OF ONE-TO-ONE FUNCTION

A function f is a **one-to-one function** if, for elements a and b from the domain of f,

$$a \neq b \quad \text{implies} \quad f(a) \neq f(b).$$

EXAMPLE 1   *Deciding Whether a Function is One-to-One*   Decide whether each of the following functions is one-to-one.

(a) $f(x) = -4x + 12$ **(b)** $f(x) = \sqrt{25 - x^2}$

SOLUTION

(a)   Suppose that $a \neq b$. Then, $-4a \neq -4b$, and $-4a + 12 \neq -4b + 12$. Thus, the fact that $a \neq b$ implies that $f(a) \neq f(b)$, so f is one-to-one.

(b)   If $a = 3$ and $b = -3$, then $3 \neq -3$, but

$$f(3) = \sqrt{25 - 3^2} = \sqrt{25 - 9} = \sqrt{16} = 4$$

and

$$f(-3) = \sqrt{25 - (-3)^2} = \sqrt{25 - 9} = 4.$$

Here, even though $3 \neq -3$, $f(3) = f(-3)$. By definition, this is not a one-to-one function.   ◆

As shown in Example 1(b), a way to show that a function is *not* one-to-one is to produce a pair of unequal numbers that leads to the same function value. There is also a useful graphical test that tells whether or not a function is one-to-one. This *horizontal line test* for one-to-one functions can be summarized as follows.

THE HORIZONTAL LINE TEST

If every horizontal line intersects the graph of a function at no more than one point, then the function is one-to-one.

Note In Example 1(b), the graph of the function is a semicircle. There are infinitely many horizontal lines that cut the graph of a semicircle in two points, so the horizontal line test shows that the function is not one-to-one.

EXAMPLE 2 *Using the Horizontal Line Test* Use the horizontal line test to determine whether the graphs in Figures 53 and 54 are graphs of one-to-one functions.

(a)
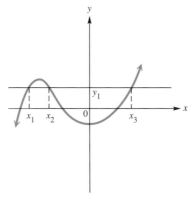

(b)
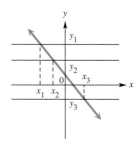

FIGURE 53   **FIGURE 54**

SOLUTION

(a) Each point where the horizontal line intersects the graph in Figure 53 has the same value of y but a different value of x. Since more than one (here, three) different values of x lead to the same value of y, the function is not one-to-one.

(b) Every horizontal line will intersect the graph in Figure 54 at exactly one point. This function is one-to-one.   ◆

For Group Discussion Based on your knowledge of the elementary functions studied so far in this text, answer the following questions. In each case, assume that the function has the largest possible domain.

1. Will a linear function (nonconstant) always be one-to-one?
2. Will an odd degree polynomial function always be one-to-one? If not, explain.
3. Will an even degree polynomial function ever be one-to-one? Why or why not?
4. Without graphing, tell whether each function is one-to-one. Then, support your answer with a graph.

    a. $f(x) = x$ **b.** $f(x) = x^n$, n odd   **c.** $f(x) = x^n$, n even

    d. $f(x) = |x|$ **e.** $f(x) = \sqrt[n]{x}$, n odd   **f.** $f(x) = \sqrt[n]{x}$, n even

    g. $f(x) = |2x + 4|$ **h.** $f(x) = \dfrac{1}{x}$ **i.** $f(x) = \dfrac{1}{x^2}$

INVERSE FUNCTIONS AND THEIR GRAPHS

Certain pairs of one-to-one functions "undo" one another. For example, if

$$f(x) = 8x + 5 \qquad\qquad \text{and} \qquad g(x) = \frac{x - 5}{8},$$

then

$$f(10) = 8 \cdot 10 + 5 = 85 \qquad \text{and} \qquad g(85) = \frac{85 - 5}{8} = 10.$$

Starting with 10, we "applied" function f and then "applied" function g to the result, which gave back the number 10. See Figure 55.

FIGURE 55

Similarly, for these same functions, check that

$$f(3) = 29 \qquad \text{and} \qquad g(29) = 3,$$
$$f(-5) = -35 \qquad \text{and} \qquad g(-35) = -5,$$
$$g(2) = -\frac{3}{8} \qquad \text{and} \qquad f\left(-\frac{3}{8}\right) = 2.$$

In particular, for these functions,

$$f[g(2)] = 2 \qquad \text{and} \qquad g[f(2)] = 2.$$

In fact for *any* value of x,

$$f[g(x)] = x \qquad \text{and} \qquad g[f(x)] = x,$$

or

$$(f \circ g)(x) = x \qquad \text{and} \qquad (g \circ f)(x) = x.$$

Because of this property, g is called the *inverse* of f.

DEFINITION OF INVERSE FUNCTION

Let f by a one-to-one function. Then, g is the **inverse function** of f if
$$(f \circ g)(x) = x \quad \text{for every } x \text{ in the domain of } g,$$
and
$$(g \circ f)(x) = x \quad \text{for every } x \text{ in the domain of } f.$$

EXAMPLE 3   *Deciding Whether Two Functions Are Inverses*   Let $f(x) = x^3 - 1$, and let $g(x) = \sqrt[3]{x + 1}$. Is g the inverse function of f?

SOLUTION   First, note that f is one-to-one. Then, use the definition to find

$$(f \circ g)(x) = f[g(x)] = \left(\sqrt[3]{x + 1}\right)^3 - 1$$
$$= x + 1 - 1 = x$$
$$(g \circ f)(x) = g[f(x)] = \sqrt[3]{(x^3 - 1) + 1} = \sqrt[3]{x^3} = x.$$

Since $(f \circ g)(x) = x$ and $(g \circ f)(x) = x$, function g is the inverse of function f. Also, f is the inverse of function g. ◆

A special notation is often used for inverse functions: if g is the inverse function of f, then g can be written as f^{-1} (read "f-inverse"). In Example 3,

$$f^{-1}(x) = \sqrt[3]{x + 1}.$$

Caution Do not confuse the -1 in f^{-1} with a negative exponent. The symbol $f^{-1}(x)$ does not represent $1/f(x)$; it represents the inverse function of f. Keep in mind that a function f can have an inverse function f^{-1} if and only if f is one-to-one.

The definition of inverse function can be used to show that the domain of f equals the range of f^{-1}, and the range of f equals the domain of f^{-1}. See Figure 56.

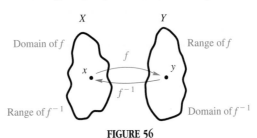

FIGURE 56

For the inverse functions f and g discussed at the beginning of this section, $f(6) = 18$ and $g(18) = 6$. Graphically, this means that the point $(6, 18)$ lies on the graph of f, and the point $(18, 6)$ lies on the graph of g.

The inverse of any one-to-one function f can be found by exchanging the components of the ordered pairs of f. The rule for the inverse of a function defined by $y = f(x)$ also is found by exchanging x and y. For example, if $f(x) = 7x - 2$, then $y = 7x - 2$. The function f is one-to-one, so that f^{-1} exists.

One method of finding the rule for f^{-1} is as follows:

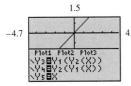

In this screen, $Y_1 = f(x) = 7x - 2$ and $Y_2 = f^{-1}(x) = \dfrac{x + 2}{7}$. Graphing both compositions (Y_3 and Y_4) and the identity function (Y_5) leads to coinciding graphs, supporting the discussion in the text.

$$y = 7x - 2 \quad \text{Given one-to-one function}$$
$$x = 7y - 2 \quad \text{Exchange } x \text{ and } y \text{ to find an equation that relates the variables in the inverse. (Step 1)}$$

$$x + 2 = 7y$$
$$\frac{x + 2}{7} = y \quad \text{Solve for } y. \text{ (Step 2)}$$

$$f^{-1}(x) = \frac{x + 2}{7}. \quad \text{Use inverse notation. (Step 3)}$$

We can now verify that $(f \circ f^{-1})(x) = (f^{-1} \circ f)(x) = x$ to complete our work.

EXAMPLE 4 *Finding the Inverse of a Function* Find the inverse, if it exists, of
$$f(x) = \frac{4x + 6}{5}.$$

SOLUTION This function is linear and is, therefore, one-to-one. Thus, it has an inverse. Let $f(x) = y$, and solve for x, getting

$$y = \frac{4x + 6}{5}$$

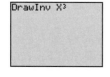

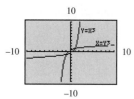

The inverses discussed in Example 4 are defined in Y_1 and Y_2. Notice that the *x*- and *y*-values in their ordered pairs are reversed.

$$x = \frac{4y + 6}{5} \qquad \text{Exchange } x \text{ and } y.$$

$$5x = 4y + 6 \qquad \text{Multiply by 5.}$$

$$5x - 6 = 4y \qquad \text{Subtract 6.}$$

$$y = \frac{5x - 6}{4} \qquad \text{Divide by 4.}$$

or

$$f^{-1}(x) = \frac{5x - 6}{4}.$$

The domain and range of both f and f^{-1} are the set of real numbers. In function f, the value of y is found by multiplying x by 4, adding 6 to the product, then dividing that sum by 5. In the equation for the inverse, x is *multiplied* by 5, then 6 is *subtracted*, and the result is *divided* by 4. This shows how an inverse function is used to "undo" what the function does to the variable x.   ◆

Suppose f and f^{-1} are inverse functions, and $f(a) = b$ for real numbers a and b. Then, by the definition of inverse, $f^{-1}(b) = a$. This shows that if a point (a, b) is on the graph of f, then (b, a) will belong to the graph of f^{-1}. As shown in Figure 57, the points (a, b) and (b, a) are reflections of one another across the line $y = x$. Thus, the graph of f^{-1} can be obtained from the graph of f by reflecting the graph of f across the line $y = x$.

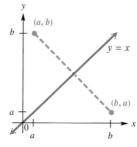

FIGURE 57

Some graphing calculators allow the user to draw inverses.

> ### GEOMETRIC RELATIONSHIP BETWEEN THE GRAPHS OF f AND f^{-1}
>
> If a function f is one-to-one, the graph of its inverse f^{-1} is a reflection of the graph of f across the line $y = x$.

For each of the following pairs of inverse functions, a calculator-generated graph is shown. See Figures 58, 59, and 60. In each case, a square window is used to give the correct perspective, and the line $y = x$ is also graphed to illustrate how the reflection appears.

$f(x) = 3x$ and $f^{-1}(x) = \frac{1}{3}x$ (functions from the section introduction)

$f(x) = x^3 - 1$ and $f^{-1}(x) = \sqrt[3]{x + 1}$ (functions from Example 3)

$f(x) = \frac{4x + 6}{5}$ and $f^{-1}(x) = \frac{5x - 6}{4}$ (functions from Example 4)

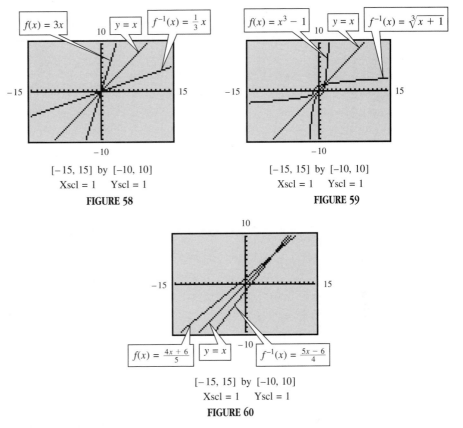

[−15, 15] by [−10, 10]
Xscl = 1   Yscl = 1
FIGURE 58

[−15, 15] by [−10, 10]
Xscl = 1   Yscl = 1
FIGURE 59

[−15, 15] by [−10, 10]
Xscl = 1   Yscl = 1
FIGURE 60

The next example considers a function with domain that is not $(-\infty, \infty)$.

EXAMPLE 5 *Finding the Inverse of a Function Whose Domain is Not* $(-\infty, \infty)$
Consider the function $f(x) = \sqrt{x + 5}$.

(a) What is the domain of f?

(b) Is f one-to-one?

(c) Find the rule for f^{-1}.

SOLUTION

(a) Using the methods described in Section 4.3, we know that for f to have real values, $x + 5$ must be nonnegative. Therefore,

$$x + 5 \geq 0$$
$$x \geq -5.$$

The domain is $[-5, \infty)$.

(b) We could use the horizontal line test, but another way to determine whether a function is one-to-one is to examine whether it is increasing over its entire domain, or whether it is decreasing over its entire domain. Figure 61 on the next page supports the first of these; thus, f is one-to-one.

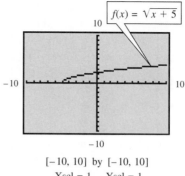

$$[-10, 10] \text{ by } [-10, 10]$$
Xscl = 1   Yscl = 1
f is a one-to-one
function.

FIGURE 61

(c) $y = \sqrt{x + 5}, \quad x \ge -5$ Replace $f(x)$ with y.

$\quad\quad x = \sqrt{y + 5}, \quad y \ge -5$ Exchange x and y.

$\quad\quad x^2 = y + 5$ Square both sides.

$\quad\quad y = x^2 - 5$ Solve for y.

The range of f^{-1} is indicated in the second line of the work above. The original function had a range of $[0, \infty)$; thus, the inverse must have this as its domain. Therefore,

$$f^{-1}(x) = x^2 - 5, \quad x \ge 0.$$

Note that the restriction on the domain of f^{-1} is necessary; otherwise, the function would not be one-to-one, as required. The graphs of f and f^{-1} are shown in Figure 62.   ◆

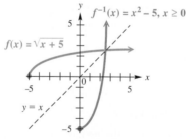

FIGURE 62

We can now summarize some important facts about inverses.

IMPORTANT FACTS ABOUT INVERSES

1. If f is one-to-one, then f^{-1} exists.
2. The domain of f is equal to the range of f^{-1}, and the range of f is equal to the domain of f^{-1}.
3. If the point (a, b) lies on the graph of f, then (b, a) lies on the graph of f^{-1}.
4. To find the rule for f^{-1}, replace $f(x)$ with y, interchange x and y, and solve for y. This gives f^{-1}.
5. The graphs of f and f^{-1} are reflections of each other across the line $y = x$.

AN APPLICATION OF INVERSE FUNCTIONS

Inverse functions can be used to send and receive coded information. The functions are usually very complicated. A simple example might use the function $f(x) = 2x + 5$. (Note that it is one-to-one.) Suppose that each letter of the alphabet is assigned a numerical value according to its position, as follows.

A	1	F	6	K	11	P	16	U	21
B	2	G	7	L	12	Q	17	V	22
C	3	H	8	M	13	R	18	W	23
D	4	I	9	N	14	S	19	X	24
E	5	J	10	O	15	T	20	Y	25
								Z	26

EXAMPLE 6   *Using a Function to Code a Message*   Use the one-to-one function $f(x) = 2x + 5$ and the numerical values above to code the word ALGEBRA.

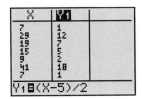

FIGURE 63

SOLUTION   The word ALGEBRA would be encoded as

$$7 \quad 29 \quad 19 \quad 15 \quad 9 \quad 41 \quad 7,$$

because $f(A) = f(1) = 2(1) + 5 = 7$, $f(L) = f(12) = 2(12) + 5 = 29$, and so on. The message would then be decoded by using the inverse of f, $f^{-1}(x) = \frac{x-5}{2}$. For example, $f^{-1}(7) = \frac{7-5}{2} = 1 = A$, $f^{-1}(29) = \frac{29-5}{2} = 12 = L$, and so on. The table feature of a graphing calculator can be very useful for this procedure. Figure 63 shows how decoding can be accomplished for the word ALGEBRA. ◆

For Group Discussion

1. You are an agent for a detective agency and know that today's function for your code is $f(x) = 4x - 5$. You receive the following message. (Read across.)

    ```
    47  95  23  67  -1  59  27  31  51  23   7  -1  43   7  79  43  -1  75
    55  67  31  71  75  27  15  23  67  15  -1  75  15  71  75  75  27  31
    51  23  71  31  51   7  15  71  43  31   7  15  11   3  67  15  -1  11
    ```

    Use the letter/number assignment described earlier to decode the message.
2. Why is a one-to-one function essential in this coding/decoding process?

4.5   EXERCISES   Tape 6

Indicate which screens suggest that Y_1 *and* Y_2 *are inverse functions.*

1.

X	Y₁
1	6
3	8
-4	1
3.2	14
0	8.2
-10	5
	-5
X=1	

X	Y₂
6	1
8	3
1	-4
14	9
8.2	3.2
5	0
-5	-10
X=6	

2.

X	Y₁
1	7
3	9
-4	2
9	15
3.2	9.2
0	6
-10	-4
X=1	

X	Y₂
7	-1
9	-3
2	4
15	-9
9.2	-3.2
6	0
-4	10
X=7	

3.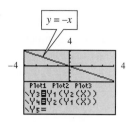

4.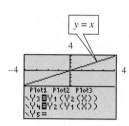

Based on your reading of the examples and exposition in this section, answer each of the following.

5. In order for a function to have an inverse, it must be _____.

6. For a function f to be of the type mentioned in Exercise 5, if $a \ne b$, then _____.

7. If f and g are inverses, then $(f \circ g)(x) =$ _____, and _____ $= x$.

8. The domain of f is equal to the _____ of f^{-1}, and the range of f is equal to the _____ of f^{-1}.

9. If the point (a, b) lies on the graph of f, and f has an inverse, then the point _____ lies on the graph of f^{-1}.

10. If the graphs of f and f^{-1} intersect, they do so at a point that satisfies what condition?

11. If a function f has an inverse, then the graph of f^{-1} may be obtained by reflecting the graph of f across the line with equation _____.

12. If a function f has an inverse, and $f(-3) = 6$, then $f^{-1}(6) =$ _____.

13. If $f(-4) = 16$ and $f(4) = 16$, then f _____ have an inverse because _____.
(does/does not)

14. If f is a function that has an inverse, and the graph of f lies completely within the second quadrant, then the graph of f^{-1} lies completely within the _____ quadrant.

Decide whether the function graphed or defined is one-to-one. For graphs, assume that a comprehensive graph is shown.

15.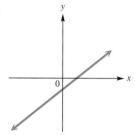

16.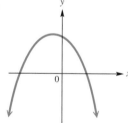

17.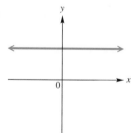

18.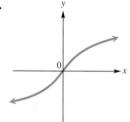

19.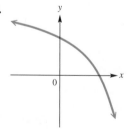

20.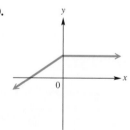

21.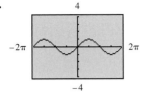

22.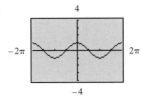

23.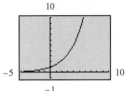

24.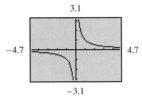

25. $y = 4x - 5$

26. $y = -x^2$

27. $y = (x - 2)^2$

28. $y = \sqrt{36 - x^2}$

29. $y = -\sqrt{100 - x^2}$

30. $y = 2x^3 + 1$

31. $y = -\sqrt[3]{x + 5}$

32. $y = \dfrac{x + 2}{x - 3}$

33. Explain why a polynomial function of even degree cannot be one-to-one.

34. Explain why a polynomial function of odd degree *may* not be one-to-one.

In Exercises 35–37, an everyday activity is described. Keeping in mind that an inverse operation "undoes" what an operation does, describe the inverse activity.

35. tying your shoelaces

36. starting a car

37. entering a room

Decide whether the functions in each pair are inverses of each other.

38.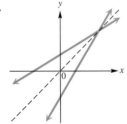

39.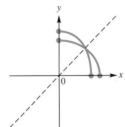

40.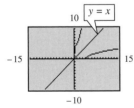

41.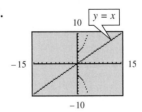

Dot Mode

42. $f(x) = -\dfrac{3}{11}x,\ g(x) = -\dfrac{11}{3}x$

43. $f(x) = 2x + 4,\ g(x) = \dfrac{1}{2}x - 2$

44. $f(x) = 5x - 5,\ g(x) = \dfrac{1}{5}x + 1$

45. $f(x) = 8x - 7,\ g(x) = \dfrac{x + 8}{7}$

46. $f(x) = \dfrac{1}{x},\ g(x) = \dfrac{1}{x}$

47. $f(x) = \dfrac{2x + 3}{x - 1},\ g(x) = \dfrac{x + 3}{x - 2}$

48. On a square sheet of paper, sketch the graph of any one-to-one function, using the entire sheet. Pattern your axes as shown in the figure near the top of the following page. Now, hold the sheet with your left hand at the left end of the *x*-axis and your right hand at the right end of the *x*-axis. Rotate the sheet 180° across the *x*-axis, so that the back of the sheet is now face up. Then, rotate the sheet 90° in a counterclockwise direction. Hold the sheet up to the light, and observe the graph from the back side of the sheet. How does it compare to your original graph?

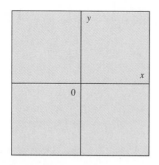

Draw by hand the graph of the inverse of each function in Exercises 49–52. Show both f and f⁻¹ in your sketch.

49.

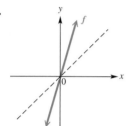

50.

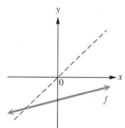

51.

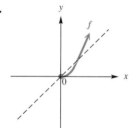

52.

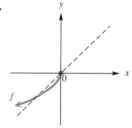

Each function f defined in Exercises 53–58 is one-to-one. Find f^{-1} analytically, and graph by hand both f and f^{-1} on the same axes.

53. $f(x) = 3x - 4$

54. $f(x) = 4x - 5$

55. $f(x) = \frac{1}{3}x$

56. $f(x) = -\frac{2}{5}x$

57. $f(x) = x^3 + 1$

58. $f(x) = \sqrt[3]{x - 5}$

59. Explain why the function $f(x) = x^4$ does not have an inverse. Give examples of ordered pairs to illustrate your explanation.

60. Some modern graphing calculators have the capability to "draw" the graph of the inverse of a function. Read your owner's manual to see if yours has this capability. If it does, use it to graph $f(x) = (x + 3)^3$ and f^{-1} on the same axes.

RELATING CONCEPTS (EXERCISES 61–64)

A graphing calculator can be used to support many of the concepts of inverse functions. Work Exercises 61–64 in order.

61. Define $f(x)$ as $y_1 = 2x - 8$ on your calculator. Find $f^{-1}(x)$ analytically, and enter it as y_2.

62. Enter y_3 as $y_1(y_2(x))$, using function notation on your calculator. Graph only y_3. What is the special name given to the function that is graphed? (Section 2.1)

63. Use a table with TblStart $= 0$ and ΔTbl $= 1$ to generate the values of $y_3 = y_1(y_2(x))$. What is the result?

64. Let $k =$ the number of letters in your last name. Evaluate $y_1(y_2(k))$ on the home screen. What is the result?

While a function may not be one-to-one when defined over its "natural" domain, it may be possible to restrict the domain in such a way that it is one-to-one and the range of the function is unchanged. For example, if we restrict the domain of the function $f(x) = x^2$ (which is not one-to-one over $(-\infty, \infty)$) to $[0, \infty)$, we obtain a one-to-one function whose range is still the same. See the accompanying figure. Notice that we could also choose to restrict the domain of $f(x) = x^2$ to $(-\infty, 0]$ and obtain the graph of a one-to-one function, except that it would be the left half of the parabola.

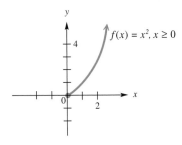

For each function defined in Exercises 65–70, decide on a suitable restriction on the domain so that the function is one-to-one and the range is not changed. You may wish to use a calculator-generated graph to help you decide.

65. $f(x) = -x^2 + 4$ **66.** $f(x) = (x - 1)^2$

67. $f(x) = |x - 6|$ **68.** $f(x) = x^4$

69. $f(x) = x^4 + x^2 - 6$ **70.** $f(x) = -\sqrt{x^2 - 16}$

Using the restrictions given here for the functions defined in Exercises 65–68, find a rule for f^{-1}.

71. $f(x) = -x^2 + 4, \quad x \geq 0$ **72.** $f(x) = (x - 1)^2, \quad x \geq 1$

73. $f(x) = |x - 6|, \quad x \geq 6$ **74.** $f(x) = x^4, \quad x \geq 0$

Use the alphabet coding assignment given in this section for Exercises 75–78. (See Example 6.)

75. The function $f(x) = 3x - 2$ was used to encode the following message:

     37 25 19 61 13 34 22 1 55 1 52 52 25 64 13 10.

Find the inverse function and decode the message.

76. The function $f(x) = 2x - 9$ was used to encode the following message:

     -5 9 5 5 9 27 15 29 -1 21 19 31 -3 27 41.

Find the inverse function and decode the message.

77. Encode the message SEND HELP, using the one-to-one function $f(x) = x^3 - 1$. Give the inverse function that the decoder would need when the message is received.

78. Encode the message SAILOR BEWARE, using the one-to-one function $f(x) = (x + 1)^3$. Give the inverse function that the decoder would need when the message is received.

CHAPTER 4   SUMMARY

Section	Important Concepts

4.1 Graphs of Rational Functions

DEFINITION OF RATIONAL FUNCTION

A function f of the form $\frac{p}{q}$ defined by

$$f(x) = \frac{p(x)}{q(x)},$$

where $p(x)$ and $q(x)$ are polynomials, is called a rational function.

THE RATIONAL FUNCTIONS $f(x) = \dfrac{1}{x}$ and $f(x) = \dfrac{1}{x^2}$

For the graph of: See page:

The reciprocal function $f(x) = \dfrac{1}{x}$ 317

$f(x) = \dfrac{1}{x^2}$ 318

GRAPHING RATIONAL FUNCTIONS

Let $f(x) = \frac{p(x)}{q(x)}$ define a function where the rational expression is written in lowest terms. To sketch its graph, follow the steps below.

1. Find any vertical asymptotes.
2. Find any horizontal or oblique asymptote.
3. Find the y-intercept by evaluating $f(0)$.
4. Find the x-intercepts, if any, by solving $f(x) = 0$. (These will be the zeros of the numerator, p.)
5. Determine whether the graph will intersect its nonvertical asymptote by solving $f(x) = k$, where k is the y-value of the nonvertical asymptote, or $f(x) = mx + b$, where $y = mx + b$ is the equation of the oblique asymptote.
6. Plot a few selected points, as necessary. Choose an x-value in each interval of the domain, as determined by the vertical asymptotes and x-intercepts.
7. Complete the sketch.

4.2 Rational Equations, Inequalities, and Applications

SOLVING EQUATIONS INVOLVING RATIONAL FUNCTIONS ANALYTICALLY

1. Determine all values for which the rational functions will be undefined.
2. Multiply both sides of the equation by the least common denominator of all rational expressions in the equation.
3. Solve the resulting equation.
4. Reject any values that were determined in Step 1.

INVERSE VARIATION

Let n be a positive real number. Then, y varies inversely as the nth power of x, or y is inversely proportional to the nth power of x, if a nonzero real number k exists such that

$$y = \frac{k}{x^n}.$$

If $n = 1$, then $y = \frac{k}{x}$, and y varies inversely as x.

JOINT VARIATION

Let m and n be real numbers. Then, y varies jointly as the nth power of x and the mth power of z, if a nonzero real number k exists such that

$$y = kx^nz^m.$$

4.3 Graphs of Root Functions

DEFINITION OF $\sqrt[n]{a}$

Suppose that n is a positive integer greater than 1. Let a be any real number.

   (i)   If $a > 0$, $\sqrt[n]{a}$ is the positive real number b such that $b^n = a$.

   (ii) If $a < 0$ and n is odd, $\sqrt[n]{a}$ is the negative real number b such that $b^n = a$.

  (iii) If $a < 0$ and n is even, then there is no real number $\sqrt[n]{a}$.

  (iv) If $a = 0$, then $\sqrt[n]{a} = 0$.

DEFINITION OF $a^{1/n}$

   (i) If n is an *even* positive integer and if $a > 0$, then $a^{1/n}$ is the *positive* real number whose nth power is a. That is, $a^{1/n} = \sqrt[n]{a}$.

   (ii) If n is an *odd* positive integer and if a is any real number, then $a^{1/n}$ is the single real number whose nth power is a. That is, $a^{1/n} = \sqrt[n]{a}$.

DEFINITION OF $a^{m/n}$

If $\frac{m}{n}$ is a rational number, where n is a positive integer greater than 1, and a is a real number such that $\sqrt[n]{a}$ is also real, then

$$a^{m/n} = \left(\sqrt[n]{a}\right)^m = \sqrt[n]{a^m}.$$

THE ROOT FUNCTIONS $f(x) = \sqrt[n]{x}$

For the graph of:	See page:
$f(x) = \sqrt[n]{x}$, n is even	352
$f(x) = \sqrt[n]{x}$, n is odd	352

4.4 Root Equations, Inequalities, and Applications

SOLVING EQUATIONS INVOLVING ROOT FUNCTIONS ANALYTICALLY

1. Isolate a term involving a root on one side of the equation.
2. Raise both sides of the equation to a power that will eliminate the radical or rational exponent.
3. Solve the resulting equation. (If a root is still present after Step 2, repeat Steps 1 and 2.)
4. Check all proposed solutions in the original equation.

4.5 Inverse Functions

DEFINITION OF A ONE-TO-ONE FUNCTION

A function f is a one-to-one function if, for elements a and b from the domain of f,

$$a \neq b \quad \text{implies} \quad f(a) \neq f(b).$$

HORIZONTAL LINE TEST

If every horizontal line intersects the graph of a function at no more than one point, then the function is one-to-one.

DEFINITION OF INVERSE FUNCTION

Let f be a one-to-one function. Then, g is the inverse function of f if

$$(f \circ g)(x) = x \quad \text{for every } x \text{ in the domain of } g,$$

and

$$(g \circ f)(x) = x \quad \text{for every } x \text{ in the domain of } f.$$

IMPORTANT FACTS ABOUT INVERSES

1. If f is one-to-one, then f^{-1} exists.
2. The domain of f is equal to the range of f^{-1}, and the range of f is equal to the domain of f^{-1}.
3. If the point (a, b) lies on the graph of f, then (b, a) lies on the graph of f^{-1}.
4. To find the rule for f^{-1}, replace $f(x)$ with y, interchange x and y, and solve for y. This gives f^{-1}.
5. The graphs of f and f^{-1} are reflections of each other across the line $y = x$.

CHAPTER 4   REVIEW EXERCISES

Use the terminology of Chapter 2 to explain how the graph of the function can be obtained from the graph of $y = \frac{1}{x}$.

1. $y = -\dfrac{2}{x + 3}$

2. $y = -\dfrac{1}{x} + 6$

3. $y = \dfrac{4}{x} - 3$

Graph each of the following functions, using a graphing calculator. You may wish first to make a rough sketch by hand, using the guidelines of Section 4.1, so that it will be easier to find an appropriate window. Your graph should be a comprehensive one. Also, give the equations of the horizontal, vertical, and oblique asymptotes.

4. $f(x) = \dfrac{4x - 3}{2x - 1}$

5. $f(x) = \dfrac{6x}{(x - 1)(x + 2)}$

6. $f(x) = \dfrac{2x}{x^2 - 1}$

7. $f(x) = \dfrac{x^2 + 4}{x + 2}$

8. $f(x) = \dfrac{x^2 - 1}{x}$

9. $f(x) = \dfrac{-2}{x^2 + 1}$

10. $f(x) = \dfrac{4x^2 - 9}{2x + 3}$

11. Under what conditions will the graph of a rational function defined by an expression reduced to lowest terms have an oblique asymptote?

12. Find an equation for the rational function graphed here.

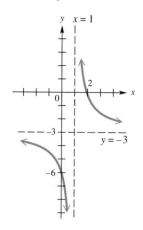

Solve the rational equation in part (a) analytically. Then, use a graph to determine the solution sets of the associated inequalities in parts (b) and (c).

13. (a) $\dfrac{3x-2}{x+1}=0$

   (b) $\dfrac{3x-2}{x+1}<0$

   (c) $\dfrac{3x-2}{x+1}>0$

14. (a) $\dfrac{5}{2x+5}=\dfrac{3}{x+2}$

   (b) $\dfrac{5}{2x+5}<\dfrac{3}{x+2}$

   (c) $\dfrac{5}{2x+5}>\dfrac{3}{x+2}$

15. (a) $\dfrac{3}{x-2}+\dfrac{1}{x+1}=\dfrac{1}{x^2-x-2}$

   (b) $\dfrac{3}{x-2}+\dfrac{1}{x+1}\le\dfrac{1}{x^2-x-2}$

   (c) $\dfrac{3}{x-2}+\dfrac{1}{x+1}\ge\dfrac{1}{x^2-x-2}$

16. (a) $1-\dfrac{5}{x}+\dfrac{6}{x^2}=0$

   (b) $1-\dfrac{5}{x}+\dfrac{6}{x^2}\le0$

   (c) $1-\dfrac{5}{x}+\dfrac{6}{x^2}\ge0$

A comprehensive graph of a rational function f is shown here. Use the figure to find the solution set of each of the following.

17. $f(x)=0$

18. $f(x)>0$

19. $f(x)<0$

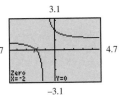

The graph has a vertical asymptote at $x=-1$.

Solve each problem involving rational functions.

20. *Antique-Car Competition*   Antique-car owners often enter their cars in a *concours d'elegance* in which a maximum of 100 points can be awarded to a particular car. Points are awarded for the general attractiveness of the car. The function defined by

$$C(x)=\dfrac{10x}{49(101-x)}$$

expresses the cost, in thousands of dollars, of restoring a car so that it will win x points.

   (a) Use a graphing calculator to graph the function in the window $[0, 101]$ by $[0, 5]$.

   (b) How much would an owner expect to pay to restore a car in order to earn 95 points?

21. *Environmental Pollution*   In situations involving environmental pollution, a cost-benefit model expresses cost as a function of the percentage of pollutant removed from the environment. Suppose a cost-benefit model is expressed as

$$C(x)=\dfrac{6.7x}{100-x},$$

where $C(x)$ is the cost, in thousands of dollars, of removing x percent of a certain pollutant.

   (a) Use a graphing calculator to graph the function in the window $[0, 100]$ by $[0, 150]$.

   (b) How much would it cost to remove 95% of the pollutant?

Solve each problem involving variation. Give approximations to the nearest hundredth.

22. Suppose r varies directly as x, and inversely as the square of y. If r is 10 when x is 5 and y is 3, find r when x is 12 and y is 4.

23. Suppose m varies jointly as n and the square of p, and inversely as q. If m is 20 when n is 5, p is 6, and q is 18, find m when n is 7, p is 11, and q is 2.

24. *Sports Arena Construction*   A sports arena requires a beam 16 m long, 24 cm wide, and 8 cm high. The maximum load of a horizontal beam that is supported at both ends varies directly as the width and square of the height, and inversely as the length between supports. If a beam of the same material 8 m long, 12 cm wide, and 15 cm high can support a maximum of 400 kg, what is the maximum load the beam in the arena will support?

25. *Weight of an Object*   The weight w of an object varies inversely as the square of the distance d between the object and the center of the earth. If a man weighs 90 kg on the surface of the earth, how much would he weigh 800 km above the surface? (The radius of the earth is about 6400 km.)

Suppose that a and b are positive numbers. Draw by hand a sketch of the general shape of each of the following functions.

26. $y=-a\sqrt{x}$

27. $y=\sqrt[3]{x+a}$

28. $y=\sqrt[3]{x-a}$

29. $y=-a\sqrt[3]{x}-b$

30. $y=\sqrt{x+a}+b$

Evaluate each of the following without the use of a calculator. Then, check your results with a calculator.

31. $-(-32)^{1/5}$ **32.** $36^{-3/2}$ **33.** $-1000^{2/3}$ **34.** $(-27)^{-4/3}$ **35.** $16^{3/4}$

Use your calculator to find each of the following. Tell whether the display represents the exact value or a rational approximation of an irrational number. Give as many digits as your display shows.

36. $\sqrt[5]{84.6}$ **37.** $\sqrt[4]{\dfrac{1}{16}}$ **38.** $12^{1/3}$ **39.** $\left(\dfrac{1}{8}\right)^{4/3}$

Consider the function $f(x) = -\sqrt{2x - 4}$ in Exercises 40–43.

40. Find the domain of f analytically.

41. Use a graph to determine the range of f.

42. Give the interval over which the function is increasing (if any).

43. Give the interval over which the function is decreasing (if any).

Consider the equation $x^2 + (y + 4)^2 = 25$ in Exercises 44–46.

44. Describe the graph of the equation.

45. Determine analytically two functions, y_1 and y_2, such that their union will give the graph of the equation.

46. Graph the equation in a square viewing window.

Suppose that $y_1 = f(x)$ is the root function graphed below, and $y_2 = g(x)$ is the absolute value function graphed on the same axes. Use the graphs to solve each equation or inequality.

47. $f(x) = 0$

48. $g(x) = 0$

49. $f(x) = g(x)$

50. $f(x) \geq 0$

51. $f(x) < g(x)$

52. $f(x) > 3$

53. $f(x) - g(x) < 0$

54. $g(x) = -1$

In Exercises 55–58, solve the radical equation in part (a) analytically. Then, use a graph to determine the solution sets of the associated inequalities in parts (b) and (c).

55. (a) $\sqrt{5 + 2x} = x + 1$

 (b) $\sqrt{5 + 2x} > x + 1$

 (c) $\sqrt{5 + 2x} < x + 1$

56. (a) $\sqrt{2x + 1} - \sqrt{x} = 1$

 (b) $\sqrt{2x + 1} - \sqrt{x} > 1$

 (c) $\sqrt{2x + 1} - \sqrt{x} < 1$

57. (a) $\sqrt[3]{6x + 2} = \sqrt[3]{4x}$

 (b) $\sqrt[3]{6x + 2} \geq \sqrt[3]{4x}$

 (c) $\sqrt[3]{6x + 2} \leq \sqrt[3]{4x}$

58. (a) $(x - 2)^{2/3} - x^{1/3} = 0$

 (b) $(x - 2)^{2/3} - x^{1/3} \geq 0$

 (c) $(x - 2)^{2/3} - x^{1/3} \leq 0$

RELATING CONCEPTS (EXERCISES 59–62)

Exercises 59–62 refer to the functions defined by

$$y_1 = \sqrt{3x + 12} - 4 \quad \text{and} \quad y_2 = \sqrt[3]{3x + 12} - 6.$$

Use a graph to help answer each problem.

59. How many solutions does the equation $y_1 = y_2$ have?

60. Based on your answer to Exercise 59, how many x-intercepts does the graph of $f(x) = y_1 - y_2$ have?

61. Describe the appearance of the graph of $-f(x) = y_2 - y_1$. (Section 2.3)

62. Describe the appearance of the graph of $y = f(-x)$. (Section 2.3)

63. *Period of a Pendulum*   The period P of a pendulum in seconds depends on its length L in feet, and is given by

$$P = 2\pi \sqrt{\frac{L}{32}}.$$

If the length of a pendulum is 8 feet, what is its period? Round to the nearest tenth.

64. *Volume of a Cylindrical Package*   A company plans to package its product in a cylinder that is open at one end. The cylinder is to have a volume of 27π cubic inches. What radius should the circular bottom of the cylinder have to minimize the cost of the material? (*Hint:* The volume of a circular cylinder is $\pi r^2 h$, where r is the radius of the circular base and h is the height; the surface area of the cylinder open at one end is $2\pi rh + \pi r^2$.)

Determine whether the function is one-to-one. Assume that graphs shown are comprehensive.

65.

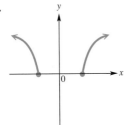

66. $f(x) = \sqrt{3x + 2}$

67.

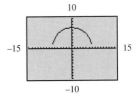

Consider the function $f(x) = \sqrt[3]{2x - 7}$ in Exercises 68–73.

68. What is the domain of f?

69. What is the range of f?

 70. Explain why f^{-1} exists.

71. Find the rule for $f^{-1}(x)$.

72. Graph both f and f^{-1} in a square viewing window, along with the line $y = x$. Describe how the graphs of f and f^{-1} are related.

73. Verify analytically that $(f \circ f^{-1})(x) = x$ and $(f^{-1} \circ f)(x) = x$.

Use the letter/number coding assignment in Section 4.5 to decode the following, using the function $f(x) = 6x - 4$.

74. 56   122   110   116   20   86   50   116

75. 2   134   26   110   86   74   26   20   122   20   26

76. Use the letter/number coding assignment in Section 4.5 and the function $f(x) = x + 10$ to code the word DANGER. What function is used to decode the word?

CHAPTER 4   TEST

1. Consider the rational function $f(x) = \frac{x^2 + x - 6}{x^2 - 3x - 4}$. Determine analytically each of the following: **(a)** the equations of the vertical asymptotes, **(b)** the equation of the horizontal asymptote, **(c)** the y-intercept, **(d)** the x-intercepts, and **(e)** the coordinates of the point where the graph of f intersects its horizontal asymptote. Then, **(f)** graph the function to obtain an accurate comprehensive graph, using a graphing calculator.

2. Consider the rational function $f(x) = \frac{x^2 - 16}{x + 4}$.

   (a) For what value of x does the graph exhibit a "hole"?

   (b) Graph the function in a window of a graphing calculator that clearly shows the "hole" in the graph.

3. Find the equation of the oblique asymptote of the graph of the rational function $f(x) = \frac{2x^2 + x - 3}{x - 2}$. Then, graph the function and its asymptote, using a graphing calculator to illustrate an accurate comprehensive graph.

4. (a) Solve the rational equation $\frac{3}{x - 2} + \frac{21}{x^2 - 4} - \frac{14}{x + 2} = 0$ analytically.

   (b) Use a graph to find the solution set of $\frac{3}{x - 2} + \frac{21}{x^2 - 4} - \frac{14}{x + 2} \geq 0$.

5. *Measure of Malnutrition*   A measure of malnutrition, called the *pelidisi*, varies directly as the cube root of a person's weight in grams, and inversely as the person's sitting height in centimeters. A person with a pelidisi below 100 is considered to be undernourished, while a pelidisi greater than 100 indicates overfeeding. A person who weighs 48,820 grams and has a sitting height of 78.7 centimeters has a pelidisi of 100. Find the pelidisi (to the nearest whole number) of a person whose weight is 54,430 grams and whose sitting height is 88.9 centimeters. Is this individual undernourished or overfed?

6. *Volume of a Cylindrical Can*   A manufacturer wants to construct a cylindrical aluminum can with a volume of 4000 cubic centimeters. If x represents the radius of the circular top and bottom of the can, the surface area S as a function of x is given by the function $S(x) = \frac{8000 + 2\pi x^3}{x}$. Use a graph to find the radius that will minimize the amount of aluminum needed, and determine what this amount will be. (*Hint:* Use the window $[-5, 42]$ by $[-1000, 10,000]$.)

7. Graph the function defined by $f(x) = -\sqrt{5 - x}$ in the standard viewing window. Then, do each of the following.

   (a) Determine the domain analytically.

   (b) Use the graph to find the range.

   (c) This function _____ over its entire domain.   (increases/decreases)

   (d) Solve the equation $f(x) = 0$ graphically.

   (e) Solve the inequality $f(x) < 0$ graphically.

8. (a) Solve the equation $\sqrt{4 - x} = x + 2$ analytically, and support the solution(s) with a graph.

   (b) Use the graph to find the solution set of $\sqrt{4 - x} > x + 2$.

   (c) Use the graph to find the solution set of $\sqrt{4 - x} \le x + 2$.

9. *Laying a Telephone Cable*   A telephone company wishes to minimize the cost of laying a cable from point P to point R. Points P and Q are directly opposite each other along the banks of a straight river. The river is 300 yards wide. Point R lies on the same side of the river as point Q, but 600 yards away. (See the figure.) If the cost per yard for the cable is \$125 per yard under the water and \$100 per yard on land, how should the company lay the cable to minimize the cost?

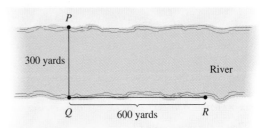

10. Consider the function $f(x) = \frac{1}{2}x - 3$.

    (a) Explain why f is one-to-one.

    (b) Find the rule for $f^{-1}(x)$ analytically.

    (c) Show that $(f \circ f^{-1})(x) = x$ and $(f^{-1} \circ f)(x) = x$.

    (d) Graph $y = f(x)$, $y = f^{-1}(x)$, and $y = x$ in a square window, and explain the geometric relationship that exists among these graphs.

CHAPTER 4   PROJECT

HOW RUGGED IS YOUR COASTLINE?*

An interesting feature of coastlines is that their ruggedness is independent of the distance from which they are viewed. From an airplane, we see irregularities as bays, peninsulas, river mouths, and so on. On foot, we see each rock outcropping and creek that makes the coastline appear more rugged. An ant sees every pebble as a mountain to be scaled. An interesting result of this phenomenon is that the total distance you travel along a coastline is dependent on the size of the steps you take. The closer (or smaller) you are, the smaller your steps will be. This means you will have more obstacles in your way, which results in a longer distance to travel. The more rugged the coastline, the longer it will be. In theory, this means that if you could take small enough steps on a rugged enough coastline, the length of the coastline would approach infinity. (If you find this topic interesting, you might want to study *fractal geometry*.)

_____

*This project is based on an idea presented by Lori Lambertson, of the Nueva School and the Exploratorium in San Francisco.

Lake Tahoe

From a mathematical perspective, we can say that the number of steps needed (y) varies inversely with the size of the steps taken (x):

$$y = \frac{k}{x^n}.$$

Each coastline will have different values for k and n, depending on its ruggedness. To find these values, you must walk the coastline. Gathering such data from a real coastline is, of course, too difficult as a classroom activity, but we can obtain similar data by using maps and "walking" the coastline with compasses (the kind used in geometry for drawing circles).

The following data were gathered by "walking" compasses along the coastline of a map of Lake Tahoe (on the California-Nevada border). For each walk, the compasses were opened to the given step size, and the legs of the compasses were walked around the lake, counting the number of steps until the journey was completed.

Step Size	Number of Steps	Total Length (step size times the number of steps)
4 inches	4.75	19 inches
2 inches	10.75	21.5 inches
1 inch	23.5	23.5 inches
$\frac{1}{2}$ inch	51	25.5 inches
$\frac{1}{4}$ inch	104	26 inches

Now we can analyze these data. First, enter them into the list editor of a graphing calculator, as shown in Figure A on the next page. Next, draw a scatterplot of the data in an appropriate viewing window, as seen in Figure B on the next page. To find an equation to model the data, we solve for k by substituting the coordinates (1, 23.5) for x and y in $y = \frac{k}{x^n}$. (We choose this point because any power of 1 is 1.) This leads to $k = 23.5$.

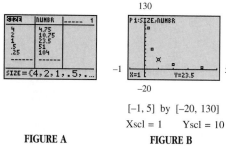

$$[-1, 5] \text{ by } [-20, 130]$$
$$\text{Xscl} = 1 \qquad \text{Yscl} = 10$$

FIGURE A   **FIGURE B**

We now know that the equation will be of the form $y = \frac{23.5}{x^n}$ for some value of n. Using the data point (4, 4.75) leads to $4.75 = \frac{23.5}{4^n}$. Solving this equation for n at this point poses a problem: we have not yet discussed how to solve such an equation analytically (this is discussed in Chapter 5). However, the multiline display and editing capabilities of a graphing calculator make it possible to estimate n. To do this, enter $23.5/4^x$ onto the home screen, and use guessing to come up with a value of x that leads to approximately 4.75. As seen in the final display of Figure C, a guess of 1.15333 works. Finally, using $n \approx 1.153$, the equation we are looking for is

$$y = \frac{23.5}{x^{1.153}}.$$

Figure D shows the graph of this equation over the scatterplot, and illustrates a remarkably good fit. Notice that the display at the bottom of the screen indicates that when $x = .5$, $y \approx 52.26$, which is very close to the experimental value of 51 seen in the original table.

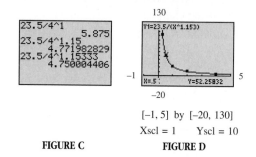

$$[-1, 5] \text{ by } [-20, 130]$$
$$\text{Xscl} = 1 \qquad \text{Yscl} = 10$$

FIGURE C   **FIGURE D**

ACTIVITIES

1. Use the equation to determine how long the coastline would be if you "walked" with step sizes of 6 inches, .1 inch, and .01 inch, respectively. Remember that the equation gives you the number of steps, so multiply that value by the step size to get the total length.

2. Gather your own data from a map of your choice. Using compasses, walk the coastline of your map and count the number of steps. Change the size of the opening of the compasses, and walk the coastline again. After you have done this with four or five different step sizes, find a mathematical model (equation) for your data, using the procedure described above.

Exponential and Logarithmic Functions

*I*n 1896, Swedish scientist Svante Arrhenius first predicted the greenhouse effect resulting from emissions of carbon dioxide by industrialized countries. In his classic calculation, he was able to estimate that a doubling of the carbon dioxide level in the atmosphere would raise the average global temperature by 7°F to 11°F. Since global warming would not be uniform, changes as small as 4.5°F in the average temperature could have drastic climatic effects, particularly on the central plains of North America. Sea levels could rise dramatically as a result of both thermal expansion and the melting of ice caps. The annual cost to the United States economy could reach $60 billion.

The burning of fossil fuels, deforestation, and changes in land use from 1850 to 1986 put approximately 312 billion tons of carbon into the atmosphere, mostly in the form of carbon dioxide. Burning of fossil fuels produces 5.4 billion tons of carbon each year which is absorbed by both the atmosphere and the oceans. A critical aspect of the accumulation of carbon dioxide in the atmosphere is that it is irreversible and its effect requires hundreds of years to disappear. In 1990, the International Panel of Climate Change (IPCC) reported that if current trends of burning of fossil fuel and deforestation continue, then future amounts of atmospheric carbon dioxide in parts per million (ppm) will increase as shown in the table.*

Year	Carbon Dioxide (ppm)
1990	353
2000	375
2075	590
2175	1090
2275	2000

Sources: Clime, W., *The Economics of Global Warming*, Institute for International Economics, Washington, D.C., 1992.

Kraljic, M. (Editor), *The Greenhouse Effect*, The H. W. Wilson Company, New York, 1992.

International Panel on Climate Change (IPCC), 1990.

Wuebbles, D., and J. Edmonds, *Primer of Greenhouse Gases*, Lewis Publishers, Inc., Chelsea, Michigan, 1991.

5.1   INTRODUCTION TO EXPONENTIAL FUNCTIONS

Preliminary Considerations ◆ Graphs of Exponential Functions ◆ Exponential Equations (Type 1) ◆ The Number *e* ◆ Compound Interest ◆ A Composite Exponential Function

PRELIMINARY CONSIDERATIONS

In Section 4.3, we studied how rational numbers are used as exponents. In particular, if r is a rational number and $r = \frac{m}{n}$, then, for appropriate values of m and n,

$$a^{m/n} = \left(\sqrt[n]{a} \right)^m.$$

For example,

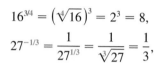
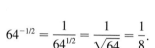

This screen supports the results in the text.

$$16^{3/4} = \left(\sqrt[4]{16} \right)^3 = 2^3 = 8,$$

$$27^{-1/3} = \frac{1}{27^{1/3}} = \frac{1}{\sqrt[3]{27}} = \frac{1}{3},$$

and

$$64^{-1/2} = \frac{1}{64^{1/2}} = \frac{1}{\sqrt{64}} = \frac{1}{8}.$$

In this section, the definition of a^r is extended to include all real (not just rational) values of the exponent r. For example, the new symbol $2^{\sqrt{3}}$ might be evaluated by approximating the exponent $\sqrt{3}$ by the numbers 1.7, 1.73, 1.732, and so on. Since these decimals approach the value of $\sqrt{3}$ more and more closely, it seems reasonable that $2^{\sqrt{3}}$ should be approximated more and more closely by the numbers $2^{1.7}$, $2^{1.73}$, $2^{1.732}$, and so on. $\left($Recall, for example, that $2^{1.7} = 2^{17/10} = \sqrt[10]{2^{17}}.\right)$ In fact, this is exactly how $2^{\sqrt{3}}$ is defined (in a more advanced course).

With this interpretation of real exponents, all rules and theorems for exponents are valid for real-number exponents as well as rational ones. In addition to the usual rules for exponents (see Chapter R), several new properties are used in this chapter. For example, if $y = 2^x$, then each value of x leads to exactly one value of y, and, therefore, $y = 2^x$ defines a function. Furthermore, if $3^x = 3^4$, then $x = 4$, and for $p > 0$, if $p^2 = 3^2$, then $p = 3$. Also,

$$4^2 < 4^3 \qquad \text{but} \qquad \left(\frac{1}{2} \right)^2 > \left(\frac{1}{2} \right)^3,$$

so that when $a > 1$, increasing the exponent on a leads to a *larger* number, but if $0 < a < 1$, increasing the exponent on a leads to a *smaller* number.

These properties are generalized below. Proofs of the properties are not given here, as they require more advanced mathematics.

   a.   If $a > 0$ and $a \neq 1$, then a^x is a unique real number for all real numbers x.
   b.   If $a > 0$ and $a \neq 1$, then $a^b = a^c$ if and only if $b = c$.
   c.   If $a > 1$ and $m < n$, then $a^m < a^n$.
   d.   If $0 < a < 1$ and $m < n$, then $a^m > a^n$.

Properties (a) and (b) require $a > 0$ so that a^x is always defined. For example, $(-6)^x$ is not a real number if $x = \frac{1}{2}$. This means that a^x will always be positive, since a is positive. In part (a), $a \neq 1$ because $1^x = 1$ for every real-number value of x, so that each value of x does not lead to a distinct real number. For Property (b) to hold, a must not equal 1 since, for example, $1^4 = 1^5$, even though $4 \neq 5$.

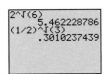

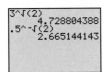

These screens show approximations for expressions having irrational exponents.

A graphing calculator can easily be used to find approximations of numbers raised to irrational powers. Using the exponentiation capabilities, we can find the following approximations:

$$2^{\sqrt{6}} \approx 5.462228786 \qquad 3^{\sqrt{2}} \approx 4.728804388$$

$$\left(\frac{1}{2}\right)^{\sqrt{3}} \approx .3010237439 \qquad .5^{-\sqrt{2}} \approx 2.665144143.$$

Later we will see how these approximations may also be found by using graphs of *exponential functions*.

GRAPHS OF EXPONENTIAL FUNCTIONS

We now define a new kind of function, the **exponential function**.

DEFINITION OF EXPONENTIAL FUNCTION

If $a > 0$, $a \neq 1$, then

$$f(x) = a^x$$

is the exponential function with base a.

Note   We do not allow 1 as a base for the exponential function because $1^x = 1$ for all real x, and thus it leads to the constant function $f(x) = 1$.

As we shall see, the behavior of the graph of an exponential function depends, in general, on the magnitude of a. Figure 1 shows the graphs of $f(x) = a^x$ for $a = 2, 3,$ and 4.

Based on our earlier discussion, the domain of $f(x) = a^x$ is $(-\infty, \infty)$. From the graphs in Figure 1, we see that the range is $(0, \infty)$, and the function is *increasing* for $a = 2, 3,$ and 4. The x-axis is the horizontal asymptote as $x \to -\infty$, and the y-intercept is 1. As a becomes larger, the graph becomes "steeper."

TECHNOLOGY NOTE

Because of the limited resolution of the graphing calculator screen, it is difficult to interpret how the graph of the exponential function $f(x) = a^x$ behaves when the curve is close to the *x*-axis, as seen in Figures 1 and 2. Remember that there is no endpoint, and that the curve approaches but never touches the *x*-axis. Tracing along the curve and observing the *y*-coordinates as $x \to -\infty$ when $a > 1$ and as $x \to \infty$ when $0 < a < 1$ will help support this fact.

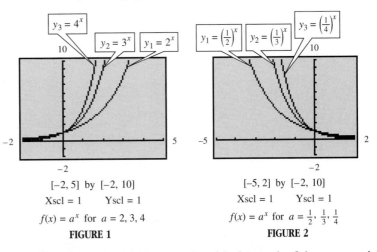

[−2, 5] by [−2, 10]
Xscl = 1   Yscl = 1
$f(x) = a^x$ for $a = 2, 3, 4$
FIGURE 1

[−5, 2] by [−2, 10]
Xscl = 1   Yscl = 1
$f(x) = a^x$ for $a = \frac{1}{2}, \frac{1}{3}, \frac{1}{4}$
FIGURE 2

If we choose the base a to be between 0 and 1, the graph of the exponential function $f(x) = a^x$ appears to be the same general type of curve, but with one important difference. Figure 2 shows the graphs of $f(x) = a^x$ for $a = \frac{1}{2}, \frac{1}{3},$ and $\frac{1}{4}$. As a gets closer to 0, the graph becomes "steeper."

Again, the domain is $(-\infty, \infty)$, the range is $(0, \infty)$, and the y-intercept is 1. However, the function is *decreasing* for $a = \frac{1}{2}, \frac{1}{3}$, and $\frac{1}{4}$, and the x-axis is the horizontal asymptote as $x \to \infty$. Our observations in Figures 1 and 2 lead to the following generalizations about the graphs of exponential functions.

EXPONENTIAL FUNCTION ($a > 1$)

$f(x) = a^x$, $a > 1$ (Figure 3)   Domain: $(-\infty, \infty)$ Range: $(0, \infty)$

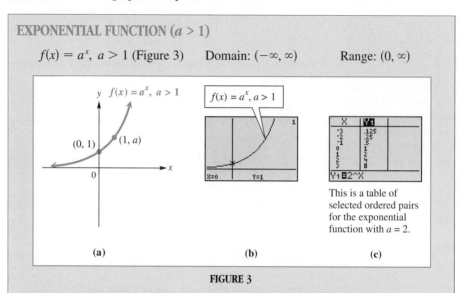

FIGURE 3

The exponential function $f(x) = a^x$, $a > 1$, increases on its entire domain, and is continuous as well. The x-axis is the horizontal asymptote as $x \to -\infty$. There is no x-intercept, and the y-intercept is 1. It is always concave up, and there are no extrema.

EXPONENTIAL FUNCTION ($0 < a < 1$)

$f(x) = a^x$, $0 < a < 1$ (Figure 4)   Domain: $(-\infty, \infty)$ Range: $(0, \infty)$

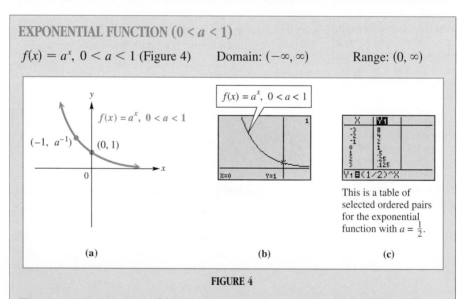

FIGURE 4

The exponential function $f(x) = a^x$, $0 < a < 1$, decreases on its entire domain, and is continuous as well. The x-axis is the horizontal asymptote as $x \to \infty$. There is no x-intercept, and the y-intercept is 1. It is always concave up, and there are no extrema.

Note The graph of $f(x) = a^{-x}$, $a > 1$, resembles the graph in Figure 4(b).

For Group Discussion

1. Using a standard viewing window, we cannot observe how the graph of the exponential function $f(x) = 2^x$ behaves for values of x less than about -3. Use the window $[-10, 0]$ by $[-.5, .5]$, and then trace to the left to see what happens.

2. Repeat Item 1 for the exponential function $f(x) = \left(\frac{1}{2}\right)^x$. Use a window of $[0, 10]$ by $[-.5, .5]$, and trace to the right.

3. Complete the following statement: For the graph of an exponential function $f(x) = a^x$, $a > 0$, $a \neq 1$, every range value appears exactly once, and thus f is a(n) _____ function. Because of this, f has a(n) _____. (*Hint*: Recall the concepts studied in Section 4.5.)

EXAMPLE 1 *Using Exponential Function Graphs to Evaluate Powers* Evaluate each of the following powers in two ways. First, use the definition of an exponent or the exponentiation capability of a calculator. Then, use a graph to support the result.

(a) $4^{3/2}$ **(b)** $2^{\sqrt{6}}$ **(c)** $.5^{-\sqrt{2}}$

SOLUTION

(a) Using the definition of a rational number as an exponent, we have

$$4^{3/2} = \left(\sqrt{4}\right)^3 = 2^3 = 8.$$

Figure 5 shows the graph of $y = 4^x$, with y evaluated for $x = \frac{3}{2} = 1.5$. Notice that when $x = 1.5$, $y = 8$.

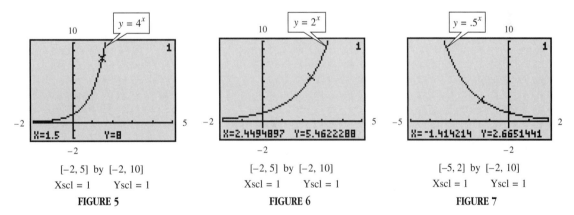

[-2, 5] by [-2, 10]	[-2, 5] by [-2, 10]	[-5, 2] by [-2, 10]
Xscl = 1 Yscl = 1	Xscl = 1 Yscl = 1	Xscl = 1 Yscl = 1
FIGURE 5	**FIGURE 6**	**FIGURE 7**

(b) Earlier in this section, we stated that the exponentiation capability of a graphing calculator will give the approximation of $2^{\sqrt{6}}$ as 5.462228786. By graphing $y = 2^x$ and letting $x = \sqrt{6} \approx 2.4494897$, we find $y \approx 5.4622288$, supporting our earlier result. See Figure 6.

(c) Again, we saw earlier that this number is approximately 2.665144143 by using the exponentiation capability of a calculator. Using the graph of $y = .5^x$, with $x = -\sqrt{2} \approx -1.414214$, we find $y \approx 2.6651441$, supporting our earlier result. See Figure 7. ◆

For Group Discussion   Repeat the procedure of Example 1 for $3^{\sqrt{2}}$ and $\left(\frac{1}{2}\right)^{\sqrt{3}}$.

EXAMPLE 2   *Comparing the Graphs of $f(x) = 2^x$ and $g(x) = \left(\frac{1}{2}\right)^x$*   Use the concepts of Section 2.3 to explain why the graph of $g(x) = \left(\frac{1}{2}\right)^x$ is a reflection across the y-axis of the graph of $f(x) = 2^x$.

SOLUTION   ANALYTIC   We must show that $g(x) = f(-x)$ to prove analytically that the graph of g is a reflection of the graph of f across the y-axis.

$$g(x) = \left(\frac{1}{2}\right)^x \qquad \text{Given}$$

$$= (2^{-1})^x \qquad \text{Definition of } a^{-1}$$

$$= 2^{-x} \qquad (a^m)^n = a^{mn}$$

$$= f(-x) \qquad \text{Because } f(x) = 2^x,\, f(-x) = 2^{-x}.$$

GRAPHICAL   Figure 8 shows traditional graphs of $f(x) = 2^x$, $g(x) = \left(\frac{1}{2}\right)^x$, and a calculator-generated graph of both on the same set of axes.   ◆

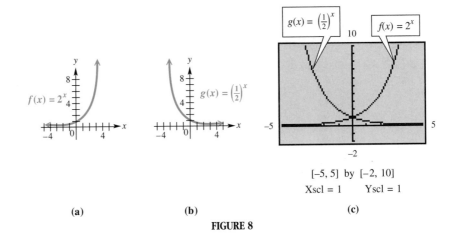

[−5, 5] by [−2, 10]
Xscl = 1   Yscl = 1

(a)   (b)   (c)

FIGURE 8

EXAMPLE 3   *Using Translations to Obtain the Graph of a Function*   Use the terminology of Chapter 2 to explain how the graph of $y = -2^x + 3$ can be obtained from the graph of $y = 2^x$. Discuss other important features of the graph.

SOLUTION   The graph of $y = -2^x$ is a reflection across the x-axis of the graph of $y = 2^x$. (Note that the base is not -2, but 2; negative numbers are not allowed as bases for exponential functions.) The $+3$ indicates that the graph is translated 3 units upward, making the horizontal asymptote the line $y = 3$ rather than the x-axis. The y-intercept is 2. The x-intercept can be approximated by using the capability of a calculator; it is approximately 1.58. (See Figure 9(b).)

The graph of $y = -2^x + 3$ is shown in both traditional and calculator-generated forms in Figure 9.   ◆

Note   We will be able to find the x-intercept of the graph in Figure 9 analytically, using concepts covered in a later section.

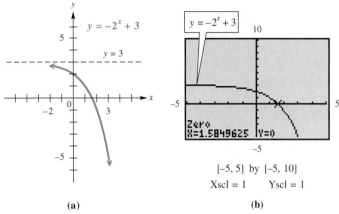

FIGURE 9

EXPONENTIAL EQUATIONS (TYPE 1)

An equation such as $\left(\frac{1}{3}\right)^x = 81$ is different from any equation studied so far in this book because the variable appears in the exponent. Notice that the base on the left side $\left(\frac{1}{3}\right)$ and the constant on the right side (81) can both easily be written as powers of a common base, 3. We shall refer to such an equation as a Type 1 exponential equation. Using Property (b) of the additional properties of exponents listed earlier in this section, we can solve Type 1 exponential equations.

EXAMPLE 4 *Solving a Type 1 Exponential Equation* Solve $\left(\frac{1}{3}\right)^x = 81$.

SOLUTION Because $\frac{1}{3} = 3^{-1}$ and $81 = 3^4$, we can solve the equation as follows.

$$\left(\frac{1}{3}\right)^x = 81$$
$$(3^{-1})^x = 3^4$$
$$3^{-x} = 3^4$$
$$-x = 4 \qquad \text{Property (b)}$$
$$x = -4$$

The solution set is $\{-4\}$. ◆

EXAMPLE 5 *Solving a Type 1 Exponential Equation* Solve $1.5^{x+1} = \left(\frac{27}{8}\right)^x$ analytically, and support the solution graphically, using the x-intercept method.

SOLUTION ANALYTIC At first glance, this equation may not seem to be of Type 1. However, notice that $1.5 = \frac{3}{2}$ and $\frac{27}{8} = \left(\frac{3}{2}\right)^3$. Thus, each base may easily be written as a power of a common base, $\frac{3}{2}$, and the method of solution used in Example 1 may be applied.

$$1.5^{x+1} = \left(\frac{27}{8}\right)^x$$
$$\left(\frac{3}{2}\right)^{x+1} = \left[\left(\frac{3}{2}\right)^3\right]^x \qquad \text{Write each base as a power of } \frac{3}{2}.$$
$$\left(\frac{3}{2}\right)^{x+1} = \left(\frac{3}{2}\right)^{3x} \qquad (a^m)^n = a^{mn}$$
$$x + 1 = 3x \qquad \text{Set exponents equal.}$$
$$1 = 2x$$
$$x = \frac{1}{2} \quad \text{or} \quad .5$$

The solution set is {.5}.

GRAPHICAL   Graphing $y = 1.5^{x+1} - \left(\frac{27}{8}\right)^x$ and using the capability of a calculator confirm our analytic solution, as shown in Figure 10. ◆

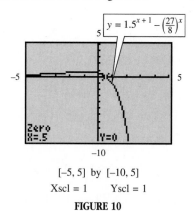

[−5, 5] by [−10, 5]

Xscl = 1   Yscl = 1

FIGURE 10

EXAMPLE 6   *Using a Graph to Solve Exponential Inequalities*   Use the graph in Figure 10 to solve each inequality.

(a) $1.5^{x+1} - \left(\frac{27}{8}\right)^x > 0$ (b) $1.5^{x+1} - \left(\frac{27}{8}\right)^x < 0$

SOLUTION   It appears that the graph of $y = 1.5^{x+1} - \left(\frac{27}{8}\right)^x$ in Figure 10 lies *above* the *x*-axis for values of *x* less than .5, and *below* the *x*-axis for values of *x* greater than .5. Tracing left and right will help to support this observation. The solution set for the inequality in part (a) is $(-\infty, .5)$, while the solution set for the inequality in part (b) is $(.5, \infty)$. ◆

THE NUMBER *e*

Perhaps the most important exponential function has the irrational number *e* as its base. The number *e* is named after the Swiss mathematician Leonhard Euler (1707–1783). There are many mathematical expressions that can be used to approximate *e*, but one of the easiest to illustrate involves the expression $\left(1 + \frac{1}{x}\right)^x$. If we let *x* take on larger and larger values (that is, approach ∞), the expression approaches *e*. The table in Figure 11 shows how this happens when we let *x* take on powers of 10.

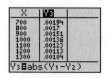

FIGURE 11

It appears that as $x \to \infty$, $\left(1 + \frac{1}{x}\right)^x$ is approaching some "limiting" number. (In fact, in calculus this number is called the *limit* of $\left(1 + \frac{1}{x}\right)^x$ as *x* approaches ∞.) The number is *e*, and to nine decimal places, the value of *e* is as follows.

$$e \approx 2.718281828$$

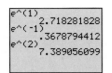

This screen supports the results in the text.

Because e is such an important base for the exponential function, calculators have the capability to find powers of e. Using the exponentiation capability, with base e, we can find the following approximations.

$$e^1 \approx 2.718281828$$
$$e^{-1} \approx .3678794412$$
$$e^2 \approx 7.389056099$$

Of course, by definition, $e^0 = 1$. A traditional graph of $f(x) = e^x$ is shown in Figure 12(a), and a calculator-generated graph is shown in Figure 12(b).

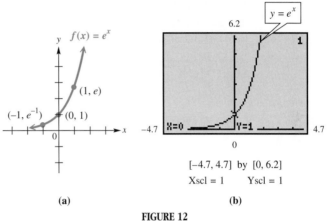

[−4.7, 4.7] by [0, 6.2]
Xscl = 1   Yscl = 1

(a)   **(b)**

FIGURE 12

We will see later in this section and in the final section of this chapter some important applications of the exponential function with base e.

COMPOUND INTEREST

The formula for *compound interest* (interest paid on both principal and interest) is an important application of exponential functions. You may recall the formula for simple interest, $I = Prt$, where P is the amount left at interest, r is the rate of interest expressed as a decimal, and t is time in years that the principal earns interest. Suppose $t = 1$ year. Then, at the end of the year, the amount has grown to

$$P + Pr = P(1 + r),$$

the original principal plus the interest. If this amount is left at the same interest rate for another year, the total amount becomes

$$[P(1 + r)] + [P(1 + r)]r = [P(1 + r)](1 + r)$$
$$= P(1 + r)^2.$$

After the third year, this will grow to

$$[P(1 + r)^2] + [P(1 + r)^2]r = [P(1 + r)^2](1 + r)$$
$$= P(1 + r)^3.$$

Continuing in this way produces the following formula for compound interest.

COMPOUND INTEREST FORMULA

Suppose that a principal of P dollars is invested at an annual interest rate r (in percent), compounded n times per year. Then, the amount A accumulated after t years is given by the formula

$$A = P\left(1 + \frac{r}{n}\right)^{nt}.$$

EXAMPLE 7   *Using the Compound Interest Formula*   Suppose that $1000 is invested at an annual rate of 8%, compounded quarterly (four times per year). Find the total amount in the account after 10 years if no withdrawals are made. Support the result with a graph.

SOLUTION   ANALYTIC   Use the compound interest formula above, with $P = 1000$, $r = .08$, $n = 4$, and $t = 10$.

$$A = P\left(1 + \frac{r}{n}\right)^{nt}$$

$$A = 1000\left(1 + \frac{.08}{4}\right)^{4 \cdot 10} \approx 2208.039664 \qquad \text{Use a calculator.}$$

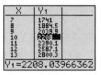

With $Y_1 = 1000\left(1 + \frac{.08}{4}\right)^{4x}$, a table can be used to determine the amount in the account described in Example 7. Compare the value of Y_1 when $x = 10$ to the one shown in the example.

To the nearest cent, there will be $2208.04 in the account after 10 years. (Note that this means that $2208.04 - $1000 = $1208.04 interest was earned.)

GRAPHICAL   Figure 13 supports this result graphically. The function $y = 1000(1.02)^x$ is graphed and evaluated for $x = 40$ (since $x = nt = 40$). The y-value is the amount in the account. (Tracing to the right on this graph gives new meaning to the phrase "watching your money grow.")   ◆

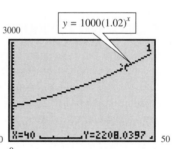

[0, 50] by [0, 3000]

Xscl = 5   Yscl = 100

FIGURE 13

The compounding formula given earlier applies if the financial institution compounds interest for a finite number of compounding periods annually. Theoretically, the number of compounding periods per year can get larger and larger (quarterly, monthly, daily, etc.), and if n is allowed to approach infinity, we say that interest is compounded *continuously*.

To derive the formula for continuous compounding, we begin with the earlier formula.

$$A = P\left(1 + \frac{r}{n}\right)^{nt}$$

Let $k = \frac{n}{r}$. Then, $n = rk$, and with these substitutions, the formula becomes

$$A = P\left(1 + \frac{1}{k}\right)^{rkt} = P\left[\left(1 + \frac{1}{k}\right)^{k}\right]^{rt}.$$

If $n \to \infty$, $k \to \infty$ as well, and the expression $\left(1 + \frac{1}{k}\right)^{k} \to e$, as discussed earlier. This leads to the formula $A = Pe^{rt}$.

CONTINUOUS COMPOUNDING FORMULA

If P dollars is deposited at a rate of interest r compounded continuously for t years, the final amount A in dollars on deposit is

$$A = Pe^{rt}.$$

EXAMPLE 8 *Solving a Continuous Compounding Problem* Suppose $5000 is deposited in an account paying 8% compounded continuously for 5 years. Find the total amount on deposit at the end of 5 years.

SOLUTION ANALYTIC Let $P = 5000$, $t = 5$, and $r = .08$. Then, $A = \mathbf{5000}\mathit{e}^{.08(5)} = 5000e^{.4}$. Using a calculator, we find that $e^{.4} \approx 1.491824698$, and, to the nearest cent, $A = 5000e^{.4} = 7459.12$, or $7459.12.

GRAPHICAL As Figure 14 shows, this result can be supported by the graph of $y = 5000e^{.08x}$. When $x = 5$, $y \approx 7459.12$. ◆

With $Y_1 = 5000e^{.08x}$, the result of Example 8 can also be supported with a table. How much would there be at the end of 8 years?

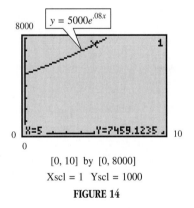

[0, 10] by [0, 8000]
Xscl = 1 Yscl = 1000
FIGURE 14

The continuous compounding formula is an example of an exponential growth function with base e. Other examples of exponential growth (and decay) will be given in Section 5.5.

A COMPOSITE EXPONENTIAL FUNCTION

If $f(x) = 2^x$ and $g(x) = -x^2$, then $(f \circ g)(x) = 2^{-x^2}$. The graph of this composite function is seen in Figure 15 on the next page. Because of the exponent on x, it is not typical of the kinds of exponential functions seen so far in this chapter. It is "bell-shaped" and is important in other branches of mathematics such as statistics.

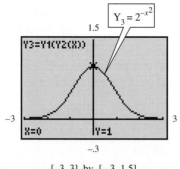

[−3, 3] by [−.3, 1.5]
Xscl = 1   Yscl = 1

The *x*-axis is the horizontal
asymptote, and the *y*-intercept is 1.
FIGURE 15

5.1 EXERCISES   Tape 6

Use the exponentiation capabilities of a calculator to find an approximation for each of the following powers. Give the maximum number of decimal places that your calculator will display.

1. $2^{\sqrt{10}}$

2. $3^{\sqrt{11}}$

3. $\left(\dfrac{1}{2}\right)^{\sqrt{2}}$

4. $\left(\dfrac{1}{3}\right)^{\sqrt{6}}$

5. $4.1^{-\sqrt{3}}$

6. $6.4^{-\sqrt{3}}$

7. $\sqrt{7}^{\sqrt{7}}$

8. $\sqrt{13}^{-\sqrt{13}}$

Use the graph of the given exponential function and the capabilities of your calculator to graphically support the result found in the specified exercise.

9. $y = 2^x$, Exercise 1

10. $y = 3^x$, Exercise 2

11. $y = \left(\dfrac{1}{2}\right)^x$, Exercise 3

12. $y = \left(\dfrac{1}{3}\right)^x$, Exercise 4

In the given figure, the graphs of $y = a^x$ for $a = 1.8, 2.3, 3.2, .4, .75$, and .31 are given. They are identified by letter, but not necessarily in the same order as the values of a just given. Use your knowledge of how the exponential function behaves for various powers of a to identify each lettered graph.

13. A

14. B

15. C

16. D

17. E

18. F

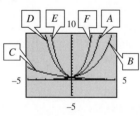

Evaluate each of the following powers in two ways. First, use the definition of an exponent. Then, use a graph to support the result.

19. $8^{2/3}$

20. $27^{-4/3}$

21. $25^{-3/2}$

22. $16^{5/4}$

In Exercises 23 and 24, the graph of an exponential function with base a is given. Follow the directions in parts (a)–(f) in each exercise.

23. (a) Is $a > 1$ or is $0 < a < 1$?

(b) Give the domain and range of f, and the equation of the asymptote.

(c) Sketch the graph of $g(x) = -a^x$.

(d) Give the domain and range of g, and the equation of the asymptote.

(e) Sketch the graph of $h(x) = a^{-x}$.

(f) Give the domain and range of h, and the equation of the asymptote.

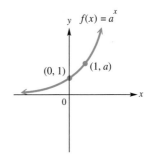

24. (a) Is $a > 1$ or is $0 < a < 1$?

(b) Give the domain and range of f, and the equation of the asymptote.

(c) Sketch the graph of $g(x) = a^x + 2$.

(d) Give the domain and range of g, and the equation of the asymptote.

(e) Sketch the graph of $h(x) = a^{x+2}$.

(f) Give the domain and range of h, and the equation of the asymptote.

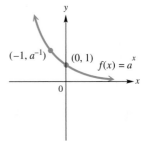

RELATING CONCEPTS (EXERCISES 25–30)

Use the terminology of Sections 2.2 and 2.3 to explain how the graph of the given function can be obtained from the graph of $y = 2^x$. $\left(\text{Recall that } \frac{1}{2} = 2^{-1}, \text{ so } \left(\frac{1}{2}\right)^x = (2^{-1})^x = 2^{-x}.\right)$

25. $y = 2^{x+5} - 3$

26. $y = 2^{x-1} + 4$

27. $y = \left(\frac{1}{2}\right)^x + 1$

28. $y = \left(\frac{1}{2}\right)^x - 6$

29. $y = -3 \cdot 2^x$

30. $y = -4 \cdot \left(\frac{1}{2}\right)^x$

In part (a) of Exercises 31–38, a Type 1 exponential equation is given. Solve the equation analytically. Use a calculator-generated graph to support your answer, applying either the intersection-of-graphs method or the x-intercept method. Then, use the graph to solve the associated inequalities in parts (b) and (c).

31. (a) $125^x = 5$

(b) $125^x > 5$

(c) $125^x < 5$

32. (a) $\left(\frac{1}{2}\right)^x = 4$

(b) $\left(\frac{1}{2}\right)^x > 4$

(c) $\left(\frac{1}{2}\right)^x < 4$

33. (a) $\left(\frac{2}{3}\right)^x = \frac{9}{4}$

(b) $\left(\frac{2}{3}\right)^x \geq \frac{9}{4}$

(c) $\left(\frac{2}{3}\right)^x \leq \frac{9}{4}$

34. (a) $\left(\frac{3}{4}\right)^x = \frac{16}{9}$

(b) $\left(\frac{3}{4}\right)^x \geq \frac{16}{9}$

(c) $\left(\frac{3}{4}\right)^x \leq \frac{16}{9}$

35. (a) $2^{3-x} = 8$

(b) $2^{3-x} > 8$

(c) $2^{3-x} < 8$

36. (a) $5^{2x+1} = 25$

(b) $5^{2x+1} > 25$

(c) $5^{2x+1} < 25$

37. (a) $27^{4x} = 9^{x+1}$

(b) $27^{4x} < 9^{x+1}$

(c) $27^{4x} > 9^{x+1}$

38. (a) $32^x = 16^{1-x}$

(b) $32^x < 16^{1-x}$

(c) $32^x > 16^{1-x}$

Use the exponentiation capabilities of a calculator to find each of the following powers of e. Then, use a calculator-generated graph of $y = e^x$ to support your result.

39. $e^{3.1}$ **40.** e^{15} **41.** $e^{-.25}$ **42.** $e^{-1.6}$ **43.** $e^{\sqrt{2}}$ **44.** $e^{-\sqrt{2}}$

Use the appropriate compound interest formula to find the amount that will be in an account, given the stated conditions.

45. $20,000 invested at 3% annual interest for 4 years compounded **(a)** annually; **(b)** semiannually

46. $35,000 invested at 4.2% annual interest for 3 years compounded **(a)** annually; **(b)** quarterly

47. $27,500 invested at 3.95% annual interest for 5 years compounded **(a)** daily ($n = 365$); **(b)** continuously

48. $15,800 invested at 4.6% annual interest for 6.5 years compounded **(a)** quarterly; **(b)** continuously

In Exercises 49 and 50, decide which of the two plans will provide a better yield.

49. Plan A: $40,000 invested for 3 years at 4.5%, compounded quarterly

Plan B: $40,000 invested for 3 years at 4.4%, compounded continuously

50. Plan A: $50,000 invested for 10 years at 4.75%, compounded daily ($n = 365$)

Plan B: $50,000 invested for 10 years at 4.7%, compounded continuously

Use the table capabilities of your calculator to work Exercises 51 and 52.

51. *Comparison of Two Accounts*   You have the choice of investing $1000.00 at an annual rate of 5%, compounded either annually or monthly. Let Y_1 represent the investment compounded annually, and let Y_2 represent the investment compounded monthly. Graph both Y_1 and Y_2, and observe the slight difference in each curve. Then, use a table to compare the graphs numerically. What is the difference between the returns on each investment after 1 year, 2 years, 5 years, 10 years, 20 years, 30 years, and 40 years?

52. *Comparison of Two Accounts*   You have the choice of investing $1000.00 at an annual rate of 7.5%, compounded daily, or at an annual rate of 7.75%, compounded annually. Let Y_1 represent the investment at 7.5% compounded daily, and let Y_2 represent the investment at 7.75% compounded annually. Graph both Y_1 and Y_2, and observe the slight difference in each curve. Then, use the table with $Y_3 = Y_1 - Y_2$ to compare the graphs numerically. What is the difference between the returns on each investment after 1 year, 2 years, 5 years, 10 years, 20 years, 30 years, and 40 years? Why does the lower interest rate yield the greater return?

Use a calculator to graph each of the following composite exponential functions. Use the window given.

53. $f(x) = \dfrac{e^x - e^{-x}}{2}$

$[-10, 10]$ by $[-10, 10]$

54. $f(x) = \dfrac{e^x + e^{-x}}{2}$

$[-10, 10]$ by $[-10, 10]$

55. $f(x) = x \cdot 2^x$

$[-5, 5]$ by $[-2, 5]$

56. $f(x) = x^2 \cdot 2^{-x}$

$[-2, 8]$ by $[-2, 5]$

In this section, we showed how Type 1 exponential equations can be solved, using analytic techniques. In Section 5.4, we will show how to solve analytically exponential equations such as $7^x = 12$, where the bases cannot easily be written as powers of the same base. We will refer to such equations as Type 2 equations. However, we can use graphical methods to solve such equations for approximate solutions, using either the x-intercept method or the intersection-of-graphs method. For example, to solve $7^x = 12$, we can graph $y_1 = 7^x$ and $y_2 = 12$ and find the x-coordinate of the point of intersection. Alternatively, we can graph $y = 7^x - 12$ and find the x-intercept. Use either of these methods to solve the equations in Exercises 57–64.

57. $7^x = 12$

58. $3^x = 5$

59. $2^{x-1} = 6$

60. $4^{x+3} = 9$

61. $e^x = 8$

62. $e^x = 5^{x+3}$

63. $10^x = 8$

64. $10^x = \left(\dfrac{1}{3}\right)^{2x+3}$

5.2   LOGARITHMS AND THEIR PROPERTIES

Introduction to Logarithms ◆ Common Logarithms ◆ Natural Logarithms ◆ Properties of Logarithms ◆ The Change-of-Base Rule

INTRODUCTION TO LOGARITHMS

In order to introduce the concept of logarithm, let us consider the exponential equation $2^3 = 8$. In this equation, 3 is the exponent to which 2 must be raised in order to obtain

8. In this context, 3 is called the logarithm to the base 2 of 8, abbreviated $3 = \log_2 8$. It is important to remember that a logarithm is an exponent and as such will possess the same properties as exponents.

DEFINITION OF LOGARITHM

For all positive numbers a, where $a \neq 1$,

$$a^k = x \qquad \text{is equivalent to} \qquad k = \log_a x.$$

A logarithm is an exponent, and $\log_a x$ is the exponent to which a must be raised in order to obtain x. The number a is called the *base* of the logarithm, and x is called the *argument* of the expression $\log_a x$. The value of x will always be positive.

The first example shows conversions between exponential and logarithmic statements.

EXAMPLE 1 *Converting between Exponential and Logarithmic Statements* The chart below shows several pairs of equivalent statements. The same statement is written in both exponential and logarithmic forms.

Exponential Form	Logarithmic Form
$2^3 = 8$	$\log_2 8 = 3$
$\left(\frac{1}{2}\right)^{-4} = 16$	$\log_{1/2} 16 = -4$
$10^5 = 100{,}000$	$\log_{10} 100{,}000 = 5$
$3^{-4} = \frac{1}{81}$	$\log_3 \left(\frac{1}{81}\right) = -4$
$5^1 = 5$	$\log_5 5 = 1$
$\left(\frac{3}{4}\right)^0 = 1$	$\log_{3/4} 1 = 0$

◆

In some cases, the method for solving Type 1 exponential equations presented in the previous section can be used to evaluate logarithms. The next example illustrates this.

EXAMPLE 2 *Finding Logarithms Analytically* Find each of the following logarithms by setting them equal to x and then solving a Type 1 exponential equation.

(a) $\log_8 4$ **(b)** $\log_{1/27} 81$

SOLUTION

(a) Let $x = \log_8 4$ and write in exponential form.

$$
\begin{aligned}
x &= \log_8 4 \\
8^x &= 4 & &\text{Exponential form} \\
(2^3)^x &= 2^2 & &\text{Write as a power of the same base, 2.} \\
2^{3x} &= 2^2 & &(a^m)^n = a^{mn} \\
3x &= 2 & &\text{Set exponents equal.} \\
x &= \frac{2}{3}
\end{aligned}
$$

Therefore, $\log_8 4 = \frac{2}{3}$.

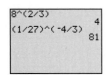

This screen uses exponential notation to support the results in Example 2.

(b)
$$x = \log_{1/27} 81$$
$$\left(\frac{1}{27}\right)^x = 81$$
$$(3^{-3})^x = 3^4$$
$$3^{-3x} = 3^4$$
$$-3x = 4$$
$$x = -\frac{4}{3}$$

Therefore, $\log_{1/27} 81 = -\frac{4}{3}$. ◆

COMMON LOGARITHMS

One of the two most important bases used for logarithms is 10. (The other is e.) Base 10 logarithms are called **common logarithms**, and the symbol that we use for the common logarithm of a positive number x is log x. Notice that when no base is indicated, the base is understood to be 10.

COMMON LOGARITHM

$\log x = \log_{10} x$ for all positive numbers x.

Remember that the argument of a common logarithm (and any base logarithm, for that matter) must be a positive number.

Graphing calculators have the capability of finding common logarithms (exact or approximate, depending upon the argument). Figure 16(a) shows a typical graphing calculator screen with several common logarithms evaluated.

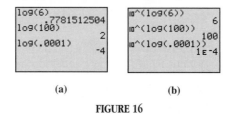

(a)   (b)

FIGURE 16

The first display indicates that .7781512504 is (approximately) the exponent to which 10 must be raised in order to obtain 6. The second says that 2 is the exponent to which 10 must be raised in order to obtain 100. This is reasonable, since $100 = 10^2$. The third display indicates that -4 is the exponent to which 10 must be raised in order to obtain .0001. Again, this is reasonable, since $.0001 = 10^{-4}$. Figure 16(b) shows how the definition of a common logarithm can be applied: raising 10 to the power log x gives x as a result.

In Figure 17, we have graphed the functions $y_1 = 10^x$ and $y_2 = 6$. Notice that the point of intersection has an x-value of .77815125, supporting our observation in Figure 16(a) for the value of log 6.

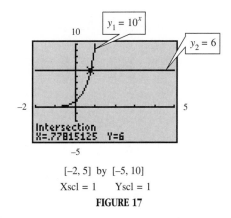

$$y_1 = 10^x$$

$$y_2 = 6$$

[−2, 5] by [−5, 10]

Xscl = 1   Yscl = 1

FIGURE 17

Common logarithms were originally used to aid mathematicians and scientists in paper-and-pencil calculations. With the advent of calculators and computers, this use is no longer necessary. However, there are still important applications of common logarithms. For example, in chemistry, the pH of a solution is defined as

$$pH = -\log[H_3O^+],$$

where $[H_3O^+]$ is the hydronium ion concentration in moles per liter.* The pH value is a measure of the acidity or alkalinity of solutions. Pure water has a pH of 7.0, substances with pH values greater than 7.0 are alkaline, and substances with pH values less than 7.0 are acidic.

EXAMPLE 3 *Finding pH and [H₃O⁺]*

(a) Find the pH of a solution with $[H_3O^+] = 2.5 \times 10^{-4}$.

(b) Find the hydronium ion concentration of a solution with pH $= 7.1$.

SOLUTION

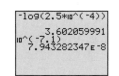

These are the calculations required in Example 3.

(a)
$$pH = -\log[H_3O^+]$$
$$pH = -\log(2.5 \times 10^{-4}) \qquad \text{Substitute.}$$
$$pH \approx 3.6 \qquad \text{Use a calculator and round to the nearest tenth.}$$

(b)
$$pH = -\log[H_3O^+]$$
$$7.1 = -\log[H_3O^+] \qquad \text{Substitute.}$$
$$-7.1 = \log[H_3O^+] \qquad \text{Multiply by } -1.$$
$$[H_3O^+] = 10^{-7.1} \qquad \text{Write in exponential form.}$$

Graphing calculators have keys marked 10^x, usually in conjunction with the log x key, that allow you to raise 10 to a power. Use this key to find that

$$[H_3O^+] = 10^{-7.1} \approx 7.9 \times 10^{-8}. \qquad \blacklozenge$$

NATURAL LOGARITHMS

In many practical applications of logarithms, the number e (introduced in the previous section) is used as the base. Logarithms to base e are called **natural logarithms.** The symbol used for the natural logarithm of a positive number x is ln x (read "el en x").

---

*A *mole* is the amount of substance that contains the same number of molecules as the number of atoms in 12 grams of carbon 12.

NATURAL LOGARITHM

$\ln x = \log_e x$ for all positive numbers x.

Natural logarithms of numbers can be found in much the same way as common logarithms, using a calculator. The natural logarithm key is usually found in conjunction with the e^x key. Figure 18 shows how the natural logarithm function and the base e exponential function can be applied.

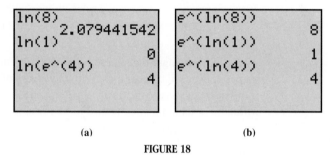

    (a)   (b)

FIGURE 18

The first display in Figure 18(a) indicates that 2.079441542 is (approximately) the exponent to which e must be raised in order to obtain 8. In Figure 19, this is supported by the fact that the graphs of $y_1 = e^x$ and $y_2 = 8$ intersect at a point with x-value 2.0794415.

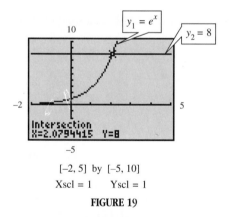

$$[-2, 5] \text{ by } [-5, 10]$$
$$\text{Xscl} = 1 \qquad \text{Yscl} = 1$$

FIGURE 19

For Group Discussion

1. The second display in Figure 18(a) indicates $\ln 1 = 0$. Use your calculator to find log 1. Discuss your results. If 1 is the argument, does it matter what base is used? Why?
2. The third display in Figure 18(a) indicates $\ln(e^4) = 4$. Use the same expression, but replace 4 with the number of letters in your last name. What is the result? Can you make a generalization?
3. Based on your answer to Item 2, predict the value of $\log(10^{14.6})$. Now verify with your calculator.

EXAMPLE 4 *Using Natural Logarithms to Solve a Continuous Compounding Problem* In Section 5.1, we learned that the formula $A = Pe^{rt}$ is used to compute the amount of money in an account earning interest compounded continuously. Suppose that $1000 is invested at 3% annual interest, compounded continuously. How long will it take for the amount to grow to $1500? Support the result graphically.

SOLUTION ANALYTIC We use the formula with $P = 1000$, $A = 1500$, and $r = .03$.

$$1500 = 1000e^{.03t} \qquad A = Pe^{rt}$$

$$1.5 = e^{.03t} \qquad \text{Divide by 1000.}$$

Because $.03t$ is the exponent to which e must be raised in order to obtain 1.5, it must equal the natural logarithm of 1.5, or ln 1.5.

$$.03t = \ln 1.5$$

$$t = \frac{\ln 1.5}{.03} \qquad \text{Divide by .03.}$$

$$t \approx 13.5 \qquad \text{Use a calculator.}$$

It will take about 13.5 years to grow to $1500.

GRAPHICAL This answer can be supported graphically by showing that the point $\left(\frac{\ln 1.5}{.03}, 1500\right)$ lies on the graph of $y = 1000e^{.03x}$. See Figure 20. ◆

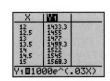

This table further supports the result of Example 4. When time (x) is 13.5 years, amount (Y_1) is $1499.3 \approx 1500$ dollars.

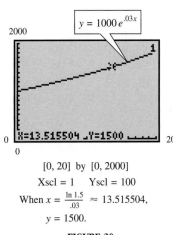

[0, 20] by [0, 2000]
Xscl = 1   Yscl = 100
When $x = \frac{\ln 1.5}{.03} \approx 13.515504$,
$y = 1500$.

FIGURE 20

PROPERTIES OF LOGARITHMS

We will now formally state several properties of logarithms.

PROPERTIES OF LOGARITHMS

1. If $a > 0$, $a \neq 1$, then $\log_a 1 = 0$.
2. If $a > 0$, $a \neq 1$, and k is a real number, then $\log_a a^k = k$.
3. If $a > 0$, $a \neq 1$, and $k > 0$, then $a^{\log_a k} = k$.

Property 1 is true because $a^0 = 1$ for any nonzero value of a. Property 2 is verified by writing the equation in exponential form, giving the identity $a^k = a^k$. Property 3 is justified by the fact that $\log_a k$ is the exponent to which a must be raised in order to obtain k. Therefore, by the definition, $a^{\log_a k}$ must equal k.

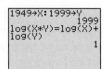

The product, quotient, and power rules can be illustrated as shown here. Letting $X = 1949$ (the birth year of one of the authors) and $Y = 1999$ (the publication year of this book), the product and quotient rules are shown to be true (since a 1 is returned). The power rule, with $R = 2$, is illustrated in a similar manner.

These rules are shown here for common logarithms, but they are valid for all bases.

Three additional rules of logarithms follow. These rules emphasize the close relationship between logarithms and exponents.

PROPERTIES OF LOGARITHMS (Continued)

For $x > 0$, $y > 0$, $a > 0$, $a \neq 1$, and any real number r,

Product Rule **4.** $\log_a xy = \log_a x + \log_a y$.
(The logarithm of the product of two numbers is equal to the sum of the logarithms of the numbers.)

Quotient Rule **5.** $\log_a \frac{x}{y} = \log_a x - \log_a y$.
(The logarithm of the quotient of two numbers is equal to the difference between the logarithms of the numbers.)

Power Rule **6.** $\log_a x^r = r \log_a x$.
(The logarithm of a number raised to a power is equal to the exponent multiplied by the logarithm of the number.)

The proof of Property 4, the product rule, follows.

PROOF Let

$$m = \log_a x \text{ and } n = \log_a y.$$

Then,

$a^m = x \quad \text{and} \quad a^n = y$	Definition of logarithm
$a^m \cdot a^n = xy$	Multiplication
$a^{m+n} = xy$	Add exponents.
$\log_a xy = m + n.$	Definition of logarithm

Since $m = \log_a x$ and $n = \log_a y$,

$$\log_a xy = \log_a x + \log_a y. \qquad \text{Substitution} \qquad \blacklozenge$$

Properties 5 and 6 are proved in a similar way. (See Exercises 95 and 96.)

EXAMPLE 5 *Using the Properties of Logarithms* Assuming that all variables represent positive real numbers, use the properties of logarithms to rewrite each of the following expressions.

(a) $\log 8x$ **(b)** $\log_9 \frac{15}{7}$ **(c)** $\log_5 \sqrt{8}$

(d) $\log_a \frac{x}{yz}$ **(e)** $\log_a \sqrt[3]{m^2}$ **(f)** $\log_b \sqrt[n]{\frac{x^3 y^5}{z^m}}$

SOLUTION

(a) $\log 8x = \log 8 + \log x$

(b) $\log_9 \frac{15}{7} = \log_9 15 - \log_9 7$

(c) $\log_5 \sqrt{8} = \log_5 8^{1/2} = \frac{1}{2} \log_5 8$

(d) $\log_a \frac{x}{yz} = \log_a x - (\log_a y + \log_a z) = \log_a x - \log_a y - \log_a z$

(e) $\log_a \sqrt[3]{m^2} = \dfrac{2}{3} \log_a m$

(f) $\log_b \sqrt[n]{\dfrac{x^3 y^5}{z^m}} = \dfrac{1}{n} \log_b \dfrac{x^3 y^5}{z^m}$

$$= \dfrac{1}{n}(\log_b x^3 + \log_b y^5 - \log_b z^m)$$

$$= \dfrac{1}{n}(3 \log_b x + 5 \log_b y - m \log_b z)$$

$$= \dfrac{3}{n} \log_b x + \dfrac{5}{n} \log_b y - \dfrac{m}{n} \log_b z$$

Notice the use of parentheses in the second step. The factor $\frac{1}{n}$ applies to each term. ◆

EXAMPLE 6 *Using the Properties of Logarithms* Use the properties of logarithms to write each of the following as a single logarithm with a coefficient of 1. Assume that all variables represent positive real numbers.

(a) $\log_3 (x + 2) + \log_3 x - \log_3 2$

(b) $2 \log_a m - 3 \log_a n$

(c) $\frac{1}{2} \log_b m + \frac{3}{2} \log_b 2n - \log_b m^2 n$

SOLUTION

(a) Using Properties 4 and 5,

$$\log_3 (x + 2) + \log_3 x - \log_3 2 = \log_3 \dfrac{(x + 2)x}{2}.$$

(b) $2 \log_a m - 3 \log_a n = \log_a m^2 - \log_a n^3 = \log_a \dfrac{m^2}{n^3}$

Here we used Property 6, then Property 5.

(c) $\dfrac{1}{2} \log_b m + \dfrac{3}{2} \log_b 2n - \log_b m^2 n$

$$= \log_b m^{1/2} + \log_b (2n)^{3/2} - \log_b m^2 n \qquad \text{Property 6}$$

$$= \log_b \dfrac{m^{1/2}(2n)^{3/2}}{m^2 n} \qquad \text{Properties 4 and 5}$$

$$= \log_b \dfrac{2^{3/2} n^{1/2}}{m^{3/2}} \qquad \text{Rules for exponents}$$

$$= \log_b \left(\dfrac{2^3 n}{m^3}\right)^{1/2} \qquad \text{Rules for exponents}$$

$$= \log_b \sqrt{\dfrac{8n}{m^3}} \qquad \text{Definition of } a^{1/n} \qquad ◆$$

Caution There is no property of logarithms to rewrite a logarithm of a *sum* or *difference*. That is why, in Example 6(a), $\log_3 (x + 2)$ was not written as $\log_3 x + \log_3 2$. Remember, $\log_3 x + \log_3 2 = \log_3 (x \cdot 2)$.

THE CHANGE-OF-BASE RULE

A natural question to ask at this point is "Can we use a calculator to find logarithms for bases other than 10 and e?" The answer is yes, but before we state the change-of-base rule, let us now return to Figure 9(b). We have the calculator-generated graph of $y = -2^x + 3$. The display shows that the x-intercept of the graph is approximately 1.5849625. If we want to find the x-intercept *analytically*, we must let $y = 0$:

$$y = -2^x + 3$$
$$0 = -2^x + 3 \qquad \text{Let } y = 0.$$
$$2^x = 3 \qquad \text{Add } 2^x \text{ to both sides.}$$
$$x = \log_2 3. \qquad \text{Rewrite in logarithmic form.}$$

Based on the earlier analysis of the graph, $\log_2 3$ must be approximately equal to 1.5849625.

Now we develop the change-of-base rule that will allow us to verify this graphical result analytically.

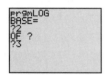

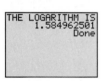

A short program that applies the change-of-base rule can easily be written to find logarithms to bases other than 10 and e. Here, we can see that $\log_2 3 \approx 1.584962501$.

> **CHANGE-OF-BASE RULE**
>
> For any positive real numbers x, a, and b, where $a \neq 1$ and $b \neq 1$,
>
> $$\log_a x = \frac{\log_b x}{\log_b a}.$$

This rule is proved by using the definition of logarithm to write $y = \log_a x$ in exponential form.

PROOF Let

$$y = \log_a x.$$
$$a^y = x \qquad \text{Change to exponential form.}$$
$$\log_b a^y = \log_b x \qquad \text{Take logarithms on both sides.*}$$
$$y \log_b a = \log_b x \qquad \text{Property 6 of logarithms}$$
$$y = \frac{\log_b x}{\log_b a} \qquad \text{Divide both sides by } \log_b a.$$
$$\log_a x = \frac{\log_b x}{\log_b a} \qquad \text{Substitute } \log_a x \text{ for } y.$$

Any positive number other than 1 can be used for base b in the change-of-base rule, but usually the only practical bases are e and 10, since calculators give logarithms only for these two bases.

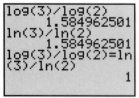

The third entry shows that the equation is true by returning a 1.

FIGURE 21

EXAMPLE 7 *Using the Change-of-Base Rule* Use the change-of-base rule to find an approximation for $\log_2 3$. Use a calculator, and apply the rule for both common and natural logarithms.

SOLUTION By the change-of-base rule,

$$\log_2 3 = \frac{\log 3}{\log 2} \quad \text{and} \quad \log_2 3 = \frac{\ln 3}{\ln 2}.$$

The display in Figure 21 shows that, in each case, $\log_2 3 \approx 1.584962501$. Notice that this supports our earlier graphical analysis. ◆

---

*This follows from the fact that the function $y = \log_a x$ is one-to-one.

For Group Discussion

1. Without using your calculator, determine the exact values of $\log_3 9$ and $\log_3 27$. Then, use the fact that 16 is between 9 and 27 to determine between what two consecutive integers $\log_3 16$ must lie. Finally, use the change-of-base rule to support your answer.
2. Without using your calculator, determine the exact values of $\log_5 \left(\frac{1}{5}\right)$ and $\log_5 1$. Then, use the fact that .68 is between $\frac{1}{5}$ and 1 to determine between what two consecutive integers $\log_5 .68$ must lie. Finally, use the change-of-base rule to support your answer.
3. Use your calculator and the change-of-base rule to support the results found analytically in Example 2: $\log_8 4 = \frac{2}{3}$ and $\log_{1/27} 81 = -\frac{4}{3}$.

5.2 EXERCISES   Tape 6

For each of the following statements, write an equivalent statement in logarithmic form.

1. $3^4 = 81$

2. $2^5 = 32$

3. $\left(\frac{1}{2}\right)^{-4} = 16$

4. $\left(\frac{2}{3}\right)^{-3} = \frac{27}{8}$

5. $10^{-4} = .0001$

6. $\left(\frac{1}{100}\right)^{-2} = 10,000$

7. $e^0 = 1$

8. $e^{1/2} = \sqrt{e}$

For each of the following statements, write an equivalent statement in exponential form.

9. $\log_6 36 = 2$

10. $\log_5 5 = 1$

11. $\log_{\sqrt{3}} 81 = 8$

12. $\log_4 \left(\frac{1}{64}\right) = -3$

13. $\log_{10} .001 = -3$

14. $\log_3 \sqrt[3]{9} = \frac{2}{3}$

15. $\log \sqrt{10} = .5$

16. $\ln e^6 = 6$

17. Explain in your own words what $\log_a x$ means.

18. In the expression $\log_a x$, why can't x be 0? Why can't x be negative?

Find each of the following logarithms using the method of Example 2. (You may wish to verify your answer by also using the change-of-base rule.)

19. $\log_5 125$

20. $\log_3 81$

21. $\log_6 \dfrac{1}{216}$

22. $\log_{1/4} 16$

23. $\log_{\sqrt{3}} 3^{12}$

24. $\log_{\sqrt[3]{5}} 25$

25. $\log_4 \dfrac{\sqrt[3]{4}}{2}$

26. $\log_9 \dfrac{\sqrt[4]{27}}{3}$

27. $\log_{1/4} \dfrac{16^2}{2^{-3}}$

28. Simplify each of the following.
 (a) $3^{\log_3 7}$
 (b) $4^{\log_4 9}$
 (c) $12^{\log_{12} 4}$
 (d) $a^{\log_a k}$ $(k > 0, a > 0, a \neq 1)$

29. Simplify each of the following.
 (a) $\log_3 3^{19}$
 (b) $\log_4 4^{17}$
 (c) $\log_{12} 12^{1/3}$
 (d) $\log_a \sqrt{a}$ $(a > 0, a \neq 1)$

30. Simplify each of the following.
 (a) $\log_3 1$
 (b) $\log_4 1$
 (c) $\log_{12} 1$
 (d) $\log_a 1$ $(a > 0, a \neq 1)$

Use a calculator to find a decimal approximation for each of the following common or natural logarithms.

31. log 43

32. log 1247

33. log .783

34. log .014

35. log 28^3

36. log (47×93)

37. ln 43

38. ln 1247

39. ln .783

40. ln .014

41. ln 28^3

42. ln (47×93)

43. **(a)** Use a calculator to find a decimal approximation of each of the following common logarithms: log 2.367, log 23.67, log 236.7, log 2367.

  (b) Write each of the following numbers in scientific notation: 2.367, 23.67, 236.7, 2367.

  (c) Compare the results in part (a) to the expressions in part (b). What similarities do you find? What differences do you find?

44. Use the definition of logarithm and a calculator to find the value that was stored in A to produce these screens.

  (a)
```
log(A)
      1.113943352
```

  (b)
```
ln(A)
         3.17805383
```

RELATING CONCEPTS (EXERCISES 45–48)

In Exercises 45–48, assume a > 1, and work them in sequential order.

45. If $f(x) = a^x$ has an inverse f^{-1}, sketch f and f^{-1} on the same set of axes. (Section 4.5)

46. If f^{-1} exists, find a rule for f^{-1} analytically. (Section 4.5)

47. If $a = e$, what is the rule for $f^{-1}(x)$?

48. If $a = 10$, what is the rule for $f^{-1}(x)$?

For each of the following substances, find the pH from the given hydronium ion $[H_3O^+]$ concentration.

49. grapefruit, 6.3×10^{-4}

50. limes, 1.6×10^{-2}

51. crackers, 3.9×10^{-9}

52. sodium hydroxide (lye), 3.2×10^{-14}

Find the hydronium ion $[H_3O^+]$ concentration for each of the following substances for the given pH.

53. soda pop, 2.7

54. wine, 3.4

55. beer, 4.8

56. drinking water, 6.5

Suppose that $2500 is invested in an account that pays interest compounded continuously. Find the amount of time that it would take for the account to grow to the given amount at the given rate. (See Example 4.)

57. $3000, at 3.75%

58. $3500, at 4.25%

59. $5000, at 5%

60. $5000, at 6%

Use the product, quotient, and power rules of logarithms to rewrite, if possible, each of the following logarithms. Assume that all variables represent positive real numbers.

61. $\log_3 \dfrac{2}{5}$

62. $\log_4 \dfrac{6}{7}$

63. $\log_2 \dfrac{6x}{y}$

64. $\log_3 \dfrac{4p}{q}$

65. $\log_5 \dfrac{5\sqrt{7}}{3}$

66. $\log_2 \dfrac{2\sqrt{3}}{5}$

67. $\log_4 (2x + 5y)$

68. $\log_6 (7m + 3q)$

69. $\log_k \dfrac{pq^2}{m}$

70. $\log_z \dfrac{x^5 y^3}{3}$

71. $\log_m \sqrt{\dfrac{5r^3}{z^5}}$

72. $\log_p \sqrt[3]{\dfrac{m^5 n^4}{t^2}}$

Use the product, quotient, and power rules of logarithms to rewrite each of the following as a single logarithm. Assume that all variables represent positive real numbers.

73. $\log_a x + \log_a y - \log_a m$

74. $(\log_b k - \log_b m) - \log_b a$

75. $2 \log_m a - 3 \log_m b^2$

76. $\dfrac{1}{2} \log_y p^3 q^4 - \dfrac{2}{3} \log_y p^4 q^3$

77. $2 \log_a (z - 1) + \log_a (3z + 2), z > 1$

78. $\log_b (2y + 5) - \dfrac{1}{2} \log_b (y + 3)$

79. $-\dfrac{2}{3} \log_5 5m^2 + \dfrac{1}{2} \log_5 25m^2$

80. $-\dfrac{3}{4} \log_3 16p^4 - \dfrac{2}{3} \log_3 8p^3$

Use the change-of-base rule to find an approximation for each of the following logarithms.

81. $\log_5 10$

82. $\log_9 12$

83. $\log_{15} 5$

84. $\log_{1/2} 3$

85. $\log_{100} 83$

86. $\log_{200} 175$

87. $\log_{2.9} 7.5$

88. $\log_{5.8} 12.7$

RELATING CONCEPTS (EXERCISES 89–94)

Work Exercises 89–94 in sequential order.

89. Use the terminology of Chapter 2 to explain how the graph of $y = -3^x + 7$ can be obtained from the graph of $y = 3^x$. (Sections 2.2 and 2.3)

90. Graph $y_1 = 3^x$ and $y_2 = -3^x + 7$ in a window of $[-5, 5]$ by $[-10, 10]$ to support your answer in Exercise 89.

91. Use the capabilities of your calculator to find an approximation for the x-intercept of the graph of y_2 in Exercise 90. (Section 1.5)

92. Solve $0 = -3^x + 7$ for x, expressing x in terms of base 3 logarithms.

93. Use the change-of-base rule to find an approximation for the solution of the equation in Exercise 92.

94. Compare your results in Exercises 91 and 93.

95. Prove Property 5 of logarithms.

96. Prove Property 6 of logarithms.

5.3 INTRODUCTION TO LOGARITHMIC FUNCTIONS

Preliminary Considerations ◆ Graphs of Logarithmic Functions ◆ Connections: Earlier Results Supported by Logarithmic Function Graphs

PRELIMINARY CONSIDERATIONS

The function $f(x) = a^x$, $a > 1$, is increasing on its entire domain. Likewise, if $0 < a < 1$, the function is decreasing on its entire domain. Therefore, for all allowable bases of the exponential function $f(x) = a^x$, the graph passes the horizontal line test and is one-to-one. As a result, f has an inverse.

We can find the rule for f^{-1} analytically, using the steps first described in Section 4.5.

$f(x) = a^x$	Exponential function
$y = a^x$	Replace $f(x)$ with y.
$x = a^y$	Exchange x and y to obtain an equation for the inverse.
$y = \log_a x$	Convert from exponential form to logarithmic form.
$f^{-1}(x) = \log_a x$	$y = f^{-1}(x)$

This final equation indicates that the logarithmic function with base a is the inverse of the exponential function with base a. To confirm this, we show that $(f \circ f^{-1})(x) = x$ and $(f^{-1} \circ f)(x) = x$:

$$(f \circ f^{-1})(x) = f[f^{-1}(x)] = a^{\log_a x} = x \quad \text{(from Section 5.2)}$$
$$(f^{-1} \circ f)(x) = f^{-1}[f(x)] = \log_a a^x = x \quad \text{(from Section 5.2)}.$$

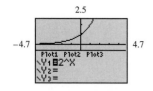

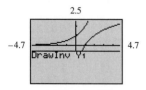

When directed to draw the inverse of $Y_1 = 2^x$, the calculator provides the graph of $y = \log_2 x$.

RELATIONSHIP BETWEEN BASE a EXPONENTIAL AND LOGARITHMIC FUNCTIONS

The functions

$$f(x) = a^x \quad \text{and} \quad g(x) = \log_a x$$

are inverses.

GRAPHS OF LOGARITHMIC FUNCTIONS

Recall from Section 4.5 that the graph of the inverse of a one-to-one function can be obtained by reflecting the graph of the function across the line $y = x$. Figure 22 shows both traditional and calculator-generated graphs of a pair of inverse functions, $y = 2^x$ and $y = \log_2 x$. These are typical shapes for such graphs where $a > 1$.

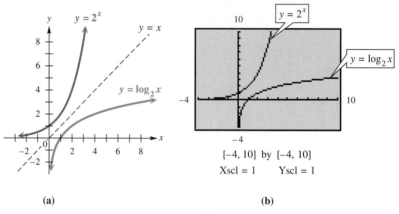

(a)   (b)

FIGURE 22

Figure 23 shows both traditional and calculator-generated graphs of $y = \left(\frac{1}{2}\right)^x$ and $y = \log_{1/2} x$. These are typical shapes for such graphs where $0 < a < 1$.

The most important logarithmic functions are $y = \ln x$ (base e) and $y = \log x$ (base 10). These are easily graphed on a graphing calculator, as seen in Figures 24 and 25.

Because e and 10 are the only logarithmic bases that appear on a graphing calculator, if we wish to graph a logarithmic function for some other base, we must use the change-of-base rule. For example, to graph $y = \log_2 x$, we may graph either

$$y = \frac{\log x}{\log 2} \qquad \text{or} \qquad y = \frac{\ln x}{\ln 2}.$$

Similarly, if we wish to graph $y = \log_{1/2} x$, we may graph either

$$y = \frac{\log x}{\log \frac{1}{2}} \qquad \text{or} \qquad y = \frac{\ln x}{\ln \frac{1}{2}}.$$

These two graphs can be seen in Figures 22(b) and 23(b).

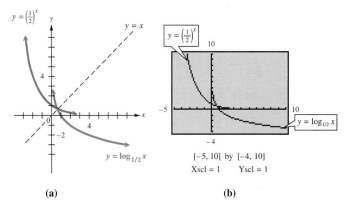

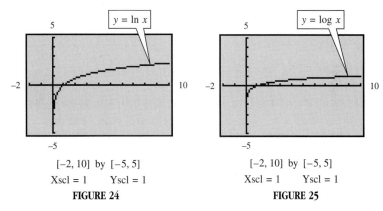

[−5, 10] by [−4, 10]
Xscl = 1 Yscl = 1

(a) (b)

FIGURE 23

TECHNOLOGY NOTE

Just as we saw in Section 5.1 that it is difficult to interpret from a calculator screen the behavior of the graph of an exponential function as it approaches its horizontal asymptote, a similar problem occurs for graphs of logarithmic functions near their vertical asymptotes. Refer to Figures 24 and 25, and be aware that there are no endpoints, and that as x approaches 0 from the right, y approaches $-\infty$.

[−2, 10] by [−5, 5]
Xscl = 1 Yscl = 1

FIGURE 24

[−2, 10] by [−5, 5]
Xscl = 1 Yscl = 1

FIGURE 25

We now summarize information about the graphs of logarithmic functions.

LOGARITHMIC FUNCTION ($a > 1$)

$f(x) = \log_a x$, $a > 1$ (Figure 26) Domain: $(0, \infty)$ Range: $(-\infty, \infty)$

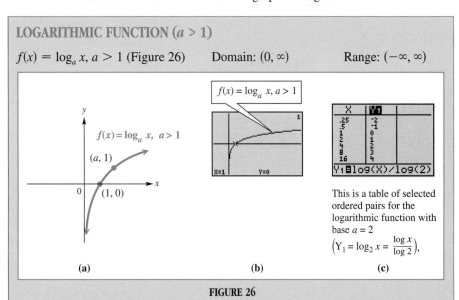

This is a table of selected ordered pairs for the logarithmic function with base $a = 2$ $\left(Y_1 = \log_2 x = \frac{\log x}{\log 2}\right)$.

(a) (b) (c)

FIGURE 26

The logarithmic function $f(x) = \log_a x$, $a > 1$, increases on its entire domain, and is continuous as well. The y-axis is a vertical asymptote as x approaches 0 from the right. The x-intercept is 1, and there is no y-intercept. It is always concave down, and there are no extrema.

LOGARITHMIC FUNCTION ($0 < a < 1$)

$f(x) = \log_a x$, $0 < a < 1$ (Figure 27)   Domain: $(0, \infty)$ Range: $(-\infty, \infty)$

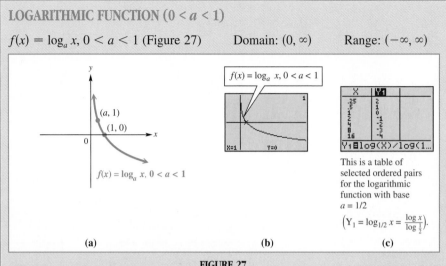

This is a table of selected ordered pairs for the logarithmic function with base $a = 1/2$

$\left(Y_1 = \log_{1/2} x = \dfrac{\log x}{\log \frac{1}{2}} \right).$

(a)   (b)   (c)

FIGURE 27

The logarithmic function $f(x) = \log_a x$, $0 < a < 1$, decreases on its entire domain, and is continuous as well. The y-axis is a vertical asymptote as x approaches 0 from the right. The x-intercept is 1, and there is no y-intercept. It is always concave up, and there are no extrema.

A function of the form $y = \log_a f(x)$ is defined only for values for which $f(x) > 0$. The first example shows how the domain of a function defined by a logarithm is determined.

EXAMPLE 1   *Determining the Domain of a Function Defined by a Logarithm*
Find the domain of each of the following functions analytically.

(a) $f(x) = \log_2 (x - 1)$ **(b)** $f(x) = \log_3 x - 1$
(c) $f(x) = \log_3 |x|$ **(d)** $f(x) = \ln (x^2 - 4)$

SOLUTION

(a)   For this function to be defined, we must have $x - 1 > 0$, or equivalently, $x > 1$. The domain is $(1, \infty)$.

(b)   Since we are interested in the logarithm to the base 3 of x (and not $x - 1$), we must have $x > 0$. The domain is $(0, \infty)$.

(c) $|x| > 0$ is true for all real numbers x except 0. Therefore, the domain is $(-\infty, 0) \cup (0, \infty)$, or equivalently, $\{x | x \neq 0\}$.

(d)   To solve $x^2 - 4 > 0$ analytically, we use a sign graph, first introduced in Section 3.3. Factor $x^2 - 4$ as $(x + 2)(x - 2)$, and then determine the signs of the factors in the intervals $(-\infty, -2)$, $(-2, 2)$, and $(2, \infty)$. See Figure 28. The product $(x + 2)(x - 2)$ is positive in the intervals $(-\infty, -2)$ and $(2, \infty)$. The domain is the union of these two intervals: $(-\infty, -2) \cup (2, \infty)$.   ◆

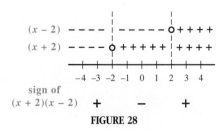

FIGURE 28

EXAMPLE 2 *Analyzing the Graph of a Function Defined by a Logarithm*
Use the terminology of Chapter 2 to explain how the graph of $y = \log_2(x - 1)$ can be obtained from the graph of $y = \log_2 x$. Discuss other important features of the graph.

SOLUTION Because the argument is $x - 1$, the graph of $y = \log_2(x - 1)$ is obtained from the graph of $y = \log_2 x$ by shifting 1 unit to the right. The vertical asymptote also moves 1 unit to the right, so its equation is $x = 1$. The x-intercept is 2. The domain is $(1, \infty)$, as found in Example 1(a), and the range is $(-\infty, \infty)$. It is always increasing. See Figure 29, which shows both traditional and calculator-generated graphs. (The graph in (b) was obtained by using $y = \frac{\log(x - 1)}{\log 2}$.) ◆

This table provides support for the result in Example 2: the domain of $y = \log_2(x - 1)$ (entered here as

$$Y_1 = \log(x - 1)/\log 2$$

consists of real numbers greater than 1.

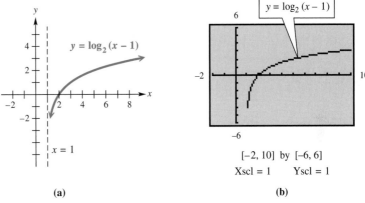

(a)   **(b)**

FIGURE 29

EXAMPLE 3 *Analyzing the Graph of a Function Defined by a Logarithm*
Repeat the procedure of Example 2 for $y = \log_3 x - 1$.

SOLUTION Here, the 1 is subtracted from $\log_3 x$, so the effect is that the graph of $y = \log_3 x$ is shifted down 1 unit. The vertical asymptote, the y-axis, is not affected. The x-intercept of $y = \log_3 x - 1$ is 3, and there is no y-intercept. The domain is $(0, \infty)$, as found in Example 1(b), and the range is $(-\infty, \infty)$. It is always increasing. See Figure 30 for both types of graphs. (The graph in (b) was obtained by using $y = \frac{\log x}{\log 3} - 1$.) ◆

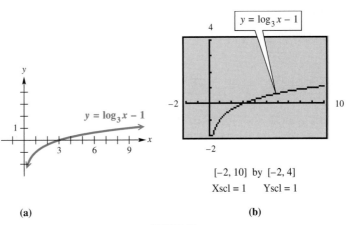

(a)   **(b)**

FIGURE 30

EXAMPLE 4   *Analyzing the Graph of a Function Defined by a Logarithm*
Discuss the symmetry exhibited by the graph of the function $f(x) = \log_3 |x|$.

SOLUTION   Recall that for all real numbers x, $|-x| = |x|$. Now, we find $f(-x)$.

$$f(-x) = \log_3 |-x| = \log_3 |x| = f(x).$$

Because $f(-x) = f(x)$, the graph is symmetric with respect to the y-axis, and f is an even function. It consists of two parts, as shown in Figure 31. The domain is $(-\infty, 0) \cup (0, \infty)$, as determined in Example 1(c), and the y-axis serves both as a vertical asymptote and the axis of symmetry.   ◆

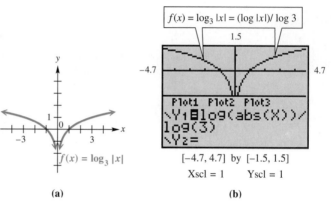

FIGURE 31

EXAMPLE 5   *Supporting an Analytic Domain Determination Graphically*
Use a graph to support the result obtained in Example 1(d): the domain of $f(x) = \ln(x^2 - 4)$ is $(-\infty, -2) \cup (2, \infty)$.

SOLUTION   Figure 32 shows both types of graphs of this function. Notice that for $[-2, 2]$, the graph does not exist, as expected.   ◆

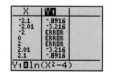

This table provides support for the result in Example 5.

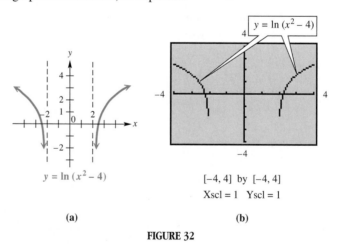

FIGURE 32

Caution   It would be easy to misinterpret the calculator-generated graphs in Figures 31(b) and 32(b), by thinking that there are "endpoints" on the two branches. In Figure 32(b), the lines $x = -2$ and $x = 2$ are vertical asymptotes, and as x approaches -2 from the left, $f(x) \to -\infty$, while as x approaches 2 from the right, $f(x) \to -\infty$. Here, we see another reason for learning the concepts. Calculator-generated graphs are limited in their resolution, and may not show important details in certain windows.

For Group Discussion

1. Use a calculator to evaluate $f(2.1)$, $f(2.01)$, $f(2.001)$, and $f(2.0001)$ for $f(x) = \ln(x^2 - 4)$. Describe what happens as x approaches 2 from the right. How does this tie in with the discussion in the preceding Caution?
2. Repeat Item 1 for $f(-2.1)$, $f(-2.01)$, $f(-2.001)$, and $f(-2.0001)$.

CONNECTIONS: EARLIER RESULTS SUPPORTED BY LOGARITHMIC FUNCTION GRAPHS

Some of the results obtained in the previous two sections of this chapter can be supported with graphs of logarithmic functions. The remaining examples in this section illustrate this.

EXAMPLE 6 *Supporting Graphically a Result Determined Analytically* In Example 2(a) of Section 5.2, we determined analytically $\log_8 4 = \frac{2}{3}$. Support this result graphically, using the graph of $y = \log_8 x$.

SOLUTION We graph $y = \log_8 x$ on a graphing calculator by using the change-of-base rule: $\log_8 x = \frac{\log x}{\log 8}$. Then, we evaluate y for $x = 4$. Figure 33 indicates that when $x = 4$, y is a decimal approximation for $\frac{2}{3}$, supporting our analytic result. ◆

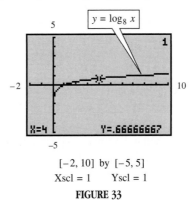

[−2, 10] by [−5, 5]
Xscl = 1   Yscl = 1
FIGURE 33

EXAMPLE 7 *Supporting Graphically a Result Determined by a Calculator Function* In Figure 16(a), we show a calculator approximation of $\log 6$: $\log 6 \approx .7781512504$. Support this result, using the graph of the common logarithm function $y = \log x$.

SOLUTION By graphing $y = \log x$ and letting $x = 6$, we find that $y \approx .77815125$, supporting the result found earlier. See Figure 34. ◆

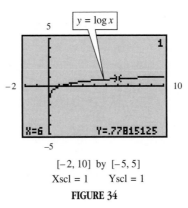

[−2, 10] by [−5, 5]
Xscl = 1   Yscl = 1
FIGURE 34

EXAMPLE 8 *Using a Property of Logarithms to Describe a Translation of a Graph* In Example 5(a) of Section 5.2, we used the product rule for logarithms to write log $8x$ as log 8 + log x. Use this result to explain how the graph of $y = \log 8x$ may be obtained from the graph of $y = \log x$ by a translation.

SOLUTION Because log $8 \approx .90309$, the graph of $y_2 = \log 8x = \log 8 + \log x$ is obtained by shifting the graph of $y_1 = \log x$ up log $8 \approx .90309$ unit. Figure 35 shows the two graphs. To support our answer, we choose an arbitrary x-value, say 5, and determine the corresponding y-value for each graph. If we subtract log 5 from log $8(5) = $ log 40, using the common logarithm function, we get the same result, approximately .903089987, shown in Figure 36. ◆

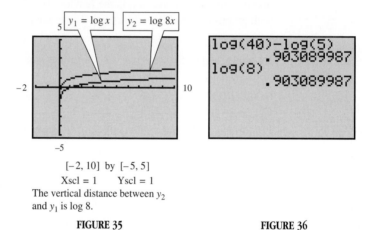

[−2, 10] by [−5, 5]
Xscl = 1   Yscl = 1
The vertical distance between y_2
and y_1 is log 8.

FIGURE 35   **FIGURE 36**

EXAMPLE 9 *Reinforcing the Inverse Relationship between Exponential and Logarithmic Functions* The one-to-one function $f(x) = -2^x + 3$ is discussed and graphed in Example 3 of Section 5.1 and the corresponding figure (Figure 9). Find f^{-1} analytically, graph both f and f^{-1} in the same window, and discuss the inverse relationships exhibited by the functions and their graphs.

SOLUTION To find f^{-1}, we use the analytic method.

$f(x) = -2^x + 3$	Given
$y = -2^x + 3$	Replace $f(x)$ with y.
$x = -2^y + 3$	Exchange x and y to obtain the inverse relationship.
$2^y = -x + 3$	Add 2^y and subtract x.
$y = \log_2 (-x + 3)$	Write in logarithmic form.
$f^{-1}(x) = \log_2 (-x + 3)$	Replace y with $f^{-1}(x)$.

Figure 37 shows both f and f^{-1} graphed. We now list some of the features of these inverses.

	Domain	Range	x-intercept	y-intercept	Asymptote
$f(x) = -2^x + 3$	$(-\infty, \infty)$	$(-\infty, 3)$	$\log_2 3 \approx 1.58$	2	horizontal: $y = 3$
$f^{-1}(x) = \log_2 (-x + 3)$	$(-\infty, 3)$	$(-\infty, \infty)$	2	$\log_2 3 \approx 1.58$	vertical: $x = 3$

Notice in the chart how the roles of x and y are reversed in f and f^{-1}. ◆

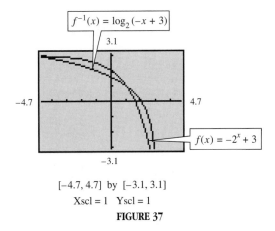

$$f^{-1}(x) = \log_2(-x + 3)$$

$$f(x) = -2^x + 3$$

[−4.7, 4.7] by [−3.1, 3.1]
Xscl = 1   Yscl = 1

FIGURE 37

WHAT WENT WRONG A student wanted to support the power rule for logarithms (Property 6 from Section 5.2): $\log_a x^r = r \log_a x$. The student defined Y_1 and Y_2, as shown in the screen on the top left, and expected the two graphs to be the same. However, the graph of Y_1, as seen in the top right screen, was different from the graph of Y_2, as seen in the bottom screen.

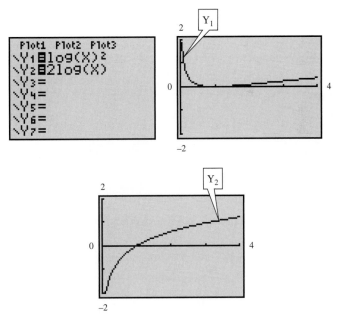

1. What went wrong?
2. How can the student change the input for Y_1 to obtain the same graph as that of Y_2?

5.3   EXERCISES  Tape 6

The graph of an exponential function f is given, with three points labeled. Sketch the graph of
f^{-1} by hand, labeling three points on its graph. For f^{-1}, also state the domain, the range,
whether it increases or decreases on its domain, and the equation of its vertical asymptote.

1.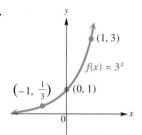

2.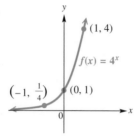

3.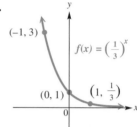

4.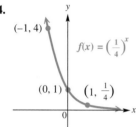

5.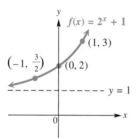

6.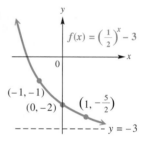

7. In Exercises 1–6, each function f is an exponential function. Therefore, each function f^{-1} is
a(n) _____ function.

8. Use the table shown, along with the display at the bottom of the screen, to give the value of
the logarithm.

(a) $\log_4 8$

(b) $\log_3 81$

(c) $\log_2 .25$

X	Y1
5	1.161
6	1.2925
7	1.4037
8	1.5
9	1.585
10	1.661
11	1.7297

Y1=log(X)/log(4)

X	Y1
78	3.9656
79	3.9772
80	3.9887
81	4
82	4.0112
83	4.0222
84	4.0331

Y1=log(X)/log(3)

X	Y1
.25	-2
.26	-1.943
.27	-1.889
.28	-1.837
.29	-1.786
.3	-1.737
.31	-1.69

Y1=log(X)/log(2)

Find the domain of each of the following logarithmic functions analytically. You may wish to
support your answer graphically.

9. $y = \log 2x$

10. $y = \log \dfrac{x}{3}$

11. $y = \log(3x + 7)$

12. $y = \log(3 - 6x)$

13. $y = \ln(x^2 + 7)$

14. $y = \ln(-x^2 - 4)$

15. $y = \log_4(x^2 - 4x - 21)$

16. $y = \log_6(2x^2 - 7x - 4)$

17. $y = \log(x^2 + x - 1)$

18. $y = \ln(x^2 + x + 1)$

19. $y = \log(x^3 - x)$

20. $y = \log\left(\dfrac{x + 3}{x - 4}\right)$

In Exercises 21–28, match the correct graph to the given equation. You should do this by your
knowledge of graphs, and not by generating your own graph on your calculator.

21. $y = e^x + 3$

22. $y = e^x - 3$

23. $y = e^{x+3}$

24. $y = e^{x-3}$

25. $y = \ln x + 3$

26. $y = \ln x - 3$

27. $y = \ln (x - 3)$

28. $y = \ln (x + 3)$

A.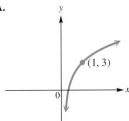

B.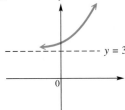

C.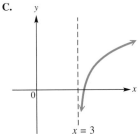

D.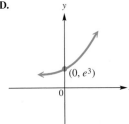

E.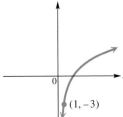

F.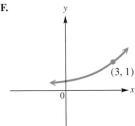

G.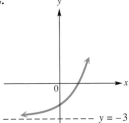

H.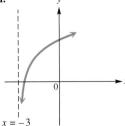

Use the terminology of Chapter 2 to explain how the graph of the given function can be obtained from the graph of $y = \log_2 x$.

29. $y = \log_2 (x + 4)$

30. $y = \log_2 (x - 6)$

31. $y = 3 \log_2 x + 1$

32. $y = -4 \log_2 x - 8$

33. $y = \log_2 (-x) + 1$

34. $y = -\log_2 (-x)$

RELATING CONCEPTS (EXERCISES 35–40)

Work Exercises 35–40 in order. They apply to the function $f(x) = \log_4 (2x^2 - x)$.

35. Determine analytically the domain of f. (Section 3.3)

36. Use the change-of-base rule to express, in terms of common logarithms, the equation you would use to graph f on your calculator. (Section 5.3)

37. Determine analytically the x-intercepts of the graph of f. (Section 3.3)

38. Give the equations of the vertical asymptotes of the graph of f. (Section 5.3)

39. Explain why the graph of f has no y-intercept. (Section 5.3)

40. Graph f in the window $[-2.5, 2.5]$ by $[-5, 2.5]$. Based on this graph, give the solution sets of $f(x) = 0$, $f(x) < 0$, and $f(x) > 0$. (Sections 1.5, 1.6)

41. Graph $y = \log x^2$ and $y = 2 \log x$ on separate viewing screens. It would seem, at first glance, that by applying the power rule for logarithms, these graphs should be the same. Are they? If not, why not? (*Hint:* Consider the domain in each case.)

42. Graph $f(x) = \log_3 |x|$ in the window $[-4, 4]$ by $[-4, 4]$, and compare it to the traditional graph in Figure 31(a). How might one easily misinterpret the domain of the function simply by observing the calculator-generated graph? What is the domain of this function?

*In Exercises 43–46, evaluate the logarithm in three ways: (**a**) Use the method of Example 2 of Section 5.2 to find the exact value analytically. (**b**) Support the result of (a) by using the change-of-base rule and common logarithms on your calculator. (**c**) Support the result of (a) by locating the appropriate point on the graph of the function $y = \log_a x$.*

43. $\log_9 27$ **44.** $\log_4\left(\dfrac{1}{8}\right)$ **45.** $\log_{16}\left(\dfrac{1}{8}\right)$ **46.** $\log_2\sqrt{8}$

Find approximations for the following common and natural logarithms by using a graph or a table. Then, support your answer by using the common or natural logarithm key of your calculator.

47. $\log 7$ **48.** $\log 9$ **49.** $\ln 7$

50. $\ln 9$ **51.** $\log 14.7$ **52.** $\ln 14.7$

In Exercises 53–56, a one-to-one exponential function f is given. Find f^{-1} analytically. Graph both f and f^{-1} in the same viewing window, and discuss the inverse relationship exhibited by the functions and their graphs. (See Example 9.)

53. $f(x) = 4^x - 3$ **54.** $f(x) = \left(\dfrac{1}{2}\right)^x - 5$ **55.** $f(x) = -10^x + 4$ **56.** $f(x) = -e^x + 6$

57. The graph of $y = 10^x$ is shown with the coordinates of a point displayed at the bottom of the screen. Write the *logarithmic* equation associated with the display.

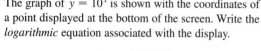
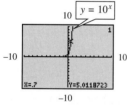

58. The graph of $y = e^x$ is shown with the coordinates of a point displayed at the bottom of the screen. Write the *logarithmic* equation associated with the display.

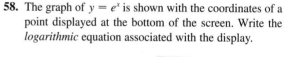
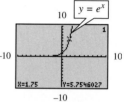

59. Why do the values for Y_1 show ERROR for *x*-values greater than or equal to 4?

60. Based on the table, give an approximation for $\log_7 24$.

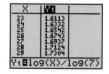

5.4 EXPONENTIAL AND LOGARITHMIC EQUATIONS AND INEQUALITIES

Preliminary Considerations ◆ Exponential Equations and Inequalities (Type 2) ◆ Logarithmic Equations and Inequalities ◆ Equations and Inequalities Involving Both Exponentials and Logarithms ◆ Formulas Involving Exponentials and Logarithms

PRELIMINARY CONSIDERATIONS

In this section, we will examine methods of solving equations involving exponential and logarithmic expressions. We have solved some types of these in the earlier sections. For example, in Section 5.1 we learned how to solve Type 1 exponential equations—

those involving expressions that could easily be written as powers of the same base. We also saw how to solve exponential equations involving bases of e and 10.

General methods for solving exponential and logarithmic equations depend on the following properties. These properties follow from the fact that exponential and logarithmic functions are, in general, one-to-one. Property 1 was used in Section 5.1 to solve Type 1 exponential equations.

PROPERTIES OF LOGARITHMIC AND EXPONENTIAL FUNCTIONS

For $b > 0$ and $b \neq 1$:

1. $b^x = b^y$ if and only if $x = y$.
2. If $x > 0$ and $y > 0$,
$$\log_b x = \log_b y \quad \text{if and only if} \quad x = y.$$

EXPONENTIAL EQUATIONS AND INEQUALITIES (TYPE 2)

A Type 2 exponential equation or inequality is one in which the exponential expressions cannot easily be written as powers of the same base. Examples of Type 2 equations are $7^x = 12$ and $2^{3x+1} = 3^{4-x}$. The general strategy in solving these equations is to use Property 2 by taking the same base logarithm of both sides (usually either common or natural) and then applying the power rule for logarithms to eliminate the variable exponents. Then, the equation is solved using familiar algebraic techniques.

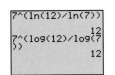

Raising 7 to the power $\frac{\ln 12}{\ln 7}$ or to the power $\frac{\log 12}{\log 7}$ gives a result of 12. See Example 1.

An equation-solving routine supports the result of Example 1. (Some graphing calculators offer more than one option for such routines.)

EXAMPLE 1 *Solving a Type 2 Exponential Equation* Solve the equation $7^x = 12$ analytically. Support the result with a graph.

SOLUTION ANALYTIC In Section 5.1, we saw that Property 1 cannot be used to solve this equation, so we apply Property 2. While any appropriate base b can be used to apply Property 2, the best practical base to use is base 10 or base e. Taking base e (natural) logarithms of both sides gives

$$7^x = 12$$
$$\ln 7^x = \ln 12$$
$$x \ln 7 = \ln 12 \qquad \text{Property 6 of logarithms}$$
$$x = \frac{\ln 12}{\ln 7}. \qquad \text{Divide by ln 7.}$$

The expression $\frac{\ln 12}{\ln 7}$ is the *exact* solution of $7^x = 12$. Had we used common logarithms instead, the solution would have the form $\frac{\log 12}{\log 7}$. In either case, we can use a calculator to find a decimal approximation for the exact solution. We find that to the nearest thousandth,

$$\frac{\ln 12}{\ln 7} = \frac{\log 12}{\log 7} \approx 1.277.$$

The solution set can be expressed with the exact solution as

$$\left\{ \frac{\ln 12}{\ln 7} \right\} \qquad \text{or} \qquad \left\{ \frac{\log 12}{\log 7} \right\},$$

while it is expressed as $\{1.277\}$ with an approximate solution.

GRAPHICAL The solution can be supported graphically by graphing $y_1 = 7^x$ and $y_2 = 12$, and then using the intersection-of-graphs method. The x-coordinate of the point of intersection is approximately 1.277, as seen in Figure 38(a). Figure 38(b) illustrates the *x*-intercept method of solution. The *x*-intercept of $y = 7^x - 12$ is also approximately 1.277. ◆

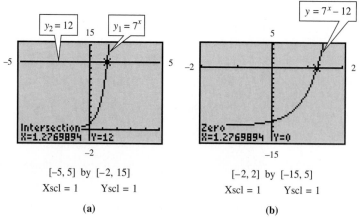

[-5, 5] by [-2, 15]
Xscl = 1   Yscl = 1
(a)

[-2, 2] by [-15, 5]
Xscl = 1   Yscl = 1
(b)

FIGURE 38

EXAMPLE 2   *Solving a Type 2 Exponential Inequality Graphically*

(a)   Use Figure 38(a) to solve $7^x < 12$.   **(b)**   Use Figure 38(b) to solve $7^x > 12$.

SOLUTION

(a)   Because the graph of $y_1 = 7^x$ is below the graph of $y_2 = 12$ for all *x*-values less than $\frac{\ln 12}{\ln 7} \approx 1.277$, we can express the solution set as

$$\left(-\infty, \frac{\ln 12}{\ln 7}\right) \quad \text{or} \quad (-\infty, 1.277).$$

(b)   This inequality is equivalent to $7^x - 12 > 0$. Because the graph of $y = 7^x - 12$ is above the *x*-axis for all values of *x* greater than $\frac{\ln 12}{\ln 7} \approx 1.277$, we can express the solution set as

$$\left(\frac{\ln 12}{\ln 7}, \infty\right) \quad \text{or} \quad (1.277, \infty). \quad ◆$$

EXAMPLE 3   *Solving a Type 2 Exponential Equation*   Solve the equation $2^{3x+1} = 3^{4-x}$ analytically, and support the solution graphically.

SOLUTION   ANALYTIC   To solve this equation analytically, we take logarithms on both sides, use the power rule, and solve the resulting linear equation.

$$2^{3x+1} = 3^{4-x} \qquad \text{Given equation}$$

$$\log 2^{3x+1} = \log 3^{4-x} \qquad \text{Take common logarithms on both sides.}$$

$$(3x + 1) \log 2 = (4 - x) \log 3 \qquad \text{Use the power rule of logarithms.}$$

$$3x \log 2 + \log 2 = 4 \log 3 - x \log 3 \qquad \text{Distributive property}$$

$$3x \log 2 + x \log 3 = 4 \log 3 - \log 2 \qquad \text{Transform so that all } x\text{-terms are on one side.}$$

$$x(3 \log 2 + \log 3) = 4 \log 3 - \log 2 \qquad \text{Factor out } x.$$

$$x = \frac{4 \log 3 - \log 2}{3 \log 2 + \log 3} \qquad \text{Divide by } 3 \log 2 + \log 3.$$

This expression is the exact solution in terms of common logarithms. A more compact form of this exact solution can be found by using properties of logarithms.

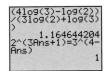

See if you can determine how this screen supports the result in Example 3.

$$x = \frac{\log 3^4 - \log 2}{\log 2^3 + \log 3}$$ Power rule

$$x = \frac{\log 81 - \log 2}{\log 8 + \log 3}$$ $3^4 = 81; \, 2^3 = 8$

$$x = \frac{\log \frac{81}{2}}{\log 24}$$ Quotient and product rules

The solution set using the exact value is $\left\{ \frac{\log(81/2)}{\log 24} \right\}$, while a calculator approximation of the solution is 1.165, giving the solution set $\{1.165\}$.

GRAPHICAL   This solution can be supported graphically by locating the point of intersection of the graphs of $y_1 = 2^{3x+1}$ and $y_2 = 3^{4-x}$, and finding the x-coordinate of the point. As seen in Figure 39, the x-coordinate is approximately 1.165, supporting our analytic solution.   ◆

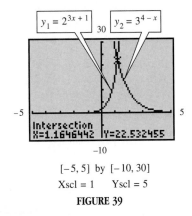

$[-5, 5]$ by $[-10, 30]$
Xscl = 1   Yscl = 5
FIGURE 39

For Group Discussion

1. Use Figure 39 to determine the solution sets of
$$2^{3x+1} < 3^{4-x} \quad \text{and} \quad 2^{3x+1} > 3^{4-x}.$$

2. Discuss the difference between the exact solution and an approximate solution of an exponential equation. In general, can you find an exact solution from a graph?

LOGARITHMIC EQUATIONS AND INEQUALITIES

The next examples show how to solve logarithmic equations and associated inequalities. The properties of logarithms given in Section 5.2 are useful here, as is Property 2 from this section. As we shall see, it is important to note the domain of the variable in the original form of a logarithmic equation, so that extraneous values can be rejected if they appear.

EXAMPLE 4   *Solving a Logarithmic Equation*   Solve
$$\log_3 (x + 6) - \log_3 (x + 2) = \log_3 x$$
analytically. Support the result graphically.

SOLUTION   ANALYTIC   First, we note that x must be positive, since any nonpositive value of x would cause $\log_3 x$ to be undefined. Using the quotient property of logarithms, we can rewrite the equation as
$$\log_3 \frac{x + 6}{x + 2} = \log_3 x.$$

Now the equation is in the proper form to use Property 2.

$$\frac{x + 6}{x + 2} = x \qquad \text{Property 2}$$

$$x + 6 = x(x + 2) \qquad \text{Multiply by } x + 2.$$

$$x + 6 = x^2 + 2x \qquad \text{Distributive property}$$

$$x^2 + x - 6 = 0 \qquad \text{Get 0 on one side.}$$

$$(x + 3)(x - 2) = 0 \qquad \text{Use the zero-product property.}$$

$$x = -3 \qquad \text{or} \qquad x = 2$$

The negative solution ($x = -3$) cannot be used because it is not in the domain of $\log_3 x$ in the original equation. For this reason, the only valid solution is the positive number 2, giving the solution set $\{2\}$.

GRAPHICAL   Figure 40 shows that the x-coordinate of the point of intersection of the graphs of $y_1 = \log_3 (x + 6) - \log_3 (x + 2)$ and $y_2 = \log_3 x$ is 2, supporting our analytic solution. Notice that the graphs do not intersect when $x = -3$, further supporting our conclusion that -3 is an extraneous value. ◆

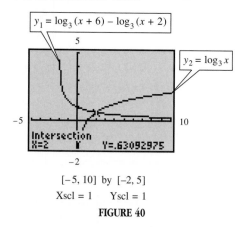

$[-5, 10]$ by $[-2, 5]$
Xscl $= 1$ Yscl $= 1$
FIGURE 40

Note   The equation in Example 4 could also have been solved by transforming so that all logarithmic terms appeared on one side. Then, by applying the properties of logarithms from Section 5.2 and rewriting in exponential form, the same solution set would result.

EXAMPLE 5   *Solving a Logarithmic Equation*   Solve

$$\log(3x + 2) + \log(x - 1) = 1$$

analytically, and support the solution graphically.

SOLUTION   ANALYTIC   Recall from Section 5.2 that $\log x$ means $\log_{10} x$.

$$\log(3x + 2) + \log(x - 1) = 1 \qquad \text{Given equation}$$

$$\log(3x + 2)(x - 1) = 1 \qquad \text{Product rule}$$

$$(3x + 2)(x - 1) = 10^1 \qquad \text{Exponential form}$$

$$3x^2 - x - 2 = 10$$

$$3x^2 - x - 12 = 0 \qquad \text{Standard form}$$

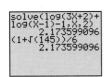

The approximation for $\frac{1 + \sqrt{145}}{6}$ and the solution obtained by the equation solver agree, further supporting the result of Example 5.

Now, use the quadratic formula with $a = 3$, $b = -1$, and $c = -12$ to get

$$x = \frac{1 \pm \sqrt{1 + 144}}{6}.$$

If $x = (1 - \sqrt{145})/6$, then $x - 1 < 0$; therefore, $\log(x - 1)$ is not acceptable, and this proposed solution must be discarded, giving the solution set

$$\left\{ \frac{1 + \sqrt{145}}{6} \right\}.$$

GRAPHICAL To support this solution graphically, we first find that $(1 + \sqrt{145})/6 \approx 2.174$ by using the arithmetic and square root functions of a calculator. Then we can choose to use the x-intercept method of solution, and graph $y = \log(3x + 2) + \log(x - 1) - 1$. As seen in Figure 41, the x-intercept agrees with the earlier approximation. ◆

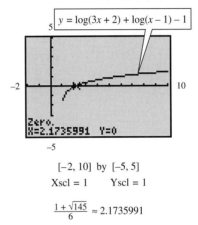

[-2, 10] by [-5, 5]
Xscl = 1   Yscl = 1

$$\frac{1 + \sqrt{145}}{6} \approx 2.1735991$$

FIGURE 41

For Group Discussion   Use Figure 41 to answer each of the following items.

1. Give the exact solution set of $\log(3x + 2) + \log(x - 1) - 1 \geq 0$.
2. Give the exact solution set of $\log(3x + 2) + \log(x - 1) - 1 \leq 0$.
(*Hint:* Pay attention to the domain.)

EQUATIONS AND INEQUALITIES INVOLVING BOTH EXPONENTIALS AND LOGARITHMS

We now look at equations that involve both exponentials and logarithms.

EXAMPLE 6 *Solving a Composite Exponential Equation*   Solve $e^{-2\ln x} = \frac{1}{16}$ analytically. Support the solution graphically.

SOLUTION ANALYTIC   Use a property of logarithms to rewrite the exponent on the left side of the equation.

$$e^{-2\ln x} = \frac{1}{16}$$

$$e^{\ln x^{-2}} = \frac{1}{16} \qquad \text{Property 6 of logarithms}$$

$$x^{-2} = \frac{1}{16} \qquad a^{\log_a x} = x$$

$$x^{-2} = 4^{-2} \qquad \tfrac{1}{16} = \tfrac{1}{4^2} = 4^{-2}; -4 \text{ is not a valid base, since } -4 < 0,$$
$$\text{and } x > 0.$$

$$x = 4 \qquad \text{Property 1}$$

An analytic check shows that 4 is indeed a solution.

GRAPHICAL   To support graphically, we show that $y = e^{-2\ln x} - \frac{1}{16}$ has x-intercept 4. See Figure 42. The solution set is $\{4\}$. The associated inequalities may be solved graphically, using Figure 42. Exercise 15 requires these solutions.   ◆

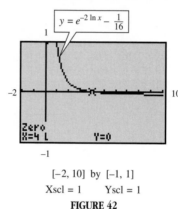

$$[-2, 10] \text{ by } [-1, 1]$$
$$\text{Xscl} = 1 \qquad \text{Yscl} = 1$$
FIGURE 42

EXAMPLE 7   *Solving a Composite Logarithmic Equation*   Solve
$$\ln e^{\ln x} - \ln (x - 3) = \ln 2$$
analytically. Give an analytic check.

SOLUTION   On the left, $\ln e^{\ln x} = \ln x$. The equation becomes
$$\ln x - \ln (x - 3) = \ln 2$$

$$\ln \frac{x}{x-3} = \ln 2 \qquad \text{Quotient rule}$$

$$\frac{x}{x-3} = 2 \qquad \text{Property 2}$$

$$x = 2x - 6 \qquad \text{Multiply by } x - 3.$$

$$6 = x.$$

To check, let $x = 6$ in the original equation.

$$\ln e^{\ln x} - \ln (x - 3) = \ln 2 \qquad \text{Given equation}$$
$$\ln e^{\ln 6} - \ln (6 - 3) = \ln 2 \qquad ? \quad \text{Let } x = 6.$$
$$\ln 6 - \ln 3 = \ln 2 \qquad ?$$
$$\ln \left(\frac{6}{3} \right) = \ln 2 \qquad ?$$
$$\ln 2 = \ln 2 \qquad \text{True}$$

The solution set is {6}. In Exercise 16, you are asked to support the solution graphically, and solve the associated inequalities. ◆

A summary of the methods used for solving equations in this section follows.

SOLVING EXPONENTIAL AND LOGARITHMIC EQUATIONS

An exponential or logarithmic equation may be solved by changing the equation into one of the following forms, where a and b are real numbers, $a > 0$, and $a \neq 1$.

1. $a^{f(x)} = b$
   Solve by taking logarithms of each side. (Natural logarithms are the best choice if $a = e$.)
2. $\log_a f(x) = \log_a g(x)$
   From the given equation, $f(x) = g(x)$, which is solved analytically.
3. $\log_a f(x) = b$
   Solve by using the definition of logarithm to write the expression in exponential form as $f(x) = a^b$.

FORMULAS INVOLVING EXPONENTIALS AND LOGARITHMS

In the next section, we will examine many applications of exponential and logarithmic functions. We now present two examples that illustrate how formulas based on such functions may be solved for one variable in terms of the other.

EXAMPLE 8 *Solving an Exponential Formula from Psychology for a Particular Variable*   The strength of a habit is a function of the number of times the habit is repeated. If N is the number of repetitions and H is the strength of the habit, then, according to psychologist C. L. Hull,

$$H = 1000(1 - e^{-kN}),$$

where k is a constant. Solve this equation for k.

SOLUTION   We must first solve the equation for e^{-kN}.

$$\frac{H}{1000} = 1 - e^{-kN} \qquad \text{Divide by 1000.}$$

$$\frac{H}{1000} - 1 = -e^{-kN} \qquad \text{Subtract 1.}$$

$$e^{-kN} = 1 - \frac{H}{1000} \qquad \text{Multiply by } -1.$$

Now, solve for k. As shown earlier, we take logarithms on each side of the equation and use the fact that $\ln e^x = x$.

$$\ln e^{-kN} = \ln\left(1 - \frac{H}{1000}\right)$$

$$-kN = \ln\left(1 - \frac{H}{1000}\right) \qquad \ln e^x = x$$

$$k = -\frac{1}{N}\ln\left(1 - \frac{H}{1000}\right) \qquad \text{Multiply by } -\tfrac{1}{N}.$$

With the last equation, if one pair of values for H and N is known, k can be found, and the equation can then be used to find either H or N for given values of the other variable. ◆

EXAMPLE 9 *Solving a Logarithmic Formula from Biology for a Particular Variable* The formula

$$S = a \ln \left(1 + \frac{n}{a} \right)$$

gives the number of species in a sample, where n is the number of individuals in the sample, and a is a constant indicating the diversity of species in the community. Solve the equation for n.

SOLUTION We begin by solving for $\ln\left(1 + \frac{n}{a}\right)$. Then we can change to exponential form and solve the resulting equation for n.

$$\frac{S}{a} = \ln \left(1 + \frac{n}{a} \right) \qquad \text{Divide by } a.$$

$$e^{S/a} = 1 + \frac{n}{a} \qquad \text{Write in exponential form.}$$

$$e^{S/a} - 1 = \frac{n}{a} \qquad \text{Subtract 1.}$$

$$n = a(e^{S/a} - 1) \qquad \text{Multiply by } a.$$

Using this equation and given values of S and a, we can find the number of individuals in a sample. ◆

5.4 EXERCISES   Tape 7

In Exercises 1–4, a logarithmic or exponential function f is graphed. Use the graph to find the solution set of (a) $f(x) = 0$ and (b) $f(x) > 0$.

1.

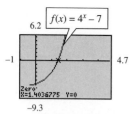

2.

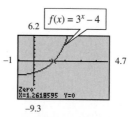

3.

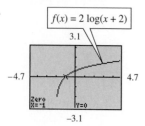

4.

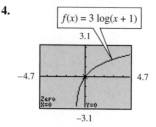

In Exercises 5–10, solve the equation in part (a) analytically, and express solutions in exact *form. If logarithms are required to express solutions, use common logarithms. Then, use a graph to support your solutions. Solve the associated inequalities in parts (b) and (c) by referring to the solutions in part (a) and the graph. Express endpoints of intervals as exact values, again using common logarithms if necessary.*

5. (a) $3^x = 10$ **6. (a)** $.5^x = .3$ **7. (a)** $2^{x+3} = 5^x$

  (b) $3^x > 10$ **(b)** $.5^x > .3$ **(b)** $2^{x+3} \geq 5^x$

  (c) $3^x < 10$ **(c)** $.5^x < .3$ **(c)** $2^{x+3} \leq 5^x$

8. (a) $6^{x+3} = 4^x$ **9. (a)** $\log(x - 3) = 1 - \log x$ **10. (a)** $\log(x - 6) = 2 - \log(x + 15)$

  (b) $6^{x+3} \geq 4^x$ **(b)** $\log(x - 3) < 1 - \log x$ **(b)** $\log(x - 6) < 2 - \log(x + 15)$

  (c) $6^{x+3} \leq 4^x$ **(c)** $\log(x - 3) > 1 - \log x$ **(c)** $\log(x - 6) > 2 - \log(x + 15)$

Solve each equation analytically. You may wish to support your result graphically.

11. $\ln (4x - 2) = \ln 4 - \ln (x - 2)$ **12.** $\ln (5 + 4x) - \ln (3 + x) = \ln 3$

13. $\log_5 (x + 2) + \log_5 (x - 2) = 1$ **14.** $\log_2 x + \log_2 (x - 7) = 3$

15. Use the result of Example 6 of this section and the corresponding figure (Figure 42) to solve the following inequalities.

  (a) $e^{-2 \ln x} > \dfrac{1}{16}$ **(b)** $e^{-2 \ln x} < \dfrac{1}{16}$

16. Use the result of Example 7 of this section to do each of the following.

  (a) Support the solution of the equation using a calculator-generated graph.

  (b) Solve the inequality $\ln e^{\ln x} - \ln (x - 3) > \ln 2$, using the graph from part (a).

  (c) Solve the inequality $\ln e^{\ln x} - \ln (x - 3) < \ln 2$, using the graph from part (a).

17. Use the fact that $\log 10 = 1$ to rewrite the equation in Example 5 with a logarithm on the right side. Then, solve the equation using the method of Example 4.

18. For what x-values shown in this table is $Y_1 \geq 0$?

In Exercises 19–24, follow the same directions as those for Exercises 5–10, but, if necessary, write solutions and endpoints as decimals rounded to the nearest thousandth.

19. (a) $3^{x+2} = 5$ **20. (a)** $5^{2-x} = 12$ **21. (a)** $2e^{5x+2} = 8$

  (b) $3^{x+2} > 5$ **(b)** $5^{2-x} > 12$ **(b)** $2e^{5x+2} < 8$

  (c) $3^{x+2} < 5$ **(c)** $5^{2-x} < 12$ **(c)** $2e^{5x+2} > 8$

22. (a) $10e^{3x-7} = 5$ **23. (a)** $\log_4 x + \log_4 (x + 2) = 1$ **24. (a)** $\log_5 (x + 4) + \log_5 (x - 1) = 0$

  (b) $10e^{3x-7} < 5$ **(b)** $\log_4 x + \log_4 (x + 2) \leq 1$ **(b)** $\log_5 (x + 4) + \log_5 (x - 1) \leq 0$

  (c) $10e^{3x-7} > 5$ **(c)** $\log_4 x + \log_4 (x + 2) \geq 1$ **(c)** $\log_5 (x + 4) + \log_5 (x - 1) \geq 0$

25. Use a graph to explain why the equation $\ln x - \ln (x + 1) = \ln 5$ has no solution.

26. Use a graph to explain why the inequality $\ln x + 1 \geq \ln (x - 4)$ has solution set $(4, \infty)$.

Use an equation-solving routine of a graphing calculator to find the solution of each of the following equations. Round your result to the nearest thousandth.

27. $1.5^{\log x} = e^{.5}$ **28.** $1.5^{\ln x} = 10^{.5}$

Each of the following formulas comes from an application of exponential or logarithmic functions. Solve the formula for the indicated variable.

29. $r = p - k \ln t$,   for t

30. $p = a + \dfrac{k}{\ln x}$,   for x

31. $T = T_0 + (T_1 - T_0)10^{-kt}$,   for t

32. $A = \dfrac{Pi}{1 - (1 + i)^{-n}}$,   for n

33. $A = T_0 + Ce^{-kt}$,   for k

34. $y = \dfrac{K}{1 + ae^{-bx}}$,   for b

35. $y = A + B(1 - e^{-Cx})$,   for x

36. $m = 6 - 2.5 \log\left(\dfrac{M}{M_0}\right)$,   for M

37. $\log A = \log B - C \log x$,   for A

38. $d = 10 \log\left(\dfrac{I}{I_0}\right)$,   for I

RELATING CONCEPTS (EXERCISES 39–44)

In Chapter 3, we introduced methods of solving quadratic equations. These methods can be applied to equations that may not necessarily be quadratic, but may be quadratic in form. Consider the equation

$$e^{2x} - 4e^x + 3 = 0,$$

and work Exercises 39–44 in order.

39. The expression e^{2x} is equivalent to $(e^x)^2$. Explain why this is so. (Section R.1)

40. The given equation is equivalent to $(e^x)^2 - 4e^x + 3 = 0$. Factor the left side of this equation. (Section R.2)

41. Solve the equation in Exercise 40 by the zero-product property. Give exact values. (Section 3.2)

42. Support your solution(s) in Exercise 41, using a calculator-generated graph of $y = e^{2x} - 4e^x + 3$. (Section 1.5)

43. Use the graph from Exercise 42 to solve the inequality $e^{2x} - 4e^x + 3 > 0$. (Section 1.6)

44. Use the graph from Exercise 42 to solve the inequality $e^{2x} - 4e^x + 3 < 0$. (Section 1.6)

In general, it is not possible to find exact solutions analytically for equations that involve exponential or logarithmic functions together with polynomial, radical, and rational functions. However, it is possible to solve them graphically with a calculator, using either the intersection-of-graphs method or the x-intercept method. Solve the following equations, using a purely graphical method, and express solutions to the nearest thousandth if an approximation is appropriate.

45. $x^2 = 2^x$

46. $x^2 - 4 = e^{x-4} + 4$

47. $\log x = x^2 - 8x + 14$

48. $\ln x = -\sqrt[3]{x + 3}$

49. $e^x = \dfrac{1}{x + 2}$

50. $3^{-x} = \sqrt{x + 5}$

Use any method (analytic or graphical) to solve each of the following equations. If appropriate, round your solution to the nearest thousandth.

51. $100(1 + .02)^{3+x} = 150$

52. $500(1 + .05)^{x/4} = 200$

53. $\log_2 \sqrt{2x^2} - 1 = .5$

54. $\log_2 (\log_2 x) = 1$

55. $\log x = \sqrt{\log x}$

56. $\log x^2 = (\log x)^2$

57. $\ln (\ln e^{-x}) = \ln 3$

58. $10^{5 \log x} = 32$

59. $e^x - 6 = -\dfrac{8}{e^x}$

60. $e^{x + \ln 3} = 4e^x$

5.5 APPLICATIONS AND MODELING WITH EXPONENTIAL AND LOGARITHMIC FUNCTIONS

Physical Science Applications ◆ Financial Applications ◆ Biological and Medical Applications ◆ Modeling Data with Exponential or Logarithmic Functions

In the first two sections of this chapter, we saw how exponential functions can be used in computing interest and how logarithms are used to determine pH levels of substances in chemistry. These are just two of many types of applications of exponential and logarithmic functions. In this section, we will examine other applications in various fields of study.

PHYSICAL SCIENCE APPLICATIONS

The formula $A = Pe^{rt}$, introduced in Section 5.1 in conjunction with continuous compounding of interest, is an example of an exponential growth function. A function of the form $A(t) = A_0 e^{kt}$, where A_0 represents the initial quantity present, t represents time elapsed, $k > 0$ represents the growth constant associated with the quantity, and $A(t)$ represents the amount present at time t, is called an **exponential growth function.** As we would expect, this is an increasing function, because $e > 1$ and $k > 0$. On the other hand, a function of the form $A(t) = A_0 e^{-kt}$ is an **exponential decay function.** It is a decreasing function because

$$e^{-k} = (e^{-1})^k = \left(\frac{1}{e}\right)^k,$$

and $0 < \frac{1}{e} < 1$. In both cases, we usually restrict t to be nonnegative, giving a domain of $[0, \infty)$. (Why do you think this is so?) Figure 43 shows graphs of typical growth and decay functions.

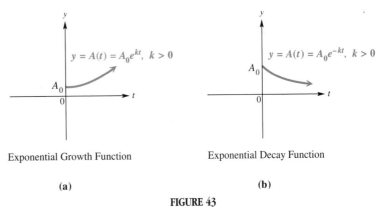

Exponential Growth Function

(a)

Exponential Decay Function

(b)

FIGURE 43

If a quantity decays exponentially, the amount of time that it takes to become one-half its original amount is called the **half-life.** The first example uses this idea.

EXAMPLE 1 *Analyzing an Exponential Decay Function Involving Radioactive Isotopes* Nuclear energy derived from radioactive isotopes can be used to supply power to space vehicles. Suppose that the output of the radioactive power supply for a certain satellite is given by the function

$$y = 40e^{-.004t},$$

where y is measured in watts and t is the time in days.

(a) What is the initial output of the power supply?

(b) After how many days will the output be reduced to 35 watts?

(c) After how many days will the output be half of its initial amount? (That is, what is its half-life?) Support this result with a graph.

SOLUTION

(a) Let $t = 0$ in the equation.

$$y = 40e^{-.004t} \qquad \text{Given function}$$

$$y = 40e^{-.004(0)} \qquad \text{Let } t = 0.$$

$$y = 40e^0$$

$$y = 40 \qquad\qquad e^0 = 1$$

The initial output is 40 watts.

(b) Let $y = 35$, and solve for t.

$$35 = 40e^{-.004t}$$

$$\frac{35}{40} = e^{-.004t} \qquad \text{Divide by 40.}$$

$$\ln\left(\frac{35}{40}\right) = \ln e^{-.004t} \qquad \text{Take the natural logarithm.}$$

$$\ln\left(\frac{35}{40}\right) = -.004t \qquad \ln e^k = k$$

$$t = \frac{\ln\left(\dfrac{35}{40}\right)}{-.004} \qquad \text{Divide by } -.004.$$

Using a calculator, we find that $t \approx 33.4$. It will take about 33.4 days for the output to be reduced to 35 watts.

(c) ANALYTIC   Because the initial amount is 40, we must find the value of t for which $y = \frac{1}{2}(40) = 20$.

$$20 = 40e^{-.004t}$$

$$.5 = e^{-.004t} \qquad \text{Divide by 40.}$$

$$\ln .5 = \ln e^{-.004t} \qquad \text{Take the natural logarithm.}$$

$$\ln .5 = -.004t \qquad \ln e^k = k$$

$$t \approx 173 \qquad \text{Divide by } -.004.$$

The half-life is approximately 173 days.

 GRAPHICAL   The graph in Figure 44 supports the result that when $x = t = 173$, $y \approx 20 = \frac{1}{2}(40)$.   ◆

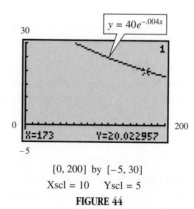

$y = 40e^{-.004x}$

X=173   Y=20.022957

[0, 200] by [−5, 30]

Xscl = 10   Yscl = 5

FIGURE 44

EXAMPLE 2 *Using an Exponential Function to Find the Age of a Fossil*
Carbon-14 is a radioactive form of carbon that is found in all living plants and animals. After a plant or animal dies, the radiocarbon disintegrates. Scientists determine the age of the remains by comparing the amount of carbon-14 present with the amount found in living plants and animals. The amount of carbon-14 present after t years is given by the exponential equation $A(t) = A_0 e^{-kt}$, with $k \approx (\ln 2)\left(\frac{1}{5700}\right)$.

(a) Find the half-life.

(b) Charcoal from an ancient fire pit on Java contained $\frac{1}{4}$ the carbon-14 of a living sample of the source material of the same size. Estimate the age of the charcoal.*

SOLUTION

(a) Let $A(t) = \left(\frac{1}{2}\right)A_0$ and $k = (\ln 2)\left(\frac{1}{5700}\right)$.

$$\frac{1}{2} A_0 = A_0 e^{-(\ln 2)(1/5700)t}$$

$$\frac{1}{2} = e^{-(\ln 2)(1/5700)t} \qquad \text{Divide by } A_0.$$

$$\ln \frac{1}{2} = \ln e^{-(\ln 2)(1/5700)t} \qquad \text{Take logarithms on both sides.}$$

$$\ln \frac{1}{2} = -\frac{\ln 2}{5700} t \qquad \ln e^x = x$$

$$-\frac{5700}{\ln 2} \ln \frac{1}{2} = t \qquad \text{Multiply by } -\frac{5700}{\ln 2}.$$

$$-\frac{5700}{\ln 2} (\ln 1 - \ln 2) = t \qquad \text{Quotient rule for logarithms}$$

$$-\frac{5700}{\ln 2} (-\ln 2) = t \qquad \ln 1 = 0$$

$$5700 = t$$

The half-life is 5700 years.

(b) Let $A(t) = \frac{1}{4} A_0$ and $k = (\ln 2)\left(\frac{1}{5700}\right)$.

$$\frac{1}{4} A_0 = A_0 e^{-(\ln 2)(1/5700)t}$$

$$\frac{1}{4} = e^{-(\ln 2)(1/5700)t}$$

$$\ln \frac{1}{4} = \ln e^{-(\ln 2)(1/5700)t}$$

$$\ln \frac{1}{4} = -\frac{\ln 2}{5700} t$$

$$-\frac{5700}{\ln 2} \ln \frac{1}{4} = t$$

$$t = 11{,}400$$

The charcoal is about 11,400 years old.   ◆

---

*Adapted from *A Sourcebook of Applications of School Mathematics* by Donald Bushaw et al. Copyright © 1980 by The Mathematical Association of America. Reprinted by permission.

EXAMPLE 3   *Measuring Sound Intensity*   The loudness of sounds is measured in a unit called a *decibel*. To measure with this unit, we first assign an intensity of I_0 to a very faint sound, called the *threshold sound*. If a particular sound has intensity I, then the decibel rating of this louder sound is

$$d = 10 \log \frac{I}{I_0}.$$

Find the decibel rating of a sound with intensity $10{,}000I_0$.

SOLUTION   Let $I = 10{,}000I_0$ and find d.

$$d = 10 \log \frac{10{,}000I_0}{I_0}$$

$$= 10 \log 10{,}000$$

$$= 10(4) \qquad\qquad \log 10{,}000 = 4$$

$$= 40$$

The sound has a decibel rating of 40.   ◆

EXAMPLE 4   *Measuring Age Using an "Atomic Clock"*   Geologists sometimes measure the age of rocks by using "atomic clocks." By measuring the amounts of potassium-40 and argon-40 in a rock, the age t of the specimen in years is found with the formula

$$t = (1.26 \times 10^9) \frac{\ln[1 + 8.33(A/K)]}{\ln 2}.$$

A and K are respectively the numbers of atoms of argon-40 and potassium-40 in the specimen. The ratio $\frac{A}{K}$ for a sample of granite from New Hampshire is .212. How old is the sample?

SOLUTION   Since $\frac{A}{K}$ is .212, we have

$$t = (1.26 \times 10^9) \frac{\ln[1 + 8.33(.212)]}{\ln 2} \approx 1.85 \times 10^9.$$

The granite is about 1.85 billion years old.*   ◆

FINANCIAL APPLICATIONS

The formulas

$$A = P\left(1 + \frac{r}{n}\right)^{nt} \qquad \text{and} \qquad A = Pe^{rt}$$

were introduced in Section 5.1. They apply to compound interest: the first, to interest compounded a finite number of times annually, and the second to interest compounded continuously. We can use logarithms to determine how long it will take for a particular investment to grow to a desired amount.

---

*Adapted from *A Sourcebook of Applications of School Mathematics* by Donald Bushaw et al. Copyright © 1980 by the Mathematical Association of America. Reprinted by permission.

EXAMPLE 5 *Using Compound Interest Formulas to Determine Time Required*

(a) How long will it take $1000 invested at 6%, compounded quarterly, to grow to $2700?

(b) How long will it take for the money in an account that is compounded continuously at 8% interest to double?

SOLUTION

(a) Here, we must find t such that $A = 2700$, $P = 1000$, $r = .06$, and $n = 4$.

$$A = P\left(1 + \frac{r}{n}\right)^{nt} \qquad \text{Given formula}$$

$$2700 = 1000\left(1 + \frac{.06}{4}\right)^{4t} \qquad \text{Substitute.}$$

$$2700 = 1000(1.015)^{4t}$$

$$2.7 = 1.015^{4t} \qquad \text{Divide by 1000.}$$

A table supports the result of Example 5(a). Notice that Y_1 = 2700 for some x-value between 16 and 17, closer to 17. We found analytically that this value is approximately 16.678.

To solve for t, we may use either common or natural logarithms. Choosing common logarithms, we obtain

$$\log 2.7 = \log 1.015^{4t}$$

$$\log 2.7 = 4t \log 1.015 \qquad \text{Power rule}$$

$$\frac{\log 2.7}{4 \log 1.015} = t \qquad \text{Divide by 4 log 1.015.}$$

$$t \approx 16.678. \qquad \text{Use a calculator.}$$

It will take about $16\frac{2}{3}$ years for the initial amount to grow to $2700.

(b) Use the formula for continuous compounding, $A = Pe^{rt}$, to find the time t that makes $A = 2P$. Substitute $2P$ for A and .08 for r; then, solve for t.

$$A = Pe^{rt}$$

$$2P = Pe^{.08t} \qquad \text{Substitute.}$$

$$2 = e^{.08t} \qquad \text{Divide by } P.$$

Taking natural logarithms on both sides gives

$$\ln 2 = \ln e^{.08t}.$$

Use the property $\ln e^x = x$ to get $\ln e^{.08t} = .08t$.

$$\ln 2 = .08t$$

$$\frac{\ln 2}{.08} = t \qquad \text{Divide by .08.}$$

$$8.664 \approx t$$

It will take about $8\frac{2}{3}$ years for the amount to double. ◆

If a quantity grows exponentially, the amount of time that it takes to become twice its original amount is called the **doubling time**. This is analogous to half-life for quantities that decay exponentially. In both cases, the actual amount present initially does not affect the doubling time or half-life.

If a loan is taken out from a lending institution, it is customary to pay the loan back in installments on a regular basis at a certain rate of interest for a length of time called the *term* of the loan. This process is called **amortization**.

AMORTIZATION OF A LOAN

If a principal of P dollars is amortized over a period of t years, and payments are made each $\frac{1}{n}$th of a year, with an annual interest rate of r (as a decimal), the payment p that must be made each period is given by the formula

$$p = \frac{Pr}{n\left[1 - \left(1 + \dfrac{r}{n}\right)^{-nt}\right]}.$$

The total interest I that will be paid during the term of the loan is given by the formula

$$I = npt - P.$$

EXAMPLE 6   *Using the Amortization Formulas for Automobile Financing*
After going through the tried-and-true American tradition of "haggling" with the used-car manager at your local dealer's showroom, you decide on a price of $14,200 for a beautiful 1994 Cadillac Sedan DeVille. Your down payment is $4000, and you are able to find an institution that will lend you the balance of the price at 8.5%. The term of the loan is to be 3 years, and payments are to be made monthly. How much will your monthly payment be, and how much interest will you have paid when the loan is completely paid off?

SOLUTION   To find the monthly payment p, use the first amortization formula with $P = 14,200 - 4000 = 10,200$, $r = .085$, $n = 12$, and $t = 3$.

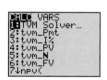

Modern graphing calculators such as the TI-83 are capable of financial calculations. Consult the owner's manual of your particular model.

$$p = \frac{10,200(.085)}{12\left[1 - \left(1 + \dfrac{.085}{12}\right)^{-12\cdot3}\right]} \approx 321.99$$

The monthly payment will be $321.99.

The total amount of interest (I) paid over the life of the loan is found by using the second amortization formula, with $n = 12$, $p = 321.99$, $t = 3$, and $P = 10,200$.

$$I = 12(321.99)(3) - 10,200 \approx 1391.64$$

The total amount of interest paid will be $1391.64.   ◆

BIOLOGICAL AND MEDICAL APPLICATIONS

Base two logarithms are used in a formula that measures the diversity of species in a community, as seen in the next example.

EXAMPLE 7   *Measuring the Diversity of Species*   One measure, H, of the diversity of species in an ecological community is given by the formula

$$H = -[P_1 \log_2 P_1 + P_2 \log_2 P_2 + \cdots + P_n \log_2 P_n],$$

where $P_1, P_2, \ldots, P_n$ are the proportions of a sample belonging to each of n species found in the sample. Suppose that a community has two species, where there are 90 of one and 10 of the other.

(a) Find P_1 and P_2.

(b) Find H for the community.

SOLUTION (a) Since there are $90 + 10 = 100$ total,

$$P_1 = \frac{90}{100} = .9 \quad \text{and} \quad P_2 = \frac{10}{100} = .1.$$

(b) $H = -[.9 \log_2 .9 + .1 \log_2 .1]$.

By the change-of-base rule,

$$\log_2 .9 = \frac{\log .9}{\log 2} \approx -.152$$

and

$$\log_2 .1 = \frac{\log .1}{\log 2} \approx -3.322.$$

Using these in the formula for H above, we find $H \approx .469$. If the number in each species is the same, the measure of diversity is 1, representing "perfect" diversity. In a community with little diversity, H is close to 0. In this example, since $H \approx .5$, there is neither great nor little diversity. ◆

The following example illustrates how the amount of medication left in the body after a certain period of time can be determined.

EXAMPLE 8 *Determining the Amount of Medication Remaining* When physicians prescribe medication, they must consider how the drug's effectiveness decreases over time. If, each hour, a drug is only 90% as effective as the previous hour, at some point the patient will not be receiving enough medication and must receive another dose. This situation can be modeled with a geometric sequence (see Section 3 in the final chapter). If the initial dose was 200 mg and the drug was administered 3 hours ago, the expression $200(.90)^2$ represents the amount of effective medication still available. Thus, $200(.90)^2 = 162$ mg are still in the system. (The exponent is equal to the number of hours since the drug was administered, less one.) How long will it take for this initial dose to reach the dangerously low level of 50 mg? Support the answer with a graph.

SOLUTION ANALYTIC We must solve the equation $200(.90)^x = 50$.

$$200(.90)^x = 50$$
$$(.90)^x = .25$$
$$\log (.90)^x = \log .25$$
$$x \log .90 = \log .25$$
$$x = \frac{\log .25}{\log .90} \approx 13.16$$

Since x represents one *less than* the number of hours since the drug was administered, the drug will reach a level of 50 mg in about 14 hours.

GRAPHICAL The graph in Figure 45 on the following page supports the result that when $x \approx 13.16$, $y = 50$. ◆

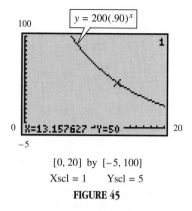

[0, 20] by [−5, 100]
Xscl = 1   Yscl = 5
FIGURE 45

For Group Discussion   Reproduce the graph shown in Figure 45 and trace from the left to the right. Discuss how the word "decreasing" applies to this activity in both mathematical and real-life contexts.

MODELING DATA WITH EXPONENTIAL OR LOGARITHMIC FUNCTIONS

Some data can be modeled quite well using exponential or logarithmic functions. Modern graphing calculators have the capability of determining such models for appropriate data. The final two examples illustrate these types of models.

EXAMPLE 9   *Modeling Data for Atmospheric Pressure Using an Exponential Function*   The atmospheric pressure (in millibars) at a given altitude (in meters) is shown in the table that follows.*

TECHNOLOGY NOTE

Some models of graphing calculators will automatically set an appropriate window for data points. Consult your owner's manual.

Altitude (x)	Pressure (P)
0	1013
1000	899
2000	795
3000	701
4000	617
5000	541
6000	472
7000	411
8000	357
9000	308
10,000	265

(a)   Use a graphing calculator to plot the data in the window $[-1000, 11{,}000]$ by $[0, 1200]$, and discuss the appearance of the graph.

(b)   Use the points $(0, 1013)$ and $(10{,}000, 265)$ to find values for a and r in the equation $P(x) = ae^{-rx}$ and determine an exponential model equation for the data. Plot the graph of this equation over the data points.

(c)   Use the capability of a graphing calculator to find an exponential function that models all the data, and graph it over the data points.

---

*Source: Miller, A., and J. Thompson, Elements of Meteorology, Charles E. Merrill Publishing Company, Columbus, Ohio, 1975.

(d) Use the functions in parts (b) and (c) to predict the pressure at 1500 meters, and compare the results to the actual value of 846 millibars.

SOLUTION

(a) Figure 46 shows the plot of these data. The curve resembles that of a decreasing exponential function.

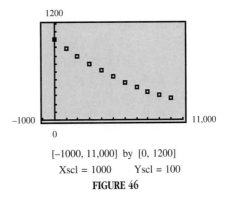

[−1000, 11,000] by [0, 1200]
Xscl = 1000   Yscl = 100
FIGURE 46

(b) We first use the point (0, 1013).

$$P(x) = ae^{-rx}$$
$$1013 = ae^{-r(0)} \qquad x = 0, P(0) = 1013$$
$$1013 = a(1) \qquad e^0 = 1$$
$$a = 1013$$

Next, use the point (10,000, 265), with $a = 1013$, to find r.

$$P(10,000) = 1013e^{-10,000r}$$
$$265 = 1013e^{-10,000r}$$
$$\frac{265}{1013} = e^{-10,000r}$$
$$-10,000r = \ln\left(\tfrac{265}{1013}\right)$$
$$r = \left(\ln\left(\tfrac{265}{1013}\right)\right)/(-10,000) \approx .0001341$$

Thus, we have $P(x) = 1013e^{-.0001341x}$. The graph is shown in Figure 47.

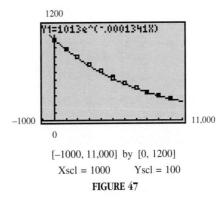

[−1000, 11,000] by [0, 1200]
Xscl = 1000   Yscl = 100
FIGURE 47

(c) Using the general form $y = a * b^x$, a graphing calculator indicates that a model for the data is $y = 1038 * (.999866)^x$. See Figure 48 for the graph of this model.

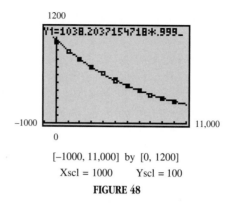

[−1000, 11,000] by [0, 1200]
Xscl = 1000   Yscl = 100

FIGURE 48

(d) For part (b), we have $P(1500) = 1013e^{-.0001341(1500)} \approx 828$ millibars, and for part (c) we have $y = 1038 * (.999866)^{1500} \approx 849$ millibars. Both are very close to the actual value of 846 millibars, especially the function from part (c).   ◆

For Group Discussion   Discuss the reasons for the model in part (c) of Example 9 giving a better prediction than the model in part (b).

EXAMPLE 10   *Using a Calculator to Find a Logarithmic Model for Interest Rates*   The following table lists the interest rates (in percent) for various U.S. securities in January, 1996.*

Time (x)	Yield (y)
3 months (.25 year)	5.71
6 months (.5 year)	6.37
1 year	6.87
2 years	7.34
3 years	7.52
5 years	7.63
10 years	7.68

These data are entered into a graphing calculator and plotted, with the results shown in Figure 49(a). Notice how the appearance of the points suggests a logarithmic curve. The calculator provides a model of the form $y = a + b \ln x$ for these data, as shown in Figure 49(b), and the graph of this model appears with the data points in Figure 49(c).

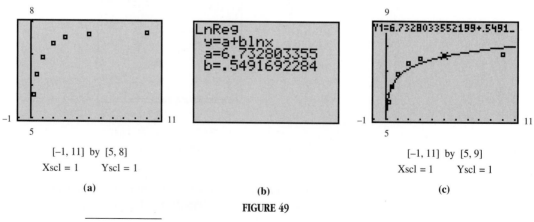

[−1, 11] by [5, 8]
Xscl = 1   Yscl = 1

(a)

[−1, 11] by [5, 9]
Xscl = 1   Yscl = 1

(c)

(b)

FIGURE 49

Source: Reuters.

Use this model to predict the yield for a 4-year time period.

SOLUTION We will use the graph and let $x = 4$ to find that $y \approx 7.49$ (percent). See the display at the bottom of the graph in Figure 50. ◆

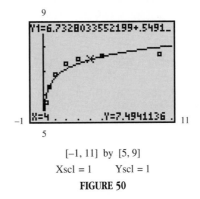

[−1, 11] by [5, 9]
Xscl = 1   Yscl = 1
FIGURE 50

5.5 EXERCISES

    Tape 7

In Exercises 1–26, solve each of the problems from the physical sciences.

1. *Carbon Dioxide Levels* The International Panel on Climate Change (IPCC) in 1990 published its finding that if current trends of burning fossil fuel and deforestation continue, then future amounts of atmospheric carbon dioxide in parts per million (ppm) will increase as shown in the table.

Year	1990	2000	2075	2175	2275
Carbon Dioxide	353	375	590	1090	2000

   (a) Plot the data in the window [1975, 2300] by [300, 2100]. Do the carbon dioxide levels appear to grow exponentially?

   (b) The function defined by $y = 353e^{.0060857(t-1990)}$ is a good model for the data. A graph of the function shows that it is very close to the data points. From the graph, estimate when future levels of carbon dioxide will double and triple over the preindustrial level of 280 ppm.

2. *Radiative Forcing* (Refer to Exercise 1.) Carbon dioxide in the atmosphere traps heat from the sun. Presently, the net incoming solar radiation reaching the earth's surface is 240 watts per square meter (w/m^2). The relationship between additional watts per square meter of heat trapped by the increased carbon dioxide R and the average rise in global temperature T (in °F) increases proportionately. This additional solar radiation trapped by carbon dioxide is called *radiative forcing*. It is measured in watts per square meter.

   (a) Is T a linear or exponential function of R?

   (b) Let T represent the temperature increase resulting from an additional radiative forcing of R w/m^2. Use the graph from Exercise 1 to write T as a function of R.

   (c) Find the global temperature increase when $R = 5$ w/m^2.

3. *Radiative Forcing* In 1896, the Swedish scientist Svante Arrhenius estimated the radiative forcing R caused by additional atmospheric carbon dioxide, using the logarithmic equation $R = k \ln(C/C_0)$, where C_0 is the preindustrial amount of carbon dioxide, C is the current carbon dioxide level, and k is a constant. Arrhenius determined that $10 \le k \le 16$ when $C = 2C_0$. (*Source:* Clime, W., *The Economics of Global Warming,* Institute for International Economics, Washington, D.C., 1992.)

   (a) Let $C = 2C_0$. Is the relationship between R and k linear or logarithmic?

   (b) The average global temperature increase T (in °F) is given by $T(R) = 1.03R$. (See Exercise 2.) Write T as a function of k.

   (c) Use $T(k)$ to find the range of the rise in global temperature T (rounded to the nearest degree) that Arrhenius predicted.

4. *Radiative Forcing* Using computer models, the International Panel on Climate Change (IPCC) in 1990 estimated k to be 6.3 in the radiative forcing equation $R = k \ln(C/C_0)$, where C_0 is the preindustrial amount of carbon dioxide and C is the current level. (*Source:* Clime, W., *The Economics of Global Warming,* Institute for International Economics, Washington, D.C., 1992.)

(a) What radiative forcing R (in w/m^2) is expected by the IPCC if the carbon dioxide level in the atmosphere doubles from its preindustrial level?

(b) Determine the global temperature increase predicted by the IPCC if the carbon dioxide levels were to double. Read the chapter-opening application and discuss this result.

5. *Global Temperatures* (Refer to Exercise 3.) According to the IPCC, if present trends continue, future increases in average global temperatures (in °F) can be modeled by $T = 6.489 \cdot \ln(C/280)$, where C is the concentration of atmospheric carbon dioxide (in ppm). C can be modeled by the function $C(x) = 353(1.006)^{x-1990}$, where x is the year. (*Source:* International Panel on Climate Change (IPCC), 1990.)

(a) Write T as a function of x.

(b) With a graphing calculator, graph $C(x)$ and $T(x)$ on the interval [1990, 2275], using different coordinate axes. Describe each function's graph. How are C and T related?

(c) Approximate the slope of the graph of T. What does this slope represent?

(d) Use graphing to estimate x and C when $T = 10°F$.

6. *Decay of Lead* A sample of 500 grams of radioactive lead-210 decays to polonium-210 according to the function $A(t) = 500e^{-.032t}$, where t is time in years. Find the amount of the sample remaining after each of the following.

(a) 4 years

(b) 8 years

(c) 20 years

(d) Find the half-life.

(e) Graph $y = A(t)$ in the window [0, 300] by [0, 750].

7. *Decay of Plutonium* Repeat Exercise 6 for 500 grams of plutonium-241, which decays according to the function $A(t) = A_0 e^{-.053t}$, where t is time in years.

8. *Decay of Radium* Find the half-life of radium-226, which decays according to the function $A(t) = A_0 e^{-.00043t}$, where t is time in years.

9. *Decibel Levels* (See Example 3.) Find the decibel ratings of the following sounds, having intensities as given. Round each answer to the nearest whole number.

(a) whisper, $115I_0$

(b) busy street, $9,500,000I_0$

(c) heavy truck, 20 m away, $1,200,000,000I_0$

(d) rock music, $895,000,000,000I_0$

(e) jetliner at takeoff, $109,000,000,000,000I_0$

The intensity of an earthquake, measured on the Richter scale, is given by $\log\left(\frac{I}{I_0}\right)$, where I_0 is the intensity of an earthquake of a certain small size. Use this information in Exercises 10–12.

10. *Earthquake Intensity* Find the Richter scale magnitude of an earthquake that has each of the following intensities.

(a) $1000I_0$

(b) $1,000,000I_0$

(c) $100,000,000I_0$

11. *Earthquake Intensity* On July 14, 1991, Peshawar, Pakistan, was shaken by an earthquake that measured 6.6 on the Richter scale.

(a) Express this reading in terms of I_0.

(b) In February of the same year, a quake measuring 6.5 on the Richter scale killed about 900 people in the mountains of Pakistan and Afghanistan. Express the intensity of a 6.5 reading in terms of I_0.

(c) Compare this with the force of the earthquake with a measure of 6.6.

12. *Earthquake Intensity* The San Francisco earthquake of 1906 had a Richter scale rating of 8.3.

(a) Express the intensity of this earthquake as a multiple of I_0.

(b) In 1989, the San Francisco region experienced an earthquake with a Richter scale rating of 7.1. Express the intensity of this earthquake as a multiple of I_0.

(c) Compare the intensity of the two San Francisco earthquakes discussed above.

13. *Magnitude of a Star* The magnitude M of a star is defined by the equation $M = 6 - \frac{5}{2}\log\frac{I}{I_0}$, where I_0 is the measure of a just-visible star, and I is the actual intensity of the star being measured. The dimmest stars are of magnitude 6, and the brightest are of magnitude 1. Determine the ratio of light intensities between a star of magnitude 1 and a star of magnitude 3.

14. *Snow Levels* In the central Sierra Nevada mountains of California, the percent of moisture that falls as snow rather than rain is approximated reasonably well by the function $p(h) = 86.3 \ln h - 680$, where h is the altitude in feet, and $p(h)$ is the percent of snow. (This model is valid for $h \geq 3000$.) Find the percent of snow that falls at the following altitudes.

(a) 3000 feet   (b) 4000 feet   (c) 7000 feet

15. *Air Pressure* The air pressure in pounds per square inch h feet above sea level is given by the function $P(h) = 14.7e^{-.0000385h}$. At approximately what height is the pressure 10% of the pressure at sea level?

16. *CFC-11 Levels* Chlorofluorocarbons (CFCs) are greenhouse gases that were produced by humans after 1930. CFCs are found in refrigeration units, foaming agents, and aerosols. They have great potential for destroying the ozone layer. As a result, governments have agreed to phase out their production by the year 2000. CFC-11 is an example of a CFC that has increased faster than any other greenhouse gas. The following graph displays approximate concentrations of atmospheric CFC-11 in parts per billion (ppb) since 1950.

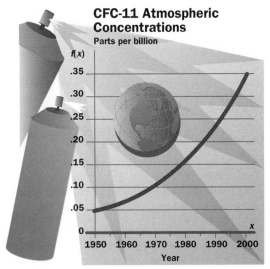

(*Source:* Nilsson, A., *Greenhouse Earth*, John Wiley & Sons, New York, 1992.)

(a) Use the graph to find an exponential function with $f(x) = A_0 a^{x-1950}$ that models the concentration of CFC-11 in the atmosphere from 1950 to 2000, where x is the year.

(b) Approximate the average annual percent increase of CFC-11 during this time period.

17. *Cost-Benefit Equation for CO_2 Emissions* One action that government could take to reduce carbon emissions into the atmosphere is to place a tax on fossil fuel. This tax would be based on the amount of carbon dioxide emitted into the air when the fuel is burned. The *cost-benefit equation* $\ln(1 - P) = -.0034 - .0053T$ describes the approximate relationship between a tax of T dollars per ton of carbon and the corresponding percent reduction P (in decimal form) of emissions of carbon dioxide. (*Source:* Nordhause, W., "To Slow or Not to Slow: The Economics of the Greenhouse Effect," Yale University, New Haven, Connecticut.)

(a) Write P as a function of T.

(b) Graph P for $0 \leq T \leq 1000$.

(c) Determine P when $T = \$60$, and interpret this result.

(d) What value of T will give a 50% reduction in carbon emissions?

The information in Example 2 allows us to use the function $A(t) = A_0 e^{-.0001216t}$ to approximate the amount of carbon-14 remaining in a sample, where t is in years. Use this function in Exercises 18–21. $\left(\text{Note: } -.0001216 \approx \dfrac{\ln 2}{-5700}. \right)$

18. *Carbon-14 Dating* Suppose an Egyptian mummy is discovered in which the amount of carbon-14 present is only about one-third the amount found in the atmosphere. About how long ago did the Egyptian die?

19. *Carbon-14 Dating* A sample from a refuse deposit near the Strait of Magellan had 60% of the carbon-14 of a contemporary living sample. How old was the sample?

20. *Carbon-14 Dating* Paint from the Lascaux caves of France contains 15% of the normal amount of carbon-14. Estimate the age of the caves.

21. *Carbon-14 Dating* Estimate the age of a specimen that contains 20% of the carbon-14 of a comparable living specimen.

Newton's law of cooling says that the rate at which a body cools is proportional to the difference in temperature between the body and the environment into which it is introduced. The temperature $f(t)$ of the body at time t in appropriate units after being introduced into an environment having constant temperature T_0 is $f(t) = T_0 + Ce^{-kt}$, where C and k are constants. Use this result in Exercises 22 and 23.

22. *Newton's Law of Cooling* Boiling water, at 100°C, is placed in a freezer at 0°C. The temperature of the water is 50°C after 24 minutes. Find the temperature of the water after 96 minutes.

23. *Newton's Law of Cooling* A piece of metal is heated to 300°C and then placed in a cooling liquid at 50°C. After 4 minutes, the metal has cooled to 175°C. Find its temperature after 12 minutes.

24. *Rock Sample Age* Use the function defined by

$$t = T \frac{\ln[1 + 8.33(A/K)]}{\ln 2}$$

to estimate the age of a rock sample, if tests show that A/K is .103 for the sample. Let $T = 1.26 \times 10^9$.

25. *Visitors to U.S. National Parks* The heights in the bar graph on the following page represent the number of visitors (in millions) to U.S. national parks from 1950 to 1994. Suppose x represents the number of years since 1900—thus, 1950 is represented by 50, 1960 is represented by 60, and so on. The logarithmic function

defined by $f(x) = -266 + 72 \ln x$ closely approximates the data. Use this function to estimate the number of visitors in the year 2000. What assumption must we make to estimate the number of visitors in years beyond 1993?

Visitors to National Parks

in millions

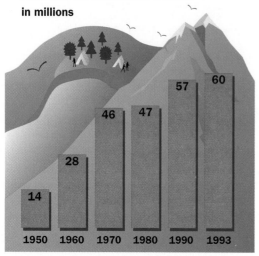

(*Source: Statistical Abstract of the United States 1995.*)

26. *Planets' Distance from the Sun and Period of Revolution*   The following table contains the planets' average distances D from the sun and their periods P of revolution around the sun in years. The distances have been normalized so that Earth is one unit away from the sun. For example, since Jupiter's distance is 5.2, its distance from the sun is 5.2 times farther than Earth's. (*Source:* Ronan, C., *The Natural History of the Universe,* Macmillan Publishing Co., New York, 1991.)

Planet	D	P
Mercury	.39	.24
Venus	.72	.62
Earth	1	1
Mars	1.52	1.89
Jupiter	5.2	11.9
Saturn	9.54	29.5
Uranus	19.2	84.0
Neptune	30.1	164.8

(a) Plot the point $(\ln D, \ln P)$ for each planet on the *xy*-coordinate axes, using a graphing calculator. Do the data points appear to be linear?

(b) Determine a linear equation that approximates the data points. Graph your line and the data on the same coordinate axes.

(c) Use this linear equation to predict the period of the planet Pluto if its distance is 39.5. Compare your answer to the true value of 248.5 years.

In Exercises 27–44, solve each of the financial applications.

27. *Interest on an Account*   How long will it take for $1000 to grow to $5000 at an interest rate of 3.5% if interest is compounded **(a)** quarterly **(b)** continuously?

28. *Interest on an Account*   How long will it take for $5000 to grow to $8400 at an interest rate of 6% if interest is compounded **(a)** semiannually **(b)** continuously?

29. *Interest on an Account*   George Duda wants to buy a $30,000 car. He has saved $27,000. Find the number of years (to the nearest tenth) it will take for his $27,000 to grow to $30,000 at 6% interest, compounded quarterly.

30. *Doubling Time*   Find the doubling time of an investment earning 2.5% interest if interest is compounded **(a)** quarterly **(b)** continuously.

31. *Comparison of Investments*   Karen Guardino, who is self-employed, wants to invest $60,000 in a pension plan. One investment offers 7%, compounded quarterly. Another offers 6.75%, compounded continuously. Which investment will earn more interest in 5 years? How much more will the better plan earn?

32. *Growth of an Account*   If Karen (see Exercise 31) chooses the plan with continuous compounding, how long will it take for her $60,000 to grow to $80,000?

*The interest rate stated by a financial institution is sometimes called the **nominal rate.** If interest is compounded, the actual rate is, in general, higher than the nominal rate, and is called the **effective rate.** If r is the nominal rate and n is the number of times interest is compounded annually, then*

$$R = \left(1 + \frac{r}{n}\right)^n - 1$$

is the effective rate. Here, R represents the annual rate that the investment would earn if simple interest were paid. Use this formula in Exercises 33 and 34.

33. *Effective Rate*   Find the effective rate if the nominal rate is 6% and interest is compounded quarterly.

34. *Effective Rate*   Find the effective rate if the nominal rate is 4.5% and interest is compounded daily ($n = 365$).

*In the formula $A = P\left(1 + \frac{r}{n}\right)^{nt}$, we can interpret P as the **present value** of A dollars t years from now, earning annual interest r compounded n times per year. In this context, A is called the **future value.** If we solve the formula for P, we obtain*

$$P = A\left(1 + \frac{r}{n}\right)^{-nt}.$$

Use this formula in Exercises 35–38.

35. *Present Value*   Find the present value of $10,000 five years from now, if interest is compounded semiannually at 12%.

36. *Present Value* Find the present value of $25,000 2.75 years from now, if interest is compounded quarterly at 6%.

37. *Future Value* Find the interest rate necessary for a present value of $25,000 to grow to a future value of $31,360, if interest is compounded annually for 2 years.

38. *Future Value* Find the interest rate necessary for a present value of $1200 to grow to a future value of $1780, if interest is compounded quarterly for 5 years.

*In Exercises 39–42, use the amortization formulas given in this section to find **(a)** the monthly ($n = 12$) payment on a loan with the given conditions and **(b)** the total interest that will be paid during the term of the loan.*

39. *Amortization* $8500 is amortized over 4 years with an interest rate of 7.5%

40. *Amortization* $9600 is amortized over 5 years with an interest rate of 9.2%

41. *Amortization* $55,000 is amortized over 15 years with an interest rate of 6.25%

42. *Amortization* $125,000 is amortized over 30 years with an interest rate of 7.25%

43. *Growth of an Account* **(a)** Use the table feature of your graphing calculator to find how long it will take $1500 invested at 5.75%, compounded daily, to triple in value. Zoom in on the solution by systematically decreasing ΔTbl. Find the answer to the nearest day. (Find your answer to the nearest day by eventually letting ΔTbl = 1/365. The decimal part of the solution can be multiplied by 365 to determine the number of days greater than the nearest year. For example, if the solution is determined to be 16.2027 years, then multiply .2027 by 365 to get 73.9855. The solution is then, to the nearest day, 16 years and 74 days.) Confirm your answer analytically.

(b) Use the table feature to find how long it will take $2000 invested at 8%, compounded daily, to be worth $5000.

44. *Growth of an Account* If interest is compounded continuously and the interest rate is tripled, what effect will this have on the time required for an investment to double?

In Exercises 45–52, solve each of the problems from the biological sciences and medicine.

45. *Growth of Bacteria* *Escherichia coli* is a strain of bacteria that occurs naturally in many organisms. Under certain conditions, the number of bacteria present in a colony is approximated by the function $A(t) = A_0 e^{.023t}$, where t is in minutes. If $A_0 = 2,400,000$, find the number of bacteria at the following times.

(a) 5 minutes   **(b)** 10 minutes   **(c)** 60 minutes

46. *Growth of Bacteria* The growth of bacteria in food products makes it necessary to time-date some products (such as milk) so that they will be sold and consumed before the bacteria count becomes too high. Suppose for a certain product that the number of bacteria present is given by

$$f(t) = 500e^{.1t},$$

under certain storage conditions, where t is time in days after packing of the product and the value of $f(t)$ is in millions.

(a) If the product cannot be safely eaten after the bacteria count reaches 3,000,000,000, how long will this take?

(b) If $t = 0$ corresponds to January 1, what date should be placed on the product?

A large cloud of radioactive debris from a nuclear explosion has floated over the Pacific Northwest, contaminating much of the hay supply. Consequently, farmers in the area are concerned that the cows who eat this hay will give contaminated milk. (The tolerance level for radioactive iodine in milk is 0.) The percent of the initial amount of radioactive iodine still present in the hay after t days is approximated by $P(t) = 100e^{-.1t}$. Use this information in Exercises 47 and 48.

47. *Radioactive Iodine Level* Some scientists feel that the hay is safe after the percent of radioactive iodine has declined to 10% of the original amount. Find the number of days before the hay can be used.

48. *Radioactive Iodine Level* Other scientists believe that the hay is not safe until the level of radioactive iodine has declined to only 1% of the original level. Find the number of days this would take.

For Exercises 49 and 50, refer to Example 8.

49. *Drug Level* If 250 mg of a drug are administered, and the drug is only 75% as effective each subsequent hour, how much effective medicine will remain in the person's system after 6 hours?

50. *Drug Level* A new drug has been introduced which is 80% as effective each hour as the previous hour. A minimum of 20 mg must remain in the patient's bloodstream during the course of treatment. If 100 mg are administered, how many hours may elapse before another dose is necessary?

51. *Cesarean Section Births* The number of Cesarean section deliveries in the United States has increased over the years. Between the years 1980 and 1989, the number of such births, in thousands, can be approximated by the function $f(t) = 625e^{.0516t}$, where $t = 1$ corresponds to the year 1980. Based on this function, what would be the approximate number of Cesarean section deliveries in 1996? (*Source:* U.S. National Center for Health Statistics.)

52. *Outpatient Surgery Growth*   The growth of outpatient surgery as a percent of total surgeries at hospitals is shown in the accompanying graph. Connecting the tops of the bars with a continuous curve would give a graph that indicates logarithmic growth. The function with $f(x) = -1317 + 304 \ln x$, where x represents the number of years since 1900 and $f(x)$ is the percent, approximates the curve reasonably well. What does this function predict for the percent of outpatient surgeries in 1998?

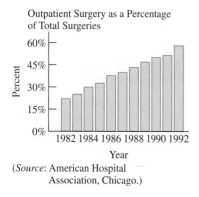

Outpatient Surgery as a Percentage of Total Surgeries

(Source: American Hospital Association, Chicago.)

In Exercises 53–60, solve the problems from the social sciences.

53. *Internet Use*   The number of Internet users was estimated to be 1.6 million in October of 1989 and 39 million in October of 1994. This number has grown exponentially. The function $f(x) = a(b^x)$ can be used to model the number of users, where a and b are constants and $x = 0$ (in months) corresponds to October of 1989. (*Source:* Genesis Corp.)

(a) Write the information in the first sentence as two ordered pairs of the form $(x, f(x))$, and substitute these values into the function to get a system of two equations with two variables. (See Chapter 7.)

(b) Solve the system from part (a) to approximate the values of a and b to four decimal places, and determine $f(x)$.

(c) Use $f(x)$ to estimate the number of users in October of 1992. Compare your estimate with the actual value of 10.3 million.

(d) Graphically determine the month and year when there were 30 million users.

54. *Racial Mix in the United States*   In 1995, the U.S. racial mix was 75.3% white, 9.0% Hispanic, 12.0% African American, 2.9% Asian/Pacific Islanders, and .8% Native American. Their respective percentages of executives, managers, and administrators in private-industry communication firms were 84.0%, 9.5%, 3.4%, 2.0%, and 0%, respectively. Assume these percents remain constant. If the number of African American executives, managers, and administrators increases

at a rate of 5% per year, determine the year when their representation will reach 12.0% of the total number of executives, managers, and administrators in communications. (*Source:* U.S. Labor Department's Glass Ceiling Commission.)

55. *Racial Mix in the United States*   (Refer to Exercise 54.) In private-industry retail trade, the percentages in 1995 were 80.8% white, 4.8% Hispanic, 4.9% African American, 5.2% Asian/Pacific Islanders, and 0% Native American. Assume these percents remain constant. If the number of Hispanic executives, managers, and administrators increases at a rate of 3% per year, determine the year when their representation will reach 9.0% of the total number in retail trade. (*Source:* U.S. Labor Department's Glass Ceiling Commission.)

56. *Social Security Tax*   The maximum Social Security tax in 1985 was $2791.80 and in 1995 it was $4681.80. Use an exponential growth function to approximate the annual percentage increase. Then, compare it to consumer prices which rose at an annual rate of 3.56%. (*Source:* Social Security Administration.)

57. *Recreation Expenditures*   Personal consumption expenditures for recreation in billions of dollars in the United States during the years 1984 through 1990 can be approximated by the function $A(t) = 185.4e^{.0587t}$, where $t = 0$ corresponds to the year 1984. Based on this model, how much would personal consumption expenditures be in 1996? (*Source:* U.S. Bureau of Economic Analysis.)

58. *Sale of Books in the United States*   The number of books, in millions, sold per year in the United States between 1985 and 1990 can be approximated by the function $A(t) = 1757e^{.0264t}$, where $t = 0$ corresponds to the year 1985. Based on this model, how many books would be sold in 1996? (*Source:* Book Industry Study Group.)

59. *Measure of Living Standards*   One measure of living standards in the United States is given by $L = 9 + 2e^{.15t}$, where t is the number of years since 1982. Find L for the following years.

(a) 1982   (b) 1986   (c) 1992

(d) Graph L in the window [0, 10] by [0, 30].

60. *Population Decline*   A midwestern city finds its residents moving to the suburbs. Its population is declining according to the relationship $P = P_0e^{-.04t}$, where t is time measured in years and P_0 is the population at time $t = 0$. Assume that $P_0 = 1,000,000$.

(a) Find the population at time $t = 1$.

(b) Estimate the time it will take for the population to be reduced to 750,000.

(c) How long will it take for the population to decline to half the initial number?

Many environmental situations place effective limits on the growth of the number of an organism in an area. Many such limited growth situations are described by the **logistic function,** defined by

$$G(t) = \frac{MG_0}{G_0 + (M - G_0)e^{-kMt}},$$

where G_0 is the initial number present, M is the maximum possible size of the population, and k is a positive constant. The screen shown here illustrates a typical logistic function graph. Graphing calculators have the capability of determining logistic function models for appropriate data.

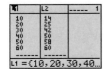 

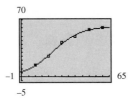

61. *Logistic Function* Assume that $G_0 = 100$, $M = 2500$, $k = .0004$, and $t = $ time in decades (10-year periods).

(a) Use a calculator to graph the S-shaped function, using $0 \le t \le 8$, $0 \le y \le 2500$.

(b) Estimate the value of $G(2)$ from the graph. Then, evaluate $G(2)$ analytically to find the population after 20 years.

(c) Find the t-coordinate of the intersection of the curve with the horizontal line $Y = 1000$ to estimate the number of decades required for the population to reach 1000. Then, solve $G(t) = 1000$ analytically to obtain the exact value of t.

62. *Logistic Function for Fatherless Children* The graph shows that the percent y of U.S. children growing up without a father has increased rapidly since 1950. If x represents the number of years since 1900, the function defined by

$$f(t) = \frac{25}{1 + 1364.3e^{-x/9.316}}$$

models the data fairly well.

(a) From the graph, in what year were 20% of U.S. children living without a father?

(b) Based on the function f, what was the percent in 1985?

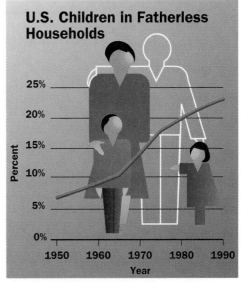

(*Sources:* National Longitudinal Survey of Youth; U.S. Department of Commerce; Bureau of the Census.)

CHAPTER 5   SUMMARY

Section	Important Concepts
5.1 Introduction to Exponential Functions	**DEFINITION OF EXPONENTIAL FUNCTION** If $a > 0$, $a \ne 1$, then $$f(x) = a^x$$ is the exponential function with base a.

THE EXPONENTIAL FUNCTIONS $f(x) = a^x, 0 < a < 1$
$f(x) = a^x, a > 1$

For the graph of: See page:
$f(x) = a^x, 0 < a < 1$ 398
$f(x) = a^x, a > 1$ 398

COMPOUNDING FORMULAS

$$A = P\left(1 + \frac{r}{n}\right)^{nt} \qquad A = Pe^{rt}$$

To solve a Type 1 exponential equation such as $4^x = 8^{2x-3}$, write each base as a power of the same base, apply the power rule for exponents, set exponents equal, and solve the resulting equation.

5.2 Logarithms and Their Properties

DEFINITION OF LOGARITHM
For all positive numbers a, where $a \neq 1$,

$$a^k = x \qquad \text{is equivalent to} \qquad k = \log_a x.$$

A logarithm is an exponent, and $\log_a x$ is the exponent to which a must be raised in order to obtain x. The number a is called the *base* of the logarithm, and x is called the *argument* of the expression $\log_a x$. The value of x will always be positive.

COMMON LOGARITHM
$\log x = \log_{10} x$ for all positive numbers x.

NATURAL LOGARITHM
$\ln x = \log_e x$ for all positive numbers x.

PROPERTIES OF LOGARITHMS
   1. If $a > 0$, $a \neq 1$, then $\log_a 1 = 0$.
   2. If $a > 0$, $a \neq 1$, and k is a real number, then $\log_a a^k = k$.
   3. If $a > 0$, $a \neq 1$, and $k > 0$, then $a^{\log_a k} = k$.

For $x > 0$, $y > 0$, $a > 0$, $a \neq 1$, and any real number r,

Product Rule   **4.** $\log_a xy = \log_a x + \log_a y$.
   (The logarithm of the product of two numbers is equal to the sum of the logarithms of the numbers.)

Quotient Rule   **5.** $\log_a \frac{x}{y} = \log_a x - \log_a y$.
   (The logarithm of the quotient of two numbers is equal to the difference between the logarithms of the numbers.)

Power Rule   **6.** $\log_a x^r = r \log_a x$.
   (The logarithm of a number raised to a power is equal to the exponent multiplied by the logarithm of the number.)

CHANGE-OF-BASE RULE
For any positive real numbers x, a, and b, where $a \neq 1$ and $b \neq 1$,

$$\log_a x = \frac{\log_b x}{\log_b a}.$$

5.3 Introduction to Logarithmic Functions

The functions

$$f(x) = a^x \quad \text{and} \quad g(x) = \log_a x$$

are inverses.

THE LOGARITHMIC FUNCTIONS $f(x) = \log_a x, \, 0 < a < 1$
$$f(x) = \log_a x, \, a > 1$$

For the graph of: See page:
$f(x) = \log_a x, \, 0 < a < 1$ 422
$f(x) = \log_a x, \, a > 1$ 421

5.4 Exponential and Logarithmic Equations and Inequalities

PROPERTIES OF LOGARITHMIC AND EXPONENTIAL FUNCTIONS
For $b > 0$ and $b \neq 1$,

  1. $b^x = b^y$ if and only if $x = y$.
  2.   If $x > 0$ and $y > 0$,

$$\log_b x = \log_b y \quad \text{if and only if} \quad x = y.$$

To solve a Type 2 exponential equation such as $2^x = 3^{x+1}$, take the same base logarithm on both sides, apply the power rule for logarithms so that the variables are no longer in the exponents, and solve the resulting equation.

SOLVING EXPONENTIAL AND LOGARITHMIC EQUATIONS
An exponential or logarithmic equation may be solved by changing the equation into one of the following forms, where a and b are real numbers, $a > 0$, and $a \neq 1$.

  1. $a^{f(x)} = b$
      Solve by taking logarithms of each side. (Natural logarithms are the best choice if $a = e$.)
  2. $\log_a f(x) = \log_a g(x)$
      From the given equation, $f(x) = g(x)$, which is solved analytically.
  3. $\log_a f(x) = b$
      Solve by using the definition of logarithm to write the expression in exponential form as $f(x) = a^b$.

5.5 Applications and Modeling with Exponential and Logarithmic Functions

EXPONENTIAL GROWTH AND DECAY FUNCTIONS

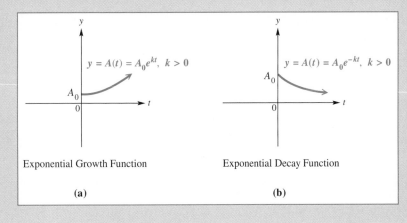

Exponential Growth Function   Exponential Decay Function

(a)   **(b)**

CHAPTER 5   REVIEW EXERCISES

Match each equation with the graph that most closely resembles its graph. Assume that a > 1.

1. $y = a^{x+2}$

2. $y = a^x + 2$

3. $y = -a^x + 2$

4. $y = a^{-x} + 2$

A.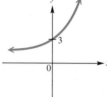

B.

C.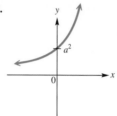

D.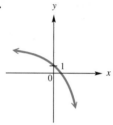

Consider the exponential function $y = f(x) = a^x$ graphed here. Answer the following based on the graph.

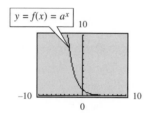

5. What is true about the value of a in comparison to 1?

6. What is the domain of f?

7. What is the range of f?

8. What is the value of $f(0)$?

9. Sketch the graph of $y = f^{-1}(x)$ by hand.

10. What is the expression that defines $f^{-1}(x)$?

Solve the equation in part (a) analytically. Then, solve the inequalities in parts (b) and (c) by using a graph.

11. (a) $\left(\dfrac{1}{8}\right)^{-2x} = 2^{x+3}$

(b) $\left(\dfrac{1}{8}\right)^{-2x} \geq 2^{x+3}$

(c) $\left(\dfrac{1}{8}\right)^{-2x} \leq 2^{x+3}$

12. (a) $3^{-x} = \left(\dfrac{1}{27}\right)^{1-2x}$

(b) $3^{-x} > \left(\dfrac{1}{27}\right)^{1-2x}$

(c) $3^{-x} < \left(\dfrac{1}{27}\right)^{1-2x}$

13. (a) $.5^{-x} = .25^{x+1}$

(b) $.5^{-x} < .25^{x+1}$

(c) $.5^{-x} > .25^{x+1}$

14. (a) $.4^x = 2.5^{1-x}$

(b) $.4^x < 2.5^{1-x}$

(c) $.4^x > 2.5^{1-x}$

15. The graphs of $y = x^2$ and $y = 2^x$ have the points (2, 4) and (4, 16) in common. There is a third point in common to the graphs whose coordinates can be approximated by using a graphing calculator. Find the coordinates, giving as many decimal places as your calculator will display.

16. A calculator-generated graph of $y = \log_2 x$ is shown with the values of the ordered pair with $x = 5$. What does the value of y represent?

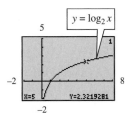

Use a calculator to find an approximation for each logarithm. Give the maximum number of digits possible on your calculator.

17. log 58.3

18. log .00233

19. ln 58.3

20. ln .00233

Evaluate each of the following, giving exact or approximate values as directed. In the case of approximations, give as many decimal places as your calculator shows.

21. $\log_{13} 1$ (exact)

22. $\ln e^{\sqrt{6}}$ (exact)

23. $\log_5 5^{12}$ (exact)

24. $7^{\log_7 13}$ (exact)

25. $\log_4 9$ (approximate)

26. x, if $3^x = 5$ (approximate)

In Exercises 27–32, identify the corresponding graph for each function.

27. $f(x) = \log_2 x$

28. $f(x) = \log_2 (2x)$

29. $f(x) = \log_2 \left(\dfrac{1}{x}\right)$

30. $f(x) = \log_2 \left(\dfrac{x}{2}\right)$

31. $f(x) = \log_2 (x - 1)$

32. $f(x) = \log_2 (-x)$

A.   **B.**

C.   **D.**

E.   **F.**

33. What is the base of the logarithmic function whose graph contains the point (81, 4)?

34. What is the base of the exponential function whose graph contains the point $(-4, 1/16)$?

Use properties of logarithms, if possible, to write each of the following logarithms as a sum, difference, or product of logarithms.

35. $\log_3 \left(\dfrac{mn}{5r}\right)$

36. $\log_2 \left(\dfrac{\sqrt{7}}{15}\right)$

37. $\log_5 \left(x^2 y^4 \sqrt[5]{m^3 p}\right)$

38. $\log_7 (7k + 5r^2)$

Solve the equation in part (a) analytically. Then, solve the inequalities in parts (b) and (c) by using a graph.

39. (a) $\log (x + 3) + \log x = 1$

   (b) $\log (x + 3) + \log x > 1$

   (c) $\log (x + 3) + \log x < 1$

40. (a) $\ln e^{\ln x} - \ln (x - 4) = \ln 3$

   (b) $\ln e^{\ln x} - \ln (x - 4) \geq \ln 3$

   (c) $\ln e^{\ln x} - \ln (x - 4) \leq \ln 3$

41. (a) $\ln e^{\ln 2} - \ln (x - 1) = \ln 5$

   (b) $\ln e^{\ln 2} - \ln (x - 1) \geq \ln 5$

   (c) $\ln e^{\ln 2} - \ln (x - 1) \leq \ln 5$

Use one of the methods described in this chapter to solve each equation. Round to the nearest thousandth if necessary.

42. $8^x = 32$

43. $\dfrac{8}{27} = x^{-3}$

44. $10^{2r-3} = 17$

45. $e^{x+1} = 10$

46. $\log_{64} x = \dfrac{1}{3}$

47. $\ln (6x) - \ln (x + 1) = \ln 4$

48. $\log_{16} \sqrt{x + 1} = \dfrac{1}{4}$

49. $\ln x + 3 \ln 2 = \ln \dfrac{2}{x}$

50. $\ln \left[\ln (e^{-x}) \right] = \ln 3$

51. Solve $N = a + b \ln \dfrac{c}{d}$ for c.

Use the x-intercept method to estimate the solution(s) of each equation.

52. $\log_{10} x = x - 2$

53. $2^{-x} = \log_{10} x$

54. $x^2 - 3 = \log x$

Solve each application of exponential or logarithmic functions.

55. *Interest Rate*   What annual interest rate, to the nearest tenth, will produce $8780 if $3500 is left at interest for 10 years?

56. *Growth of an Account*   Find the number of years (to the nearest tenth) needed for $48,000 to become $58,344 at 5% interest, compounded semiannually.

57. *Growth of an Account*   Manuel deposits $10,000 for 12 years in an account paying 12%, compounded annually. He then puts this total amount on deposit in another account paying 10%, compounded semiannually, for another 9 years. Find the total amount on deposit after the entire 21-year period.

58. *Growth of an Account*   Anne Kelly deposits $12,000 for 8 years in an account paying 5%, compounded annually. She then leaves the money alone with no further deposits at 6%, compounded annually, for an additional 6 years. Find the total amount on deposit after the entire 14-year period.

59. *Growth of an Account*   Suppose that $2000 is invested in an account that pays 3% annually and then is left untouched for 5 years.

   (a) How much will be in the account if interest is compounded quarterly (4 times per year)?

   (b) How much will be in the account if interest is compounded continuously?

   (c) To the nearest tenth of a year, how long will it take the $2000 to triple if interest is compounded continuously?

60. *Gross National Product*   Suppose the gross national product (GNP) of a small country (in millions of dollars) is approximated by $G(t) = 15 + 2 \log t$, where t is time in years, for $1 \leq t \leq 6$. Find the GNP at the following times.

   (a) 1 year   **(b)** 2 years   **(c)** 5 years

61. *Pollutant Concentration*   The concentration of pollutants, in grams per liter, in the east fork of the Big Weasel River is approximated by $P(x) = .04e^{-4x}$, where x is the number of miles downstream from a paper mill that the measurement is taken. Find each of the following.

   (a) $P(.5)$

   (b) $P(1)$

   (c) The concentration of pollutants 2 miles downstream

   (d) The number of miles downstream where the concentration of pollutants is .002 gram per liter

62. *Repetitive Skills*   A person learning certain skills involving repetition tends to learn quickly at first. Then, learning tapers off and approaches some upper limit. Suppose the number of symbols per minute that a textbook typesetter can produce is given by $p(x) = 250 - 120(2.8)^{-5x}$, where x is the number of months the typesetter has been in training. Find each of the following.

   (a) $p(2)$

   (b) $p(4)$

   (c) $p(10)$

   (d) Graph $y = p(x)$ in the window [0, 10] by [0, 300], and support the answer to part (a).

63. *Decay by Radioactivity*   The amount of radioactive material, in grams, present after t days is given by $A(t) = 600e^{-.05t}$.

   (a) Find the amount present after 12 days.

   (b) Find the half-life of the material.

64. *Free Fall of a Skydiver*   A skydiver in free fall travels at the speed of $v(t) = 176(1 - e^{-.18t})$ ft per sec after t sec. How long will it take for the skydiver to attain the speed of 147 ft per sec (100 mph)?

65. *Atmospheric Pressure*   (Refer to Example 9 in Section 5.5.) The atmospheric pressure (in millibars) at a given altitude (in meters) is listed in the table.

   (a) Plot the points $(x, \ln P)$ on the coordinate axes. Is there a linear relationship between x and $\ln P$?

   (b) If $P = Ce^{kx}$ with constants C and k, explain why there is a linear relationship between x and $\ln P$.

Altitude (x)	Pressure (P)
0	1013
1000	899
2000	795
3000	701
4000	617
5000	541
6000	472
7000	411
8000	357
9000	308
10,000	265

CHAPTER 5   TEST

1. Match each equation with its graph.

   (a) $y = \log_{1/2} x$ **(b)** $y = e^x$

   (c) $y = \ln x$ **(d)** $y = \left(\frac{1}{2}\right)^x$

   A.
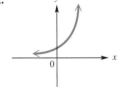

   B.
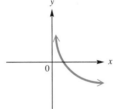

   C.
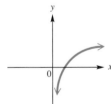

   D.
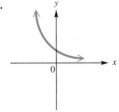

2. Consider the function $f(x) = -2^{x-1} + 8$.

   (a) Graph it in the standard viewing window of your calculator.

   (b) Use the terminology of Chapter 2 to explain how the graph of f can be obtained from the graph of $y = 2^x$.

   (c) Give the domain and the range of f.

   (d) Does the graph have an asymptote? If so, is it vertical or horizontal, and what is its equation?

   (e) Find the x- and y-intercepts analytically, and use the graph from part (a) to support your answers graphically.

3. (a) Solve the equation $\left(\frac{1}{8}\right)^{2x-3} = 16^{x+1}$ analytically.

   (b) Use a graph and the result of part (a) to solve the inequality $\left(\frac{1}{8}\right)^{2x-3} > 16^{x+1}$.

   (c) Use a graph and the result of part (a) to solve the inequality $\left(\frac{1}{8}\right)^{2x-3} < 16^{x+1}$.

4. *Growth of an Account*   Suppose that $10,000 is invested at 5.5% for 4 years. Find the total amount present at the end of this time period if interest is compounded **(a)** quarterly and **(b)** continuously.

5. One of your friends is taking another mathematics course and tells you, "I have no idea what an expression like $\log_5 27$ really means." Write an explanation of what it means, and tell how you can find an approximation for it with a calculator.

6. Use a calculator to find an approximation of each logarithm to the nearest thousandth.

   (a) $\log 45.6$ **(b)** $\ln 470$ **(c)** $\log_3 769$

7. Use the power, quotient, and product properties of logarithms to write $\log \frac{m^3 n}{\sqrt{y}}$ as an equivalent expression.

8. Consider the equation $\log_2 x + \log_2 (x + 2) = 3$.

   (a) Solve the equation analytically. If there is an extraneous solution, what is it?

(b) To support the solution in part (a), we may graph $y_1 = \log_2 x + \log_2 (x + 2) - 3$ and find the x-intercept. Write an expression for y_1, using the change-of-base rule with base 10, and graph the function to support the solution from part (a).

(c) Use the graph to solve the inequality $\log_2 x + \log_2 (x + 2) > 3$.

9. Solve the equation $6^{2-x} = 2^{3x+1}$. Give the solution set
(a) with an exact value, using common logarithms and
(b) with an approximation to the nearest thousandth.

10. A population is increasing according to the growth law $y = 2e^{.02t}$, where y is in millions and t is in years. Match each question with one of the solutions A, B, C, or D.

(a) How long will it take for the population to triple?

(b) When will the population reach 3 million?

(c) How large will the population be in 3 years?

(d) How large will the population be in 4 months?

A Evaluate $2e^{.02(1/3)}$.   **B** Solve $2e^{.02t} = 3 \cdot 2$ for t.

C Evaluate $2e^{.02(3)}$.   **D** Solve $2e^{.02t} = 3$ for t.

11. *Drug Level in the Bloodstream*   After a medical drug is injected directly into the bloodstream, it is gradually eliminated from the body. Graph each of the following functions on the interval [0, 10]. Use [0, 500] for the range of $A(t)$. Use a graphing calculator to determine the function that best describes the amount $A(t)$ (in milligrams) of a drug remaining in the body after t hours if 350 milligrams were initially injected.

(a) $A(t) = t^2 - t + 350$

(b) $A(t) = 350 \log(t + 1)$

(c) $A(t) = 350(.75)^t$

(d) $A(t) = 100(.95)^t$

12. *Transistors on Computer Chips*   Computing power of personal computers has increased dramatically as a result of the ability to place an increasing number of transistors on a single processor chip. The table lists the number of transistors on some popular computer chips made by Intel. (*Source:* Intel.)

Year	Chip	Transistors
1971	4004	2300
1986	386DX	275,000
1989	486DX	1,200,000
1993	Pentium	3,300,000
1995	P6	5,500,000

(a) Plot the data. Let the x-value represent the year, where $x = 0$ corresponds to 1971, and let the y-value represent the number of transistors.

 (b) Discuss which type of function $y = f(x)$ describes the data best, where a and b are constants.

(i) $f(x) = ax + b$ (linear)

(ii) $f(x) = a \ln b(x + 1)$ (logarithmic)

(iii) $f(x) = ae^{bx}$ (exponential)

(c) Determine a function f that approximates these data. Plot f and the data on the same coordinate system.

(d) Assuming that the present trend continues, use f to predict the number of transistors on a chip in the year 2000.

CHAPTER 5   PROJECT

CALIFORNIA'S PRISON POPULATION
(with a lesson about careless use of mathematical models)

In 1995, the California Department of Finance and the California Department of Corrections reported the following information on the state population and the prison population in the state.

Year	State Population	Prison Population
1980	23,668,049	23,511
1994	31,960,600	125,000
1999 (projected)	35,824,000	245,000

It should be noted that *no methods were given to justify the projected 1999 figures.*

In this project, let's proceed under the assumption that these populations, like many, will grow exponentially over time. Assuming the growth rate does not change, a function of the form $f(x) = a \cdot b^x$ can be used to model each population. In these models, *x* will represent the year, *f*(*x*) will be the population in year *x*, the constant *a* will be the initial population (that is, when $x = 0$), and the constant *b* will be the growth rate.

ACTIVITIES

1. Use the data for 1980 and 1994 to find two exponential functions to model the data. Use $x = 0$ for 1980 and $x = 14$ for 1994. Construct the functions so that the function value gives the population in thousands. Use the two data points in each case to find *a* and *b*.

2. Use your models to determine the rate of increase in each function. Express in terms of percent.

3. Use your graphing calculator to draw a scatterplot of each data set. Then graph your model function over the data points. (In the figures that follow, we illustrate how a TI-83 calculator is used to enter the data points for the state prison data and draw a scatterplot, and how a model function is graphed over the scatterplot.)

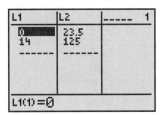

L1 represents year and L2 represents prison population data (in thousands).

FIGURE A

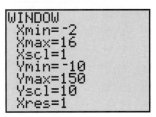

This window will allow the user to view the two data points for the prison population.

FIGURE B

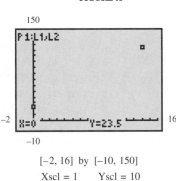

[−2, 16] by [−10, 150]
Xscl = 1   Yscl = 10

This is a scatterplot of the two prison data points.

FIGURE C

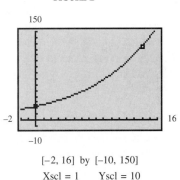

[−2, 16] by [−10, 150]
Xscl = 1   Yscl = 10

The exponential function graph determined as described in Activity 1 passes through the two data points.

FIGURE D

4. Use your models to project the 1999 populations. How close are your projections to the ones provided?

5. Use your models to predict when the state population in California will reach 100 million, and when the prison population will reach 1 million.

6. Graph both functions in the same window, so that a comprehensive graph of each is visible. Use the graphs to predict what will happen near the middle of the twenty-first century. What is obviously wrong with the prediction? What is the cause of this error? Comment on the following statement: "A model is valid only within the domain of the given data."

The Conic Sections and Parametric Equations

Since the beginning of civilization, people have been fascinated by and compelled to understand the universe they live in. The Greeks together with the early Christian astronomers believed that earth was the center of the universe and the sun and planets traveled in circular orbits around earth. The circle was regarded as the perfect geometric shape. Later, in the sixteenth century, the greatest observational astronomer of the age, Tycho Brahe, recorded precise data on planetary movement in the sky. With Brahe's data, Johannes Kepler in 1619 empirically determined that the planets do not move in circular orbits, but rather in elliptical orbits around the sun. In 1686, Newton used Kepler's work to show analytically that elliptical orbits were a result of his famous theory of gravitation. Edmund Halley determined that comets also followed elliptical orbits around the sun and accurately predicted the return of Halley's comet. People now know that both celestial objects and satellites move through space in one of three types of paths: elliptical, parabolic, or hyperbolic. These curves are called *conic sections* and were discovered in 200 B.C. by the Greek geometer Apollonius. Today, scientists are searching the sky for information about the beginning of the universe and for signs of life elsewhere. Enormous radio telescopes with parabolic dishes—much like television satellite dishes—search the sky continuously for new information.

Throughout history, parabolas, ellipses, and hyperbolas have played a central role in our understanding of the universe. In this chapter, we will learn about these age-old curves that have had such a profound influence on our understanding of who we are and the cosmos we live in.*

---

*Sources: Boorse, H., L. Motz, and J. Weaver, *The Atomic Scientists*, John Wiley & Sons, Inc., 1989; National Council of Teachers of Mathematics, *Historical Topics for the Mathematics Classroom*, Thirty-first Yearbook, Washington, D.C., 1969; Sagan, C., *Cosmos*, Random House, New York, 1980.

6.1 CIRCLES AND PARABOLAS

Introduction to the Conic Sections ◆ Equations and Graphs of Circles ◆ An Application of Circles ◆ Equations and Graphs of Parabolas ◆ An Application of Parabolas

INTRODUCTION TO THE CONIC SECTIONS

In the first two sections of this chapter, we will expand upon some ideas introduced earlier in the text. In Chapter 3, we saw that the graph of a quadratic function is a parabola, and in Section 4.3 we saw how horizontal parabolas can be graphed by using the union of root functions. In Section 4.3, we also saw that the graph of $x^2 + y^2 = r^2$ is a circle with center at the origin and radius r. By treating a circle as the union of two root functions, we can use a graphing calculator to graph the circle. We will now examine parabolas and circles in terms of their actual definitions, based on the distance formula (Section 1.4). Along with ellipses and hyperbolas, they form a group of curves known as the **conic sections**.

By intersecting a cone with a plane, we may obtain the conic sections. Figure 1 illustrates these curves.

These curves can be defined mathematically by using the distance formula: the distance between the points $A(x_1, y_1)$ and $B(x_2, y_2)$, symbolized $d(A, B)$, is given by the expression $\sqrt{(x_2 - x_1)^2 + (y_2 - y_1)^2}$.

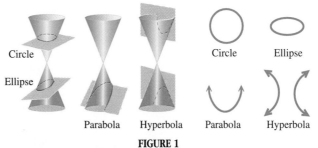

Circle   Ellipse   Parabola   Hyperbola   Circle   Ellipse   Parabola   Hyperbola

FIGURE 1

EQUATIONS AND GRAPHS OF CIRCLES

We begin with the definition of a circle.

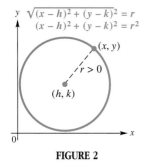

y $\sqrt{(x-h)^2 + (y-k)^2} = r$
$(x-h)^2 + (y-k)^2 = r^2$
(x, y)
$r > 0$
(h, k)
x

FIGURE 2

DEFINITION OF A CIRCLE

A **circle** is a set of points in a plane, each of which is equidistant from a fixed point. The distance is called the **radius** of the circle, and the fixed point is called the **center**.

Suppose that a circle has center (h, k) and radius $r > 0$, as shown in Figure 2. Then, by the distance formula, if (x, y) is any point on the circle, $\sqrt{(x - h)^2 + (y - k)^2} = r$. Squaring both sides of this equation gives us the center-radius form of the equation of the circle.

CENTER-RADIUS FORM OF THE EQUATION OF A CIRCLE

The circle with center (h, k) and radius r has equation $(x - h)^2 + (y - k)^2 = r^2$, the **center-radius form** of the equation of a circle.

Notice that a circle is an example of a mathematical relation, but is not the graph of a function, as it does not pass the vertical line test (Section 1.2).

EXAMPLE 1 *Finding the Equation of a Circle* Find the center-radius form of the equation of a circle with radius 6 and center at $(-3, 4)$. Graph the circle by hand, and give the domain and the range of the relation.

SOLUTION Using the center-radius form with $h = -3$, $k = 4$, and $r = 6$, we find that the equation of the circle is

$$(x - (-3))^2 + (y - 4)^2 = 6^2$$

or

$$(x + 3)^2 + (y - 4)^2 = 36.$$

Its graph is shown in Figure 3. As seen there, the domain is $[-9, 3]$ and the range is $[-2, 10]$. (We will see later how this circle can be graphed, using a graphing calculator.) ◆

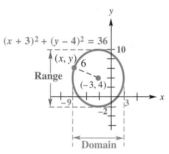

FIGURE 3

If a circle has center at the origin $(0, 0)$, then its equation is found by using $h = 0$ and $k = 0$ in the center-radius form.

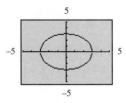

FIGURE 4

EQUATION OF A CIRCLE WITH CENTER AT THE ORIGIN

A circle with center $(0, 0)$ and radius r has equation $x^2 + y^2 = r^2$.

Figure 4 shows the graph of a circle with center at the origin and radius r.

EXAMPLE 2 *Finding the Equation of a Circle with Center at the Origin* Find the equation of a circle with center at the origin and radius 3. Give a traditional graph, and state the domain and the range of the relation.

SOLUTION Using the form $x^2 + y^2 = r^2$ with $r = 3$, we find that the equation of the circle is $x^2 + y^2 = 9$. The graph is shown in Figure 5. Both the domain and the range are $[-3, 3]$. ◆

A graphing calculator in the function mode cannot directly graph a circle. In order to do so, we must first solve the equation of the circle for y, obtaining two functions y_1 and y_2. The union of these two graphs will be the graph of the entire circle.

5

−5 ┤ 5

−5

[−5, 5] by [−5, 5]
Xscl = 1   Yscl = 1

The graph of $x^2 + y^2 = 9$ on a rectangular screen looks like an ellipse.

Note In order to obtain an undistorted graph on a graphing calculator screen, a *square* screen must be used. See your instruction manual if necessary.

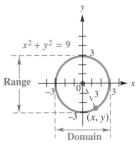

FIGURE 5

EXAMPLE 3 *Graphing a Circle Using a Graphing Calculator* For each of the following circles, solve the equation for y, and then use a graphing calculator to graph the circle in a square viewing window.

(a) $x^2 + y^2 = 9$

(b) $(x + 3)^2 + (y - 4)^2 = 36$

SOLUTION In both cases, we must solve for y. Recall that if $k > 0$, $y^2 = k$ has two real solutions, $\sqrt{k}$ and $-\sqrt{k}$.

(a) $x^2 + y^2 = 9$

$$y^2 = 9 - x^2 \qquad \text{Subtract } x^2.$$

$$y = \pm\sqrt{9 - x^2} \qquad \text{Take square roots.}$$

We graph two functions, $y_1 = \sqrt{9 - x^2}$ and $y_2 = -\sqrt{9 - x^2}$. See Figure 6(a), and compare to the traditional graph in Figure 5.

(b) $(x + 3)^2 + (y - 4)^2 = 36$

$$(y - 4)^2 = 36 - (x + 3)^2 \qquad \text{Subtract } (x + 3)^2.$$

$$y - 4 = \pm\sqrt{36 - (x + 3)^2} \qquad \text{Take square roots.}$$

$$y = 4 \pm \sqrt{36 - (x + 3)^2} \qquad \text{Add 4.}$$

Here, the two functions to be graphed are

$$y_1 = 4 + \sqrt{36 - (x + 3)^2} \quad \text{and} \quad y_2 = 4 - \sqrt{36 - (x + 3)^2}.$$

See Figure 6(b), and compare to Figure 3. ◆

TECHNOLOGY NOTE

When entering expressions like those required for graphing the functions graphed in Figures 6(a) and 6(b), you must be careful to include the expressions under the radicals within parentheses. For example, the function y_1 in Example 3(b) must be entered as $4 + \sqrt{(36 - (x + 3)^2)}$.

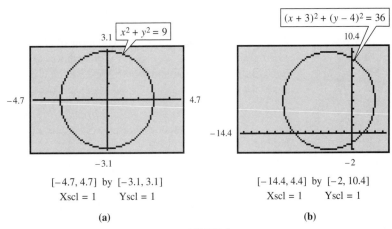

$[-4.7, 4.7]$ by $[-3.1, 3.1]$
Xscl = 1   Yscl = 1
(a)

$[-14.4, 4.4]$ by $[-2, 10.4]$
Xscl = 1   Yscl = 1
(b)

FIGURE 6

For Group Discussion

1. Notice in Example 3(b) that the final expressions for y_1 and y_2 have radicand* $36 - (x + 3)^2$. If this were simplified analytically, the result would be $27 - x^2 - 6x$. Use your calculator to replace the given radicands with this simplified result. Do you get the same graphs? Does it matter to the calculator which form you use? What would be a possible drawback of attempting to use the simplified form? (*Hint*: Nobody's perfect.)

2. Suppose the functions y_1 and y_2 from Example 3(a) were graphed in a standard window. What might be a possible misinterpretation by a student who has not studied the mathematical theory along with the technological approach?

The table feature of a graphing calculator suggests the domain of the two functions used to get the graph of the circle $x^2 + y^2 = 9$. Figure 7 shows two views of the table. You can scroll through the table by using the up arrow and the down arrow. Notice that the table shows ERROR for Y_1 and Y_2 when $x < -3$ and when $x > 3$. Y_1 and Y_2 are evaluated only when $-3 \le x \le 3$. This suggests that the domain of both Y_1 and Y_2 is the closed interval $[-3, 3]$. This result can be supported graphically in a window that evaluates the endpoints of each semicircle (a decimal window will usually work).

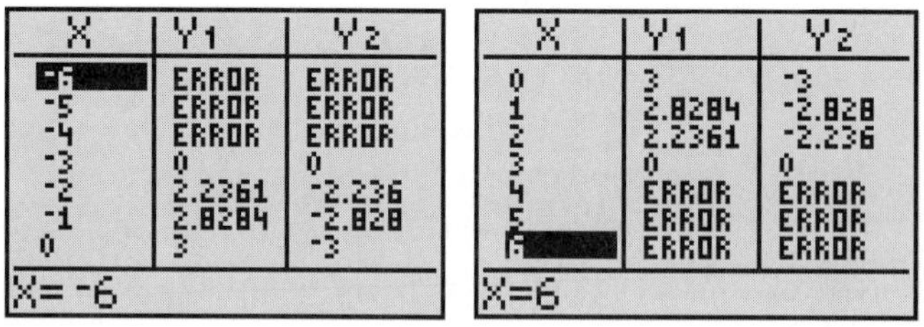

X	Y1	Y2
-6	ERROR	ERROR
-5	ERROR	ERROR
-4	ERROR	ERROR
-3	0	0
-2	2.2361	-2.236
-1	2.8284	-2.828
0	3	-3

X= -6

X	Y1	Y2
0	3	-3
1	2.8284	-2.828
2	2.2361	-2.236
3	0	0
4	ERROR	ERROR
5	ERROR	ERROR
6	ERROR	ERROR

X=6

(a)   (b)

FIGURE 7

Starting with the center-radius form of the equation of a circle, $(x - h)^2 + (y - k)^2 = r^2$, and squaring $x - h$ and $y - k$ gives an equation of the form

$$x^2 + y^2 + cx + dy + e = 0, \tag{*}$$

where c, d, and e are real numbers. This result is the **general form of the equation of a circle**. Also, starting with an equation in the form of (*), the process of completing the square can be used to get an equation of the form $(x - h)^2 + (y - k)^2 = m$ for some number m. If $m > 0$, then $r^2 = m$, and the equation represents a circle with radius $\sqrt{m}$. If $m = 0$, then the equation represents the single point (h, k). If $m < 0$, no points satisfy the equation.

EXAMPLE 4   *Finding the Center and Radius by Completing the Square*   Find the center and the radius of the circle with equation

$$x^2 - 6x + y^2 + 4y - 3 = 0.$$

Then, graph the circle, using a graphing calculator.

SOLUTION   ANALYTIC   Our goal is to obtain an equivalent equation of the form $(x - h)^2 + (y - k)^2 = r^2$. To do this, first write the equation with the constant on the right.

$$x^2 - 6x + y^2 + 4y = 3$$

*The **radicand** is the expression under the radical symbol.

Now we complete the square in both x and y. To complete the square in x, we add $\left[\frac{1}{2}(-6)\right]^2 = 9$ to both sides, and to complete the square in y, we add $\left[\frac{1}{2}(4)\right]^2 = 4$ to both sides. Insert parentheses as shown.

$$(x^2 - 6x + 9) + (y^2 + 4y + 4) = 3 + 9 + 4$$

Now, factor on the left and add on the right.

$$(x - 3)^2 + (y + 2)^2 = 16 \qquad\qquad\qquad \textbf{(*)}$$

$$(x - 3)^2 + (y - (-2))^2 = 4^2 \qquad \text{Write } +2 \text{ as } -(-2) \text{ and 16 as } 4^2.$$

The circle has its center at $(3, -2)$ and its radius is 4. A traditional graph is shown in Figure 8(a).

GRAPHICAL   To graph it using a graphing calculator, use the equation in the line marked (*), and solve for y to get $y = -2 \pm \sqrt{16 - (x - 3)^2}$. Let y_1 and y_2 define these two expressions (one with $+$ and the other with $-$) to obtain the calculator-generated graph shown in Figure 8(b).   ◆

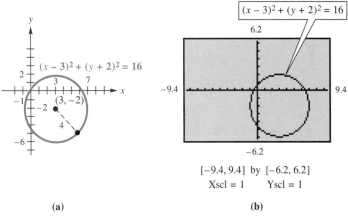

(a)   (b)

FIGURE 8

AN APPLICATION OF CIRCLES

Seismologists can locate the epicenter of an earthquake by determining the intersection of three circles. The radii of these circles represent the distances from the epicenter to each of three receiving stations. The centers of the circles represent the receiving stations.

EXAMPLE 5   *Using Circles to Locate the Epicenter of an Earthquake*   Suppose that an earthquake is recorded by three receiving stations, A, B, and C, located on a coordinate plane 2, 5, and 4 units, respectively, from the epicenter. Use Figure 9 to determine the location of the epicenter with respect to the coordinate plane.

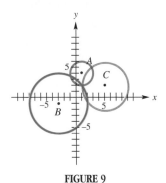

FIGURE 9

SOLUTION   Graphically, it appears that the epicenter is located at $(1, 2)$. To check this algebraically, determine the equation for each circle, and substitute $x = 1$ and $y = 2$.

Station A:

$$(x - 1)^2 + (y - 4)^2 = 4 \qquad ?$$
$$(1 - 1)^2 + (2 - 4)^2 = 4 \qquad ?$$
$$0 + 4 = 4 \qquad ?$$
$$4 = 4 \qquad \text{True}$$

Station B:

$$(x + 3)^2 + (y + 1)^2 = 25 \qquad ?$$
$$(1 + 3)^2 + (2 + 1)^2 = 25 \qquad ?$$
$$16 + 9 = 25 \qquad ?$$
$$25 = 25 \qquad \text{True}$$

Station C:

$$(x - 5)^2 + (y - 2)^2 = 16 \qquad ?$$
$$(1 - 5)^2 + (2 - 2)^2 = 16 \qquad ?$$
$$16 + 0 = 16 \qquad ?$$
$$16 = 16 \qquad \text{True}$$

Thus, we can be sure that the epicenter lies at $(1, 2)$.   ◆

EQUATIONS AND GRAPHS OF PARABOLAS

The definition of a parabola is also based on distance.

DEFINITION OF A PARABOLA

A **parabola** is the set of points in a plane equidistant from a fixed point and a fixed line. The fixed point is called the **focus**, and the fixed line, the **directrix**, of the parabola.

An equation of a parabola can be found from the definition as follows. Let the directrix be the line $y = -c$ and the focus be the point F with coordinates $(0, c)$, as shown in Figure 10. To get the equation of the set of points that are the same distance from the line $y = -c$ and the point $(0, c)$, choose one such point P and give it coordinates (x, y). Then, since $d(P, F)$ and $d(P, D)$ must have the same length, using the distance formula gives

$$d(P, F) = d(P, D)$$
$$\sqrt{(x - 0)^2 + (y - c)^2} = \sqrt{(x - x)^2 + (y - (-c))^2}$$
$$\sqrt{x^2 + (y - c)^2} = \sqrt{(y + c)^2}$$
$$x^2 + y^2 - 2yc + c^2 = y^2 + 2yc + c^2$$
$$x^2 = 4cy$$
$$y = \frac{1}{4c} x^2.$$

This discussion is summarized as follows.

PARABOLA WITH A VERTICAL AXIS

The parabola with focus at $(0, c)$ and directrix $y = -c$ has equation $y = \frac{1}{4c}x^2$. The parabola has a vertical axis, opens upward if $c > 0$, and opens downward if $c < 0$.

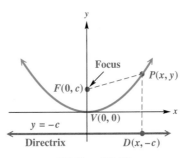

$$d(P, F) = d(P, D)$$
for all P on the curve.

FIGURE 10

If the directrix is the line $x = -c$, and the focus is at $(c, 0)$, using the definition of a parabola and the distance formula leads to the equation of a parabola with a horizontal axis. (See Exercise 103.)

PARABOLA WITH A HORIZONTAL AXIS

The parabola with focus at $(c, 0)$ and directrix $x = -c$ has equation $x = \frac{1}{4c}y^2$. The parabola opens to the right if $c > 0$, to the left if $c < 0$, and has a horizontal axis.

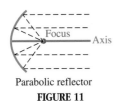

Parabolic reflector
FIGURE 11

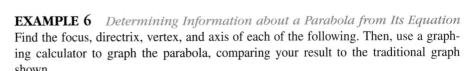

The graphs of the equations $y_1 = 2\sqrt{x}$ and $y_2 = -2\sqrt{x}$ form a horizontal parabola. Here $c = 1 > 0$, so $x \geq 0$.

The geometric properties of parabolas lead to many practical applications. For example, if a light source is placed at the focus of a parabolic reflector, as in Figure 11, light rays reflect parallel to the axis, making a spotlight or flashlight. The process also works in reverse. Light rays from a distant source come in parallel to the axis and are reflected to a point at the focus. (If such a reflector is aimed at the sun, a temperature of several thousand degrees may be obtained.) This use of parabolic reflection is seen in the satellite dishes used to pick up signals from communications satellites.

Note   A parabola with a horizontal axis is not the graph of a function. However, since the equation $x = \frac{1}{4c}y^2$ is equivalent to $y_1 = 2\sqrt{cx}$ or $y_2 = -2\sqrt{cx}$, such a parabola can be graphed with a graphing calculator, using the same general procedure discussed for circles in Example 3(a). If $c > 0$, then $x \geq 0$, and if $c < 0$, then $x \leq 0$.

EXAMPLE 6   *Determining Information about a Parabola from Its Equation*
Find the focus, directrix, vertex, and axis of each of the following. Then, use a graphing calculator to graph the parabola, comparing your result to the traditional graph shown.

(a) $y = \frac{1}{8}x^2$ **(b)** $x = -\frac{1}{28}y^2$

SOLUTION

(a)   The equation indicates that this is a vertical parabola. Because $4c = 8$, $c = 2$. Therefore, the focus is at $(0, 2)$, and the directrix has the equation $y = -2$. The vertex is at $(0, 0)$, and the axis is the y-axis. To graph this parabola (which defines a function), we simply enter y_1 as $\frac{1}{8}x^2$. Figure 12 shows both a traditional graph and a calculator-generated graph of this function.

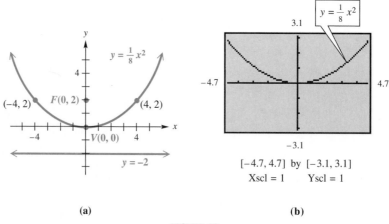

(a)   (b)

FIGURE 12

(b) ANALYTIC   Here we must think of the negative sign as being in the denominator to solve for c. Since $4c = -28$, $c = -7$. The parabola is horizontal (and is thus not the graph of a function), with focus $(-7, 0)$, directrix $x = 7$, vertex $(0, 0)$, and the x-axis is the axis of the parabola. Because c is negative, the graph opens to the left, as shown in Figure 13(a).

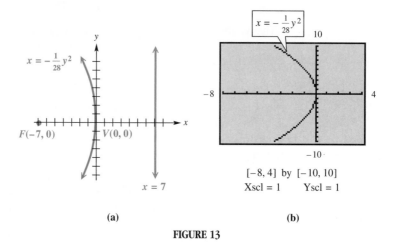

(a)   (b)

FIGURE 13

GRAPHICAL   To graph this parabola with a graphing calculator, we must first transform it analytically into the union of two functions.

$$x = -\frac{1}{28}y^2$$
$$-28x = y^2$$
$$y = \pm\sqrt{-28x}$$
$$y_1 = \sqrt{-28x} \quad \text{or} \quad y_2 = -\sqrt{-28x}$$

Note from the equations for y_1 and y_2 that x must be nonpositive for y to represent a real number. See Figure 13(b) for the calculator-generated graph.   ◆

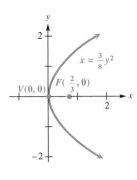

FIGURE 14

EXAMPLE 7 *Determining the Equation of a Parabola from Given Information*   Write an equation for each parabola described.

(a)   focus $\left(\frac{2}{3}, 0\right)$ and vertex at the origin

(b)   vertical axis, vertex at the origin, through the point $(-2, 12)$

SOLUTION

(a)   Since the focus $\left(\frac{2}{3}, 0\right)$ is on the *x*-axis, the parabola is horizontal and opens to the right because $c = \frac{2}{3}$ is positive. The equation is of the form $x = \frac{1}{4c}y^2$, so

$$x = \frac{1}{4\left(\frac{2}{3}\right)}\, y^2 = \frac{1}{\left(\frac{8}{3}\right)}\, y^2 = \frac{3}{8}\, y^2$$

is the equation of the parabola. It is graphed in Figure 14.

(b)   The parabola will have an equation of the form $y = \frac{1}{4c}x^2$ because the axis is vertical. Since the point $(-2, 12)$ is on the graph, it must satisfy the equation. We substitute -2 for *x* and 12 for *y* to find *c*.

$$12 = \frac{1}{4c}(-2)^2 = \frac{1}{4c}(4) = \frac{1}{c}$$

$$c = \frac{1}{12}$$

The equation of the parabola is $y = \dfrac{1}{4\left(\frac{1}{12}\right)}\, x^2$ or $y = 3x^2$.   ◆

In Chapter 3, we showed that the graph of the quadratic function defined by the equation $y = a(x - h)^2 + k$ is a parabola with vertex at (h, k) and the line $x = h$ as its axis. The relation defined by $x = a(y - k)^2 + h$ also has a parabola as its graph, but since *x* and *y* are interchanged, the graph of this new relation is symmetric to the graph of $y = a(x - h)^2 + k$ with respect to the line $y = x$. This reflection changes the vertical axis to a horizontal axis; however, the vertex is still at (h, k).

TRANSLATION OF A HORIZONTAL PARABOLA

The parabola with vertex at (h, k) and the horizontal line $y = k$ as axis has an equation of the form $x = a(y - k)^2 + h$. The parabola opens to the right if $a > 0$ and to the left if $a < 0$.

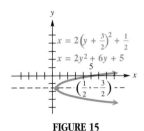

FIGURE 15

Figure 15 shows the graph of $x = 2\left(y + \frac{3}{2}\right)^2 + \frac{1}{2}$. It is a horizontal parabola opening to the right, with vertex at $\left(\frac{1}{2}, -\frac{3}{2}\right)$. If you refer to Example 8 in Section 4.3, you will see that we obtained a calculator-generated graph of this same relation. We started with the equation $x = 2y^2 + 6y + 5$ in that example. Notice that this latter equation is equivalent to the one given above.

AN APPLICATION OF PARABOLAS

EXAMPLE 8 *Applying the Reflective Property of Parabolas to a Radio Telescope*   The Parkes radio telescope has a parabolic dish shape with a diameter of 210 feet and a depth of 32 feet. Because of this parabolic shape, distant rays hitting the dish are reflected directly toward the focus. A cross-section of the dish is shown in Figure 16 on the next page. (*Source:* Mar, J., and H. Liebowitz, *Structure Technology for Large Radio and Radar Telescope Systems*, The MIT Press, Massachusetts Institute of Technology, Cambridge, MA, 1969.)

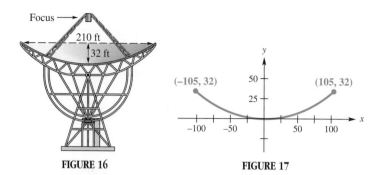

FIGURE 16   **FIGURE 17**

(a) Determine an equation describing this cross-section by placing the vertex at the origin with the parabola opening upward.

(b) The receiver must be placed at the focus of the parabola. How far from the vertex of the parabolic dish should the receiver be located?

SOLUTION

(a) Locate the vertex at the origin, as shown in Figure 17. Then, the form of the parabola will be $y = ax^2$. The parabola must pass through the point $\left(\frac{210}{2}, 32\right) = (105, 32)$. Thus,

$$32 = a(105)^2$$

$$a = \frac{32}{105^2} = \frac{32}{11{,}025},$$

so the cross-section can be described by

$$y = \frac{32}{11{,}025}x^2.$$

(b) Since $y = \frac{32}{11{,}025}x^2$,

$$4c = \frac{1}{a} = \frac{11{,}025}{32}$$

$$c = \frac{11{,}025}{128} \approx 86.1.$$

The receiver should be located at $(0, 86.1)$, or 86.1 feet above the vertex.   ◆

WHAT WENT WRONG   A student graphed the relation $x^2 + y^2 = 9$ in the window $[-6, 6]$ by $[-4, 4]$, using the two functions defined as $y_1 = \sqrt{9 - x^2}$ and $y_2 = -\sqrt{9 - x^2}$. He was surprised to see that the two functions did not meet at the points $(-3, 0)$ and $(3, 0)$ as he expected.

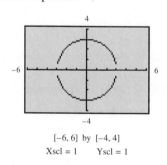

$[-6, 6]$ by $[-4, 4]$
Xscl = 1   Yscl = 1

1. What went wrong?

2. How could the problem be corrected?

?

6.1 EXERCISES   Tape 7

Match each equation with its calculator-generated graph in Exercises 1–10. Do this first with-out actually using your calculator. Then, check your answer by generating a calculator graph of your own. (Every window has Xscl = Yscl = 1.)

1. $y = x^2$

2. $x = y^2$

3. $x = 2(y + 3)^2 - 4$

4. $y = 2(x + 3)^2 - 4$

5. $y = -\frac{1}{3}x^2$

6. $x = -\frac{1}{3}y^2$

7. $x^2 + y^2 = 25$

8. $(x - 3)^2 + (y + 4)^2 = 25$

9. $(x + 3)^2 + (y - 4)^2 = 25$

10. $x^2 + y^2 = -4$

A.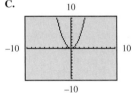

B.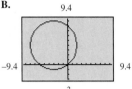

C.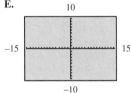

D.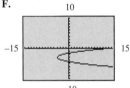

E.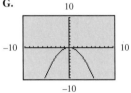

F.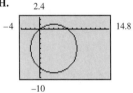

G.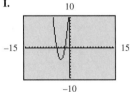

H.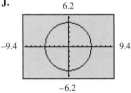

I. (graph)

J. (graph)

Find the center-radius form for each circle satisfying the given conditions.

11. Center $(1, 4)$, radius 3

12. Center $(-2, 5)$, radius 4

13. Center $(0, 0)$, radius 1

14. Center $(0, 0)$, radius 5

15. Center $\left(\frac{2}{3}, -\frac{4}{5}\right)$, radius $\frac{3}{7}$

16. Center $\left(-\frac{1}{2}, -\frac{1}{4}\right)$, radius $\frac{12}{5}$

17. Center $(-1, 2)$, passing through $(2, 6)$

18. Center $(2, -7)$, passing through $(-2, -4)$

19. Center $(-3, -2)$, tangent to the x-axis (*Hint: Tangent to* means touching at one point.)

20. Center $(5, -1)$, tangent to the y-axis

RELATING CONCEPTS (EXERCISES 21–24)

The figure shows a circle and a diameter of the circle. The endpoints of the diameter are $(-1, \ 3)$ *and* $(5, \ -9)$.

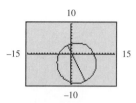

21. Find the coordinates of the center of the circle. (Section 1.4)

22. Find the radius of the circle. (Section 1.4)

23. Find the center-radius form of the equation of the circle.

24. Read the owner's manual of your calculator to see if your model has a DRAW capability. If it does, use it to duplicate the figure shown for Exercises 21–23.

25. Find the center-radius form of the equation of the circle with endpoints of a diameter having coordinates $(3, \ -5)$ and $(-7, 3)$.

26. Suppose that a circle is tangent to both axes, is completely within the third quadrant, and has a radius of $\sqrt{2}$. Find the center-radius form of the equation of the circle.

Graph each of the following circles by hand (that is, without using a graphing calculator). Give the domain and the range.

27. $x^2 + y^2 = 36$ **28.** $(x - 2)^2 + y^2 = 36$

29. $(x + 2)^2 + (y - 5)^2 = 16$ **30.** $(x - 5)^2 + (y + 4)^2 = 49$

Graph each of the following circles, using a graphing calculator. Follow the procedure explained in Example 3. Use a square viewing window.

31. $x^2 + y^2 = 81$ **32.** $x^2 + (y + 3)^2 = 49$

33. $(x - 4)^2 + (y - 3)^2 = 25$ **34.** $(x + 3)^2 + (y + 2)^2 = 36$

35. Describe the graph of the equation $(x - 3)^2 + (y - 3)^2 = 0$.

36. Describe the graph of the equation $(x - 3)^2 + (y - 3)^2 = -1$.

Use the table of your graphing calculator to find the points of intersection of the two half-graphs. Analytically determine the domain of each relation. Support your answers graphically in a decimal window.

37. $x^2 + y^2 = 36$ **38.** $(x + 4)^2 + y^2 = 81$ **39.** $(x - 5)^2 + (y - 2)^2 = 16$

40. $(x + 1)^2 + (y + 3)^2 = 25$ **41.** $x = (y - 1)^2 + 2$ **42.** $x = -(y + 2)^2 - 3$

Find the center and the radius of each of the following circles.

43. $x^2 + 6x + y^2 + 8y = -9$ **44.** $x^2 - 4x + y^2 + 12y + 4 = 0$ **45.** $x^2 - 12x + y^2 + 10y + 25 = 0$

46. $x^2 + 8x + y^2 - 6y = -16$ **47.** $x^2 + y^2 = 2y + 48$ **48.** $x^2 + 4x + y^2 = 21$

Refer to Example 5 to solve the problems in Exercises 49 and 50.

49. Show analytically that if three receiving stations at $(1, 4)$, $(-6, 0)$, and $(5, -2)$ record distances to an earthquake epicenter of 4 units, 5 units, and 10 units, respectively, that the epicenter would lie at $(-3, 4)$.

50. Three receiving stations record the presence of an earthquake. The location of the receiving center and the distance to the epicenter are contained in the following three equations: $(x - 2)^2 + (y - 1)^2 = 25$, $(x + 2)^2 + (y - 2)^2 = 16$, and $(x - 1)^2 + (y + 2)^2 = 9$. Determine the location of the earthquake epicenter.

51. A new type of ski, the super sidecut, is quickly becoming popular. When tipped on an edge, these skis carve a circle with a smaller turn radius than traditional skis. Give equations for the three circles with arcs pictured below, assuming a center at the origin.

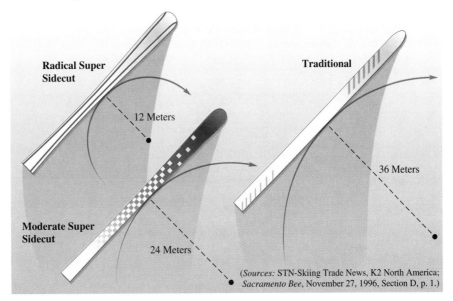

Radical Super Sidecut

12 Meters

Traditional

36 Meters

Moderate Super Sidecut

24 Meters

(*Sources:* STN-Skiing Trade News, K2 North America; *Sacramento Bee*, November 27, 1996, Section D, p. 1.)

Each equation in Exercises 52–59 defines a parabola. Without actually graphing, match the description given in the column on the right with the equation.

52. $y = (x - 4)^2 - 2$ **A.** vertex $(2, -4)$, opens down

53. $y = (x - 2)^2 - 4$ **B.** vertex $(2, -4)$, opens up

54. $y = -(x - 4)^2 - 2$ **C.** vertex $(4, -2)$, opens down

55. $y = -(x - 2)^2 - 4$ **D.** vertex $(4, -2)$, opens up

56. $x = (y - 4)^2 - 2$ **E.** vertex $(-2, 4)$, opens left

57. $x = (y - 2)^2 - 4$ **F.** vertex $(-2, 4)$, opens right

58. $x = -(y - 4)^2 - 2$ **G.** vertex $(-4, 2)$, opens left

59. $x = -(y - 2)^2 - 4$ **H.** vertex $(-4, 2)$, opens right

60. For the graph of $y = a(x - h)^2 + k$, in what quadrant is the vertex if:

     (a) $h < 0, k < 0$;   **(b)** $h < 0, k > 0$;   **(c)** $h > 0, k < 0$;   **(d)** $h > 0, k > 0$?

61. Repeat parts (a)–(d) of Exercise 60 for the graph of $x = a(y - k)^2 + h$.

Give the coordinates of the focus, the equation of the directrix, and the axis of each of the following parabolas.

62. $y = \dfrac{1}{16}x^2$ **63.** $y = \dfrac{1}{4}x^2$ **64.** $y = -2x^2$ **65.** $y = 9x^2$

66. $x = 16y^2$ **67.** $x = -32y^2$ **68.** $x = -\dfrac{1}{16}y^2$ **69.** $x = -\dfrac{1}{4}y^2$

Write an equation for each of the following parabolas with vertex at the origin.

70. focus $(0, -2)$ **71.** focus $(5, 0)$ **72.** focus $\left(-\dfrac{1}{2}, 0\right)$ **73.** focus $\left(0, \dfrac{1}{4}\right)$

74. through $(2, -2\sqrt{2})$, opening to the right

75. through $(\sqrt{3}, 3)$, opening upward

76. through $(\sqrt{10}, -5)$, opening downward

77. through $(-3, 3)$, opening to the left

78. through $(2, -4)$, symmetric with respect to the y-axis

79. through $(3, 2)$, symmetric with respect to the x-axis

Graph each of the following parabolas either by hand or using a graphing calculator. Give the coordinates of the vertex, the axis, the domain, and the range. (Hint: Each of these is a parabola with a vertical axis and is thus a function. Therefore, each may be graphed directly with a graphing calculator in the function mode.)

80. $y = (x + 3)^2 - 4$

81. $y = (x - 5)^2 - 4$

82. $y = -2(x + 3)^2 + 2$

83. $y = \frac{2}{3}(x - 2)^2 - 1$

84. $y = x^2 - 2x + 3$

85. $y = x^2 + 6x + 5$

86. $y = 2x^2 - 4x + 5$

87. $y = -3x^2 + 24x - 46$

Graph each of the following parabolas either by hand or using a graphing calculator. Give the coordinates of the vertex, the axis, the domain, and the range. (Hint: Each of these is a parabola with a horizontal axis and is thus not a function. To graph with a graphing calculator in function mode, you must first rewrite the equation in terms of two functions, y_1 and y_2. See Example 6(b), and refer to Example 8 of Section 4.3.)

88. $x = y^2 + 2$

89. $x = (y + 1)^2$

90. $x = (y - 3)^2$

91. $x = (y + 2)^2 - 1$

92. $x = (y - 4)^2 + 2$

93. $x = -2(y + 3)^2$

94. $x = \frac{2}{3}(y - 3)^2 + 2$

95. $x = y^2 + 2y - 8$

96. $x = -4y^2 - 4y - 3$

Solve each problem.

97. *Path of an Object on a Planet* When an object moves under the influence of a constant force (without air resistance), its path is parabolic. This would occur if a ball is thrown near the surface of a planet or other celestial object. The graphing calculator can be used to simulate something that would be impossible to view in real life. Suppose two balls are simultaneously thrown upward at a 45° angle on two different planets. If their initial velocities are both 30 miles per hour, then their xy-coordinates in feet at time x in seconds can be expressed by the equation $y = x - (g/1922)x^2$, where g is the acceleration due to gravity. The value of g will vary depending on the mass and size of the planet. (*Source:* Zeilik, M., S. Gregory, and E. Smith, *Introductory Astronomy and Astrophysics,* Saunders College Publishers, 1992.)

(a) On Earth, $g = 32.2$; on Mars, $g = 12.6$. Find the two equations, and graph on the same coordinate axes the paths of the two balls thrown on Earth and Mars. Use the interval $[0, 180]$ for x. (*Hint:* If possible, set the mode on your graphing calculator to simultaneous.)

(b) Determine the difference in the horizontal distances traveled by the two balls.

98. *Path of an Object on a Planet* (Refer to Exercise 97.) Suppose the two balls are now thrown upward at a 60° angle on Mars and the moon. If their initial velocity is 60 miles per hour, then their xy-coordinates in feet at time x in seconds can be expressed by the equation

$y = \frac{19}{11}x - \frac{g}{3872}x^2$. (*Source:* Zeilik, M., S. Gregory, and E. Smith, *Introductory Astronomy and Astrophysics,* Saunders College Publishers, 1992.)

(a) On the same coordinate axes, graph the paths of the balls if $g = 5.2$ for the moon.

(b) Determine the maximum height of each ball to the nearest foot.

99. *Design of a Radio Telescope* The U.S. Naval Research Laboratory designed a giant radio telescope weighing 3450 tons. Its parabolic dish had a diameter of 300 feet with a focal length (the distance from the focus to the parabolic surface) of 128.5 feet. Determine the maximum depth of the 300-foot dish. (*Source:* Mar, J., and H. Liebowitz, *Structure Technology for Large Radio and Radar Telescope Systems,* The MIT Press, Cambridge, MA, 1969.)

100. *Path of an Alpha Particle* When an alpha particle is moving in a horizontal path along the positive x-axis and passes between charged plates, it is deflected in a parabolic path. If the plate is charged with 2000 volts and is .4 meter long, an alpha particle's path can be described by the equation $y = (-k/(2v_0))x^2$, where $k = 5 \times 10^{-9}$ is constant and v_0 is the initial velocity of the alpha particle. If $v_0 = 10^7$ meters/sec, what is the deflection of the alpha particle's path in the y-direction when $x = .4$ meter? (*Source:* Semat, H., and J. Albright, *Introduction to Atomic and Nuclear Physics,* Holt, Rinehart, and Winston, Inc., 1972.)

101. *Height of a Bridge's Cable Supports* The cable in the center portion of a bridge is supported as shown in the figure to form a parabola. The center support is 10 feet high, the tallest supports are 210 feet high, and the distance between the two tallest supports is 400 feet. Find the height of the remaining supports, if the supports are evenly spaced. (Ignore the width of the supports.)

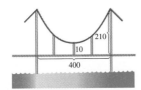

102. *Parabolic Arch*   An arch in the shape of a parabola has the dimensions shown in the figure in the next column. How wide is the arch 9 feet up?

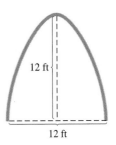

103. Prove that the parabola with focus $(c, 0)$ and directrix $x = -c$ has the equation $x = \frac{1}{4c} y^2$.

104. Use the definition of a parabola to find an equation of the parabola with vertex at (h, k) and a vertical axis. Let the distance from the vertex to the focus and the distance from the vertex to the directrix be c, where $c > 0$.

6.2 ELLIPSES AND HYPERBOLAS

Equations and Graphs of Ellipses   ◆   An Application of Ellipses   ◆   Equations and Graphs of Hyperbolas

EQUATIONS AND GRAPHS OF ELLIPSES

We have studied two types of second-degree relations thus far: parabolas and circles. We now look at another type, the *ellipse*. The definition of an ellipse is also based on distance.

> **DEFINITION OF ELLIPSE**
>
> An **ellipse** is the set of all points in a plane such that the sum of their distances from two fixed points is always the same (constant). The two fixed points are called the **foci** (plural of *focus*) of the ellipse.

For example, the ellipse in Figure 18 has foci at points F and F'. By the definition, the ellipse is made up of all points P such that the sum $d(P, F) + d(P, F')$ is constant. This ellipse has its **center** at the origin. Points V and V' are the **vertices** of the ellipse, and the line segment connecting V and V' is the **major axis**. The foci always lie on the major axis. The line segment from B to B' is the **minor axis**. The major axis has length $2a$, and the minor axis has length $2b$.

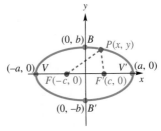

FIGURE 18

If the foci are chosen to be on the *x*-axis (or *y*-axis), with the center of the ellipse at the origin, then the distance formula and the definition of an ellipse can be used to obtain the following results. (See Exercise 70.)

STANDARD FORMS OF EQUATIONS FOR ELLIPSES

The ellipse with center at the origin and equation

$$\frac{x^2}{a^2} + \frac{y^2}{b^2} = 1$$

has vertices $(\pm a, 0)$, endpoints of the minor axis $(0, \pm b)$, and foci $(\pm c, 0)$, where $c^2 = a^2 - b^2$. The ellipse with center at the origin and equation

$$\frac{y^2}{a^2} + \frac{x^2}{b^2} = 1$$

has vertices $(0, \pm a)$, endpoints of the minor axis $(\pm b, 0)$, and foci $(0, \pm c)$, where $c^2 = a^2 - b^2$.

Do not be confused by the two standard forms—in one case a^2 is associated with x^2; in the other case a^2 is associated with y^2. However, when graphing an ellipse in a traditional manner, it is necessary only to find the intercepts of the graph—if the positive x-intercept is larger than the positive y-intercept, the major axis is horizontal, and otherwise it is vertical. When using the relationship $a^2 - b^2 = c^2$, choose a^2 and b^2 so that a^2 is larger than b^2.

In an equation of an ellipse, the coefficients of x^2 and y^2 must be different positive numbers. (What happens if the coefficients are equal?)

An ellipse is the graph of a relation. As suggested by Figure 18, if the ellipse has equation $\frac{x^2}{a^2} + \frac{y^2}{b^2} = 1$, the domain is $[-a, a]$ and the range is $[-b, b]$. Notice that the ellipse in Figure 18 is symmetric with respect to the x-axis, the y-axis, and the origin.

EXAMPLE 1 *Graphing an Ellipse Centered at the Origin* Transform the equation $4x^2 + 9y^2 = 36$ into the standard form for an ellipse. Graph the ellipse in a traditional manner by finding intercepts. Then, solve for y and graph the ellipse, using a graphing calculator.

SOLUTION ANALYTIC Begin by dividing both sides by 36.

$$\frac{x^2}{9} + \frac{y^2}{4} = 1$$

This ellipse is centered at the origin, with x-intercepts 3 and -3, and y-intercepts 2 and -2. The domain of this relation is $[-3, 3]$, and the range is $[-2, 2]$. See Figure 19(a).

GRAPHICAL Solving the equation for y gives the two functions

$$y_1 = 2\sqrt{1 - \frac{x^2}{9}} \quad \text{and} \quad y_2 = -2\sqrt{1 - \frac{x^2}{9}}.$$

Graphing these in a square window gives the graph shown in Figure 19(b). ◆

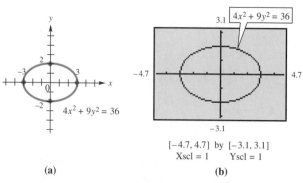

$[-4.7, 4.7]$ by $[-3.1, 3.1]$
Xscl = 1   Yscl = 1
(a)   **(b)**

FIGURE 19

EXAMPLE 2 *Finding Foci of an Ellipse*   Find the coordinates of the foci of the ellipse in Example 1.

SOLUTION   Since $9 > 4$, we can find the foci by letting $a^2 = 9$ and $b^2 = 4$ in the equation $c^2 = a^2 - b^2$. Then, solve for c:

$$c^2 = a^2 - b^2 = 9 - 4 = 5$$
$$c = \sqrt{5}.$$

(By definition, $c > 0$. See Figure 18.) The major axis is along the x-axis, so the foci have coordinates $\left(-\sqrt{5}, 0\right)$ and $\left(\sqrt{5}, 0\right)$.   ◆

EXAMPLE 3 *Finding the Equation of an Ellipse*   Find the equation of the ellipse having center at the origin, foci at $(0, 3)$ and $(0, -3)$, and major axis of length 8 units.

SOLUTION   Since the major axis is 8 units long, $2a = 8$ or $a = 4$. Use the relationship $a^2 - b^2 = c^2$ to find b^2. Here $a = 4$ and $c = 3$.

$$a^2 - b^2 = c^2$$
$$4^2 - b^2 = 3^2 \qquad \text{Substitute for } a \text{ and } c.$$
$$16 - b^2 = 9$$
$$b^2 = 7$$

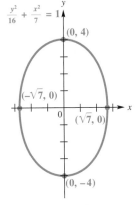

$\frac{y^2}{16} + \frac{x^2}{7} = 1$

$(0, 4)$

$(-\sqrt{7}, 0)$

$(\sqrt{7}, 0)$

$(0, -4)$

FIGURE 20

Since the foci are on the y-axis, the larger intercept, a, is used to find the denominator for y^2, giving the equation in standard form as $\frac{y^2}{16} + \frac{x^2}{7} = 1$. A traditional graph of this ellipse is shown in Figure 20.   ◆

Just as a circle need not have its center at the origin, an ellipse may also have its center translated away from the origin.

ELLIPSE CENTERED AT (h, k)

An ellipse centered at (h, k) with horizontal major axis of length $2a$ and vertical minor axis of length $2b$ has equation

$$\frac{(x - h)^2}{a^2} + \frac{(y - k)^2}{b^2} = 1.$$

There is a similar result for ellipses having a vertical major axis.

The definition of an ellipse can be used to prove the statement above.

EXAMPLE 4 *Graphing an Ellipse Translated away from the Origin (Traditional Approach)*   Use a traditional approach to graph $\frac{(y + 1)^2}{16} + \frac{(x - 2)^2}{9} = 1$.

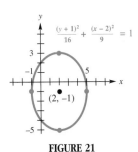

$\frac{(y + 1)^2}{16} + \frac{(x - 2)^2}{9} = 1$

$(2, -1)$

FIGURE 21

SOLUTION   The graph of this equation is an ellipse centered at $(2, -1)$. As mentioned earlier, ellipses always have $a > b$. For this ellipse, then, $a = 4$ and $b = 3$. Since $a = 4$ is associated with y^2, the vertices of the ellipse are on the vertical line through $(2, -1)$. Find the vertices by locating two points on the vertical line through $(2, -1)$, one 4 units up from $(2, -1)$ and one 4 units down. The vertices are $(2, 3)$ and $(2, -5)$. Locate two other points on the ellipse by locating points on a horizontal line through $(2, -1)$, one 3 units to the right and one 3 units to the left. The graph is shown in Figure 21. As the graph suggests, the domain is $[-1, 5]$ and the range is $[-5, 3]$.   ◆

EXAMPLE 5   *Graphing an Ellipse Translated away from the Origin (Calculator Approach)*   Solve the equation in Example 4 for *y*, and give a calculator-generated graph of the ellipse.

SOLUTION

$$\frac{(y + 1)^2}{16} + \frac{(x - 2)^2}{9} = 1 \qquad \text{Given equation}$$

$$\frac{(y + 1)^2}{16} = 1 - \frac{(x - 2)^2}{9} \qquad \text{Subtract } \tfrac{(x - 2)^2}{9}.$$

$$(y + 1)^2 = 16 - \frac{16(x - 2)^2}{9} \qquad \text{Multiply by 16.}$$

$$y + 1 = \pm\sqrt{16 - \frac{16(x - 2)^2}{9}} \qquad \text{Take square roots.}$$

$$y = -1 \pm \sqrt{16 - \frac{16(x - 2)^2}{9}} \qquad \text{Subtract 1.}$$

This final equation indicates that the graph may be obtained by graphing

$$y_1 = -1 + \sqrt{16 - \frac{16(x - 2)^2}{9}} \quad \text{and} \quad y_2 = -1 - \sqrt{16 - \frac{16(x - 2)^2}{9}}.$$

Using a square window gives the graph shown in Figure 22.   ◆

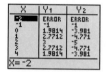

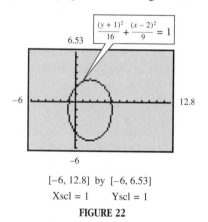

$$\frac{(y + 1)^2}{16} + \frac{(x - 2)^2}{9} = 1$$

[−6, 12.8] by [−6, 6.53]
Xscl = 1   Yscl = 1
FIGURE 22

The tables show that the points of intersection of the two halves of the ellipse are $(-1, -1)$ and $(5, -1)$. The tables also suggest that the domain of the relation in Example 5 is $[-1, 5]$.

Caution   Entering the expressions for y_1 and y_2 in Example 5 requires that parentheses be used with extreme care.

AN APPLICATION OF ELLIPSES

Ellipses have many useful applications. As the earth makes its year-long journey around the sun, it traces an ellipse. Spacecraft travel around the earth in elliptical orbits, and planets make elliptical orbits around the sun. An interesting recent application is the use of an elliptical tub in the nonsurgical removal of kidney stones. In this procedure, the reflective property of the ellipse is used. If a beam is projected from one focus onto the ellipse, it will reflect to the other focus. This feature has helped scientists develop the lithotripter, a machine that uses shock waves to crush kidney stones. The waves originate at one focus and are reflected to hit the kidney stone, which is positioned at the second focus.

EXAMPLE 6 *Applying Properties of an Ellipse to a Lithotripter* If a lithotripter is based on the ellipse $\frac{x^2}{36} + \frac{y^2}{27} = 1$, determine how many units the kidney stone and the wave source must be placed from the center of the ellipse.

SOLUTION Since 36 appears in the denominator of the term involving x^2, and $36 > 27$, the major axis will lie along the x-axis, with $a^2 = 36$. Since $b^2 = 27$, $c^2 = 36 - 27 = 9$. Therefore, $c = 3$, and the foci will be at $(-3, 0)$ and $(3, 0)$. Thus, the kidney stone and the wave source must each be 3 units from the center of the ellipse on the longer axis. See Figure 23. ◆

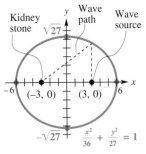

FIGURE 23

For Group Discussion What is a "whispering gallery"? How does its special feature correspond to the foci of an ellipse?

EQUATIONS AND GRAPHS OF HYPERBOLAS

An ellipse was defined as the set of all points in a plane the sum of whose distances from two fixed points is a constant. A *hyperbola* is defined similarly.

DEFINITION OF HYPERBOLA

A **hyperbola** is the set of all points in a plane the *difference* of whose distances from two fixed points is constant. The two fixed points are called the **foci** of the hyperbola.

Suppose a hyperbola has center at the origin and foci at $F'(-c, 0)$ and $F(c, 0)$. The midpoint of the segment $F'F$ is the **center** of the hyperbola, and the points $V'(-a, 0)$ and $V(a, 0)$ are the **vertices** of the hyperbola. The line segment $V'V$ is the **transverse axis** of the hyperbola. See Figure 24.

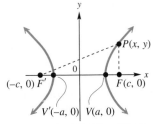

For this hyperbola, $d(P, F') - d(P, F) = 2a$.

FIGURE 24

Using the distance formula in conjunction with the definition of a hyperbola, you can verify the following standard forms of the equations for hyperbolas with center at the origin. (See Exercise 71.)

STANDARD FORMS OF EQUATIONS FOR HYPERBOLAS

The hyperbola with center at the origin and equation

$$\frac{x^2}{a^2} - \frac{y^2}{b^2} = 1$$

has vertices $(\pm a, 0)$ and foci $(\pm c, 0)$, where $c^2 = a^2 + b^2$. The hyperbola with center at the origin and equation

$$\frac{y^2}{a^2} - \frac{x^2}{b^2} = 1$$

has vertices $(0, \pm a)$ and foci $(0, \pm c)$, where $c^2 = a^2 + b^2$.

Starting with the equation for a hyperbola $\frac{x^2}{a^2} - \frac{y^2}{b^2} = 1$ and solving for y gives

$$\frac{x^2}{a^2} - 1 = \frac{y^2}{b^2}$$

$$\frac{x^2 - a^2}{a^2} = \frac{y^2}{b^2}$$

or

$$y = \pm\frac{b}{a}\sqrt{x^2 - a^2}. \qquad (*)$$

If x^2 is very large in comparison to a^2, the difference $x^2 - a^2$ would be very close to x^2. If this happens, then the points satisfying equation (*) above would be very close to one of the lines

$$y = \pm\frac{b}{a}x.$$

Thus, as $|x|$ gets larger and larger, the points of the hyperbola $\frac{x^2}{a^2} - \frac{y^2}{b^2} = 1$ come closer to the lines $y = \pm\frac{b}{a}x$. These lines, called the **asymptotes** of the hyperbola, are very helpful when graphing the hyperbola with traditional graphing methods. The lines are the extended diagonals of the rectangle whose vertices are (a, b), $(-a, b)$, $(a, -b)$, and $(-a, -b)$. This rectangle is called the **fundamental rectangle** of the hyperbola.

Results similar to those above hold for a hyperbola of the form $\frac{y^2}{a^2} - \frac{x^2}{b^2} = 1$.

EXAMPLE 7 *Graphing a Hyperbola Centered at the Origin (Traditional Approach)* Graph $\frac{x^2}{25} - \frac{y^2}{49} = 1$, using traditional graphing methods.

SOLUTION For this hyperbola, $a = 5$ and $b = 7$. With these values, $y = \pm\frac{b}{a}x$ becomes $y = \pm\frac{7}{5}x$. The four points, $(5, 7)$, $(5, -7)$, $(-5, 7)$, and $(-5, -7)$, lead to the rectangle shown in Figure 25. The extended diagonals of this rectangle are the asymptotes of the hyperbola. The hyperbola has x-intercepts 5 and -5. The domain is $(-\infty, -5] \cup [5, \infty)$, and the range is $(-\infty, \infty)$. The final graph is shown in Figure 25. ◆

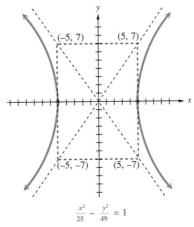

$$\frac{x^2}{25} - \frac{y^2}{49} = 1$$

FIGURE 25

EXAMPLE 8 *Graphing a Hyperbola Centered at the Origin (Calculator Approach)* Solve the equation in Example 7 for *y*, and give a calculator-generated graph of the hyperbola.

SOLUTION We solve for *y* as follows.

$$\frac{x^2}{25} - \frac{y^2}{49} = 1 \qquad \text{Given equation}$$

$$-\frac{y^2}{49} = 1 - \frac{x^2}{25} \qquad \text{Subtract } \tfrac{x^2}{25}.$$

$$\frac{y^2}{49} = \frac{x^2}{25} - 1 \qquad \text{Multiply by } -1.$$

$$\frac{y}{7} = \pm\sqrt{\frac{x^2}{25} - 1} \qquad \text{Take square roots.}$$

$$y = \pm 7\sqrt{\frac{x^2}{25} - 1} \qquad \text{Multiply by 7.}$$

The final equation indicates that the hyperbola is composed of the union of two functions, defined as

$$y_1 = 7\sqrt{\frac{x^2}{25} - 1} \quad \text{and} \quad y_2 = -7\sqrt{\frac{x^2}{25} - 1}.$$

Figure 26 shows the calculator-generated graph of this hyperbola. Again, a square window gives the proper perspective. Compare it to the graph in Figure 25. ◆

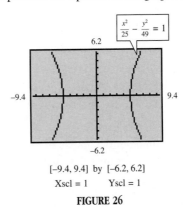

[−9.4, 9.4] by [−6.2, 6.2]
Xscl = 1   Yscl = 1
FIGURE 26

For Group Discussion   Determine the equation of the asymptote with positive slope for the hyperbola in Example 8. Then, graph that asymptote, along with y_1 as defined in the example. Use various viewing windows to "watch" the hyperbola approach the asymptote.

EXAMPLE 9   *Graphing a Hyperbola Centered at the Origin (Both Approaches)*   Graph $25y^2 - 4x^2 = 9$, using both traditional and graphing calculator approaches.

SOLUTION   ANALYTIC   First, let us use a traditional approach. We divide each side by 9 to get $\frac{25y^2}{9} - \frac{4x^2}{9} = 1$. To determine the values of a and b, write the equation as

$$\frac{y^2}{\dfrac{9}{25}} - \frac{x^2}{\dfrac{9}{4}} = 1.$$

This hyperbola is centered at the origin, has foci on the y-axis, and has y-intercepts $-\frac{3}{5}$ and $\frac{3}{5}$. Use the points $\left(\frac{3}{2}, \frac{3}{5}\right)$, $\left(-\frac{3}{2}, \frac{3}{5}\right)$, $\left(\frac{3}{2}, -\frac{3}{5}\right)$, and $\left(-\frac{3}{2}, -\frac{3}{5}\right)$ to get the fundamental rectangle shown in Figure 27(a). Use the diagonals of this rectangle to determine the asymptotes for the graph. The domain is $(-\infty, \infty)$, and the range is $\left(-\infty, -\frac{3}{5}\right] \cup \left[\frac{3}{5}, \infty\right)$.

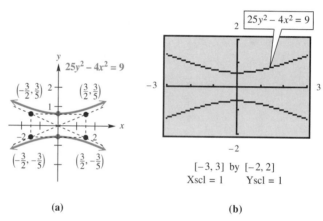

(a)   (b)

FIGURE 27

GRAPHICAL   To graph this hyperbola using a graphing calculator, we go back to the original form of the equation and solve for y.

$$25y^2 - 4x^2 = 9 \qquad \text{Given equation}$$
$$25y^2 = 9 + 4x^2 \qquad \text{Add } 4x^2.$$
$$5y = \pm\sqrt{9 + 4x^2} \qquad \text{Take square roots.}$$
$$y = \pm\frac{1}{5}\sqrt{9 + 4x^2} \qquad \text{Multiply by } \tfrac{1}{5}.$$

Therefore, we graph

$$y_1 = \frac{1}{5}\sqrt{9 + 4x^2} \qquad \text{and} \qquad y_2 = -\frac{1}{5}\sqrt{9 + 4x^2}.$$

The union of the graphs of these two functions is the graph of the hyperbola. See Figure 27(b) for the calculator-generated graph.   ◆

Earlier we saw how ellipses can be translated away from the origin. This same translation can be made with hyperbolas, as shown in the next example.

EXAMPLE 10   *Graphing a Hyperbola Translated away from the Origin (Traditional Approach)*   Graph $\frac{(y + 2)^2}{9} - \frac{(x + 3)^2}{4} = 1$.

SOLUTION   This hyperbola has the same graph as $\frac{y^2}{9} - \frac{x^2}{4} = 1$, except that it is centered at $(-3, -2)$. See Figure 28.   ◆

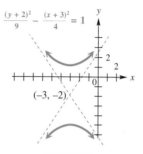

FIGURE 28

For Group Discussion   How would you go about graphing the hyperbola in Example 10 on your graphing calculator? See if you can duplicate the calculator-generated graph of that hyperbola as seen in Figure 29.

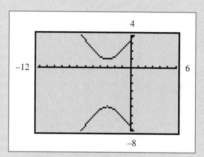

FIGURE 29

The two branches of a hyperbola are reflections about two different axes, and also about a point. What are the axes and the point for the hyperbola in Figure 29? (The reflecting property of the hyperbola was used on the Hubble space telescope.)

6.2   **EXERCISES**   Tape 7

Match each equation with its calculator-generated graph in Exercises 1–10 on the following page. Do this first without actually using your calculator. Then, check your answer by generating a calculator graph of your own. (Every window has Xscl = Yscl = 1.)

1. $\dfrac{y^2}{16} + \dfrac{x^2}{4} = 1$

2. $\dfrac{x^2}{16} + \dfrac{y^2}{4} = 1$

3. $\dfrac{x^2}{4} - \dfrac{y^2}{16} = 1$

4. $\dfrac{y^2}{4} - \dfrac{x^2}{16} = 1$

5. $\dfrac{(y-4)^2}{25} + \dfrac{(x+2)^2}{9} = 1$

6. $\dfrac{(y+4)^2}{25} + \dfrac{(x-2)^2}{9} = 1$

7. $\dfrac{(x+2)^2}{9} - \dfrac{(y-4)^2}{25} = 1$

8. $\dfrac{(x-2)^2}{9} - \dfrac{(y+4)^2}{25} = 1$

9. $36x^2 + 4y^2 = 144$

10. $9x^2 - 4y^2 = 36$

A.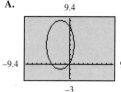

B.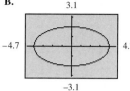

C.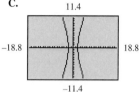

D.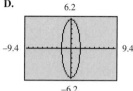

E.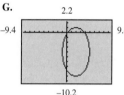

F.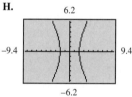

G.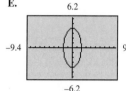

H.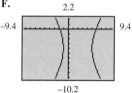

I.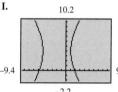

J.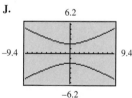

 11. Explain how a circle can be interpreted as a special case of an ellipse.

12. If an ellipse has vertices at $(-3, 0)$, $(3, 0)$, $(0, 5)$, and $(0, -5)$, what is its domain? What is its range?

Graph each of the following ellipses, using either a traditional method or a graphing calculator. Give the domain and the range.

13. $\dfrac{x^2}{9} + \dfrac{y^2}{4} = 1$

14. $\dfrac{y^2}{36} + \dfrac{x^2}{16} = 1$

15. $9x^2 + 6y^2 = 54$

16. $12x^2 + 8y^2 = 96$

17. $\dfrac{25y^2}{36} + \dfrac{64x^2}{9} = 1$

18. $\dfrac{16y^2}{9} + \dfrac{121x^2}{25} = 1$

19. $\dfrac{(y+3)^2}{25} + \dfrac{(x-1)^2}{9} = 1$

20. $\dfrac{(y-2)^2}{36} + \dfrac{(x+3)^2}{16} = 1$

21. $\dfrac{(x-2)^2}{16} + \dfrac{(y-1)^2}{9} = 1$

22. $\dfrac{(y+2)^2}{36} + \dfrac{(x+3)^2}{25} = 1$

23. Discuss the symmetries exhibited by a hyperbola centered at the origin.

24. The ellipse $\frac{(y+1)^2}{16} + \frac{(x-2)^2}{9} = 1$ is graphed in Figure 21. What is the equation of its horizontal axis of symmetry? What is the equation of its vertical axis of symmetry?

RELATING CONCEPTS (EXERCISES 25–30)

In Example 5, we show how to graph the ellipse $\dfrac{(y+1)^2}{16} + \dfrac{(x-2)^2}{9} = 1$ *with a graphing calculator by solving for y. Its graph is the union of the graphs of the two functions*

$$y_1 = -1 + \sqrt{16 - \frac{16(x-2)^2}{9}} \quad \text{and} \quad y_2 = -1 - \sqrt{16 - \frac{16(x-2)^2}{9}}.$$

The domain of this relation is $[-1, 5]$*. Work Exercises 25–30 in order.*

25. The relation is defined only when the radicand in y_1 and y_2 is greater than or equal to 0. Write the inequality that would need to be solved in order to find the domain analytically. (Section 4.4)

26. Let y represent the expression in x found in the radicand. What conic section is the graph of this function?

27. Graph the function defined by the radicand with a graphing calculator in the window $[-10, 10]$ by $[-10, 20]$.

28. Use the graph in Exercise 27 to solve the inequality of Exercise 25. (Section 4.4)

29. Explain how the solution set from Exercise 28 confirms what was found earlier by using the graph of the original ellipse.

30. Solve the inequality of Exercise 25 analytically, using a sign graph. (Section 3.3)

Graph each of the following hyperbolas by using either a traditional method or a graphing calculator. Give the domain and the range.

31. $\dfrac{x^2}{16} - \dfrac{y^2}{9} = 1$

32. $\dfrac{y^2}{9} - \dfrac{x^2}{9} = 1$

33. $49y^2 - 36x^2 = 1764$

34. $144x^2 - 49y^2 = 7056$

35. $\dfrac{4x^2}{9} - \dfrac{25y^2}{16} = 1$

36. $x^2 - y^2 = 1$

37. $\dfrac{(x-1)^2}{9} - \dfrac{(y+3)^2}{25} = 1$

38. $\dfrac{(x+3)^2}{16} - \dfrac{(y-2)^2}{36} = 1$

39. $\dfrac{(y-5)^2}{4} - \dfrac{(x+1)^2}{9} = 1$

40. $\dfrac{(y+1)^2}{25} - \dfrac{(x-3)^2}{36} = 1$

RELATING CONCEPTS (EXERCISES 41–42)

41. In Example 8, we show how to graph the hyperbola $\frac{x^2}{25} - \frac{y^2}{49} = 1$ on a graphing calculator by considering the union of the graphs of the two functions

$$y_1 = 7\sqrt{\frac{x^2}{25} - 1} \quad \text{and} \quad y_2 = -7\sqrt{\frac{x^2}{25} - 1}.$$

If we graph $y = \frac{x^2}{25} - 1$, the function defined by the radicand of y_1 and y_2, our graph is a parabola. Show how the solution set of $\frac{x^2}{25} - 1 \geq 0$ can be determined graphically, and explain how it relates to the domain of the given hyperbola. (*Hint:* Use the window $[-10, 10]$ by $[-2, 10]$.)

42. In Example 9, we show how to graph the hyperbola $25y^2 - 4x^2 = 9$ with a graphing calculator by considering the union of the graphs of the two functions

$$y_1 = \frac{1}{5}\sqrt{9 + 4x^2} \quad \text{and} \quad y_2 = -\frac{1}{5}\sqrt{9 + 4x^2}.$$

The function defined by the radicand, $y = 9 + 4x^2$, is a translation of the graph of the squaring function.

(a) Use the terminology of Chapter 2 to explain how the graph of $y = 9 + 4x^2$ can be obtained by a translation of the graph of $y = x^2$. (Section 2.2)

(b) Which one of the following graphs most closely resembles the graph of $y = 9 + 4x^2$?

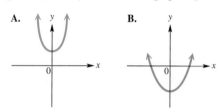

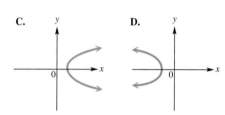

📄 **(c)** Explain how the domain of the graph of your choice in part (b) supports the conclusion concerning the domain of the original hyperbola found in Example 9.

43. Use a table to find and compare the domains of

$$\frac{x^2}{4} + \frac{y^2}{25} = 1 \quad \text{and} \quad \frac{x^2}{4} - \frac{y^2}{25} = 1.$$

In particular, note when each relation returns an error message. Support your answer graphically.

44. Use a table to find and compare the domains of

$$\frac{x^2}{36} - \frac{y^2}{9} = 1 \quad \text{and} \quad \frac{y^2}{36} - \frac{x^2}{9} = 1.$$

Support your answer graphically.

Find an equation for each ellipse described.

45. x-intercepts ± 4; foci at $(-2, 0)$ and $(2, 0)$

46. y-intercepts ± 3; foci at $(0, \sqrt{3}), (0, -\sqrt{3})$

47. endpoints of major axis at $(6, 0), (-6, 0)$; $c = 4$

48. vertices $(0, 5), (0, -5)$; $b = 2$

49. center $(3, -2)$; $a = 5, c = 3$; major axis vertical

50. center $(2, 0)$; minor axis of length 6; major axis horizontal and of length 9

Find an equation for each hyperbola described.

51. x-intercepts ± 3; foci at $(-4, 0), (4, 0)$

52. y-intercepts ± 5; foci at $(0, 3\sqrt{3}), (0, -3\sqrt{3})$

53. asymptotes $y = \pm\frac{3}{5}x$; y-intercepts 3 and -3

54. y-intercept -2; center at origin; passing through $(2, 3)$

RELATING CONCEPTS (EXERCISES 55–60)

Consider the ellipse $\dfrac{x^2}{16} + \dfrac{y^2}{12} = 1$ *and the hyperbola*

$\dfrac{x^2}{4} - \dfrac{y^2}{12} = 1.$

55. Find the foci of the ellipse. Call them F_1 and F_2.

56. Graph the ellipse with your calculator and trace to find the coordinates of several points on the ellipse.

57. For each of the points P, verify that

[Distance of P from F_1] + [Distance of P from F_2] = 8.

58. Repeat Exercises 55 and 56 for the hyperbola.

59. For each of the points P from Exercise 58, verify that

[Distance of P from F_1] − [Distance of P from F_2] = 4.

60. How do Exercises 57 and 59 relate to the definitions of the ellipse and the hyperbola we have given in this section?

Solve each application of ellipses or hyperbolas.

61. *Shape of a Lithotripter*   A patient is placed 12 units away from the source of the shock waves of a lithotripter. The lithotripter is based on an ellipse with a minor axis that measures 16 units. Find an equation of an ellipse that would satisfy this situation.

62. *Orbit of Venus*   The orbit of Venus is an ellipse, with the sun at one focus. An approximate equation for the orbit is $\dfrac{x^2}{5013} + \dfrac{y^2}{4970} = 1$, where x and y are measured in millions of miles.

   (a) Find the length of the major axis.

   (b) Find the length of the minor axis.

63. *The Roman Coliseum*   The Roman Coliseum is an ellipse with major axis 620 feet and minor axis 513 feet. Find the distance between the foci of this ellipse.

64. *The Roman Coliseum*   A formula for the approximate circumference of an ellipse is

$$C \approx 2\pi \sqrt{\dfrac{a^2 + b^2}{2}},$$

where a and b are the lengths, as shown in the figure. Use this formula to find the approximate circumference of the Roman Coliseum (see Exercise 63).

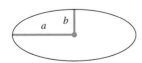

65. *Height of an Overpass*   A one-way road passes under an overpass in the form of half of an ellipse, 15 feet high at the center and 20 feet wide. Assuming a truck is 12 feet wide, what is the height of the tallest truck that can pass under the overpass?

66. *A Location-Finding System*   Ships and planes often use a location-finding system called LORAN. With this system, a radio transmitter at M on the figure sends out a series of pulses. When each pulse is received at transmitter S, it then sends out a pulse. A ship at P receives pulses from both M and S. A receiver on the ship measures the difference in the arrival times of the pulses. The navigator then consults a special map, showing certain curves according to the differences in arrival times. In this way, the ship can be located as lying on a portion of what type of curve?

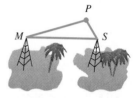

67. *Orbit of a Satellite*   The coordinates in miles for the orbit of the artificial satellite *Explorer VII* can be described by the equation $\dfrac{x^2}{a^2} + \dfrac{y^2}{b^2} = 1$, where $a = 4465$ and $b = 4462$. Earth's center is located at one focus of its elliptical orbit. (*Sources:* Loh, W., *Dynamics and Thermodynamics of Planetary Entry*, Prentice Hall, Inc., Englewood Cliffs, NJ, 1963; Thomson, W., *Introduction to Space Dynamics*, John Wiley & Sons, Inc., New York, 1961.)

   (a) Graph both the orbit of *Explorer VII* and of Earth on the same coordinate axes if the average radius of Earth is 3960 miles. Use the window [−6750, 6750] by [−4500, 4500].

   (b) Determine the maximum and minimum heights of the satellite above Earth's surface.

68. *Structure of an Atom*   In 1911, Ernest Rutherford discovered the basic structure of the atom by "shooting" positively charged alpha particles with a speed of 10^7 m/sec at a piece of gold foil 6×10^{-7} m thick. Only a small percentage of the alpha particles struck a gold nucleus head-on and were deflected directly back toward their source. The rest of the particles often followed a hyperbolic trajectory because they were repelled by positively charged gold nuclei. Thus, Rutherford proposed that the atom was composed of mostly empty space with a small and dense nucleus. The figure on the following page shows an alpha particle A initially approaching a gold nucleus N and being deflected at an angle $\theta = 90°$. N is located at a focus of the hyperbola, and the trajectory of A passes through a vertex of the hyperbola. (*Source:* Semat, H., and J. Albright, *Introduction to Atomic and Nuclear Physics*, Holt, Rinehart, and Winston, Inc., 1972.)

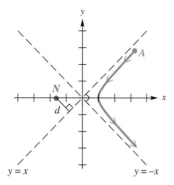

(a) Determine the equation of the trajectory of the alpha particle if $d = 5 \times 10^{-14}$ m.

(b) What was the minimum distance between the centers of the alpha particle and the gold nucleus?

69. *Sound Detection*   Microphones are placed at points $(-c, 0)$ and $(c, 0)$. An explosion occurs at point $P(x, y)$ having positive x-coordinate. (See the figure.) The sound is detected at the closer microphone t sec before being detected at the farther microphone. Assume that sound travels at a speed of 330 m per sec, and show that P must be on the hyperbola

$$\frac{x^2}{330^2 t^2} - \frac{y^2}{4c^2 - 330^2 t^2} = \frac{1}{4}.$$

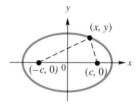

70. Suppose that $(c, 0)$ and $(-c, 0)$ are the foci of an ellipse. Suppose also that the sum of the distances from any point (x, y) of the ellipse to the two foci is $2a$. See the accompanying figure.

(a) Use the distance formula to show that the equation of the resulting ellipse is $\frac{x^2}{a^2} + \frac{y^2}{a^2 - c^2} = 1$.

(b) Show that a and $-a$ are the x-intercepts.

(c) Let $b^2 = a^2 - c^2$, and show that b and $-b$ are the y-intercepts.

71. Suppose a hyperbola has center at the origin, foci at $F'(-c, 0)$ and $F(c, 0)$, and equation $d(P, F') - d(P, F) = 2a$. Let $b^2 = c^2 - a^2$, and show that an equation of the hyperbola is $\frac{x^2}{a^2} - \frac{y^2}{b^2} = 1$.

72. (a) Use the result of Exercise 70(a) to find an equation of an ellipse with foci $(3, 0)$ and $(-3, 0)$, where the sum of the distances from any point of the ellipse to the two foci is 10.

(b) Use the result of Exercise 71 to find an equation of a hyperbola with center at the origin, foci at $(-2, 0)$ and $(2, 0)$, and the absolute value of the difference of the distances from any point of the hyperbola to the two foci equal to 2.

6.3 SUMMARY OF THE CONIC SECTIONS

Characteristics of the Various Conic Sections ◆ Identification of Equations of Conic Sections ◆ Rotated Conics ◆ Eccentricity

CHARACTERISTICS OF THE VARIOUS CONIC SECTIONS

The conic sections presented so far in this chapter all have equations that can be written in the form

$$Ax^2 + Dx + Cy^2 + Ey + F = 0,$$

where either A or C must be nonzero. The special characteristics of each of the conic sections are summarized below.

Equations of Conic Sections

Conic Section	Characteristic	Example
Parabola	Either $A = 0$ or $C = 0$, but not both.	$y = x^2$
		$x = 3y^2 + 2y - 4$
Circle	$A = C \neq 0$	$x^2 + y^2 = 16$
Ellipse	$A \neq C$, $AC > 0$	$\frac{x^2}{16} + \frac{y^2}{25} = 1$
Hyperbola	$AC < 0$	$x^2 - y^2 = 1$

The following chart summarizes our work with conic sections.

Summary of the Conic Sections

Equation and Graph	Description	Identification
$(x - h)^2 + (y - k)^2 = r^2$ *Circle*	Center is at (h, k), and radius is r.	x^2- and y^2-terms have the same positive coefficient.
$y = a(x - h)^2 + k$ *Parabola*	Opens upward if $a > 0$, downward if $a < 0$. Vertex is at (h, k).	x^2-term y is not squared.
$x = a(y - k)^2 + h$ *Parabola*	Opens to right if $a > 0$, to left if $a < 0$. Vertex is at (h, k).	y^2-term x is not squared.
$\dfrac{x^2}{a^2} + \dfrac{y^2}{b^2} = 1$ *Ellipse*	x-intercepts are a and $-a$. y-intercepts are b and $-b$.	x^2- and y^2-terms have different positive coefficients.
$\dfrac{x^2}{a^2} - \dfrac{y^2}{b^2} = 1$ *Hyperbola*	x-intercepts are a and $-a$. Asymptotes found from (a, b), $(a, -b)$, $(-a, -b)$, and $(-a, b)$.	x^2 has a positive coefficient. y^2 has a negative coefficient.
$\dfrac{y^2}{a^2} - \dfrac{x^2}{b^2} = 1$ *Hyperbola*	y-intercepts are a and $-a$. Asymptotes found from (b, a), $(b, -a)$, $(-b, -a)$, and $(-b, a)$.	y^2 has a positive coefficient. x^2 has a negative coefficient.

IDENTIFICATION OF EQUATIONS OF CONIC SECTIONS

In order to recognize the type of graph that a given conic section has, it is sometimes necessary to transform the equation into a more familiar form, as shown in the next examples.

EXAMPLE 1   *Determining the Type of a Conic Section from Its Equation*
Decide on the type of conic section represented by each of the following equations, and give each graph.

(a) $25y^2 - 4x^2 = 100$

(b) $x^2 = 25 + 5y^2$

(c) $4x^2 - 16x + 9y^2 + 54y = -61$

(d) $x^2 - 8x + y^2 + 10y = -41$

(e) $x^2 - 6x + 8y - 7 = 0$

SOLUTION

(a)   Divide each side by 100 to get $\dfrac{y^2}{4} - \dfrac{x^2}{25} = 1$. This is a hyperbola centered at the origin, with foci on the *y*-axis, and *y*-intercepts 2 and -2. The points $(5, 2)$, $(5, -2)$, $(-5, 2)$, and $(-5, -2)$ determine the fundamental rectangle. The diagonals of the rectangle are the asymptotes, and their equations are $y = \pm\frac{2}{5}x$. The graph is shown in Figure 30. Both traditional and calculator-generated graphs are shown.

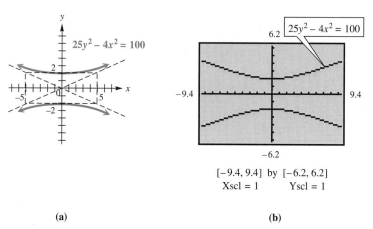

(a)

[−9.4, 9.4] by [−6.2, 6.2]
Xscl = 1   Yscl = 1

(b)

FIGURE 30

(b)   Rewriting the equation as

$$x^2 - 5y^2 = 25 \quad \text{or} \quad \frac{x^2}{25} - \frac{y^2}{5} = 1$$

shows that the equation represents a hyperbola centered at the origin, with asymptotes $y = \frac{\pm\sqrt{5}}{5}x$. The *x*-intercepts are ±5; both types of graphs are shown in Figure 31.

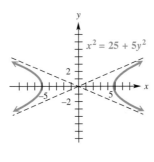

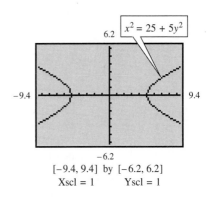

(a)   (b)

FIGURE 31

(c) Since the coefficients of the x^2- and y^2-terms of $4x^2 - 16x + 9y^2 + 54y = -61$ are unequal and both positive, this equation might represent an ellipse. (It might also represent a single point or no points at all.) To find out, complete the square on x and y.

$$4(x^2 - 4x \quad) + 9(y^2 + 6y \quad) = -61 \qquad \text{Factor out a 4;}$$
$$\text{factor out a 9.}$$

$$4(x^2 - 4x + 4 - 4) + 9(y^2 + 6y + 9 - 9) = -61 \qquad \text{Add and subtract the same quantity.}$$

$$4(x^2 - 4x + 4) - 16 + 9(y^2 + 6y + 9) - 81 = -61 \qquad \text{Regroup and distribute.}$$

$$4(x - 2)^2 + 9(y + 3)^2 = 36 \qquad \text{Add 97 and factor.}$$

$$\frac{(x - 2)^2}{9} + \frac{(y + 3)^2}{4} = 1 \qquad \text{Divide by 36.}$$

This equation represents an ellipse having center $(2, -3)$ and graph as shown in Figure 32. (See if you can duplicate the calculator-generated graph.)

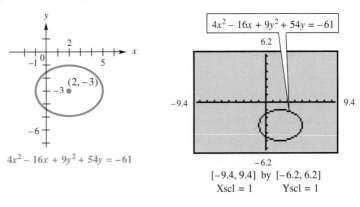

$$4x^2 - 16x + 9y^2 + 54y = -61$$

(a)   (b)

FIGURE 32

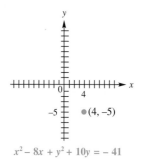

$x^2 - 8x + y^2 + 10y = -41$

FIGURE 33

(d) Complete the square on both x and y, as follows:

$$x^2 - 8x + y^2 + 10y = -41$$
$$(x^2 - 8x + 16) + (y^2 + 10y + 25) = -41 + 16 + 25$$
$$(x - 4)^2 + (y + 5)^2 = 0.$$

This result shows that the equation is that of a circle of radius 0; that is, the point $(4, -5)$. See Figure 33. Had a negative number been obtained on the right (instead of 0), the equation would have represented no points at all, and there would be no graph.

(e) Since only one variable of $x^2 - 6x + 8y - 7 = 0$ is squared (x, and not y), the equation represents a parabola. Rearrange the terms with y (the variable that is not squared) alone on one side. Then, complete the square on the other side of the equation.

$$8y = -x^2 + 6x + 7$$
$$8y = -(x^2 - 6x \quad) + 7 \qquad \text{Regroup and factor out } -1.$$
$$8y = -(x^2 - 6x + 9) + 7 + 9 \qquad \text{Add 0 in the form } -9 + 9.$$
$$8y = -(x - 3)^2 + 16 \qquad \text{Factor.}$$
$$y = -\frac{1}{8}(x - 3)^2 + 2 \qquad \text{Multiply both sides by } \tfrac{1}{8}.$$

The parabola has a vertex at $(3, 2)$ and opens downward, as shown in both graphs in Figure 34. ◆

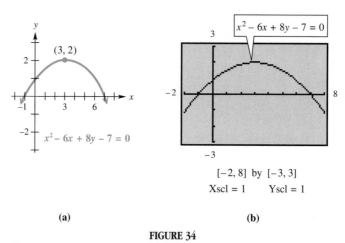

(a)   (b)

FIGURE 34

> *For Group Discussion*   Use an integer window and the equation in Example 1(d) to graph the single point $(4, -5)$ on your calculator screen.

ROTATED CONICS

We saw at the beginning of this section that the equation $Ax^2 + Cy^2 + Dx + Ey + F = 0$ represents a conic section that is translated horizontally if $D \neq 0$ and vertically if $E \neq 0$. The letter B, which was not used in the equation, is reserved for the one other possible second-degree term, Bxy. If $B \neq 0$, the equation represents a conic section that is *rotated* with respect to the *xy*-coordinate axes. For example, the graph of the conic with equation $x^2 + xy = 4$, which can be solved for *y* to get

$$y = \frac{4 - x^2}{x},$$

is the rotated hyperbola shown in Figure 35.

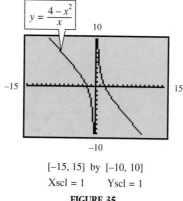

$[-15, 15]$ by $[-10, 10]$
Xscl = 1   Yscl = 1
FIGURE 35

TECHNOLOGY NOTE

The graphs of the two functions that form a rotated conic may not quite meet. (See, for example, Exercise 37 in this section.) This is a limitation of the calculator, and again we see the importance of understanding the mathematics.

In Section 4.3, Example 8, we showed how to graph a rotated conic that had an equation with a y^2-term as well as a *y*-term. As shown in that example, it is necessary to complete the square to solve the equation.

ECCENTRICITY

In Sections 6.1 and 6.2, we introduced the various definitions of the conic sections. The conic sections (or conics) can all be characterized by one general definition.

DEFINITION OF A CONIC

A **conic** is the set of all points $P(x, y)$ in a plane such that the ratio of the distance from *P* to a fixed point and the distance from *P* to a fixed line is constant.

As with parabolas, the fixed line is the **directrix**, and the fixed point is the **focus**. In Figure 36, the focus is $F(c, 0)$, and the directrix is the line $x = -c$. The constant ratio is called the **eccentricity** of the conic, written as *e*. (This is not the same *e* as the base of natural logarithms.)

By definition, the distances $d(P, F)$ and $d(P, D)$ in Figure 36 are equal if the conic is a parabola. Thus, a parabola always has eccentricity $e = 1$.

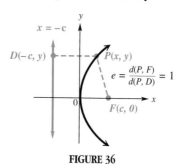

FIGURE 36

For the ellipse and hyperbola, it can be shown that the constant ratio of the definition is $e = \frac{c}{a}$, where c is the distance from the center of the figure to a focus, and a is the distance from the center to a vertex. By the definition of an ellipse, $a^2 > b^2$ and $c = \sqrt{a^2 - b^2}$. Thus, for the ellipse,

$$0 < c < a$$

$$0 < \frac{c}{a} < 1$$

$$0 < e < 1.$$

If a is constant, letting c approach 0 would force the ratio $\frac{c}{a}$ to approach 0, which also forces b to approach a (so that $\sqrt{a^2 - b^2} = c$ would approach 0). Since b leads to the y-intercepts, this means that the x- and y-intercepts are almost the same, producing an ellipse very close in shape to a circle when e is very close to 0. In a similar manner, if e approaches 1, then b will approach 0. The path of the earth around the sun is an ellipse that is very nearly circular. In fact, for this ellipse, $e \approx .017$. On the other hand, the path of Halley's comet is a very flat ellipse, with $e \approx .98$.

EXAMPLE 2   *Finding Eccentricity from the Equation of an Ellipse*   Find the eccentricity of each ellipse.

(a) $\dfrac{x^2}{9} + \dfrac{y^2}{16} = 1$ **(b)** $5x^2 + 10y^2 = 50$

SOLUTION

(a)   Since $16 > 9$, let $a^2 = 16$, which gives $a = 4$. Also,

$$c = \sqrt{a^2 - b^2} = \sqrt{16 - 9} = \sqrt{7}.$$

Finally, $e = \dfrac{\sqrt{7}}{4} \approx .66.$

(b)   Divide by 50 to get $\dfrac{x^2}{10} + \dfrac{y^2}{5} = 1$. Here, $a^2 = 10$, with $a = \sqrt{10}$. Now, find c.

$$c = \sqrt{10 - 5} = \sqrt{5}$$

and

$$e = \frac{\sqrt{5}}{\sqrt{10}} = \frac{1}{\sqrt{2}} \approx .71 \qquad \blacklozenge$$

As mentioned above, the hyperbola

$$\frac{x^2}{a^2} - \frac{y^2}{b^2} = 1 \quad \text{or} \quad \frac{y^2}{a^2} - \frac{x^2}{b^2} = 1,$$

where $c = \sqrt{a^2 + b^2}$, also has eccentricity $e = \frac{c}{a}$. By definition, $c = \sqrt{a^2 + b^2} > a$, so that $\frac{c}{a} > 1$, and so for a hyperbola, $e > 1$.

EXAMPLE 3   *Finding Eccentricity from the Equation of a Hyperbola*   Find the eccentricity of the hyperbola

$$\frac{x^2}{9} - \frac{y^2}{4} = 1.$$

SOLUTION   Here $a^2 = 9$; thus, $a = 3$, $c = \sqrt{9 + 4} = \sqrt{13}$, and

$$e = \frac{c}{a} = \frac{\sqrt{13}}{3} \approx 1.2. \qquad \blacklozenge$$

The following chart summarizes this discussion of eccentricity.

ECCENTRICITY OF CONICS

Conic	Eccentricity		
Parabola	$e = 1$		
Ellipse	$e = \frac{c}{a}$	and	$0 < e < 1$
Hyperbola	$e = \frac{c}{a}$	and	$e > 1$

For Group Discussion   Answer the following questions in order.

1. As the two foci of an ellipse move closer and closer together, what familiar shape does the ellipse begin to resemble?
2. For a circle with center at the origin, how do the values of a and b compare?
3. What is the value of c as defined above for a circle?
4. What is the eccentricity of a circle?

EXAMPLE 4   *Finding the Equation of a Conic Using Eccentricity*   Find an equation for the following conics with centers at the origin.

(a)   focus at $(3, 0)$ and eccentricity 2

(b)   vertex at $(0, -8)$ and $e = \frac{1}{2}$

SOLUTION

(a)   Since $e = 2$, which is greater than 1, the conic is a hyperbola with $c = 3$. From $e = \frac{c}{a}$, find a by substituting $e = 2$ and $c = 3$.

$$e = \frac{c}{a}$$

$$2 = \frac{3}{a}$$

$$a = \frac{3}{2}$$

Now, find b.

$$b^2 = c^2 - a^2 = 9 - \frac{9}{4} = \frac{27}{4}$$

The given focus is on the x-axis, so the x^2-term is positive, and the equation is

$$\frac{x^2}{\frac{9}{4}} - \frac{y^2}{\frac{27}{4}} = 1, \quad \text{or} \quad \frac{4x^2}{9} - \frac{4y^2}{27} = 1.$$

(b)   The graph of the conic with vertex at $(0, -8)$ and $e = \frac{1}{2}$ is an ellipse because $e = \frac{1}{2} < 1$. From the given vertex, we know that the vertices are on the y-axis and $a = 8$. Use $e = \frac{c}{a}$ to find c.

$$e = \frac{c}{a}$$

$$\frac{1}{2} = \frac{c}{8}$$

$$c = 4$$

Since $b^2 = a^2 - c^2$,

$$b^2 = 64 - 16 = 48,$$

and the equation is $\dfrac{y^2}{64} + \dfrac{x^2}{48} = 1$.   ◆

EXAMPLE 5   *Applying an Ellipse to the Orbit of a Planet*   The orbit of the planet Mars is an ellipse with the sun at one focus. The eccentricity of the ellipse is .0935, and the closest distance that Mars comes to the sun is 128.5 million miles. (*Source: The World Almanac and Book of Facts*, 1995.) Find the maximum distance of Mars from the sun.

SOLUTION   Figure 37 shows the orbit of Mars with the origin at the center of the ellipse and the sun at one focus. Mars is closest to the sun when Mars is at the right endpoint of the major axis and farthest from the sun when Mars is at the left endpoint. Therefore, the smallest distance is $a - c$, and the greatest distance is $a + c$.

Since $a - c = 128.5$, $c = a - 128.5$. Therefore,

$$e = \frac{c}{a} = \frac{a - 128.5}{a} = .0935$$

$$a - 128.5 = .0935a$$

$$.9065a = 128.5$$

$$a \approx 141.8$$

$$c = 141.8 - 128.5 = 13.3$$

$$a + c = 141.8 + 13.3 = 155.1.$$

Thus, the maximum distance of Mars from the sun is about 155.1 million miles.

◆

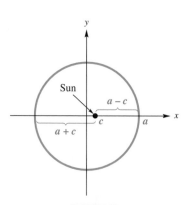

FIGURE 37

6.3 EXERCISES   Tape 7

In Exercises 1–12, the equation of a conic section is given in a familiar form. Identify the type of graph that the equation has, without actually graphing.

1. $x^2 + y^2 = 144$

2. $(x - 2)^2 + (y + 3)^2 = 25$

3. $y = 2x^2 + 3x - 4$

4. $x = 3y^2 + 5y - 6$

5. $x = -3(y - 4)^2 + 1$

6. $\dfrac{x^2}{25} + \dfrac{y^2}{36} = 1$

7. $\dfrac{x^2}{49} + \dfrac{y^2}{100} = 1$

8. $x^2 - y^2 = 1$

9. $\dfrac{x^2}{4} - \dfrac{y^2}{16} = 1$

10. $\dfrac{(x + 2)^2}{9} + \dfrac{(y - 4)^2}{16} = 1$

11. $\dfrac{x^2}{25} - \dfrac{y^2}{25} = 1$

12. $y = 4(x + 3)^2 - 7$

For each of the following equations that has a graph, identify the corresponding graph. It may be necessary to transform the equation.

13. $\dfrac{x^2}{4} = 1 - \dfrac{y^2}{9}$

14. $\dfrac{x^2}{4} = 1 + \dfrac{y^2}{9}$

15. $\dfrac{x^2}{4} + \dfrac{y^2}{4} = -1$

16. $x^2 = 25 + y^2$

17. $x^2 = 25 - y^2$

18. $9x^2 + 36y^2 = 36$

19. $x^2 = 4y - 8$

20. $\dfrac{(x + 3)^2}{16} + \dfrac{(y - 2)^2}{16} = 1$

21. $\dfrac{(x - 4)^2}{8} + \dfrac{(y + 1)^2}{2} = 0$

22. $y^2 - 4y = x + 4$

23. $(x + 7)^2 + (y - 5)^2 + 4 = 0$

24. $4(x - 3)^2 + 3(y + 4)^2 = 0$

25. $3x^2 + 6x + 3y^2 - 12y = 12$

26. $2x^2 - 8x + 2y^2 + 20y = 12$

27. $x^2 - 6x + y = 0$

28. $x - 4y^2 - 8y = 0$

29. $4x^2 - 8x - y^2 - 6y = 6$

30. $x^2 + 2x = x^2 - 4y - 2$

31. $4x^2 - 8x + 9y^2 + 54y = -84$

32. $3x^2 + 12x + 3y^2 = -11$

33. $6x^2 - 12x + 6y^2 - 18y + 25 = 0$

34. $4x^2 - 24x + 5y^2 + 10y + 41 = 0$

Use your calculator to graph each equation.

35. $5x^2 + 12xy - 10 = 0$

36. $3xy + x - 5 = 0$

37. $x^2 - xy + y^2 = 6$

38. $x^2 + xy + y^2 = 3$

39. Suppose that both A and C are zero in the equation $Ax^2 + Bx + Cy^2 + Dy + E = 0$. What kind of graph does this equation have?

40. How can the graph of the equation discussed in Exercise 39 be obtained by a plane intersecting a cone?

Find the eccentricity of each of the following ellipses and hyperbolas.

41. $12x^2 + 9y^2 = 36$

42. $8x^2 - y^2 = 16$

43. $x^2 - y^2 = 4$

44. $x^2 + 2y^2 = 8$

45. $4x^2 + 7y^2 = 28$

46. $9x^2 - y^2 = 1$

47. $x^2 - 9y^2 = 18$

48. $x^2 + 10y^2 = 10$

Write an equation for each of the following conics. Each parabola has vertex at the origin, and each ellipse or hyperbola is centered at the origin.

49. focus at $(0, 8)$ and $e = 1$

50. focus at $(-2, 0)$ and $e = 1$

51. focus at $(3, 0)$ and $e = \frac{1}{2}$

52. focus at $(0, -2)$ and $e = \frac{2}{3}$

53. vertex at $(-6, 0)$ and $e = 2$

54. vertex at $(0, 4)$ and $e = \frac{5}{3}$

55. focus at $(0, -1)$ and $e = 1$

56. focus at $(2, 0)$ and $e = \frac{6}{5}$

57. vertical major axis of length 6 and $e = \frac{4}{5}$

58. y-intercepts -4 and 4 and $e = \frac{7}{3}$

59. Calculator-generated graphs are shown in Figures A–D. Arrange the figures in order so that the first in the list has the smallest eccentricity and the rest have eccentricities in increasing order.

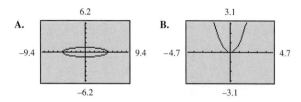

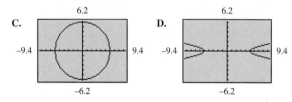

60. *Orbit of Mars*   The orbit of Mars around the sun is an ellipse with equation $\dfrac{x^2}{5013} + \dfrac{y^2}{4970} = 1$, where x and y are measured in millions of miles. Find the eccentricity of this ellipse.

61. *Orbits of Neptune and Pluto*   Neptune and Pluto both have elliptical orbits with the sun at one focus. Neptune's orbit has a semimajor axis of $a = 30.1$ astronomical units (AU) with an eccentricity of $e = .009$, whereas Pluto's orbit has $a = 39.4$ and $e = .249$. (1 AU is equal to the average distance from Earth to the sun and is approximately 149,600,000 kilometers.) (*Source:* Zeilik, M., S. Gregory, and E. Smith, *Introductory Astronomy and Astrophysics*, Saunders College Publishers, 1992.)

  (a) Position the sun at the origin and determine equations for each orbit.

  (b) Graph both equations on the same coordinate axes. Use the window $[-60, 60]$ by $[-40, 40]$.

62. *Velocity of a Planet in Orbit*   The maximum and minimum velocities in kilometers per second of a planet moving in an elliptical orbit can be calculated with the equations

$$v_{max} = \frac{2\pi a}{P} \sqrt{\frac{1+e}{1-e}}$$

and

$$v_{min} = \frac{2\pi a}{P} \sqrt{\frac{1-e}{1+e}},$$

where a is the semimajor axis in kilometers, P is its orbital period in seconds, and e is the eccentricity of the orbit. (*Source:* Zeilik, M., S. Gregory, and E. Smith, *Introductory Astronomy and Astrophysics*, Saunders College Publishers, 1992.)

  (a) Calculate v_{max} and v_{min} for Earth if $a = 1.496 \times 10^8$ kilometers and $e = .0167$.

  (b) If an object has a circular orbit, what can be said about its orbital velocity?

  (c) Kepler showed that the sun is located at a focus of a planet's elliptical orbit. He also showed that a planet's minimum velocity occurs when its distance from the sun is maximum and a planet's maximum velocity occurs when its distance from the sun is minimum. Where do the maximum and minimum velocities occur in an elliptical orbit?

63. *Distance between Halley's Comet and the Sun*   Halley's comet has an elliptical orbit of eccentricity .9673 with the sun at one of the foci. The greatest distance of the comet from the sun is 3281 million miles. (*Source: The World Almanac and Book of Facts*, 1995.) Find the shortest distance between Halley's comet and the sun.

64. *Orbit of Earth*   The orbit of Earth is an ellipse with the sun at one focus. The distance between Earth and the sun ranges from 91.4 to 94.6 million miles. (*Source: The World Almanac and Book of Facts*, 1995.) Find the eccentricity of Earth's orbit.

6.4   PARAMETRIC EQUATIONS

Basic Concepts ◆ Graphs of Parametric Equations and Their Rectangular Equivalents ◆ Alternative Forms of Parametric Equations ◆ An Application

BASIC CONCEPTS

Up to now in this text, we have graphed sets of ordered pairs of real numbers that corresponded to a function of the form $y = f(x)$ or a relation of the form $Ax^2 + Bxy + Cy^2 + Dx + Ey + F = 0$. Another way to determine a set of ordered pairs involves two functions f and g defined by $x = f(t)$ and $y = g(t)$, where t is a real number in some interval I. Each value of t leads to a corresponding x-value and a corresponding y-value, and thus to an ordered pair (x, y).

DEFINITION OF PARAMETRIC EQUATIONS OF A PLANE CURVE

A **plane curve** is a set of points (x, y) such that $x = f(t)$, $y = g(t)$, and f and g are both defined on an interval I. The equations $x = f(t)$ and $y = g(t)$ are **parametric equations** with **parameter t**.

Note   Just as graphing calculators are capable of graphing rectangular and polar equations, they are also capable of graphing plane curves defined by parametric equations. You should familiarize yourself with how your particular model handles them.

GRAPHS OF PARAMETRIC EQUATIONS AND THEIR RECTANGULAR EQUIVALENTS

EXAMPLE 1   *Graphing a Plane Curve Defined Parametrically*   For the plane curve defined by the parametric equations

$$x = t^2, \quad y = 2t + 3, \quad \text{for } t \text{ in the interval } [-3, 3],$$

graph the curve with a graphing calculator, and then find an equivalent rectangular equation.

SOLUTION   Figure 38 shows the graph of the curve, which appears to be a horizontal parabola.

The table shows values of $X_{1T} = T^2$, $Y_{1T} = 2T + 3$ for T in $[-3, 3]$.

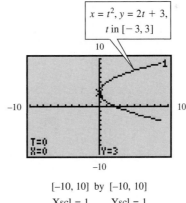

$x = t^2, y = 2t + 3,$
$t \text{ in } [-3, 3]$

$[-10, 10]$ by $[-10, 10]$
Xscl $= 1$ Yscl $= 1$

When $t = 0$, $x = 0$, and $y = 3$.

FIGURE 38

To find an equivalent rectangular equation, we analytically eliminate the parameter t. For this curve, we will solve for t in the second equation, $y = 2t + 3$, to begin.

$$y = 2t + 3$$
$$2t = y - 3$$
$$t = \frac{y - 3}{2}$$

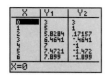

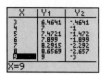

Solving the rectangular equation in Example 1 for y gives $Y_1 = 3 + 2\sqrt{x}$ or $Y_2 = 3 - 2\sqrt{x}$. Compare the ordered pairs from the first table (X_{1T}, Y_{1T}) with the ordered pairs (X, Y_1) and (X, Y_2) from this table.

Now, substitute the result in the first equation to get

$$x = t^2 = \left(\frac{y-3}{2}\right)^2 = \frac{(y-3)^2}{4} = \frac{1}{4}(y-3)^2.$$

This is indeed an equation of a parabola. It has a horizontal axis and opens to the right. Because t is in $[-3, 3]$, x is in $[0, 9]$ and y is in $[-3, 9]$. The rectangular equation must be given with its restricted domain as

$$x = \frac{1}{4}(y-3)^2, \quad \text{for } x \text{ in } [0, 9]. \qquad \blacklozenge$$

For Group Discussion   The display at the bottom of Figure 38 indicates particular values of t, x, and y. Discuss how these relate to both the parametric form and the rectangular form of the equation of this parabola.

EXAMPLE 2   *Graphing a Plane Curve Defined Parametrically*   Graph the plane curve defined by

$$x = 2t + 5, \quad y = \sqrt{4 - t^2}, \quad \text{for } t \text{ in } [-2, 2],$$

and then find an equivalent rectangular equation.

SOLUTION   The graph with both x and y in $[-10, 10]$ is shown in Figure 39(a). It appears to be a semiellipse. Changing to a square window gives the graph in Figure 39(b), which gives a truer representation of the figure.

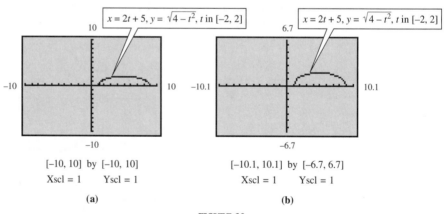

FIGURE 39

To get an equivalent rectangular equation, solve the first equation for t, then substitute into the second equation. Since t is in $[-2, 2]$, x is in the interval $[1, 9]$.

$$2t + 5 = x$$
$$2t = x - 5$$
$$t = \frac{x-5}{2}$$

The rectangular equation is

$$y = \sqrt{4 - t^2} = \sqrt{4 - \left(\frac{x-5}{2}\right)^2}, \quad \text{for } x \text{ in } [1, 9].$$

By squaring both sides and rearranging terms, we get the equation

$$y^2 = 4 - \left(\frac{x-5}{2}\right)^2$$
$$4y^2 = 16 - (x-5)^2$$
$$4y^2 + (x-5)^2 = 16$$
$$\frac{y^2}{4} + \frac{(x-5)^2}{16} = 1,$$

which represents the complete ellipse. (Why is the parametric graph only half the ellipse?) ◆

ALTERNATIVE FORMS OF PARAMETRIC EQUATIONS

Parametric representations of a curve are not unique. In fact, there are infinitely many parametric representations of a given curve. If the curve can be described by a rectangular equation $y = f(x)$, with domain X, then one simple parametric representation is

$$x = t, \quad y = f(t), \quad \text{for } t \text{ in } X.$$

The next example shows how one plane curve has alternative parametric equation forms.

EXAMPLE 3 *Finding Alternative Parametric Equation Forms*   Give two parametric representations for the parabola $y = (x-2)^2 + 1$.

SOLUTION   The simplest choice is to let

$$x = t, \quad y = (t-2)^2 + 1, \quad \text{for } t \text{ in } (-\infty, \infty).$$

Another choice that leads to a simpler equation for y is

$$x = t + 2, \quad y = t^2 + 1, \quad \text{for } t \text{ in } (-\infty, \infty). \quad ◆$$

> *For Group Discussion*   Consider the rectangular equation $y - 1 = (x-2)^2$. Why are the parametric equations $x = (t-2)^{1/2}$, $y = t + 1$, for t in $[2, \infty)$ not equivalent to the rectangular equation?

AN APPLICATION

Of the many applications of parametric equations, one of the most useful allows us to determine the path of a moving object whose position is given by the function $x = f(t)$, $y = g(t)$, where t represents time. The parametric equations give the position of the object at any time t.

EXAMPLE 4 *Examining Parametric Equations Defining the Position of an Object in Motion*   The motion of a projectile moving in a direction at a 45° angle with the horizontal (neglecting air resistance) is given by

$$x = v_0 \frac{\sqrt{2}}{2} t, \quad y = v_0 \frac{\sqrt{2}}{2} t - 16t^2, \quad \text{for } t \text{ in } [0, k],$$

where t is time in seconds, v_0 is the initial speed of the projectile, x and y are in feet, and k is a positive real number. See Figure 40 on the following page. Find the rectangular form of the equation.

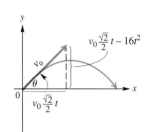

FIGURE 40

SOLUTION   Solving the first equation for t and substituting the result into the second equation gives (after simplification)

$$y = x - \frac{32}{v_0^2} x^2,$$

the equation of a vertical parabola opening downward, as shown in Figure 40.   ◆

6.4   EXERCISES   Tape 8

For each plane curve, use a graphing calculator to generate the curve over the interval for the parameter t, using the window specified. Then, find a rectangular equation for the curve.

1. $x = 2t$, $y = t + 1$,   for t in $[-2, 3]$
Window: $[-8, 8]$ by $[-8, 8]$

2. $x = t + 2$, $y = t^2$,   for t in $[-1, 1]$
Window: $[0, 4]$ by $[-2, 2]$

3. $x = \sqrt{t}$, $y = 3t - 4$,   for t in $[0, 4]$
Window: $[-6, 6]$ by $[-6, 10]$

4. $x = t^2$, $y = \sqrt{t}$,   for t in $[0, 4]$
Window: $[-2, 20]$ by $[0, 4]$

5. $x = t^3 + 1$, $y = t^3 - 1$,   for t in $[-3, 3]$
Window: $[-30, 30]$ by $[-30, 30]$

6. $x = 2t - 1$, $y = t^2 + 2$,   for t in $[-10, 10]$
Window: $[-20, 20]$ by $[0, 120]$

7. $x = 2^t$, $y = \sqrt{3t - 1}$,   for t in $\left[\frac{1}{3}, 4\right]$
Window: $[-2, 30]$ by $[-2, 10]$

8. $x = \ln(t - 1)$, $y = 2t - 1$,   for t in $(1, 10]$
Window: $[-5, 5]$ by $[-2, 20]$

9. $x = t + 2$, $y = -\frac{1}{2}\sqrt{9 - t^2}$,   for t in $[-3, 3]$
Window: $[-6, 6]$ by $[-4, 4]$

10. $x = t$, $y = \sqrt{4 - t^2}$,   for t in $[-2, 2]$
Window: $[-6, 6]$ by $[-4, 4]$

11. $x = t$, $y = \frac{1}{t}$,   for t in $(-\infty, 0) \cup (0, \infty)$
Window: $[-6, 6]$ by $[-4, 4]$

12. $x = 2t - 1$, $y = \frac{1}{t}$,   for t in $(-\infty, 0) \cup (0, \infty)$
Window: $[-6, 6]$ by $[-4, 4]$

For each plane curve, find a rectangular equation.

13. $x = 3t^2$, $y = 4t^3$,   for t in $(-\infty, \infty)$

14. $x = 2t^3$, $y = -t^2$,   for t in $(-\infty, \infty)$

15. $x = t$, $y = \sqrt{t^2 + 2}$,   for t in $(-\infty, \infty)$

16. $x = \sqrt{t}$, $y = t^2 - 1$,   for t in $[0, \infty)$

17. $x = e^t$, $y = e^{-t}$,   for t in $(-\infty, \infty)$

18. $x = e^{2t}$, $y = e^t$,   for t in $(-\infty, \infty)$

19. $x = \frac{1}{\sqrt{t + 2}}$, $y = \frac{t}{t + 2}$,   for t in $(-2, \infty)$

20. $x = \frac{t}{t - 1}$, $y = \frac{1}{\sqrt{t - 1}}$,   for t in $(1, \infty)$

21. $x = t + 2$, $y = \frac{1}{t + 2}$,   for $t \neq -2$

22. $x = t - 3$, $y = \frac{2}{t - 3}$,   for $t \neq 3$

23. $x = t^2$, $y = 2 \ln t$,   for t in $(0, \infty)$

24. $x = \ln t$, $y = 3 \ln t$,   for t in $(0, \infty)$

Give two parametric representations for the plane curves with the following rectangular equations. Use your calculator to verify your results.

25. $y = 2x + 3$

26. $y = \frac{3}{2}x - 4$

27. $y = \sqrt{3x + 2}$, x in $\left[-\frac{2}{3}, \infty\right)$

28. $y = (x + 1)^2 + 1$

29. *Firing a Projectile*   A projectile is fired with an initial velocity of 400 feet per second at an angle of 45° with the horizontal. Find each of the following. (See Example 4.)

 (a) the time when it strikes the ground

 (b) the range (horizontal distance covered)

 (c) the maximum altitude

30. *Firing a Projectile*   If a projectile is fired at an angle of 30° with the horizontal, the parametric equations describing its motion are $x = v_0 \dfrac{\sqrt{3}}{2}t$, $y = \dfrac{v_0}{2}t - 16t^2$, for t in $[0, \infty)$. Repeat Exercise 29 if the projectile is fired at 800 feet per second.

31. Show that the rectangular equation for the curve defined by $x = v_0 \dfrac{\sqrt{2}}{2}t$, $y = v_0 \dfrac{\sqrt{2}}{2}t - 16t^2$, for t in $[0, k]$, is $y = x - \dfrac{32}{v_0^2}x^2$. (See Example 4.)

32. Find the vertex of the parabola given by the rectangular equation of Exercise 31.

33. *Path of a Projectile*   A projectile moves so that its position at any time t is given by the equations $x = 60t$ and $y = 80t - 16t^2$. Graph the path of the projectile, and find the equivalent rectangular equation. Use the window $[0, 300]$ by $[0, 200]$.

34. *Path of a Projectile*   Repeat Exercise 33 if the path is given by the equations $x = t^2$ and $y = -16t + 64\sqrt{t}$. Use the window $[0, 300]$ by $[0, 200]$.

35. Give two parametric representations of the line through the point (x_1, y_1) with slope m.

36. Give two parametric representations of the parabola $y = a(x - h)^2 + k$.

CHAPTER 6   SUMMARY

Section	Important Concepts
6.1 Circles and Parabolas	**DEFINITION OF A CIRCLE** A circle is a set of points in a plane, each of which is equidistant from a fixed point. The distance is called the radius of the circle, and the fixed point is called the center. **CENTER-RADIUS FORM OF THE EQUATION OF A CIRCLE** The circle with center (h, k) and radius r has equation $(x - h)^2 + (y - k)^2 = r^2$. A circle with center $(0, 0)$ and radius r has equation $x^2 + y^2 = r^2$. **GENERAL FORM OF THE EQUATION OF A CIRCLE** For real numbers c, d, and e, the general form of the equation of a circle is $$x^2 + y^2 + cx + dy + e = 0.$$ **DEFINITION OF A PARABOLA** A parabola is the set of points in a plane equidistant from a fixed point and a fixed line. The fixed point is called the focus, and the fixed line, the directrix, of the parabola. **PARABOLA WITH A VERTICAL AXIS** The parabola with focus at $(0, c)$ and directrix $y = -c$ has equation $y = \frac{1}{4c}x^2$. The parabola has a vertical axis, opens upward if $c > 0$, and opens downward if $c < 0$. **PARABOLA WITH A HORIZONTAL AXIS** The parabola with focus at $(c, 0)$ and directrix $x = -c$ has equation $x = \frac{1}{4c}y^2$. The parabola opens to the right if $c > 0$, to the left if $c < 0$, and has a horizontal axis. **TRANSLATION OF A HORIZONTAL PARABOLA** The parabola with vertex at (h, k) and the horizontal line $y = k$ as axis has an equation of the form $x = a(y - k)^2 + h$. The parabola opens to the right if $a > 0$ and to the left if $a < 0$.

6.2 Ellipses and Hyperbolas

DEFINITION OF AN ELLIPSE

An ellipse is the set of all points in a plane such that the sum of their distances from two fixed points is always the same (constant). The two fixed points are called the foci (plural of *focus*) of the ellipse.

STANDARD FORMS OF EQUATIONS FOR ELLIPSES

The ellipse with center at the origin and equation $\frac{x^2}{a^2} + \frac{y^2}{b^2} = 1$ has vertices $(\pm a, 0)$, endpoints of the minor axis $(0, \pm b)$, and foci $(\pm c, 0)$, where $c^2 = a^2 - b^2$. The ellipse with center at the origin and equation $\frac{y^2}{a^2} + \frac{x^2}{b^2} = 1$ has vertices $(0, \pm a)$, endpoints of the minor axis $(\pm b, 0)$, and foci $(0, \pm c)$, where $c^2 = a^2 - b^2$.

DEFINITION OF A HYPERBOLA

A hyperbola is the set of all points in a plane the *difference* of whose distances from two fixed points is constant. The two fixed points are called the foci of the hyperbola.

STANDARD FORMS OF EQUATIONS FOR HYPERBOLAS

The hyperbola with center at the origin and equation $\frac{x^2}{a^2} - \frac{y^2}{b^2} = 1$ has vertices $(\pm a, 0)$ and foci $(\pm c, 0)$, where $c^2 = a^2 + b^2$. The hyperbola with center at the origin and equation $\frac{y^2}{a^2} - \frac{x^2}{b^2} = 1$ has vertices $(0, \pm a)$ and foci $(0, \pm c)$, where $c^2 = a^2 + b^2$.

6.3 Summary of the Conic Sections

Equations of Conic Sections

Conic Section	Characteristic	Example
Parabola	In the equation $Ax^2 + Cy^2 + Dx + Ey + F = 0$, Either $A = 0$ or $C = 0$, but not both.	$y = x^2$ $x = 3y^2 + 2y - 4$
Circle	$A = C \neq 0$.	$x^2 + y^2 = 16$
Ellipse	$A \neq C, AC > 0$.	$\frac{x^2}{16} + \frac{y^2}{25} = 1$
Hyperbola	$AC < 0$.	$x^2 - y^2 = 1$

DEFINITION OF A CONIC

A conic is the set of all points $P(x, y)$ in a plane such that the ratio of the distance from P to a fixed point and the distance from P to a fixed line is constant.

Eccentricity of Conics

Conic	Eccentricity
Parabola	$e = 1$
Ellipse	$e = \frac{c}{a}$ and $0 < e < 1$
Hyperbola	$e = \frac{c}{a}$ and $e > 1$

6.4 Parametric Equations

DEFINITION OF PARAMETRIC EQUATIONS OF A PLANE CURVE

A plane curve is a set of points (x, y) such that $x = f(t)$, $y = g(t)$, and f and g are both defined on an interval I. The equations $x = f(t)$ and $y = g(t)$ are parametric equations with parameter t.

CHAPTER 6   REVIEW EXERCISES

Write an equation for the circle satisfying the given conditions.

1. center $(-2, 3)$, radius 5

2. center $\left(\sqrt{5}, -\sqrt{7}\right)$, radius $\sqrt{3}$

3. center $(-8, 1)$, passing through $(0, 16)$

4. center $(3, -6)$, tangent to the x-axis

Find the center and radius of the following circles.

5. $x^2 - 4x + y^2 + 6y + 12 = 0$

6. $x^2 - 6x + y^2 - 10y + 30 = 0$

7. $2x^2 + 14x + 2y^2 + 6y = -2$

8. $3x^2 + 3y^2 + 33x - 15y = 0$

 9. Describe the graph of $(x - 4)^2 + (y - 5)^2 = 0$.

Give the focus, directrix, and axis for the parabola, and graph it by hand.

10. $x = -\dfrac{3}{2}y^2$ **11.** $x = \dfrac{1}{2}y^2$ **12.** $3x^2 = y$ **13.** $x^2 + 2y = 0$

Write an equation for the parabola with vertex at the origin that satisfies the given conditions.

14. focus $(4, 0)$

15. through $(2, 5)$, opening to the right

16. through $(3, -4)$, opening downward

Write an equation for each parabola.

17. vertex $(-5, 6)$, focus $(2, 6)$

18. vertex $(4, 3)$, focus $(4, 5)$

Graph the following ellipses and hyperbolas by hand, and give the coordinates of the vertices.

19. $\dfrac{y^2}{9} + \dfrac{x^2}{5} = 1$ **20.** $\dfrac{x^2}{16} + \dfrac{y^2}{4} = 1$ **21.** $\dfrac{x^2}{64} - \dfrac{y^2}{36} = 1$

22. $\dfrac{y^2}{25} - \dfrac{x^2}{9} = 1$ **23.** $\dfrac{(x - 3)^2}{4} + (y + 1)^2 = 1$ **24.** $\dfrac{(x - 2)^2}{9} + \dfrac{(y + 3)^2}{4} = 1$

25. $\dfrac{(y + 2)^2}{4} - \dfrac{(x + 3)^2}{9} = 1$ **26.** $\dfrac{(x + 1)^2}{16} - \dfrac{(y - 2)^2}{4} = 1$

Write equations for the following conic sections with centers at the origin.

27. ellipse; vertex at $(0, 4)$, focus at $(0, 2)$

28. ellipse; x-intercept 6, focus at $(-2, 0)$

29. hyperbola; focus at $(0, -5)$, y-intercepts -4 and 4

30. hyperbola; y-intercept -2, passing through $(2, 3)$

31. focus at $(0, -3)$ and $e = \frac{2}{3}$

32. focus at $(5, 0)$ and $e = \frac{5}{2}$

33. Consider the circle with equation $x^2 + y^2 + 2x + 6y - 15 = 0$.

   (a) What are the coordinates of the center?

   (b) What is the radius?

   (c) What two functions must be graphed to graph this circle with your calculator?

Match each equation with its graph.

34. $4x^2 + y^2 = 36$

35. $x = 2y^2 + 3$

36. $(x - 1)^2 + (y + 2)^2 = 36$

37. $\dfrac{x^2}{36} + \dfrac{y^2}{9} = 1$

38. $(y - 1)^2 - (x - 2)^2 = 36$

39. $y^2 = 36 + 4x^2$

A.
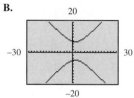

B.
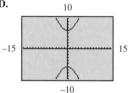

C.
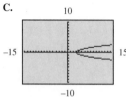

D.
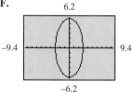

E.
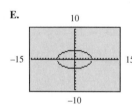

F.

Find the eccentricity of each ellipse or hyperbola.

40. $9x^2 + 25y^2 = 225$

41. $4x^2 + 9y^2 = 36$

42. $9x^2 - y^2 = 9$

Use a graphing calculator to graph each conic.

43. $5x^2 + 8xy + 5y^2 = 9$

44. $24xy - 7y^2 + 36 = 0$

45. $x^2 - xy + 2x - 3y = 0$

Use a graphing calculator to graph the plane curve defined by the parametric equations. Use the window specified.

46. $x = 4t - 3$, $y = t^2$, for t in $[-3, 4]$
Window: $[-20, 20]$ by $[-20, 20]$

47. $x = t^2$, $y = t^3$, for t in $[-2, 2]$
Window: $[-15, 15]$ by $[-10, 10]$

48. $x = t + \ln t$, $y = t + e^t$, for t in $(0, 2]$
Window: $[-10, 5]$ by $[0, 10]$

49. Without graphing, predict the type of graph that the parametric equations $x = t$, $y = t^2$ will have for t in $[-6, 6]$.

Find a rectangular equation for each plane curve with the given parametric equations.

50. $x = 3t + 2$, $y = t - 1$, for t in $[-5, 5]$

51. $x = \sqrt{t - 1}$, $y = \sqrt{t}$, for t in $[1, \infty)$

52. $x = \dfrac{1}{t + 3}$, $y = t + 3$, for $t \neq -3$

Solve each problem.

53. *Orbit of Venus* The orbit of Venus is an ellipse with the sun at one of the foci. The eccentricity of the orbit is $e = .006775$, and the major axis has length 134.5 million miles. (*Source: The World Almanac and Book of Facts, 1995.*) Find the smallest and greatest distances of Venus from the sun.

54. *Orbit of the Comet Swift-Tuttle* Comet Swift-Tuttle has an elliptical orbit of eccentricity $e = .964$ with the sun at one of the foci. Find the equation of the comet given that the closest it comes to the sun is 89 million miles.

Trajectory of a Satellite *When a satellite is near Earth, its orbital trajectory may trace out a hyperbola, parabola, or ellipse. The type of trajectory depends on the satellite's velocity V in meters per second. It will be hyperbolic if $V > k/\sqrt{D}$, parabolic if $V = k/\sqrt{D}$, and elliptic if $V < k/\sqrt{D}$, where $k = 2.82 \times 10^7$ is a constant and D is the distance in meters from the satellite to the center of Earth.* (Sources: *Loh, W., Dynamics and Thermodynamics of Planetary Entry, Prentice Hall, Inc., Englewood Cliffs, NJ, 1963; Thomson, W., Introduction to Space Dynamics, John Wiley & Sons, Inc., New York, 1961.*)

55. When the artificial satellite *Explorer IV* was at a maximum distance D of 42.5×10^6 m from Earth's center, it had a velocity V of 2090 m/sec. Determine the shape of its trajectory.

56. If a satellite is scheduled to leave Earth's gravitational influence, its velocity must be increased so that its trajectory changes from elliptic to hyperbolic. Determine the minimum increase in velocity necessary for *Explorer IV* to escape Earth's gravitational influence when $D = 42.5 \times 10^6$ m.

57. Explain why it is easier to change a satellite's trajectory from an ellipse to a hyperbola when D is maximum rather than minimum.

58. If $Ax^2 + Cy^2 + Dx + Ey + F = 0$ is the general equation of an ellipse, find its center point by completing the square.

59. Graph the ellipse $\dfrac{x^2}{16} + \dfrac{y^2}{12} = 1$ with a graphing calculator. Use tracing to find the coordinates of several points on the ellipse. For each of these points P, verify that
[Distance of P from (2, 0)]
$$= \frac{1}{2} \text{[Distance of } P \text{ from the line } x = 8].$$

60. Graph the hyperbola $\dfrac{x^2}{4} - \dfrac{y^2}{12} = 1$ with a graphing calculator. Use tracing to find the coordinates of several points on the hyperbola. For each of these points P, verify that
[Distance of P from (4, 0)]
$$= 2\text{[Distance of } P \text{ from the line } x = 1].$$

CHAPTER 6 TEST

1. Match each equation with its graph.
(a) $4(x + 3)^2 - (y + 2)^2 = 16$
(b) $(x - 3)^2 + (y - 2)^2 = 16$
(c) $(x + 3)^2 + (y - 2)^2 = 16$
(d) $(x + 3)^2 + y = 4$
(e) $x + (y - 2)^2 = 4$
(f) $4(x + 3)^2 + (y + 2)^2 = 16$

A.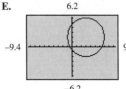

B.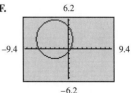

C.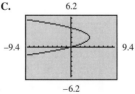

D.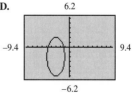

E.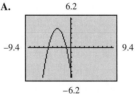

F.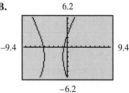

2. Give the coordinates of the focus and the equation of the directrix for the parabola with equation $x = 8y^2$.

3. Graph $y = -\sqrt{1 - \dfrac{x^2}{36}}$. Is the graph that of a function?

4. The screen shown here gives the graph of $\dfrac{x^2}{25} - \dfrac{y^2}{49} = 1$ as generated by a graphing calcula-

tor. What two functions were used to obtain the graph?

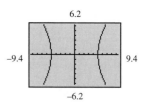

Describe the graph, if any, of each equation. Give the location of the radius, center, vertices, and foci, as applicable.

5. $\dfrac{y^2}{4} - \dfrac{x^2}{9} = 1$

6. $x^2 + 4y^2 + 2x - 16y + 17 = 0$

7. $x = t - 1, \quad y = 4 - \sqrt{8 + 2t - t^2}, \quad$ for t in $[-2, 4]$

8. $y^2 - 2x - 8y + 22 = 0$

9. Write an equation for each of the following.

    (a) The conic with center at the origin, focus at $(0, -2)$, and $e = 1$

    (b) The conic centered at the origin with vertical major axis of length 6 and $e = \frac{5}{6}$

10. *Shape of a Bridge's Arch* An arch of a bridge has the shape of the top half of an ellipse. The arch is 40 feet wide and 12 feet high at the center. Find the equation of the complete ellipse. Find the height of the arch 10 feet from the center at the bottom.

CHAPTER 6   PROJECT

MODELING THE PATH OF A BOUNCING BALL, USING A PIECEWISE-DEFINED FUNCTION

The height of each bounce of a bouncing ball varies quadratically with time. This means that the graph of each bounce will be a parabola. Figure A is a scatterplot of a bouncing ball, captured with a CBR (Calculator Based Ranger) connected to a TI-83. The CBR measured the height of the ball (in feet) at a fixed time interval (in seconds), generating two lists of data that were then downloaded into the STAT editor of the TI-83. The tick marks on the x-axis represent 1-second intervals, and the tick marks on the y-axis represent 1-foot intervals.

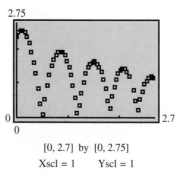

[0, 2.7] by [0, 2.75]
Xscl = 1   Yscl = 1

FIGURE A

Since each bounce can be modeled with a parabola, the entire path of the ball can be modeled with a piecewise-defined function. The graph consists of all or part of five bounces; therefore, the function will have five pieces. Find this function, using the data lists that generated the scatterplot.

Time	Height	Time	Height	Time	Height
0	2.37249	.934912	1.80222	1.869694	.60089
.042496	2.48284	.977404	1.66124	1.912184	.88197
.084992	2.536	1.019894	1.47971	1.954674	1.10494
.127488	2.53149	1.062384	1.23962	1.997164	1.27115
.169984	2.46122	1.104874	.92341	2.039654	1.37565
.21248	2.34186	1.147364	.55945	2.082144	1.41574
.254976	2.16168	1.189854	.12928	2.124634	1.40223
.297472	1.92384	1.232344	.37297	2.167124	1.33602
.339968	1.61799	1.274834	.73107	2.209614	1.20809
.382464	1.27476	1.317324	1.02881	2.252104	1.02476
.42496	.85089	1.359814	1.26845	2.294594	.78512
.467456	.37882	1.402304	1.44863	2.337084	.48648
.509952	.10045	1.444794	1.54592	2.379574	.20991
.552448	.53423	1.487284	1.60808	2.422064	.5108
.594944	.9153	1.529774	1.6225	2.464554	.76125
.63744	1.23286	1.572264	1.57295	2.507044	.9635
.679936	1.48782	1.614754	1.47836	2.549534	1.08467
.722432	1.6752	1.657244	1.32701	2.592024	1.1689
.764928	1.81754	1.699734	1.10854	2.634514	1.19683
.807424	1.89321	1.742224	.83332	2.677004	1.168
.84992	1.91889	1.784714	.49234	2.719494	1.0671
.892416	1.88826	1.827204	.2509	2.761984	.95134

We will find a piecewise-defined function that models the graph of the bouncing ball in Figure A. First, we must determine the endpoints of each parabola or partial parabola by looking through the data list. For example, to find the endpoints of the second parabola, look in the data list for the first height that is close to zero. We find the first near-zero height of .10045 occurs at time .509952, and the next near-zero height of .12928 occurs at 1.189854. So, the domain of the second parabola is the interval (.51, 1.19). Its vertex is at time .84992, which corresponds to the largest height in the list between these endpoints, 1.91889.

The equation of each parabola should be in the form $y = a(x - h)^2 + k$. Since the ball is a projectile in motion, a is the force of gravity, -16 feet per second per second. The values of h and k are the coordinates of each vertex, (.84992, 1.91889) for the second parabola. The resolution of the calculator screen is such that rounding to hundredths will not result in the loss of detail. Thus, the equation and domain of the second parabola are $y = -16(x - .85)^2 + 1.92$, on the interval (.51, 1.19).

ACTIVITIES

1. Find the equations and domains of the four remaining parabolas. Write a piecewise-defined function for the five parabolas and graph it. (Refer to the Chapter 2 Project to recall how to write a piecewise-defined function.) Prepare a report that contains the piecewise function, the graph, and a discussion comparing the graph with the scatterplot in Figure A.
2. The TI-83 calculator has a graph style called "Path" in which a circular cursor traces the leading edge of the graph and draws a path. If you are using this model, set the graph style to Path and watch the ball bounce. (*Note:* In reality, the ball bounces straight up and down.)
3. Working in groups of 3 or 4, gather your own original data. Write a piecewise function to model the path of the ball. (Different groups should use different balls.)

CHAPTER

Matrices and Systems of Equations and Inequalities

*I*n order to make predictions and forecasts about the future, professionals in many fields attempt to determine relationships between different factors. These relationships often result in equations containing more than one variable. When quantities are interrelated, *systems of equations* in several variables are used to describe their relationship. As early as 4000 B.C. in Mesopotamia, people were able to solve up to 10 equations having 10 variables. In 1940, John Atanasoff, a physicist from Iowa State University, needed to solve a system of equations containing 29 equations and 29 variables. This need to solve a large *linear system* led him to invent the first fully electronic digital computer. Modern supercomputers are capable of performing billions of calculations in a single second and solving more than 600,000 equations simultaneously. This capability has enabled people to predict the weather, design modern aircraft, and develop faster computer chips.

In this chapter, we will see how systems of equations are used to make predictions and forecasts in areas such as the environment and business. Whether one is predicting the number of newborn antelope in Wyoming, forecasting gross sales for a business, or analyzing data related to the greenhouse effect, systems of equations are involved. Many of the techniques presented in this chapter are currently used by professionals throughout the world to solve important applications.

7.1 SYSTEMS OF EQUATIONS

Introduction ◆ The Substitution Method ◆ The Addition (or Elimination) Method ◆ Connections: Relating Systems of Equations and the Intersection-of-Graphs Method ◆ Nonlinear Systems ◆ Applications of Systems

INTRODUCTION

Many mathematics applications require the simultaneous solution of a large number of equations having many variables. Such a set of equations is called a **system of equations.** The **solutions** of a system of equations must satisfy every equation in the system. In this chapter, we discuss several methods of solving systems of equations or inequalities. The matrix techniques we explain later in the chapter have gained particular importance with the increasing availability of computers and graphing calculators.

The definition of a linear equation given in an earlier chapter can be extended to more variables; any equation of the form $a_1x_1 + a_2x_2 + \cdots + a_nx_n = b$ for real numbers $a_1, a_2, \ldots, a_n$ (not all of which are 0), and b, is a **linear equation** or a **first-degree equation in n unknowns.**

The solution set of a linear equation in two variables is an infinite set of ordered pairs. Since the graph of such an equation is a straight line, there are three possibilities for the solution set of a system of two linear equations in two variables. An example of each possibility is shown in Figure 1.

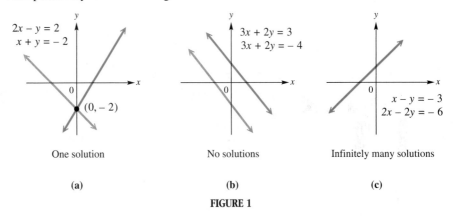

One solution   No solutions   Infinitely many solutions

(a)   (b)   (c)

FIGURE 1

The possible graphs are as follows.

1. The graphs of the two equations intersect in a single point. The coordinates of this point give the solution of the system. This is the most common case. See Figure 1(a).
2. The graphs are distinct parallel lines. In this case, the system is said to be **inconsistent.** That is, there is no solution common to both equations. See Figure 1(b).
3. The graphs are the same line. In this case, the equations are said to be **dependent,** and any solution of one equation is also a solution of the other. Thus, there are infinitely many solutions. See Figure 1(c).

In earlier courses, you have had experience in solving systems of two linear equations, using the substitution or addition (or elimination) methods. The first three examples illustrate these methods.

THE SUBSTITUTION METHOD

Although the *number* of solutions of a linear system can often be seen from the graph of the equations of the system, it is usually difficult to determine an exact solution from the graph. The substitution method is the most general analytic method of solving a system of equations. In a system of two equations with two variables, the **substitution method** involves using one equation to find an expression for one variable in terms of the other, then substituting into the other equation of the system.

EXAMPLE 1 *Solving a System by Substitution* Solve the system

$$3x + 2y = 11 \tag{1}$$
$$-x + y = 3 \tag{2}$$

by substitution. Support the solution graphically.

SOLUTION ANALYTIC While there are several ways to approach this solution, one way is to solve equation (2) for y to get $y = x + 3$. We substitute $x + 3$ for y in equation (1) to get

$$3x + 2(x + 3) = 11. \tag{3}$$

Notice the careful use of parentheses. Now we solve equation (3) for x.

$$3x + 2x + 6 = 11$$
$$5x + 6 = 11$$
$$5x = 5$$
$$x = 1$$

Replace x with 1 in $y = x + 3$ to find that $y = 1 + 3 = 4$. The solution set for this system is $\{(1, 4)\}$.

GRAPHICAL To support this solution graphically, we must graph equations (1) and (2) in an appropriate viewing window, and determine the coordinates of the point of intersection. First, we solve equation (1) for y to get $y_1 = -1.5x + 5.5$ and equation (2) for y to get $y_2 = x + 3$. Graphing these in the standard viewing window (which is appropriate for this system) and using the capability of the calculator for finding the coordinates of a point of intersection of two graphs, we find that the point is indeed $(1, 4)$ as seen in Figure 2. ◆

The table also shows that the two graphs intersect at $(1, 4)$.

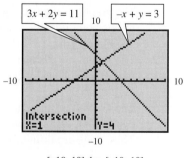

[−10, 10] by [−10, 10]
Xscl = 1   Yscl = 1
FIGURE 2

THE ADDITION (OR ELIMINATION) METHOD

Another method of solving systems of two equations is the **addition method.** With this method, we first multiply the equations on both sides by suitable numbers, so that when they are added, one variable is eliminated. (Thus, this method is also called the *elimination method.*) The result is an equation in one variable that can be solved by methods from Chapter 1. The solution is then substituted into one of the original equations, making it possible to solve for the other variable. In this process, the given system is replaced by new systems that have the same solution set as the original system. Systems that have the same solution set are called **equivalent systems.**

EXAMPLE 2   *Solving a System by Addition*   Solve the system

$$3x - 4y = 1 \tag{4}$$
$$2x + 3y = 12 \tag{5}$$

by addition, and support the solution graphically.

SOLUTION   ANALYTIC   To eliminate x, multiply both sides of equation (4) by -2 and both sides of equation (5) by 3 to get equations (6) and (7).

$$-6x + 8y = -2 \tag{6}$$
$$6x + 9y = 36 \tag{7}$$

Although this new system is not the same as the given system, it will have the same solution set.

Now, add the two equations to eliminate x, and then solve the result for y.

$$\begin{array}{r} -6x + 8y = -2 \\ \underline{6x + 9y = 36} \\ 17y = 34 \\ y = 2 \end{array}$$

Substitute 2 for y in equation (4) or (5). Choosing equation (4) gives

$$3x - 4(2) = 1$$
$$3x = 9$$
$$x = 3.$$

The solution set of the given system is $\{(3, 2)\}$, which can be checked analytically by substituting 3 for x and 2 for y in equation (5).

GRAPHICAL   To support this solution graphically, we must solve equations (4) and (5) for y, and graph both in an appropriate viewing window.

$$\text{Equation (4): } 3x - 4y = 1$$
$$-4y = -3x + 1$$
$$y = \frac{3}{4}x - \frac{1}{4}$$

$$\text{Equation (5): } 2x + 3y = 12$$
$$3y = -2x + 12$$
$$y = -\frac{2}{3}x + 4$$

As seen in Figure 3 on the next page, the graphs of $y_1 = \frac{3}{4}x - \frac{1}{4}$ and $y_2 = -\frac{2}{3}x + 4$ intersect at the point $(3, 2)$, supporting our earlier analytic result.   ◆

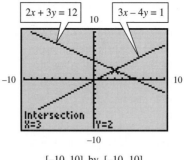

$$[-10, 10] \text{ by } [-10, 10]$$
$$\text{Xscl} = 1 \qquad \text{Yscl} = 1$$

The line with positive slope is $y_1 = \frac{3}{4}x - \frac{1}{4}$,

and the line with negative slope is $y_2 = -\frac{2}{3}x + 4$.

FIGURE 3

CONNECTIONS: RELATING SYSTEMS OF EQUATIONS AND THE INTERSECTION-OF-GRAPHS METHOD

In Chapter 1, we learned how to support an analytic solution of a linear equation by two methods: the intersection-of-graphs method and the x-intercept method. The former is closely related to the concept of solving systems of linear equations. Suppose, for example, we wish to solve the linear equation $\frac{3}{4}x - \frac{1}{4} = -\frac{2}{3}x + 4$ analytically. We can begin by multiplying both sides of the equation by 12, to get

$$12\left(\frac{3}{4}x - \frac{1}{4}\right) = 12\left(-\frac{2}{3}x + 4\right)$$
$$9x - 3 = -8x + 48.$$

Then, we complete the solution in the usual manner.

$$9x + 8x = 3 + 48$$
$$17x = 51$$
$$x = 3$$

To check this solution analytically, we substitute 3 for x in the original equation.

$$\frac{3}{4}(3) - \frac{1}{4} = -\frac{2}{3}(3) + 4 \qquad ?$$

$$\frac{9}{4} - \frac{1}{4} = -2 + 4 \qquad ?$$

$$\frac{8}{4} = 2 \qquad\qquad ?$$

$$2 = 2 \qquad\qquad \text{True}$$

Since both sides of the original equation yield 2 when $x = 3$, our analytic work is correct, and the solution set of the original equation is {3}.

Now, compare this work with the system in Example 2 and the graphs in Figure 3. The linear equation we just solved had $y_1 = \frac{3}{4}x - \frac{1}{4}$ as its original left side and $y_2 = -\frac{2}{3}x + 4$ as its original right side. Its solution set, {3}, consists of the x-coordinate of the point of intersection of the two lines, while the y-coordinate of the point of intersection, 2, is the result obtained when 3 is substituted for x in the equation.

We can see from the preceding discussion that there is a close connection between the solution of a linear equation in one variable and the solution of a linear system in two variables. The former has a solution set consisting of only domain values, while the latter has a solution set consisting of both domain and range values, in the form of ordered pairs. Yet the connection is apparent.

For Group Discussion   Suppose that we are given the following:

$$y_1 = f(x) = -2.5x - .5$$

and

$$y_2 = g(x) = 2x - 5.$$

Figure 4 shows the graphs of y_1 and y_2 along with the coordinates of the point of intersection of the lines.

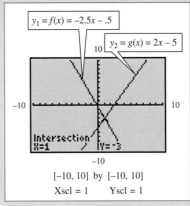

$$[-10, 10] \text{ by } [-10, 10]$$
$$Xscl = 1 \qquad Yscl = 1$$

FIGURE 4

1. What is the solution set of the linear equation $f(x) = g(x)$; that is, $-2.5x - .5 = 2x - 5$?
2. What is the solution set of the system of linear equations below?

$$y = -2.5x - .5$$
$$y = 2x - 5$$

3. Discuss the connections between your answers to Items 1 and 2.

The examples presented thus far in this section have all been systems with a single solution. The next example illustrates the analytic procedure used to determine whether a system is inconsistent or has dependent equations.

EXAMPLE 3   *Solving an Inconsistent System and a System with Dependent Equations*   Solve each system, using the addition method, and support the solution graphically.

(a) $3x - 2y = 4$ **(b)** $-4x + y = 2$

         $-6x + 4y = 7$ $8x - 2y = -4$

SOLUTION

(a)   ANALYTIC   The variable x can be eliminated by multiplying both sides of the first equation by 2 and then adding.

$$\begin{array}{r} 6x - 4y = 8 \\ -6x + 4y = 7 \\ \hline 0 = 15 \end{array} \qquad \text{False}$$

Both variables were eliminated here, leaving the false statement $0 = 15$, a signal that these two equations have no solutions in common. The system is inconsistent, and the solution set is $\emptyset$.

GRAPHICAL In slope-intercept form, the first equation is written $y = 1.5x - 2$ and the second is written $y = 1.5x + 1.75$. Since the lines have the same slope but different y-intercepts, they are parallel, and have no point of intersection. This supports our analytic conclusion. See Figure 5.

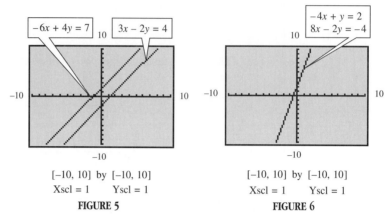

[−10, 10] by [−10, 10]
Xscl = 1   Yscl = 1
FIGURE 5

[−10, 10] by [−10, 10]
Xscl = 1   Yscl = 1
FIGURE 6

(b) ANALYTIC We can eliminate x by multiplying both sides of the first equation by 2 and then adding it to the second equation.

$$-8x + 2y = 4$$
$$\underline{8x - 2y = -4}$$
$$0 = 0 \quad \text{True}$$

This true statement, $0 = 0$, indicates that a solution of one equation is also a solution of the other, so the solution set is an infinite set of ordered pairs.

We will write the solution of a system of dependent equations as an ordered pair by expressing x in terms of y as follows. Choose either equation and solve for x. Choosing the first equation gives

$$-4x + y = 2$$
$$x = \frac{2 - y}{-4} = \frac{y - 2}{4}.$$

We write the solution set as $\left\{\left(\frac{y-2}{4}, y\right)\right\}$. By selecting values for y and calculating the corresponding values for x, individual ordered pairs of the solution set can be found. For example, if $y = -2$, $x = \frac{-2-2}{4} = -1$, and the ordered pair $(-1, -2)$ is a solution. Verify that $(0, 2)$ and $(6, 26)$ are also solutions.

GRAPHICAL The graphs of the equations are the same line, since both equations are equivalent to $y = 4x + 2$. See Figure 6. The two equations are dependent. ◆

Note In Example 3(b), we wrote the solution set in a form with the variable y arbitrary. However, it would be acceptable to write the ordered pair with x arbitrary. In this case, the solution set would be written $\{(x, 4x + 2)\}$. By selecting values for x and solving for y in this ordered pair, individual solutions can be found. Verify again that $(-1, -2)$ is a solution.

NONLINEAR SYSTEMS

A **nonlinear system of equations** is one in which at least one of the equations is not a linear equation.

EXAMPLE 4 *Solving a Nonlinear System by Substitution* Use the substitution method to solve the system

$$3x^2 - 2y = 5 \tag{8}$$
$$x + 3y = -4, \tag{9}$$

and support graphically.

SOLUTION ANALYTIC The graph of equation (8) is a parabola, and that of equation (9) is a line. There may be 0, 1, or 2 points of intersection of these graphs.

Although either equation could be solved for either variable, it is best to use the linear equation. Here, it is simpler to solve equation (9) for y, since x is squared in equation (8).

$$x + 3y = -4$$
$$3y = -4 - x$$
$$y = \frac{-4 - x}{3} \tag{10}$$

Substituting this value of y into equation (8) gives

$$3x^2 - 2\left(\frac{-4 - x}{3}\right) = 5.$$
$$9x^2 - 2(-4 - x) = 15 \qquad \text{Multiply by 3.}$$
$$9x^2 + 8 + 2x = 15$$
$$9x^2 + 2x - 7 = 0$$
$$(9x - 7)(x + 1) = 0$$
$$x = \frac{7}{9} \quad \text{or} \quad x = -1$$

Substitute both values of x into equation (10) to find the corresponding y-values.

$$y = \frac{-4 - \left(\frac{7}{9}\right)}{3} \qquad \text{or} \qquad y = \frac{-4 - (-1)}{3}$$
$$y = -\frac{43}{27} \qquad\qquad\qquad y = -1$$

Check in the original system that the solution set is $\left\{\left(\frac{7}{9}, -\frac{43}{27}\right), (-1, -1)\right\}$.

TECHNOLOGY NOTE

The decimal values for x and y displayed at the bottom of Figure 7(b) represent the repeating decimal forms for the fractions 7/9 and $-43/27$, respectively.

GRAPHICAL To support our solution graphically, we graph equation (8) as $y_1 = 1.5x^2 - 2.5$ and equation (9) as $y_2 = -\frac{1}{3}x - \frac{4}{3}$. As Figure 7 indicates, the points of intersection are $(-1, -1)$ and $(.\overline{7}, -1.592)$. confirming our analytic solution. ◆

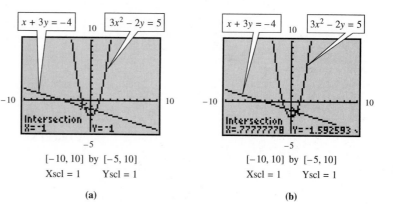

$$[-10, 10] \text{ by } [-5, 10]$$
$$\text{Xscl} = 1 \qquad \text{Yscl} = 1$$
$$\text{(a)}$$

$$[-10, 10] \text{ by } [-5, 10]$$
$$\text{Xscl} = 1 \qquad \text{Yscl} = 1$$
$$\text{(b)}$$

FIGURE 7

Some nonlinear systems can be solved by the addition method. This method works well if we can eliminate completely one variable from the system, as shown in the following example.

EXAMPLE 5 *Solving a Nonlinear System by Addition* Solve the system

$$x^2 + y^2 = 4 \tag{11}$$
$$2x^2 - y^2 = 8 \tag{12}$$

by addition, and support graphically.

SOLUTION ANALYTIC The graph of equation (11) is a circle and that of equation (12) is a hyperbola. Visualizing these suggests that there may be 0, 1, 2, 3, or 4 points of intersection.

Add the two equations to eliminate y^2.

$$
\begin{array}{rl}
x^2 + y^2 = & 4 \\
2x^2 - y^2 = & 8 \\
\hline
3x^2 = & 12 \\
x^2 = & 4
\end{array}
$$

$$x = 2 \quad \text{or} \quad x = -2$$

Substituting into equation (11) gives the corresponding values of y.

$$
\begin{array}{ccc}
2^2 + y^2 = 4 & \text{or} & (-2)^2 + y^2 = 4 \\
y^2 = 0 & & y^2 = 0 \\
y = 0 & & y = 0
\end{array}
$$

The solution set of the system is $\{(2, 0), (-2, 0)\}$.

GRAPHICAL To support the solution graphically, we graph equation (11) as the union of $y_1 = \sqrt{4 - x^2}$ and $y_2 = -\sqrt{4 - x^2}$, and equation (12) as the union of $y_3 = \sqrt{2x^2 - 8}$ and $y_4 = -\sqrt{2x^2 - 8}$. As suggested in Figure 8, the two points of intersection are indeed $(2, 0)$ and $(-2, 0)$. ◆

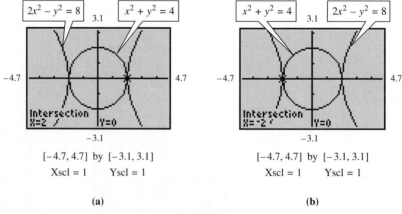

$[-4.7, 4.7]$ by $[-3.1, 3.1]$
Xscl = 1   Yscl = 1

(a)

$[-4.7, 4.7]$ by $[-3.1, 3.1]$
Xscl = 1   Yscl = 1

(b)

FIGURE 8

Graphing calculators will often allow us to solve systems using purely graphical methods. Many systems are difficult or even impossible to solve using strictly analytic methods.

EXAMPLE 6 *Solving a Nonlinear System Graphically* Solve the system

$$y = 2^x \tag{13}$$

$$|x + 2| - y = 0, \tag{14}$$

using graphical methods.

SOLUTION We enter equation (13) as $y_1 = 2^x$ and equation (14) as $y_2 = |x + 2|$. As seen in Figure 9, the various windows indicate that there are three points of intersection of the graphs, and thus three solutions. Using the capabilities of the calculator, we find that $(2, 4)$ is an exact solution, and $(-2.22, .22)$ and $(-1.69, .31)$ are approximate solutions. Therefore, the solution set is

$$\{(2, 4), (-2.22, .22), (-1.69, .31)\}. \quad \blacklozenge$$

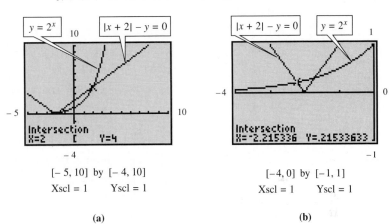

[-5, 10] by [-4, 10]
Xscl = 1   Yscl = 1

(a)

[-4, 0] by [-1, 1]
Xscl = 1   Yscl = 1

(b)

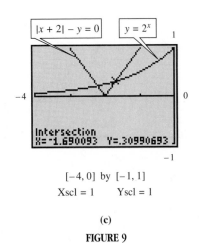

[-4, 0] by [-1, 1]
Xscl = 1   Yscl = 1

(c)

FIGURE 9

APPLICATIONS OF SYSTEMS

Usually, as the price of an item increases, demand for the item decreases and the supply of the item increases. The price of gasoline in early 1998 illustrates this phenomenon. The price where supply and demand are equal is called the *equilibrium price,* and the resulting supply or demand is called the *equilibrium supply* or *equilibrium demand.*

EXAMPLE 7   *Solving a Supply and Demand Problem*

(a)   Suppose that the supply of a product is related to its price by the equation $p = \frac{2}{3}q$, where p is price in dollars and q is supply in appropriate units. (Here, q stands for quantity.) Find the price for the supply levels $q = 9$ and $q = 18$.

(b)   Suppose demand and price for the same product are related by $p = -\frac{1}{3}q + 18$, where p is price and q is demand. Find the price for the demand levels $q = 6$ and $q = 18$.

(c)   Using $x = q$ and $y = p$, graph both functions on the same axes.

(d)   Solve the system analytically to find the equilibrium price, supply, and demand.

SOLUTION

(a)   When $q = 9$, $p = \frac{2}{3}q = \frac{2}{3}(9) = 6$. When $q = 18$, $p = \frac{2}{3}q = \frac{2}{3}(18) = 12$.

(b)   When $q = 6$, $p = -\frac{1}{3}q + 18 = -\frac{1}{3}(6) + 18 = 16$, and when $q = 18$, $p = -\frac{1}{3}q + 18 = -\frac{1}{3}(18) + 18 = 12$.

(c)   Enter $y_1 = \frac{2}{3}x$ and $y_2 = -\frac{1}{3}x + 18$, and graph in an appropriate window. See Figure 10.

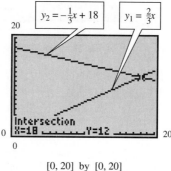

[0, 20] by [0, 20]

Xscl = 1   Yscl = 1

The point (18, 12) is the equilibrium point.

FIGURE 10

(d)   Solve the system

$$p = \frac{2}{3}q$$

$$p = -\frac{1}{3}q + 18,$$

using substitution.

$$\frac{2}{3}q = -\frac{1}{3}q + 18 \qquad \text{Substitute } \tfrac{2}{3}q \text{ for } p.$$

$$q = 18$$

This gives 18 units as the equilibrium supply or demand. Find the equilibrium price by substituting 18 for q in either equation. Using $p = \frac{2}{3}q$ gives

$$p = \frac{2}{3}(18) = 12,$$

or \$12, the equilibrium price. The point (18, 12) that gives the equilibrium values is shown in Figure 10. ◆

EXAMPLE 8 *Solving a Distance, Rate, and Time Problem Graphically* Los Angeles and Chico, California, are 480 miles apart. Stuart leaves Chico at noon and heads for Los Angeles at 65 mph. John leaves Los Angeles, also at noon, and heads for Chico at 55 mph. If x represents the time traveled, how far from Los Angeles will they be when they pass, and what time will it be?

SOLUTION Figure 11 shows how Stuart's distance from Los Angeles can be described as Y_1, and John's distance from Los Angeles can be described as Y_2. When they pass, $Y_1 = Y_2$. Solving analytically, we have

$$480 - 65x = 55x$$
$$480 = 120x$$
$$x = 4.$$

Substituting 4 for x in either Y_1 or Y_2 gives 220. Stuart and John will meet at 4 P.M., 220 miles from Los Angeles. Figures 12(a) and 12(b) support this analytic solution with both a table and a graph. ◆

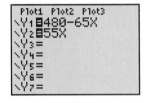

FIGURE 11

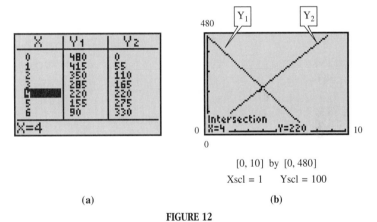

(a)

(b)

FIGURE 12

[0, 10] by [0, 480]
Xscl = 1   Yscl = 100

Sometimes, applications lead to nonlinear systems, as shown in the next example. A knowledge of the theory of polynomial functions, introduced in Chapter 3, is often useful in solving such problems.

EXAMPLE 9 *Using a Nonlinear System to Find the Dimensions of a Box* A box with an open top has a square base and four sides of equal height. The volume of the box is 75 cubic inches, and the surface area is 85 square inches. What are the dimensions of the box?

SOLUTION If each side of the square base measures x inches and the height measures y inches, then the volume is

$$x^2y = 75 \qquad \text{(Volume formula)}$$

and the surface area is

$$x^2 + 4xy = 85. \qquad \text{(Sum of areas of base and sides)}$$

Solve the first equation for y to get $y = \frac{75}{x^2}$, and substitute this into the second equation.

$$x^2 + 4x\left(\frac{75}{x^2}\right) = 85$$

$$x^2 + \frac{300}{x} = 85$$

$$x^3 + 300 = 85x \qquad \text{Multiply by } x.$$

$$x^3 - 85x + 300 = 0$$

The solutions of this equation are x-intercepts of the graph of $y = x^3 - 85x + 300$. A comprehensive graph of this function indicates that there are three real solutions. However, one of them is negative and must be rejected. As Figure 13 indicates, one positive solution is 5 and the other positive solution is approximately 5.64. (In Exercise 89, you are asked to show that the exact value of this latter solution is $\frac{-5 + \sqrt{265}}{2}$.) By substituting back into the first equation, we find that when $x = 5$, $y = 3$, and when $x \approx 5.64$, $y \approx 2.36$. Therefore, this problem has two solutions: the box may have a base 5 inches by 5 inches and a height 3 inches, or it may have a base 5.64 inches by 5.64 inches and a height 2.36 inches.   ◆

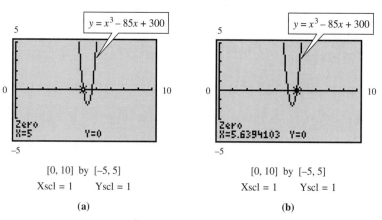

[0, 10] by [−5, 5]
Xscl = 1   Yscl = 1

(a)

[0, 10] by [−5, 5]
Xscl = 1   Yscl = 1

(b)

The positive roots of $x^3 - 85x + 300 = 0$ are 5 and approximately 5.64.

FIGURE 13

WHAT WENT WRONG ❓   To support the solution found in Example 2, a student entered the functions shown on the left and got the graph shown on the right.

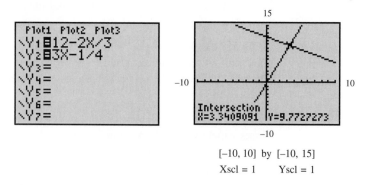

[−10, 10] by [−10, 15]
Xscl = 1   Yscl = 1

1.   What went wrong?
2.   What can be done to correct it?

7.1   EXERCISES      Tape 8

In Exercises 1–8, a system of equations is given, along with a graph indicating the coordinates of one point of intersection of the graphs of the equations in the system. Show by direct substitution, using the displayed values of x and y, that the indicated point is indeed a solution of the system.

1. $4x - y = 3$
     $-2x + 3y = 1$

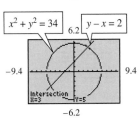

2. $5x + 3y = 1$
     $-3x - 4y = 6$

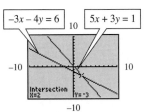

3. $4x + 2y = 3$
     $-3x - 3y = 0$

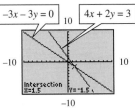

4. $.5x + y = -.2$
     $10x + 5y = .5$

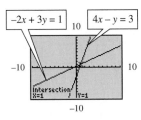

5. $y - x = 2$
     $x^2 + y^2 = 34$

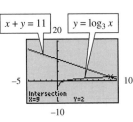

6. $x^2 - y^2 = 12$
     $x^2 + y^2 = 20$

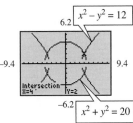

7. $y = \log_3 x$
     $x + y = 11$

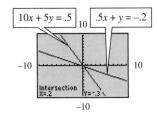

8. $y = |x + 3|$
     $y = 4x^2$

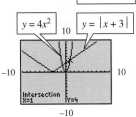

Without graphing, decide whether the system has a single solution, no solution, or infinitely many solutions.

9. One line has positive slope and one line has negative slope.

10. One line has slope 0 and one line has undefined slope.

11. Both lines have slope 0 and have the same y-intercept.

12. Both lines have undefined slope and have the same x-intercept.

13. $x + y = 4$
      $kx + ky = 4k \quad (k \neq 0)$

14. $x + y = 10$
      $kx + ky = 9k \quad (k \neq 0)$

Use the substitution method to solve each system. Check your answers. Support your answers graphically if you wish.

15. $y = 2x + 3$
$3x + 4y = 78$

16. $y = 4x - 6$
$2x + 5y = -8$

17. $3x - 2y = 12$
$5x = 4 - 2y$

18. $8x + 3y = 2$
$5x = 17 + 6y$

19. Refer to Example 2 in this section. If we began solving the system by eliminating y, by what numbers might we have multiplied equations (4) and (5)?

20. Explain how one can determine whether a system is inconsistent or has dependent equations when using the substitution or elimination method.

Use the addition method to solve each system. Support your answers graphically if you wish.

21. $4x - y = 9$
$-8x + 2y = -18$

22. $9x - 5y = 1$
$-18x + 10y = 1$

23. $5x + 7y = 6$
$10x - 3y = 46$

24. $12x - 5y = 9$
$3x - 8y = -18$

25. $\dfrac{x}{2} + \dfrac{y}{3} = 8$

$\dfrac{2x}{3} + \dfrac{3y}{2} = 17$

26. $\dfrac{x}{5} + 3y = 31$

$2x - \dfrac{y}{5} = 8$

RELATING CONCEPTS (EXERCISES 27–32)

The system

$$\frac{5}{x} + \frac{15}{y} = 16$$

$$\frac{5}{x} + \frac{4}{y} = 5$$

is not a linear system, because the variables appear in the denominator. However, it can be solved in a manner similar to the method for solving a linear system by using a substitution-of-variable technique. Let $t = \frac{1}{x}$ and let $u = \frac{1}{y}$.

27. Write a system of equations in t and u by making the appropriate substitutions.

28. Solve the system in Exercise 27 for t and u.

29. Solve the given system for x and y by using the equations relating t and x, and u and y. (Section 4.2)

30. Refer to the first equation in the given system, and solve for y in terms of x to obtain a rational function. (Section 4.2)

31. Repeat Exercise 30 for the second equation in the given system. (Section 4.2)

32. Using a viewing window of $[0, 10]$ by $[0, 2]$, show that the point of intersection of the graphs of the functions in Exercises 30 and 31 has the same x- and y-values as found in Exercise 29.

Use the substitution-of-variable technique to solve the systems in Exercises 33–34 analytically.

33. $\dfrac{2}{x} + \dfrac{1}{y} = \dfrac{3}{2}$

$\dfrac{3}{x} - \dfrac{1}{y} = 1$

34. $\dfrac{2}{x} + \dfrac{1}{y} = 11$

$\dfrac{3}{x} - \dfrac{5}{y} = 10$

Use only a graphical approach to solve the systems in Exercises 35–36. Express solutions to the nearest hundredth. In Exercise 36, use the e and π keys on your calculator.

35. $\sqrt{5}x + \sqrt[3]{6}y = 9$
$\sqrt{2}x + \sqrt[5]{9}y = 12$

36. $\pi x + ey = 3$
$ex + \pi y = 4$

37. Determine visually the other solution of the system in Exercise 5. Then, support your answer graphically.

38. Use symmetry to determine the other three solutions of the system in Exercise 6.

In Exercises 39–48, draw a sketch of the two graphs described, with the indicated number of points of intersection. (There may be more than one way to do this in some cases.)

39. a line and a circle; no points

40. a line and a circle; one point

41. a line and a circle; two points

42. a line and a hyperbola; no points

43. a line and a hyperbola; one point

44. a line and a hyperbola; two points

45. a circle and an ellipse; four points

46. a parabola and an ellipse; one point

47. a parabola and a hyperbola; four points

48. a circle and a hyperbola; two points

Give all solutions of the following nonlinear systems of equations. Use analytic methods, and support your solutions graphically.

49. $y = -x^2 + 2$
$x - y = 0$

50. $y = (x - 1)^2$
$x - 3y = -1$

51. $3x^2 + 2y^2 = 5$
$x - y = -2$

52. $x^2 + y^2 = 5$
$-3x + 4y = 2$

53. $x^2 + y^2 = 10$
$2x^2 - y^2 = 17$

54. $x^2 + y^2 = 4$
$2x^2 - 3y^2 = -12$

55. $x^2 + 2y^2 = 9$
$3x^2 - 4y^2 = 27$

56. $2x^2 + 3y^2 = 5$
$3x^2 - 4y^2 = -1$

57. $2x^2 + 2y^2 = 20$
$3x^2 + 3y^2 = 30$

58. $x^2 + y^2 = 4$
$5x^2 + 5y^2 = 28$

59. $x^2 - xy + y^2 = 5$
$2x^2 + xy - y^2 = 10$

60. $3x^2 + 2xy - y^2 = 9$
$x^2 - xy + y^2 = 9$

61. $y = |x - 1|$
$y = x^2 - 4$

62. $2x^2 - y^2 = 4$
$|x| = |y|$

Use a purely graphical method to solve each system in Exercises 63–66. Give x- and y-coordinates correct to the nearest hundredth.

63. $y = \log(x + 5)$
$y = x^2$

64. $y = 5^x$
$xy = 1$

65. $y = e^{x+1}$
$2x + y = 3$

66. $y = \sqrt[3]{x - 4}$
$x^2 + y^2 = 6$

Use a system of equations to solve each problem.

67. *Business Loans* To start a new business, Shannon d'Hemecourt borrowed money from two financial institutions. One loan was at 7% interest and the other was for one-third as much money at 8% interest. How much did she borrow at each rate if the total amount of annual interest was $1160?

68. *Shipment Costs* A manufacturer of portable compact disc players shipped 200 of the players to its two Quebec warehouses. It costs $3 per unit to ship to Warehouse A, and $2.50 per unit to ship to Warehouse B. If the total shipping cost was $537.50, how many were shipped to each warehouse?

69. *Octane Ratings of Gasoline* Octane ratings show the percent of isooctane in gasoline. An octane rating of 98, for example, indicates a gasoline that is 98% isooctane.

How many gallons of 98-octane gasoline should be mixed with 92-octane gasoline to produce 40 gallons of 94-octane gasoline?

70. *Mixing Solutions* A chemist needs 10 liters of a 24% alcohol solution. She has on hand a 30% alcohol solution and an 18% alcohol solution. How many liters of each should be mixed to get the required solution?

71. *Investment Decisions* Kathy Rodgers plans to invest $30,000 she won in a lottery. With part of the money she buys a mutual fund, paying 4.5% a year. The rest she invests in utility bonds, paying 5% per year. The first year her investments bring a return of $1410. How much is invested at each rate?

72. *Mixing Milk* How much milk that is 3% butterfat should be mixed with milk that is 18% butterfat to get 25 gallons of milk that is 4.8% butterfat?

Finding the Break-Even Point The break-even point for a company is the point where its costs equal its revenues. If both cost and revenue are expressed as linear equations, the break-even point is the solution of a linear system. In each of the following exercises, C represents the cost in dollars to produce x items, and R represents the revenue in dollars from the sale of x items. Use the substitution method to find the break-even point in each case; that is, the point where C = R. Then, find the value of C and R at that point.

73. $C = 20x + 10,000$
$R = 30x - 11,000$

74. $C = 4x + 125$
$R = 9x - 200$

Equilibrium Supply and Demand In each of the following exercises, p is the price of an item in dollars, while q represents the supply in one equation and the demand in the other. Find the equilibrium price and the equilibrium supply/demand.

75. $p = 80 - \dfrac{3}{5}q$

$p = \dfrac{2}{5}q$

76. $p = 630 - \dfrac{3}{4}q$

$p = \dfrac{3}{4}q$

77. Let the supply and demand equations for banana smoothies be

$$\text{supply: } p = \frac{3}{2}q \quad \text{and} \quad \text{demand: } p = 81 - \frac{3}{4}q.$$

(a) Graph these on the same axes. Let $x = q$, $y = p$, and use the window [0, 120] by [0, 100]. Find the point of intersection.

(b) Find the equilibrium demand analytically.

(c) Find the equilibrium price analytically.

78. *Unmarried Single Parents* The figure shows a graph that accompanied a recent news article in *The Sacramento Bee* newspaper. The graph indicates that the percent of children living with a never-married parent has increased and appears about to overtake the percent of children living with a single divorced parent. (*Source: The Sacramento Bee*, July 20, 1994. Copyright, The Sacramento Bee, 1995. Reprinted by permission.)

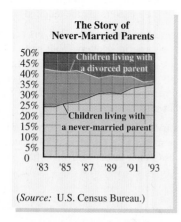

(*Source:* U.S. Census Bureau.)

(a) From the graph, estimate the percents in 1983 and 1993 for each curve. Let 1983 be represented by 0, and find two data points for each curve.

(b) Use these data points to write a linear equation that approximates each curve.

(c) Use a calculator to graph the lines and find the point of intersection. Interpret your answer.

Solve each problem by using a system of equations in two variables.

79. Find the equation of the straight line through (2, 4) that touches the parabola $y = x^2$ at only one point. (*Note:* Recall that a quadratic equation has a unique solution when the discriminant is 0.)

80. For what values of b will the line $x + 2y = b$ touch the circle $x^2 + y^2 = 9$ in only one point?

81. For what nonzero values of a do the graphs of $x^2 + y^2 = 25$ and $\dfrac{x^2}{a^2} + \dfrac{y^2}{25} = 1$ have exactly two points in common?

82. Find the equation of the line passing through the points of intersection of the graphs of $y = x^2$ and $x^2 + y^2 = 90$.

83. Suppose that you are given the equations of two circles that are known to intersect in exactly two points. Explain how you would find the equation of the only chord common to these circles.

Solve the following applications of systems.

84. *Atmospheric Carbon Emissions* The emissions of carbon into the atmosphere from 1950 to 1990 are modeled in the graph for both Western Europe and Eastern Europe together with the former USSR. This carbon combines with oxygen to form carbon dioxide, which is believed to contribute to the greenhouse effect.

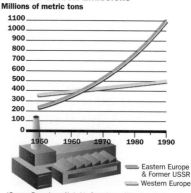

(*Source:* Rosenberg, N. (ed.), *Greenhouse Warming: Abatement and Adaptation*, Resources for the Future, Washington, D.C., 1995.)

(a) Interpret this graph. How are emissions changing with time?

(b) Use the graph to estimate the year and the amount when the carbon emissions were equal.

(c) The equation $W = 375(1.008)^{(t-1950)}$ models the emissions in Western Europe, while the equation $E = 260(1.038)^{(t-1950)}$ models the emissions from Eastern Europe and the former USSR. Use these equations to determine the year and emission levels when $W = E$.

85. *Circuit Gain* In electronics, circuit gain is given by

$$G = \frac{Bt}{R + R_t},$$

where R is the value of a resistor, t is temperature, and B is a constant. The sensitivity of the circuit to temperature is given by

$$S = \frac{BR}{(R + R_t)^2}.$$

If $B = 3.7$ and t is 90K (Kelvin), find the values of R and R_t that will make $G = .4$ and $S = .001$.

86. *Travel Time* (Refer to Example 8.) John leaves Slidell, Louisiana, at 12 noon, driving north. He maintains an average speed of 60 mph. Stuart leaves Slidell at 1:30 P.M., also heading north. Stuart averages 75 mph. Assuming Stuart does not get stopped for speeding, at what time will he overtake John, and how far will they be from Slidell at that time? Solve numerically, using a table. Support graphically, and confirm analytically.

RELATING CONCEPTS (EXERCISES 87–92)

In Example 9, we solved a nonlinear system of equations by solving the related polynomial equation $x^3 - 85x + 300 = 0$.

87. Use synthetic division to show that a real solution of this equation is 5. (Section 3.6)

88. Factor $x^3 - 85x + 300$, using the results of Exercise 87. (Section 3.6)

89. From Exercise 88, show that the other positive solution has an exact value of $\frac{-5 + \sqrt{265}}{2}$. (Section 3.3)

90. Use the graph of $y = x^3 - 85x + 300$ to show that the negative solution of the polynomial equation is approximately -10.64.

91. What is the exact value of the negative solution? (Section 3.3)

92. Why was the negative solution rejected in the problem solved in Example 9?

7.2 SOLUTION OF LINEAR SYSTEMS BY THE ECHELON METHOD

Geometric Considerations ◆ Analytic Solution of Systems in Three Variables ◆ Applications of Systems

GEOMETRIC CONSIDERATIONS

Our work with systems of equations and inequalities so far has dealt strictly with systems in two variables. We can extend the ideas of systems of equations to linear equations of the form $Ax + By + Cz = D$. A solution of such an equation is called an **ordered triple,** and is denoted (x, y, z). For example, $(1, 2, -4)$ is a solution of $2x + 5y - 3z = 24$. The solution set of such an equation is an infinite set of ordered triples. In geometry, the graph of a linear equation in three variables is a plane in three-dimensional space. Considering the possible intersections of the planes representing three equations in three unknowns shows that the solution set of such a system may be either a single ordered triple (x, y, z), an infinite set of ordered triples (dependent equations), or the empty set (an inconsistent system). See Figure 14 on the next page.

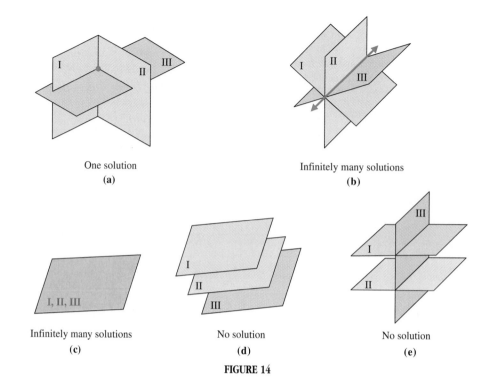

One solution

(a)

Infinitely many solutions

(b)

Infinitely many solutions

(c)

No solution

(d)

No solution

(e)

FIGURE 14

ANALYTIC SOLUTION OF SYSTEMS IN THREE VARIABLES

A linear system of equations can be solved by repeated use of the addition method, using the following transformations.

TRANSFORMATIONS OF A SYSTEM

The following transformations can be applied to a system of linear equations to get an equivalent system:

1. exchanging any two equations;
2. multiplying both sides of an equation by any nonzero real number;
3. replacing any equation by a nonzero multiple of that equation plus a nonzero multiple of any other equation.

A systematic approach for using the three transformations to solve a system is called the **echelon method.** The goal of the echelon method is to use the transformations to rewrite the equations of the system until the system has a triangular form. For a system of three equations in three variables, for example, the system should have the form

$$x + ay + bz = c$$
$$y + dz = e$$
$$z = f,$$

where a, b, c, d, e, and f are constants. Then the value of z from the third equation can be substituted back into the second equation to find y, and the values of y and z can be substituted into the first equation to find x. This is called **back-substitution.**

EXAMPLE 1 *Solving a System of Three Equations by the Echelon Method*
Solve the system

$$2x + y - z = 2 \tag{1}$$
$$x + 3y + 2z = 1 \tag{2}$$
$$x + y + z = 2. \tag{3}$$

SOLUTION As shown above, in the first equation x should have a coefficient of 1. We can use transformation 1 to exchange equations (1) and (2) to get the system

$$x + 3y + 2z = 1 \tag{2}$$
$$2x + y - z = 2 \tag{1}$$
$$x + y + z = 2. \tag{3}$$

Next we use transformation 3 to eliminate the x-term in equations (1) and (3). Multiply equation (2) by -2 and add the results to equation (1) to get an equivalent system. We will indicate this process by the notation $-2R_1 + R_2 \rightarrow R_2$. ($R$ stands for row.) Similarly, multiplying equation (2) by -1 and adding the results to equation (3) eliminates the x-term there.

$$\begin{array}{lll} & x + 3y + 2z = 1 & \textbf{(2)} \\ -2R_1 + R_2 \rightarrow R_2 & -5y - 5z = 0 & \textbf{(4)} \\ -1R_1 + R_3 \rightarrow R_3 & -2y - z = 1 & \textbf{(5)} \end{array}$$

Use transformation 2 to get a coefficient of 1 for y in equation (4); multiply by $-1/5$.

$$\begin{array}{lll} & x + 3y + 2z = 1 & \textbf{(2)} \\ (-1/5)R_2 \rightarrow R_2 & y + z = 0 & \textbf{(6)} \\ & -2y - z = 1 & \textbf{(5)} \end{array}$$

To eliminate the y-term in equation (5), multiply equation (6) by 2 and add the results to equation (5).

$$\begin{array}{lll} & x + 3y + 2z = 1 & \textbf{(2)} \\ & y + z = 0 & \textbf{(6)} \\ 2R_2 + R_3 \rightarrow R_3 & z = 1 & \end{array}$$

The system is now in triangular form. Substitute 1 for z in equation (6) to get $y = -1$. Finally, substitute 1 for z and -1 for y in equation (2) to get $x = 2$. The solution set of the system is $\{(2, -1, 1)\}$. ◆

Although the echelon method may seem confusing at first, it has the advantages of being systematic and leading to the matrix method we discuss in the next section.

EXAMPLE 2 *Solving a System Having Two Equations with Three Variables*
Solve the system

$$x + 2y + z = 4 \tag{7}$$
$$3x - y - 4z = -9. \tag{8}$$

SOLUTION Geometrically, the solution is the intersection of the two planes given by equations (7) and (8). The intersection of two different nonparallel planes is a line. Thus, there will be an infinite number of ordered triples in the solution set, representing the points on the line of intersection. To describe these ordered triples, use the echelon method to get the system in triangular form as much as possible.

$$x + 2y + z = 4 \qquad \textbf{(7)}$$
$$-3R_1 + R_2 \rightarrow R_2 \qquad -7y - 7z = -21$$

$$x + 2y + z = 4 \qquad \textbf{(7)}$$
$$(-1/7)R_2 \rightarrow R_2 \qquad y + z = 3$$

This is as far as we can go with the echelon method. Solve $y + z = 3$ for y to get $y = 3 - z$. Now, express x also in terms of z by solving equation (7) for x and substituting $3 - z$ for y.

$$x + 2y + z = 4$$
$$x = -2y - z + 4$$
$$x = -2(3 - z) - z + 4 \qquad \text{Let } y = 3 - z.$$
$$x = z - 2$$

The solution set is written $\{(z - 2, 3 - z, z)\}$. The system has an infinite number of solutions. For any arbitrary value of z, $y = 3 - z$ and $x = z - 2$. For example, if $z = 1, y = 3 - 1 = 2$ and $x = 1 - 2 = -1$, giving the solution $(-1, 2, 1)$. Verify that another solution is $(0, 1, 2)$.   ◆

A system like the one in Example 2 occurs when two of the equations in a system of three equations with three variables are dependent. In such a case, there are really only two equations in three variables, and Example 2 illustrates the method of solution. On the other hand, an inconsistent system is indicated by a false statement at some point in the solution, as in Example 3(a) of Section 7.1.

APPLICATIONS OF SYSTEMS

Many applied problems involve more than one unknown quantity. To solve such a problem using a system, determine the unknown quantities you are asked to find, and let different variables represent each of these quantities. Then, write a system of equations, one for each variable, to get a unique solution, and solve it, using one of the methods shown in this chapter. Be sure that you answer the question(s) posed in the problem, and check to see that your answer is reasonable.

EXAMPLE 3   *Solving a System of Three Equations to Satisfy Feed Requirements*   An animal feed is made from three ingredients: corn, soybeans, and cottonseed. One unit of each ingredient provides units of protein, fat, and fiber, as shown in the table below. How many units of each ingredient should be used to make a feed that contains 22 units of protein, 28 units of fat, and 18 units of fiber?

	Corn	**Soybeans**	**Cottonseed**	**Total**
Protein	.25	.4	.2	22
Fat	.4	.2	.3	28
Fiber	.3	.2	.1	18

SOLUTION   Let x represent the number of units of corn, y, the number of units of soybeans, and z, the number of units of cottonseed that are required. Since the total amount of protein is to be 22 units,

$$.25x + .4y + .2z = 22.$$

Also, for the 28 units of fat,

$$.4x + .2y + .3z = 28,$$

and, for the 18 units of fiber,

$$.3x + .2y + .1z = 18.$$

Multiply the first equation on both sides by 100, and the second and third equations by 10 to get the system

$$25x + 40y + 20z = 2200$$
$$4x + 2y + 3z = 280$$
$$3x + 2y + z = 180.$$

Using the methods described in this section, we can show that $x = 40$, $y = 15$, and $z = 30$. The feed should contain 40 units of corn, 15 units of soybeans, and 30 units of cottonseed to fulfill the given requirements.   ◆

Note The table shown in Example 3 is useful in setting up the equations of the system, since the coefficients in each equation can be read from left to right. This idea is extended in the next section, where we introduce solution of systems by matrices.

EXAMPLE 4 *Solving a Feed Requirements Application with Fewer Equations Than Variables* In Example 3, suppose that only the fat and fiber content of the feed are of interest. How would the solution be changed?

SOLUTION We would need to solve the system

$$4x + 2y + 3z = 280$$
$$3x + 2y + z = 180,$$

which does not have a single, unique solution. Following the procedure outlined in Example 2, we find the solution set of this system is $\{(100 - 2z, 2.5z - 60, z)\}$. In this applied problem, however, all three variables must be nonnegative, so z must satisfy the conditions

$$100 - 2z \geq 0, \qquad 2.5z - 60 \geq 0, \qquad z \geq 0.$$

From the first inequality, $z \leq 50$; from the second inequality, $z \geq 24$; thus, $24 \leq z \leq 50$. Only solutions with z in this range are usable.   ◆

Three noncollinear points lie on the graph of a parabola with an equation of the form $y = ax^2 + bx + c$. The procedure for finding a curve to fit certain points (called *data points*) is called **curve-fitting,** and it is studied extensively in statistics. The following example illustrates how curve-fitting is accomplished by solving a system of equations.

EXAMPLE 5 *Using a System to Fit a Parabola to Three Data Points* Find the equation of the parabola $y = ax^2 + bx + c$ that passes through $(2, 4)$, $(-1, 1)$, and $(-2, 5)$.

SOLUTION Since the three points lie on the graph of the equation $y = ax^2 + bx + c$, they must satisfy the equation. Substituting each ordered pair into the equation gives three equations with three variables.

$$4 = a(2)^2 + b(2) + c \qquad \text{or} \qquad 4 = 4a + 2b + c \qquad \textbf{(9)}$$
$$1 = a(-1)^2 + b(-1) + c \qquad \text{or} \qquad 1 = a - b + c \qquad \textbf{(10)}$$
$$5 = a(-2)^2 + b(-2) + c \qquad \text{or} \qquad 5 = 4a - 2b + c \qquad \textbf{(11)}$$

This system can be solved by the addition method. First, eliminate c, using equations (9) and (10).

$$\begin{array}{ll} 4 = 4a + 2b + c & \\ \underline{-1 = -a + b - c} & \text{-1 times equation (10)} \\ 3 = 3a + 3b & \qquad\qquad\qquad\textbf{(12)} \end{array}$$

Now, use equations (10) and (11) to also eliminate c.

$$\begin{array}{r} 1 = a - b + c \\ -5 = -4a + 2b - c \\ \hline -4 = -3a + b \end{array}$$ -1 times equation (11)

(13)

Solve the system of equations (12) and (13) in two variables by eliminating a.

$$\begin{array}{r} 3 = 3a + 3b \\ -4 = -3a + b \\ \hline -1 = 4b \end{array}$$

$$-\frac{1}{4} = b$$

Find a by substituting $-\frac{1}{4}$ for b in equation (12), which is equivalent to $1 = a + b$.

$$1 = a + b \qquad \text{Equation (12) divided by 3}$$

$$1 = a - \frac{1}{4} \qquad \text{Let } b = -\tfrac{1}{4}.$$

$$\frac{5}{4} = a$$

Finally, find c by substituting $a = \frac{5}{4}$ and $b = -\frac{1}{4}$ in equation (10).

$$1 = a - b + c$$

$$1 = \frac{5}{4} - \left(-\frac{1}{4}\right) + c \qquad a = \tfrac{5}{4}, b = -\tfrac{1}{4}$$

$$1 = \frac{6}{4} + c$$

$$-\frac{1}{2} = c$$

An equation of the parabola is $y = \frac{5}{4}x^2 - \frac{1}{4}x - \frac{1}{2}$, or equivalently, $y = 1.25x^2 - .25x - .5$.

This result may be supported graphically by graphing $y = 1.25x^2 - .25x - .5$ and showing that the points $(2, 4)$, $(-1, 1)$, and $(-2, 5)$ do indeed lie on the parabola. See Figure 15. ◆

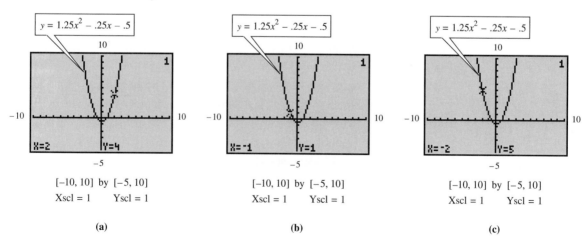

The points $(2, 4)$, $(-1, 1)$, and $(-2, 5)$ all lie on the graph of $y = 1.25x^2 - .25x - .5$.

FIGURE 15

For Group Discussion   Suppose you want to find a polynomial function of the form $y = P(x)$ with a graph that goes through four given ordered pairs. How would you proceed? What kind of polynomial would P be? If you were given five pairs of numbers, what kind of polynomial would P be? In a practical problem of this type with a set of at least five points, how could you decide which kind of polynomial would fit the data best?

7.2   EXERCISES   Tape 8

Verify that the given ordered triple is a solution of the system by substituting the given values of x, y, and z into each equation.

1. $(-3, 6, 1)$
$$2x + y - z = -1$$
$$x - y + 3z = -6$$
$$-4x + y + z = 19$$

2. $\left(\dfrac{1}{2}, -\dfrac{3}{4}, \dfrac{1}{6}\right)$
$$2x + 8y - 6z = -6$$
$$x + y + z = -\dfrac{1}{12}$$
$$x + 3z = 1$$

3. $(-.2, .4, .5)$
$$5x - y + 2z = -.4$$
$$x + 4z = 1.8$$
$$-3y + z = -.7$$

4. Why is the graphing capability of a graphing calculator not used to solve a system of equations with more than two variables?

Solve each of the following systems analytically. (Hint: In Exercises 19–22, let $t = \frac{1}{x}$, $u = \frac{1}{y}$, and $v = \frac{1}{z}$. Solve for t, u, and v, and then solve for x, y, and z.)

5. $\quad x + y + z = 2$
$\quad 2x + y - z = 5$
$\quad x - y + z = -2$

6. $\quad 2x + y + z = 9$
$\quad -x - y + z = 1$
$\quad 3x - y + z = 9$

7. $\quad x + 3y + 4z = 14$
$\quad 2x - 3y + 2z = 10$
$\quad 3x - y + z = 9$

8. $\quad 4x - y + 3z = -2$
$\quad 3x + 5y - z = 15$
$\quad -2x + y + 4z = 14$

9. $\quad x + 2y + 3z = 8$
$\quad 3x - y + 2z = 5$
$\quad -2x - 4y - 6z = 5$

10. $3x - 2y - 8z = 1$
$\quad 9x - 6y - 24z = -2$
$\quad x - y + z = 1$

11. $\quad x + 4y - z = 6$
$\quad 2x - y + z = 3$
$\quad 3x + 2y + 3z = 16$

12. $\quad 4x - 3y + z = 9$
$\quad 3x + 2y - 2z = 4$
$\quad x - y + 3z = 5$

13. $\quad 5x + y - 3z = -6$
$\quad 2x + 3y + z = 5$
$\quad -3x - 2y + 4z = 3$

14. $2x - 5y + 4z = -35$
$\quad 5x + 3y - z = 1$
$\quad x + y + z = 1$

15. $\quad x - 3y - 2z = -3$
$\quad 3x + 2y - z = 12$
$\quad -x - y + 4z = 3$

16. $\quad x + y + z = 3$
$\quad 3x - 3y - 4z = -1$
$\quad x + y + 3z = 11$

17. $2x + 6y - z = 6$
$\quad 4x - 3y + 5z = -5$
$\quad 6x + 9y - 2z = 11$

18. $\quad 8x - 3y + 6z = -2$
$\quad 4x + 9y + 4z = 18$
$\quad 12x - 3y + 8z = -2$

19. $\dfrac{1}{x} + \dfrac{1}{y} - \dfrac{1}{z} = \dfrac{1}{4}$
$\dfrac{2}{x} - \dfrac{1}{y} + \dfrac{3}{z} = \dfrac{9}{4}$
$-\dfrac{1}{x} - \dfrac{2}{y} + \dfrac{4}{z} = 1$

20.
$$\frac{3}{x} + \frac{2}{y} - \frac{1}{z} = \frac{11}{6}$$
$$\frac{1}{x} - \frac{1}{y} + \frac{3}{z} = -\frac{11}{12}$$
$$\frac{2}{x} + \frac{1}{y} + \frac{1}{z} = \frac{7}{12}$$

21.
$$\frac{2}{x} - \frac{2}{y} + \frac{1}{z} = -1$$
$$\frac{4}{x} + \frac{1}{y} - \frac{2}{z} = -9$$
$$\frac{1}{x} + \frac{1}{y} - \frac{3}{z} = -9$$

22.
$$\frac{5}{x} - \frac{1}{y} - \frac{2}{z} = -6$$
$$-\frac{1}{x} + \frac{3}{y} - \frac{3}{z} = -12$$
$$\frac{2}{x} - \frac{1}{y} - \frac{1}{z} = 6$$

23. Consider the linear equation in three variables $x + y + z = 4$. Find a pair of linear equations that, when considered together with the given equation, will form a system having the following.

    (a) exactly one solution

    (b) no solution

    (c) infinitely many solutions

24. Refer to Example 2 in this section. Write the solution set with x arbitrary.

25. Give an example, using your immediate surroundings, of three planes that intersect in a single point.

26. Give an example, using your immediate surroundings, of three planes that intersect in a line.

Solve each of the following systems in terms of the arbitrary variable z.

27. $x - 2y + 3z = 6$
         $2x - y + 2z = 5$

28. $3x + 4y - z = 13$
         $x + y + 2z = 15$

29. $5x - 4y + z = 9$
         $x + y = 15$

30. $x - y + z = -6$
         $4x + y + z = 7$

31. $3x - 5y - 4z = -7$
         $y - z = -13$

32. $3x - 2y + z = 15$
         $x + 4y - z = 11$

Use a system of equations in three variables to solve the following problems.

33. *Coin Collecting*   A coin collection contains a total of 29 coins, made up of cents, nickels, and quarters. The number of quarters is 8 less than the number of cents. The total face value of the coins is $1.77. How many of each denomination are there?

34. *Mixing Waters*   A sparkling water distributor wants to make up 300 gallons of sparkling water to sell for $6.00 per gallon. She wishes to mix three grades of water selling for $9.00, $3.00, and $4.50 per gallon, respectively. She must use twice as much of the $4.50 water as the $3.00 water. How many gallons of each should she use?

35. *Mixing Glue*   A glue company needs to make some glue that it can sell for $120 per barrel. It wants to use 150 barrels of glue worth $100 per barrel, along with some glue worth $150 per barrel, and glue worth $190 per barrel. It must use the same number of barrels of $150 and $190 glue. How much of the $150 and $190 glue will be needed? How many barrels of $120 glue will be produced?

36. *Pricing Concert Tickets*   Frank Capek and his Generation Gap group sells three kinds of concert tickets: "up close," "middle," and "farther back." "Up close" tickets cost $6 more than "middle" tickets, while "middle" tickets cost $3 more than "farther back" tickets. Twice the cost of an "up close" ticket is $3 more than 3 times the cost of a "farther back" seat. Find the price of each kind of ticket.

37. *Triangle Dimensions*   The perimeter of a triangle is 59 inches. The longest side is 11 inches longer than the medium side, and the medium side is 3 inches more than the shortest side. Find the length of each side of the triangle.

38. *Triangle Dimensions*   The sum of the measures of the angles of any triangle is 180°. In a certain triangle, the largest angle measures 55° less than twice the medium angle, and the smallest measures 25° less than the medium angle. Find the measures of each of the three angles.

39. *Investment Decisions*   Tom Accardo wins $100,000 in the Louisiana state lottery. He invests part of the money in real estate with an annual return of 5% and another part in a money market account at 4.5% interest. He invests the rest, which amounts to $20,000 less than the sum of the other two parts, in certificates of deposit that pay 3.75%. If the total annual interest on the money is $4450, how much was invested at each rate?

40. *Investment Decisions*   Jane Ann invests $10,000 received in an inheritance in three parts. With one part she buys mutual funds which offer a return of 4% per year. The second part, which amounts to twice the first, is used to buy government bonds paying 4.5% per year. She puts the rest into a savings account that pays 2.5% annual interest. During the first year, the total interest is $415. How much did she invest at each rate?

41. *Scheduling Production*   Felsted Furniture makes dining room furniture. A buffet requires 30 hours for construction and 10 hours for finishing. A chair requires 10 hours for construction and 10 hours for finishing. A table requires 10 hours for construction and 30 hours for finishing. The construction department has 350 hours of labor and the finishing department has 150 hours of labor available each week. How many pieces of each type of furniture should be produced each week if the factory is to run at full capacity?

42. *Scheduling Deliveries*   Kelley Karpet Kleaners sells rug cleaning machines. The EZ model weighs 10

pounds and comes in a 10-cubic-foot box. The compact model weighs 20 pounds and comes in an 8-cubic-foot box. The commercial model weighs 60 pounds and comes in a 28-cubic-foot box. Each of their delivery vans has 248 cubic feet of space and can hold a maximum of 440 pounds. In order for a van to be fully loaded, how many of each model should it carry?

Use the method of Example 5 in Exercises 43–48.

43. Find a, b, and c so that the graph of the equation $y = ax^2 + bx + c$ passes through the points $(2, 3)$, $(-1, 0)$, and $(-2, 2)$.

44. Find a, b, and c so that $(2, 14)$, $(0, 0)$, and $(-1, -1)$ lie on the graph of $y = ax^2 + bx + c$.

Find the equation of the parabola shown. In each exercise, three views of the same curve are given.

45.

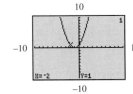

46.

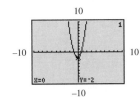

47. *Atmospheric Carbon Dioxide*   Determining the amount of carbon dioxide in the atmosphere is important because carbon dioxide is known to be a greenhouse gas. Carbon dioxide concentrations (in parts per million) have been measured at Mauna Loa, Hawaii, over the past 30 years. This concentration has increased quadratically. The table lists readings for three years.

Year	CO_2
1958	315
1973	325
1988	352

(*Source:* Nilsson, A., *Greenhouse Earth*, John Wiley & Sons, New York, 1992.)

(a) If the quadratic relationship between the carbon dioxide concentration C and the year t is expressed as $C = at^2 + bt + c$, where $t = 0$ corresponds to 1958, use a linear system of equations to determine the constants a, b, and c, and give the equation for these data.

(b) Predict the year when the amounts of carbon dioxide in the atmosphere will double from its 1958 level.

48. *Aircraft Altitude and Speed*   For certain aircraft there exists a quadratic relationship between an airplane's maximum speed S (in knots) and its ceiling C, or highest altitude possible (in thousands of feet). The table lists three airplanes that conform to this relationship.

Airplane	Max. Speed	Ceiling
Hawkeye	320	33
Corsair	600	40
Tomcat	1283	50

(*Source:* Sanders, D., *Statistics: A First Course*, Fifth Edition, McGraw-Hill, Inc., 1995.)

(a) If the quadratic relationship between C and S is written as $C = aS^2 + bS + c$, use a linear system of equations to determine the constants a, b, and c, and give the equation for these data.

(b) A new aircraft of this type has a ceiling of 45,000 feet. Predict its top speed.

Given three noncollinear points, there is one and only one circle that passes through them. Knowing that the equation of a circle may be written in the form

$$x^2 + y^2 + ax + by + c = 0,$$

find the equation of the circle described or graphed in Exercises 49 and 50.

49. passing through the points $(2, 1)$, $(-1, 0)$, and $(3, 3)$

50.

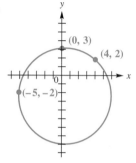

Determining the Position of a Particle   *Suppose that the position of a particle moving along a straight line is given by the function* $s(t) = at^2 + bt + c$, *where t is time in seconds and a, b, and c are real numbers.*

51. If $s(0) = 5$, $s(1) = 23$, and $s(2) = 37$, find a, b, and c. Then, find $s(8)$.

52. If $s(0) = -10$, $s(1) = 6$, and $s(2) = 30$, find a, b, and c. Then, find $s(10)$.

7.3   SOLUTION OF LINEAR SYSTEMS BY ROW TRANSFORMATIONS

Matrices and Technology   ◆   Solving Systems by Matrix Row Transformations   ◆   Special Cases   ◆   An Application

MATRICES AND TECHNOLOGY

TECHNOLOGY NOTE

The manner in which matrices are entered and displayed varies greatly among different manufacturers, and sometimes even differs among the various models manufactured by the same corporation. As always, we suggest that you refer to your owner's manual when necessary.

In this section and Section 7.6, we show two methods for solving linear systems with matrices. Matrix methods are particularly suitable for calculator or computer solutions of large systems of equations that have many unknowns. Graphing calculators are programmed to do the necessary computations when the appropriate commands are entered. We believe that a knowledge of the concepts is essential in understanding how matrices are used; simply viewing results on a screen without the knowledge of how they were computed is not an acceptable method of using the matrix capabilities of graphing calculators. Examples and exercises involving matrices are presented in a manner similar to that of a traditional algebra text. However, we urge you to learn how to use your particular model by having it available at all times. Sample screens are shown in the margins. In many cases, intermediate steps shown in print will not be necessary when using a calculator, but they are included for completeness.

For Group Discussion   Refer to the owner's manual of your particular make and model calculator to find the pages that describe how to use the matrix capabilities. Mark these pages with a bookmark or some other method so that you can readily refer to them while studying this chapter.

SOLVING SYSTEMS BY MATRIX ROW TRANSFORMATIONS

The echelon method used to solve linear systems of equations in the previous section can be streamlined to a systematic method by using *matrices* (singular: *matrix*). To begin, consider a system of three equations and three unknowns such as

$$a_1x + b_1y + c_1z = d_1$$
$$a_2x + b_2y + c_2z = d_2$$
$$a_3x + b_3y + \,\cdot\, \cdot = d_3.$$

This system can be written in an abbreviated form as

$$\begin{bmatrix} a_1 & b_1 & c_1 & d_1 \\ a_2 & b_2 & c_2 & d_2 \\ a_3 & b_3 & c_3 & d_3 \end{bmatrix}.$$

TECHNOLOGY NOTE

It is likely that the vertical bar separating the coefficients of the system from the constants will not appear on the graphing calculator screen.

Such a rectangular array of numbers enclosed by brackets is called a **matrix.** Each number in the array is an **element** or **entry.** The constants in the last column of the matrix can be set apart from the coefficients of the variables with a vertical line, as shown in the following **augmented matrix.** (Because the matrix of coefficients has an extra column determined by the constants of the system, the coefficient matrix is *augmented.*)

Rows $\begin{bmatrix} a_1 & b_1 & c_1 & d_1 \\ a_2 & b_2 & c_2 & d_2 \\ a_3 & b_3 & c_3 & d_3 \end{bmatrix}$

Columns

As an example, the system

$$x + 3y + 2z = 1$$
$$2x + y - z = 2$$
$$x + y + z = 2$$

has the augmented matrix

$$\begin{bmatrix} 1 & 3 & 2 & 1 \\ 2 & 1 & -1 & 2 \\ 1 & 1 & 1 & 2 \end{bmatrix}.$$

This matrix has 3 rows (horizontal) and 4 columns (vertical). To refer to a number in the matrix, use its row and column numbers. For example, the number 3 is in the first row, second column position.

The rows of this augmented matrix can be treated just like the equations of the system of linear equations. Since the augmented matrix is nothing more than a short form of the system, any transformation of the matrix that results in an equivalent system of equations can be performed. Operations that produce such transformations are given below.

TECHNOLOGY NOTE

Refer to your owner's manual to see how these transformations are accomplished with your model.

MATRIX ROW TRANSFORMATIONS

For any augmented matrix of a system of linear equations, the following row transformations will result in the matrix of an equivalent system.

1. Any two rows may be interchanged.
2. The elements of any row may be multiplied by a nonzero real number.
3. Any row may be changed by adding to its elements a multiple of the corresponding elements of another row.

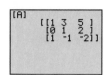

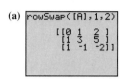

EXAMPLE 1   *Using the Row Transformations*   We show the results of using the three row transformations below. The given matrix and calculator screens that correspond to parts (a)–(c) are shown in the margin.

SOLUTION

(a)   The first row transformation is used to change the matrix

$$\begin{bmatrix} 1 & 3 & 5 \\ 0 & 1 & 2 \\ 1 & -1 & -2 \end{bmatrix} \text{ to } \begin{bmatrix} 0 & 1 & 2 \\ 1 & 3 & 5 \\ 1 & -1 & -2 \end{bmatrix}$$

by interchanging the first two rows.

(b)   Using the second row transformation with $k = -2$ changes

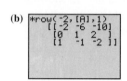

$$\begin{bmatrix} 1 & 3 & 5 \\ 0 & 1 & 2 \\ 1 & -1 & -2 \end{bmatrix} \text{ to } \begin{bmatrix} -2 & -6 & -10 \\ 0 & 1 & 2 \\ 1 & -1 & -2 \end{bmatrix},$$

where the elements of the first row of the original matrix were multiplied by -2.

(c)   The third row transformation is used to change

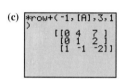

$$\begin{bmatrix} 1 & 3 & 5 \\ 0 & 1 & 2 \\ 1 & -1 & -2 \end{bmatrix} \text{ to } \begin{bmatrix} 0 & 4 & 7 \\ 0 & 1 & 2 \\ 1 & -1 & -2 \end{bmatrix},$$

by multiplying each element in the third row of the original matrix by -1 and adding the results to the corresponding elements in the first row of that matrix. That is, the elements in the new first row were found as follows.

These screens show how the TI-83 accomplishes the transformations in Example 1.

$$\begin{bmatrix} 1 + 1(-1) & 3 + (-1)(-1) & 5 + (-2)(-1) \\ 0 & 1 & 2 \\ 1 & -1 & -2 \end{bmatrix} = \begin{bmatrix} 0 & 4 & 7 \\ 0 & 1 & 2 \\ 1 & -1 & -2 \end{bmatrix}$$

Rows two and three were left unchanged.   ◆

The matrix row transformations are used to transform the augmented matrix of a system into one that is in **triangular** (or **echelon**) form. When in triangular form, the matrix will have 1s down the diagonal from upper left to lower right and 0s below each 1, as shown here.

$$\begin{bmatrix} 1 & 5 & 3 & | & 7 \\ 0 & 1 & 2 & | & 9 \\ 0 & 0 & 1 & | & 4 \end{bmatrix}$$

The 1s lie on the **main diagonal** of the matrix.

Before using matrices to solve a linear system, we must arrange the system in the proper form, with variable terms on the left side of the equation and constant terms on the right. The variable terms must be in the same order in each of the equations.

The method of using matrices to solve a system of linear equations, to be developed in this section, is called the **Gaussian reduction method,** after the mathematician Karl Friedrich Gauss (1777–1855). The following example illustrates this matrix method and compares it with the echelon method discussed in the previous section.

EXAMPLE 2   *Comparing the Elimination and Gaussian Reduction Methods*
Solve the system

$$3x - 4y = 1$$

$$5x + 2y = 19.$$

SOLUTION   The procedure for matrix solution is parallel to the echelon method used in the previous section, except for the last step. First, write the augmented matrix for the system. Here the system is already in the proper form.

Echelon Method	Gaussian Reduction Method
$3x - 4y = 1$ **(1)** $\quad$ $5x + 2y = 19$ **(2)**	$\begin{bmatrix} 3 & -4 & \vert & 1 \\ 5 & 2 & \vert & 19 \end{bmatrix}$
Divide both sides of equation (1) by 3 so that x has a coefficient of 1.	Using row transformation (2), multiply each element of row 1 by $\frac{1}{3}$.
$x - \frac{4}{3}y = \frac{1}{3}$ **(3)** $\quad$ $5x + 2y = 19$	$\begin{bmatrix} 1 & -\frac{4}{3} & \vert & \frac{1}{3} \\ 5 & 2 & \vert & 19 \end{bmatrix}$
Eliminate x from equation (2) by adding -5 times equation (3) to equation (2).	Using row transformation (3), add -5 times the elements of row 1 to the elements of row 2.
$x - \frac{4}{3}y = \frac{1}{3}$ $\quad\quad$ $\frac{26}{3}y = \frac{52}{3}$ **(4)**	$\begin{bmatrix} 1 & -\frac{4}{3} & \vert & \frac{1}{3} \\ 0 & \frac{26}{3} & \vert & \frac{52}{3} \end{bmatrix}$
Multiply both sides of equation (4) by $\frac{3}{26}$ to get $y = 2$.	Multiply the elements of row 2 by $\frac{3}{26}$, using row transformation (2).
$x - \frac{4}{3}y = \frac{1}{3}$ $\quad\quad$ $y = 2$	$\begin{bmatrix} 1 & -\frac{4}{3} & \vert & \frac{1}{3} \\ 0 & 1 & \vert & 2 \end{bmatrix}$
The system is in triangular form.	Write the corresponding equations. $x - \frac{4}{3}y = \frac{1}{3}$ $\quad\quad$ $y = 2$

Finish the solution in either method by substituting 2 for y in the first equation to get $x = 3$. The solution of the system is (3, 2) with solution set $\{(3, 2)\}$.   ◆

EXAMPLE 3   *Using the Gaussian Reduction Method*   Use the Gaussian reduction method to solve the linear system

$$2x + 6y = 28$$

$$4x - 3y = -19.$$

SOLUTION   We begin with the augmented matrix

$$\begin{bmatrix} 2 & 6 & \vert & 28 \\ 4 & -3 & \vert & -19 \end{bmatrix}.$$

It is best to work vertically, in columns, beginning in each column with the element which is to become 1. This is the same order used in Section 7.2 to arrange a system of equations in triangular form. The augmented matrix has a 2 in the row 1, column 1 position. To get 1 in this position, use the second transformation and multiply each entry in row 1 by $\frac{1}{2}$. This is indicated below with the notation $\frac{1}{2}R_1$ next to the new row 1.

$$\begin{bmatrix} 1 & 3 & | & 14 \\ 4 & -3 & | & -19 \end{bmatrix} \quad \tfrac{1}{2}\,\text{R}_1$$

To get 0 in row 2, column 1, add -4 times row 1 to row 2.

$$\begin{bmatrix} 1 & 3 & | & 14 \\ 0 & -15 & | & -75 \end{bmatrix} \quad -4\text{R}_1 + \text{R}_2$$

To get 1 in row 2, column 2, multiply each element of row 2 by $-\frac{1}{15}$, which gives

$$\begin{bmatrix} 1 & 3 & | & 14 \\ 0 & 1 & | & 5 \end{bmatrix}. \quad -\tfrac{1}{15}\,\text{R}_2$$

This matrix corresponds to the system

$$x + 3y = 14 \tag{1}$$
$$y = 5. \tag{2}$$

Substitute 5 for y in equation (1) to get $x = -1$. The solution set of the system is thus $\{(-1, 5)\}$. ◆

The Gaussian reduction method can be extended to larger systems. The final matrix always will have 0s below the diagonal of 1s on the left of the vertical bar. To transform the matrix, it is best to work column by column from upper left to lower right. For each column, first perform the step that gives the 1s, then get the 0s below the 1 in that column.

EXAMPLE 4 *Using the Gaussian Reduction Method* Use the Gaussian reduction method to solve the system below, and support with a calculator.

$$\begin{aligned} x - y + 5z &= -6 \\ 3x + 3y - z &= 10 \\ x + 3y + 2z &= 5 \end{aligned}$$

SOLUTION ANALYTIC We begin by writing the augmented matrix of the linear system.

$$\begin{bmatrix} 1 & -1 & 5 & | & -6 \\ 3 & 3 & -1 & | & 10 \\ 1 & 3 & 2 & | & 5 \end{bmatrix}$$

There is already a 1 in row 1, column 1. The next thing to do is get 0s in the rest of column 1. First, add to row 2 the results of multiplying row 1 by -3.

$$\begin{bmatrix} 1 & -1 & 5 & | & -6 \\ 0 & 6 & -16 & | & 28 \\ 1 & 3 & 2 & | & 5 \end{bmatrix} \quad -3\text{R}_1 + \text{R}_2$$

Now, add to row 3 the results of multiplying row 1 by -1.

$$\begin{bmatrix} 1 & -1 & 5 & | & -6 \\ 0 & 6 & -16 & | & 28 \\ 0 & 4 & -3 & | & 11 \end{bmatrix} \quad -1\text{R}_1 + \text{R}_3$$

To get 1 in row 2, column 2, multiply row 2 by $\frac{1}{6}$.

$$\begin{bmatrix} 1 & -1 & 5 & | & -6 \\ 0 & 1 & -\frac{8}{3} & | & \frac{14}{3} \\ 0 & 4 & -3 & | & 11 \end{bmatrix} \quad \tfrac{1}{6}\,\text{R}_2$$

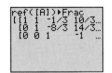

Some calculators will return the triangular (echelon) form of a matrix with one step. The screen shows the echelon form of the matrix in Example 4. The dots at the right indicate that more of the matrix can be seen by scrolling to the right. This can be avoided by limiting the number of decimal places used.

Notice that the entries in row 1, column 3 and row 1, columns 2 and 4 are different than the corresponding entries in the echelon form shown in Example 4. This occurs because the steps were performed in an alternative way. The final result is the same.

Next, to get 0 in row 3, column 2, add to row 3 the results of multiplying row 2 by -4.

$$\begin{bmatrix} 1 & -1 & 5 & | & -6 \\ 0 & 1 & -\frac{8}{3} & | & \frac{14}{3} \\ 0 & 0 & \frac{23}{3} & | & -\frac{23}{3} \end{bmatrix} \quad -4R_2 + R_3$$

Now, multiply the last row by $\frac{3}{23}$ to get 1 in row 3, column 3.

$$\begin{bmatrix} 1 & -1 & 5 & | & -6 \\ 0 & 1 & -\frac{8}{3} & | & \frac{14}{3} \\ 0 & 0 & 1 & | & -1 \end{bmatrix} \quad \frac{3}{23}R_3$$

While not absolutely necessary, we may obtain all integer elements by multiplying row 2 by 3.

$$\begin{bmatrix} 1 & -1 & 5 & | & -6 \\ 0 & 3 & -8 & | & 14 \\ 0 & 0 & 1 & | & -1 \end{bmatrix} \quad 3R_2$$

The final matrix corresponds to the system of equations

$$x - y + 5z = -6 \tag{1}$$
$$3y - 8z = 14 \tag{2}$$
$$z = -1. \tag{3}$$

We know that $z = -1$ from equation (3). Using back-substitution into equation (2), we find that $y = 2$, and then substituting into equation (1) we find that $x = 1$. The solution set of the system is $\{(1, 2, -1)\}$.

GRAPHICAL   A calculator solution is shown in the margin.   ◆

Note   There is usually more than one way to approach the solution of a system with Gaussian reduction.

SPECIAL CASES

The next two examples show how to recognize inconsistent systems or systems with dependent equations when solving such systems by Gaussian reduction.

EXAMPLE 5   *Using the Gaussian Reduction Method (No Solutions)*   Solve the system

$$x + y = 2$$
$$2x + 2y = 5.$$

SOLUTION   We start with the augmented matrix.

$$\begin{bmatrix} 1 & 1 & | & 2 \\ 2 & 2 & | & 5 \end{bmatrix}$$

Next, add to row 2 the results of multiplying row 1 by -2.

$$\begin{bmatrix} 1 & 1 & | & 2 \\ 0 & 0 & | & 1 \end{bmatrix} \quad -2R_1 + R_2$$

This matrix gives the system of equations

$$x + y = 2$$
$$0 = 1,$$

an inconsistent system with no solution. The solution set is $\emptyset$.   ◆

Whenever a row of the augmented matrix is of the form

$$0 \ 0 \ 0 \ \dots \ | \ a, \quad \text{where } a \neq 0,$$

the system is inconsistent and there will be no solution, since this row corresponds to the equation $0 = a$. A row of the matrix of a linear system in the form

$$0 \ \ 0 \ \ 0 \ \dots \ \big| \ 0$$

indicates that the equations of the system are *dependent*.

EXAMPLE 6   *Using the Gaussian Reduction Method (Infinitely Many Solutions)*   Use the Gaussian reduction method to solve the system, and support with a calculator.

$$2x - 5y + 3z = 1$$
$$x + 4y - 2z = 8$$

SOLUTION   ANALYTIC   Recall from Section 7.2 (Example 2) that a system with two equations and three variables may have an infinite number of solutions. The Gaussian reduction method can be used to indicate the solution with one arbitrary variable. Start with the augmented matrix

$$\begin{bmatrix} 2 & -5 & 3 & | & 1 \\ 1 & 4 & -2 & | & 8 \end{bmatrix}.$$

Exchange rows to get a 1 in the row 1, column 1 position.

$$\begin{bmatrix} 1 & 4 & -2 & | & 8 \\ 2 & -5 & 3 & | & 1 \end{bmatrix}$$

Now, multiply each element in row 1 by -2 and add to the corresponding element in row 2.

$$\begin{bmatrix} 1 & 4 & -2 & | & 8 \\ 0 & -13 & 7 & | & -15 \end{bmatrix} \quad -2R_1 + R_2$$

Multiply each element in row 2 by $-\frac{1}{13}$.

$$\begin{bmatrix} 1 & 4 & -2 & | & 8 \\ 0 & 1 & -\frac{7}{13} & | & \frac{15}{13} \end{bmatrix} \quad -\frac{1}{13}R_2$$

This is as far as we can go with the Gaussian reduction method. The equations which correspond to the final matrix are

$$x + 4y - 2z = 8 \qquad \text{and} \qquad y - \frac{7}{13}z = \frac{15}{13}.$$

Solve the second equation for y: $y = \frac{15}{13} + \frac{7}{13}z$. Now, substitute this result for y in the first equation and solve for x.

$$x + 4y - 2z = 8$$
$$x + 4\left(\frac{15}{13} + \frac{7}{13}z\right) - 2z = 8$$
$$x + \frac{60}{13} + \frac{28}{13}z - 2z = 8$$
$$x + \frac{60}{13} + \frac{2}{13}z = 8$$
$$x = 8 - \frac{60}{13} - \frac{2}{13}z$$
$$x = \frac{44}{13} - \frac{2}{13}z$$

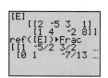

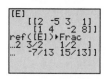

The two screens show the results of using row transformations for the system in Example 6. (In the second screen, we have shifted the display to the right to show the rest of the matrix.) Notice that the top row of this matrix is different than the one we found when we worked the problem by hand. Verify that this matrix gives the same values for x and y as we found. This shows again how technology may produce the same results in a different way.

The solution set can now be written with z arbitrary as

$$\left\{ \left(\frac{44 - 2z}{13}, \frac{15 + 7z}{13}, z \right) \right\}.$$

GRAPHICAL   See the margin on the previous page for a calculator solution.   ◆

AN APPLICATION

If an applied problem leads to a system of equations, the system may often be solved using Gaussian reduction.

EXAMPLE 7   *Solving a Machine Scheduling Problem*   A manufacturer produces chairs in 3 styles: 604, 610, and 618. Each chair requires time on 3 machines, as shown in the following chart. The total available time per week for each machine is also shown in the chart. All available time must be used each week. How many of each style chair can be made under these conditions each week?

Style	Number Made Each Week	Hours on Machine A	Hours on Machine B	Hours on Machine C
604	x	1	3	2
610	y	2	2	3
618	z	1	2	1
Total hours per week		30	56	44

SOLUTION   The total number of hours per week on machine A for the three styles is $x + 2y + z$, which must equal 30, giving the equation

$$x + 2y + z = 30.$$

Similarly, the total hours on machine B leads to the equation

$$3x + 2y + 2z = 56,$$

and the total hours on machine C produces the equation

$$2x + 3y + z = 44.$$

These three equations form a system with the augmented matrix

$$\begin{bmatrix} 1 & 2 & 1 & | & 30 \\ 3 & 2 & 2 & | & 56 \\ 2 & 3 & 1 & | & 44 \end{bmatrix}.$$

Use the Gaussian reduction method on this matrix to get a matrix that gives the solution of the system, $(8, 6, 10)$. The manufacturer can produce 8 style 604 chairs, 6 style 610 chairs, and 10 style 618 chairs each week.   ◆

For Group Discussion   Figure 16 on the next page shows three views of the graph of the same quadratic function $f(x) = ax^2 + bx + c$. Discuss how the values of a, b, and c can be found. (See Exercises 55–62 for a follow-up to this discussion.)

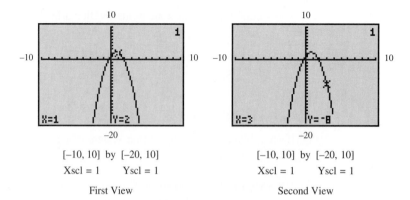

[−10, 10] by [−20, 10]

Xscl = 1   Yscl = 1

First View

[−10, 10] by [−20, 10]

Xscl = 1   Yscl = 1

Second View

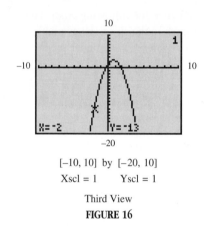

[−10, 10] by [−20, 10]

Xscl = 1   Yscl = 1

Third View

FIGURE 16

7.3 EXERCISES   Tape 8

Use the row transformation described to transform each of the following matrices as indicated.

1. $\begin{bmatrix} 2 & 4 \\ 4 & 7 \end{bmatrix}$; −2 times row 1

2. $\begin{bmatrix} -1 & 4 \\ 7 & 0 \end{bmatrix}$; 7 times row 1

3. $\begin{bmatrix} 1 & 5 & 6 \\ -2 & 3 & -1 \\ 4 & 7 & 0 \end{bmatrix}$; row 1 added to row 2

4. $\begin{bmatrix} 2 & 5 & 6 \\ 4 & -1 & 2 \\ 3 & 7 & 1 \end{bmatrix}$; row 3 added to row 1

5. $\begin{bmatrix} -3 & 1 & -4 \\ 2 & 1 & 3 \\ -7 & 5 & 2 \end{bmatrix}$; −5 times row 2 added to row 3

6. $\begin{bmatrix} 4 & 10 & -8 \\ 7 & 4 & 3 \\ -1 & 1 & 0 \end{bmatrix}$; −4 times row 3 added to row 2

Write the augmented matrix for each of the following systems. Do not solve the system.

7. $2x + 3y = 11$
$\quad\ x + 2y = 8$

8. $3x + 5y = -13$
$\quad 2x + 3y = -9$

9. $x + 5y = 6$
$\quad x + 2y = 8$

10. $2x + 7y = 1$
$\qquad\quad 5x = -15$

11. $2x + y + z = 3$
$\quad 3x - 4y + 2z = -7$
$\quad\ x + y + z = 2$

12. $4x - 2y + 3z = 4$
$\quad 3x + 5y + z = 7$
$\quad 5x - y + 4z = 7$

13. $\quad x + y = 2$
$\quad 2y + z = -4$
$\qquad\quad z = 2$

14. $\qquad\quad x = 6$
$\quad y + 2z = 2$
$\quad x - 3z = 6$

Write the system of equations associated with each of the following augmented matrices. Do not try to solve.

15. $\begin{bmatrix} 2 & 1 & | & 1 \\ 3 & -2 & | & -9 \end{bmatrix}$

16. $\begin{bmatrix} 1 & -5 & | & -18 \\ 6 & 2 & | & 20 \end{bmatrix}$

17. $\begin{bmatrix} 1 & 0 & 0 & | & 2 \\ 0 & 1 & 0 & | & 3 \\ 0 & 0 & 1 & | & -2 \end{bmatrix}$

18. $\begin{bmatrix} 1 & 0 & 1 & | & 4 \\ 0 & 1 & 0 & | & 2 \\ 0 & 0 & 1 & | & 3 \end{bmatrix}$

19. $\begin{bmatrix} 3 & 2 & 1 & | & 1 \\ 0 & 2 & 4 & | & 22 \\ -1 & -2 & 3 & | & 15 \end{bmatrix}$

20. $\begin{bmatrix} 2 & 1 & 3 & | & 12 \\ 4 & -3 & 0 & | & 10 \\ 5 & 0 & -4 & | & -11 \end{bmatrix}$

Use the Gaussian reduction method to solve each of the following systems of equations.

21. $x + y = 5$
$x - y = -1$

22. $x + 2y = 5$
$2x + y = -2$

23. $x + y = -3$
$2x - 5y = -6$

24. $3x - 2y = 4$
$3x + y = -2$

25. $2x - 3y = 10$
$2x + 2y = 5$

26. $4x + y = 5$
$2x + y = 3$

27. $2x - 3y = 2$
$4x - 6y = 1$

28. $x + 2y = 1$
$2x + 4y = 3$

29. $6x - 3y = 1$
$-12x + 6y = -2$

30. $x - y = 1$
$-x + y = -1$

31. $x + y = -1$
$y + z = 4$
$x + z = 1$

32. $x - z = -3$
$y + z = 9$
$x + z = 7$

33. $x + y - z = 6$
$2x - y + z = -9$
$x - 2y + 3z = 1$

34. $x + 3y - 6z = 7$
$2x - y + 2z = 0$
$x + y + 2z = -1$

35. $-x + y = -1$
$y - z = 6$
$x + z = -1$

36. $x + y = 1$
$2x - z = 0$
$y + 2z = -2$

37. $2x - y + 3z = 0$
$x + 2y - z = 5$
$2y + z = 1$

38. $4x + 2y - 3z = 6$
$x - 4y + z = -4$
$-x + 2z = 2$

39. Compare the use of an augmented matrix as a shorthand way of writing a system of linear equations and the use of synthetic division as a shorthand way to divide polynomials.

40. Compare the use of the third row transformation on a matrix and the elimination method of solving a system of linear equations.

Solve each of the systems in Exercises 41–46 by Gaussian reduction. Let z be the arbitrary variable if necessary.

41. $x - 3y + 2z = 10$
$2x - y - z = 8$

42. $3x + y - z = 12$
$x + 2y + z = 10$

43. $x + 2y - z = 0$
$3x - y + z = 6$
$-2x - 4y + 2z = 0$

44. $3x + 5y - z = 0$
$4x - y + 2z = 1$
$-6x - 10y + 2z = 0$

45. $x - 2y + z = 5$
$-2x + 4y - 2z = 2$
$2x + y - z = 2$

46. $3x + 6y - 3z = 12$
$-x - 2y + z = 16$
$x + y - 2z = 20$

Solve each applied problem by writing a system of equations and then solving the system, using Gaussian reduction.

47. *Distributing Income* A working couple earned a total of $4352. The wife earned $64 per day; the husband earned $8 per day less. Find the number of days each worked if the total number of days worked by both was 72.

48. *Scheduling Production* Caltek Computer Company makes two products: computer monitors and printers. Both require time on two machines—monitors: 1 hr on machine *A* and 2 hr on machine *B*; printers: 3 hr on machine *A* and 1 hr on machine *B*. Both machines operate 15 hr per day. How many of each product can be produced in a day under these conditions?

49. *Scheduling Production* A company produces two models of bicycles, model 201 and model 301. Model 201 requires 2 hr of assembly time, and model 301 requires 3 hr of assembly time. The parts for model 201 cost $25 per bike; those for model 301 cost $30 per bike. If the company has a total of 34 hr of assembly time and $365 available per day for these two models, how many of each can be made in a day?

50. *Investment Decisions* Jeff Balius deposits some money in a bank account paying 3% per year. He uses some additional money, amounting to $\frac{1}{3}$ the amount placed in the bank, to buy bonds paying 4% per year. With the balance of his funds, he buys a 4.5% certificate of deposit. The first year, his investments bring a return of $400. If the total of the investments is $10,000, how much is invested at each rate?

51. *Financing Expansion*   To get necessary funds for a planned expansion, a small company took out three loans totaling $25,000. The company was able to borrow some of the money at 8%. It borrowed $2000 more than $\frac{1}{2}$ the amount of the 8% loan at 10%, and the rest at 9%. The total annual interest was $2220. How much did the company borrow at each rate?

52. *Mixing Salt Solutions*   Rebecca Isaac-Fahey, a biologist, has three salt solutions: some 5% solution, some 15% solution, and some 25% solution. She needs to mix some of each to get 50 liters of 20% solution. She wants to use twice as much of the 5% solution as the 15% solution. How much of each solution should she use?

We have seen how Gaussian reduction can be used to solve systems with up to three equations in three unknowns. It can be extended to larger systems as well. Use Gaussian reduction to solve each system in Exercises 53 and 54, and express your solutions in the form (x, y, z, w).

53.
$$x + 3y - 2z - w = 9$$
$$4x + y + z + 2w = 2$$
$$-3x - y + z - w = -5$$
$$x - y - 3z - 2w = 2$$

54.
$$3x + 2y - w = 0$$
$$2x + z + 2w = 5$$
$$x + 2y - z = -2$$
$$2x - y + z + w = 2$$

RELATING CONCEPTS (EXERCISES 55–62)

After reading the group discussion exercise and observing Figure 16 at the end of this section, you should have reached the conclusion that the values a, b, and c can be found by substituting the corresponding x- and y-values in each case into the equation $y = ax^2 + bx + c$ to obtain equations in a, b, and c. Work Exercises 55–62 in order.

55. Give the equation obtained when $x = 1$ and $y = 2$.

56. Give the equation obtained when $x = -2$ and $y = -13$.

57. Give the equation obtained when $x = 3$ and $y = -8$.

58. Before solving the system formed by the equations in Exercises 55–57, what do you know from the graph about the *sign* of a? Why is this so? (Section 3.2)

59. What do you know about the *sign* of c? (Section 3.2)

60. Use Gaussian reduction to solve the system formed by the equations in Exercises 55–57. Give the values of a, b, and c.

61. If $f(x) = ax^2 + bx + c$, determine $f(-1.5)$ analytically for the function found in Exercise 60. (Section 3.2)

62. Support your result for Exercise 61 graphically. (Section 3.2)

63. Suppose that $g(x) = ax^3 + bx^2 + cx + d$, with $g(-2) = -38$, $g(0) = -2$, $g(1) = 1$, and $g(3) = 37$. Find the values of a, b, c, and d, and then find $g(4)$ both analytically and graphically.

64. Solve the system
$$3x^2 + y^2 = 4$$
$$-5x^2 + 2y^2 = -3$$
for x^2 and y^2, using Gaussian reduction. Then, solve for x and y, using methods described in Section 7.1. (*Hint:* Start by letting $x^2 = t$ and $y^2 = s$.)

RELATING CONCEPTS (EXERCISES 65–70)

In this section, we used the Gaussian reduction method to solve a linear system by putting it in triangular form. Another procedure for solving linear systems, the Gauss-Jordan method, goes further by using row transformations to get zeros above (as well as below) the main diagonal of 1s.

65. Refer to Example 4. Begin with the next to last matrix given there,
$$\begin{bmatrix} 1 & -1 & 5 & | & -6 \\ 0 & 1 & -\frac{8}{3} & | & \frac{14}{3} \\ 0 & 0 & 1 & | & -1 \end{bmatrix},$$
and use row transformation (3) to get a 0 as the second element in row 1.

66. Using your answer from Exercise 65, use row transformation (3) to get a 0 as the third element in row 1.

67. Using your answer from Exercise 66, use row transformation (3) to get a 0 as the third element in row 2.

68. Write the system associated with your answer from Exercise 67.

69. What is the solution of the system? Does it agree with the solution we found in Example 4?

70. Some graphing calculators will give results in this form. If your calculator has a command for the Gauss-Jordan method (reduced row-echelon form), use it to work Example 4 with one command.

71. *Determining the Number of Fawns* To model the spring fawn count F from the adult population A, the precipitation P, and the severity of the winter W, environmentalists have found that the equation $F = a + bA + cP + dW$ can be used, where the coefficients a, b, c, and d are constants that must be determined before one can use the equation. (*Sources:* Brase, C., and C. Brase, *Understandable Statistics*, Lexington, MA: D.C. Heath and Company, 1995; Bureau of Land Management.) The following table lists the results of four different years.

Fawns	Adults	Precip.	Winter
239	871	11.5	3
234	847	12.2	2
192	685	10.6	5
343	969	14.2	1

(a) Substitute the values for F, W, A, and P from the table for each of the four years into the equation $F = a + bA + cP + dW$, and obtain four linear equations involving a, b, c, and d.

(b) Write an augmented matrix representing the system and solve for a, b, c, and d.

(c) Write the equation for F, using these values for the coefficients.

(d) If a winter has a severity of 3, the adult antelope population is 960, and the precipitation is 12.6 inches, predict the spring fawn count. (Compare this with the actual count of 320.)

7.4 PROPERTIES OF MATRICES

Introduction ◆ Terminology of Matrices ◆ Operations on Matrices ◆ An Application of Matrix Algebra

INTRODUCTION

The use of matrix notation in solving a system of linear equations was illustrated in the previous section. In this section, we will examine matrices as mathematical entities with properties analogous to those of real numbers. To motivate the concept of a matrix, consider the following. Suppose you are the manager of a health food store and you receive the following shipments of vitamins from two suppliers: from Dexter, 2 cartons of vitamin A pills, 7 cartons of vitamin E pills, and 5 cartons of vitamin K pills; from Sullivan, 4 cartons of vitamin A pills, 6 cartons of vitamin E pills, and 9 cartons of vitamin K pills.

It might be helpful to rewrite the information in a chart to make it more understandable.

	Cartons of Vitamins		
Manufacturer	A	E	K
Dexter	2	7	5
Sullivan	4	6	9

The information is clearer when presented this way. In fact, as long as you remember what each number represents, you can remove all the labels and write the numbers as a matrix.

$$\begin{bmatrix} 2 & 7 & 5 \\ 4 & 6 & 9 \end{bmatrix}$$

This array of numbers gives all the information needed.

TERMINOLOGY OF MATRICES

Matrices are classified by their **dimension,** that is, by the number of rows and columns that they contain. For example, the matrix

$$\begin{bmatrix} 2 & 7 & -5 \\ 3 & -6 & 0 \end{bmatrix}$$

has two rows and three columns, with dimension 2×3; a matrix with m rows and n columns has dimension $m \times n$. The number of rows is always given first.

Certain matrices have special names: an $n \times n$ matrix is a **square matrix** of dimension $n \times n$. Also, a matrix with just one row is a **row matrix,** and a matrix with just one column is a **column matrix.**

Two matrices are **equal** if they have the same dimension and if each pair of corresponding elements, position by position, is equal. Using this definition, the matrices

$$\begin{bmatrix} 2 & 1 \\ 3 & -5 \end{bmatrix} \quad \text{and} \quad \begin{bmatrix} 1 & 2 \\ -5 & 3 \end{bmatrix}$$

are *not* equal (even though they contain the same elements and have the same dimension), because the corresponding elements differ.

TECHNOLOGY NOTE

The TEST function of some graphing calculators will allow you to determine whether two matrices are equal, returning a 1 if they are or a 0 if they are not. Refer to your owner's manual to see if your model is capable of this.

EXAMPLE 1 *Classifying Matrices by Dimension* Find the dimension of each matrix, and determine any special characteristics.

(a) $\begin{bmatrix} 6 & 5 \\ 3 & 4 \\ 5 & -1 \end{bmatrix}$ **(b)** $\begin{bmatrix} 5 & 8 & 9 \\ 0 & 5 & -3 \\ -4 & 0 & 5 \end{bmatrix}$ **(c)** $[1 \quad 6 \quad 5 \quad -2 \quad 5]$ **(d)** $\begin{bmatrix} 3 \\ -5 \\ 0 \\ 2 \end{bmatrix}$

SOLUTION

(a) The matrix $\begin{bmatrix} 6 & 5 \\ 3 & 4 \\ 5 & -1 \end{bmatrix}$ is a 3×2 matrix, because it has 3 rows and 2 columns.

(b) $\begin{bmatrix} 5 & 8 & 9 \\ 0 & 5 & -3 \\ -4 & 0 & 5 \end{bmatrix}$ is a 3×3 matrix. It is also a square matrix.

(c) $[1 \quad 6 \quad 5 \quad -2 \quad 5]$ is a 1×5 matrix. It is an example of a row matrix.

(d) $\begin{bmatrix} 3 \\ -5 \\ 0 \\ 2 \end{bmatrix}$ is a 4×1 column matrix. ◆

EXAMPLE 2 *Determining Equality of Matrices*

(a) If $A = \begin{bmatrix} 2 & 1 \\ p & q \end{bmatrix}$ and $B = \begin{bmatrix} x & y \\ -1 & 0 \end{bmatrix}$, find the values of x, y, p, and q such that $A = B$.

(b) Find the values of x and y, if possible, such that

$$\begin{bmatrix} x \\ y \end{bmatrix} = \begin{bmatrix} 1 \\ 4 \\ 0 \end{bmatrix}.$$

SOLUTION

(a) From the definition of equality given above, the only way that the statement

$$\begin{bmatrix} 2 & 1 \\ p & q \end{bmatrix} = \begin{bmatrix} x & y \\ -1 & 0 \end{bmatrix}$$

can be true is if $2 = x$, $1 = y$, $p = -1$, and $q = 0$.

(b) The statement

$$\begin{bmatrix} x \\ y \end{bmatrix} = \begin{bmatrix} 1 \\ 4 \\ 0 \end{bmatrix}$$

can never be true, since the two matrices have different dimensions. (One is 2×1 and the other is 3×1.) ◆

TECHNOLOGY NOTE

The remaining material in this section discusses the addition, subtraction, and multiplication of matrices. As explained in the text, these operations are defined only when there are appropriate restrictions on the dimensions of the matrices being added, subtracted, or multiplied. Graphing calculators will perform these operations provided the dimensions are compatible for the operation. A dimension error message will occur if the operation cannot be performed.

OPERATIONS ON MATRICES

At the beginning of this section, we used the matrix

$$\begin{bmatrix} 2 & 7 & 5 \\ 4 & 6 & 9 \end{bmatrix},$$

where the columns represent the numbers of cartons of three different types of vitamins (A, E, and K, respectively), and the rows represent two different manufacturers (Dexter and Sullivan, respectively). For example, the element 7 represents 7 cartons of vitamin E pills from Dexter, and so on. Suppose another shipment from these two suppliers is described by the following matrix.

$$\begin{bmatrix} 3 & 12 & 10 \\ 15 & 11 & 8 \end{bmatrix}$$

Here, for example, 8 cartons of vitamin K pills arrived from Sullivan. The total number of cartons of each kind of pill that were received from these two shipments can be found from these two matrices.

In the first shipment, 2 cartons of vitamin A pills were received from Dexter, and in the second shipment, 3 cartons of vitamin A pills were received from Dexter. Altogether, $2 + 3$, or 5, cartons of these pills were received. Corresponding elements can be added to find the total number of cartons of each type of pill received.

$$\begin{bmatrix} 2 & 7 & 5 \\ 4 & 6 & 9 \end{bmatrix} + \begin{bmatrix} 3 & 12 & 10 \\ 15 & 11 & 8 \end{bmatrix} = \begin{bmatrix} 2+3 & 7+12 & 5+10 \\ 4+15 & 6+11 & 9+8 \end{bmatrix}$$

$$= \begin{bmatrix} 5 & 19 & 15 \\ 19 & 17 & 17 \end{bmatrix}$$

The last matrix gives the total number of cartons of each type of pill that were received. For example, 15 cartons of vitamin K pills were received from Dexter. Generalizing from this example leads to the following definition.

```
[C]
     [[2 7 5]
      [4 6 9]]
[D]
     [[3 12 10]
      [15 11 8 ]]
```

```
[C]+[D]
     [[5  19 15]
      [19 17 17]]
```

These screens show the sum of the two matrices added in the text.

> **DEFINITION OF MATRIX ADDITION**
>
> The sum of two $m \times n$ matrices A and B is the $m \times n$ matrix $A + B$ in which each element is the sum of the corresponding elements of A and B.

Caution   Only matrices with the same dimension can be added.

EXAMPLE 3   *Adding Matrices*   Find the following sums.

(a) $\begin{bmatrix} 5 & -6 \\ 8 & 9 \end{bmatrix} + \begin{bmatrix} -4 & 6 \\ 8 & -3 \end{bmatrix}$

(b) $\begin{bmatrix} 2 \\ 5 \\ 8 \end{bmatrix} + \begin{bmatrix} -6 \\ 3 \\ 12 \end{bmatrix}$

(c) $A + B$, if $A = \begin{bmatrix} 5 & 8 \\ 6 & 2 \end{bmatrix}$ and $B = \begin{bmatrix} 3 & 9 & 1 \\ 4 & 2 & 5 \end{bmatrix}$

SOLUTION

(a) $\begin{bmatrix} 5 & -6 \\ 8 & 9 \end{bmatrix} + \begin{bmatrix} -4 & 6 \\ 8 & -3 \end{bmatrix} = \begin{bmatrix} 5 + (-4) & -6 + 6 \\ 8 + 8 & 9 + (-3) \end{bmatrix} = \begin{bmatrix} 1 & 0 \\ 16 & 6 \end{bmatrix}$

(b) $\begin{bmatrix} 2 \\ 5 \\ 8 \end{bmatrix} + \begin{bmatrix} -6 \\ 3 \\ 12 \end{bmatrix} = \begin{bmatrix} -4 \\ 8 \\ 20 \end{bmatrix}$

(c)   The matrices

$$A = \begin{bmatrix} 5 & 8 \\ 6 & 2 \end{bmatrix} \quad \text{and} \quad B = \begin{bmatrix} 3 & 9 & 1 \\ 4 & 2 & 5 \end{bmatrix}$$

have different dimensions. Therefore, the sum $A + B$ does not exist.   ◆

A matrix containing only zero elements is called a **zero matrix.** For example, $[0 \quad 0 \quad 0]$ is the 1×3 zero matrix, while

$$\begin{bmatrix} 0 & 0 & 0 \\ 0 & 0 & 0 \end{bmatrix}$$

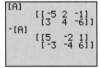

This screen supports the discussion regarding additive inverses of matrices.

is the 2×3 zero matrix. A zero matrix can be written with any dimension.

The additive inverse of a real number a is the unique real number $-a$ such that $a + (-a) = 0$ and $-a + a = 0$. Given a matrix A, a matrix $-A$ can be found such that $A + (-A) = O$, where O is the appropriate zero matrix, and $-A + A = O$. For example, if

$$A = \begin{bmatrix} -5 & 2 & -1 \\ 3 & 4 & -6 \end{bmatrix},$$

then the elements of matrix $-A$ are the additive inverses of the corresponding elements of A. (Remember that each element of A is a real number and thus has an additive inverse.)

$$-A = \begin{bmatrix} 5 & -2 & 1 \\ -3 & -4 & 6 \end{bmatrix}$$

To check, first test that $A + (-A)$ equals O, the appropriate zero matrix.

$$A + (-A) = \begin{bmatrix} -5 & 2 & -1 \\ 3 & 4 & -6 \end{bmatrix} + \begin{bmatrix} 5 & -2 & 1 \\ -3 & -4 & 6 \end{bmatrix}$$

$$= \begin{bmatrix} 0 & 0 & 0 \\ 0 & 0 & 0 \end{bmatrix} = O$$

Then, test that $-A + A$ is also O. Matrix $-A$ is the **additive inverse,** or **negative,** of matrix A. Every matrix has a unique additive inverse.

Just as subtraction of real numbers is defined in terms of the additive inverse, subtraction of matrices is defined in the same way.

DEFINITION OF MATRIX SUBTRACTION

If A and B are matrices with the same dimension, then

$$A - B = A + (-B).$$

EXAMPLE 4 *Subtracting Matrices* Find the following differences.

(a) $\begin{bmatrix} -5 & 6 \\ 2 & 4 \end{bmatrix} - \begin{bmatrix} -3 & 2 \\ 5 & -8 \end{bmatrix}$

(b) $\begin{bmatrix} 8 & 6 & -4 \end{bmatrix} - \begin{bmatrix} 3 & 5 & -8 \end{bmatrix}$

(c) $\begin{bmatrix} -2 & 5 \\ 0 & 1 \end{bmatrix} - \begin{bmatrix} 3 \\ 5 \end{bmatrix}$

SOLUTION

(a) $\begin{bmatrix} -5 & 6 \\ 2 & 4 \end{bmatrix} - \begin{bmatrix} -3 & 2 \\ 5 & -8 \end{bmatrix} = \begin{bmatrix} -5 & 6 \\ 2 & 4 \end{bmatrix} + \begin{bmatrix} 3 & -2 \\ -5 & 8 \end{bmatrix} = \begin{bmatrix} -2 & 4 \\ -3 & 12 \end{bmatrix}$

(b) $\begin{bmatrix} 8 & 6 & -4 \end{bmatrix} - \begin{bmatrix} 3 & 5 & -8 \end{bmatrix} = \begin{bmatrix} 5 & 1 & 4 \end{bmatrix}$

(c) The matrices

$$\begin{bmatrix} -2 & 5 \\ 0 & 1 \end{bmatrix} \quad \text{and} \quad \begin{bmatrix} 3 \\ 5 \end{bmatrix}$$

have different dimensions and cannot be subtracted. ◆

[C]
      [[-5 6]
       [2 4]]
[D]
      [[-3 2]
       [5 -8]]

[C]-[D]
      [[-2 4]
       [-3 12]]

These screens show the subtraction in Example 4(a).

If a matrix A is added to itself, each element in the sum is twice as large as the corresponding element of A. For example,

$$\begin{bmatrix} 2 & 5 \\ 1 & 3 \\ 4 & 6 \end{bmatrix} + \begin{bmatrix} 2 & 5 \\ 1 & 3 \\ 4 & 6 \end{bmatrix} = \begin{bmatrix} 4 & 10 \\ 2 & 6 \\ 8 & 12 \end{bmatrix} = 2\begin{bmatrix} 2 & 5 \\ 1 & 3 \\ 4 & 6 \end{bmatrix}.$$

In the last expression, the number 2 in front of the matrix is called a **scalar** to distinguish it from a matrix. (A scalar is just a special name for a real number.) The example above suggests the following definition of multiplication of a matrix by a scalar.

DEFINITION OF MULTIPLICATION OF A MATRIX BY A SCALAR

The product of a scalar k and a matrix X is the matrix kX, each of whose elements is k times the corresponding element of X.

EXAMPLE 5 *Multiplying a Matrix by a Scalar* Perform the following multiplications.

(a) $5\begin{bmatrix} 2 & -3 \\ 0 & 4 \end{bmatrix}$ **(b)** $\frac{3}{4}\begin{bmatrix} 20 & 36 \\ 12 & -16 \end{bmatrix}$

SOLUTION

(a) $5\begin{bmatrix} 2 & -3 \\ 0 & 4 \end{bmatrix} = \begin{bmatrix} 10 & -15 \\ 0 & 20 \end{bmatrix}$ **(b)** $\frac{3}{4}\begin{bmatrix} 20 & 36 \\ 12 & -16 \end{bmatrix} = \begin{bmatrix} 15 & 27 \\ 9 & -12 \end{bmatrix}$ ◆

5*[C]
[[10 -15]
[0 20]]

This screen supports the result of Example 5(a), where matrix C is $\begin{bmatrix} 2 & -3 \\ 0 & 4 \end{bmatrix}$.

In order to motivate the definition of multiplying a matrix by another matrix, let us return to the vitamin pills example. The matrix below shows the number of cartons of each type of vitamin received from Dexter and Sullivan, respectively.

$$\begin{bmatrix} 2 & 7 & 5 \\ 4 & 6 & 9 \end{bmatrix}$$

Now, suppose that each carton of vitamin A pills costs the store $12, each carton of vitamin E pills costs $18, and each carton of vitamin K pills costs $9. To find the total cost of the pills from Dexter, we multiply as follows.

Vitamin	Number of Cartons	Cost per Carton	Total Cost
A	2	$12	$ 24
E	7	$18	$126
K	5	$ 9	$ 45
			$195 Total from Dexter

The Dexter pills cost a total of $195.

This result is the sum of three products: $2(\$12) + 7(\$18) + 5(\$9) = \$195$. In the same way, using the second row of the matrix and the three costs shows us that the total cost of the Sullivan pills is $4(\$12) + 6(\$18) + 9(\$9) = \$237$. The costs, $12, $18, and $9, can be written as a column matrix.

$$\begin{bmatrix} 12 \\ 18 \\ 9 \end{bmatrix}$$

The total costs for each supplier, $195 and $237, also can be written as a column matrix.

$$\begin{bmatrix} 195 \\ 237 \end{bmatrix}$$

The product of the matrices

$$\begin{bmatrix} 2 & 7 & 5 \\ 4 & 6 & 9 \end{bmatrix} \quad \text{and} \quad \begin{bmatrix} 12 \\ 18 \\ 9 \end{bmatrix}$$

can be written as follows.

$$\begin{bmatrix} 2 & 7 & 5 \\ 4 & 6 & 9 \end{bmatrix}\begin{bmatrix} 12 \\ 18 \\ 9 \end{bmatrix} = \begin{bmatrix} 2 \cdot 12 + 7 \cdot 18 + 5 \cdot 9 \\ 4 \cdot 12 + 6 \cdot 18 + 9 \cdot 9 \end{bmatrix} = \begin{bmatrix} 195 \\ 237 \end{bmatrix}$$

Each element of the product was found by multiplying the elements of the *rows* of the matrix on the left and the corresponding elements of the *columns* of the matrix on the right, and then finding the sum of these products. Notice that the product of a 2×3 matrix and a 3×1 matrix is a 2×1 matrix.

Generalizing from this example gives the following definition of matrix multiplication.

DEFINITION OF MATRIX MULTIPLICATION

The product AB of an $m \times n$ matrix A and an $n \times k$ matrix B is found as follows. To get the ith row, jth column element of AB, multiply each element in the ith row of A by the corresponding element in the jth column of B. The sum of these products will give the element of row i, column j of AB. The dimension of AB is $m \times k$.

This definition requires the following restriction on matrix multiplication.

RESTRICTION ON MATRIX MULTIPLICATION

The product AB of two matrices A and B can be found only if the number of *columns* of A is the same as the number of *rows* of B. The final product will have as many rows as A and as many columns as B.

The next example illustrates how matrices are multiplied.

EXAMPLE 6 *Finding the Product of Two Matrices* Find the product.

$$\begin{bmatrix} -3 & 4 & 2 \\ 5 & 0 & 4 \end{bmatrix} \begin{bmatrix} -6 & 4 \\ 2 & 3 \\ 3 & -2 \end{bmatrix}$$

SOLUTION

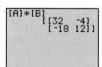

These screens support the result of Example 6.

Step 1 $\begin{bmatrix} -3 & 4 & 2 \\ 5 & 0 & 4 \end{bmatrix} \begin{bmatrix} -6 & 4 \\ 2 & 3 \\ 3 & -2 \end{bmatrix}$ $(-3)(-6) + 4(2) + 2(3) = 32$

Step 2 $\begin{bmatrix} -3 & 4 & 2 \\ 5 & 0 & 4 \end{bmatrix} \begin{bmatrix} -6 & 4 \\ 2 & 3 \\ 3 & -2 \end{bmatrix}$ $(-3)(4) + 4(3) + 2(-2) = -4$

Step 3 $\begin{bmatrix} -3 & 4 & 2 \\ 5 & 0 & 4 \end{bmatrix} \begin{bmatrix} -6 & 4 \\ 2 & 3 \\ 3 & -2 \end{bmatrix}$ $5(-6) + 0(2) + 4(3) = -18$

Step 4 $\begin{bmatrix} -3 & 4 & 2 \\ 5 & 0 & 4 \end{bmatrix} \begin{bmatrix} -6 & 4 \\ 2 & 3 \\ 3 & -2 \end{bmatrix}$ $5(4) + 0(3) + 4(-2) = 12$

Step 5 Write the product.

$$\begin{bmatrix} -3 & 4 & 2 \\ 5 & 0 & 4 \end{bmatrix} \begin{bmatrix} -6 & 4 \\ 2 & 3 \\ 3 & -2 \end{bmatrix} = \begin{bmatrix} 32 & -4 \\ -18 & 12 \end{bmatrix}$$

As this example shows, the product of a 2×3 matrix and a 3×2 matrix is a 2×2 matrix. ◆

EXAMPLE 7   *Deciding Whether Two Matrices Can be Multiplied*   Suppose matrix A is 2×2, while matrix B is 2×4. Can the product AB be calculated? What is the dimension of the product?

SOLUTION   The following diagram helps answer these questions.

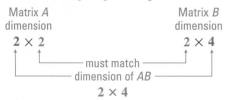

The product AB can be calculated because A has two columns and B has two rows. The dimension of the product is 2×4. (However, the product BA could not be found.)   ◆

EXAMPLE 8   *Multiplying Matrices in Different Orders*   If possible, find AB and BA, where

$$A = \begin{bmatrix} 1 & -3 \\ 7 & 2 \end{bmatrix} \quad \text{and} \quad B = \begin{bmatrix} 1 & 0 & -1 & 2 \\ 3 & 1 & 4 & -1 \end{bmatrix}.$$

SOLUTION   Use the definition of matrix multiplication to find AB.

$$\begin{aligned} AB &= \begin{bmatrix} 1 & -3 \\ 7 & 2 \end{bmatrix} \begin{bmatrix} 1 & 0 & -1 & 2 \\ 3 & 1 & 4 & -1 \end{bmatrix} \\ &= \begin{bmatrix} 1(1) + (-3)3 & 1(0) + (-3)1 & 1(-1) + (-3)4 & 1(2) + (-3)(-1) \\ 7(1) + 2(3) & 7(0) + 2(1) & 7(-1) + 2(4) & 7(2) + 2(-1) \end{bmatrix} \\ &= \begin{bmatrix} -8 & -3 & -13 & 5 \\ 13 & 2 & 1 & 12 \end{bmatrix} \end{aligned}$$

Since B is a 2×4 matrix, and A is a 2×2 matrix, the product BA cannot be found.   ◆

EXAMPLE 9   *Multiplying Square Matrices in Different Orders*   Find MN and NM, given

$$M = \begin{bmatrix} 1 & 3 \\ -2 & 4 \end{bmatrix} \quad \text{and} \quad N = \begin{bmatrix} 2 & 5 \\ 10 & -3 \end{bmatrix}.$$

SOLUTION   By the definition of matrix multiplication,

$$\begin{aligned} MN &= \begin{bmatrix} 1 & 3 \\ -2 & 4 \end{bmatrix} \begin{bmatrix} 2 & 5 \\ 10 & -3 \end{bmatrix} \\ &= \begin{bmatrix} 2 + 30 & 5 - 9 \\ -4 + 40 & -10 - 12 \end{bmatrix} \\ &= \begin{bmatrix} 32 & -4 \\ 36 & -22 \end{bmatrix}. \end{aligned}$$

Similarly,

$$\begin{aligned} NM &= \begin{bmatrix} 2 & 5 \\ 10 & -3 \end{bmatrix} \begin{bmatrix} 1 & 3 \\ -2 & 4 \end{bmatrix} \\ &= \begin{bmatrix} 2 - 10 & 6 + 20 \\ 10 + 6 & 30 - 12 \end{bmatrix} \\ &= \begin{bmatrix} -8 & 26 \\ 16 & 18 \end{bmatrix}. \end{aligned}$$ ◆

In Example 8, the product AB could be found, but not BA. In Example 9, although both MN and NM could be found, they were not equal, showing that multiplication of matrices is not commutative. This fact distinguishes matrix arithmetic from the arithmetic of real numbers.

Caution Since multiplication of square matrices is, in general, not commutative, be careful about the order when you multiply matrices.

After reading Examples 3–9, you may wish to experiment with your calculator to see if you can duplicate the results obtained.

AN APPLICATION OF MATRIX ALGEBRA

EXAMPLE 10 *Using Matrix Multiplication to Determine Costs and Amounts of Materials* A contractor builds three kinds of houses, models A, B, and C, with a choice of two styles, colonial or ranch. Matrix P below shows the number of each kind of house the contractor is planning to build for a new 100-home subdivision. The amounts for each of the main materials used depend on the style of the house. These amounts are shown in matrix Q below, while matrix R gives the cost in dollars for each kind of material. Concrete is measured here in cubic yards, lumber in 1000 board feet, brick in 1000s, and shingles in 100 square feet.

$$
\begin{array}{c}
\phantom{\text{Model A}} \\[2pt]
\text{Model A} \\
\text{Model B} \\
\text{Model C}
\end{array}
\begin{array}{c}
\text{Colonial} \quad \text{Ranch} \\
\begin{bmatrix} 0 & 30 \\ 10 & 20 \\ 20 & 20 \end{bmatrix} = P
\end{array}
$$

$$
\begin{array}{c}
 \\
\text{Colonial} \\
\text{Ranch}
\end{array}
\begin{array}{c}
\text{Concrete} \quad \text{Lumber} \quad \text{Brick} \quad \text{Shingles} \\
\begin{bmatrix} 10 & 2 & 0 & 2 \\ 50 & 1 & 20 & 2 \end{bmatrix} = Q
\end{array}
\qquad
\begin{array}{c}
\text{Cost per} \\
\text{Unit} \\[2pt]
\begin{array}{c}
\text{Concrete} \\
\text{Lumber} \\
\text{Brick} \\
\text{Shingles}
\end{array}
\begin{bmatrix} 20 \\ 180 \\ 60 \\ 25 \end{bmatrix} = R
\end{array}
$$

(a) What is the total cost of materials for all houses of each model?

(b) How much of each of the four kinds of material must be ordered?

(c) What is the total cost of the materials?

SOLUTION

(a) To find the materials cost for each model, first find matrix PQ, which will show the total amount of each material needed for all houses of each model.

$$
PQ = \begin{bmatrix} 0 & 30 \\ 10 & 20 \\ 20 & 20 \end{bmatrix} \begin{bmatrix} 10 & 2 & 0 & 2 \\ 50 & 1 & 20 & 2 \end{bmatrix}
$$

$$
= \begin{array}{c}
\text{Concrete} \quad \text{Lumber} \quad \text{Brick} \quad \text{Shingles} \\
\begin{bmatrix} 1500 & 30 & 600 & 60 \\ 1100 & 40 & 400 & 60 \\ 1200 & 60 & 400 & 80 \end{bmatrix}
\begin{array}{l} \text{Model A} \\ \text{Model B} \\ \text{Model C} \end{array}
\end{array}
$$

Multiplying PQ and the cost matrix R gives the total cost of materials for each model.

$$
(PQ)R = \begin{bmatrix} 1500 & 30 & 600 & 60 \\ 1100 & 40 & 400 & 60 \\ 1200 & 60 & 400 & 80 \end{bmatrix} \begin{bmatrix} 20 \\ 180 \\ 60 \\ 25 \end{bmatrix} = \begin{bmatrix} 72{,}900 \\ 54{,}700 \\ 60{,}800 \end{bmatrix} \begin{array}{l} \text{Model A} \\ \text{Model B} \\ \text{Model C} \end{array}
$$

(b) The totals of the columns of matrix PQ will give a matrix whose elements represent the total amounts of each material needed for the subdivision. Call this matrix T, and write it as a row matrix.

$$T = [3800 \quad 130 \quad 1400 \quad 200]$$

(c) The total cost of all the materials is given by the product of matrix R, the cost matrix, and matrix T, the total amounts matrix. To multiply these and get a 1×1 matrix, representing the total cost, requires multiplying a 1×4 matrix and a 4×1 matrix. This is why in (b) a row matrix was written rather than a column matrix. The total materials cost is given by TR, so

$$TR = [3800 \quad 130 \quad 1400 \quad 200] \begin{bmatrix} 20 \\ 180 \\ 60 \\ 25 \end{bmatrix} = [188,400].$$

The total cost of the materials is $188,400. ◆

7.4 EXERCISES   Tape 8

1. What is the maximum dimension for a matrix allowed by your particular calculator?

2. What condition is necessary for a matrix to be a square matrix? A zero matrix?

Find the dimension of each of the following matrices. Identify any square, column, or row matrices.

3. $\begin{bmatrix} -4 & 8 \\ 2 & 3 \end{bmatrix}$

4. $\begin{bmatrix} -9 & 6 & 2 \\ 4 & 1 & 8 \end{bmatrix}$

5. $\begin{bmatrix} -6 & 8 & 0 & 0 \\ 4 & 1 & 9 & 2 \\ 3 & -5 & 7 & 1 \end{bmatrix}$

6. $[8 \quad -2 \quad 4 \quad 6 \quad 3]$

7. $\begin{bmatrix} 2 \\ 4 \end{bmatrix}$

8. $[-9]$

9. $\begin{bmatrix} -4 & 2 & 3 \\ -8 & 2 & 1 \\ 4 & 6 & 8 \end{bmatrix}$

10. $\begin{bmatrix} -4 & 2 \\ 3 & 5 \end{bmatrix}$

Find the values of the variables in each of the following matrices.

11. $\begin{bmatrix} x+6 & y+2 \\ 8 & 3 \end{bmatrix} = \begin{bmatrix} -9 & 7 \\ 8 & k \end{bmatrix}$

12. $\begin{bmatrix} 9 & 7 \\ r & 0 \end{bmatrix} = \begin{bmatrix} m-3 & n+5 \\ 8 & 0 \end{bmatrix}$

13. $\begin{bmatrix} -7+z & 4r & 8s \\ 6p & 2 & 5 \end{bmatrix} + \begin{bmatrix} -9 & 8r & 3 \\ 2 & 5 & 4 \end{bmatrix} = \begin{bmatrix} 2 & 36 & 27 \\ 20 & 7 & 12a \end{bmatrix}$

14. $\begin{bmatrix} a+2 & 3z+1 & 5m \\ 4k & 0 & 3 \end{bmatrix} + \begin{bmatrix} 3a & 2z & 5m \\ 2k & 5 & 6 \end{bmatrix} = \begin{bmatrix} 10 & -14 & 80 \\ 10 & 5 & 9 \end{bmatrix}$

15. Your friend missed the lecture on adding matrices. In your own words, explain to him how to add two matrices.

16. Explain to a friend in your own words how to subtract two matrices.

Perform each of the following operations, whenever possible.

17. $3\begin{bmatrix} 6 & -1 & 4 \\ 2 & 8 & -3 \\ -4 & 5 & 6 \end{bmatrix} + 5\begin{bmatrix} -2 & -8 & -6 \\ 4 & 1 & 3 \\ 2 & -1 & 5 \end{bmatrix}$

18. $4\begin{bmatrix} 1 & -4 \\ 2 & -3 \\ -8 & 4 \end{bmatrix} - 3\begin{bmatrix} -6 & 9 \\ -2 & 5 \\ -7 & -12 \end{bmatrix}$

19. $\begin{bmatrix} -8 & 4 & 0 \\ 2 & 5 & 0 \end{bmatrix} + \begin{bmatrix} 6 & 3 \\ 8 & 9 \end{bmatrix}$

20. $\begin{bmatrix} 2 \\ 3 \end{bmatrix} - \begin{bmatrix} 8 & 1 \\ 9 & 4 \end{bmatrix}$

21. $\begin{bmatrix} 9 & 4 & 1 & -2 \\ 5 & -6 & 3 & 4 \\ 2 & -5 & 1 & 2 \end{bmatrix} - \begin{bmatrix} -2 & 5 & 1 & 3 \\ 0 & 1 & 0 & 2 \\ -8 & 3 & 2 & 1 \end{bmatrix} + \begin{bmatrix} 2 & 4 & 0 & 3 \\ 4 & -5 & 1 & 6 \\ 2 & -3 & 0 & 8 \end{bmatrix}$

22. $\begin{bmatrix} 6 & -2 & 4 \\ -2 & 5 & 8 \\ 1 & 0 & 2 \end{bmatrix} + \begin{bmatrix} 3 & 0 & 8 \\ 1 & -2 & 4 \\ 6 & 9 & -2 \end{bmatrix} - \begin{bmatrix} -4 & 2 & 1 \\ 0 & 3 & -2 \\ 4 & 2 & 0 \end{bmatrix}$

23. $\begin{bmatrix} -4x + 2y & -3x + y \\ 6x - 3y & 2x - 5y \end{bmatrix} + \begin{bmatrix} -8x + 6y & 2x \\ 3y - 5x & 6x + 4y \end{bmatrix}$

24. $\begin{bmatrix} 4k - 8y \\ 6z - 3x \\ 2k + 5a \\ -4m + 2n \end{bmatrix} - \begin{bmatrix} 5k + 6y \\ 2z + 5x \\ 4k + 6a \\ 4m - 2n \end{bmatrix}$

Solve each problem.

25. *Stock Investments* David Cundiff bought 7 shares of Shell stock, 9 shares of Texaco stock, and 8 shares of BP stock. The following month, he bought 2 shares of Shell stock, no Texaco, and 6 shares of BP. Write this information first as a 3×2 matrix and then as a 2×3 matrix.

26. *Computer Sales* Margie Bezzone works in a computer store. The first week she sold 5 computers, 3 printers, 4 disk drives, and 6 monitors. The next week she sold 4 computers, 2 printers, 6 disk drives, and 5 monitors. Write this information first as a 2×4 matrix and then as a 4×2 matrix.

27. *Comparing Miles Driven by Men and Women* A study revealed that the average number of miles driven in the United States has been increasing steadily, with miles driven by women increasing much faster than miles driven by men. In 1969, women drove 5411 thousand miles and men drove 11,352 thousand miles. In 1990, women drove 9371 thousand miles and men drove 15,956 thousand miles. Write this information as a 2×2 matrix in two ways.

28. *Chinese Population Policies* The proportion of the population of China living in cities was slowed for a time by government-imposed birth-control policies. The urban population proportion increased again in later years. In 1952, the proportion was 12.5%; in 1960, 19.7%; in 1975, 12.1%; and in 1985, 19.7%. Write this information as a row matrix and as a column matrix.

Let $A = \begin{bmatrix} -2 & 4 \\ 0 & 3 \end{bmatrix}$ *and* $B = \begin{bmatrix} -6 & 2 \\ 4 & 0 \end{bmatrix}$. *Perform the operations to find the following matrices.*

29. $2A$

30. $-3B$

31. $-4A$

32. $5B$

33. $2A - B$

34. $-4A + 5B$

35. $3A - 11B$

36. $-2A + 4B$

37. Based on the screen shown here, what is matrix A?

```
[B]
     [[4  6  -5]
      [-6  3  2 ]]
[A]+[B]
     [[6   12  0 ]
      [-10 -4 11]]
```

38. Based on the screen shown here, what is matrix B?

```
[A]
     [[3  6  5]
      [-2 1  4]]
[A]-[B]
     [[9  0  -5]
      [-4 6  -3]]
```

The dimensions of matrices A and B are given. Find the dimensions of the product AB and of the product BA if the products are defined. If they are not defined, say so.

39. A is 4×2, B is 2×4

40. A is 3×1, B is 1×3

41. A is 3×5, B is 5×2

42. A is 7×3, B is 2×7

43. A is 4×3, B is 2×5

44. A is 1×6, B is 2×4

45. The product MN of two matrices can be found only if the number of _____ of M equals the number of _____ of N.

46. True or false: For matrices A and B, if AB can be found, then BA can always be found, too.

47. To find the product AB of matrices A and B, the first row, second column entry is found by multiplying the _____ elements in A and the _____ elements in B and then _____ these products.

48. If a matrix is multiplied by a zero matrix of appropriate dimension, what must the product be?

Find each of the following matrix products, whenever possible.

49. $\begin{bmatrix} p & q \\ r & s \end{bmatrix} \begin{bmatrix} a & c \\ b & d \end{bmatrix}$

50. $\begin{bmatrix} a & b & c \\ d & e & f \\ g & h & i \end{bmatrix} \begin{bmatrix} x \\ y \\ z \end{bmatrix}$

51. $\begin{bmatrix} 3 & -4 & 1 \\ 5 & 0 & 2 \end{bmatrix} \begin{bmatrix} -1 \\ 4 \\ 2 \end{bmatrix}$

52. $\begin{bmatrix} -6 & 3 & 5 \\ 2 & 9 & 1 \end{bmatrix} \begin{bmatrix} -2 \\ 0 \\ 3 \end{bmatrix}$

53. $\begin{bmatrix} 5 & 2 \\ -1 & 4 \end{bmatrix} \begin{bmatrix} 3 & -2 \\ 1 & 0 \end{bmatrix}$

54. $\begin{bmatrix} -4 & 0 \\ 1 & 3 \end{bmatrix} \begin{bmatrix} -2 & 4 \\ 0 & 1 \end{bmatrix}$

55. $\begin{bmatrix} 2 & 2 & -1 \\ 3 & 0 & 1 \end{bmatrix} \begin{bmatrix} 0 & 2 \\ -1 & 4 \\ 0 & 2 \end{bmatrix}$

56. $\begin{bmatrix} -9 & 2 & 1 \\ 3 & 0 & 0 \end{bmatrix} \begin{bmatrix} 2 \\ -1 \\ 4 \end{bmatrix}$

57. $\begin{bmatrix} -1 & 2 & 0 \\ 0 & 3 & 2 \\ 0 & 1 & 4 \end{bmatrix} \begin{bmatrix} 2 & -1 & 2 \\ 0 & 2 & 1 \\ 3 & 0 & -1 \end{bmatrix}$

58. $\begin{bmatrix} -2 & -3 & -4 \\ 2 & -1 & 0 \\ 4 & -2 & 3 \end{bmatrix} \begin{bmatrix} 0 & 1 & 4 \\ 1 & 2 & -1 \\ 3 & 2 & -2 \end{bmatrix}$

59. $[-2 \ \ 4 \ \ 1] \begin{bmatrix} 3 & -2 & 4 \\ 2 & 1 & 0 \\ 0 & -1 & 4 \end{bmatrix}$

60. $[0 \ \ 3 \ \ -4] \begin{bmatrix} -2 & 6 & 3 \\ 0 & 4 & 2 \\ -1 & 1 & 4 \end{bmatrix}$

61. $\begin{bmatrix} -2 & 1 & 4 \\ 0 & 1 & 2 \end{bmatrix} \begin{bmatrix} -2 & 1 & 0 \\ 0 & -2 & 0 \\ 4 & 1 & 2 \end{bmatrix}$

62. $\begin{bmatrix} -1 & 0 & 0 \\ 0 & 1 & 4 \\ 2 & 1 & 4 \end{bmatrix} \begin{bmatrix} 4 & -2 & 5 \\ 0 & 1 & 4 \\ 2 & -9 & 0 \end{bmatrix}$

63. $\begin{bmatrix} -3 & 0 & 2 & 1 \\ 4 & 0 & 2 & 6 \end{bmatrix} \begin{bmatrix} -4 & 2 \\ 0 & 1 \end{bmatrix}$

64. $\begin{bmatrix} -1 & 2 & 4 & 1 \\ 0 & 2 & -3 & 5 \end{bmatrix} \begin{bmatrix} 1 & 2 & 4 \\ -2 & 5 & 1 \end{bmatrix}$

65. $[-2 \ \ 4 \ \ 6] \begin{bmatrix} 3 \\ -2 \\ 1 \end{bmatrix}$

66. $[4 \ \ 0 \ \ 2] \begin{bmatrix} -5 \\ 1 \\ 6 \end{bmatrix}$

67. $\begin{bmatrix} 3 \\ -2 \\ 1 \end{bmatrix} [-2 \ \ 4 \ \ 6]$

68. $\begin{bmatrix} -5 \\ 1 \\ 6 \end{bmatrix} [4 \ \ 0 \ \ 2]$

Let $A = \begin{bmatrix} -2 & 4 \\ 1 & 3 \end{bmatrix}$, $B = \begin{bmatrix} -2 & 1 \\ 3 & 6 \end{bmatrix}$, *and* $C = \begin{bmatrix} 5 & -2 & 1 \\ 0 & 3 & 7 \end{bmatrix}$. *Find each of the following products.*

69. AB **70.** BA **71.** AC **72.** CA

73. Did you get the same answer in Exercises 69 and 70? What about Exercises 71 and 72? Do you think that matrix multiplication is commutative?

74. For any matrices P and Q, what must be true for both PQ and QP to exist?

Solve each problem.

75. *Income from Sales* Yummy Yogurt sells three types of yogurt: nonfat, regular, and super creamy at three locations. Location I sells 50 gallons of nonfat, 100 gallons of regular, and 30 gallons of super creamy each day. Location II sells 10 gallons of nonfat, and Location III sells 60 gallons of nonfat each day. Daily sales of regular yogurt are 90 gallons at Location II and 120 gallons at Location III. At Location II, 50 gallons of super creamy are sold each day, and 40 gallons of super creamy are sold each day at Location III.

(a) Write a 3×3 matrix that shows the sales figures for the three locations.

(b) The income per gallon for nonfat, regular, and super creamy is $12, $10, and $15, respectively. Write a 1×3 or 3×1 matrix displaying the income.

(c) Find a matrix product that gives the daily income at each of the three locations.

(d) What is Yummy Yogurt's total daily income from the three locations?

76. *Purchasing Costs* The Bread Box, a small neighborhood bakery, sells four main items: sweet rolls, bread, cakes, and pies. The amount of each ingredient (in cups, except for eggs) required for these items is given by matrix A.

	Eggs	Flour	Sugar	Shortening	Milk
Rolls (doz)	1	4	$\frac{1}{4}$	$\frac{1}{4}$	1
Bread (loaves)	0	3	0	$\frac{1}{4}$	0
Cakes	4	3	2	1	1
Pies (crust)	0	1	0	$\frac{1}{3}$	0

$= A$

The cost (in cents) for each ingredient when purchased in large lots or small lots is given in matrix B.

	Large lot	Small lot
Eggs	5	5
Flour	8	10
Sugar	10	12
Shortening	12	15
Milk	5	6

$= B$

(a) Use matrix multiplication to find a matrix giving the comparative cost per item for the two purchase options.

(b) Suppose a day's orders consist of 20 dozen sweet rolls, 200 loaves of bread, 50 cakes, and 60 pies. Write the orders as a 1×4 matrix and, using matrix multiplication, write as a matrix the amount of each ingredient needed to fill the day's orders.

(c) Use matrix multiplication to find a matrix giving the costs under the two purchase options to fill the day's orders.

For the following exercises, let $A = \begin{bmatrix} a & b \\ c & d \end{bmatrix}$, $B = \begin{bmatrix} e & f \\ g & h \end{bmatrix}$, and $C = \begin{bmatrix} j & m \\ k & n \end{bmatrix}$. Decide which of the following statements are true for these three matrices. Then, make a conjecture as to whether a similar property holds for any square matrices of the same order.

77. $(AB)C = A(BC)$ (associative property)

78. $A(B + C) = AB + AC$ (distributive property)

79. $k(A + B) = kA + kB$ for any real number k

80. $(k + p)A = kA + pA$ for any real numbers k and p

7.5 DETERMINANTS AND CRAMER'S RULE

Determinants of Matrices of Dimension 2×2 ◆ Determinants of Larger Matrices ◆ Derivation of Cramer's Rule ◆ Using Cramer's Rule to Solve Systems

It is often convenient to use subscript notation to name the elements of a matrix as in the following matrix A.

$$A = \begin{bmatrix} a_{11} & a_{12} & a_{13} & \cdots & a_{1n} \\ a_{21} & a_{22} & a_{23} & \cdots & a_{2n} \\ a_{31} & a_{32} & a_{33} & \cdots & a_{3n} \\ \vdots & \vdots & \vdots & & \vdots \\ a_{m1} & a_{m2} & a_{m3} & \cdots & a_{mn} \end{bmatrix}$$

With this notation, the row 1, column 1 element is a_{11}; the row 2, column 3 element is a_{23}; and, in general, the row i, column j element is a_{ij}. Subscript notation is used in this section to define determinants.

DETERMINANTS OF MATRICES OF DIMENSION 2×2

Associated with every square matrix A is a real number called the **determinant** of A. There are several symbols used to represent the determinant of A, including $|A|$, $\delta(A)$, and det A. In this text, we will use det A.

DEFINITION OF THE DETERMINANT OF A 2 × 2 MATRIX

The **determinant of a 2 × 2 matrix** A,

$$A = \begin{bmatrix} a_{11} & a_{12} \\ a_{21} & a_{22} \end{bmatrix},$$

is defined as

$$\det A = a_{11}a_{22} - a_{21}a_{12}.$$

Note   Be able to distinguish between a matrix and its determinant. A matrix is an array of numbers, while a determinant is a real number associated with a square matrix.

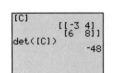

In this screen, the symbol

det [C]

represents the determinant of matrix C. With a graphing calculator, we can define a matrix and then use the capability of the calculator to find the determinant of the matrix. Compare the result here to the one in Example 1.

EXAMPLE 1   *Evaluating the Determinant of a Dimension 2 × 2 Matrix*   If $C = \begin{bmatrix} -3 & 4 \\ 6 & 8 \end{bmatrix}$, find det C.

SOLUTION $\det C = \det \begin{bmatrix} -3 & 4 \\ 6 & 8 \end{bmatrix} = -3(8) - 6(4) = -48.$ ◆

EXAMPLE 2   *Solving an Equation Involving a Determinant*   If $A = \begin{bmatrix} x & 3 \\ -1 & 5 \end{bmatrix}$ and det $A = 33$, find the value of x.

SOLUTION   Since $\det A = 5x - (-3)$, or $5x + 3$, we have

$$5x + 3 = 33$$
$$5x = 30$$
$$x = 6.$$

Check to see that $\det \begin{bmatrix} 6 & 3 \\ -1 & 5 \end{bmatrix} = 33.$ ◆

DETERMINANTS OF LARGER MATRICES

We now investigate determinants of larger matrices, beginning with dimension 3 × 3.

DEFINITION OF THE DETERMINANT OF A 3 × 3 MATRIX

The **determinant of a square matrix** A of dimension 3 × 3,

$$A = \begin{bmatrix} a_{11} & a_{12} & a_{13} \\ a_{21} & a_{22} & a_{23} \\ a_{31} & a_{32} & a_{33} \end{bmatrix},$$

is defined as

$$\det A = \det \begin{bmatrix} a_{11} & a_{12} & a_{13} \\ a_{21} & a_{22} & a_{23} \\ a_{31} & a_{32} & a_{33} \end{bmatrix} = (a_{11}a_{22}a_{33} + a_{12}a_{23}a_{31} + a_{13}a_{21}a_{32})$$
$$- (a_{31}a_{22}a_{13} + a_{32}a_{23}a_{11} + a_{33}a_{21}a_{12}).$$

An easy method for calculating dimension 3×3 determinants is found by rearranging and factoring the terms given above to get

$$\det\begin{bmatrix} a_{11} & a_{12} & a_{13} \\ a_{21} & a_{22} & a_{23} \\ a_{31} & a_{32} & a_{33} \end{bmatrix} = a_{11}(a_{22}a_{33} - a_{32}a_{23}) - a_{21}(a_{12}a_{33} - a_{32}a_{13})$$

$$+ a_{31}(a_{12}a_{23} - a_{22}a_{13}).$$

Each of the quantities in parentheses represents the determinant of a dimension 2×2 matrix that is the part of the dimension 3×3 matrix remaining when the row and column of the multiplier are eliminated, as shown below.

$$a_{11}(a_{22}a_{33} - a_{32}a_{23}) \quad \begin{bmatrix} a_{11} & a_{12} & a_{13} \\ a_{21} & a_{22} & a_{23} \\ a_{31} & a_{32} & a_{33} \end{bmatrix}$$

$$a_{21}(a_{12}a_{33} - a_{32}a_{13}) \quad \begin{bmatrix} a_{11} & a_{12} & a_{13} \\ a_{21} & a_{22} & a_{23} \\ a_{31} & a_{32} & a_{33} \end{bmatrix}$$

$$a_{31}(a_{12}a_{23} - a_{22}a_{13}) \quad \begin{bmatrix} a_{11} & a_{12} & a_{13} \\ a_{21} & a_{22} & a_{23} \\ a_{31} & a_{32} & a_{33} \end{bmatrix}$$

These determinants of the dimension 3×3 matrices are called **minors** of an element in the dimension 3×3 matrix. The symbol M_{ij} represents the minor that results when row i and column j are eliminated. The following list gives some of the minors from the matrix above.

Element	Minor	Element	Minor
a_{11}	$M_{11} = \det\begin{bmatrix} a_{22} & a_{23} \\ a_{32} & a_{33} \end{bmatrix}$	a_{22}	$M_{22} = \det\begin{bmatrix} a_{11} & a_{13} \\ a_{31} & a_{33} \end{bmatrix}$
a_{21}	$M_{21} = \det\begin{bmatrix} a_{12} & a_{13} \\ a_{32} & a_{33} \end{bmatrix}$	a_{23}	$M_{23} = \det\begin{bmatrix} a_{11} & a_{12} \\ a_{31} & a_{32} \end{bmatrix}$
a_{31}	$M_{31} = \det\begin{bmatrix} a_{12} & a_{13} \\ a_{22} & a_{23} \end{bmatrix}$	a_{33}	$M_{33} = \det\begin{bmatrix} a_{11} & a_{12} \\ a_{21} & a_{22} \end{bmatrix}$

In a matrix of dimension 4×4, the minors are determinants of dimension 3×3 matrices. Similarly, a dimension $n \times n$ matrix has minors that are determinants of dimension $(n - 1) \times (n - 1)$ matrices.

To find the determinant of a dimension 3×3 or larger square matrix, first choose any row or column. Then, the minor of each element in that row or column must be multiplied by $+1$ or -1, depending on whether the sum of the row numbers and column numbers is even or odd. The product of a minor and the number $+1$ or -1 is called a *cofactor*.

DEFINITION OF COFACTOR

Let M_{ij} be the minor for element a_{ij} in a dimension $n \times n$ matrix. The **cofactor** of a_{ij}, written A_{ij}, is

$$A_{ij} = (-1)^{i+j} \cdot M_{ij}.$$

Finally, the determinant of a dimension $n \times n$ matrix is found as follows.

> ### FINDING THE DETERMINANT OF A MATRIX
>
> Multiply each element in any row or column of the matrix by its cofactor. The sum of these products gives the value of the determinant.

The process of forming this sum of products is called **expansion by a given row or column.**

EXAMPLE 3 *Finding the Cofactor of an Element*   For the matrix

$$\begin{bmatrix} 6 & 2 & 4 \\ 8 & 9 & 3 \\ 1 & 2 & 0 \end{bmatrix},$$

find the cofactor of each of the following elements.

(a) 6   **(b)** 3   **(c)** 8

SOLUTION

(a)   Since 6 is in the first row and first column of the matrix, $i = 1$ and $j = 1$.

$$M_{11} = \det \begin{bmatrix} 9 & 3 \\ 2 & 0 \end{bmatrix} = -6.$$ The cofactor is $(-1)^{1+1} \cdot -6 = 1 \cdot -6 = -6$.

(b)   Here $i = 2$ and $j = 3$. $M_{23} = \det \begin{bmatrix} 6 & 2 \\ 1 & 2 \end{bmatrix} = 10$. The cofactor is $(-1)^{2+3} \cdot 10 = -1 \cdot 10 = -10$.

(c)   We have $i = 2$ and $j = 1$. $M_{21} = \det \begin{bmatrix} 2 & 4 \\ 2 & 0 \end{bmatrix} = -8$. The cofactor here is $(-1)^{2+1} \cdot -8 = -1 \cdot -8 = 8$.   ◆

EXAMPLE 4 *Evaluating the Determinant of a Dimension 3 × 3 Matrix*

Evaluate $\det \begin{bmatrix} 2 & -3 & -2 \\ -1 & -4 & -3 \\ -1 & 0 & 2 \end{bmatrix}$, expanding by the second column.

SOLUTION   To find this determinant, first get the minors of each element in the second column.

$$M_{12} = \det \begin{bmatrix} -1 & -3 \\ -1 & 2 \end{bmatrix} = -1(2) - (-1)(-3) = -5$$

$$M_{22} = \det \begin{bmatrix} 2 & -2 \\ -1 & 2 \end{bmatrix} = 2(2) - (-1)(-2) = 2$$

$$M_{32} = \det \begin{bmatrix} 2 & -2 \\ -1 & -3 \end{bmatrix} = 2(-3) - (-1)(-2) = -8$$

Now, find the cofactor of each of these minors.

$$A_{12} = (-1)^{1+2} \cdot M_{12} = (-1)^3 \cdot (-5) = (-1)(-5) = 5$$
$$A_{22} = (-1)^{2+2} \cdot M_{22} = (-1)^4 \cdot 2 = 1 \cdot 2 = 2$$
$$A_{32} = (-1)^{3+2} \cdot M_{32} = (-1)^5 \cdot (-8) = (-1)(-8) = 8$$

The determinant is found by multiplying each cofactor by its corresponding element in the matrix and finding the sum of these products.

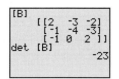

The result in Example 4 is supported with a graphing calculator.

$$\det \begin{bmatrix} 2 & -3 & -2 \\ -1 & -4 & -3 \\ -1 & 0 & 2 \end{bmatrix} = a_{12} \cdot A_{12} + a_{22} \cdot A_{22} + a_{32} \cdot A_{32}$$

$$= -3(5) + (-4)(2) + (0)(8)$$

$$= -15 + (-8) + 0 = -23 \qquad \blacklozenge$$

Caution Be very careful to keep track of all negative signs when evaluating determinants by hand. Work carefully, writing down each step as in the examples. Skipping steps frequently leads to errors in these computations.

Exactly the same answer would be found using any row or column of the matrix. One reason that column 2 was used in Example 4 is that it contains a 0 element, so that it was not really necessary to calculate M_{32} and A_{32} above. One learns quickly that zeros can be very useful in working with determinants.

Instead of calculating $(-1)^{i+j}$ for a given element, the following sign checker-boards can be used.

For Dimension 3 × 3 Matrices	For Dimension 4 × 4 Matrices
+ − + − + − + − +	+ − + − − + − + + − + − − + − +

The signs alternate for each row and column, beginning with + in the first row, first column position. Thus, these arrays of signs can be reproduced as needed. If we expand a square matrix of dimension 3 × 3 about row 3, for example, the first minor would have a + sign associated with it, the second minor a − sign, and the third minor a + sign. These arrays of signs can be extended in this way for determinants of square matrices of dimension 5 × 5 and greater.

EXAMPLE 5 *Evaluating the Determinant of a Dimension 4 × 4 Square Matrix* Evaluate

$$\det \begin{bmatrix} -1 & -2 & 3 & 2 \\ 0 & 1 & 4 & -2 \\ 3 & -1 & 4 & 0 \\ 2 & 1 & 0 & 3 \end{bmatrix}.$$

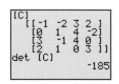

The result in Example 5 is supported with a graphing calculator.

SOLUTION Expanding by minors about the fourth row gives

$$-2 \det \begin{bmatrix} -2 & 3 & 2 \\ 1 & 4 & -2 \\ -1 & 4 & 0 \end{bmatrix} + 1 \det \begin{bmatrix} -1 & 3 & 2 \\ 0 & 4 & -2 \\ 3 & 4 & 0 \end{bmatrix}$$

$$-0 \det \begin{bmatrix} -1 & -2 & 2 \\ 0 & 1 & -2 \\ 3 & -1 & 0 \end{bmatrix} + 3 \det \begin{bmatrix} -1 & -2 & 3 \\ 0 & 1 & 4 \\ 3 & -1 & 4 \end{bmatrix}$$

$$= -2(6) + 1(-50) - 0 + 3(-41)$$

$$= -185. \qquad \blacklozenge$$

Note As seen in these examples, the computation involved in finding determinants of dimension 3 × 3 and larger square matrices is quite tedious. For this reason, it is customary to use graphing calculators with matrix capabilities (or, in some cases, computers) to evaluate such determinants. We have presented the theory behind these processes so that you can see how such determinants are defined mathematically.

DERIVATION OF CRAMER'S RULE

Cramer's rule is a method of solving a linear system of equations using determinants. The determinant of the matrix of coefficients allows us to first decide whether a single solution exists before we actually solve the system.

To derive Cramer's rule, we use the elimination method to solve the general system of two equations with two variables,

$$a_1x + b_1y = c_1 \tag{1}$$
$$a_2x + b_2y = c_2. \tag{2}$$

To begin, eliminate y and solve for x by first multiplying both sides of equation (1) by b_2 and both sides of equation (2) by $-b_1$. Then, add these results and solve for x.

$$a_1b_2x + b_1b_2y = c_1b_2$$
$$\underline{-a_2b_1x - b_1b_2y = -c_2b_1}$$
$$(a_1b_2 - a_2b_1)x = c_1b_2 - c_2b_1$$
$$x = \frac{c_1b_2 - c_2b_1}{a_1b_2 - a_2b_1}, \text{ if } a_1b_2 - a_2b_1 \neq 0$$

Solve for y by multiplying both sides of equation (1) by $-a_2$ and equation (2) by a_1 and then adding the two equations.

$$-a_1a_2x - a_2b_1y = -a_2c_1$$
$$\underline{a_1a_2x + a_1b_2y = a_1c_2}$$
$$(a_1b_2 - a_2b_1)y = a_1c_2 - a_2c_1$$
$$y = \frac{a_1c_2 - a_2c_1}{a_1b_2 - a_2b_1}, \text{ if } a_1b_2 - a_2b_1 \neq 0$$

Both numerators and the common denominator of these values for x and y can be written as determinants, since

$$c_1b_2 - c_2b_1 = \det\begin{bmatrix} c_1 & b_1 \\ c_2 & b_2 \end{bmatrix}, \qquad a_1c_2 - a_2c_1 = \det\begin{bmatrix} a_1 & c_1 \\ a_2 & c_2 \end{bmatrix},$$

and

$$a_1b_2 - a_2b_1 = \det\begin{bmatrix} a_1 & b_1 \\ a_2 & b_2 \end{bmatrix}.$$

Using these determinants, the solutions for x and y become

$$x = \frac{\det\begin{bmatrix} c_1 & b_1 \\ c_2 & b_2 \end{bmatrix}}{\det\begin{bmatrix} a_1 & b_1 \\ a_2 & b_2 \end{bmatrix}} \quad \text{and} \quad y = \frac{\det\begin{bmatrix} a_1 & c_1 \\ a_2 & c_2 \end{bmatrix}}{\det\begin{bmatrix} a_1 & b_1 \\ a_2 & b_2 \end{bmatrix}}, \text{ if } \det\begin{bmatrix} a_1 & b_1 \\ a_2 & b_2 \end{bmatrix} \neq 0.$$

CRAMER'S RULE FOR 2 × 2 SYSTEMS

For the system

$$a_1x + b_1y = c_1$$
$$a_2x + b_2y = c_2,$$

$$x = \frac{D_x}{D} \quad \text{and} \quad y = \frac{D_y}{D},$$

where

$$D_x = \det\begin{bmatrix} c_1 & b_1 \\ c_2 & b_2 \end{bmatrix}, \quad D_y = \det\begin{bmatrix} a_1 & c_1 \\ a_2 & c_2 \end{bmatrix}, \quad \text{and} \quad D = \det\begin{bmatrix} a_1 & b_1 \\ a_2 & b_2 \end{bmatrix} \neq 0.$$

Although this theorem is well-known as Cramer's rule, it was probably first discovered by Colin Maclaurin (1698–1746) as early as 1729 and was published under his name in 1748, two years earlier than Cramer's first publication of the rule in 1750.

Cramer's rule is used to solve a system of linear equations by evaluating the three determinants D, D_x, and D_y and then writing the appropriate quotients for x and y.

Caution   As indicated above, Cramer's rule does not apply if $D = 0$. When $D = 0$, the system is inconsistent or has dependent equations. For this reason, it is a good idea to evaluate D first.

USING CRAMER'S RULE TO SOLVE SYSTEMS

The next example shows how to apply Cramer's rule to a system of linear equations in two unknowns.

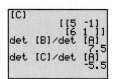

By defining matrices and using a graphing calculator to compute their determinants, you can apply Cramer's rule to solve a system of equations. These two screens show how the system in Example 6 can be solved with a graphing calculator.

EXAMPLE 6   *Applying Cramer's Rule to a System with Two Equations*   Use Cramer's rule to solve the system both analytically and graphically.

$$5x + 7y = -1$$
$$6x + 8y = 1$$

SOLUTION   ANALYTIC   By Cramer's rule, $x = \frac{D_x}{D}$ and $y = \frac{D_y}{D}$. As mentioned above, it is a good idea to find D first, because if $D = 0$, Cramer's rule does not apply. If $D \neq 0$, then find D_x and D_y.

$$D = \det\begin{bmatrix} 5 & 7 \\ 6 & 8 \end{bmatrix} = 5(8) - 6(7) = -2$$

$$D_x = \det\begin{bmatrix} -1 & 7 \\ 1 & 8 \end{bmatrix} = (-1)(8) - (1)(7) = -15$$

$$D_y = \det\begin{bmatrix} 5 & -1 \\ 6 & 1 \end{bmatrix} = 5(1) - (6)(-1) = 11$$

From Cramer's rule,

$$x = \frac{D_x}{D} = \frac{-15}{-2} = \frac{15}{2} \qquad \text{and} \qquad y = \frac{D_y}{D} = \frac{11}{-2} = -\frac{11}{2}.$$

The solution set is $\left\{\left(\frac{15}{2}, -\frac{11}{2}\right)\right\}$, as can be verified by substituting in the given system.

GRAPHICAL   See the margin for graphical support.   ◆

Cramer's rule can be generalized to systems of three equations in three variables (or n equations in n variables).

CRAMER'S RULE FOR 3 × 3 SYSTEMS

For the system

$$a_1x + b_1y + c_1z = d_1$$
$$a_2x + b_2y + c_2z = d_2$$
$$a_3x + b_3y + c_3z = d_3,$$

$$x = \frac{D_x}{D}, \qquad y = \frac{D_y}{D}, \qquad \text{and} \qquad z = \frac{D_z}{D},$$

where

$$D_x = \det\begin{bmatrix} d_1 & b_1 & c_1 \\ d_2 & b_2 & c_2 \\ d_3 & b_3 & c_3 \end{bmatrix}, \qquad D_y = \det\begin{bmatrix} a_1 & d_1 & c_1 \\ a_2 & d_2 & c_2 \\ a_3 & d_3 & c_3 \end{bmatrix},$$

$$D_z = \det\begin{bmatrix} a_1 & b_1 & d_1 \\ a_2 & b_2 & d_2 \\ a_3 & b_3 & d_3 \end{bmatrix}, \qquad \text{and} \qquad D = \det\begin{bmatrix} a_1 & b_1 & c_1 \\ a_2 & b_2 & c_2 \\ a_3 & b_3 & c_3 \end{bmatrix} \neq 0.$$

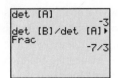

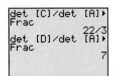

When using Cramer's rule, be sure to check that the determinant that appears in the denominator is not zero, as shown in the first display (it is -3). Refer to Example 7. The values of x, y, and z are shown in their fraction forms: $-7/3$, $22/3$, and 7.

EXAMPLE 7   *Applying Cramer's Rule to a System with Three Equations*   Use Cramer's rule to solve the system

$$x + y - z = -2$$
$$2x - y + z = -5$$
$$x - 2y + 3z = 4.$$

SOLUTION   Verify that the required determinants are as follows:

$$D = \det \begin{bmatrix} 1 & 1 & -1 \\ 2 & -1 & 1 \\ 1 & -2 & 3 \end{bmatrix} = -3, \qquad D_x = \det \begin{bmatrix} -2 & 1 & -1 \\ -5 & -1 & 1 \\ 4 & -2 & 3 \end{bmatrix} = 7,$$

$$D_y = \det \begin{bmatrix} 1 & -2 & -1 \\ 2 & -5 & 1 \\ 1 & 4 & 3 \end{bmatrix} = -22, \qquad D_z = \det \begin{bmatrix} 1 & 1 & -2 \\ 2 & -1 & -5 \\ 1 & -2 & 4 \end{bmatrix} = -21.$$

Thus,

$$x = \frac{D_x}{D} = \frac{7}{-3} = -\frac{7}{3}, \quad y = \frac{D_y}{D} = \frac{-22}{-3} = \frac{22}{3}, \quad z = \frac{D_z}{D} = \frac{-21}{-3} = 7,$$

so the solution set is $\left\{ \left(-\frac{7}{3}, \frac{22}{3}, 7 \right) \right\}$.   ◆

EXAMPLE 8   *Verifying That Cramer's Rule Does Not Apply*   Show why Cramer's rule does not apply to the system

$$2x - 3y + 4z = 10$$
$$6x - 9y + 12z = 24$$
$$x + 2y - 3z = 5.$$

SOLUTION   First, find D by expanding about column 1.

$$D = \det \begin{bmatrix} 2 & -3 & 4 \\ 6 & -9 & 12 \\ 1 & 2 & -3 \end{bmatrix} = 2 \det \begin{bmatrix} -9 & 12 \\ 2 & -3 \end{bmatrix} - 6 \det \begin{bmatrix} -3 & 4 \\ 2 & -3 \end{bmatrix}$$
$$+ 1 \det \begin{bmatrix} -3 & 4 \\ -9 & 12 \end{bmatrix}$$
$$= 2(3) - 6(1) + 1(0)$$
$$= 0$$

As mentioned above, Cramer's rule does not apply if $D = 0$. When $D = 0$, the system either is inconsistent or contains dependent equations. Use the elimination method or Gaussian reduction to tell which is the case. Verify that this system is inconsistent.   ◆

7.5 EXERCISES   Tape 9

Find the determinant of each of the following matrices.

1. $\begin{bmatrix} 5 & 8 \\ 2 & -4 \end{bmatrix}$ **2.** $\begin{bmatrix} -3 & 0 \\ 0 & 9 \end{bmatrix}$ **3.** $\begin{bmatrix} -1 & -2 \\ 5 & 3 \end{bmatrix}$ **4.** $\begin{bmatrix} 6 & -4 \\ 0 & -1 \end{bmatrix}$

5. $\begin{bmatrix} 9 & 3 \\ -3 & -1 \end{bmatrix}$ **6.** $\begin{bmatrix} 0 & 2 \\ 1 & 5 \end{bmatrix}$ **7.** $\begin{bmatrix} 3 & 4 \\ 5 & -2 \end{bmatrix}$ **8.** $\begin{bmatrix} -9 & 7 \\ 2 & 6 \end{bmatrix}$

9. $\begin{bmatrix} -2 & 0 & 0 \\ 4 & 0 & 1 \\ 3 & 4 & 2 \end{bmatrix}$ **10.** $\begin{bmatrix} 3 & -2 & 0 \\ 0 & -1 & 1 \\ 4 & 0 & 2 \end{bmatrix}$ **11.** $\begin{bmatrix} 1 & 2 & 0 \\ -1 & 2 & -1 \\ 0 & 1 & 4 \end{bmatrix}$ **12.** $\begin{bmatrix} 2 & 1 & -1 \\ 4 & 7 & -2 \\ 2 & 4 & 0 \end{bmatrix}$

13. $\begin{bmatrix} 10 & 2 & 1 \\ -1 & 4 & 3 \\ -3 & 8 & 10 \end{bmatrix}$ **14.** $\begin{bmatrix} 7 & -1 & 1 \\ 1 & -7 & 2 \\ -2 & 1 & 1 \end{bmatrix}$ **15.** $\begin{bmatrix} 1 & -2 & 3 \\ 0 & 0 & 0 \\ 1 & 10 & -12 \end{bmatrix}$ **16.** $\begin{bmatrix} 2 & 3 & 0 \\ 1 & 9 & 0 \\ -1 & -2 & 0 \end{bmatrix}$

17. $\begin{bmatrix} 3 & 3 & -1 \\ 2 & 6 & 0 \\ -6 & -6 & 2 \end{bmatrix}$ **18.** $\begin{bmatrix} 5 & -3 & 2 \\ -5 & 3 & -2 \\ 1 & 0 & 1 \end{bmatrix}$ **19.** $\begin{bmatrix} 4 & 0 & 0 & 2 \\ -1 & 0 & 3 & 0 \\ 2 & 4 & 0 & 1 \\ 0 & 0 & 1 & 2 \end{bmatrix}$ **20.** $\begin{bmatrix} -2 & 0 & 4 & 2 \\ 3 & 6 & 0 & 4 \\ 0 & 0 & 0 & 3 \\ 9 & 0 & 2 & -1 \end{bmatrix}$

21. Explain why a matrix with a row or column of all zeros must have 0 as the value of its determinant.

RELATING CONCEPTS (EXERCISES 22–25)

The equation $\det\begin{bmatrix} x & 2 & 1 \\ -1 & x & 4 \\ -2 & 0 & 5 \end{bmatrix} = 45$ *is an example of a determinant equation. It can be solved by finding an expression in x for the determinant and then solving the resulting equation. Work Exercises 22–25 in order, to see how to solve this equation.*

22. Use one of the methods described in this section to write the determinant as a polynomial in x.

23. Replace the determinant with the expression you found in Exercise 22. What kind of equation is this (based on the degree of the polynomial)? (Section 3.2)

24. Solve the equation found in Exercise 23. (Section 3.3)

25. Verify that when the solutions are substituted for x in the original determinant, the equation is satisfied.

Solve each of the following equations for x.

26. $A = \begin{bmatrix} 5 & x \\ -3 & 2 \end{bmatrix}$ and $\det A = 6$

27. $A = \begin{bmatrix} -.5 & 2 \\ x & x \end{bmatrix}$ and $\det A = 0$

28. $\det\begin{bmatrix} x & 3 \\ x & x \end{bmatrix} = 4$

29. $\det\begin{bmatrix} 2x & x \\ 11 & x \end{bmatrix} = 6$

30. $\det\begin{bmatrix} -2 & 0 & 1 \\ -1 & 3 & x \\ 5 & -2 & 0 \end{bmatrix} = 3$

31. $\det\begin{bmatrix} 4 & 3 & 0 \\ 2 & 0 & 1 \\ -3 & x & -1 \end{bmatrix} = 5$

32. $\det\begin{bmatrix} 5 & 3x & -3 \\ 0 & 2 & -1 \\ 4 & -1 & x \end{bmatrix} = -7$

33. $\det\begin{bmatrix} 2x & 1 & -1 \\ 0 & 4 & x \\ 3 & 0 & 2 \end{bmatrix} = x$

A formal study of determinants usually examines the following properties.

1. *If every element in a row or column of a matrix is 0, then the determinant equals 0.*

2. *If corresponding rows and columns of a matrix are interchanged, the determinant is not changed.*

3. *Interchanging two rows (or columns) of a matrix reverses the sign of the determinant.*

4. *If every element of a row (or column) of a matrix is multiplied by the real number k, then the determinant of the new matrix is k times the determinant of the original matrix.*

5. *The determinant of a matrix with two identical rows (or columns) equals 0.*

6. *The determinant of a matrix is unchanged if a multiple of a row (or column) of the matrix is added to the corresponding elements of another row (or column).*

Refer to these properties in Exercises 34–40.

34. Use Property 1 to evaluate $\det\begin{bmatrix} 0 & 0 & 0 \\ 2 & 1 & 4 \\ -3 & 8 & 6 \end{bmatrix}$. Then, verify your result, using expansion by minors or a calculator.

35. Repeat Exercise 34 for $\det\begin{bmatrix} .5 & 0 & \sqrt{2} \\ 6 & 0 & -3 \\ -1 & 0 & 5 \end{bmatrix}$.

36. Use Property 2 to complete the following statement: The value of $\det \begin{bmatrix} 2 & 1 & 6 \\ 3 & 0 & 5 \\ -4 & 6 & 9 \end{bmatrix}$ is 1. Therefore, the value of

$\det \begin{bmatrix} 2 & 3 & -4 \\ 1 & 0 & 6 \\ 6 & 5 & 9 \end{bmatrix}$ is _____, because _____.

37. Use Property 3 and the information given in Exercise 36 to complete the following statement: The value of

$\det \begin{bmatrix} 3 & 0 & 5 \\ 2 & 1 & 6 \\ -4 & 6 & 9 \end{bmatrix}$ is _____, because _____.

38. Use Property 4 to complete the given statement: The value of $\det \begin{bmatrix} 2 & -3 \\ 4 & 1 \end{bmatrix}$ is 14. Therefore, $\det \begin{bmatrix} 2 & -3 \\ -20 & -5 \end{bmatrix} =$

_____, because _____.

39. Use Property 5 to find $\det \begin{bmatrix} -4 & 2 & 3 \\ 0 & 1 & 6 \\ -4 & 2 & 3 \end{bmatrix}$.

40. Multiply each element of the first column of the matrix $A = \begin{bmatrix} -2 & 4 & 1 \\ 2 & 1 & 5 \\ 3 & 0 & 2 \end{bmatrix}$ by 3, and add the results to the third column

to get a new matrix B.

(a) What is the new matrix B?

(b) $\det A = 37$, therefore $\det B =$ _____.

RELATING CONCEPTS (EXERCISES 41–44)

Determinants can be used to find the equation of a line passing through two given points. Exercises 41–44 show how this is done.

41. Expand the determinant in the equation $\det \begin{bmatrix} x & y & 1 \\ 2 & 3 & 1 \\ -1 & 4 & 1 \end{bmatrix} = 0$.

42. Find the equation of the line through $(2, 3)$ and $(-1, 4)$. How does your answer compare with the answer to Exercise 41? (Section 1.4)

43. Write the equation of the line through the points (x_1, y_1) and (x_2, y_2), using the point-slope formula. (Section 1.4)

44. Expand the determinant below and compare the resulting equation with your answer to Exercise 43. What do you find?

$$\det \begin{bmatrix} x & y & 1 \\ x_1 & y_1 & 1 \\ x_2 & y_2 & 1 \end{bmatrix} = 0$$

Use Cramer's rule to solve each of the following systems of equations. If $D = 0$, use another method to complete the solution.

45. $x + y = 4$
$2x - y = 2$

46. $3x + 2y = -4$
$2x - y = -5$

47. $4x + 3y = -7$
$2x + 3y = -11$

48. $4x - y = 0$
$2x + 3y = 14$

49. $3x + 2y = 4$
$6x + 4y = 8$

50. $1.5x + 3y = 5$
$2x + 4y = 3$

51. $2x - 3y = -5$
$x + 5y = 17$

52. $x + 9y = -15$
$3x + 2y = 5$

53. $4x - y + 3z = -3$
$3x + y + z = 0$
$2x - y + 4z = 0$

54. $5x + 2y + z = 15$
$2x - y + z = 9$
$4x + 3y + 2z = 13$

55. $2x - y + 4z = -2$
$3x + 2y - z = -3$
$x + 4y + 2z = 17$

56. $x + y + z = 4$
$2x - y + 3z = 4$
$4x + 2y - z = -15$

57. $4x - 3y + z = -1$
$5x + 7y + 2z = -2$
$3x - 5y - z = 1$

58. $2x - 3y + z = 8$
$-x - 5y + z = -4$
$3x - 5y + 2z = 12$

59. $2x - y + 3z = 1$
$-2x + y - 3z = 2$
$5x - y + z = 2$

60. $-2x - 2y + 3z = 4$
$5x + 7y - z = 2$
$2x + 2y - 3z = -4$

61. $3x + 2y - w = 0$
$2x + z + 2w = 5$
$x + 2y - z = -2$
$2x - y + z + w = 2$

62. $x + 2y - z + w = 8$
$2x - y + 2w = 8$
$y + 3z = 5$
$x - z = 4$

63. In your own words, explain what it means when applying Cramer's rule if $D = 0$.

64. Describe D_x, D_y, and D_z in terms of the coefficients and constants in the given system of equations.

RELATING CONCEPTS (EXERCISES 65–66)

Determinants can be used to find the area of a triangle, given the coordinates of its vertices. Given a triangle PQR with vertices (x_1, y_1), (x_2, y_2), and (x_3, y_3), as in the figure, it can be shown that the area of the triangle is given by A, where

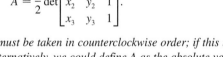

$$A = \frac{1}{2} \det \begin{bmatrix} x_1 & y_1 & 1 \\ x_2 & y_2 & 1 \\ x_3 & y_3 & 1 \end{bmatrix}.$$

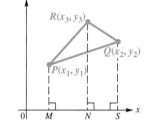

The points (x_1, y_1), (x_2, y_2), (x_3, y_3) must be taken in counterclockwise order; if this is not done, then A may have the wrong sign. Alternatively, we could define A as the absolute value of $\frac{1}{2}$ the determinant shown above. Use the formula given to find the area of the triangles in Exercises 65 and 66.

65. (a) $P(0, 0), Q(0, 2), R(1, 1)$

(b) $P(0, 1), Q(2, 0), R(1, 3)$

(c) $P(2, 5), Q(-1, 3), R(4, 0)$

66. (a) $P(2, -2), Q(0, 0), R(-3, -4)$

(b) $P(4, 7), Q(5, -2), R(1, 1)$

(c) $P(1, 2), Q(4, 3), R(3, 5)$

7.6 SOLUTION OF LINEAR SYSTEMS BY MATRIX INVERSES

Identity Matrices ◆ Multiplicative Inverses of Square Matrices ◆ Solving Linear Systems, Using Inverse Matrices

IDENTITY MATRICES

The identity property for real numbers tells us that $a \cdot 1 = a$ and $1 \cdot a = a$ for any real number a. **Square matrices** (those with the same number of rows as columns) have a similar property. If there is to be a multiplicative identity matrix I, such that

$$AI = A \qquad \text{and} \qquad IA = A,$$

for any matrix A, then A and I must be square matrices of the same dimension. Otherwise, it would not be possible to find both products. For example, let A be the 2×2

matrix $A = \begin{bmatrix} a_{11} & a_{12} \\ a_{21} & a_{22} \end{bmatrix}$, and let $I_2 = \begin{bmatrix} x_{11} & x_{12} \\ x_{21} & x_{22} \end{bmatrix}$ represent the 2 × 2 identity matrix. (The subscript 2 is used to denote the fact that it is a square matrix of dimension 2 × 2.) To find I_2, use the fact that $I_2A = A$, so

$$\begin{bmatrix} x_{11} & x_{12} \\ x_{21} & x_{22} \end{bmatrix} \begin{bmatrix} a_{11} & a_{12} \\ a_{21} & a_{22} \end{bmatrix} = \begin{bmatrix} a_{11} & a_{12} \\ a_{21} & a_{22} \end{bmatrix}.$$

Multiplying the two matrices on the left side of this equation and setting the elements of the product matrix equal to the corresponding elements of A gives the following system of equations with variables x_{11}, x_{12}, x_{21}, and x_{22}.

$$x_{11}a_{11} + x_{12}a_{21} = a_{11}$$
$$x_{11}a_{12} + x_{12}a_{22} = a_{12}$$
$$x_{21}a_{11} + x_{22}a_{21} = a_{21}$$
$$x_{21}a_{12} + x_{22}a_{22} = a_{22}$$

Notice that this is really two systems of equations in two variables. Use one of the methods introduced earlier to find the solution of this system: $x_{11} = 1$, $x_{12} = x_{21} = 0$, and $x_{22} = 1$. From the solution of the system, the 2 × 2 identity matrix is $I_2 = \begin{bmatrix} 1 & 0 \\ 0 & 1 \end{bmatrix}$. Check that with this definition of I_2, both $AI_2 = A$ and $I_2A = A$.

EXAMPLE 1   *Verifying the Identity Property*   Let $M = \begin{bmatrix} -2 & 6 \\ 3 & 5 \end{bmatrix}$. Verify that $MI_2 = M$.

SOLUTION

$$MI_2 = \begin{bmatrix} -2 & 6 \\ 3 & 5 \end{bmatrix} \begin{bmatrix} 1 & 0 \\ 0 & 1 \end{bmatrix} = \begin{bmatrix} -2 & 6 \\ 3 & 5 \end{bmatrix} = M \qquad \blacklozenge$$

For Group Discussion   Verify that the matrix product I_2M in Example 1 also equals M. Do you think it is always the case that if A is a square matrix of dimension $n \times n$, and $AI = A$, then $IA = A$? Why or why not?

The 2 × 2 identity matrix found above suggests the following generalization.

DEFINITION OF AN $n \times n$ IDENTITY MATRIX

For any value of n there is an $n \times n$ identity matrix having 1s down the main diagonal and 0s elsewhere. The **$n \times n$ identity matrix** is given by I_n, where

$$I_n = \begin{bmatrix} 1 & 0 & \cdots & 0 \\ 0 & 1 & \cdots & 0 \\ \vdots & \vdots & a_{ij} & \vdots \\ 0 & 0 & \cdots & 1 \end{bmatrix}.$$

Here $a_{ij} = 1$ when $i = j$ (the diagonal elements) and $a_{ij} = 0$ otherwise.

EXAMPLE 2 *Stating and Verifying the 3 × 3 Identity Matrix*   Let
$K = \begin{bmatrix} -2 & 4 & 0 \\ 3 & 5 & 9 \\ 0 & 8 & -6 \end{bmatrix}$. Give the 3 × 3 identity matrix I_3 and show that $KI_3 = K$.

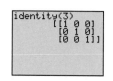

The calculator screen shows the 3 × 3 identity matrix.

SOLUTION   The 3 × 3 identity matrix is $I_3 = \begin{bmatrix} 1 & 0 & 0 \\ 0 & 1 & 0 \\ 0 & 0 & 1 \end{bmatrix}$. By the definition of matrix multiplication,

$$KI_3 = \begin{bmatrix} -2 & 4 & 0 \\ 3 & 5 & 9 \\ 0 & 8 & -6 \end{bmatrix}\begin{bmatrix} 1 & 0 & 0 \\ 0 & 1 & 0 \\ 0 & 0 & 1 \end{bmatrix} = \begin{bmatrix} -2 & 4 & 0 \\ 3 & 5 & 9 \\ 0 & 8 & -6 \end{bmatrix} = K. \qquad \blacklozenge$$

MULTIPLICATIVE INVERSES OF SQUARE MATRICES

TECHNOLOGY NOTE

As explained in the text, the multiplicative inverse exists only when the determinant is not equal to 0. A graphing calculator will return the error message **singular matrix** if you attempt to find the multiplicative inverse of a matrix that has determinant 0.

Suppose that $A = \begin{bmatrix} a & b \\ c & d \end{bmatrix}$. Is it possible to find a matrix B such that $AB = I_2$ and $BA = I_2$? If so, then B is the inverse of A.

Let $B = \begin{bmatrix} x & y \\ z & w \end{bmatrix}$. We must find x, y, z, and w so that $\begin{bmatrix} a & b \\ c & d \end{bmatrix}\begin{bmatrix} x & y \\ z & w \end{bmatrix} = \begin{bmatrix} 1 & 0 \\ 0 & 1 \end{bmatrix}$.
Multiplying the two matrices and setting the elements of the product equal to the corresponding elements in I_2 gives the following two systems of equations.

$$ax + bz = 1 \qquad ay + bw = 0$$
$$cx + dz = 0 \qquad cy + dw = 1$$

By solving these two systems, we get the values of x, y, z, and w:

$$x = \frac{d}{ad - cb}, \quad y = \frac{-b}{ad - cb}, \quad z = \frac{-c}{ad - cb}, \quad \text{and} \quad w = \frac{a}{ad - cb}.$$

Verify that $AB = I_2$ and that $BA = I_2$. As a result, we can conclude that B is the inverse of A, written A^{-1}, provided that $\det A = ad - cb \neq 0$.

DEFINITION OF THE MULTIPLICATIVE INVERSE OF A SQUARE MATRIX OF DIMENSION 2 × 2

If $A = \begin{bmatrix} a & b \\ c & d \end{bmatrix}$ and $\det A \neq 0$, then

$$A^{-1} = \frac{1}{\det A}\begin{bmatrix} d & -b \\ -c & a \end{bmatrix} \quad \text{or} \quad A^{-1} = \begin{bmatrix} \dfrac{d}{ad - cb} & \dfrac{-b}{ad - cb} \\ \dfrac{-c}{ad - cb} & \dfrac{a}{ad - cb} \end{bmatrix}.$$

EXAMPLE 3 *Finding the Multiplicative Inverse of a Square Matrix of Dimension 2 × 2*   Find A^{-1}, if it exists, for each of the following matrices.

(a) $A = \begin{bmatrix} 2 & 3 \\ 1 & -1 \end{bmatrix}$ **(b)** $A = \begin{bmatrix} 3 & -6 \\ 2 & -4 \end{bmatrix}$

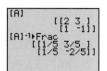

The calculator-generated matrix inverse for the matrix in Example 3(a) is shown here.

SOLUTION

(a) To find A^{-1} for $\begin{bmatrix} 2 & 3 \\ 1 & -1 \end{bmatrix}$, first we find det A and determine whether A^{-1} actually exists.

$$\det A = 2(-1) - 1(3) = -5$$

Since $-5 \neq 0$, A^{-1} exists. By the rule given earlier,

$$A^{-1} = \frac{1}{-5}\begin{bmatrix} -1 & -3 \\ -1 & 2 \end{bmatrix} = \begin{bmatrix} \frac{1}{5} & \frac{3}{5} \\ \frac{1}{5} & -\frac{2}{5} \end{bmatrix}.$$

(b) Here, A^{-1} does not exist because

$$\det A = 3(-4) - 2(-6) = 0. \quad \blacklozenge$$

As seen in Example 3, it is always necessary for the determinant to be nonzero for the inverse to exist.

For Group Discussion

1. What happens if you use your calculator to try to find A^{-1} in Example 3(b)?

2. We have seen that under certain conditions, the multiplicative inverse of a matrix does not exist. How does this compare to the multiplicative inverse property for real numbers? Does the multiplicative inverse of $\frac{x-4}{5}$ exist if $x = 4$?

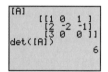

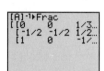

Compare these screens with the work in Example 4. Note that the right portion of the inverse matrix (shown in the lower screen) is not complete.

EXAMPLE 4 *Verifying Matrix Inverses* Consider the matrix $A = \begin{bmatrix} 1 & 0 & 1 \\ 2 & -2 & -1 \\ 3 & 0 & 0 \end{bmatrix}$.

(a) Show that det $A \neq 0$. **(b)** Verify that $A^{-1} = \begin{bmatrix} 0 & 0 & \frac{1}{3} \\ -\frac{1}{2} & -\frac{1}{2} & \frac{1}{2} \\ 1 & 0 & -\frac{1}{3} \end{bmatrix}$.

SOLUTION

(a) Expanding about the third row, we find

$$\det A = 3 \cdot \det\begin{bmatrix} 0 & 1 \\ -2 & -1 \end{bmatrix} = 3(0 + 2) = 6.$$

(Because $6 \neq 0$, A^{-1} exists.)

(b) First, show $A \cdot A^{-1} = I_3$.

$$A \cdot A^{-1} = \begin{bmatrix} 1 & 0 & 1 \\ 2 & -2 & -1 \\ 3 & 0 & 0 \end{bmatrix} \cdot \begin{bmatrix} 0 & 0 & \frac{1}{3} \\ -\frac{1}{2} & -\frac{1}{2} & \frac{1}{2} \\ 1 & 0 & -\frac{1}{3} \end{bmatrix}$$

$$= \begin{bmatrix} 0 + 0 + 1 & 0 + 0 + 0 & \frac{1}{3} + 0 - \frac{1}{3} \\ 0 + 1 - 1 & 0 + 1 + 0 & \frac{2}{3} - 1 + \frac{1}{3} \\ 0 + 0 + 0 & 0 + 0 + 0 & 1 + 0 + 0 \end{bmatrix}$$

$$= \begin{bmatrix} 1 & 0 & 0 \\ 0 & 1 & 0 \\ 0 & 0 & 1 \end{bmatrix} = I_3$$

Next, show that $A^{-1} \cdot A = I_3$. Because $A \cdot A^{-1} = A^{-1} \cdot A = I_3$, we know that the inverse of A is

$$\begin{bmatrix} 0 & 0 & \frac{1}{3} \\ -\frac{1}{2} & -\frac{1}{2} & \frac{1}{2} \\ 1 & 0 & -\frac{1}{3} \end{bmatrix}. \qquad \blacklozenge$$

While there are methods of finding inverses of square matrices of dimension 3×3 or greater by hand computation, we seldom do this because it is so tedious. Calculators and computers perform this function quickly and efficiently. Learn how to use your calculator to find a matrix inverse. You may wish to experiment to see whether you can duplicate the results in Examples 1–4. Be careful to use the correct notation.

SOLVING LINEAR SYSTEMS, USING INVERSE MATRICES

We used matrices to solve systems of linear equations by the Gaussian reduction method in Section 7.3. Another way to use matrices to solve linear systems is to write the system as a matrix equation $AX = B$, where A is the matrix of the coefficients of the variables of the system, X is the matrix of the variables, and B is the matrix of the constants. Matrix A is called the **coefficient matrix.**

To solve the matrix equation $AX = B$, first see if A^{-1} exists. Assuming A^{-1} exists, and using the facts that $A^{-1}A = I$ and $IX = X$, gives

$$AX = B$$
$$A^{-1}(AX) = A^{-1}B \qquad \text{Multiply both sides by } A^{-1}.$$
$$(A^{-1}A)X = A^{-1}B \qquad \text{Associative property}$$
$$IX = A^{-1}B \qquad \text{Multiplicative inverse property}$$
$$X = A^{-1}B. \qquad \text{Identity property}$$

Caution When multiplying by matrices on both sides of a matrix equation, be careful to multiply in the same order on both sides of the equation, since multiplication of matrices is not commutative (unlike multiplication of real numbers).

The following example illustrates the use of the inverse matrix to solve a system of three equations in three variables.

EXAMPLE 5 *Solving a System of Equations, Using a Matrix Inverse* Solve the system

$$2x + y + 3z = 1$$
$$x - 2y + z = -3$$
$$-3x + y - 2z = -4,$$

using the inverse of the coefficient matrix.

SOLUTION First, find the determinant of the coefficient matrix to be sure that it is not 0. Using the method described earlier or a calculator, we find that for

$$A = \begin{bmatrix} 2 & 1 & 3 \\ 1 & -2 & 1 \\ -3 & 1 & -2 \end{bmatrix},$$

$\det A = -10 \neq 0$. Therefore, A^{-1} exists.

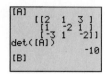

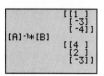

A calculator performs math operations on matrices just as it does on numbers. These screens show matrix A and det A (on top) and matrix B and the product of A^{-1} and B (at the bottom).

With

$$X = \begin{bmatrix} x \\ y \\ z \end{bmatrix} \quad \text{and} \quad B = \begin{bmatrix} 1 \\ -3 \\ -4 \end{bmatrix},$$

we use the calculator to evaluate $A^{-1}B$:

$$A^{-1}B = \begin{bmatrix} 2 & 1 & 3 \\ 1 & -2 & 1 \\ -3 & 1 & -2 \end{bmatrix}^{-1} \begin{bmatrix} 1 \\ -3 \\ -4 \end{bmatrix} = \begin{bmatrix} 4 \\ 2 \\ -3 \end{bmatrix}.$$

Since $X = A^{-1}B$, we have $x = 4$, $y = 2$, and $z = -3$. The solution set of the system is $\{(4, 2, -3)\}$.   ◆

Caution   Always evaluate the determinant of the coefficient matrix *before* using the inverse matrix method. If the determinant is 0, the system either is inconsistent or has dependent equations.

The final example of this section continues the discussion of curve-fitting introduced earlier in this chapter.

EXAMPLE 6   *Using a System to Find the Equation for a Cubic Polynomial*
Figure 17 shows four views of the graph of a polynomial function of the form $P(x) = ax^3 + bx^2 + cx + d$. Use the points indicated to write a system of four equations in the variables a, b, c, and d, and then use the inverse matrix method to solve the system. What is the equation that defines this graph?

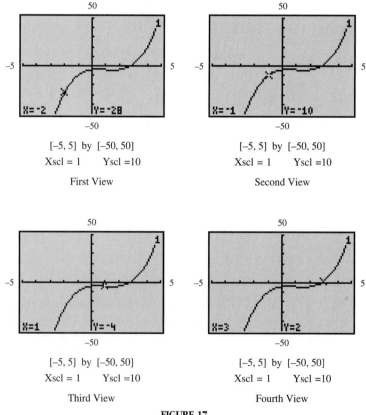

[-5, 5] by [-50, 50]
Xscl = 1   Yscl =10

First View

[-5, 5] by [-50, 50]
Xscl = 1   Yscl =10

Second View

[-5, 5] by [-50, 50]
Xscl = 1   Yscl =10

Third View

[-5, 5] by [-50, 50]
Xscl = 1   Yscl =10

Fourth View

FIGURE 17

SOLUTION   We see from the graph that $P(-2) = -28$, $P(-1) = -10$, $P(1) = -4$, and $P(3) = 2$. From the first of these, we get

$$P(-2) = a(-2)^3 + b(-2)^2 + c(-2) + d = -28$$

or, equivalently,

$$-8a + 4b - 2c + d = -28.$$

Similarly, from the others, we find the following equations.

From $P(-1) = -10$: $\quad -a + b - c + d = -10$

From $P(1) = -4$: $\qquad a + b + c + d = -4$

From $P(3) = 2$: $\qquad 27a + 9b + 3c + d = 2$

Now we must solve the system formed by these four equations:

$$-8a + 4b - 2c + d = -28$$
$$-a + b - c + d = -10$$
$$a + b + c + d = -4$$
$$27a + 9b + 3c + d = 2.$$

We will use the inverse matrix method to solve the system. Let

$$A = \begin{bmatrix} -8 & 4 & -2 & 1 \\ -1 & 1 & -1 & 1 \\ 1 & 1 & 1 & 1 \\ 27 & 9 & 3 & 1 \end{bmatrix}, \quad X = \begin{bmatrix} a \\ b \\ c \\ d \end{bmatrix}, \quad \text{and} \quad B = \begin{bmatrix} -28 \\ -10 \\ -4 \\ 2 \end{bmatrix}.$$

Because det $A = 240 \neq 0$, a unique solution exists. Based on our discussion earlier, $X = A^{-1}B$:

$$X = \begin{bmatrix} -8 & 4 & -2 & 1 \\ -1 & 1 & -1 & 1 \\ 1 & 1 & 1 & 1 \\ 27 & 9 & 3 & 1 \end{bmatrix}^{-1} \begin{bmatrix} -28 \\ -10 \\ -4 \\ 2 \end{bmatrix} = \begin{bmatrix} 1 \\ -3 \\ 2 \\ -4 \end{bmatrix}.$$

Therefore, $a = 1$, $b = -3$, $c = 2$, and $d = -4$. The polynomial is $P(x) = x^3 - 3x^2 + 2x - 4$. ◆

For Group Discussion   Use a window with dimensions $[-5, 5]$ by $[-50, 50]$ to reproduce the graph in Figure 17, based on the result of Example 6.

7.6   EXERCISES   **Tape 9**

Multiply the given pair of matrices in both directions to determine whether they are inverses of each other. Remember that the product must be the identity matrix for them to be inverses.

1. $\begin{bmatrix} 2 & 3 \\ 1 & 1 \end{bmatrix}, \begin{bmatrix} -1 & 3 \\ 1 & -2 \end{bmatrix}$
   2. $\begin{bmatrix} 5 & 7 \\ 2 & 3 \end{bmatrix} \begin{bmatrix} 3 & -7 \\ -2 & 5 \end{bmatrix}$
   3. $\begin{bmatrix} 2 & 1 \\ 3 & 2 \end{bmatrix}, \begin{bmatrix} 2 & 1 \\ -3 & 2 \end{bmatrix}$
   4. $\begin{bmatrix} -1 & 2 \\ 3 & -5 \end{bmatrix} \begin{bmatrix} -5 & -2 \\ -3 & -1 \end{bmatrix}$

5. $\begin{bmatrix} 1 & -2 & -3 \\ 2 & -2 & -5 \\ -1 & 1 & 4 \end{bmatrix}, \begin{bmatrix} -1 & \frac{5}{3} & \frac{4}{3} \\ -1 & \frac{1}{3} & -\frac{1}{3} \\ 0 & \frac{1}{3} & \frac{2}{3} \end{bmatrix}$
   6. $\begin{bmatrix} 1 & 2 & -1 \\ 2 & -1 & 3 \\ 3 & -2 & 3 \end{bmatrix}, \begin{bmatrix} \frac{3}{10} & -\frac{2}{5} & \frac{1}{2} \\ \frac{3}{10} & \frac{3}{5} & -\frac{1}{2} \\ -\frac{1}{10} & \frac{4}{5} & -\frac{1}{2} \end{bmatrix}$

7. $\begin{bmatrix} 1 & 2 & -1 \\ 0 & 1 & 3 \\ 2 & 1 & -2 \end{bmatrix}, \begin{bmatrix} 1 & 1 & 2 \\ 1 & 1 & 1 \\ 2 & 3 & 4 \end{bmatrix}$

8. $\begin{bmatrix} 2 & -1 & 4 \\ 0 & 5 & 0 \\ 3 & 2 & -1 \end{bmatrix}, \begin{bmatrix} 1 & 0 & 1 \\ 6 & 4 & 2 \\ 1 & 1 & 0 \end{bmatrix}$

9. Under what condition will the inverse of a square matrix not exist?

10. Explain why a square matrix of dimension 2×2 will not have an inverse if either a column or a row contains all zeros.

Find the inverse, if it exists, of each matrix.

11. $\begin{bmatrix} 1 & -1 \\ 2 & 0 \end{bmatrix}$

12. $\begin{bmatrix} 3 & -1 \\ -5 & 2 \end{bmatrix}$

13. $\begin{bmatrix} -6 & 4 \\ -3 & 2 \end{bmatrix}$

14. $\begin{bmatrix} -1 & 2 \\ -2 & -1 \end{bmatrix}$

15. $\begin{bmatrix} -1 & -2 \\ 3 & 4 \end{bmatrix}$

16. $\begin{bmatrix} 5 & 10 \\ -3 & -6 \end{bmatrix}$

17. $\begin{bmatrix} .6 & .2 \\ .5 & .1 \end{bmatrix}$

18. $\begin{bmatrix} .8 & -.3 \\ .5 & -.2 \end{bmatrix}$

19. $\begin{bmatrix} 1 & 0 & 0 \\ 0 & -1 & 0 \\ 1 & 0 & 1 \end{bmatrix}$

20. $\begin{bmatrix} 1 & 3 & 3 \\ -1 & 0 & 0 \\ -4 & -4 & -3 \end{bmatrix}$

21. $\begin{bmatrix} -1 & -1 & -1 \\ 4 & 5 & 0 \\ 0 & 1 & -3 \end{bmatrix}$

22. $\begin{bmatrix} 2 & 0 & 4 \\ 3 & 1 & 5 \\ -1 & 1 & -2 \end{bmatrix}$

23. $\begin{bmatrix} -.4 & .1 & .2 \\ 0 & .6 & .8 \\ .3 & 0 & -.2 \end{bmatrix}$

24. $\begin{bmatrix} .8 & .2 & .1 \\ -.2 & 0 & .3 \\ 0 & 0 & .5 \end{bmatrix}$

25. $\begin{bmatrix} 2 & 1 & 2 \\ 5 & 10 & 5 \\ 3 & 6 & 3 \end{bmatrix}$

26. $\begin{bmatrix} 5 & -3 & 2 \\ -5 & 3 & -2 \\ 1 & 0 & 1 \end{bmatrix}$

RELATING CONCEPTS (EXERCISES 27–32)

One method for finding matrix inverses analytically is developed by using a system of linear equations. To see how this is done, work Exercises 27–32 in order. Use the matrices

$$A = \begin{bmatrix} 2 & 4 \\ 1 & -1 \end{bmatrix} \quad \text{and} \quad A^{-1} = \begin{bmatrix} x & y \\ z & w \end{bmatrix}.$$

27. Find the product AA^{-1}.

28. Set $AA^{-1} = I_2$ and set corresponding elements equal to get four equations.

29. Write the equations from Exercise 28 as two systems of two equations in two variables. Solve the systems. (Section 7.1)

30. From Exercise 29, what is matrix A^{-1}?

31. Write the augmented matrix $[A|I_2]$, and use row operations to change it to the form $[I_2|B]$. (Section 7.3)

32. Compare matrix B with matrix A^{-1} from Exercise 30. What do you find? How can you use row transformations to find A^{-1} for any matrix A?

Write the matrix of coefficients A, the matrix of variables X, and the matrix of constants B for each of the following systems. Do not solve.

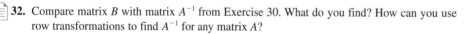

33. $\begin{aligned} x + y &= 8 \\ 2x - y &= 4 \end{aligned}$

34. $\begin{aligned} 2x + y &= 9 \\ x + 3y &= 17 \end{aligned}$

35. $\begin{aligned} 4x + 5y &= 7 \\ 2x + 3y &= 5 \end{aligned}$

36. $\begin{aligned} 2x + 3y &= -2 \\ 5x + 4y &= -12 \end{aligned}$

Solve each of the following systems by using the inverse of the coefficient matrix.

37. $\begin{aligned} 2x - y &= -8 \\ 3x + y &= -2 \end{aligned}$

38. $\begin{aligned} x + 3y &= -12 \\ 2x - y &= 11 \end{aligned}$

39. $\begin{aligned} 2x + 3y &= -10 \\ 3x + 4y &= -12 \end{aligned}$

40. $\begin{aligned} 2x - 3y &= 10 \\ 2x + 2y &= 5 \end{aligned}$

41. $\begin{aligned} 2x - 5y &= 10 \\ 4x - 5y &= 15 \end{aligned}$

42. $\begin{aligned} 2x - 3y &= 2 \\ 4x - 6y &= 1 \end{aligned}$

43. $\begin{aligned} 2x + 4z &= 14 \\ 3x + y + 5z &= 19 \\ -x + y - 2z &= -7 \end{aligned}$

44. $\begin{aligned} 3x + 6y + 3z &= 12 \\ 6x + 4y - 2z &= -4 \\ y - z &= -3 \end{aligned}$

45.
$$x + 3y + z = 2$$
$$x - 2y + 3z = -3$$
$$2x - 3y - z = 34$$

46.
$$x + y - z = 6$$
$$2x - y + z = -9$$
$$x - 2y + 3z = 1$$

47.
$$x + 3y - 2z - w = 9$$
$$4x + y + z + 2w = 2$$
$$-3x - y + z - w = -5$$
$$x - y - 3z - 2w = 2$$

48.
$$3x + 2y - w = 0$$
$$2x + z + 2w = 5$$
$$x + 2y - z = -2$$
$$2x - y + z + w = 2$$

In Exercises 49 and 50, use the method of Example 6 to find the cubic polynomial P(x) that defines the curve shown in the four figures.

49.

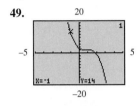
First View

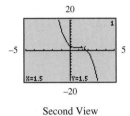
Second View

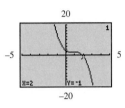
Third View

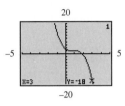
Fourth View

50.

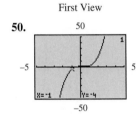
First View

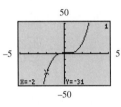
Second View

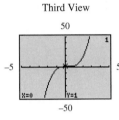
Third View

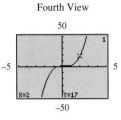
Fourth View

51. Find the fourth degree polynomial $P(x)$ satisfying the following conditions:
$$P(-2) = 13, \quad P(-1) = 2, \quad P(0) = -1, \quad P(1) = 4, \quad P(2) = 41.$$

52. Find the fifth degree polynomial $P(x)$ satisfying the following conditions:
$$P(-2) = -8, \quad P(-1) = -1, \quad P(0) = -4, \quad P(1) = -5, \quad P(2) = 8, \quad P(3) = 167.$$

Solve each application.

53. *Predicting Sales* The amount of plate-glass sales S (in millions of dollars) can be affected by the number of new building contracts B issued (in millions) and automobiles A produced (in millions). A plate-glass company in California wants to forecast future sales by using the past three years of sales. The totals for three years are given in the table.

S	A	B
602.7	5.543	37.14
656.7	6.933	41.30
778.5	7.638	45.62

In order to describe the relationship between these variables, the equation $S = a + bA + cB$ was used, where the coefficients a, b, and c are constants that must be determined. (*Source:* Makridakis, S., and S. Wheelwright, *Forecasting Methods for Management,* John Wiley & Sons, Inc., 1989.)

(a) Substitute the values for S, A, and B for each year from the table into the equation $S = a + bA + cB$ and obtain three linear equations involving a, b, and c.

(b) Solve this linear system for a, b, and c. Use matrix inverse methods.

(c) Write the equation for S, using these values for the coefficients.

(d) For the next year it is estimated that $A = 7.752$ and $B = 47.38$. Predict S. (The actual value for S was 877.6.)

(e) It is predicted that in six years $A = 8.9$ and $B = 66.25$. Find the value for S in this situation and discuss its validity.

54. *Predicting Sales* The number of automobile tire sales is dependent on several variables. In one study, the relationship between annual tire sales S (in thousands), automobile registrations R (in millions), and personal disposable income I (in millions of dollars) was investigated. The results for three years are given in the table.

S	R	I
10,170	112.9	307.5
15,305	132.9	621.63
21,289	155.2	1937.13

In order to describe the relationship between these variables, mathematicians often use the equation $S = a + bR + cI$, where the coefficients a, b, and c are constants that must be determined before the equation can be used. (*Source:* Jarrett, J., *Business Forecasting Methods,* Basil Blackwell, Ltd., Cambridge, MA, 1991.)

(a) Substitute the values for S, R, and I for each year from the table into the equation $S = a + bR + cI$ and obtain three linear equations involving a, b, and c.

(b) Solve this linear system for a, b, and c. Use matrix inverse methods.

(c) Write the equation for S, using these values for the coefficients.

(d) If $R = 117.6$ and $I = 310.73$, predict S. (The actual value for S was 11,314.)

(e) If $R = 143.8$ and $I = 829.06$, predict S. (The actual value for S was 18,481.)

7.7 SYSTEMS OF INEQUALITIES AND LINEAR PROGRAMMING

Graphs of Inequalities in Two Variables ◆ Systems of Inequalities ◆ A Graphing Calculator Approach to Inequalities and Systems of Inequalities ◆ Linear Programming

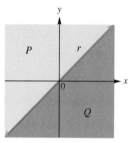

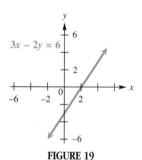

FIGURE 18

GRAPHS OF INEQUALITIES IN TWO VARIABLES

We begin our study of inequalities in two variables by considering the simplest type: linear inequalities. A line divides a plane into three sets of points: the points of the line itself and the points belonging to the two regions determined by the line. Each of these two regions is called a **half-plane.** In Figure 18, line r divides the plane into three different sets of points: line r, half-plane P, and half-plane Q. The points on r belong neither to P nor to Q. Line r is the **boundary** of each half-plane.

A **linear inequality in two variables** is an inequality of the form

$$Ax + By > C, \qquad Ax + By \geq C,$$
$$Ax + By < C, \qquad \text{or} \qquad Ax + By \leq C,$$

where A, B, and C are real numbers, with A and B not both equal to 0.

Let us now see how a linear inequality in two variables is graphed, using a traditional approach. Such a graph is a half-plane, perhaps with its boundary. For example, to graph the linear inequality $3x - 2y \leq 6$, first graph the boundary, $3x - 2y = 6$, as shown in Figure 19.

Since the points of the line $3x - 2y = 6$ satisfy $3x - 2y \leq 6$, this line is part of the solution. To decide which half-plane (the one above the line $3x - 2y = 6$ or the one below the line) is part of the solution, solve the original inequality for y.

$$3x - 2y \leq 6$$
$$-2y \leq -3x + 6$$
$$y \geq \frac{3}{2}x - 3 \qquad \text{Multiply by } -\tfrac{1}{2}; \text{ change } \leq \text{ to } \geq.$$

For a particular value of x, the inequality will be satisfied by all values of y that are *greater than* or equal to $\frac{3}{2}x - 3$. This means that the solution includes the half-plane *above* the line, as well as the line itself. The domain and range are both $(-\infty, \infty)$. See Figure 20.

There is an alternative method for deciding which side of the boundary line to shade. Choose as a test point any point not on the graph of the equation. The origin,

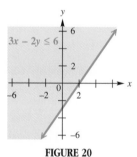

FIGURE 20

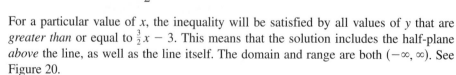

FIGURE 19

(0, 0), is often a good test point (as long as it does not lie on the boundary line). Substituting 0 for x and 0 for y in the inequality $3x - 2y \leq 6$ gives

$$3(0) - 2(0) \leq 6$$
$$0 \leq 6,$$

a true statement. Since (0, 0) leads to a true result, it is part of the solution of the inequality. Shade the side of the graph containing (0, 0), as shown in Figure 20.

EXAMPLE 1 *Graphing a Linear Inequality, Using a Traditional Approach*
Graph $x + 4y > 4$, using traditional graphing methods.

SOLUTION   The boundary here is the straight line $x + 4y = 4$. Since the points on this line do not satisfy $x + 4y > 4$, it is customary to graph the boundary as a dashed line, as in Figure 21. To decide which half-plane satisfies the inequality, use a test point. Choosing (0, 0) as a test point gives $0 + 4 \cdot 0 > 4$, or $0 > 4$, a false statement. Since (0, 0) leads to a false statement, shade the side of the graph *not* containing (0, 0), as in Figure 21.

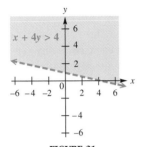

FIGURE 21

Alternatively, we can determine which half-plane should be shaded by solving the original inequality for y:

$$x + 4y > 4$$
$$4y > -x + 4$$
$$y > -\frac{1}{4}x + 1.$$

Since the inequality is true for any y-value *greater than* $-\frac{1}{4}x + 1$, we shade the half-plane *above* the boundary line. This procedure will be helpful when we study the graphing calculator approach later in this section.   ◆

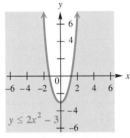

FIGURE 22

EXAMPLE 2 *Graphing a Nonlinear Inequality, Using a Traditional Approach*
Graph $y \leq 2x^2 - 3$, using traditional graphing methods.

SOLUTION   First, graph the boundary, $y = 2x^2 - 3$. It is graphed as a solid curve, since the symbol in the given inequality includes the $=$ sign. The inequality is given in a convenient form; we want values of y that are *less than* $2x^2 - 3$, so we shade *below* the boundary. See Figure 22.   ◆

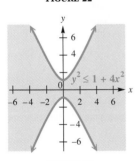

FIGURE 23

EXAMPLE 3 *Graphing Second Degree Inequalities, Using a Traditional Approach* Graph each inequality.

(a) $y^2 \leq 1 + 4x^2$ **(b)** $y^2 \geq 1 + 4x^2$

SOLUTION

(a) Write the boundary $y^2 = 1 + 4x^2$ as $y^2 - 4x^2 = 1$, a hyperbola with y-intercepts 1 and -1, as shown in Figure 23. Select any point not on the hyperbola and test it in the original inequality. Since (0, 0) satisfies this inequality, shade the area between the two branches of the hyperbola, as shown in Figure 23. The points on the hyperbola are part of the solution.

(b) The boundary is the same as in part (a). Since (0, 0) makes this inequality false, the two regions above and below the branches of the hyperbola are shaded, as shown in Figure 24. Again, the points on the hyperbola itself are included in the solution.   ◆

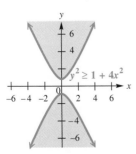

FIGURE 24

SYSTEMS OF INEQUALITIES

Just as the solution set of a system of equations is the set of all solutions common to all the equations in the system, the solution set of a system of inequalities is the set of all solutions common to all the inequalities in the system. The solution set of a system of inequalities, such as

$$x + y < 3$$
$$y > \frac{1}{2}x^2,$$

is the *intersection* of the solution sets of the individual inequalities. This is best visualized by its graph. Using a traditional graphing approach, we graph each inequality separately and then identify the region of solution by shading heavily the region common to all the graphs.

EXAMPLE 4   *Graphing a System of Inequalities, Using a Traditional Approach*   Use traditional graphing methods to graph the solution set of the system

$$x + y < 3$$
$$y > \frac{1}{2}x^2.$$

SOLUTION   The first inequality is equivalent to $y < -x + 3$, so we shade *below* the dashed line $y = -x + 3$, as seen in Figure 25. To graph $y > \frac{1}{2}x^2$, we shade *above* the dashed parabola $y = \frac{1}{2}x^2$, as seen in Figure 26. The solution set of the system is shown in Figure 27, and consists of the intersection of the two regions.   ◆

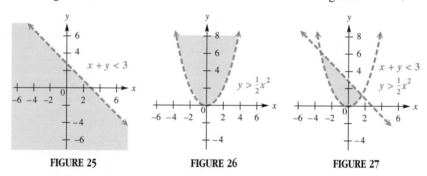

FIGURE 25   FIGURE 26   FIGURE 27

Note   While we illustrated three graphs in the solution of Example 4, in practice it is customary to give only the final graph (Figure 27). The two individual inequalities were shown simply to illustrate the procedure.

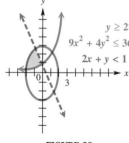

FIGURE 28

EXAMPLE 5   *Graphing a System of Inequalities, Using Traditional Methods*   Use traditional methods to graph the solution set of the system

$$y \geq 2^x$$
$$9x^2 + 4y^2 \leq 36$$
$$2x + y < 1.$$

SOLUTION   Graph the three inequalities on the same axes and shade the region common to all three, as shown in Figure 28. Two boundary lines are solid and one is dashed.   ◆

A GRAPHING CALCULATOR APPROACH TO INEQUALITIES AND SYSTEMS OF INEQUALITIES

Modern graphing calculators have the capability of shading above or below graphs of functions. Because the various makes and models differ in their shading commands, we suggest that you read your owner's manual carefully to determine how yours accomplishes shading.

EXAMPLE 6 *Graphing a Linear Inequality, Using a Graphing Calculator*
Graph $x + 4y > 4$, using a graphing calculator.

SOLUTION As seen in Example 1, this inequality is equivalent to $y > -\frac{1}{4}x + 1$. We direct the calculator to graph $y = -\frac{1}{4}x + 1$, and shade above the line. See Figure 29. Notice that we cannot tell from the graph that the boundary line is not included in the solution set. Again, this illustrates how the mathematical concepts must be understood to correctly interpret what we see on the screen. ◆

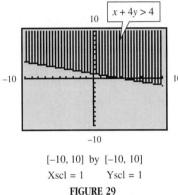

[−10, 10] by [−10, 10]
Xscl = 1   Yscl = 1
FIGURE 29

EXAMPLE 7 *Graphing a Nonlinear Inequality, Using a Graphing Calculator*
Graph $y \leq 2x^2 - 3$, using a graphing calculator.

SOLUTION The graph is that of the parabola $y = 2x^2 - 3$ and the region below the graph. Here, the parabola itself is understood to be included in the solution set. See Figure 30, and compare it to the traditional graph in Figure 22. ◆

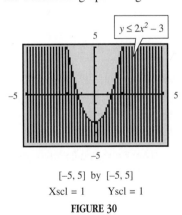

[−5, 5] by [−5, 5]
Xscl = 1   Yscl = 1
FIGURE 30

EXAMPLE 8 *Graphing a System of Inequalities, Using a Graphing Calculator* Graph the solution set of the system

$$x + y < 3$$

$$y > \frac{1}{2}x^2,$$

using a graphing calculator.

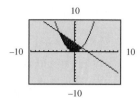

Some calculators show shaded regions like this. Compare this graph with the graph of the solution for Example 8 in Figure 31.

SOLUTION   This system was first seen in Example 4. We must direct the calculator to shade below the line $y = -x + 3$ *and* above the parabola $y = \frac{1}{2}x^2$. This is done in Figure 31. Compare to the traditional graph in Figure 27.   ◆

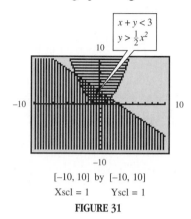

$$x + y < 3$$
$$y > \frac{1}{2}x^2$$

[−10, 10] by [−10, 10]
Xscl = 1   Yscl = 1

FIGURE 31

LINEAR PROGRAMMING

An important application of systems of inequalities is known as *linear programming*. This procedure is used to find minimum cost, maximum profit, the maximum amount of earning that can take place under given conditions, and so on. The procedure was developed by George Dantzig in 1947, when he was working on a problem of allocating supplies for the Air Force in a way that minimized total cost. The procedure involves a system of inequalities like the one seen in the next example.

EXAMPLE 9   *Graphing a System of Several Linear Inequalities, Using Traditional Methods*   Graph the solution set of the system

$$2x + 3y \geq 12$$
$$7x + 4y \geq 28$$
$$y \leq 6$$
$$x \leq 5,$$

using traditional methods.

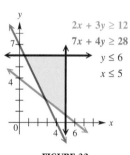

FIGURE 32

SOLUTION   The graph is obtained by graphing the four inequalities on the same axes and shading the region common to all four, as shown in Figure 32. As the graph shows, the boundary lines are all solid.   ◆

The system considered in Example 9 is typical of linear programming problems. To illustrate the method of linear programming, consider the following problem.

The Smith Company makes two products, tape decks and amplifiers. Each tape deck gives a profit of \$3, while each amplifier produces \$7 profit. The company must manufacture at least 1 tape deck per day to satisfy one of its customers, but no more than 5 because of production problems. Also, the number of amplifiers produced cannot exceed 6 per day. As a further requirement, the number of tape decks cannot exceed the number of amplifiers. How many of each should the company manufacture in order to obtain the maximum profit?

To begin, translate the statement of the problem into symbols by assuming

x = number of tape decks to be produced daily

y = number of amplifiers to be produced daily.

7.7 *Systems of Inequalities and Linear Programming*   **589**

According to the statement of the problem given above, the company must produce at least 1 tape deck (1 or more), so $x \geq 1$. Since no more than 5 tape decks may be produced, $x \leq 5$. Since no more than 6 amplifiers may be made in one day, $y \leq 6$. The requirement that the number of tape decks may not exceed the number of amplifiers translates as $x \leq y$. The number of tape decks and of amplifiers cannot be negative, so $x \geq 0$ and $y \geq 0$. These restrictions, or **constraints,** that are placed on production form the system of inequalities

$$x \geq 1, \quad x \leq 5, \quad y \leq 6, \quad x \leq y, \quad x \geq 0, \quad y \geq 0.$$

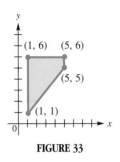

FIGURE 33

The maximum possible profit that the company can make, subject to these constraints, is found by sketching the graph of the solution of the system. See Figure 33. The only feasible values of x and y are those that satisfy all constraints. These values correspond to points that lie on the boundary or in the shaded region, called the **region of feasible solutions.**

Since each tape deck gives a profit of $3, the daily profit from the production of x tape decks is $3x$ dollars. Also, the profit from the production of y amplifiers will be $7y$ dollars per day. The total daily profit is thus given by the following **objective function:**

$$\text{Profit} = 3x + 7y.$$

The problem of the Smith Company may now be stated as follows: find values of x and y in the region of feasible solutions as shown in Figure 33 that will produce the maximum possible value of $3x + 7y$.

It can be shown that any optimum value (maximum or minimum) will always occur at a **vertex** (or **corner point**) of the region of feasible solutions. Locate the point (x, y) that gives the maximum profit by checking the coordinates of the vertex points, shown in Figure 33 and listed below. Find the profit that corresponds to each coordinate pair and choose the one that gives the maximum profit.

Point	Profit = $3x + 7y$	
(1, 1)	$3(1) + 7(1) = 10$	
(1, 6)	$3(1) + 7(6) = 45$	
(5, 6)	$3(5) + 7(6) = 57$	← Maximum
(5, 5)	$3(5) + 7(5) = 50$	

The maximum profit of $57 is obtained when 5 tape decks and 6 amplifiers are produced each day.

EXAMPLE 10   *Solving a Minimum Cost Problem, Using Linear Programming*   Margaret Westmoreland, who is very health-conscious, takes vitamin pills. Each day, she must have at least 16 units of vitamin A, at least 5 units of vitamin B_1, and at least 20 units of vitamin C. She can choose between red pills, costing 10¢ each, which contain 8 units of A, 1 of B_1, and 2 of C; and blue pills, costing 20¢ each, which contain 2 units of A, 1 of B_1, and 7 of C. How many of each pill should she buy in order to minimize her cost yet fulfill her daily requirements?

SOLUTION   Let x represent the number of red pills to buy, and let y represent the number of blue pills to buy. Then, the cost in pennies per day is given by Cost $= 10x + 20y$. Since she buys x of the 10¢ pills and y of the 20¢ pills, she gets vitamin A as follows: 8 units from each red pill and 2 units from each blue pill. Altogether, she gets $8x + 2y$ units of A per day. Since she needs at least 16 units, $8x + 2y \geq 16$. Each red pill or each blue pill supplies 1 unit of vitamin B_1. Margaret needs at least

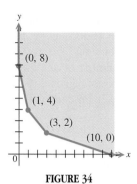

FIGURE 34

5 units per day, so $x + y \geq 5$. For vitamin C, the inequality is $2x + 7y \geq 20$. Also, $x \geq 0$ and $y \geq 0$, since Margaret cannot buy negative numbers of the pills.

The total cost of the pills can be minimized by finding the solution of the system of inequalities formed by the constraints. (See Figure 34.) The solution to this minimizing problem will also occur at a vertex point. Check the coordinates of the vertex points in the cost function to find the lowest cost.

Point	Cost = $10x + 20y$	
(10, 0)	$10(10) + 20(0) = 100$	
(3, 2)	$10(3) + 20(2) = 70$	← Minimum
(1, 4)	$10(1) + 20(4) = 90$	
(0, 8)	$10(0) + 20(8) = 160$	

Margaret's solution is to buy 3 red pills and 2 blue ones, for a total cost of 70¢ per day. She receives minimum amounts of B_1 and C, but an excess of vitamin A. Even with an excess of A, this is still the best buy.   ◆

To solve a linear programming problem in general, use the following steps.

SOLVING A LINEAR PROGRAMMING PROBLEM

1. Write the objective function and all necessary constraints.
2. Graph the feasible region.
3. Identify all vertices or corner points.
4. Find the value of the objective function at each vertex.
5. The solution is given by the vertex producing the optimum value of the objective function.

7.7   EXERCISES   **Tape 9**

Use traditional graphing methods to graph each of the following inequalities.

1. $x \leq 3$ **2.** $y \leq -2$ **3.** $x + 2y \leq 6$ **4.** $x - y \geq 2$

5. $2x + 3y \geq 4$ **6.** $4y - 3x < 5$ **7.** $3x - 5y > 6$ **8.** $x < 3 + 2y$

9. $5x \leq 4y - 2$ **10.** $2x > 3 - 4y$ **11.** $y < 3x^2 + 2$ **12.** $y \leq x^2 - 4$

13. $y > (x - 1)^2 + 2$ **14.** $y > 2(x + 3)^2 - 1$ **15.** $x^2 + (y + 3)^2 \leq 16$ **16.** $(x - 4)^2 + (y + 3)^2 \leq 9$

17. $4x^2 \leq 4 - y^2$ **18.** $x^2 + 9y^2 > 9$ **19.** $9x^2 - 16y^2 > 144$ **20.** $4x^2 \leq 36 + 9y^2$

21. Which one of the following is a description of the graph of the inequality

$$(x - 5)^2 + (y - 2)^2 < 4?$$

  (a) the region inside a circle with center $(-5, -2)$ and radius 2

  (b) the region inside a circle with center $(5, 2)$ and radius 2

  (c) the region inside a circle with center $(-5, -2)$ and radius 4

  (d) the region outside a circle with center $(5, 2)$ and radius 4

22. Without graphing, write a description of the graph of the nonlinear inequality $y > 2(x - 3)^2 + 2$.

23. Which one of the following inequalities satisfies the following description: the region outside an ellipse centered at the origin, with x-intercepts 4 and -4, and y-intercepts 9 and -9?

    (a) $\dfrac{x^2}{4} + \dfrac{y^2}{9} > 1$ **(b)** $\dfrac{x^2}{16} - \dfrac{y^2}{81} > 1$ **(c)** $\dfrac{x^2}{16} + \dfrac{y^2}{81} > 0$ **(d)** $\dfrac{x^2}{16} + \dfrac{y^2}{81} > 1$

24. Explain how it is determined whether the boundary of an inequality is a solid line or a dashed line.

In Exercises 25–28, match the inequality with the appropriate calculator-generated graph. You should not use your calculator, but rather use your knowledge of the concepts involved in graphing inequalities.

25. $y \le 3x - 6$

26. $y \ge 3x - 6$

27. $y \le -3x - 6$

28. $y \ge -3x - 6$

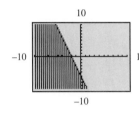
      A.

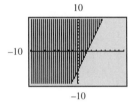
      B.

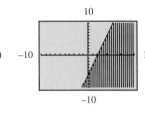
      C.

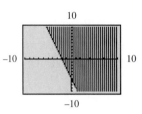
      D.

Use traditional graphing methods to graph each of the following systems of inequalities.

29. $x + y \le 4$
    $x - 2y \ge 6$

30. $2x + y > 2$
    $x - 3y < 6$

31. $4x + 3y < 12$
    $y + 4x > -4$

32. $3x + 5y \le 15$
    $x - 3y \ge 9$

33. $x + y \le 6$
    $2x + 2y \ge 12$

34. $3x + 4y < 15$
    $6x + 8y > 30$

35. $x + 2y \le 4$
    $y \ge x^2 - 1$

36. $4x - 3y \le 12$
    $y \le x^2$

37. $y \le -x^2$
    $y \ge x^2 - 6$

38. $x^2 + y^2 \le 9$
    $x \le -y^2$

39. $x^2 - y^2 < 1$
    $-1 < y < 1$

40. $x^2 + y^2 \le 36$
    $-4 \le x \le 4$

41. $2x^2 - y^2 > 4$
    $2y^2 - x^2 > 4$

42. $y \ge x^2 + 4x + 4$
    $y < -x^2$

43. $\dfrac{x^2}{16} + \dfrac{y^2}{9} \le 1$

    $\dfrac{x^2}{4} - \dfrac{y^2}{16} \ge 1$

44. $\dfrac{x^2}{36} - \dfrac{y^2}{9} \ge 1$

    $\dfrac{x^2}{81} + y^2 \le 1$

45. $\dfrac{x^2}{4} + \dfrac{y^2}{9} > 1$

    $x^2 - y^2 \ge 1$

    $-4 \le x \le 4$

46. $2x - 3y < 6$
    $4x^2 + 9y^2 < 36$
    $x \ge -1$

47. $y \ge 3^x$
    $y \ge 2$

48. $y \le \left(\dfrac{1}{2}\right)^x$

    $y \ge 4$

49. $|x| \ge 2$
    $|y| \ge 4$
    $y < x^2$

50. $|x| + 2 \ge 4$
    $|y| \le 1$
    $\dfrac{x^2}{9} + \dfrac{y^2}{16} \le 1$

51. $y \le |x + 2|$
    $\dfrac{x^2}{16} - \dfrac{y^2}{9} \le 1$

52. $y \le \log x$
    $y \ge |x - 2|$

53. Which one of the following is a description of the solution set of the system below?

$$x^2 + 4y^2 < 36$$
$$y < x$$

(a) all points outside the ellipse $x^2 + 4y^2 = 36$ and above the line $y = x$

(b) all points outside the ellipse $x^2 + 4y^2 = 36$ and below the line $y = x$

(c) all points inside the ellipse $x^2 + 4y^2 = 36$ and above the line $y = x$

(d) all points inside the ellipse $x^2 + 4y^2 = 36$ and below the line $y = x$

54. Fill in the blanks with the appropriate responses. The graph of the system

$$y > x^2 + 2$$
$$x^2 + y^2 < 16$$
$$y < 7$$

consists of all points _____ the parabola $y = x^2 + 2$, _____ the circle
$\quad\quad\quad\quad\quad\quad\quad\quad$ (above/below) $\quad\quad\quad\quad\quad\quad\quad\quad\quad\quad$ (inside/outside)
$x^2 + y^2 = 16$, and _____ the line $y = 7$.
$\quad\quad\quad\quad\quad\quad\quad$ (above/below)

In Exercises 55–58, match the system of inequalities with the appropriate calculator-generated graph. You should not use your calculator, but rather use your knowledge of the concepts involved in graphing systems of inequalities.

55. $y \geq x$
$\quad$ $y \leq 2x - 3$

56. $y \geq x^2$
$\quad$ $y < 5$

57. $x^2 + y^2 \leq 16$
$\quad$ $y \geq 0$

58. $y \leq x$
$\quad$ $y \geq 2x - 3$

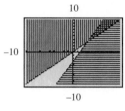
A.

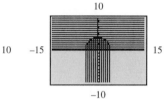
B.

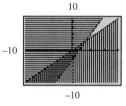
C.

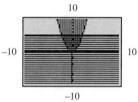
D.

TECHNOLOGY NOTE The heavy lines at $y = 0$ (the x-axis) in the graphs for Exercises 55–58 and at $y = 5$ (in choice D) are caused by the fact that the shading lines are horizontal and overlap the boundary lines at those places. Again, we must understand the mathematics to correctly interpret the technology.

Use the shading capabilities of your graphing calculator to graph each inequality or system of inequalities.

59. $3x + 2y \geq 6$

60. $y \leq x^2 + 5$

61. $x + y \geq 2$
$\quad$ $x + y \leq 6$

62. $y \geq |x + 2|$
$\quad$ $y \leq 6$

63. $y \geq 2^x$
$\quad$ $y \leq 8$

64. $y \leq x^3 + x^2 - 4x - 4$

The following systems of linear inequalities involve more than two inequalities. Use traditional graphing methods to graph the solution set of each system.

65. $3x - 2y \geq 6$
$\quad$ $x + y \leq -5$
$\quad\quad$ $y \leq 4$

66. $2x + 3y \leq 12$
$\quad$ $2x + 3y > -6$
$\quad\quad$ $3x + y < 4$
$\quad\quad\quad$ $x \geq 0$
$\quad\quad\quad$ $y \geq 0$

The graphs below represent regions of feasible solutions. Determine the maximum and minimum values of the given objective function.

67. Objective function: $3x + 5y$

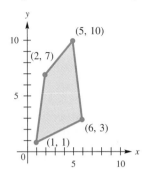

68. Objective function: $6x + y$

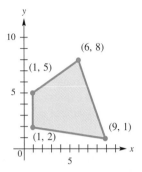

Solve each of the following linear programming problems.

69. *Maximum Profit from Farm Animals* Farmer Jones raises only pigs and geese. She wants to raise no more than 16 animals, with no more than 12 geese. She spends $50 to raise a pig and $20 to raise a goose. She has $500 available for this purpose. Find the maximum profit she can make if she makes a profit of $80 per goose and $40 per pig.

70. *Maximum Inquiries about Displayed Products* A wholesaler of party goods wishes to display her products at a convention of social secretaries in such a way that she gets the maximum number of inquiries about her whistles and hats. Her booth at the convention has 12 square meters of floor space to be used for display purposes. A display unit for hats requires 2 square meters, and for whistles, 4 square meters. Experience tells the wholesaler that she should never have more than a total of 5 units of whistles and hats on display at one time. If she receives three inquiries for each unit of hats and two inquiries for each unit of whistles on display, how many of each should she display in order to get the maximum number of inquiries?

71. *Satisfying Diet Requirements at Minimum Cost* Peggy LeBeau uses two food supplements, I and II. She can get these supplements from two different products, A and B. Product A provides 3 g per serving of supplement I and 2 g per serving of supplement II. Product B provides 2 g per serving of supplement I and 4 g per serving of supplement II. Her program director, Paulette, has recommended that she include at least 15 g of each supplement in her daily diet. If product A costs 25¢ per serving and product B costs 40¢ per serving, how can she satisfy her requirements most economically?

72. *Minimum Shipment Costs* A manufacturer of refrigerators must ship at least 100 refrigerators to its two West Coast warehouses. Each warehouse holds a maximum of 100 refrigerators. Warehouse A holds 25 refrigerators already, while warehouse B has 20 on hand. It costs $12 to ship a refrigerator to warehouse A and $10 to ship one to warehouse B. How many refrigerators should be shipped to each warehouse to minimize cost? What is the minimum cost?

73. *Maximizing Manufacturing Revenues* A machine shop manufactures two types of bolts. Each can be made on any of three groups of machines, but the time required on each group differs, as shown in the table below.

		Machine Groups		
		I	**II**	**III**
Bolts	**Type 1**	.1 min	.1 min	.1 min
	Type 2	.1 min	.4 min	.5 min

Production schedules are made up one day at a time. In a day, there are 240, 720, and 160 minutes available, respectively, on these machines. Type 1 bolts sell for 10¢ and type 2 bolts for 12¢. How many of each type of bolt should be manufactured per day to maximize revenue? What is the maximum revenue?

74. *Maximizing Manufacturing Revenues* A certain kind of manufacturing process requires that oil refineries manufacture at least 2 gallons of gasoline for each gallon of fuel oil. To meet the winter demand for fuel oil, at least 3 million gallons a day must be produced. The demand for gasoline is no more than 6.4 million gallons per day. If the price of gasoline is $1.90 and the price of fuel oil is $1.50 per gallon, how much of each should be produced to maximize revenue?

75. *Maximizing Aid to Disaster Victims* Earthquake victims in China need medical supplies and bottled water. Each medical kit measures 1 cubic foot and weighs 10 pounds. Each container of water is also 1 cubic foot but weighs 20 pounds. The plane can carry only 80,000 pounds, with a total volume of 6000 cubic feet. Each medical kit will aid 4 people, while each container of water will serve 10 people. How many of each should be sent in order to maximize the number of people aided?

76. *Maximizing Aid to Disaster Victims* If each medical kit could aid 6 people instead of 4, how would the results in Exercise 75 change?

7.8   PARTIAL FRACTIONS

Introduction ◆ Distinct Linear Factors ◆ Repeated Linear Factors ◆ Distinct Linear and Quadratic Factors ◆ Repeated Quadratic Factors

INTRODUCTION

The sums of rational expressions are found by combining two or more rational expressions into one rational expression. Here, the reverse process is considered: given one rational expression, express it as the sum of two or more rational expressions. A special type of sum of rational expressions is called the **partial fraction decomposition;** each term in the sum is a **partial fraction.** The technique of decomposing a rational expression into partial fractions is useful in calculus and other areas of mathematics.

To form a partial fraction decomposition of a rational expression, we use the following steps.

PARTIAL FRACTION DECOMPOSITION OF $\frac{f(x)}{g(x)}$

Step 1 If $f(x)/g(x)$ is not a proper fraction (a fraction with the numerator of lower degree than the denominator), divide $f(x)$ by $g(x)$. For example,

$$\frac{x^4 - 3x^3 + x^2 + 5x}{x^2 + 3} = x^2 - 3x - 2 + \frac{14x + 6}{x^2 + 3}.$$

Then, apply the following steps to the remainder, which is a proper fraction.

Step 2 Factor $g(x)$ completely into factors of the form $(ax + b)^m$ or $(cx^2 + dx + e)^n$, where $cx^2 + dx + e$ is irreducible and m and n are integers.

Step 3

(a) For each distinct linear factor $(ax + b)$, the decomposition must include the term $\frac{A}{ax + b}$.

(b) For each repeated linear factor $(ax + b)^m$, the decomposition must include the terms

$$\frac{A_1}{ax + b} + \frac{A_2}{(ax + b)^2} + \cdots + \frac{A_m}{(ax + b)^m}.$$

Step 4

(a) For each distinct quadratic factor $(cx^2 + dx + e)$, the decomposition must include the term $\frac{Bx + C}{cx^2 + dx + e}$.

(b) For each repeated quadratic factor $(cx^2 + dx + e)^n$, the decomposition must include the terms

$$\frac{B_1x + C_1}{cx^2 + dx + e} + \frac{B_2x + C_2}{(cx^2 + dx + e)^2} + \cdots + \frac{B_nx + C_n}{(cx^2 + dx + e)^n}.$$

Step 5 Use algebraic techniques to solve for the constants in the numerators of the decomposition.

To find the constants in step 5, the goal is to get a system of equations with as many equations as there are unknowns in the numerators. One method for getting these equations is to substitute values for x on both sides of the rational equation formed from steps 3 or 4.

DISTINCT LINEAR FACTORS

EXAMPLE 1 *Finding a Partial Fraction Decomposition* Find the partial fraction decomposition of

$$\frac{2x^4 - 8x^2 + 5x - 2}{x^3 - 4x}.$$

SOLUTION The given fraction is not a proper fraction; the numerator has higher degree than the denominator. Perform the division.

$$
\begin{array}{r}
2x \\
x^3 - 4x \overline{)2x^4 - 8x^2 + 5x - 2} \\
\underline{2x^4 - 8x^2} \\
5x - 2
\end{array}
$$

The quotient is $\dfrac{2x^4 - 8x^2 + 5x - 2}{x^3 - 4x} = 2x + \dfrac{5x - 2}{x^3 - 4x}.$ Now, work with the remainder fraction. Factor the denominator as $x^3 - 4x = x(x + 2)(x - 2)$. Since the factors are **distinct linear factors,** use step 3(a) to write the decomposition as

$$\frac{5x - 2}{x^3 - 4x} = \frac{A}{x} + \frac{B}{x + 2} + \frac{C}{x - 2}, \tag{1}$$

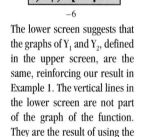

The lower screen suggests that the graphs of Y_1 and Y_2, defined in the upper screen, are the same, reinforcing our result in Example 1. The vertical lines in the lower screen are not part of the graph of the function. They are the result of using the connected graphing mode.

where A, B, and C are constants that need to be found. Multiply both sides of equation (1) by $x(x + 2)(x - 2)$, getting

$$5x - 2 = A(x + 2)(x - 2) + Bx(x - 2) + Cx(x + 2). \tag{2}$$

Equation (1) is an identity, since both sides represent the same rational expression. Thus, equation (2) is also an identity. Equation (1) holds for all values of x except 0, -2, and 2. However, equation (2) holds for all values of x. In particular, substituting 0 for x in equation (2) gives $-2 = -4A$, so that $A = \frac{1}{2}$. Similarly, choosing $x = -2$ gives $-12 = 8B$, so that $B = -\frac{3}{2}$. Finally, choosing $x = 2$, $8 = 8C$, so that $C = 1$. The remainder rational expression can be written as the following sum of partial fractions:

$$\frac{5x - 2}{x^3 - 4x} = \frac{1}{2x} + \frac{-3}{2(x + 2)} + \frac{1}{x - 2},$$

and the given rational expression can be written as

$$\frac{2x^4 - 8x^2 + 5x - 2}{x^3 - 4x} = 2x + \frac{1}{2x} + \frac{-3}{2(x + 2)} + \frac{1}{x - 2}.$$

Check the work by combining the terms on the right. ◆

REPEATED LINEAR FACTORS

EXAMPLE 2 *Finding a Partial Fraction Decomposition* Find the partial fraction decomposition of $\frac{2x}{(x - 1)^3}$.

SOLUTION This is a proper fraction. The denominator is already factored with **repeated linear factors.** Write the decomposition as shown, by using step 3(b) above.

$$\frac{2x}{(x - 1)^3} = \frac{A}{x - 1} + \frac{B}{(x - 1)^2} + \frac{C}{(x - 1)^3}$$

Clear the denominators by multiplying both sides of this equation by $(x - 1)^3$.

$$2x = A(x - 1)^2 + B(x - 1) + C$$

Substituting 1 for x leads to $C = 2$, so that

$$2x = A(x - 1)^2 + B(x - 1) + 2. \tag{1}$$

The only root has been substituted, and values for A and B still need to be found. However, *any* number can be substituted for x. For example, when we choose $x = -1$ (because it is easy to substitute), equation (1) becomes

$$-2 = 4A - 2B + 2$$
$$-4 = 4A - 2B$$
$$-2 = 2A - B. \tag{2}$$

Substituting 0 for x in equation (1) gives

$$0 = A - B + 2$$
$$2 = -A + B. \tag{3}$$

Now, solve the system of equations (2) and (3) to get $A = 0$ and $B = 2$. The partial fraction decomposition is

$$\frac{2x}{(x - 1)^3} = \frac{2}{(x - 1)^2} + \frac{2}{(x - 1)^3}.$$

Three substitutions were needed because there were three constants to evaluate, A, B, and C.

To check this result, combine the terms on the right.   ◆

DISTINCT LINEAR AND QUADRATIC FACTORS

EXAMPLE 3   *Finding a Partial Fraction Decomposition*   Find the partial fraction decomposition of $\frac{x^2 + 3x - 1}{(x + 1)(x^2 + 2)}$.

SOLUTION   This denominator has **distinct linear and quadratic factors,** where neither is repeated. Since $x^2 + 2$ cannot be factored, it is irreducible. The partial fraction decomposition is

$$\frac{x^2 + 3x - 1}{(x + 1)(x^2 + 2)} = \frac{A}{x + 1} + \frac{Bx + C}{x^2 + 2}.$$

Multiply both sides by $(x + 1)(x^2 + 2)$ to get

$$x^2 + 3x - 1 = A(x^2 + 2) + (Bx + C)(x + 1). \tag{1}$$

First, substitute -1 for x to get

$$(-1)^2 + 3(-1) - 1 = A[(-1)^2 + 2] + 0$$
$$-3 = 3A$$
$$A = -1.$$

Replace A with -1 in equation (1) and substitute any value for x. For instance, if $x = 0$,

$$0^2 + 3(0) - 1 = -1(0^2 + 2) + (B \cdot 0 + C)(0 + 1)$$
$$-1 = -2 + C$$
$$C = 1.$$

Now, letting $A = -1$ and $C = 1$, substitute again in equation (1), using another number for x. For $x = 1$,

$$3 = -3 + (B + 1)(2)$$
$$6 = 2B + 2$$
$$B = 2.$$

Using $A = -1$, $B = 2$, and $C = 1$, the partial fraction decomposition is

$$\frac{x^2 + 3x - 1}{(x + 1)(x^2 + 2)} = \frac{-1}{x + 1} + \frac{2x + 1}{x^2 + 2}.$$

Again, this work can be checked by combining terms on the right. ◆

For fractions with denominators that have quadratic factors, another method is often more convenient. The system of equations is formed by equating coefficients of like terms on both sides of the partial fraction decomposition. For instance, in Example 3, after both sides were multiplied by the common denominator, the equation was

$$x^2 + 3x - 1 = A(x^2 + 2) + (Bx + C)(x + 1).$$

Multiplying on the right and collecting like terms, we have

$$x^2 + 3x - 1 = Ax^2 + 2A + Bx^2 + Bx + Cx + C$$
$$x^2 + 3x - 1 = (A + B)x^2 + (B + C)x + (C + 2A).$$

Now, equating the coefficients of like powers of x gives the three equations

$$1 = A + B$$
$$3 = B + C$$
$$-1 = C + 2A.$$

Solving this system of equations for A, B, and C would give the partial fraction decomposition. The next example uses a combination of the two methods.

REPEATED QUADRATIC FACTORS

EXAMPLE 4 *Finding a Partial Fraction Decomposition* Find the partial fraction decomposition of $\frac{2x}{(x^2 + 1)^2(x - 1)}$.

SOLUTION This expression has both a linear factor and a **repeated quadratic factor.** By steps 3(a) and 4(b) from the beginning of this section,

$$\frac{2x}{(x^2 + 1)^2(x - 1)} = \frac{Ax + B}{x^2 + 1} + \frac{Cx + D}{(x^2 + 1)^2} + \frac{E}{x - 1}.$$

Multiplication of both sides by $(x^2 + 1)^2(x - 1)$ leads to

$$2x = (Ax + B)(x^2 + 1)(x - 1) + (Cx + D)(x - 1) + E(x^2 + 1)^2. \quad \textbf{(1)}$$

If $x = 1$, equation (1) reduces to $2 = 4E$, or $E = \frac{1}{2}$. Substituting $1/2$ for E in equation (1) and combining terms on the right gives

$$2x = (A + 1/2)x^4 + (-A + B)x^3 + (A - B + C + 1)x^2 +$$
$$(-A + B + D - C)x + (-B - D + 1/2). \quad \textbf{(2)}$$

To get additional equations involving the unknowns, equate the coefficients of like powers of x on the two sides of equation (2). Setting corresponding coefficients of x^4 equal,

$0 = A + \frac{1}{2}$ or $A = -\frac{1}{2}$. From the corresponding coefficients of x^3, $0 = -A + B$. Since $A = -1/2$, $B = -1/2$. Using the coefficients of x^2, $0 = A - B + C + 1$. Since $A = -1/2$ and $B = -1/2$, $C = -1$. Finally, from the coefficients of x, $2 = -A + B + D - C$. Substituting for A, B, and C gives $D = 1$. With $A = -1/2$, $B = -1/2$, $C = -1$, $D = 1$, and $E = 1/2$, the given fraction has the partial fraction decomposition

$$\frac{2x}{(x^2 + 1)^2(x - 1)} = \frac{-\dfrac{1}{2}x - \dfrac{1}{2}}{x^2 + 1} + \frac{-x + 1}{(x^2 + 1)^2} + \frac{\dfrac{1}{2}}{x - 1}$$

or

$$\frac{2x}{(x^2 + 1)^2(x - 1)} = \frac{-(x + 1)}{2(x^2 + 1)} + \frac{-x + 1}{(x^2 + 1)^2} + \frac{1}{2(x - 1)}. \qquad \blacklozenge$$

In summary, to solve for the constants in the numerators of a partial fraction decomposition, use either of the following methods or a combination of the two.

Method 1   For Linear Factors

1. Multiply both sides of the rational expression by the common denominator.
2. Substitute the zero of each factor in the resulting equation. For repeated linear factors, substitute as many other numbers as necessary to find all the constants in the numerators. The number of substitutions required will equal the number of constants $A, B, \ldots$.

Method 2   For Quadratic Factors

1. Multiply both sides of the rational expression by the common denominator.
2. Collect terms on the right side of the resulting equation.
3. Equate the coefficients of like terms to get a system of equations.
4. Solve the system to find the constants in the numerators.

7.8 EXERCISES   Tape 9

Find the partial fraction decomposition for the following rational expressions.

1. $\dfrac{5}{3x(2x + 1)}$ **2.** $\dfrac{3x - 1}{x(x + 1)}$ **3.** $\dfrac{4x + 2}{(x + 2)(2x - 1)}$ **4.** $\dfrac{x + 2}{(x + 1)(x - 1)}$

5. $\dfrac{x}{x^2 + 4x - 5}$ **6.** $\dfrac{5x - 3}{(x + 1)(x - 3)}$ **7.** $\dfrac{2x}{(x + 1)(x + 2)^2}$ **8.** $\dfrac{2}{x^2(x + 3)}$

9. $\dfrac{4}{x(1 - x)}$ **10.** $\dfrac{4x^2 - 4x^3}{x^2(1 - x)}$ **11.** $\dfrac{4x^2 - x - 15}{x(x + 1)(x - 1)}$ **12.** $\dfrac{2x + 1}{(x + 2)^3}$

13. $\dfrac{x^2}{x^2 + 2x + 1}$ **14.** $\dfrac{3}{x^2 + 4x + 3}$ **15.** $\dfrac{2x^5 + 3x^4 - 3x^3 - 2x^2 + x}{2x^2 + 5x + 2}$

16. $\dfrac{6x^5 + 7x^4 - x^2 + 2x}{3x^2 + 2x - 1}$ **17.** $\dfrac{x^3 + 4}{9x^3 - 4x}$ **18.** $\dfrac{x^3 + 2}{x^3 - 3x^2 + 2x}$

19. $\dfrac{-3}{x^2(x^2 + 5)}$ **20.** $\dfrac{2x + 1}{(x + 1)(x^2 + 2)}$ **21.** $\dfrac{3x - 2}{(x + 4)(3x^2 + 1)}$

22. $\dfrac{3}{x(x+1)(x^2+1)}$ **23.** $\dfrac{1}{x(2x+1)(3x^2+4)}$ **24.** $\dfrac{x^4+1}{x(x^2+1)^2}$

25. $\dfrac{3x-1}{x(2x^2+1)^2}$ **26.** $\dfrac{3x^4+x^3+5x^2-x+4}{(x-1)(x^2+1)^2}$ **27.** $\dfrac{-x^4-8x^2+3x-10}{(x+2)(x^2+4)^2}$

28. $\dfrac{x^2}{x^4-1}$ **29.** $\dfrac{5x^5+10x^4-15x^3+4x^2+13x-9}{x^3+2x^2-3x}$ **30.** $\dfrac{3x^6+3x^4+3x}{x^4+x^2}$

*Determine whether each of the following partial fraction decompositions is correct by graphing
the left side and the right side of the equation on the same coordinate system and observing
whether the graphs coincide.*

31. $\dfrac{4x^2-3x-4}{x^3+x^2-2x}=\dfrac{2}{x}+\dfrac{-1}{x-1}+\dfrac{3}{x+2}$ **32.** $\dfrac{1}{(x-1)(x+2)}=\dfrac{1}{x-1}-\dfrac{1}{x+2}$

33. $\dfrac{x^3-2x}{(x^2+2x+2)^2}=\dfrac{x-2}{x^2+2x+2}+\dfrac{2}{(x^2+2x+2)^2}$ **34.** $\dfrac{2x+4}{x^2(x-2)}=\dfrac{-2}{x}+\dfrac{-2}{x^2}+\dfrac{2}{x-2}$

CHAPTER 7   SUMMARY

Section	Important Concepts
7.1 Systems of Equations	Systems of equations in two variables may be solved by the substitution method, the elimination method, or graphically, using the intersection-of-graphs method.
7.2 Solution of Linear Systems by the Echelon Method	**TRANSFORMATIONS OF A SYSTEM** The following transformations can be applied to a system of linear equations to get an equivalent system: 1. exchanging any two equations; 2. multiplying both sides of an equation by any nonzero real number; 3. replacing any equation by a nonzero multiple of that equation plus a nonzero multiple of any other equation.
7.3 Solution of Linear Systems by Row Transformations	**MATRIX ROW TRANSFORMATIONS** For any augmented matrix of a system of linear equations, the following row transformations will result in the matrix of an equivalent system. 1. Any two rows may be interchanged. 2. The elements of any row may be multiplied by a nonzero real number. 3. Any row may be changed by adding to its elements a multiple of the corresponding elements of another row.
7.4 Properties of Matrices	**DEFINITION OF MATRIX ADDITION** The sum of two $m \times n$ matrices A and B is the $m \times n$ matrix $A+B$ in which each element is the sum of the corresponding elements of A and B. **MATRIX SUBTRACTION** If A and B are matrices with the same dimension, then $A-B=A+(-B)$.

DEFINITION OF MATRIX MULTIPLICATION

The product AB of an $m \times n$ matrix A and an $n \times k$ matrix B is found as follows. To get the ith row, jth column element of AB, multiply each element in the ith row of A by the corresponding element in the jth column of B. The sum of these products will give the element of row i, column j of AB. The dimension of AB is $m \times k$.

7.5 Determinants and Cramer's Rule

DEFINITION OF THE DETERMINANT OF A 2 × 2 MATRIX

The **determinant of a 2 × 2 matrix** A, $A = \begin{bmatrix} a_{11} & a_{12} \\ a_{21} & a_{22} \end{bmatrix}$, is defined as $\det A = a_{11}a_{22} - a_{21}a_{12}$.

DEFINITION OF COFACTOR

Let M_{ij} be the minor for element a_{ij} in a dimension $n \times n$ matrix. The cofactor of a_{ij} is $A_{ij} = (-1)^{i+j} \cdot M_{ij}$.

FINDING THE DETERMINANT OF A MATRIX

Multiply each element in any row or column of the matrix by its cofactor. The sum of these products gives the value of the determinant.

CRAMER'S RULE FOR 2 × 2 SYSTEMS

For the system

$$a_1x + b_1y = c_1$$
$$a_2x + b_2y = c_2,$$

$$x = \frac{D_x}{D} \quad \text{and} \quad y = \frac{D_y}{D},$$

where

$$D_x = \det \begin{bmatrix} c_1 & b_1 \\ c_2 & b_2 \end{bmatrix}, \quad D_y = \det \begin{bmatrix} a_1 & c_1 \\ a_2 & c_2 \end{bmatrix}, \quad \text{and} \quad D = \det \begin{bmatrix} a_1 & b_1 \\ a_2 & b_2 \end{bmatrix} \neq 0.$$

Cramer's rule can be extended to 3×3 and larger systems.

7.6 Solution of Linear Systems by Matrix Inverses

DEFINITION OF THE MULTIPLICATIVE INVERSE OF A SQUARE MATRIX OF DIMENSION 2 × 2

If $A = \begin{bmatrix} a & b \\ c & d \end{bmatrix}$ and $\det A \neq 0$, then

$$A^{-1} = \frac{1}{\det A} \begin{bmatrix} d & -b \\ -c & a \end{bmatrix}$$

or

$$A^{-1} = \begin{bmatrix} \dfrac{d}{ad-cb} & \dfrac{-b}{ad-cb} \\ \dfrac{-c}{ad-cb} & \dfrac{a}{ad-cb} \end{bmatrix}.$$

To find the multiplicative inverse of a 3×3 (or larger) matrix A, use a calculator or perform row transformations on $[A \,|\, I_n]$ to change it to $[I_n \,|\, B]$, where matrix B is A^{-1}.

7.7 Systems of Inequalities and Linear Programming

GRAPHING INEQUALITIES

1. For a function f, the graph of $y < f(x)$ consists of all the points that are below the graph of $y = f(x)$; the graph of $y > f(x)$ consists of all the points that are above the graph of $y = f(x)$.

2. If the inequality is not or cannot be solved for y, choose a test point not on the boundary. If the test point satisfies the inequality, the graph includes all points on the same side of the boundary as the test point. Otherwise, the graph includes all points on the other side of the boundary.

SOLVING A LINEAR PROGRAMMING PROBLEM

1. Write the objective function and all necessary constraints.
2. Graph the feasible region.
3. Identify all vertices or corner points.
4. Find the value of the objective function at each vertex.
5. The solution is given by the vertex producing the optimum value of the objective function.

7.8 Partial Fractions

To solve for the constants in the numerators of a partial fraction decomposition, use either of the following methods or a combination of the two.

Method 1 For Linear Factors

1. Multiply both sides of the rational expression by the common denominator.
2. Substitute the zero of each factor in the resulting equation. For repeated linear factors, substitute as many other numbers as necessary to find all the constants in the numerators. The number of substitutions required will equal the number of constants $A, B, \ldots$.

Method 2 For Quadratic Factors

1. Multiply both sides of the rational expression by the common denominator.
2. Collect terms on the right side of the resulting equation.
3. Equate the coefficients of like terms to get a system of equations.
4. Solve the system to find the constants in the numerators.

CHAPTER 7   REVIEW EXERCISES

Use the substitution method to solve each of the following systems. Identify any systems with dependent equations or any inconsistent systems.

1. $4x - 3y = -1$
   $3x + 5y = 50$

2. $\dfrac{x}{2} - \dfrac{y}{5} = \dfrac{11}{10}$
   $2x - \dfrac{4y}{5} = \dfrac{22}{5}$

3. $4x + 5y = 5$
   $3x + 7y = -6$

4. $y = x^2 - 1$
   $x + y = 1$

5. $x^2 + y^2 = 2$
   $3x + y = 4$

6. $x^2 + 2y^2 = 22$
   $2x^2 - y^2 = -1$

7. $x^2 - 4y^2 = 19$
   $x^2 + y^2 = 29$

8. $xy = 4$
   $x - 6y = 2$

9. $x^2 + 2xy + y^2 = 4$
   $x = 3y - 2$

10. Use your calculator with a window of $[-18, 18]$ by $[-12, 12]$ to answer the following.

   (a) Do the circle $x^2 + y^2 = 144$ and the line $x + 2y = 8$ have any points in common?

   (b) Approximate any intersection points to the nearest tenth.

   (c) Find the exact values of the coordinates of the points of intersection analytically.

11. Can a system of two linear equations in two variables have exactly two solutions? Explain.

12. Consider the system in Exercise 5.

  (a) To graph the first equation, what two functions must you enter into your calculator?

  (b) To graph the second equation, what function must you enter?

  (c) What would be an appropriate window in which to graph this system?

Use the addition or echelon method to solve each of the following linear systems.

13. $2x - 3y + z = -5$
  $x + 4y + 2z = 13$
  $5x + 5y + 3z = 14$

14. $x - 3y = 12$
  $2y + 5z = 1$
  $4x + z = 25$

15. $x + y - z = 5$
  $2x + y + 3z = 2$
  $4x - y + 2z = -1$

16. $5x - 3y + 2z = -5$
  $2x + y - z = 4$
  $-4x - 2y + 2z = -1$

17. Can a system consisting of two equations in three variables have a unique solution? Explain.

Use the Gaussian reduction method to solve the system.

18. $2x + 3y = 10$
  $-3x + y = 18$

19. $3x + y = -7$
  $x - y = -5$

20. $x - z = -3$
  $y + z = 6$
  $2x - 3z = -9$

21. $2x - y + 4z = -1$
  $-3x + 5y - z = 5$
  $2x + 3y + 2z = 3$

Perform each of the following operations whenever possible.

22. $\begin{bmatrix} 3 & -4 & 2 \\ 5 & -1 & 6 \end{bmatrix} + \begin{bmatrix} -3 & 2 & 5 \\ 1 & 0 & 4 \end{bmatrix}$

23. $\begin{bmatrix} 3 \\ 2 \\ 5 \end{bmatrix} - \begin{bmatrix} 8 \\ -4 \\ 6 \end{bmatrix} + \begin{bmatrix} 1 \\ 0 \\ 2 \end{bmatrix}$

24. $\begin{bmatrix} 2 & 5 & 8 \\ 1 & 9 & 2 \end{bmatrix} - \begin{bmatrix} 3 & 4 \\ 7 & 1 \end{bmatrix}$

25. $3\begin{bmatrix} 2 & 4 \\ -1 & 4 \end{bmatrix} - 2\begin{bmatrix} 5 & 8 \\ 2 & -2 \end{bmatrix}$

26. $-1\begin{bmatrix} 3 & -5 & 2 \\ 1 & 7 & -4 \end{bmatrix} + 5\begin{bmatrix} 0 & 2 \\ -1 & 3 \end{bmatrix}$

27. $10\begin{bmatrix} 2x + 3y & 4x + y \\ x - 5y & 6x + 2y \end{bmatrix} + 2\begin{bmatrix} -3x - y & x + 6y \\ 4x + 2y & 5x - y \end{bmatrix}$

28. Complete the following sentence. The sum of two $m \times n$ matrices A and B is found _____.

Find each of the following matrix products whenever possible.

29. $\begin{bmatrix} -3 & 4 \\ 2 & 8 \end{bmatrix}\begin{bmatrix} -1 & 0 \\ 2 & 5 \end{bmatrix}$

30. $\begin{bmatrix} 3 & 2 & -1 \\ 4 & 0 & 6 \end{bmatrix}\begin{bmatrix} -2 & 0 \\ 0 & 2 \\ 3 & 1 \end{bmatrix}$

31. $\begin{bmatrix} 1 & -2 & 4 & 2 \\ 0 & 1 & -1 & 8 \end{bmatrix}\begin{bmatrix} -1 \\ 2 \\ 0 \\ 1 \end{bmatrix}$

32. $\begin{bmatrix} 1 & 2 & 5 \\ -3 & 4 & 7 \\ 0 & 2 & -1 \end{bmatrix}\begin{bmatrix} 4 & 2 & 3 \\ 10 & -5 & 6 \end{bmatrix}$

33. $\begin{bmatrix} 4 & 2 & 3 \\ 10 & -5 & 6 \end{bmatrix}\begin{bmatrix} 1 & 2 & 5 \\ -3 & 4 & 7 \\ 0 & 2 & -1 \end{bmatrix}$

34. $\begin{bmatrix} 3 & -1 & 0 \end{bmatrix}\begin{bmatrix} 1 & 3 & 2 \\ 2 & -4 & 0 \\ 5 & 7 & 3 \end{bmatrix}$

35. Which of the following properties does not apply to multiplication of matrices?

  (a) commutative   **(b)** associative   **(c)** distributive   **(d)** identity

36. What must be true of two matrices to find their product?

Decide whether or not each of the following pairs of matrices are multiplicative inverses.

37. $\begin{bmatrix} 2 & -3 \\ 1 & -2 \end{bmatrix}, \begin{bmatrix} 2 & -3 \\ 1 & -2 \end{bmatrix}$

38. $\begin{bmatrix} 1 & 0 \\ 2 & -3 \end{bmatrix}, \begin{bmatrix} 1 & 0 \\ \frac{2}{3} & -\frac{1}{3} \end{bmatrix}$

39. $\begin{bmatrix} 2 & 0 & 6 \\ 0 & 1 & 0 \\ 1 & 0 & 1 \end{bmatrix}, \begin{bmatrix} -1 & 0 & \frac{3}{2} \\ 0 & 1 & 0 \\ \frac{1}{4} & 0 & -1 \end{bmatrix}$

40. $\begin{bmatrix} 1 & 0 & 2 \\ 0 & 2 & 4 \\ 0 & 0 & 1 \end{bmatrix}, \begin{bmatrix} 1 & 0 & -2 \\ 0 & \frac{1}{2} & -2 \\ 0 & 0 & 1 \end{bmatrix}$

Find the inverse, if it exists, for each of the following matrices.

41. $\begin{bmatrix} 2 & 1 \\ 5 & 3 \end{bmatrix}$

42. $\begin{bmatrix} -4 & 2 \\ 0 & 3 \end{bmatrix}$

43. $\begin{bmatrix} 2 & 0 \\ -1 & 5 \end{bmatrix}$

44. $\begin{bmatrix} 2 & 0 & 4 \\ 1 & -1 & 0 \\ 0 & 1 & -2 \end{bmatrix}$

45. $\begin{bmatrix} 2 & -1 & 0 \\ 1 & 0 & 1 \\ 1 & -2 & 0 \end{bmatrix}$

46. $\begin{bmatrix} 2 & 3 & 5 \\ -2 & -3 & -5 \\ 1 & 4 & 2 \end{bmatrix}$

Use matrix inverses to solve each of the following. Identify any inconsistent systems or systems with dependent equations.

47. $x + y = 4$
$2x + 3y = 10$

48. $5x - 3y = -2$
$2x + 7y = -9$

49. $2x + y = 5$
$3x - 2y = 4$

50. $x - 2y = 7$
$3x + y = 7$

51. $3x - 2y + 4z = 1$
$4x + y - 5z = 2$
$-6x + 4y - 8z = -2$

52. $x + 2y = -1$
$3y - z = -5$
$x + 2y - z = -3$

53. $x + y + z = 1$
$2x - y = -2$
$3y + z = 2$

54. $x = -3$
$y + z = 6$
$2x - 3z = -9$

Evaluate each of the following determinants.

55. $\det \begin{bmatrix} -1 & 8 \\ 2 & 9 \end{bmatrix}$

56. $\det \begin{bmatrix} -2 & 4 \\ 0 & 3 \end{bmatrix}$

57. $\det \begin{bmatrix} -2 & 4 & 1 \\ 3 & 0 & 2 \\ -1 & 0 & 3 \end{bmatrix}$

58. $\det \begin{bmatrix} -1 & 2 & 3 \\ 4 & 0 & 3 \\ 5 & -1 & 2 \end{bmatrix}$

Solve each of the following determinant equations for x.

59. $\det \begin{bmatrix} -3 & 2 \\ 1 & x \end{bmatrix} = 5$

60. $\det \begin{bmatrix} 3x & 7 \\ -x & 4 \end{bmatrix} = 8$

61. $\det \begin{bmatrix} 2 & 5 & 0 \\ 1 & 3x & -1 \\ 0 & 2 & 0 \end{bmatrix} = 4$

62. $\det \begin{bmatrix} 6x & 2 & 0 \\ 1 & 5 & 3 \\ x & 2 & -1 \end{bmatrix} = 2x$

Exercises 63 and 64 refer to the system below.

$$3x - y = 28$$
$$2x + y = 2$$

63. Suppose you are asked to solve this system using Cramer's rule.

(a) What is the value of D?   (b) What is the value of D_x?

(c) What is the value of D_y?   (d) Find x and y, using Cramer's rule.

64. Suppose you are asked to solve this system by using the inverse matrix method. (a) What is A? (b) What is B? (c) Explain how you would go about solving the system, using these matrices.

65. Cramer's rule has the condition that $D \neq 0$. Why is this necessary? What is true of the system when $D = 0$?

Solve each of the following systems by Cramer's rule. Identify any systems with dependent equations or any inconsistent systems.

66. $3x + y = -1$
$5x + 4y = 10$

67. $3x + 7y = 2$
$5x - y = -22$

68. $2x - 5y = 8$
$3x + 4y = 10$

69. $3x + 2y + z = 2$
$4x - y + 3z = -16$
$x + 3y - z = 12$

70. $5x - 2y - z = 8$
$-5x + 2y + z = -8$
$x - 4y - 2z = 0$

71. $-x + 3y - 4z = 2$
$2x + 4y + z = 3$
$3x - z = 9$

Write a system of linear equations, and then use the system to solve the problem.

72. *Determining the Contents of a Meal* A cup of uncooked rice contains 15 grams of protein and 810 calories. A cup of uncooked soybeans contains 22.5 grams of protein and 270 calories. How many cups of each should be used for a meal containing 9.5 grams of protein and 324 calories?

73. *Determining Order Quantities* A company sells $3\frac{1}{2}''$ diskettes for 40¢ each and sells $5\frac{1}{4}''$ diskettes for 30¢ each. The company receives $38 for an order of 100 diskettes. However, the customer neglected to specify how many of each size to send. Determine the number of each size of diskette that should be sent.

74. *Mixing Teas* Three kinds of tea worth $4.60, $5.75, and $6.50 per pound are to be mixed to get 20 pounds of tea worth $5.25 per pound. The amount of $4.60 tea used is to be equal to the total amount of the other two kinds together. How many pounds of each tea should be used?

75. *Mixing Solutions of a Drug* A 5% solution of a drug is to be mixed with some 15% solution and some 10% solution to get 20 milliliters of 8% solution. The amount of 5% solution used must be 2 milliliters more than the sum of the other two solutions. How many ml of each solution should be used?

76. Find the equation of the quadratic polynomial $P(x)$ that defines the curve shown in the figures.

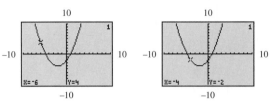

First View   Second View

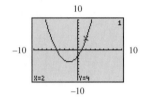

Third View

Graph the solution of each of the following systems of inequalities.

77. $x + y \leq 6$
    $2x - y \geq 3$

78. $x - 3y \geq 6$
    $y^2 \leq 16 - x^2$

79. $9x^2 + 16y^2 \geq 144$
    $x^2 - y^2 \geq 16$

80. Find $x \geq 0$ and $y \geq 0$ such that

$$3x + 2y \leq 12$$
$$5x + y \geq 5$$

and $2x + 4y$ is maximized.

81. Find $x \geq 0$ and $y \geq 0$ such that

$$x + y \leq 50$$
$$2x + y \geq 20$$
$$x + 2y \geq 30$$

and $4x + 2y$ is minimized.

Time Management   Set up the following linear programming problem to be solved by writing a system of inequalities to express the constraints and the equation of the objective function.

82. A bakery makes both cakes and cookies. Each batch of cakes requires 2 hours in the oven and 3 hours in the decorating room. Each batch of cookies needs $1\frac{1}{2}$ hours in the oven and $\frac{2}{3}$ hour in the decorating room. The oven is available no more than 15 hours per day, and the decorating room can be used no more than 13 hours per day. Set up a system of inequalities, and then graph the solution of the system. How many batches of cakes and cookies should the bakery make in order to maximize profits if cookies produce a profit of $20 per batch and cakes produce a profit of $30 per batch?

Find the partial fraction decomposition of each rational expression.

83. $\dfrac{5x - 2}{x^2 - 4}$

84. $\dfrac{x + 2}{x^3 + 2x^2 + x}$

CHAPTER 7   TEST

1. Consider the system of equations

$$x^2 - 4y^2 = -15$$
$$3x + y = 1.$$

(a) What type of graph does each equation have?

(b) How many points of intersection of the two graphs are possible?

(c) Solve the system.

(d) What does your answer to part (c) tell you about the actual number of points of intersection of the graphs?

2. Consider the linear system of equations

$$2x + y + z = 3$$
$$x + 2y - z = 3$$
$$3x - y + z = 5.$$

(a) Solve the system, using Gaussian reduction (row transformations).

(b) Compare the steps used to solve this system by the echelon method with the steps used with the Gaussian method. How are they similar? How do they differ?

3. Perform the indicated matrix operations if possible.

(a) $3\begin{bmatrix} 2 & 3 \\ 1 & -4 \\ 5 & 9 \end{bmatrix} - \begin{bmatrix} -2 & 6 \\ 3 & -1 \\ 0 & 8 \end{bmatrix}$

(b) $\begin{bmatrix} 1 \\ 2 \end{bmatrix} + \begin{bmatrix} 4 \\ -6 \end{bmatrix} + \begin{bmatrix} 2 & 8 \\ -7 & 5 \end{bmatrix}$

(c) $\begin{bmatrix} 2 & 1 & -3 \\ 4 & 0 & 5 \end{bmatrix} \begin{bmatrix} 1 & 3 \\ 2 & 4 \\ 3 & -2 \end{bmatrix}$

4. Suppose A and B are both $n \times n$ matrices.

(a) Can AB be found?

(b) Can BA be found?

(c) Does $AB = BA$? Explain why or why not.

(d) If A is $n \times n$ and B is $m \times n$, can either AB or BA be found?

5. Evaluate each determinant.

(a) $\det \begin{bmatrix} 6 & 8 \\ 2 & -7 \end{bmatrix}$

(b) $\det \begin{bmatrix} 2 & 0 & 8 \\ -1 & 7 & 9 \\ 12 & 5 & -3 \end{bmatrix}$

6. Solve the system by Cramer's rule.

$$2x - 3y = -33$$
$$4x + 5y = 11$$

7. Consider the system of equations

$$x + y - z = -4$$
$$2x - 3y - z = 5$$
$$x + 2y + 2z = 3.$$

(a) Write the matrix of coefficients A, the matrix of variables X, and the matrix of constants B for this system.

(b) Find A^{-1}.

(c) Use the matrix inverse to solve the system.

(d) If the first equation in the system above is replaced by $.5x + y + z = 1.5$, the system cannot be solved by using the matrix inverse. Explain why.

8. The solution set of a system of inequalities is shown below. Which of the following systems is it?

(a) $x - 3y \le 6$
$y^2 \le 16 - x^2$

(b) $x - 3y \ge 6$
$y^2 \le 16 - x^2$

(c) $x - 3y \le 6$
$y^2 \ge 16 - x^2$

(d) $x - 3y \ge 6$
$y^2 \ge 16 - x^2$

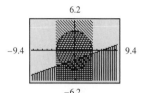

9. *Vehicle Thefts in the United States*   After rising for several years, the number of vehicle thefts in the United States began decreasing in 1991. The table lists the number of thefts (in millions) in several recent years, with 1988 represented by year 0, 1990 by year 2, and so on.

Year	Number of Thefts
1988	1.41
1990	1.65
1992	1.66
1994	1.58
1996	1.40

(a) Plot the data points on a graph. What type of function would provide a good fit?

(b) Write a system of equations that can be used to find the equation of the function you gave in part (a), using the data points (0, 1.41), (4, 1.66), and (8, 1.40).

(c) Solve the system and give the equation of the function. Verify your answer by graphing the function with the data points.

(d) Use the equation from part (c) to predict the number of vehicle thefts in 1998.

10. *Traffic Congestion at Intersections*   During rush hours, substantial traffic congestion is encountered at the traffic intersections shown in the figure on the next page. (All streets are one-way.) To improve the flow of traffic, the traffic engineers gather data. The numbers and variables in the figure give the number of cars per hour that enter and leave each intersection. For example, 700 cars per hour come down M Street to intersection A, and 300 cars per hour come to intersection A on 10th Street. A total of x_1 of these cars leave A on M Street, while x_4 cars leave A on 10th Street. The number of cars that enter A must equal the number that leave, so $x_1 + x_4 = 700 + 300$.

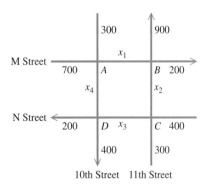

(a) Write four equations that represent the traffic in and out of the four intersections.

(b) Set up an augmented matrix and use the Gaussian method to reduce the system to triangular form.

(c) Determine the restrictions on the variables so that all four equations are satisfied and all variables are nonnegative.

(d) Does the system have a unique solution? If so, what is it? If not, give one solution that satisfies all the constraints.

CHAPTER 7   PROJECT

FINDING A POLYNOMIAL THAT GOES THROUGH ANY NUMBER OF GIVEN POINTS USING A MATRIX

In the Chapter 3 Project, we found ninth degree polynomials with zeros that depended on the digits of our Social Security numbers. In this project, we'll learn to find a polynomial that goes through a given set of points in the plane. Recall that three points define a second degree polynomial, four points define a third degree polynomial, five points define a fourth degree polynomial, and so on. The only restriction on the points is that they must each have a different x-coordinate. Since polynomials define functions, no two distinct points can have the same x-coordinate.

In the following example, we will use the same Social Security number (539-58-0954) that we used in the Chapter 3 Project to find an eighth degree polynomial that lies on the nine points with x-coordinates from 1 to 9 and with y-coordinates that are the digits of the Social Security number. The nine ordered pairs are

$$(1, 5), (2, 3), (3, 9), (4, 5), (5, 8), (6, 0), (7, 9), (8, 5), \text{ and } (9, 4).$$

These nine ordered pairs are entered in the lists shown in Figure A.

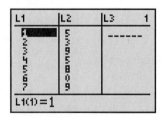 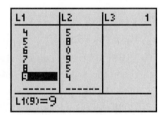

FIGURE A

To get the scatterplot of the data in Figure C, we enter the window settings shown in Figure B.

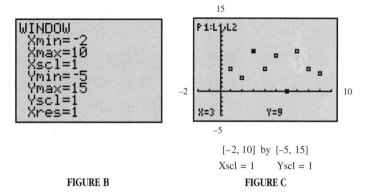

[-2, 10] by [-5, 15]
Xscl = 1   Yscl = 1

FIGURE B   **FIGURE C**

The nine ordered pairs are now used to write a system of nine equations with nine unknowns by substituting each pair of x- and y-values into the general polynomial equation

$$ax^8 + bx^7 + cx^6 + dx^5 + ex^4 + fx^3 + gx^2 + hx + i = y. \qquad (*)$$

$$a1^8 + b1^7 + c1^6 + d1^5 + e1^4 + f1^3 + g1^2 + h1^1 + i = 5$$
$$a2^8 + b2^7 + c2^6 + d2^5 + e2^4 + f2^3 + g2^2 + h2^1 + i = 3$$
$$a3^8 + b3^7 + c3^6 + d3^5 + e3^4 + f3^3 + g3^2 + h3^1 + i = 9$$
$$a4^8 + b4^7 + c4^6 + d4^5 + e4^4 + f4^3 + g4^2 + h4^1 + i = 5$$
$$a5^8 + b5^7 + c5^6 + d5^5 + e5^4 + f5^3 + g5^2 + h5^1 + i = 8$$
$$a6^8 + b6^7 + c6^6 + d6^5 + e6^4 + f6^3 + g6^2 + h6^1 + i = 0$$
$$a7^8 + b7^7 + c7^6 + d7^5 + e7^4 + f7^3 + g7^2 + h7^1 + i = 9$$
$$a8^8 + b8^7 + c8^6 + d8^5 + e8^4 + f8^3 + g8^2 + h8^1 + i = 5$$
$$a9^8 + b9^7 + c9^6 + d9^5 + e9^4 + f9^3 + g9^2 + h9^1 + i = 4$$

Next, we write an equivalent matrix equation $[A][X] = [B]$ as follows.

$$
\begin{bmatrix}
1 & 1 & 1 & 1 & 1 & 1 & 1 & 1 & 1 \\
2^8 & 2^7 & 2^6 & 2^5 & 2^4 & 2^3 & 2^2 & 2 & 1 \\
3^8 & 3^7 & 3^6 & 3^5 & 3^4 & 3^3 & 3^2 & 3 & 1 \\
4^8 & 4^7 & 4^6 & 4^5 & 4^4 & 4^3 & 4^2 & 4 & 1 \\
5^8 & 5^7 & 5^6 & 5^5 & 5^4 & 5^3 & 5^2 & 5 & 1 \\
6^8 & 6^7 & 6^6 & 6^5 & 6^4 & 6^3 & 6^2 & 6 & 1 \\
7^8 & 7^7 & 7^6 & 7^5 & 7^4 & 7^3 & 7^2 & 7 & 1 \\
8^8 & 8^7 & 8^6 & 8^5 & 8^4 & 8^3 & 8^2 & 8 & 1 \\
9^8 & 9^7 & 9^6 & 9^5 & 9^4 & 9^3 & 9^2 & 9 & 1
\end{bmatrix}
\begin{bmatrix} a \\ b \\ c \\ d \\ e \\ f \\ g \\ h \\ i \end{bmatrix}
=
\begin{bmatrix} 5 \\ 3 \\ 9 \\ 5 \\ 8 \\ 0 \\ 9 \\ 5 \\ 4 \end{bmatrix}
$$

To solve the matrix equation for the matrix of variables, enter $[A]$ and $[B]$ into the matrix editor of your graphing calculator. (See Figures D and E.) Entries can be typed in exponential form; the calculator will evaluate each entry. Although the calculator displays large entries in scientific notation, it stores the exact value as shown in Figure D, for the 6th row 1st column entry of the matrix. (The screen is not large enough to display the entire matrix at one time.)

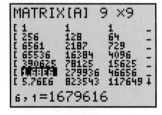

FIGURE D

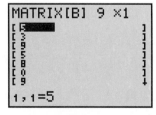

FIGURE E

Solve the system by multiplying the inverse of matrix [A] times [B], storing the solution as matrix [C], as shown in Figures F and G. (You will need to scroll down to see the entire matrix [C].)

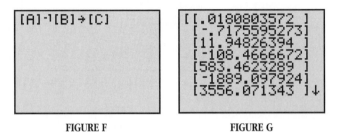

| FIGURE F | FIGURE G |

The entries in matrix [C] are *a*, *b*, *c*, and so on, the coefficients of the polynomial equation (*). We can use the rows of matrix [C] to enter the equation of the polynomial in the Y= menu, using the notation in Figure H. Be sure to use the matrix menu to enter [C] for each coordinate [C] (*m*, *n*). Figure I shows the graph of the polynomial in the same window as the scatterplot in Figure C. In Figure J, the window setting has been changed to show a comprehensive graph, including all local minima and maxima.

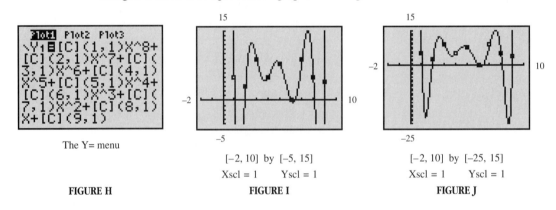

The Y= menu

FIGURE H

[−2, 10] by [−5, 15]
Xscl = 1   Yscl = 1
FIGURE I

[−2, 10] by [−25, 15]
Xscl = 1   Yscl = 1
FIGURE J

ACTIVITIES

1. Use the process outlined above to construct a second degree polynomial with three ordered pairs. Use the digits 1, 2, and 3 as the *x*-coordinates and the first three digits of your Social Security number as the *y*-coordinates of the ordered pairs.
2. Construct the eighth degree polynomial $f(x)$ that lies on the nine points with *x*-coordinates from 1 to 9 and *y*-coordinates that are the digits of your Social Security number.
3. Using your graphing calculator, find a comprehensive graph of $f(x)$, including all local extreme points, all intercepts, and clear end behavior. Sketch the graph on paper or print it using appropriate computer software. Label the extreme points with their coordinates.

Further Topics in Algebra

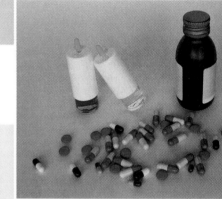

*L*argely due to the frequent use of antibiotics in society, many strains of bacteria are becoming resistant to certain drugs. This phenomenon is of interest to both the medical community and genetic engineers. Many types of bacteria contain genetic material called plasmids. Plasmids are inherited during cell division. The genetic material of bacteria can sometimes contain two different plasmids, causing drug resistance to both the antibiotics ampicillin and tetracycline. When the bacteria reproduce, the type of plasmids passed on to each new cell is random. Genetic engineers want to predict what will happen to these bacteria after many generations. Will the bacteria remain resistant to both antibiotics indefinitely?*

In order to answer this and other questions related to the spread of disease, further topics in algebra are needed.

Bacterial growth can be modeled and simulated, using sequences. The behavior of the bacteria after a long period of time often involves an element of chance. To analyze the phenomenon of chance, the study of probability is required. Solving real-world applications usually requires knowledge from several areas of mathematics. The solution to this genetic problem will require not only the new concepts of sequences and probability, but also previously learned concepts such as matrices. As our mathematical ability has increased during this mathematics course, so has the complexity of the applications that we can solve.

Source: Hoppensteadt, F., and C. Peskin, *Mathematics in Medicine and the Life Sciences*, Springer-Verlag, 1992.

8.1   SEQUENCES AND SERIES

Sequences ◆ Series and Summation Notation ◆ Summation Properties

SEQUENCES

Defined informally, a sequence is a list of numbers. We are most interested in lists of numbers that satisfy some pattern. For example, 2, 4, 6, 8, 10, . . . is a list of the natural-number multiples of 2. This can be written $2n$, where n is a natural number, so a sequence may be defined (as is a function) by a variable expression. More formally, a sequence is defined as follows.

DEFINITION OF SEQUENCE

A **sequence** is a function that has a set of natural numbers as its domain.

TECHNOLOGY NOTE

Some graphing calculators have a designated sequence mode (similar to the modes function, polar, and parametric). This mode allows the user to investigate and graph sequences defined in terms of n, where n is a natural number. You should consult your owner's manual for specific instructions on how to use this feature.

Instead of using $f(x)$ notation to indicate a sequence, it is customary to use a_n, where n represents an element in the domain of the sequence. Thus, $a_n = f(n)$. The letter n is used instead of x as a reminder that n represents a *natural number*. The elements in the range of a sequence, called the **terms** of the sequence, are $a_1, a_2, a_3, \ldots$. The elements of both the domain and the range of a sequence are *ordered*. The first term (range element) is found by letting $n = 1$, the second term is found by letting $n = 2$, and so on. The **general term**, or **nth term**, of the sequence is a_n.

Figure 1 shows graphs of $f(x) = 2x$ and $a_n = 2n$. Notice that $f(x)$ defines a "continuous" function, while a_n is "discontinuous."

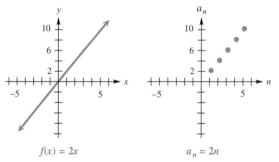

$$f(x) = 2x \qquad\qquad a_n = 2n$$

FIGURE 1

A graphing calculator provides a convenient way to list the terms in a sequence. Methods may vary, so you should refer to your owner's manual. Using the sequence mode to list the first 10 terms of the sequence with general term $n + (1/n)$ produces the result shown in Figure 2(a). Additional terms of the sequence can be seen by scrolling to the right. Sequences can also be graphed by using the sequence mode. In Figure 2(b), we show a calculator screen with the graph of $a_n = n + (1/n)$. Notice that for $n = 5$, the term is $5 + 1/5 = 5.2$.

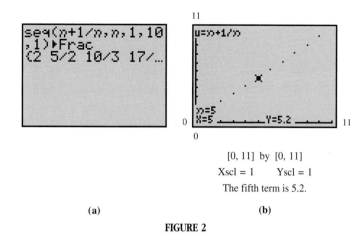

[0, 11] by [0, 11]
Xscl = 1   Yscl = 1
The fifth term is 5.2.

(a)   **(b)**

FIGURE 2

EXAMPLE 1   *Finding Terms of a Sequence from the General Term*   Write the first five terms for each of the following sequences.

(a) $a_n = \dfrac{n+1}{n+2}$ **(b)** $a_n = (-1)^n \cdot n$ **(c)** $b_n = \dfrac{(-1)^n}{2^n}$

SOLUTION

(a)   Replacing n, in turn, with 1, 2, 3, 4, and 5 gives

$$\frac{2}{3}, \frac{3}{4}, \frac{4}{5}, \frac{5}{6}, \frac{6}{7}.$$

(b)   Replace n with 1, 2, 3, 4, and 5 to get

$$n = 1: a_1 = (-1)^1 \cdot 1 = -1$$
$$n = 2: a_2 = (-1)^2 \cdot 2 = 2$$
$$n = 3: a_3 = (-1)^3 \cdot 3 = -3$$
$$n = 4: a_4 = (-1)^4 \cdot 4 = 4$$
$$n = 5: a_5 = (-1)^5 \cdot 5 = -5.$$

(c)   Here, we have $b_1 = -1/2$, $b_2 = 1/4$, $b_3 = -1/8$, $b_4 = 1/16$, and $b_5 = -1/32$. ◆

A sequence is a **finite sequence** if the domain is the set $\{1, 2, 3, 4, \ldots, n\}$, where n is a natural number. An **infinite sequence** has the set of all natural numbers as its domain.

EXAMPLE 2   *Distinguishing between Finite and Infinite Sequences*   The sequence of natural-number multiples of 2,

$$2, 4, 6, 8, 10, 12, 14, \ldots,$$

is infinite, but the sequence of days in June is finite:

$$1, 2, 3, 4, \ldots, 29, 30. \quad ◆$$

If the terms of an infinite sequence get closer and closer to some real number, the sequence is said to be **convergent** and to **converge** to that real number. Graphs of sequences illustrate this property. The sequence in Example 1(a) is graphed in Figure 3 on the next page, using both traditional and calculator methods. What number do you think this sequence converges to? A sequence that does not converge to some number is **divergent**.

Some sequences are defined by a **recursive definition**, one in which each term is defined as an expression involving the previous term. On the other hand, the sequences

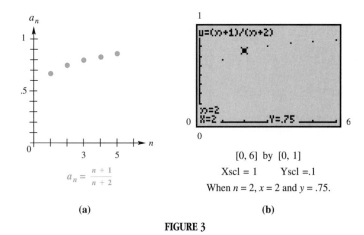

$$a_n = \frac{n+1}{n+2}$$

(a)

[0, 6] by [0, 1]
Xscl = 1   Yscl = .1
When $n = 2$, $x = 2$ and $y = .75$.

(b)

FIGURE 3

in Example 1 were defined *explicitly*, with a formula for a_n that does not depend on a previous term.

EXAMPLE 3 *Using a Recursion Formula* Find the first four terms for the sequences defined as follows.

(a) $a_1 = 4$; for $n > 1, a_n = 2 \cdot a_{n-1} + 1$

(b) $a_1 = 2$; for $n > 1, a_n = a_{n-1} + n - 1$

SOLUTION

(a) This is an example of a recursive definition. We know $a_1 = 4$. Since $a_n = 2 \cdot a_{n-1} + 1$,

$$a_2 = 2 \cdot a_1 + 1 = 2 \cdot 4 + 1 = 9$$
$$a_3 = 2 \cdot a_2 + 1 = 2 \cdot 9 + 1 = 19$$
$$a_4 = 2 \cdot a_3 + 1 = 2 \cdot 19 + 1 = 39.$$

(b)
$$a_1 = 2$$
$$a_2 = a_1 + 2 - 1 = 2 + 1 = 3$$
$$a_3 = a_2 + 3 - 1 = 3 + 2 = 5$$
$$a_4 = a_3 + 4 - 1 = 5 + 3 = 8$$ ◆

For Group Discussion One of the most famous sequences in mathematics is the **Fibonacci sequence:**

$$1, 1, 2, 3, 5, 8, 13, 21, 34, 55, \ldots.$$

This sequence is named for the Italian mathematician Leonardo of Pisa (1170–1250), who was also known as Fibonacci. The Fibonacci sequence is found in numerous places in nature. For example, male honeybees hatch from eggs that have not been fertilized, so a male bee has only one parent, a female. On the other hand, female honeybees hatch from fertilized eggs, so a female has two parents, one male and one female. The number of ancestors in consecutive generations of bees follows the Fibonacci sequence. Successive terms in the sequence also appear in plants: in the daisy head, the pineapple, and the pine cone, for instance.

1. See if you can discover the pattern in the Fibonacci sequence.
2. Using the description given above, draw a tree diagram that shows the number of ancestors of a male bee in each generation.

SERIES AND SUMMATION NOTATION

Suppose a sequence has terms $a_1, a_2, a_3, \ldots$. Then S_n is defined as the sum of the first n terms. That is, $S_n = a_1 + a_2 + a_3 + \cdots + a_n$. The sum of the first n terms of a sequence is called a **series**. Special notation is used to represent a series. The symbol Σ, the Greek capital letter *sigma*, is used to indicate a sum.

DEFINITION OF A SERIES

A **finite series** is an expression of the form

$$S_n = a_1 + a_2 + a_3 + \cdots + a_n = \sum_{i=1}^{n} a_i,$$

and an **infinite series** is an expression of the form

$$S_n = a_1 + a_2 + a_3 + \cdots + a_n + \cdots = \sum_{i=1}^{\infty} a_i.$$

The letter i is called the **index of summation**.

Caution   Do not confuse this use of i with the use of i to represent an imaginary number. Other letters may be used for the index of summation.

EXAMPLE 4   *Using Summation Notation*   Evaluate the series $\displaystyle\sum_{k=1}^{6} (2^k + 1)$. Use a graphing calculator to support the result.

SOLUTION   ANALYTIC   Write out each of the six terms, then evaluate the sum.

$$\sum_{k=1}^{6} (2^k + 1) = (2^1 + 1) + (2^2 + 1) + (2^3 + 1) + (2^4 + 1) +$$
$$(2^5 + 1) + (2^6 + 1)$$
$$= (2 + 1) + (4 + 1) + (8 + 1) + (16 + 1) +$$
$$(32 + 1) + (64 + 1)$$
$$= 3 + 5 + 9 + 17 + 33 + 65 = 132$$

GRAPHICAL   The screen in Figure 4 supports the result found above.   ◆

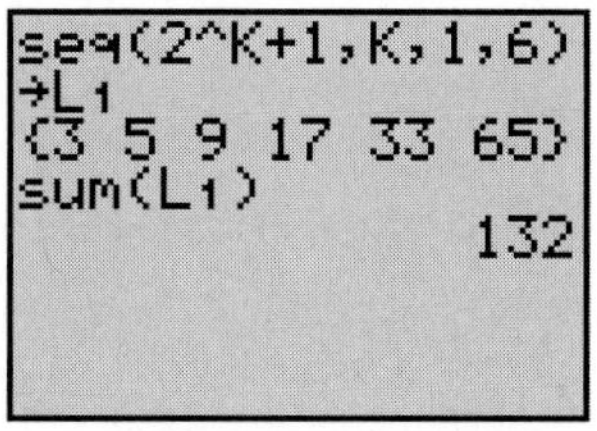

The result from Example 4 is supported in this screen. The calculator stores the sequence into a list and then computes the sum of the terms.

FIGURE 4

EXAMPLE 5   *Using Summation Notation with Subscripts*   Write out the terms for each of the following series. Evaluate each sum, if possible.

(a) $\displaystyle\sum_{j=3}^{6} a_j$

(b) $\displaystyle\sum_{i=1}^{3} (6x_i - 2)$ if $x_1 = 2$, $x_2 = 4$, $x_3 = 6$

(c) $\displaystyle\sum_{i=1}^{4} f(x_i)\,\Delta x$ if $f(x) = x^2$, $x_1 = 0$, $x_2 = 2$, $x_3 = 4$, $x_4 = 6$, and $\Delta x = 2$

SOLUTION

(a) $\displaystyle\sum_{j=3}^{6} a_j = a_3 + a_4 + a_5 + a_6$

(b) Let $i = 1$, 2, and 3, respectively, to get

$$\sum_{i=1}^{3} (6x_i - 2) = (6x_1 - 2) + (6x_2 - 2) + (6x_3 - 2).$$

Now, substitute the given values for x_1, x_2, and x_3.

$$\sum_{i=1}^{3} (6x_i - 2) = (6 \cdot 2 - 2) + (6 \cdot 4 - 2) + (6 \cdot 6 - 2)$$
$$= 10 + 22 + 34 = 66$$

(c) $\displaystyle\sum_{i=1}^{4} f(x_i)\,\Delta x = f(x_1)\,\Delta x + f(x_2)\,\Delta x + f(x_3)\,\Delta x + f(x_4)\,\Delta x$

$$= x_1^2\,\Delta x + x_2^2\,\Delta x + x_3^2\,\Delta x + x_4^2\,\Delta x$$
$$= 0^2(2) + 2^2(2) + 4^2(2) + 6^2(2)$$
$$= 0 + 8 + 32 + 72 = 112 \qquad \blacklozenge$$

EXAMPLE 6   *Representing a Series with Different Summations*   Use summation notation to rewrite each series with the index of summation starting at the indicated number.

(a) $\displaystyle\sum_{i=1}^{8} (3i - 4); 0$ (b) $\displaystyle\sum_{i=2}^{10} i^2; -1$

SOLUTION

(a) Let the new index be j. Since the new index is to start at 0, which is $1 - 1$, $j = i - 1$, or $i = j + 1$. Substitute $j + 1$ for i in the summation.

$$\sum_{i=1}^{8} (3i - 4) = \sum_{j+1=1}^{j+1=8} [3(j + 1) - 4]$$

$$= \sum_{j=0}^{j=7} (3j - 1) \quad \text{or} \quad \sum_{j=0}^{7} (3j - 1)$$

(b) Here, if the new index is j, then $i = j + 3$ and

$$\sum_{i=2}^{10} i^2 = \sum_{j+3=2}^{j+3=10} (j + 3)^2$$

$$= \sum_{j=-1}^{j=7} (j + 3)^2 \quad \text{or} \quad \sum_{j=-1}^{7} (j + 3)^2. \qquad \blacklozenge$$

SUMMATION PROPERTIES

Polynomial functions, defined by expressions of the form $f(x) = a_n x^n + a_{n-1} x^{n-1} + \cdots + a_1 x + a_0$, can be written in compact form, using summation notation, as

$$f(x) = \sum_{i=0}^{n} a_i x^i.$$

Several properties of summation are given below. These provide useful shortcuts for evaluating series.

SUMMATION PROPERTIES

If $a_1, a_2, a_3, \ldots, a_n$ and $b_1, b_2, b_3, \ldots, b_n$ are two sequences, and c is a constant, then for every positive integer n,

(a) $\displaystyle\sum_{i=1}^{n} c = nc$

(b) $\displaystyle\sum_{i=1}^{n} c a_i = c \sum_{i=1}^{n} a_i$

(c) $\displaystyle\sum_{i=1}^{n} (a_i + b_i) = \sum_{i=1}^{n} a_i + \sum_{i=1}^{n} b_i$

(d) $\displaystyle\sum_{i=1}^{n} (a_i - b_i) = \sum_{i=1}^{n} a_i - \sum_{i=1}^{n} b_i.$

To prove Property (a), expand the series to get

$$c + c + c + c + \cdots + c,$$

where there are n terms of c, so the sum is nc.

Property (c) also can be proved by first expanding the series:

$$\sum_{i=1}^{n} (a_i + b_i) = (a_1 + b_1) + (a_2 + b_2) + \cdots + (a_n + b_n).$$

Now, use the commutative and associative properties to rearrange the terms.

$$\sum_{i=1}^{n} (a_i + b_i) = (a_1 + a_2 + \cdots + a_n) + (b_1 + b_2 + \cdots + b_n)$$

$$= \sum_{i=1}^{n} a_i + \sum_{i=1}^{n} b_i$$

Proofs of the other two properties are similar.

The following results can be proved, using the methods of Section 5 of this chapter.

SUMMATION RULES

$$\sum_{i=1}^{n} i^2 = 1^2 + 2^2 + \cdots + n^2 = \frac{n(n+1)(2n+1)}{6}$$

and

$$\sum_{i=1}^{n} i = 1 + 2 + \cdots + n = \frac{n(n+1)}{2}$$

These summations are used in the next example.

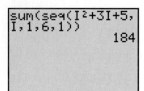

FIGURE 5

EXAMPLE 7 *Using the Summation Properties* Use the properties of series to evaluate $\sum_{i=1}^{6} (i^2 + 3i + 5)$. Support the result with a graphing calculator.

SOLUTION

ANALYTIC

$$\sum_{i=1}^{6} (i^2 + 3i + 5) = \sum_{i=1}^{6} i^2 + \sum_{i=1}^{6} 3i + \sum_{i=1}^{6} 5 \qquad \text{Property (c)}$$

$$= \sum_{i=1}^{6} i^2 + 3 \sum_{i=1}^{6} i + \sum_{i=1}^{6} 5 \qquad \text{Property (b)}$$

$$= \sum_{i=1}^{6} i^2 + 3 \sum_{i=1}^{6} i + 6(5) \qquad \text{Property (a)}$$

By substituting the results given just before this example, we get

$$= \frac{6(6 + 1)(2 \cdot 6 + 1)}{6} + 3\left[\frac{6(6 + 1)}{2}\right] + 6(5)$$

$$= 91 + 3(21) + 6(5) = 184.$$

GRAPHICAL Figure 5 supports this result. ◆

8.1 EXERCISES Tape 14

Write the first five terms of each sequence.

1. $a_n = 4n + 10$ **2.** $a_n = 6n - 3$ **3.** $a_n = 2^{n-1}$ **4.** $a_n = -3^n$

5. $a_n = \left(\dfrac{1}{3}\right)^n (n - 1)$ **6.** $a_n = (-2)^n(n)$ **7.** $a_n = (-1)^n(2n)$ **8.** $a_n = (-1)^{n-1}(n + 1)$

9. $a_n = \dfrac{4n - 1}{n^2 + 2}$ **10.** $a_n = \dfrac{n^2 - 1}{n^2 + 1}$

11. Your friend does not understand what is meant by the nth term or general term of a sequence. How would you explain this idea?

12. How are sequences related to functions?

Decide whether the given sequence is finite or infinite.

13. the sequence of days of the week   **14.** the sequence of dates in the month of November

15. 1, 2, 3, 4   **16.** $-1, -2, -3, -4$

17. 1, 2, 3, 4, . . .   **18.** $-1, -2, -3, -4, . . .$

19. $a_1 = 3$;   for $2 \le n \le 10, a_n = 3 \cdot a_{n-1}$ **20.** $a_1 = 1$; $a_2 = 3$;   for $n \ge 3, a_n = a_{n-1} + a_{n-2}$

Find the first four terms for each sequence.

21. $a_1 = -2$, $a_n = a_{n-1} + 3$,   for $n > 1$

22. $a_1 = -1$, $a_n = a_{n-1} - 4$,   for $n > 1$

23. $a_1 = 1$, $a_2 = 1$, $a_n = a_{n-1} + a_{n-2}$,   for $n \ge 3$ (the Fibonacci sequence)

24. $a_1 = 2$, $a_n = n \cdot a_{n-1}$,   for $n > 1$

Evaluate each series.

25. $\displaystyle\sum_{i=1}^{5} (2i + 1)$

26. $\displaystyle\sum_{i=1}^{6} (3i - 2)$

27. $\displaystyle\sum_{j=1}^{4} \frac{1}{j}$

28. $\displaystyle\sum_{i=1}^{5} (i + 1)^{-1}$

29. $\displaystyle\sum_{i=1}^{4} i^i$

30. $\displaystyle\sum_{k=1}^{4} (k + 1)^2$

31. $\displaystyle\sum_{k=1}^{6} (-1)^k \cdot k$

32. $\displaystyle\sum_{i=1}^{7} (-1)^{i+1} \cdot i^2$

Evaluate the terms for each sum, where $x_1 = -2$, $x_2 = -1$, $x_3 = 0$, $x_4 = 1$, and $x_5 = 2$.

33. $\displaystyle\sum_{i=1}^{5} (2x_i + 3)$

34. $\displaystyle\sum_{i=1}^{4} x_i^2$

35. $\displaystyle\sum_{i=1}^{3} (3x_i - x_i^2)$

36. $\displaystyle\sum_{i=1}^{3} (x_i^2 + 1)$

37. $\displaystyle\sum_{i=2}^{5} \frac{x_i + 1}{x_i + 2}$

38. $\displaystyle\sum_{i=1}^{5} \frac{x_i}{x_i + 3}$

Evaluate the terms of $\sum_{i=1}^{4} f(x_i) \Delta x$, with $x_1 = 0$, $x_2 = 2$, $x_3 = 4$, $x_4 = 6$, and $\Delta x = .5$, for the functions defined.

39. $f(x) = 4x - 7$

40. $f(x) = 6 + 2x$

41. $f(x) = 2x^2$

42. $f(x) = x^2 - 1$

43. $f(x) = \dfrac{-2}{x + 1}$

44. $f(x) = \dfrac{5}{2x - 1}$

Use summation notation to rewrite each series with the index of summation starting at the indicated number.

45. $\displaystyle\sum_{i=1}^{5} (6 - 3i)$; 3

46. $\displaystyle\sum_{i=1}^{7} (5i + 2)$; -2

47. $\displaystyle\sum_{i=1}^{10} 2(3)^i$; 0

48. $\displaystyle\sum_{i=-1}^{6} 5(2)^i$; 3

49. $\displaystyle\sum_{i=-1}^{9} (i^2 - 2i)$; 0

50. $\displaystyle\sum_{i=3}^{11} (2i^2 + 1)$; 0

Use the summation properties to evaluate each series. The following sums may be needed.

$$\sum_{i=1}^{n} i = \frac{n(n + 1)}{2}$$

$$\sum_{i=1}^{n} i^2 = \frac{n(n + 1)(2n + 1)}{6}$$

$$\sum_{i=1}^{n} i^3 = \frac{n^2(n + 1)^2}{4}$$

51. $\displaystyle\sum_{i=1}^{5} (5i + 3)$

52. $\displaystyle\sum_{i=1}^{5} (8i - 1)$

53. $\displaystyle\sum_{i=1}^{5} (4i^2 - 2i + 6)$

54. $\displaystyle\sum_{i=1}^{6} (2 + i - i^2)$

55. $\displaystyle\sum_{i=1}^{4} (3i^3 + 2i - 4)$

56. $\displaystyle\sum_{i=1}^{6} (i^2 + 2i^3)$

Use the sequence graphing capability of a graphing calculator to graph the first ten terms of each sequence as defined. Use the graph to make a conjecture as to whether the sequence converges or diverges. If you think it converges, determine the number to which it converges.

57. $a_n = \dfrac{n + 4}{2n}$

58. $a_n = \dfrac{1 + 4n}{2n}$

59. $a_n = 2e^n$

60. $a_n = n(n + 2)$

61. $a_n = \left(1 + \dfrac{1}{n}\right)^n$

62. $a_n = (1 + n)^{1/n}$

Solve these problems involving sequences.

63. *Bacterial Growth* Certain strains of bacteria cannot produce an amino acid called histidine. This amino acid is necessary for them to produce proteins and reproduce. If these bacteria are cultured in a medium with sufficient histidine, they will double in size and then divide every 40 minutes. Let N_1 be the initial number of bacteria cells, N_2 the number after 40 minutes, N_3 the number after 80 minutes, and N_j the number after $40(j - 1)$ minutes. (*Source:* Hoppensteadt, F., and C. Peskin, *Mathematics in Medicine and the Life Sciences*, Springer-Verlag, 1992.)

(a) Write N_{j+1} in terms of N_j for $j \geq 1$.

(b) Determine the number of bacteria after two hours if $N_1 = 230$.

(c) Graph the sequence N_j for $j = 1, 2, 3, \ldots, 7$. Use the window $[0, 10]$ by $[0, 15{,}000]$.

(d) Describe the growth of these bacteria when there are unlimited nutrients.

64. *Verhulst's Model for Bacteria Growth* Refer to Exercise 63. If the bacteria are not cultured in a medium with sufficient nutrients, competition will ensue and the growth will slow. According to Verhulst's model, the number of bacteria N_j at time $40(j - 1)$ in minutes can be determined by the sequence $N_{j+1} = \left[\dfrac{2}{1 + (N_j/K)}\right]N_j$, where K is a constant and $j \geq 1$.

(*Source:* Hoppensteadt, F., and C. Peskin, *Mathematics in Medicine and the Life Sciences*, Springer-Verlag, 1992.)

(a) If $N_1 = 230$ and $K = 5000$, make a table of N_j for $j = 1, 2, 3, \ldots, 20$. Round values in the table to the nearest integer.

(b) Graph the sequence N_j for $j = 1, 2, 3, \ldots, 20$. Use the window $[0, 20]$ by $[0, 6000]$.

(c) Describe the growth of these bacteria when there are limited nutrients.

(d) Make a conjecture as to why K is called the *saturation constant*. Test your conjecture.

8.2 ARITHMETIC SEQUENCES AND SERIES

Arithmetic Sequences ◆ Arithmetic Series

ARITHMETIC SEQUENCES

A sequence in which each term after the first is obtained by adding a fixed number to the previous term is an **arithmetic sequence** (or **arithmetic progression**). The fixed number that is added is the **common difference**. The sequence

$$5, 9, 13, 17, 21, \ldots$$

is an arithmetic sequence because each term after the first is obtained by adding 4 to the previous term. That is,

$$9 = 5 + 4$$
$$13 = 9 + 4$$
$$17 = 13 + 4$$
$$21 = 17 + 4,$$

and so on. The common difference is 4.

If the common difference of an arithmetic sequence is d, then by the definition of an arithmetic sequence,

$$d = a_{n+1} - a_n$$

for every positive integer n in its domain.

EXAMPLE 1 *Finding the Common Difference* Find the common difference d for the arithmetic sequence $-9, -7, -5, -3, -1, \ldots$.

SOLUTION Since this sequence is arithmetic, d can be found by choosing any two adjacent terms and subtracting the first from the second. Choosing -7 and -5 gives

$d = -5 - (-7) = 2$. Choosing -9 and -7 would give $d = -7 - (-9) = 2$, the same result. ◆

If a_1 and d are known, then all the terms of an arithmetic sequence can be found.

EXAMPLE 2 *Finding the Terms Given a_1 and d* Write the first five terms for each of the following arithmetic sequences.

(a) The first term is 7, and the common difference is -3.

(b) $a_1 = -12$, $d = 5$

SOLUTION

(a) Here,

$$a_1 = 7 \quad \text{and} \quad d = -3.$$
$$a_2 = 7 + (-3) = 4,$$
$$a_3 = 4 + (-3) = 1,$$
$$a_4 = 1 + (-3) = -2,$$
$$a_5 = -2 + (-3) = -5.$$

(b) Starting with a_1, add d to each term to get the next term.

$$a_1 = -12$$
$$a_2 = -12 + d = -12 + 5 = -7$$
$$a_3 = -7 + d = -7 + 5 = -2$$
$$a_4 = -2 + d = -2 + 5 = 3$$
$$a_5 = 3 + d = 3 + 5 = 8 \quad ◆$$

If a_1 is the first term of an arithmetic sequence and d is the common difference, then the terms of the sequence are given by

$$a_1 = a_1$$
$$a_2 = a_1 + d$$
$$a_3 = a_2 + d = a_1 + d + d = a_1 + 2d$$
$$a_4 = a_3 + d = a_1 + 2d + d = a_1 + 3d$$
$$a_5 = a_1 + 4d$$
$$a_6 = a_1 + 5d,$$

and, by this pattern, $a_n = a_1 + (n-1)d$. This result can be proven by mathematical induction (see Section 5 of this chapter); a summary is given below.

FORMULA FOR THE nTH TERM OF AN ARITHMETIC SEQUENCE

In an arithmetic sequence with first term a_1 and common difference d, the nth term, a_n, is given by
$$a_n = a_1 + (n-1)d.$$

EXAMPLE 3 *Using the Formula for the nth Term* Find a_{13} and a_n for the arithmetic sequence

$$-3, 1, 5, 9, \ldots.$$

SOLUTION   Here, $a_1 = -3$ and $d = 1 - (-3) = 4$. To find a_{13}, substitute 13 for n in the preceding formula.

$$a_{13} = a_1 + (13 - 1)d$$
$$a_{13} = -3 + (12)4$$
$$a_{13} = -3 + 48$$
$$a_{13} = 45$$

Find a_n by substituting values for a_1 and d in the formula for a_n.

$$a_n = -3 + (n - 1) \cdot 4$$
$$a_n = -3 + 4n - 4 \qquad \text{Distributive property}$$
$$a_n = 4n - 7 \qquad \blacklozenge$$

EXAMPLE 4   *Using the Formula for the nth Term*   Find a_{18} and a_n for the arithmetic sequence having $a_2 = 9$ and $a_3 = 15$.

SOLUTION   Find d first; $d = a_3 - a_2 = 15 - 9 = 6$.

Since $a_2 = a_1 + d,$

$$9 = a_1 + 6 \qquad \text{Let } a_2 = 9, \, d = 6.$$
$$a_1 = 3.$$

Then, $a_{18} = 3 + (18 - 1) \cdot 6 \qquad \text{Formula for } a_n; \quad n = 18$

$$a_{18} = 105,$$

and $a_n = 3 + (n - 1) \cdot 6$

$$a_n = 3 + 6n - 6 \qquad \text{Distributive property}$$
$$a_n = 6n - 3. \qquad \blacklozenge$$

EXAMPLE 5   *Using the Formula for the nth Term*   Suppose that an arithmetic sequence has $a_8 = -16$ and $a_{16} = -40$. Find a_1.

SOLUTION   We must find d first. Since $a_8 = a_1 + (8 - 1)d$, replacing a_8 with -16 gives $-16 = a_1 + 7d$ or $a_1 = -16 - 7d$. Similarly, $-40 = a_1 + 15d$ or $a_1 = -40 - 15d$. From these two equations, using the substitution method from Chapter 7,

$$-16 - 7d = -40 - 15d.$$

Verify that $d = -3$. To find a_1, substitute -3 for d in $-16 = a_1 + 7d$.

$$-16 = a_1 + 7d$$
$$-16 = a_1 + 7(-3) \qquad \text{Let } d = -3.$$
$$a_1 = 5 \qquad \blacklozenge$$

ARITHMETIC SERIES

The indicated sum of the terms of an arithmetic sequence is an **arithmetic series**. To illustrate when it is necessary to find the sum of the terms of an arithmetic sequence, suppose that a person borrows $3000 and agrees to pay $100 per month plus interest of 1% per month on the unpaid balance until the loan is paid off. The first month, $100 is paid to reduce the loan, plus interest of $(.01)3000 = 30$ dollars. The second month, another $100 is paid toward the loan and $(.01)2900 = 29$ dollars is paid for interest.

Since the loan is reduced by \$100 each month, interest payments decrease by $(.01)100 = 1$ dollar each month, forming the arithmetic sequence

$$30, 29, 28, \ldots, 3, 2, 1.$$

The total amount of interest paid is given by the sum S_n of the terms of this sequence. A formula will be developed here to find this sum without adding all 30 numbers directly.

Since the sequence is arithmetic, the sum of the first n terms can be written as follows.

$$S_n = a_1 + [a_1 + d] + [a_1 + 2d] + \cdots + [a_1 + (n - 1)d]$$

The formula for the general term was used in the last expression. Now, write the same sum in reverse order, beginning with a_n and *subtracting d.*

$$S_n = a_n + [a_n - d] + [a_n - 2d] + \cdots + [a_n - (n - 1)d]$$

Adding respective sides of these two equations term by term gives

$$S_n + S_n = (a_1 + a_n) + (a_1 + a_n) + \cdots + (a_1 + a_n)$$

or

$$2S_n = n(a_1 + a_n),$$

since there are n terms of $a_1 + a_n$ on the right. Now, solve for S_n to get

$$S_n = \frac{n}{2}(a_1 + a_n).$$

Using the formula $a_n = a_1 + (n - 1)d$, this result for S_n can also be written as

$$S_n = \frac{n}{2}[a_1 + a_1 + (n - 1)d]$$

or

$$S_n = \frac{n}{2}[2a_1 + (n - 1)d],$$

an alternative formula for the sum of the first n terms of an arithmetic sequence.

A summary of this work with sums of arithmetic sequences follows.

FORMULAS FOR THE SUM OF THE FIRST *n* TERMS OF AN ARITHMETIC SEQUENCE

If an arithmetic sequence has first term a_1 and common difference d, then the sum of the first n terms is given by

$$S_n = \frac{n}{2}(a_1 + a_n)$$

or

$$S_n = \frac{n}{2}[2a_1 + (n - 1)d].$$

The first formula is used when the first and last terms are known; otherwise, the second formula is used.

Either one of these formulas can be used to find the total interest on the $3000 loan discussed above. In the sequence of interest payments, $a_1 = 30$, $d = -1$, $n = 30$, and $a_n = 1$. Choosing the first formula,

$$S_n = \frac{n}{2}(a_1 + a_n),$$

gives

$$S_{30} = \frac{30}{2}(30 + 1) = 15(31) = 465,$$

so a total of $465 interest will be paid over the 30 months.

EXAMPLE 6   *Using the Sum Formulas*

(a)   Evaluate S_{12} for the arithmetic sequence $-9, -5, -1, 3, 7, \dots$.

(b)   Use a formula for S_n to evaluate the sum of the first 60 positive integers.

SOLUTION

(a)   We want the sum of the first 12 terms. Using $a_1 = -9$, $n = 12$, and $d = 4$ in the second formula,

$$S_n = \frac{n}{2}[2a_1 + (n - 1)d]$$

$$S_{12} = \frac{12}{2}[2(-9) + 11(4)] = 156.$$

(b)   ANALYTIC   In this example, $n = 60$, $a_1 = 1$, and $a_{60} = 60$, so it is convenient to use the first of the two formulas.

$$S_n = \frac{n}{2}(a_1 + a_n)$$

$$S_{60} = \frac{60}{2}(1 + 60) = 1830$$

GRAPHICAL   See the screen in the margin.   ◆

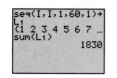

seq(I,I,1,60,1)→
L₁
{1 2 3 4 5 6 7 …
sum(L₁)
                1830

The result of Example 6(b) is supported in this screen.

EXAMPLE 7   *Using the Sum Formulas*   The sum of the first 17 terms of an arithmetic sequence is 187. If $a_{17} = -13$, find a_1 and d.

SOLUTION   Use the first formula for S_n, with $n = 17$, to find a_1.

$$S_{17} = \frac{17}{2}(a_1 + a_{17}) \qquad \text{Let } n = 17.$$

$$187 = \frac{17}{2}(a_1 - 13) \qquad \text{Let } S_{17} = 187, a_{17} = -13.$$

$$22 = a_1 - 13 \qquad \text{Multiply by } \tfrac{2}{17}.$$

$$a_1 = 35$$

Since $a_{17} = a_1 + (17 - 1)d$,

$$-13 = 35 + 16d \qquad \text{Let } a_{17} = -13, a_1 = 35.$$

$$-48 = 16d$$

$$d = -3. \qquad\qquad ◆$$

Any sum of the form

$$\sum_{i=1}^{n} (mi + p),$$

where m and p are real numbers, represents the sum of the terms of an arithmetic sequence having first term $a_1 = m(1) + p = m + p$ and common difference $d = m$. These sums can be evaluated by the formulas in this section, as shown by the next example.

EXAMPLE 8 *Using Summation Notation* Evaluate the following sums. Support the results with a graphing calculator.

(a) $\displaystyle\sum_{i=1}^{10} (4i + 8)$

(b) $\displaystyle\sum_{k=3}^{9} (4 - 3k)$

SOLUTION

(a) ANALYTIC This sum represents the sum of the first ten terms of the arithmetic sequence having $a_1 = 4 \cdot 1 + 8 = 12$, $n = 10$, and $a_n = a_{10} = 4 \cdot 10 + 8 = 48$. Thus,

$$\sum_{i=1}^{10} (4i + 8) = S_{10} = \frac{10}{2}(12 + 48) = 5(60) = 300.$$

GRAPHICAL See Figure 6(a).

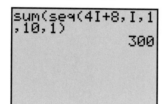
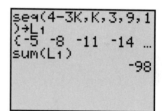

Here, the calculator stores the terms of the sequence in a list and then evaluates the sum of the terms in the list.

(a)   (b)

FIGURE 6

(b) ANALYTIC The first few terms are

$$[4 - 3(3)] + [4 - 3(4)] + [4 - 3(5)] + \cdots$$
$$= -5 + (-8) + (-11) + \cdots.$$

Thus, $a_1 = -5$ and $d = -3$. If the sequence started with $k = 1$, there would be nine terms. Since it starts at 3, two of those terms are missing, so there are seven terms and $n = 7$.

$$\sum_{k=3}^{9} (4 - 3k) = \frac{7}{2}[2(-5) + 6(-3)] = -98$$

GRAPHICAL See Figure 6(b). ◆

8.2 EXERCISES   Tape 14

Find the common difference d for each arithmetic sequence.

1. $2, 5, 8, 11, \ldots$

2. $4, 10, 16, 22, \ldots$

3. $3, -2, -7, -12, \ldots$

4. $-8, -12, -16, -20, \ldots$

5. $x + 3y, 2x + 5y, 3x + 7y, \ldots$

6. $t^2 + q, -4t^2 + 2q, -9t^2 + 3q, \ldots$

Write the first five terms for each arithmetic sequence.

7. The first term is 8, and the common difference is 6.

8. The first term is -2, and the common difference is 12.

9. $a_1 = 5, d = -2$

10. $a_1 = 4, d = 3$

11. $a_3 = 10, d = -2$

12. $a_1 = 3 - \sqrt{2}, a_2 = 3$

Find a_8 and a_n for each arithmetic sequence.

13. $a_1 = 5, d = 2$

14. $a_1 = -3, d = -4$

15. $a_3 = 2, d = 1$

16. $a_4 = 5, d = -2$

17. $a_1 = 8, a_2 = 6$

18. $a_1 = 6, a_2 = 3$

19. $a_{10} = 6, a_{12} = 15$

20. $a_{15} = 8, a_{17} = 2$

21. $a_1 = x, a_2 = x + 3$

22. $a_2 = y + 1, d = -3$

Find a_1 for each arithmetic sequence.

23. $a_5 = 27, a_{15} = 87$

24. $a_{12} = 60, a_{20} = 84$

25. $S_{16} = -160, a_{16} = -25$

26. $S_{28} = 2926, a_{28} = 199$

Find the sum of the first ten terms for each arithmetic sequence.

27. $a_1 = 8, d = 3$

28. $a_1 = -9, d = 4$

29. $a_3 = 5, a_4 = 8$

30. $a_2 = 9, a_4 = 13$

31. $5, 9, 13, \ldots$

32. $8, 6, 4, \ldots$

33. $a_1 = 10, a_{10} = 5.5$

34. $a_1 = -8, a_{10} = -1.25$

Find a_1 and d for each arithmetic sequence.

35. $S_{20} = 1090, a_{20} = 102$

36. $S_{31} = 5580, a_{31} = 360$

37. $S_{12} = -108, a_{12} = -19$

38. $S_{25} = 650, a_{25} = 62$

Evaluate each sum.

39. $\displaystyle\sum_{i=1}^{3} (i + 4)$

40. $\displaystyle\sum_{i=1}^{5} (i - 8)$

41. $\displaystyle\sum_{j=1}^{10} (2j + 3)$

42. $\displaystyle\sum_{j=1}^{15} (5j - 9)$

43. $\displaystyle\sum_{i=1}^{12} (-5 - 8i)$

44. $\displaystyle\sum_{k=1}^{19} (-3 - 4k)$

45. $\displaystyle\sum_{i=1}^{1000} i$

46. $\displaystyle\sum_{k=1}^{2000} k$

RELATING CONCEPTS (EXERCISES 47–50)

Let $f(x) = mx + b$. Work Exercises 47–50 in order.

47. Find $f(1), f(2)$, and $f(3)$. (Section 1.2)

48. Consider the sequence $f(1), f(2), f(3), \ldots$. Is it an arithmetic sequence?

49. If the sequence is arithmetic, what is the common difference?

50. What is a_n for the sequence described in Exercise 48?

Use the sequence feature of a graphing calculator to evaluate the sum of the first ten terms of the arithmetic sequence. In Exercises 53 and 54, round to the nearest thousandth.

51. $a_n = 4.2n + 9.73$

52. $a_n = 8.42n + 36.18$

53. $a_n = \sqrt{8}n + \sqrt{3}$

54. $a_n = -\sqrt[3]{4}n + \sqrt{7}$

Solve each problem involving arithmetic sequences.

55. Find the sum of all the integers from 51 to 71.

56. Find the sum of all the integers from -8 to 30.

57. *Clock Chimes* If a clock strikes the proper number of chimes each hour on the hour, how many times will it chime in a month of 30 days?

58. *Telephone Pole Stack* A stack of telephone poles has 30 in the bottom row, 29 in the next, and so on, with one pole in the top row. How many poles are in the stack?

59. *City Population Growth* The population of a city was 49,000 five years ago. Each year the zoning commission permits an increase of 580 in the population. What will the maximum population be five years from now?

60. *Supports on a Slide* A super slide of uniform slope is to be built on a level piece of land. There are to be 20 equally spaced supports, with the longest support 15 meters long and the shortest 2 meters long. Find the total length of all the supports.

61. *Rungs of a Ladder* How much material would be needed for the rungs of a ladder of 31 rungs, if the rungs taper uniformly from 18 inches to 28 inches?

62. *Growth Pattern for Children* The normal growth pattern for children aged 3–11 follows that of an arithmetic sequence. An increase in height of about 6 centimeters per year is expected. Thus, 6 would be the common dif-

ference of the sequence. A child who measures 96 centimeters at age 3 would have his expected height in subsequent years represented by the sequence 102, 108, 114, 120, 126, 132, 138, 144. Each term differs from the adjacent terms by the common difference, 6.

(a) If a child measures 98.2 centimeters at age 3 and 109.8 centimeters at age 5, what would be the common difference of the arithmetic sequence describing her yearly height?

(b) What would we expect her height to be at age 8?

63. Find all arithmetic sequences $a_1, a_2, a_3, \ldots$, such that $a_1^2, a_2^2, a_3^2, \ldots$, is also an arithmetic sequence.

64. Suppose that $a_1, a_2, a_3, \ldots$ and $b_1, b_2, b_3, \ldots$ are both arithmetic sequences. Let $d_n = a_n + c \cdot b_n$, for any real number c and every positive integer n. Show that $d_1, d_2, d_3, \ldots$ is an arithmetic sequence.

65. Suppose that $a_1, a_2, a_3, a_4, a_5, \ldots$ is an arithmetic sequence. Is $a_1, a_3, a_5, \ldots$ an arithmetic sequence?

66. Explain why the sequence log 2, log 4, log 8, log 16, . . . is arithmetic.

8.3 GEOMETRIC SEQUENCES AND SERIES

Geometric Sequences ◆ Geometric Series ◆ Infinite Geometric Series ◆ Annuities

GEOMETRIC SEQUENCES

Suppose you agreed to work for 1¢ the first day, 2¢ the second day, 4¢ the third day, 8¢ the fourth day, and so on, with your wages doubling each day. How much will you earn on day 20, after working 5 days a week for a month? How much will you have earned altogether in 20 days? These questions will be answered in this section.

A **geometric sequence** (or **geometric progression**) is a sequence in which each term after the first is obtained by multiplying the preceding term by a constant nonzero real number, called the **common ratio**. The sequence discussed above,

$$1, 2, 4, 8, 16, \ldots,$$

is an example of a geometric sequence in which the first term is 1 and the common ratio is 2.

If the common ratio of a geometric sequence is r, then by the definition of a geometric sequence,

$$r = \frac{a_{n+1}}{a_n}$$

for every positive integer n. Therefore, the common ratio can be found by choosing any term except the first and dividing it by the preceding term.

In the geometric sequence

$$2, 8, 32, 128, \ldots,$$

$r = 4$. Notice that

$$8 = 2 \cdot 4$$
$$32 = 8 \cdot 4 = (2 \cdot 4) \cdot 4 = 2 \cdot 4^2$$
$$128 = 32 \cdot 4 = (2 \cdot 4^2) \cdot 4 = 2 \cdot 4^3.$$

To generalize this, assume that a geometric sequence has first term a_1 and common ratio r. The second term can be written as $a_2 = a_1 r$, the third as $a_3 = a_2 r = (a_1 r)r = a_1 r^2$, and so on. Following this pattern, the nth term is $a_n = a_1 r^{n-1}$. Again, this result is proven by mathematical induction. (See Section 5 of this chapter.)

FORMULA FOR THE nTH TERM OF A GEOMETRIC SEQUENCE

In the geometric sequence with first term a_1 and common ratio r, the nth term is

$$a_n = a_1 r^{n-1}.$$

EXAMPLE 1   *Finding the nth Term of a Geometric Sequence of Wages*   The formula for the nth term of a geometric sequence can be used to answer the first question posed at the beginning of this section. How much will be earned on day 20 if daily wages follow the sequence

$$1, 2, 4, 8, 16, \ldots,$$

with $a_1 = 1$ and $r = 2$?

SOLUTION   To answer the question, find a_{20}.

$$a_{20} = a_1 r^{19} = 1(2)^{19} = 524{,}288 \text{ cents, or } \$5242.88 \qquad \blacklozenge$$

EXAMPLE 2   *Using the Formula for the nth Term*   Find a_5 and a_n for the following geometric sequence.

$$4, 12, 36, 108, \ldots$$

SOLUTION   The first term, a_1, is 4. Find r by choosing any term except the first and dividing it by the preceding term. For example,

$$r = \frac{36}{12} = 3.$$

Since $a_4 = 108$, $a_5 = 3 \cdot 108 = 324$. The fifth term also could be found by using the formula for a_n, $a_n = a_1 r^{n-1}$, and replacing n with 5, r with 3, and a_1 with 4.

$$a_5 = 4 \cdot (3)^{5-1} = 4 \cdot 3^4 = 324$$

By the formula,

$$a_n = 4 \cdot 3^{n-1}. \qquad \blacklozenge$$

EXAMPLE 3   *Using the Formula for the nth Term*   Find a_1 and r for the geometric sequence with third term 20 and sixth term 160.

SOLUTION   Use the formula for the nth term of a geometric sequence.

$$\text{For } n = 3,\ a_3 = a_1 r^2 = 20;$$
$$\text{for } n = 6,\ a_6 = a_1 r^5 = 160.$$

We have $a_1 r^2 = 20$, so $a_1 = 20/r^2$. Substituting this in the second equation gives

$$a_1 r^5 = 160$$

$$\left(\frac{20}{r^2}\right) r^5 = 160$$

$$20 r^3 = 160$$

$$r^3 = 8$$

$$r = 2.$$

Since $a_1 r^2 = 20$ and $r = 2$, we know $a_1 = 5$.   ◆

EXAMPLE 4   *Solving a Geometric Growth Problem for a Population of Fruit Flies*   A population of fruit flies is growing in such a way that each generation is 1.5 times as large as the last generation. Suppose there were 100 insects in the first generation. How many would there be in the fourth generation?

SOLUTION   The population of each generation can be written as a geometric sequence, with a_1 as the first-generation population, a_2 the second-generation population, and so on. Then the fourth-generation population will be a_4. Using the formula for a_n, with $n = 4$, $r = 1.5$, and $a_1 = 100$, gives

$$a_4 = a_1 r^3 = 100(1.5)^3 = 100(3.375) = 337.5.$$

In the fourth generation, the population will number about 338 insects.   ◆

GEOMETRIC SERIES

A **geometric series** is the indicated sum of the terms of a geometric sequence. In applications, it may be necessary to find the sum of the terms of such a sequence. For example, a scientist might want to know the total number of insects in four generations of the population discussed in Example 4.

To find a formula for the sum of the first n terms of a geometric sequence, S_n, first write the sum as

$$S_n = a_1 + a_2 + a_3 + \cdots + a_n$$

or

$$S_n = a_1 + a_1 r + a_1 r^2 + \cdots + a_1 r^{n-1}. \tag{1}$$

If $r = 1$, $S_n = n a_1$, which is a correct formula for this case. If $r \neq 1$, multiply both sides of equation (1) by r, obtaining

$$rS_n = a_1 r + a_1 r^2 + a_1 r^3 + \cdots + a_1 r^n. \tag{2}$$

If (2) is subtracted from (1),

$$\begin{array}{rl} S_n = & a_1 + a_1 r + a_1 r^2 + \cdots + a_1 r^{n-1} \\ rS_n = & a_1 r + a_1 r^2 + \cdots + a_1 r^{n-1} + a_1 r^n \\ \hline S_n - rS_n = & a_1 \phantom{+ a_1 r + a_1 r^2 + \cdots + a_1 r^{n-1}} - a_1 r^n \end{array}$$ Subtract.

or

$$S_n(1 - r) = a_1(1 - r^n),$$ Factor.

which finally gives

$$S_n = \frac{a_1(1 - r^n)}{1 - r}, \quad \text{where } r \neq 1.$$ Divide by $1 - r$.

This discussion is summarized below.

FORMULA FOR THE SUM OF THE FIRST n TERMS OF A GEOMETRIC SEQUENCE

If a geometric sequence has first term a_1 and common ratio r, then the sum of the first n terms is given by

$$S_n = \frac{a_1(1 - r^n)}{1 - r}, \quad \text{where } r \neq 1.$$

This formula can be used to find the total fruit fly population in Example 4 over the four-generation period. With $n = 4$, $a_1 = 100$, and $r = 1.5$,

$$S_4 = \frac{100(1 - 1.5^4)}{1 - 1.5} = \frac{100(1 - 5.0625)}{-.5} = 812.5,$$

so the total population for the four generations will amount to about 813 insects.

EXAMPLE 5   *Applying the Sum of the First n Terms*   At the beginning of this section, we posed the following question: How much will you have earned altogether after 20 days? Answer this question, and support the answer with a graphing calculator.

SOLUTION   ANALYTIC   To answer the second question posed at the beginning of this section, we must find the total amount earned in 20 days with daily wages of

$$1, 2, 4, 8, \ldots$$

cents. Since $a_1 = 1$ and $r = 2$,

$$S_{20} = \frac{1(1 - 2^{20})}{1 - 2} = 1{,}048{,}575 \text{ cents,}$$

or \$10,485.75. Not bad for 20 days of work!

GRAPHICAL   See Figure 7.   ◆

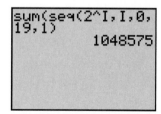   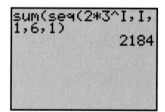

This screen supports the result in Example 5.

FIGURE 7

This screen supports the result in Example 6.

FIGURE 8

EXAMPLE 6   *Finding the Sum of the First n Terms*   Find $\sum\limits_{i=1}^{6} 2 \cdot 3^i$. Support the answer with a graphing calculator.

SOLUTION   ANALYTIC   This sum is the sum of the first six terms of a geometric sequence having $a_1 = 2 \cdot 3^1 = 6$ and $r = 3$. From the formula for S_n,

$$\sum_{i=1}^{6} 2 \cdot 3^i = S_6 = \frac{6(1 - 3^6)}{1 - 3}$$

$$= \frac{6(1 - 729)}{-2} = \frac{6(-728)}{-2} = 2184.$$

GRAPHICAL   See Figure 8.   ◆

INFINITE GEOMETRIC SERIES

Now the discussion of sums of sequences will be extended to include infinite geometric sequences such as the infinite sequence

$$2, 1, \frac{1}{2}, \frac{1}{4}, \frac{1}{8}, \frac{1}{16}, \ldots$$

with first term 2 and common ratio $1/2$. Using the earlier formula gives the following sequence.

$$S_1 = 2, \quad S_2 = 3, \quad S_3 = 3.5, \quad S_4 = 3.75, \quad S_5 = 3.875, \quad S_6 = 3.9375$$

With a calculator in function mode, we define Y_1 as S_n.

$$Y_1 = \frac{2\left(1 - \left(\frac{1}{2}\right)^x\right)}{1 - \frac{1}{2}}$$

Now, observe the first few terms as x takes on the values $1, 2, 3, \ldots$, using the table seen in Figure 9. As the table suggests, these terms, which represent the partial sums of the sequence, seem to be getting closer and closer to the number 4.

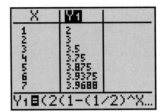

FIGURE 9

For no value of n is $S_n = 4$. However, if n is large enough, then S_n is as close to 4 as desired.* As mentioned earlier, we say the sequence converges to 4. This is expressed as

$$\lim_{n \to \infty} S_n = 4.$$

(Read: "the limit of S_n as n increases without bound is 4.") Since

$$\lim_{n \to \infty} S_n = 4,$$

the number 4 is called the **sum** of the infinite geometric sequence

$$2, 1, \frac{1}{2}, \frac{1}{4}, \ldots$$

and we say

$$2 + 1 + \frac{1}{2} + \frac{1}{4} + \cdots = 4.$$

EXAMPLE 7   *Finding the Sum of the Terms of an Infinite Geometric Sequence*

Find the sum $1 + \frac{1}{3} + \frac{1}{9} + \frac{1}{27} + \cdots$.

SOLUTION   Use the formula for the first n terms of a geometric sequence to get

$$S_1 = 1, \quad S_2 = \frac{4}{3}, \quad S_3 = \frac{13}{9}, \quad S_4 = \frac{40}{27},$$

---

*The phrases "large enough" and "as close as desired" are not nearly precise enough for mathematicians; much of a standard calculus course is devoted to making them more precise.

and in general

$$S_n = \frac{1\left[1 - \left(\frac{1}{3}\right)^n\right]}{1 - \frac{1}{3}}. \qquad \text{Let } a_1 = 1, \ r = \frac{1}{3}.$$

The chart below shows the value of $(1/3)^n$ for larger and larger values of n.

n	1	10	100	200
$\left(\frac{1}{3}\right)^n$	$\frac{1}{3}$	1.69×10^{-5}	1.94×10^{-48}	3.76×10^{-96}

As n gets larger and larger, $(1/3)^n$ gets closer and closer to 0. That is,

$$\lim_{n \to \infty} \left(\frac{1}{3}\right)^n = 0,$$

making it reasonable that

$$\lim_{n \to \infty} S_n = \lim_{n \to \infty} \frac{1\left[1 - \left(\frac{1}{3}\right)^n\right]}{1 - \frac{1}{3}} = \frac{1(1 - 0)}{1 - \frac{1}{3}} = \frac{1}{\frac{2}{3}} = \frac{3}{2}.$$

Hence,

$$1 + \frac{1}{3} + \frac{1}{9} + \frac{1}{27} + \cdots = \frac{3}{2}. \qquad \blacklozenge$$

If a geometric sequence has a first term a_1 and a common ratio r, then

$$S_n = \frac{a_1(1 - r^n)}{1 - r} \quad (r \neq 1)$$

for every positive integer n. If $-1 < r < 1$, then $\lim_{n \to \infty} r^n = 0$, and

$$\lim_{n \to \infty} S_n = \frac{a_1(1 - 0)}{1 - r} = \frac{a_1}{1 - r}.$$

This quotient, $a_1/(1 - r)$, is called the **sum of the terms of an infinite geometric sequence.** The limit $\lim_{n \to \infty} S_n$ is often expressed as S_∞ or $\sum_{i=1}^{\infty} a_i$. These results lead to the following definition.

FORMULA FOR THE SUM OF THE TERMS OF AN INFINITE GEOMETRIC SEQUENCE

The sum of an infinite geometric sequence with first term a_1 and common ratio r, where $-1 < r < 1$, is given by

$$S_\infty = \frac{a_1}{1 - r}.$$

If $|r| > 1$, the terms get larger and larger in absolute value, so there is no limit as $n \to \infty$. Hence, the sequence will not have a sum.

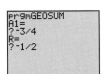

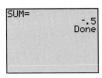

A short program supports the result of Example 8(a).

EXAMPLE 8 *Finding the Sum of the Terms of an Infinite Geometric Sequence*
Find each sum.

(a) $\displaystyle\sum_{i=1}^{\infty}\left(-\frac{3}{4}\right)\left(-\frac{1}{2}\right)^{i-1}$ (b) $\displaystyle\sum_{i=1}^{\infty}\left(\frac{3}{5}\right)^{i}$

SOLUTION

(a) ANALYTIC Here, $a_1 = -3/4$ and $r = -1/2$. Since $-1 < r < 1$, the formula above applies, and

$$S_{\infty} = \frac{a_1}{1-r} = \frac{-\dfrac{3}{4}}{1-\left(-\dfrac{1}{2}\right)} = -\frac{1}{2}.$$

GRAPHICAL See the screens in the margin.

(b) $\displaystyle\sum_{i=1}^{\infty}\left(\frac{3}{5}\right)^{i} = \sum_{i=1}^{\infty}\left(\frac{3}{5}\right)\left(\frac{3}{5}\right)^{i-1} = \frac{\dfrac{3}{5}}{1-\dfrac{3}{5}} = \frac{3}{2}$ ◆

ANNUITIES

Geometric sequences and series are very important in the mathematics of finance. An example is a sequence of equal payments made at equal periods of time, such as car payments or house payments, called an **annuity**. If the payments are accumulated in an account (with no withdrawals), the sum of the payments and interest on the payments is called the **future value** of the annuity.

EXAMPLE 9 *Finding the Future Value of an Annuity* To save money for a trip to Europe, Paula Story deposited $1000 at the *end* of each year for four years in an account paying 6% interest, compounded annually. Find the future value of this annuity. Support the answer with a graphing calculator.

SOLUTION ANALYTIC To find the future value, we use the formula for compound interest, $A = P(1 + r)^t$. The first payment earns interest for three years, the second payment for two years, and the third payment for one year. The last payment earns no interest. The total amount is

$$1000(1.06)^3 + 1000(1.06)^2 + 1000(1.06) + 1000.$$

This is the sum of a geometric sequence with first term (starting at the end of the sum as written above) $a_1 = 1000$ and common ratio $r = 1.06$. Using the formula for S_4, the sum of four terms, gives

$$S_4 = \frac{1000[1 - (1.06)^4]}{1 - 1.06} \approx 4374.62.$$

The future value of the annuity is $4374.62.

GRAPHICAL See Figure 10. ◆

The screen supports the result of the computation in Example 9.

FIGURE 10

WHAT WENT WRONG ?   The discussion preceding Example 7 justified the statement that the sum of the terms $2, 1, \frac{1}{2}, \frac{1}{4}, \frac{1}{8}, \frac{1}{16}, \ldots$ gets closer and closer to 4 by taking more and more terms of the sequence. However, this sum will always differ from 4 by some small amount. Now, look at the figure below, which is an extension of the table in Figure 9. According to the calculator, the sum *is* 4 when $x \geq 17$.

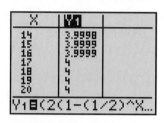

1.   What went wrong (if anything)?
2.   Discuss the limitations of technology, basing your comments on this example.

?

8.3   EXERCISES   ◆   ▣   ◀▶ Tape 14

Write out the terms of the geometric sequence that satisfies the given conditions.

1. $a_1 = \dfrac{5}{3}$, $r = 3$, $n = 4$ **2.** $a_1 = -\dfrac{3}{4}$, $r = \dfrac{2}{3}$, $n = 4$ **3.** $a_4 = 5$, $a_5 = 10$, $n = 5$ **4.** $a_3 = 16$, $a_4 = 8$, $n = 5$

Find a_5 and a_n for each geometric sequence.

5. $a_1 = 5$, $r = -2$ **6.** $a_1 = 8$, $r = -5$ **7.** $a_2 = -4$, $r = 3$

8. $a_3 = -2$, $r = 4$ **9.** $a_4 = 243$, $r = -3$ **10.** $a_4 = 18$, $r = 2$

11. $-4, -12, -36, -108, \ldots$ **12.** $-2, 6, -18, 54, \ldots$ **13.** $\dfrac{4}{5}, 2, 5, \dfrac{25}{2}, \ldots$

14. $\dfrac{1}{2}, \dfrac{2}{3}, \dfrac{8}{9}, \dfrac{32}{27}, \ldots$

RELATING CONCEPTS (EXERCISES 15–18)

Using the definition of difference *for an arithmetic sequence and* ratio *for a geometric sequence, we can find the appropriate middle term in a group of three terms so that the resulting sequence is the type desired. For example, consider the three terms*

$$5, x, .6,$$

where x is to be determined. Work Exercises 15–18 in order.

15. For these terms to form an arithmetic sequence, the difference $x - 5$ must be the same as the difference $.6 - x$. Write an equation that makes this statement. (Section 1.5)

16. Solve the equation of Exercise 15, and write the three terms of the arithmetic sequence. (Section 1.5)

17. For these terms to form a geometric sequence, the ratio $x/5$ must be the same as the ratio $.6/x$. Write an equation that makes this statement. (Section 1.8)

18. Solve the equation of Exercise 17 for its positive solution, and write the three terms of the geometric sequence. (Section 3.3)

Find a_1 and r for each geometric sequence.

19. $a_3 = 5, a_8 = \dfrac{1}{625}$ **20.** $a_2 = -6, a_7 = -192$ **21.** $a_4 = -\dfrac{1}{4}, a_9 = -\dfrac{1}{128}$ **22.** $a_3 = 50, a_7 = .005$

Use the formula for S_n to find the sum of the first five terms for each geometric sequence. Round the answers for Exercises 27 and 28 to the nearest hundredth.

23. 2, 8, 32, 128, ...   **24.** 4, 16, 64, 256, ...   **25.** $18, -9, \dfrac{9}{2}, -\dfrac{9}{4}, \ldots$

26. $12, -4, \dfrac{4}{3}, -\dfrac{4}{9}, \ldots$ **27.** $a_1 = 8.423, r = 2.859$ **28.** $a_1 = -3.772, r = -1.553$

Find each sum.

29. $\displaystyle\sum_{i=1}^{5} 3^i$ **30.** $\displaystyle\sum_{i=1}^{4} (-2)^i$ **31.** $\displaystyle\sum_{j=1}^{6} 48\left(\dfrac{1}{2}\right)^j$

32. $\displaystyle\sum_{j=1}^{5} 243\left(\dfrac{2}{3}\right)^j$ **33.** $\displaystyle\sum_{k=4}^{10} 2^k$ **34.** $\displaystyle\sum_{k=3}^{9} (-3)^k$

35. Under what conditions does the sum of an infinite geometric series exist?

36. The number .999 ... can be written as the sum of the terms of an infinite geometric sequence: $.9 + .09 + .009 + \ldots$. Here we have $a_1 = .9$ and $r = .1$. Use the formula for S_∞ to find this sum. Does your intuition indicate that your answer is correct?

Find r for each infinite geometric sequence. Identify any whose sum would not converge.

37. 12, 24, 48, 96, ...   **38.** 625, 125, 25, 5, ...   **39.** $-48, -24, -12, -6, \ldots$ **40.** $2, -10, 50, -250, \ldots$

Find each sum that converges by using the formula from this section where it applies.

41. $16 + 2 + \dfrac{1}{4} + \dfrac{1}{32} + \cdots$ **42.** $18 + 6 + 2 + \dfrac{2}{3} + \cdots$ **43.** $100 + 10 + 1 + \cdots$

44. $128 + 64 + 32 + \cdots$ **45.** $\dfrac{4}{3} + \dfrac{2}{3} + \dfrac{1}{3} + \cdots$ **46.** $\dfrac{1}{4} - \dfrac{1}{6} + \dfrac{1}{9} - \dfrac{2}{27} + \cdots$

47. $\displaystyle\sum_{i=1}^{\infty} 3\left(\dfrac{1}{4}\right)^{i-1}$ **48.** $\displaystyle\sum_{i=1}^{\infty} 5\left(-\dfrac{1}{4}\right)^{i-1}$ **49.** $\displaystyle\sum_{k=1}^{\infty} (.3)^k$ **50.** $\displaystyle\sum_{k=1}^{\infty} 10^{-k}$

RELATING CONCEPTS (EXERCISES 51–54)

Let $g(x) = ab^x$. Work Exercises 51–54 in order.

51. Find $g(1)$, $g(2)$, and $g(3)$. (Section 1.2)

52. Consider the sequence $g(1)$, $g(2)$, $g(3)$, ... Is it a geometric sequence? If so, what is the common ratio?

53. What is the general term of the sequence in Exercise 52?

54. Explain how geometric sequences are related to exponential functions. (Section 5.1)

Use the sequence feature of a graphing calculator to evaluate each of the following sums. Round to the nearest thousandth.

55. $\displaystyle\sum_{i=1}^{10} (1.4)^i$ **56.** $\displaystyle\sum_{j=1}^{6} -(3.6)^j$ **57.** $\displaystyle\sum_{j=3}^{8} 2(.4)^j$ **58.** $\displaystyle\sum_{i=4}^{9} 3(.25)^i$

Annuity Values *Find the future value of each annuity.*

59. Payments of $1000 at the end of each year for 9 years at 8% interest, compounded annually

60. Payments of $800 at the end of each year for 12 years at 7% interest, compounded annually

61. Payments of $2430 at the end of each year for 10 years at 6% interest, compounded annually

62. Payments of $1500 at the end of each year for 6 years at 5% interest, compounded annually

Solve each problem involving geometric sequences.

63. *Bacterial Growth* The strain of bacteria described in Exercise 63 in Section 8.1 will double in size and then divide every 40 minutes. Let a_1 be the initial number of bacteria cells, a_2 the number after 40 minutes, and a_n the number after $40(n-1)$ minutes. (*Source:* Hoppensteadt, F., and C. Peskin, *Mathematics in Medicine and the Life Sciences*, Springer-Verlag, 1992.)

(a) Write a formula for the nth term a_n of the geometric sequence $a_1, a_2, a_3, \ldots, a_n, \ldots$.

(b) Determine the first n where $a_n > 1{,}000{,}000$, when $a_1 = 100$.

(c) How long does it take for the number of bacteria to exceed one million?

64. *Photo Processing* The final step in processing a black-and-white photographic print is to immerse the print in a chemical called "fixer." The print is then washed in running water. Under certain conditions, 98% of the fixer in a print will be removed with 15 minutes of washing. How much of the original fixer would be left after 1 hour of washing?

65. *Chemical Mixture* A scientist has a vat containing 100 liters of a pure chemical. Twenty liters are drained and replaced with water. After complete mixing, 20 liters of the mixture are drained and replaced with water. What will be the strength of the mixture after 9 such drainings?

66. *Half-Life of a Radioactive Substance* The half-life of a radioactive substance is the time it takes for half the substance to decay. Suppose the half-life of a substance is 3 years, and 10^{15} molecules of the substance are present initially. How many molecules will be present after 15 years?

67. *Depreciation in Value* Each year a machine loses 20% of the value it had at the beginning of the year.

Find the value of the machine at the end of 6 years if it cost $100,000 new.

68. *Sugar Processing* A sugar factory receives an order for 1000 units of sugar. The production manager thus orders production of 1000 units of sugar. He forgets, however, that the production of sugar requires some sugar (to prime the machines, for example), and so he ends up with only 900 units of sugar. He then orders an additional 100 units, and receives only 90 units. A further order for 10 units produces 9 units. Finally seeing he is wrong, the manager decides to try mathematics. He views the production process as an infinite geometric progression with $a_1 = 1000$ and $r = .1$. Using this, find the number of units of sugar that he should have ordered originally.

69. *Swing of a Pendulum* A pendulum bob swings through an arc 40 centimeters long on its first swing. Each swing thereafter, it swings only 80% as far as on the previous swing. How far will it swing altogether before coming to a complete stop?

70. *Height of a Dropped Ball* Mitzi drops a ball from a height of 10 meters and notices that on each bounce the ball returns to about 3/4 of its previous height. About how far will the ball travel before it comes to rest? (*Hint:* Consider the sum of two sequences.)

71. *Number of Ancestors* Each person has two parents, four grandparents, eight great-grandparents, and so on. What is the total number of ancestors a person has, going back five generations? Ten generations?

72. *Drug Dosage* Certain medical conditions are treated with a fixed dose of a drug administered at regular intervals. Suppose a person is given 2 milligrams of a drug each day, and that during each 24-hour period the body utilizes 40% of the amount of drug that was present at the beginning of the period.

(a) Show that the amount of the drug present in the body at the end of n days is $\sum_{i=1}^{n} 2(.6)^i$.

(b) What will be the approximate quantity of the drug in the body at the end of each day after the treatment has been administered for a long period of time?

73. *Side Length of a Triangle* A sequence of equilateral triangles is constructed. The first triangle has sides 2 meters in length. To get the second triangle, midpoints of the sides of the original triangle are connected. What is the length of the side of the eighth such triangle? See the figure below.

74. *Side Lengths and Areas of Triangles*

   (a) In Exercise 73, if the process could be continued indefinitely, what would be the total perimeter of all the triangles?

   (b) What would be the total area of all the triangles, disregarding the overlapping?

75. Let $a_1, a_2, a_3, \ldots$ and $b_1, b_2, b_3, \ldots$ be geometric sequences. Let $d_n = c \cdot a_n \cdot b_n$ for any real number c and every positive integer n. Show that $d_1, d_2, d_3, \ldots$ is a geometric sequence.

76. Explain why the sequence log 6, log 36, log 1296, log 1,679,616, . . . is geometric.

8.4   THE BINOMIAL THEOREM

A Binomial Expansion Pattern ◆ Pascal's Triangle ◆ *n*-Factorial ◆ The Binomial Coefficient ◆ The Binomial Theorem ◆ *r*th Term of a Binomial Expansion

A BINOMIAL EXPANSION PATTERN

In this section, we introduce a method for writing out the terms of expressions of the form $(x + y)^n$, where n is a natural number. The formula for writing out the powers of a binomial $(x + y)^n$ as a polynomial is called the **binomial theorem**. This theorem is important when working with probability and statistics. Some expansions of $(x + y)^n$, for various nonnegative integer values of n, are given below.

$$(x + y)^0 = 1$$
$$(x + y)^1 = x + y$$
$$(x + y)^2 = x^2 + 2xy + y^2$$
$$(x + y)^3 = x^3 + 3x^2y + 3xy^2 + y^3$$
$$(x + y)^4 = x^4 + 4x^3y + 6x^2y^2 + 4xy^3 + y^4$$
$$(x + y)^5 = x^5 + 5x^4y + 10x^3y^2 + 10x^2y^3 + 5xy^4 + y^5$$

Studying these results reveals a pattern that can be used to write a general expression for $(x + y)^n$. First, notice that after the special case $(x + y)^0 = 1$, each expression begins with x raised to the same power as the binomial itself. That is, the expansion of $(x + y)^1$ has a first term of x^1, $(x + y)^2$ has a first term of x^2, $(x + y)^3$ has a first term of x^3, and so on. Also, the last term in each expansion is y to the same power as the binomial. Thus, the expansion of $(x + y)^n$ should begin with the term x^n and end with the term y^n.

Also, the exponents on x decrease by one in each term after the first, while the exponents on y, beginning with y in the second term, increase by one in each succeeding term. That is, the *variables* in the terms of the expansion of $(x + y)^n$ have the following pattern.

$$x^n, \ x^{n-1}y, \ x^{n-2}y^2, \ x^{n-3}y^3, \ \ldots, \ xy^{n-1}, \ y^n$$

This pattern suggests that the sum of the exponents on x and y in each term is n. For example, in the third term in the list above, the variable is $x^{n-2}y^2$, and the sum of the exponents is $n - 2 + 2 = n$.

PASCAL'S TRIANGLE

Now, examine the *coefficients* in the terms of the expansions shown above. Writing the coefficients alone gives the following pattern.

TECHNOLOGY NOTE

The rows of Pascal's triangle can be generated by a graphing calculator. See the screens in the margin on page 638.

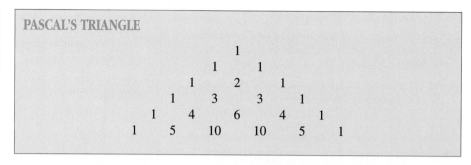

PASCAL'S TRIANGLE

With the coefficients arranged in this way, we can see that each number in the triangle is the sum of the two numbers directly above it (one to the right and one to the left). For example, if we number the rows starting with row 0, in row four, 1 is the sum of 1, the only number above it, 4 is the sum of 1 and 3, 6 is the sum of 3 and 3, and so on. This triangular array of numbers is called **Pascal's triangle**, in honor of the seventeenth-century mathematician Blaise Pascal (1623–1662), one of the first to use it extensively.

n-FACTORIAL

To get the coefficients for $(x + y)^6$, we add row six to the array of numbers given above. Adding adjacent numbers, we find that row six is

$$1 \quad 6 \quad 15 \quad 20 \quad 15 \quad 6 \quad 1.$$

Using these coefficients, the expansion of $(x + y)^6$ is

$$(x + y)^6 = x^6 + 6x^5y + 15x^4y^2 + 20x^3y^3 + 15x^2y^4 + 6xy^5 + y^6.$$

Although it is possible to use Pascal's triangle to find the coefficients of $(x + y)^n$ for any positive integer value of *n*, this becomes impractical for large values of *n* because of the need to write out all the preceding rows. A more efficient way of finding these coefficients uses **factorial notation**. The number *n*! (read "*n*-factorial") is defined as follows.

DEFINITION OF *n*-FACTORIAL

For any positive integer *n*,

$$n! = n(n - 1)(n - 2) \cdots (3)(2)(1),$$

and $0! = 1.$

EXAMPLE 1   *Evaluating Factorials*   Evaluate each factorial and support the results with a graphing calculator.

(a) 5!   **(b)** 7!   **(c)** 2!   **(d)** 1!   **(e)** 0!

SOLUTION   ANALYTIC

(a) $5! = 5 \cdot 4 \cdot 3 \cdot 2 \cdot 1 = 120$ **(b)** $7! = 7 \cdot 6 \cdot 5 \cdot 4 \cdot 3 \cdot 2 \cdot 1 = 5040$

(c) $2! = 2 \cdot 1 = 2$ **(d)** $1! = 1$

(e) $0! = 1$

GRAPHICAL   See Figure 11 for the calculator support.   ◆

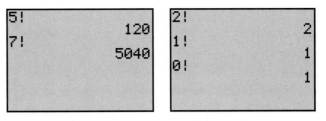

This screen supports the results in
Example 1, parts (a) and (b).

This screen supports the results in
Example 1, parts (c), (d), and (e).

(a)   (b)

FIGURE 11

THE BINOMIAL COEFFICIENT

Now, look at the coefficients of the expression

$$(x + y)^5 = x^5 + 5x^4y + 10x^3y^2 + 10x^2y^3 + 5xy^4 + y^5.$$

The coefficient of the second term, $5x^4y$, is 5, and the exponents on the variables are 4 and 1. Note that

$$5 = \frac{5!}{4!\,1!}.$$

The coefficient of the third term is 10, with exponents of 3 and 2, and

$$10 = \frac{5!}{3!\,2!}.$$

The last term (the sixth term) can be written as $y^5 = 1x^0y^5$, with coefficient 1, and exponents of 0 and 5. Since $0! = 1$, check that

$$1 = \frac{5!}{0!\,5!}.$$

Generalizing from these examples, we find that the coefficient for the term of the expansion of $(x + y)^n$ in which the variable part is x^ry^{n-r} (where $r \leq n$) will be

$$\frac{n!}{r!(n - r)!}.$$

This number, called a **binomial coefficient**, is often written as $\binom{n}{r}$ (read "n choose r") or $_nC_r$. We will use both notations.

DEFINITION OF THE BINOMIAL COEFFICIENT

For nonnegative integers n and r, with $r \leq n$, the symbol $\binom{n}{r}$ is defined as

$$\binom{n}{r} = \frac{n!}{r!(n - r)!}.$$

These binomial coefficients are just numbers from Pascal's triangle. For example, $\binom{3}{0}$ is the first number in the row that begins 1, 3, and $\binom{7}{4}$ is the fifth number in the row that begins 1, 7. Graphing calculators are able to compute binomial coefficients with the *combinations* function, denoted $_nC_r$.

EXAMPLE 2   *Evaluating Binomial Coefficients*   Evaluate each binomial coefficient. Support the results with a graphing calculator.

(a) $\begin{pmatrix} 6 \\ 2 \end{pmatrix}$ (b) $\begin{pmatrix} 8 \\ 0 \end{pmatrix}$ (c) $\begin{pmatrix} 10 \\ 10 \end{pmatrix}$

SOLUTION   ANALYTIC

(a) $\begin{pmatrix} 6 \\ 2 \end{pmatrix} = \dfrac{6!}{2!\,(6-2)!} = \dfrac{6!}{2!\,4!} = \dfrac{6 \cdot 5 \cdot 4 \cdot 3 \cdot 2 \cdot 1}{2 \cdot 1 \cdot 4 \cdot 3 \cdot 2 \cdot 1} = 15$

(b) $\begin{pmatrix} 8 \\ 0 \end{pmatrix} = \dfrac{8!}{0!\,(8-0)!} = \dfrac{8!}{0!\,8!} = \dfrac{8!}{1 \cdot 8!} = 1$

(c) $\begin{pmatrix} 10 \\ 10 \end{pmatrix} = \dfrac{10!}{10!\,(10-10)!} = \dfrac{10!}{10!\,0!} = 1$

GRAPHICAL   See Figure 12.   ◆

```
6 nCr 2
              15
8 nCr 0
               1
10 nCr 10
               1
```

FIGURE 12

For Group Discussion

1. Have members of the class compute pairs of binomial coefficients of the form $_nC_r$ and $_nC_{n-r}$; for example, $_{10}C_3$ and $_{10}C_7$.
2. Compare your results with Pascal's triangle.
3. Discuss your results and express them as a generalization.

Refer again to Pascal's triangle. Notice the symmetry in each row. This suggests that the binomial coefficients should have the same property. That is,

$$\begin{pmatrix} n \\ r \end{pmatrix} = \begin{pmatrix} n \\ n-r \end{pmatrix}.$$

```
seq(5 nCr X,X,0,
5,1)
  (1 5 10 10 5 1)
```

This is true, since

$$\begin{pmatrix} n \\ r \end{pmatrix} = \dfrac{n!}{r!\,(n-r)!} \qquad \text{and} \qquad \begin{pmatrix} n \\ n-r \end{pmatrix} = \dfrac{n!}{(n-r)!\,r!}.$$

```
 X  | Y1
----+----
 0  | 1
 1  | 6
 2  | 15
 3  | 20
 4  | 15
 5  | 6
 6  | 1
----+----
Y1 ▉6 nCr X
```

The two screens above illustrate how the sequence and table capabilities of a graphing calculator can generate rows of Pascal's triangle.

THE BINOMIAL THEOREM

Our observations about the expansion of $(x + y)^n$ are summarized as follows.

1. There are $n + 1$ terms in the expansion.
2. The first term is x^n, and the last term is y^n.
3. The exponent on x decreases by 1 and the exponent on y increases by 1 in each succeeding term.
4. The sum of the exponents on x and y in any term is n.
5. The coefficient of $x^r y^{n-r}$ is $\begin{pmatrix} n \\ r \end{pmatrix}$.

These observations about the expansion of $(x + y)^n$ for any positive integer value of n suggest the **binomial theorem**.

BINOMIAL THEOREM

For any positive integer n,

$$(x + y)^n = x^n + \binom{n}{n-1}x^{n-1}y + \binom{n}{n-2}x^{n-2}y^2 + \binom{n}{n-3}x^{n-3}y^3$$

$$+ \cdots + \binom{n}{n-r}x^{n-r}y^r + \cdots + \binom{n}{1}xy^{n-1} + y^n.$$

As stated above, the binomial theorem is a conjecture, determined inductively by looking at $(x + y)^n$ for several values of n. A proof of the binomial theorem that uses *mathematical induction* is given in Section 5 of this chapter.

Note   The binomial theorem looks much more manageable written in summation notation. The theorem can be summarized as follows:

$$(x + y)^n = \sum_{r=0}^{n} \binom{n}{r}x^{n-r}y^r.$$

EXAMPLE 3   *Applying the Binomial Theorem*   Write out the binomial expansion of $(x + y)^9$.

SOLUTION   Using the binomial theorem,

$$(x + y)^9 = x^9 + \binom{9}{1}x^8y + \binom{9}{2}x^7y^2 + \binom{9}{3}x^6y^3 + \binom{9}{4}x^5y^4 + \binom{9}{5}x^4y^5$$

$$+ \binom{9}{6}x^3y^6 + \binom{9}{7}x^2y^7 + \binom{9}{8}xy^8 + y^9.$$

Now, evaluate each of the binomial coefficients.

$$(x + y)^9 = x^9 + \frac{9!}{1!\,8!}x^8y + \frac{9!}{2!\,7!}x^7y^2 + \frac{9!}{3!\,6!}x^6y^3 + \frac{9!}{4!\,5!}x^5y^4$$

$$+ \frac{9!}{5!\,4!}x^4y^5 + \frac{9!}{6!\,3!}x^3y^6 + \frac{9!}{7!\,2!}x^2y^7 + \frac{9!}{8!\,1!}xy^8 + y^9$$

$$= x^9 + 9x^8y + 36x^7y^2 + 84x^6y^3 + 126x^5y^4 + 126x^4y^5$$

$$+ 84x^3y^6 + 36x^2y^7 + 9xy^8 + y^9 \qquad \blacklozenge$$

EXAMPLE 4   *Applying the Binomial Theorem*   Expand $\left(a - \dfrac{b}{2}\right)^5$.

SOLUTION   Write the binomial as follows.

$$\left(a - \frac{b}{2}\right)^5 = \left(a + \left(-\frac{b}{2}\right)\right)^5$$

Now, use the binomial theorem with $x = a$, $y = -b/2$, and $n = 5$ to get

$$\left(a - \frac{b}{2}\right)^5 = a^5 + \binom{5}{1}a^4\left(-\frac{b}{2}\right) + \binom{5}{2}a^3\left(-\frac{b}{2}\right)^2 + \binom{5}{3}a^2\left(-\frac{b}{2}\right)^3 + \binom{5}{4}a\left(-\frac{b}{2}\right)^4 + \left(-\frac{b}{2}\right)^5$$

$$= a^5 + 5a^4\left(-\frac{b}{2}\right) + 10a^3\left(-\frac{b}{2}\right)^2 + 10a^2\left(-\frac{b}{2}\right)^3 + 5a\left(-\frac{b}{2}\right)^4 + \left(-\frac{b}{2}\right)^5$$

$$= a^5 - \frac{5}{2}a^4b + \frac{5}{2}a^3b^2 - \frac{5}{4}a^2b^3 + \frac{5}{16}ab^4 - \frac{1}{32}b^5. \quad \blacklozenge$$

Note As Example 4 illustrates, any expansion of the *difference* of two terms has alternating signs.

rTH TERM OF A BINOMIAL EXPANSION

Any single term of a binomial expansion can be determined without writing out the whole expansion. For example, the seventh term of $(x + y)^9$ has y raised to the sixth power (since y has the power 1 in the second term, the power 2 in the third term, and so on). The exponents on x and y in each term must have a sum of 9, so the exponent on x in the seventh term is $9 - 6 = 3$. Thus, writing the coefficient as given in the binomial theorem, the seventh term should be

$$\frac{9!}{3!\ 6!}x^3y^6.$$

This is in fact the seventh term of $(x + y)^9$ found in Example 3. This discussion suggests the next theorem.

EXPRESSION FOR THE rTH TERM OF THE BINOMIAL EXPANSION

The rth term of the binomial expansion of $(x + y)^n$, where $n \geq r - 1$, is

$$\binom{n}{n - (r - 1)}x^{n-(r-1)}y^{r-1}.$$

EXAMPLE 5 *Finding a Specific Term of a Binomial Expansion* Find the fourth term of $(a + 2b)^{10}$.

SOLUTION In the fourth term, $2b$ has an exponent of 3, while a has an exponent of $10 - 3$, or 7. Using $n = 10$, $r = 4$, $x = a$, and $y = 2b$ in the formula above, we find that the fourth term is

$$\binom{10}{7}a^7(2b)^3 = 120a^7(8b^3) = 960a^7b^3. \quad \blacklozenge$$

8.4 EXERCISES   Tape 14

Evaluate the following.

1. $\dfrac{6!}{3!\ 3!}$ 2. $\dfrac{5!}{2!\ 3!}$ 3. $\dfrac{7!}{3!\ 4!}$ 4. $\dfrac{8!}{5!\ 3!}$ 5. $\dbinom{8}{3}$ 6. $\dbinom{7}{4}$

7. $\dbinom{10}{8}$ 8. $\dbinom{9}{6}$ 9. $\dbinom{13}{13}$ 10. $\dbinom{12}{12}$ 11. $\dbinom{n}{n-1}$ 12. $\dbinom{n}{n-2}$

13. How many terms are there in the expansion of $(x + y)^8$?

14. How many terms are there in the expansion of $(x + y)^{10}$?

Write out the binomial expansion for the following.

15. $(x + y)^6$

16. $(m + n)^4$

17. $(p - q)^5$

18. $(a - b)^7$

19. $(r^2 + s)^5$

20. $(m + n^2)^4$

21. $(p + 2q)^4$

22. $(3r - s)^6$

23. $(7p + 2q)^4$

24. $(4a - 5b)^5$

25. $(3x - 2y)^6$

26. $(7k - 9j)^4$

27. $\left(\dfrac{m}{2} - 1\right)^6$

28. $\left(3 + \dfrac{y}{3}\right)^5$

29. $\left(\sqrt{2}r + \dfrac{1}{m}\right)^4$

30. $\left(\dfrac{1}{k} - \sqrt{3}p\right)^3$

Write the indicated term of the binomial expansion.

31. Sixth term of $(4h - j)^8$

32. Eighth term of $(2c - 3d)^{14}$

33. Fifteenth term of $(a^2 + b)^{22}$

34. Twelfth term of $(2x + y^2)^{16}$

35. Fifteenth term of $(x - y^3)^{20}$

36. Tenth term of $(a^3 + 3b)^{11}$

Use the concepts of this section to work Exercises 37–40.

37. Find the middle term of $(3x^7 + 2y^3)^8$.

38. Find the two middle terms of $(-2m^{-1} + 3n^{-2})^{11}$.

39. Find the value of n for which the coefficients of the fifth and eighth terms in the expansion of $(x + y)^n$ are the same.

40. Find the term in the expansion of $(3 + \sqrt{x})^{11}$ that contains x^4.

RELATING CONCEPTS (EXERCISES 41–44)

In this section, we saw how the factorial of a positive integer n can be computed as a product: $n! = 1 \cdot 2 \cdot 3 \cdot \ldots \cdot n$. Calculators and computers are capable of evaluating factorials very quickly. Before the days of technology, mathematicians developed a formula for approximating large factorials. Interestingly enough, the formula involves the irrational numbers π and e. It is called **Stirling's formula:**

$$n! \approx \sqrt{2\pi n} \cdot n^n \cdot e^{-n}.$$

As an example, for a small value of n, we observe that the exact value of 5! is 120, while Stirling's formula gives the approximation as 118.019160 with a graphing calculator. This is "off" by less than 2, an error of only 1.65%.

Work Exercises 41–44 in order.

41. Use a calculator to find the exact value of 10! and the approximation, using Stirling's formula.

42. Subtract the smaller value from the larger value in Exercise 41. Divide it by 10! and convert to a percent. What is the percent error?

43. Repeat Exercises 41 and 42 for $n = 12$.

44. Repeat Exercises 41 and 42 for $n = 13$. What seems to happen as n gets larger?

In later courses, it is shown that

$$(1 + x)^n = 1 + nx + \frac{n(n - 1)}{2!}x^2 + \frac{n(n - 1)(n - 2)}{3!}x^3 + \cdots$$

for any real number n (not just positive integer values) and any real number x, where $|x| < 1$. This result, a generalized binomial theorem, may be used to find approximate values of powers and roots. For example,

$$(1.008)^{1/4} = (1 + .008)^{1/4}$$

$$= 1 + \frac{1}{4}(.008) + \frac{(1/4)(-3/4)}{2!}(.008)^2 + \frac{(1/4)(-3/4)(-7/4)}{3!}(.008)^3 + \cdots$$

$$\approx 1.002.$$

Use this result to approximate the quantities in Exercises 45–48 to the nearest thousandth.

45. $(1.02)^{-3}$

46. $\dfrac{1}{1.04^5}$

47. $(1.01)^{3/2}$

48. $(1.03)^2$

8.5 MATHEMATICAL INDUCTION

Proof by Mathematical Induction   ◆   Proving Equality Statements   ◆   Proving Inequality Statements   ◆   Proof of the Binomial Theorem

PROOF BY MATHEMATICAL INDUCTION

Many results in mathematics are claimed true for any positive integer. Any of these results could be checked for $n = 1$, $n = 2$, $n = 3$, and so on, but since the set of positive integers is infinite, it would be impossible to check every possible case. For example, let S_n represent the statement that the sum of the first n positive integers is $\frac{n(n + 1)}{2}$,

$$S_n: 1 + 2 + 3 + \cdots + n = \frac{n(n + 1)}{2}.$$

The truth of this statement can be checked quickly for the first few values of n.

If $n = 1$, S_1 is $\qquad\qquad 1 = \dfrac{1(1 + 1)}{2}$, a true statement, since $1 = 1$.

If $n = 2$, S_2 is $\qquad 1 + 2 = \dfrac{2(2 + 1)}{2}$, a true statement, since $3 = 3$.

If $n = 3$, S_3 is $\quad 1 + 2 + 3 = \dfrac{3(3 + 1)}{2}$, a true statement, since $6 = 6$.

If $n = 4$, S_4 is $1 + 2 + 3 + 4 = \dfrac{4(4 + 1)}{2}$, a true statement, since $10 = 10$.

Since the statement is true for $n = 1, 2, 3$, and 4, and so on, can we conclude that the statement is true for all positive integers by observing this finite number of examples? The answer is no. However, we have an idea that it *may* be true for all positive integers.

To prove that such a statement is true for every positive integer, we use the following principle.

PRINCIPLE OF MATHEMATICAL INDUCTION

Let S_n be a statement concerning the positive integer n. Suppose that

1. S_1 is true;
2. for any positive integer k, $k \le n$, if S_k is true, then S_{k+1} is also true.

Then, S_n is true for every positive integer value of n.

A proof by mathematical induction can be explained as follows. By assumption (1) above, the statement is true when $n = 1$. By (2) above, the fact that the statement is true for $n = 1$ implies that it is true for $n = 1 + 1 = 2$. Using (2) again, the statement is thus true for $2 + 1 = 3$, for $3 + 1 = 4$, for $4 + 1 = 5$, and so on. Continuing in this way shows that the statement must be true for *every* positive integer, no matter how large.

The situation is similar to that of an infinite number of dominoes lined up, as suggested in Figure 13. If the first domino is pushed over, it pushes the next, which pushes the next, and so on, indefinitely.

FIGURE 13

Another example of the principle of mathematical induction might be an infinite ladder. Suppose the rungs are spaced so that, whenever you are on a rung, you know you can move to the next rung. Then *if* you can get to the first rung, you can go as high up the ladder as you wish.

Two separate steps are required for a proof by mathematical induction.

PROOF BY MATHEMATICAL INDUCTION

Step 1 Prove that the statement is true for $n = 1$.

Step 2 Show that for any positive integer k, $k \le n$, if S_k is true, then S_{k+1} is also true.

PROVING EQUALITY STATEMENTS

Mathematical induction is used in the next example to prove the statement discussed above: $S_n = 1 + 2 + 3 + \cdots + n = \frac{n(n+1)}{2}$.

EXAMPLE 1 *Proving an Equality Statement by Mathematical Induction*
Let S_n represent the statement

$$1 + 2 + 3 + \cdots + n = \frac{n(n+1)}{2}.$$

Prove that S_n is true for every positive integer n.

PROOF The proof by mathematical induction is as follows.

Step 1 Show that the statement is true when $n = 1$. If $n = 1$, S_1 becomes

$$1 = \frac{1(1+1)}{2},$$

which is true.

Step 2 Show that S_k implies S_{k+1}, where S_k is the statement

$$1 + 2 + 3 + \cdots + k = \frac{k(k+1)}{2},$$

and S_{k+1} is the statement

$$1 + 2 + 3 + \cdots + k + (k+1) = \frac{(k+1)[(k+1)+1]}{2}.$$

Start with S_k.

$$1 + 2 + 3 + \cdots + k = \frac{k(k+1)}{2}$$

Add $k + 1$ to both sides of this equation.

$$1 + 2 + 3 + \cdots + k + (k + 1) = \frac{k(k + 1)}{2} + (k + 1)$$

Now, factor out the common factor $k + 1$ on the right to get

$$= (k + 1)\left(\frac{k}{2} + 1\right)$$

$$= (k + 1)\left(\frac{k + 2}{2}\right)$$

$$1 + 2 + 3 + \cdots + k + (k + 1) = \frac{(k + 1)[(k + 1) + 1]}{2}.$$

This final result is the statement for $n = k + 1$; it has been shown that if S_k is true, then S_{k+1} is also true. The two steps required for a proof by mathematical induction have now been completed, so the statement S_n is true for every positive integer value of n. ◆

Caution   Notice that the left side of the statement always includes *all* the terms up to the nth term, as well as the nth term.

EXAMPLE 2   *Proving an Equality Statement by Mathematical Induction*

Prove that $4 + 7 + 10 + \cdots + (3n + 1) = \dfrac{n(3n + 5)}{2}$ is true for all positive integers n.

PROOF

Step 1   Show that the statement is true for S_1. S_1 is

$$4 = \frac{1(3 \cdot 1 + 5)}{2}.$$

Since the right side equals 4, S_1 is a true statement.

Step 2   Show that if S_k is true, then S_{k+1} is true, where S_k is

$$4 + 7 + 10 + \cdots + (3k + 1) = \frac{k(3k + 5)}{2},$$

and S_{k+1} is

$$4 + 7 + 10 + \cdots + (3k + 1) + [3(k + 1) + 1]$$
$$= \frac{(k + 1)[3(k + 1) + 5]}{2}.$$

Start with S_k.

$$4 + 7 + 10 + \cdots + (3k + 1) = \frac{k(3k + 5)}{2}$$

To get the left side of the equation S_k to be the left side of the equation S_{k+1}, we must add the $(k + 1)$th term. Now we try to algebraically change the right side to look like the right side of S_{k+1}. Adding $[3(k + 1) + 1]$ to both sides of S_k gives

$$4 + 7 + 10 + \cdots + (3k + 1) + [3(k + 1) + 1]$$

$$= \frac{k(3k + 5)}{2} + [3(k + 1) + 1].$$

Clear the parentheses in the new term on the right side of the equals sign and simplify.

$$= \frac{k(3k + 5)}{2} + 3k + 3 + 1$$

$$= \frac{k(3k + 5)}{2} + 3k + 4$$

Now, combine the two terms on the right.

$$= \frac{k(3k + 5)}{2} + \frac{2(3k + 4)}{2} = \frac{k(3k + 5) + 2(3k + 4)}{2}$$

$$= \frac{3k^2 + 5k + 6k + 8}{2} = \frac{3k^2 + 11k + 8}{2}$$

$$= \frac{(k + 1)(3k + 8)}{2}$$

Since $3k + 8$ can be written as $3(k + 1) + 5$,

$$4 + 7 + 10 + \cdots + (3k + 1) + [3(k + 1) + 1]$$

$$= \frac{(k + 1)[3(k + 1) + 5]}{2}.$$

The final result is the statement S_{k+1}. Therefore, if S_k is true, then S_{k+1} is true. The two steps required for a proof by mathematical induction are completed, so the general statement S_n is true for every positive integer value of n. ◆

PROVING INEQUALITY STATEMENTS

EXAMPLE 3 *Proving an Inequality Statement by Mathematical Induction*
Prove that if x is a real number between 0 and 1, then for every positive integer n,

$$0 < x^n < 1.$$

PROOF   Here S_1 is the statement

if $0 < x < 1$, then $0 < x^1 < 1$,

which is true. S_k is the statement

if $0 < x < 1$, then $0 < x^k < 1$.

To show that this implies that S_{k+1} is true, multiply all three parts of $0 < x^k < 1$ by x to get

$$x \cdot 0 < x \cdot x^k < x \cdot 1.$$

(Here the fact that $0 < x$ is used.) Simplify to get

$$0 < x^{k+1} < x.$$

Since $x < 1$,

$$x^{k+1} < x < 1$$

and

$$0 < x^{k+1} < 1.$$

By this work, if S_k is true, then S_{k+1} is true. Since both conditions for a proof have been satisfied, the given statement is true for every positive integer n. ◆

PROOF OF THE BINOMIAL THEOREM

The binomial theorem, introduced in Section 4 of this chapter, can be proved by mathematical induction.

EXAMPLE 4   *Proving the Binomial Theorem*   Prove that for any positive integer n and any complex numbers x and y,

$$(x + y)^n = x^n + \binom{n}{1}x^{n-1}y + \binom{n}{2}x^{n-2}y^2 + \binom{n}{3}x^{n-3}y^3$$

$$+ \cdots + \binom{n}{r}x^{n-r}y^r + \cdots + \binom{n}{n-1}xy^{n-1} + y^n. \tag{1}$$

PROOF   Let S_n be statement (1) above. Begin by verifying S_n for $n = 1$,

$$S_1\colon\ (x + y)^1 = x^1 + y^1,$$

which is true.

Now, assume that S_n is true for the positive integer k. Statement S_k becomes (using the definition of the binomial coefficient)

$$S_k\colon\ (x + y)^k = x^k + \frac{k!}{1!\,(k-1)!}x^{k-1}y + \frac{k!}{2!\,(k-2)!}x^{k-2}y^2$$

$$+ \cdots + \frac{k!}{(k-1)!\,1!}xy^{k-1} + y^k. \tag{2}$$

Multiply both sides of equation (2) by $x + y$.

$$(x + y)^k \cdot (x + y)$$

$$= \left[x \cdot x^k + \frac{k!}{1!\,(k-1)!}x^k y + \frac{k!}{2!\,(k-2)!}x^{k-1}y^2 + \cdots + \frac{k!}{(k-1)!\,1!}x^2 y^{k-1} + xy^k \right]$$

$$+ \left[x^k \cdot y + \frac{k!}{1!\,(k-1)!}x^{k-1}y^2 + \cdots + \frac{k!}{(k-1)!\,1!}xy^k + y \cdot y^k \right]$$

Rearrange terms to get

$$(x + y)^{k+1} = x^{k+1} + \left[\frac{k!}{1!\,(k-1)!} + 1 \right]x^k y + \left[\frac{k!}{2!\,(k-2)!} + \frac{k!}{1!\,(k-1)!} \right]x^{k-1}y^2$$

$$+ \cdots + \left[1 + \frac{k!}{(k-1)!\,1!} \right]xy^k + y^{k+1}. \tag{3}$$

The first expression in brackets in equation (3) simplifies to $\binom{k+1}{1}$. To see this, note that

$$\binom{k + 1}{1} = \frac{(k+1)(k)(k-1)(k-2)\cdots 1}{1 \cdot (k)(k-1)(k-2)\cdots 1} = k + 1.$$

Also,

$$\frac{k!}{1!\,(k-1)!} + 1 = \frac{k(k-1)!}{1(k-1)!} + 1 = k + 1.$$

The second expression becomes $\binom{k+1}{2}$, the last $\binom{k+1}{k}$, and so on. The result of equation (3) is just equation (2) with every k replaced by $k + 1$. Thus, the truth of S_n when $n = k$ implies the truth of S_n for $n = k + 1$, which completes the proof of the theorem by mathematical induction.   ◆

8.5 EXERCISES   Tape 14

1. A proof by mathematical induction allows us to prove that a statement is true for all _____.

2. Suppose that step 2 in a proof by mathematical induction can be satisfied, but step 1 cannot. May we conclude that the proof is complete? Explain.

3. For which natural numbers is the statement $2^n > 2n$ *not true*?

4. Write out in full and verify the statements S_1, S_2, S_3, S_4, and S_5 for the following. Then, use mathematical induction to prove that the statement is true for every positive integer n.

$$2 + 4 + 6 + \cdots + 2n = n(n + 1)$$

Use the method of mathematical induction to prove the following statements. Assume that n is a positive integer.

5. $1 + 3 + 5 + \cdots + (2n - 1) = n^2$

6. $3 + 6 + 9 + \cdots + 3n = \dfrac{3n(n + 1)}{2}$

7. $5 + 10 + 15 + \cdots + 5n = \dfrac{5n(n + 1)}{2}$

8. $2 + 4 + 8 + \cdots + 2^n = 2^{n+1} - 2$

9. $3 + 3^2 + 3^3 + \cdots + 3^n = \dfrac{3(3^n - 1)}{2}$

10. $1^2 + 2^2 + 3^2 + \cdots + n^2 = \dfrac{n(n + 1)(2n + 1)}{6}$

11. $1^3 + 2^3 + 3^3 + \cdots + n^3 = \dfrac{n^2(n + 1)^2}{4}$

12. $5 \cdot 6 + 5 \cdot 6^2 + 5 \cdot 6^3 + \cdots + 5 \cdot 6^n = 6(6^n - 1)$

13. $7 \cdot 8 + 7 \cdot 8^2 + 7 \cdot 8^3 + \cdots + 7 \cdot 8^n = 8(8^n - 1)$

14. $\dfrac{1}{1 \cdot 2} + \dfrac{1}{2 \cdot 3} + \dfrac{1}{3 \cdot 4} + \cdots + \dfrac{1}{n(n + 1)} = \dfrac{n}{n + 1}$

15. $\dfrac{1}{1 \cdot 4} + \dfrac{1}{4 \cdot 7} + \dfrac{1}{7 \cdot 10} + \cdots + \dfrac{1}{(3n - 2)(3n + 1)} = \dfrac{n}{3n + 1}$

16. $\dfrac{1}{2} + \dfrac{1}{2^2} + \dfrac{1}{2^3} + \cdots + \dfrac{1}{2^n} = 1 - \dfrac{1}{2^n}$

17. $\dfrac{4}{5} + \dfrac{4}{5^2} + \dfrac{4}{5^3} + \cdots + \dfrac{4}{5^n} = 1 - \dfrac{1}{5^n}$

18. $x^{2n} + x^{2n-1}y + \cdots + xy^{2n-1} + y^{2n} = \dfrac{x^{2n+1} - y^{2n+1}}{x - y}$

19. $x^{2n-1} + x^{2n-2}y + \cdots + xy^{2n-2} + y^{2n-1} = \dfrac{x^{2n} - y^{2n}}{x - y}$

20. $(a^m)^n = a^{mn}$ (Assume that a and m are constant.)

21. $(ab)^n = a^n b^n$ (Assume that a and b are constant.)

22. If $a > 1$, then $a^n > 1$.

23. If $a > 1$, then $a^n > a^{n-1}$.

24. If $0 < a < 1$, then $a^n < a^{n-1}$.

RELATING CONCEPTS (EXERCISES 25–29)

Many of the statements you are asked to prove in these exercises are generalizations of properties of algebra.

25. In the statement given in Exercise 18, replace n with 1.

26. Of what factorization is this a rearrangement? (Section R.2)

27. Write a statement similar to the one in Exercise 18 for

$$\dfrac{x^{2n+1} + y^{2n+1}}{x + y}.$$

28. The statement in Exercise 20 is the _____ rule for _____. (Section R.1)

29. The statement in Exercise 21 is one of the _____ rules for _____. (Section R.1)

8.6 COUNTING THEORY

The Fundamental Principle of Counting ◆ Factorials ◆ Permutations ◆ Combinations ◆ Distinguishing between Permutations and Combinations

THE FUNDAMENTAL PRINCIPLE OF COUNTING

If there are 3 roads from Albany to Baker and 2 roads from Baker to Creswich, in how many ways can one travel from Albany to Creswich by way of Baker? For each of the 3 roads from Albany to Baker, there are 2 different roads from Baker to Creswich. Hence, there are $3 \cdot 2 = 6$ different ways to make the trip, as shown in the **tree diagram** in Figure 14.

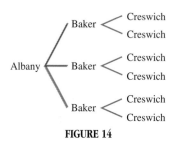

FIGURE 14

Two events are **independent events** if neither influences the outcome of the other. The opening example illustrates the fundamental principle of counting with independent events.

> **THE FUNDAMENTAL PRINCIPLE OF COUNTING**
>
> If n independent events occur, with
>
> $$m_1 \text{ ways for event 1 to occur,}$$
> $$m_2 \text{ ways for event 2 to occur,}$$
> $$\cdot$$
> $$\cdot$$
> $$\cdot$$
>
> and $\qquad m_n \text{ ways for event } n \text{ to occur,}$
>
> then there are
>
> $$m_1 \cdot m_2 \cdot \ldots \cdot m_n$$
>
> different ways for all n events to occur.

EXAMPLE 1 *Using the Fundamental Principle of Counting* A restaurant offers a choice of 3 salads, 5 main dishes, and 2 desserts. Use the fundamental principle of counting to find the number of different 3-course meals that can be selected.

SOLUTION Three events are involved: selecting a salad, selecting a main dish, and selecting a dessert. The first event can occur in 3 ways, the second event can occur in 5 ways, and the third event can occur in 2 ways; thus, there are

$$3 \cdot 5 \cdot 2 = 30 \text{ possible meals.} \qquad ◆$$

EXAMPLE 2   *Using the Fundamental Principle of Counting*   A teacher has 5 different books that he wishes to arrange in a row. How many different arrangements are possible?

SOLUTION   Five events are involved: selecting a book for the first spot, selecting a book for the second spot, and so on. For the first spot, the teacher has 5 choices. After a choice has been made, the teacher has 4 choices for the second spot. Continuing in this manner, there are 3 choices for the third spot, 2 for the fourth spot, and 1 for the fifth spot. By the fundamental principle of counting, there are

$$5 \cdot 4 \cdot 3 \cdot 2 \cdot 1 \text{ or } 120 \text{ different arrangements.} \qquad \blacklozenge$$

EXAMPLE 3   *Arranging r of n Items (r < n)*   Suppose the teacher in Example 2 wishes to place only 3 of the 5 books in a row. How many arrangements of 3 books are possible?

SOLUTION   The teacher still has 5 ways to fill the first spot, 4 ways to fill the second spot, and 3 ways to fill the third. Since only 3 books will be used, there are only 3 spots to be filled (3 events) instead of 5, with

$$5 \cdot 4 \cdot 3 = 60 \text{ arrangements.} \qquad \blacklozenge$$

FACTORIALS

As seen in Section 4 of this chapter, products such as $5 \cdot 4 \cdot 3 \cdot 2 \cdot 1$ occur often. We repeat the definition of *n*-factorial first seen there.

DEFINITION OF THE FACTORIAL OF A NATURAL NUMBER

For any natural number n,

$$n! = n(n-1)(n-2)(n-3) \ldots (2)(1).$$

By the definition of $n!$, $n[(n-1)!] = n!$ for all natural numbers $n \geq 2$. It is convenient to have this relation hold also for $n = 1$, so, by definition, we have the following.

DEFINITION OF ZERO FACTORIAL

$$0! = 1$$

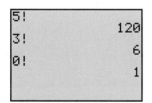

FIGURE 15

To illustrate these definitions, $5! = 5 \cdot 4 \cdot 3 \cdot 2 \cdot 1 = 120$ and $3! = 3 \cdot 2 \cdot 1 = 6$. Graphing calculators are capable of finding factorials, as seen in the screen in Figure 15.

PERMUTATIONS

Refer again to Example 3. Since each ordering of 3 books is considered a different *arrangement*, the number 60 in that example is called the number of permutations of 5 things taken 3 at a time, written $P(5, 3) = 60$. The number of ways of arranging 5 elements from a set of 5 elements, written $P(5, 5) = 120$, was found in Example 2. A **permutation** of n elements taken r at a time is one of the *arrangements* of r elements from a set of n elements. Generalizing from the examples above, the number of permutations of n elements, taken r at a time, denoted by $P(n, r)$, is

$$P(n, r) = n(n - 1)(n - 2) \cdots (n - r + 1)$$

$$= \frac{n(n - 1)(n - 2) \cdots (n - r + 1)(n - r)(n - r - 1) \cdots (2)(1)}{(n - r)(n - r - 1) \cdots (2)(1)}$$

$$= \frac{n!}{(n - r)!}.$$

In summary, we have the following result.

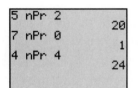

This screen supports the following permutations calculations:

$P(5, 2) = \frac{5!}{(5 - 2)!} = \frac{5!}{3!} = 4 \cdot 5 = 20$

$P(7, 0) = \frac{7!}{(7 - 0)!} = \frac{7!}{7!} = 1$

$P(4, 4) = \frac{4!}{(4 - 4)!} = \frac{4!}{0!} = 24$

FIGURE 16

FORMULA FOR PERMUTATIONS OF n ELEMENTS TAKEN r AT A TIME

If $P(n, r)$ denotes the number of permutations of n elements taken r at a time, with $r \le n$, then

$$P(n, r) = \frac{n!}{(n - r)!}.$$

Alternative notations for $P(n, r)$ are P_r^n and $_nP_r$.

Graphing calculators are capable of evaluating permutation expressions. The screen in Figure 16 shows how the calculator evaluates $P(5, 2)$, $P(7, 0)$, and $P(4, 4)$.

EXAMPLE 4 *Using the Permutations Formula* Find the following, and support the calculations with a graphing calculator.

(a) The number of permutations of the letters L, M, and N

(b) The number of permutations of 2 of the 3 letters M, N, and L

TECHNOLOGY NOTE

As seen in Figure 16, the TI-83 uses the notation

$n \quad _nP_r \quad r$

for permutations. A similar notation is used for combinations.

SOLUTION

(a) ANALYTIC By the formula for $P(n, r)$, with $n = 3$ and $r = 3$,

$$P(3, 3) = \frac{3!}{(3 - 3)!}$$

$$= \frac{3!}{0!} = \frac{3!}{1} = 3 \cdot 2 \cdot 1 = 6.$$

As shown in the tree diagram in Figure 17, the 6 permutations are LMN, LNM, MLN, MNL, NLM, NML.

GRAPHICAL See Figure 18.

(b) ANALYTIC Find $P(3, 2)$.

$$P(3, 2) = \frac{3!}{(3 - 2)!}$$

$$= \frac{3!}{1!} = \frac{3!}{1} = 6$$

This result is the same as the answer in part (a) because after the first 2 choices are made, the third is already determined, since only one letter is left.

GRAPHICAL See Figure 18. ◆

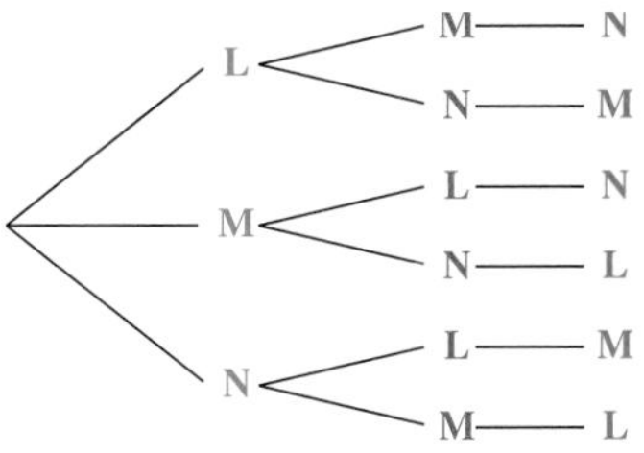

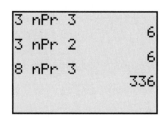

The results in Examples 4 and 5 are supported here.

FIGURE 17   FIGURE 18

EXAMPLE 5   *Using the Permutations Formula*   Suppose 8 people enter an event in a swim meet. In how many ways could the gold, silver, and bronze prizes be awarded? Support the calculation with a graphing calculator.

SOLUTION   ANALYTIC   Using the fundamental principle of counting, there are 3 choices to be made, giving $8 \cdot 7 \cdot 6 = 336$. However, the formula for $P(n, r)$ also can be used to get the same result.

$$P(8, 3) = \frac{8!}{5!} = \frac{8 \cdot 7 \cdot 6 \cdot 5 \cdot 4 \cdot 3 \cdot 2 \cdot 1}{5 \cdot 4 \cdot 3 \cdot 2 \cdot 1} = 336$$

GRAPHICAL   See Figure 18 for calculator support of this computation.   ◆

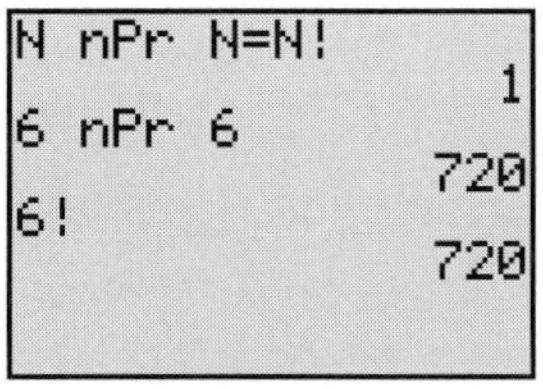

EXAMPLE 6   *Using the Permutations Formula*   In how many ways can 6 students be seated in a row of 6 desks?

SOLUTION   Use $P(n, n)$ with $n = 6$ to get

$$P(6, 6) = 6! = 6 \cdot 5 \cdot 4 \cdot 3 \cdot 2 \cdot 1 = 720.$$ ◆

Note   In Example 6, if N represents the number of students and desks, the result is N! for *any* value of N. The screen in Figure 19 illustrates this for N = 6.

FIGURE 19

COMBINATIONS

Earlier, we saw that there are 60 ways that a teacher can arrange 3 of 5 different books in a row. That is, there are 60 permutations of 5 things taken 3 at a time. Suppose now that the teacher does not wish to arrange the books in a row, but rather wishes to choose, without regard to order, any 3 of the 5 books to donate to a book sale to raise money for the school. In how many ways can this be done?

At first glance, we might say 60 again, but this is incorrect. The number 60 counts all possible *arrangements* of 3 books chosen from 5. The following 6 arrangements, however, would all lead to the same set of 3 books being given to the book sale.

mystery-biography-textbook	biography-textbook-mystery
mystery-textbook-biography	textbook-biography-mystery
biography-mystery-textbook	textbook-mystery-biography

The list shows 6 different *arrangements* of 3 books but only one *set* of 3 books. A subset of items selected *without regard to order* is called a **combination**. The number of combinations of 5 things taken 3 at a time is written $\binom{5}{3}$, or $C(5, 3)$.

Note   This combinations notation also represents the binomial coefficient defined in Section 4 of this chapter. That is, binomial coefficients are the combinations of *n* elements chosen *r* at a time.

To evaluate $\binom{5}{3}$ or $C(5, 3)$, start with the $5 \cdot 4 \cdot 3$ *permutations* of 5 things taken 3 at a time. Since order doesn't matter, and each subset of 3 items from the set of 5 items can have its elements rearranged in $3 \cdot 2 \cdot 1 = 3!$ ways, $\binom{5}{3}$ can be found by dividing the number of permutations by $3!$, or

$$\binom{5}{3} = \frac{5 \cdot 4 \cdot 3}{3!} = \frac{5 \cdot 4 \cdot 3}{3 \cdot 2 \cdot 1} = 10.$$

There are 10 ways that the teacher can choose 3 books for the book sale.

Generalizing this discussion gives the following formula for the number of combinations of n elements taken r at a time:

$$C(n, r) = \binom{n}{r} = \frac{P(n, r)}{r!}.$$

A more useful version of this formula is found as follows.

$$C(n, r) = \binom{n}{r} = \frac{P(n, r)}{r!} = \frac{n!}{(n-r)!} \cdot \frac{1}{r!} = \frac{n!}{(n-r)! \, r!}$$

This last version is the most useful for calculation and is the one we used earlier to calculate binomial coefficients. This discussion is summarized in the following result.

FORMULA FOR COMBINATIONS OF n ELEMENTS TAKEN r AT A TIME

If $\binom{n}{r}$ or $C(n, r)$ represents the number of combinations of n things taken r at a time, with $r \leq n$, then

$$\binom{n}{r} = C(n, r) = \frac{n!}{(n-r)! \, r!}.$$

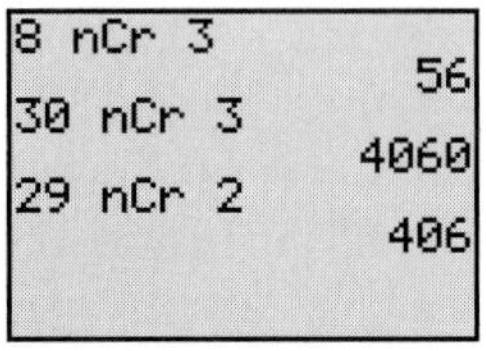

```
8 nCr 3
                56
30 nCr 3
              4060
29 nCr 2
               406
```

The results in Examples 7 and 8 are supported here.

FIGURE 20

EXAMPLE 7   *Using the Combinations Formula to Determine Possible Committees*   How many different committees of 3 people can be chosen from a group of 8 people? Use a calculator to support the calculation.

SOLUTION   ANALYTIC   Since a committee is an unordered set, use combinations to get

$$C(8, 3) = \binom{8}{3} = \frac{8!}{5! \, 3!} = \frac{8 \cdot 7 \cdot 6 \cdot 5 \cdot 4 \cdot 3 \cdot 2 \cdot 1}{5 \cdot 4 \cdot 3 \cdot 2 \cdot 1 \cdot 3 \cdot 2 \cdot 1} = 56.$$

GRAPHICAL   See Figure 20.   ◆

EXAMPLE 8   *Using the Combinations Formula to Determine Possible Selections*   Three lawyers are to be selected from a group of 30 to work on a special project.

(a)   In how many different ways can the lawyers be selected?

(b)   In how many ways can the group of 3 be selected if a certain lawyer must work on the project?

Support the calculations.

SOLUTION

(a)   ANALYTIC   Here we wish to know the number of 3-element combinations that can be formed from a set of 30 elements. (We want combinations, not permutations, since order within the group of 3 does not matter.)

$$C(30, 3) = \binom{30}{3} = \frac{30!}{27!\,3!} = \frac{30 \cdot 29 \cdot 28 \cdot 27!}{27! \cdot 3 \cdot 2 \cdot 1} = 4060$$

There are 4060 ways to select the project group.

    GRAPHICAL   See Figure 20.

(b) ANALYTIC   Since 1 lawyer already has been selected for the project, the problem is reduced to selecting 2 more from the remaining 29 lawyers.

$$C(29, 2) = \binom{29}{2} = \frac{29!}{27!\,2!} = \frac{29 \cdot 28 \cdot 27!}{27! \cdot 2 \cdot 1} = \frac{29 \cdot 28}{2 \cdot 1} = 29 \cdot 14 = 406$$

In this case, the project group can be selected in 406 ways.

    GRAPHICAL   Again, see Figure 20.   ◆

DISTINGUISHING BETWEEN PERMUTATIONS AND COMBINATIONS

The formulas for permutations and combinations given in this section will be very useful in solving probability problems in the next section. Any difficulty in using these formulas usually comes from being unable to differentiate between them. Both permutations and combinations give the number of ways to choose r objects from a set of n objects. The differences between permutations and combinations are outlined in the following chart.

Permutations	*Combinations*
Different orderings or arrangements of the r objects are different permutations.	Each choice or subset of r objects gives one combination. Order within the group of r objects does not matter.
$$P(n, r) = \frac{n!}{(n-r)!}$$	$$C(n, r) = \binom{n}{r} = \frac{n!}{(n-r)!\,r!}$$
Clue words: arrangement, schedule, order	Clue words: group, committee, sample, selection

    In the next example, concentrate on recognizing which formula should be applied.

EXAMPLE 9 *Distinguishing between Combinations and Permutations in a Selection Application*   A sales representative has 10 accounts in a certain city.

(a) In how many ways can 3 accounts be selected to call on?

(b) In how many ways can calls be scheduled for 3 of the 10 accounts?

SOLUTION

(a) Within a selection of 3 accounts, the arrangement of the visits is not important, so there are

$$C(10, 3) = \binom{10}{3} = \frac{10!}{7!\,3!} = \frac{10 \cdot 9 \cdot 8}{3 \cdot 2 \cdot 1} = 120$$

ways to select 3 accounts.

(b) To schedule calls, the sales representative must *order* each selection of 3 accounts. Use permutations here, since order is important.

$$P(10, 3) = \frac{10!}{(10-3)!} = \frac{10!}{7!} = 10 \cdot 9 \cdot 8 = 720$$

There are 720 different orders in which to call on 3 of the accounts.   ◆

EXAMPLE 10   *Distinguishing between Permutations and Combinations*   To illustrate the differences between permutations and combinations in another way, suppose 2 cans of soup are to be selected from 4 cans on a shelf: noodle (N), bean (B), mushroom (M), and tomato (T). As shown in Figure 21(a), there are 12 ways to select 2 cans from the 4 cans if the order matters (if noodle first and bean second is considered different from bean, then noodle, for example). On the other hand, if order is unimportant, then there are 6 ways to choose 2 cans of soup from the 4, as illustrated in Figure 21(b).   ◆

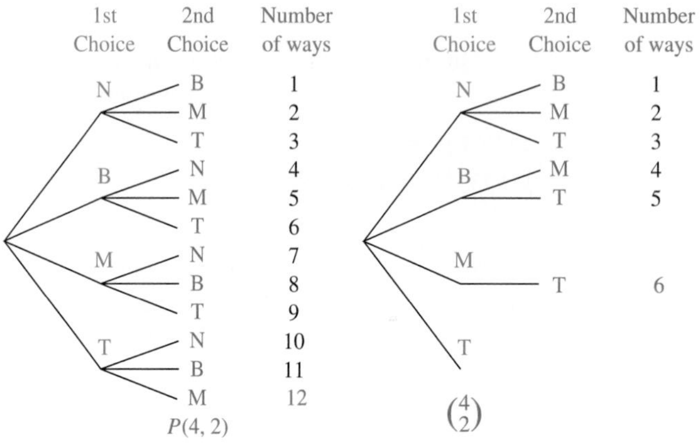

1st Choice	2nd Choice	Number of ways
N	B	1
	M	2
	T	3
B	N	4
	M	5
	T	6
M	N	7
	B	8
	T	9
T	N	10
	B	11
	M	12

$P(4, 2)$

1st Choice	2nd Choice	Number of ways
N	B	1
	M	2
	T	3
B	M	4
	T	5
M	T	6

$\binom{4}{2}$

(a)   (b)

FIGURE 21

Caution   It should be stressed that not all counting problems lend themselves to either permutations or combinations. Whenever a tree diagram or the multiplication principle can be used directly, as in Example 10, then use it.

8.6 EXERCISES   ◆   Tape 15

Evaluate the following.

1. $P(12, 8)$ **2.** $P(5, 5)$ **3.** $P(9, 2)$ **4.** $P(10, 9)$ **5.** $P(5, 1)$ **6.** $P(6, 0)$

7. $C(4, 2)$ **8.** $C(9, 3)$ **9.** $C(6, 0)$ **10.** $C(8, 1)$ **11.** $\binom{12}{4}$ **12.** $\binom{16}{3}$

13. Decide whether the situation described involves a permutation or a combination of objects. See Examples 9 and 10.

(a) a telephone number

(b) a Social Security number

(c) a hand of cards in poker

(d) a committee of politicians

(e) the "combination" on a combination lock

(f) a lottery choice of six numbers where the order does not matter

(g) an automobile license plate

14. Explain the difference between a permutation and a combination. What should you look for in a problem to decide which of these is an appropriate method of solution?

Use the fundamental principle of counting to solve each problem.

15. *Home Plan Choices* How many different types of homes are available if a builder offers a choice of 5 basic plans, 3 roof styles, and 2 exterior finishes?

16. *Auto Varieties* An auto manufacturer produces 7 models, each available in 6 different colors, with 4 different upholstery fabrics, and 5 interior colors. How many varieties of the auto are available?

17. *Radio Station Call Letters* How many different 4-letter radio-station call letters can be made

  (a) if the first letter must be K or W and no letter may be repeated;

  (b) if repeats are allowed (but the first letter is K or W)?

  (c) How many of the 4-letter call letters (starting with K or W) with no repeats end in R?

18. *Meal Choices* A menu offers a choice of 3 salads, 8 main dishes, and 5 desserts. How many different 3-course meals (salad, main dish, dessert) are possible?

19. *Names for a Baby* A couple has narrowed down the choice of a name for their new baby to 3 first names and 5 middle names. How many different first- and middle-name arrangements are possible?

20. *Concert Program Arrangement* A concert to raise money for an economics prize is to consist of 5 works: 2 overtures, 2 sonatas, and a piano concerto. In how many ways can a program with these 5 works be arranged?

21. *License Plates*   For many years, the state of California used 3 letters followed by 3 digits on its automobile license plates.

  (a) How many different license plates are possible with this arrangement?

  (b) When the state ran out of new plates, the order was reversed to 3 digits followed by 3 letters. How many additional plates were then possible?

  (c) Several years ago, the plates described in (b) were also used up. The state then issued plates with 1 letter followed by 3 digits and then 3 letters. How many plates does this scheme provide?

22. *Telephone Numbers* How many 7-digit telephone numbers are possible if the first digit cannot be zero and

  (a) only odd digits may be used;

  (b) the telephone number must be a multiple of 10 (that is, it must end in zero);

  (c) the telephone number must be a multiple of 100;

  (d) the first 3 digits are 481;

  (e) no repetitions are allowed?

Solve the following problems involving permutations.

23. *Seating People in a Row* In an experiment on social interaction, 6 people will sit in 6 seats in a row. In how many ways can this be done?

24. *Genetics Experiment* In how many ways can 7 of 10 monkeys be arranged in a row for a genetics experiment?

25. *Course Schedule Arrangement* A business school offers courses in typing, shorthand, transcription, business English, technical writing, and accounting. In how many ways can a student arrange a schedule if 3 courses are taken?

26. *Course Schedule Arrangement* If your college offers 400 courses, 20 of which are in mathematics, and your counselor arranges your schedule of 4 courses by random selection, how many schedules are possible that do not include a math course?

27. *Club Officer Choices* In a club with 15 members, how many ways can a slate of 3 officers consisting of president, vice-president, and secretary/treasurer be chosen?

28. *Batting Orders* A baseball team has 20 players. How many 9-player batting orders are possible?

29. *Basketball Positions* In how many ways can 5 players be assigned to the 5 positions on a basketball team, assuming that any player can play any position? In how many ways can 10 players be assigned to the 5 positions?

30. *Letter Arrangement* How many ways can all the letters of the word TOUGH be arranged?

Solve the following problems involving combinations.

31. *Committee Choices* A club has 30 members. If a committee of 4 is selected at random, how many committees are possible?

32. *Apple Samples* How many different samples of 3 apples can be drawn from a crate of 25 apples?

33. *Hamburger Choices* Howard Hardee's Hamburger Heaven sells hamburgers with cheese, relish, lettuce, tomato, mustard, or ketchup. How many different hamburgers can be made that use any 3 of the extras?

34. *Student Groupings* Three students are to be selected from a group of 12 students to participate in a special class. In how many ways can this be done? In how many ways can the group that will not participate be selected?

35. *Card Combinations* Five cards are marked with the numbers 1, 2, 3, 4, and 5, shuffled, and 2 cards are then drawn. How many different 2-card combinations are possible?

36. *Marble Samples* If a bag contains 15 marbles, how many samples of 2 marbles can be drawn from it? How many samples of 4 marbles?

37. *Marble Samples* In Exercise 36, if the bag contains 3 yellow, 4 white, and 8 blue marbles, how many samples of 2 can be drawn in which both marbles are blue?

38. *Apple Samples* In Exercise 32, if it is known that there are 5 rotten apples in the crate,

   (a) how many samples of 3 could be drawn in which all 3 are rotten;

   (b) how many samples of 3 could be drawn in which there are 1 rotten apple and 2 good apples?

39. *Convention Delegation Choices* A city council is composed of 5 liberals and 4 conservatives. Three members are to be selected randomly as delegates to a convention.

   (a) How many delegations are possible?

   (b) How many delegations could have all liberals?

   (c) How many delegations could have 2 liberals and 1 conservative?

   (d) If 1 member of the council serves as mayor, how many delegations are possible that include the mayor?

40. *Delegation Choices* Seven workers decide to send a delegation of 2 to their supervisor to discuss their grievances.

   (a) How many different delegations are possible?

   (b) If it is decided that a certain employee must be in the delegation, how many different delegations are possible?

   (c) If there are 2 women and 5 men in the group, how many delegations would include at least 1 woman?

Use any or all of the methods described in this section to solve the following problems.

41. *Course Schedule* If Dwight Johnston has 8 courses to choose from, how many ways can he arrange his schedule if he must pick 4 of them?

Prove each statement for positive integers n and r, with r ≤ n.

51. $P(n, n - 1) = P(n, n)$ **52.** $P(n, 1) = n$

55. $\binom{n}{0} = 1$ **56.** $\binom{n}{n - 1} = n$

42. *Pineapple Samples* How many samples of 3 pineapples can be drawn from a crate of 12?

43. *Soup Ingredients* Velma specializes in making different vegetable soups with carrots, celery, beans, peas, mushrooms, and potatoes. How many different soups can she make with any 4 ingredients?

44. *Secretary/Manager Assignments* From a pool of 7 secretaries, 3 are selected to be assigned to 3 managers, 1 secretary to each manager. In how many ways can this be done?

45. *Musical Chairs Seatings* In a game of musical chairs, 12 children will sit in 11 chairs (1 will be left out). How many seatings are possible?

46. *Plant Samples* In an experiment on plant hardiness, a researcher gathers 6 wheat plants, 3 barley plants, and 2 rye plants. She wishes to select 4 plants at random.

   (a) In how many ways can this be done?

   (b) In how many ways can this be done if exactly 2 wheat plants must be included?

47. *Committee Choices* In a club with 8 men and 11 women members, how many 5-member committees can be chosen that have the following?

   (a) all men

   (b) all women

   (c) 3 men and 2 women

   (d) no more than 3 women

48. *Committee Choices* From 10 names on a ballot, 4 will be elected to a political party committee. In how many ways can the committee of 4 be formed if each person will have a different responsibility?

49. *Plant Arrangement* In how many ways can 5 out of 9 plants be arranged in a row on a window sill?

50. *Letter Arrangement* In how many ways can all the letters of CHAMBERPOT be arranged?

53. $P(n, 0) = 1$ **54.** $\binom{n}{n} = 1$

57. $\binom{n}{n - r} = \binom{n}{r}$

58. Explain why the restriction $r \le n$ is needed in the formula for $P(n, r)$.

Series are often used in mathematics and science to make approximations. Large values of factorials often occur in counting theory. The value of n! can quickly become too large for most calculators to evaluate. To estimate n! for large values of n, one can use the property of logarithms that says

$$\log n! = \log(1 \times 2 \times 3 \times \cdots \times n)$$
$$= \log 1 + \log 2 + \log 3 + \cdots + \log n.$$

Using sum and sequence features on a calculator, one can then determine an r such that $n! \approx 10^r$ *since* $r = \log n!$. *For example, the screen shown here illustrates that a calculator will give the same approximation for* 30!, *using the factorial function and the formula just discussed.*

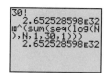

59. Use the technique described above to approximate each of the factorials. Then, try to compute the value directly on your calculator.

   (a) 50!   **(b)** 60!   **(c)** 65!

60. Use the technique described above to approximate each permutation $P(n, r)$.

   (a) $P(47, 13)$ **(b)** $P(50, 4)$ **(c)** $P(29, 21)$

8.7 PROBABILITY

Basic Concepts ◆ Complements and Venn Diagrams ◆ Odds ◆ The Union of Two Events ◆ Binomial Probability

BASIC CONCEPTS

The study of probability has become increasingly popular because it has a wide range of practical applications. The basic ideas of probability are introduced in this section.

Consider an experiment that has one or more possible **outcomes**, each of which is equally likely to occur. For example, the experiment of tossing a fair coin has two equally likely possible outcomes: landing heads up (H) or landing tails up (T). Also, the experiment of rolling a fair die has 6 equally likely outcomes: landing so the face that is up shows 1, 2, 3, 4, 5, or 6 points.

The set S of all possible outcomes of a given experiment is called the **sample space** for the experiment. (In this text, all sample spaces are finite.) One sample space for the experiment of tossing a coin could consist of the outcomes H and T. This sample space can be written in set notation as

$$S = \{H, T\}.$$

Similarly, a sample space for the experiment of rolling a single die might be

$$S = \{1, 2, 3, 4, 5, 6\}.$$

Any subset of the sample space is called an **event**. In the experiment with the die, for example, "the number showing is a 3" is an event, say E_1, such that $E_1 = \{3\}$. "The number showing is greater than 3" is also an event, say E_2, such that $E_2 = \{4, 5, 6\}$. To represent the number of outcomes that belong to event E, the notation $n(E)$ is used. Then, $n(E_1) = 1$ and $n(E_2) = 3$.

The notation $P(E)$ is used for the *probability* of an event E. If the outcomes in the sample space for an experiment are equally likely, then the probability of event E occurring is found as follows.

DEFINITION OF PROBABILITY OF EVENT E

In a sample space with equally likely outcomes, the **probability** of an event E, written $P(E)$, is the ratio of the number of outcomes in sample space S that belong to event E, $n(E)$, to the total number of outcomes in sample space S, $n(S)$. That is,

$$P(E) = \frac{n(E)}{n(S)}.$$

To use this definition to find the probability of the event E_1 given above, start with the sample space for the experiment, $S = \{1, 2, 3, 4, 5, 6\}$, and the desired event, $E_1 = \{3\}$. Since $n(E_1) = 1$, and since there are 6 equally likely outcomes in the sample space,

$$P(E_1) = \frac{n(E_1)}{n(S)} = \frac{1}{6}.$$

EXAMPLE 1   *Finding Probabilities of Events*   A single die is rolled. Write the following events in set notation and give the probability for each event.

(a) E_3: the number showing is even

(b) E_4: the number showing is greater than 4

(c) E_5: the number showing is less than 7

(d) E_6: the number showing is 7

SOLUTION

(a)   Use the definition above. Since $E_3 = \{2, 4, 6\}$, $n(E_3) = 3$. As shown above, $n(S) = 6$, so

$$P(E_3) = \frac{3}{6} = \frac{1}{2}.$$

(b)   Again, $n(S) = 6$. Event $E_4 = \{5, 6\}$, with $n(E_4) = 2$. By the definition,

$$P(E_4) = \frac{2}{6} = \frac{1}{3}.$$

(c) $E_5 = \{1, 2, 3, 4, 5, 6\}$ and $P(E_5) = \frac{6}{6} = 1$

(d) $E_6 = \emptyset$ and $P(E_6) = \frac{0}{6} = 0$ ◆

In Example 1(c), $E_5 = S$. Therefore, the event E_5 is certain to occur every time the experiment is performed. An event that is certain to occur, such as E_5, always has a probability of 1. On the other hand, $E_6 = \emptyset$ and $P(E_6)$ is 0. The probability of an impossible event, such as E_6, is always 0, since none of the outcomes in the sample space satisfies the event. For any event E, $P(E)$ is between 0 and 1 inclusive.

COMPLEMENTS AND VENN DIAGRAMS

The set of all outcomes in the sample space that do *not* belong to event E is called the **complement of E**, written E'. For example, in the experiment of drawing a single card from a standard deck of 52 cards, let E be the event "the card is an ace." Then, E' is the event "the card is not an ace." From the definition of E', for an event E,

$$E \cup E' = S   \text{and}   E \cap E' = \emptyset.*$$

---

*The **union** of two sets A and B is the set $A \cup B$ made up of all the elements from either A or B, or both. The **intersection** of sets A and B, written $A \cap B$, is made up of all the elements that belong to both sets.

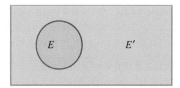

FIGURE 22

Probability concepts can be illustrated with **Venn diagrams**, as shown in Figure 22. The rectangle in Figure 22 represents the sample space in an experiment. The area inside the circle represents event E, while the area inside the rectangle, but outside the circle, represents event E'.

A standard deck of 52 cards has 4 suits: hearts, clubs, diamonds, and spades, with 13 cards of each suit. Each suit has an ace, king, queen, jack, and cards numbered from 2 to 10. The hearts and diamonds are red and the spades and clubs are black. We will refer to this standard deck of cards in the examples that follow.

EXAMPLE 2   *Using the Complement in a Probability Problem*   In the experiment of drawing a card from a well-shuffled deck, find the probability of event E, the card is an ace, and event E'.

SOLUTION   Since there are 4 aces in the deck of 52 cards, $n(E) = 4$ and $n(S) = 52$. Therefore,

$$P(E) = \frac{n(E)}{n(S)} = \frac{4}{52} = \frac{1}{13}.$$

Of the 52 cards, 48 are not aces, so

$$P(E') = \frac{n(E')}{n(S)} = \frac{48}{52} = \frac{12}{13}. \qquad \blacklozenge$$

In Example 2, $P(E) + P(E') = (1/13) + (12/13) = 1$. This is always true for any event E and its complement E'. That is,

$$P(E) + P(E') = 1.$$

This can be restated as

$$P(E) = 1 - P(E') \qquad \text{or} \qquad P(E') = 1 - P(E).$$

These two equations suggest an alternative way to compute the probability of an event. For example, if it is known that $P(E) = 1/10$, then $P(E') = 1 - 1/10 = 9/10$.

ODDS

Sometimes probability statements are expressed in terms of odds, a comparison of $P(E)$ with $P(E')$. The **odds** in favor of an event E are expressed as the ratio of $P(E)$ to $P(E')$ or as the fraction $P(E)/P(E')$. For example, if the probability of rain can be established as $1/3$, the odds that it will rain are

$$P(\text{rain}) \text{ to } P(\text{no rain}) \ = \frac{1}{3} \text{ to } \frac{2}{3} = \frac{1/3}{2/3} = \frac{1}{2} \qquad \text{or} \qquad 1 \text{ to } 2.$$

On the other hand, the odds that it will not rain are 2 to 1 (or $2/3$ to $1/3$). If the odds in favor of an event are, say, 3 to 5, then the probability of the event is $3/8$, while the probability of the complement of the event is $5/8$. If the odds favoring event E are m to n, then

$$P(E) = \frac{m}{m + n} \qquad \text{and} \qquad P(E') = \frac{n}{m + n}.$$

EXAMPLE 3 *Finding Odds in Favor of an Event*   A shirt is selected at random from a dark closet containing 6 blue shirts and 4 shirts that are not blue. Find the odds in favor of a blue shirt being selected.

SOLUTION   Let E represent "a blue shirt is selected." Then, $P(E) = 6/10$ or $3/5$. Also, $P(E') = 1 - (3/5) = 2/5$. Therefore, the odds in favor of a blue shirt being selected are

$$P(E) \text{ to } P(E') = \frac{3}{5} \text{ to } \frac{2}{5} = \frac{3/5}{2/5} = \frac{3}{2} \quad \text{or} \quad 3 \text{ to } 2. \quad \blacklozenge$$

THE UNION OF TWO EVENTS

We now extend the rules for probability to more complex events. Since events are sets, we can use set operations to find the union of two events. (The *union* of sets A and B includes all elements of set A in addition to all elements of set B.)

Suppose a fair die is tossed. Let H be the event "the result is a 3," and K the event "the result is an even number." From the results earlier in this section,

$$H = \{3\} \qquad K = \{2, 4, 6\} \qquad H \cup K = \{2, 3, 4, 6\}$$

$$P(H) = \frac{1}{6} \qquad P(K) = \frac{3}{6} = \frac{1}{2} \qquad P(H \cup K) = \frac{4}{6} = \frac{2}{3}.$$

Notice that $P(H) + P(K) = P(H \cup K)$.

Before assuming that this relationship is true in general, consider another event for this experiment, "the result is a 2," event G.

$$G = \{2\} \qquad K = \{2, 4, 6\} \qquad K \cup G = \{2, 4, 6\}$$

$$P(G) = \frac{1}{6} \qquad P(K) = \frac{3}{6} = \frac{1}{2} \qquad P(K \cup G) = \frac{3}{6} = \frac{1}{2}$$

In this case, $P(K) + P(G) \neq P(K \cup G)$. See Figure 23.

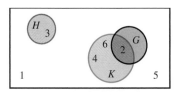

FIGURE 23

As Figure 23 suggests, the difference in the two examples above comes from the fact that events H and K cannot occur simultaneously. Such events are called **mutually exclusive events**. In fact, $H \cap K = \emptyset$, which is true for any two mutually exclusive events. Events K and G, however, can occur simultaneously. Both are satisfied if the result of the roll is a 2, the element in their intersection ($K \cap G = \{2\}$). This example suggests the following property.

FORMULA FOR THE PROBABILITY OF THE UNION OF TWO EVENTS

For any events E and F,
$$P(E \text{ or } F) = P(E \cup F) = P(E) + P(F) - P(E \cap F).$$

EXAMPLE 4 *Finding the Probability of a Union*   One card is drawn from a well-shuffled deck of 52 cards. What is the probability of the following outcomes?

(a) The card is an ace or a spade.   **(b)** The card is a 3 or a king.

SOLUTION

(a) The events "drawing an ace" and "drawing a spade" are not mutually exclusive since it is possible to draw the ace of spades, an outcome satisfying both events. The probability is

$$P(\text{ace or spade}) = P(\text{ace}) + P(\text{spade}) - P(\text{ace and spade})$$

$$= \frac{4}{52} + \frac{13}{52} - \frac{1}{52} = \frac{16}{52} = \frac{4}{13}.$$

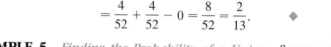

(b) "Drawing a 3" and "drawing a king" are mutually exclusive events because it is impossible to draw one card that is both a 3 and a king. Using the rule given above,

$$P(3 \text{ or } K) = P(3) + P(K) - P(3 \text{ and } K)$$

$$= \frac{4}{52} + \frac{4}{52} - 0 = \frac{8}{52} = \frac{2}{13}. \blacklozenge$$

The arithmetic in Examples 4(a) and 5(a) is easily accomplished with a calculator.

EXAMPLE 5 *Finding the Probability of a Union*   Suppose two fair dice are rolled. Find each of the following probabilities.

(a) The first die shows a 2, or the sum of the two dice is 6 or 7.

(b) The sum of the points showing is at most 4.

SOLUTION

(a) Think of the two dice as being distinguishable, one red and one green for example. (Actually, the sample space is the same even if they are not apparently distinguishable.) A sample space with equally likely outcomes is shown in Figure 24, where (1, 1) represents the event "the first (red) die shows a 1 and the second die (green) shows a 1," (1, 2) represents "the first die shows a 1 and the second die shows a 2," and so on. Let A represent the event "the first die shows a 2," and B represent the event "the sum of the results is 6 or 7." These events are indicated in Figure 24. From the diagram, event A has 6 elements, B has 11 elements, and the sample space has 36 elements. Thus,

$$P(A) = \frac{6}{36}, \quad P(B) = \frac{11}{36}, \quad \text{and} \quad P(A \cap B) = \frac{2}{36}.$$

By the union rule,

$$P(A \cup B) = P(A) + P(B) - P(A \cap B)$$

$$P(A \cup B) = \frac{6}{36} + \frac{11}{36} - \frac{2}{36} = \frac{15}{36} = \frac{5}{12}.$$

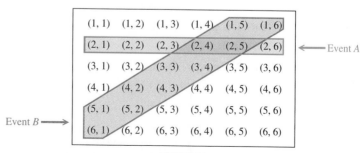

FIGURE 24

(b)   "At most 4" can be written as "2 or 3 or 4." (A sum of 1 is meaningless here.)
Then,

$$P(\text{at most } 4) = P(2 \text{ or } 3 \text{ or } 4) = P(2) + P(3) + P(4), \qquad (*)$$

since the events represented by "2," "3," and "4" are mutually exclusive.

The sample space for this experiment includes the 36 possible pairs of numbers shown in Figure 24. The pair (1, 1) is the only one with a sum of 2, so $P(2) = 1/36$. Also, $P(3) = 2/36$ since both (1, 2) and (2, 1) give a sum of 3. The pairs (1, 3), (2, 2), and (3, 1) have a sum of 4, so $P(4) = 3/36$. Substituting into equation (*) above gives

$$P(\text{at most } 4) = \frac{1}{36} + \frac{2}{36} + \frac{3}{36} = \frac{1}{6}. \qquad \blacklozenge$$

The properties of probability discussed in this section are summarized as follows.

PROPERTIES OF PROBABILITY

For any events E and F,

1. $0 \leq P(E) \leq 1$;
2. $P(\text{a certain event}) = 1$;
3. $P(\text{an impossible event}) = 0$;
4. $P(E') = 1 - P(E)$;
5. $P(E \text{ or } F) = P(E \cup F) = P(E) + P(F) - P(E \cap F)$.

Caution   When finding the probability of a union, don't forget to subtract the probability of the intersection from the sum of the probabilities of the individual events.

BINOMIAL PROBABILITY

If an experiment consists of repeated independent trials with only two outcomes in each trial, success or failure, it is called a **binomial experiment**. Let the probability of success in one trial be p. Then the probability of failure is $1 - p$, and the probability of r successes in n trials is given by

$$\binom{n}{r} p^r (1 - p)^{n-r}.$$

Notice that this expression is equivalent to the general term of the binomial expansion given in Section 4 of this chapter. Thus, the terms of the binomial expansion give the probabilities of r successes in n trials, for $0 \leq r \leq n$, in a binomial experiment.

EXAMPLE 6   *Finding Probabilities, Using a Binomial Experiment*   An experiment consists of rolling a die 10 times. Find the following probabilities.

(a)   The probability that exactly 4 of the tosses result in a 3.

(b)   The probability that in 9 of the 10 tosses, the result is not a 3.

SOLUTION

(a)   The probability of a 3 on one roll is $p = 1/6$. The required probability is

$$\binom{10}{4}\left(\frac{1}{6}\right)^4\left(1 - \frac{1}{6}\right)^{10-4} = 210\left(\frac{1}{6}\right)^4\left(\frac{5}{6}\right)^6 \approx .054.$$

(b)   This probability is

$$\binom{10}{1}\left(\frac{1}{6}\right)^1\left(\frac{5}{6}\right)^9 \approx .323. \qquad \blacklozenge$$

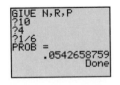

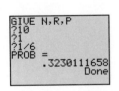

A short program for a graphing calculator can be written to compute binomial probabilities. The two screens above support the results in Example 6.

8.7   EXERCISES   Tape 15

Write a sample space with equally likely outcomes for each experiment.

1. A two-headed coin is tossed once.

2. Two ordinary coins are tossed.

3. Three ordinary coins are tossed.

4. Slips of paper marked with the numbers 1, 2, 3, 4, and 5 are placed in a box. After mixing well, two slips are drawn.

5. An unprepared student takes a 3-question true/false quiz in which he guesses the answers to all 3 questions.

6. A die is rolled and then a coin is tossed.

Write the event in set notation and give the probability of the event.

7. In Exercise 1,

   (a) the result of the toss is heads;

   (b) the result of the toss is tails.

8. In Exercise 2,

   (a) both coins show the same face;

   (b) at least one coin turns up heads.

9. In Exercise 5,

   (a) the student gets all three answers correct;

   (b) he gets all three answers wrong;

   (c) he gets exactly two answers correct;

   (d) he gets at least one answer correct.

10. In Exercise 4,

   (a) both slips are marked with even numbers;

   (b) both slips are marked with odd numbers;

   (c) both slips are marked with the same number;

   (d) one slip is marked with an odd number, the other with an even number.

11. A student gives the answer to a probability problem as 6/5. Explain why this answer must be incorrect.

12. If the probability of an event is .857, what is the probability that the event will not occur?

Work each problem.

13. *Drawing a Marble*   A marble is drawn at random from a box containing 3 yellow, 4 white, and 8 blue marbles. Find the probabilities in parts (a)–(c).

   (a) A yellow marble is drawn.

   (b) A black marble is drawn.

   (c) The marble is yellow or white.

   (d) What are the odds in favor of drawing a yellow marble?

   (e) What are the odds against drawing a blue marble?

14. *Batting Average*   A baseball player with a batting average of .300 comes to bat. What are the odds in favor of his getting a hit?

15. *Drawing Slips of Paper*   In Exercise 4, what are the odds that the sum of the numbers on the two slips of paper is 5?

16. *Probability of Rain*   If the odds that it will rain are 4 to 5, what is the probability of rain?

17. *Probability of a Candidate Losing*   If the odds that a candidate will win an election are 3 to 2, what is the probability that the candidate will lose?

18. *Drawing a Card*   A card is drawn from a well-shuffled deck of 52 cards. Find the probability that the card is

   (a) a 9;   **(b)** black;   **(c)** a black 9;

   (d) a heart;

   (e) a face card (K, Q, J of any suit);

   (f) red or a 3;

   (g) less than a 4 (consider aces as 1s).

19. *Guest Arrival at a Party*   Mrs. Schmulen invites 10 relatives to a party: her mother, 2 uncles, 3 brothers, and 4 cousins. If the chances of any one guest arriving first are equally likely, find the following probabilities.

   (a) The first guest is an uncle or a brother.

   (b) The first guest is a brother or cousin.

   (c) The first guest is a brother or her mother.

20. *Dice Rolls*   Two dice are rolled. Find the probability of the following events.

   (a) The sum of the points is at least 10.

   (b) The sum of the points is either 7 or at least 10.

   (c) The sum of the points is 2 or the dice both show the same number.

21. *Consumer Purchases*   The table shows the probability that a customer of a department store will make a purchase in the indicated price range.

Cost	Probability
Below $5	.25
$5–$19.99	.37
$20–$39.99	.11
$40–$69.99	.09
$70–$99.99	.07
$100–$149.99	.08
$150 or more	.03

Find the probability that a customer makes a purchase that is

(a) less than $20; **(b)** $40 or more;

(c) more than $99.99; **(d)** less than $100.

22. *State Lottery* One game in a state lottery requires you to pick 1 heart, 1 club, 1 diamond, and 1 spade, in that order, from the 13 cards in each suit. What is the probability of getting all four picks correct and winning $5000?

23. *State Lottery* If three of the four selections in Exercise 22 are correct, the player wins $200. Find the probability of this outcome.

24. *Partner Selection* The law firm of Alam, Bartolini, Chinn, Dickinson, and Ellsberg has two senior partners, Alam and Bartolini. Two of the attorneys are to be selected to attend a conference. Assuming that all are equally likely to be selected, find the following probabilities.

(a) Chinn is selected.

(b) Alam and Dickinson are selected.

(c) At least one senior partner is selected.

25. *Opinion Survey* The management of a firm wishes to survey the opinions of its workers, classified as follows for the purpose of an interview: 30% have worked for the company more than 5 years; 28% are female; 65% contribute to a voluntary retirement plan; and $\frac{1}{2}$ of the female workers contribute to the retirement plan. Find the following probabilities.

(a) A male worker is selected.

(b) A worker is selected who has worked for the company less than 5 years.

(c) A worker is selected who contributes to the retirement plan or is female.

26. Explain in your own words why the probability of an event must be a number between and inclusive of 0 and 1.

Gender Makeup of a Family Suppose a family has 5 children. Also, suppose that the probability of having a girl is 1/2. Find the probability that the family has the following children.

27. Exactly 2 girls and 3 boys

28. Exactly 3 girls and 2 boys

29. No girls

30. No boys

31. At least 3 boys

32. No more than 4 girls

A die is rolled 12 *times. Find the probability of rolling the following.*

33. Exactly 12 ones **34.** Exactly 6 ones

35. No more than 3 ones **36.** No more than 1 one

The screens shown here illustrate how the table feature of a graphing calculator can be used to find the probabilities of having 0, 1, 2, 3, or 4 girls in a family of 4 children. (Note that 0 appears for X = 5 and X = 6. Why is this so?) Use this approach to work Exercises 37 and 38.

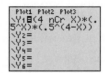

37. Find the probabilities of having 0, 1, 2, or 3 boys in a family of 3 children.

38. Find the probabilities of having 0, 1, 2, 3, 4, 5, or 6 girls in a family of 6 children.

Solve the following probability applications.

39. *Disease Infection* What will happen when an infectious disease is introduced into a family? Suppose a family has *I* infected members and *S* members who are not infected but are susceptible to contracting the disease. The probability *P* of *k* people not contracting the disease during a one-week period can be calculated by the formula $P = \binom{S}{k} q^k (1-q)^{S-k}$, where $q = (1-p)^I$ and *p* is the probability that a susceptible person contracts the disease from an infectious person. For example, if $p = .5$, then there is a 50% chance that a susceptible person exposed to one infectious person for one week will contract the disease. (*Source:* Hoppensteadt, F., and C. Peskin, *Mathematics in Medicine and the Life Sciences*, Springer-Verlag, 1992.)

(a) Compute the probability *P* of 3 family members not becoming infected within one week if there are currently 2 infected and 4 susceptible members. Assume that $p = .1$.

(b) A highly infectious disease can have $p = .5$. Repeat part (a) with this value of *p*.

(c) Determine the probability that everyone would become sick in a large family if initially $I = 1$, $S = 9$, and $p = .5$. Discuss the results.

40. *Disease Infection* (Refer to Exercise 39.) Suppose that in a family $I = 2$ and $S = 4$. If the probability *P* is .25 of there being $k = 2$ uninfected members after one week, estimate graphically the possible values of *p*. (*Hint:* Write *P* as a function of *p*.)

41. *Drug Resistance* Some bacteria, called *haploid organisms*, contain genetic material called *plasmids*.

Plasmids can cause bacteria to become resistant to certain types of antibiotic drugs. Suppose the bacteria carry two plasmids that cause drug resistance. Plasmid R_1 is resistant to the antibiotic ampicillin, whereas plasmid R_2 is resistant to the antibiotic tetracycline. If a bacterium has both plasmids, it will be resistant to both antibiotics. R_1 and R_2 are passed in cell division to daughter cells. The type of plasmids inherited by a daughter cell is random. It can have 0, 1, or 2 plasmids of type R_1. The probability $P_{i,j}$ that a mother cell with i plasmids of type R_1 produces a daughter cell with j plasmids of type R_1 can be calculated by the formula

$$P_{i,j} = \frac{\binom{2i}{j}\binom{4-2i}{2-j}}{\binom{4}{2}}.$$

(*Source:* Hoppensteadt, F., and C. Peskin, *Mathematics in Medicine and the Life Sciences*, Springer-Verlag, 1992.)

(a) Compute the nine values of $P_{i,j}$ for $0 \le i \le 2$, $0 \le j \le 2$. Assume that $\binom{0}{0} = 1$ and $\binom{i}{j} = 0$ whenever $i < j$.

(b) Write your results from part (a) in the matrix

$$P = \begin{bmatrix} P_{0,0} & P_{0,1} & P_{0,2} \\ P_{1,0} & P_{1,1} & P_{1,2} \\ P_{2,0} & P_{2,1} & P_{2,2} \end{bmatrix}.$$

(c) Describe the matrix. Where are the probabilities the greatest?

42. *Drug Resistance* (Continuation of Exercise 41) The genetic makeup of future generations of the haploid bacteria can be modeled with matrices. Let $A = [a_1 \ \ a_2 \ \ a_3]$ be a 1×3 matrix containing three probabilities. a_1 is the probability that a cell has two R_1 plasmids, a_2 is the probability that it has one R_1 plasmid and one R_2 plasmid, and a_3 is the probability that a cell has two R_2 plasmids. If the current generation of bacteria only has one plasmid of each type, then $A_1 = [0 \ \ 1 \ \ 0]$. The probabilities A_{i+1} for plasmids R_1 and R_2 in each future generation can be calculated from the equation $A_{i+1} = A_i P$, where $i \ge 1$ and the 3×3 matrix P was determined in the previous exercise. The phenomenon that results from this sequence of calculations was not well understood until fairly recently. It is now used in the genetic engineering of plasmids.

(a) If all bacteria initially have both plasmids, make a conjecture as to the types of plasmids future generations of bacteria will have.

(b) Test your conjecture by repeatedly computing with your calculator the matrix product $A_{i+1} = A_i P$, with $A_1 = [0 \ \ 1 \ \ 0]$ for $i = 1, 2, 3, \ldots, 12$. Interpret the result. It may surprise you.

CHAPTER 8   SUMMARY

Section	Important Concepts
8.1 Sequences and Series	**DEFINITION OF SEQUENCE** A sequence is a function that has a set of natural numbers as its domain. **DEFINITION OF A SERIES** A finite series is an expression of the form $$S_n = a_1 + a_2 + a_3 + \cdots + a_n = \sum_{i=1}^{n} a_i,$$ and an infinite series is an expression of the form $$S_n = a_1 + a_2 + a_3 + \cdots + a_n + \cdots = \sum_{i=1}^{\infty} a_i.$$

SUMMATION PROPERTIES

If $a_1, a_2, a_3, \ldots, a_n$ and $b_1, b_2, b_3, \ldots, b_n$ are two sequences, and c is a constant, then for every positive integer n,

(a) $\displaystyle\sum_{i=1}^{n} c = nc$
(b) $\displaystyle\sum_{i=1}^{n} ca_i = c \sum_{i=1}^{n} a_i$

(c) $\displaystyle\sum_{i=1}^{n} (a_i + b_i) = \sum_{i=1}^{n} a_i + \sum_{i=1}^{n} b_i$
(d) $\displaystyle\sum_{i=1}^{n} (a_i - b_i) = \sum_{i=1}^{n} a_i - \sum_{i=1}^{n} b_i.$

8.2 Arithmetic
Sequences and Series

DEFINITION OF ARITHMETIC SEQUENCE

A sequence in which each term after the first is obtained by adding a fixed number to the previous term is an arithmetic sequence (or arithmetic progression). The fixed number that is added is the common difference.

FORMULA FOR THE nTH TERM OF AN ARITHMETIC SEQUENCE

In an arithmetic sequence with first term a_1 and common difference d, the nth term, a_n, is given by

$$a_n = a_1 + (n - 1)d.$$

FORMULAS FOR THE SUM OF THE FIRST n TERMS OF AN ARITHMETIC SEQUENCE

If an arithmetic sequence has first term a_1 and common difference d, then the sum of the first n terms is given by

$$S_n = \frac{n}{2}(a_1 + a_n)$$

or

$$S_n = \frac{n}{2}[2a_1 + (n - 1)d].$$

8.3 Geometric
Sequences and Series

DEFINITION OF GEOMETRIC SEQUENCE

A geometric sequence (or geometric progression) is a sequence in which each term after the first is obtained by multiplying the preceding term by a constant nonzero real number, called the common ratio.

FORMULA FOR THE nTH TERM OF A GEOMETRIC SEQUENCE

In the geometric sequence with first term a_1 and common ratio r, the nth term is given by

$$a_n = a_1 r^{n-1}.$$

FORMULA FOR THE SUM OF THE FIRST n TERMS OF A GEOMETRIC SEQUENCE

If a geometric sequence has first term a_1 and common ratio r, then the sum of the first n terms is given by

$$S_n = \frac{a_1(1 - r^n)}{1 - r}, \quad \text{where } r \neq 1.$$

FORMULA FOR THE SUM OF THE TERMS OF AN INFINITE GEOMETRIC SEQUENCE

The sum of an infinite geometric sequence with first term a_1 and common ratio r, where $-1 < r < 1$, is given by

$$S_\infty = \frac{a_1}{1 - r}.$$

8.4 The Binomial
Theorem

PASCAL'S TRIANGLE

$$
\begin{array}{ccccccccccc}
 & & & & & 1 & & & & & \\
 & & & & 1 & & 1 & & & & \\
 & & & 1 & & 2 & & 1 & & & \\
 & & 1 & & 3 & & 3 & & 1 & & \\
 & 1 & & 4 & & 6 & & 4 & & 1 & \\
1 & & 5 & & 10 & & 10 & & 5 & & 1
\end{array}
$$

DEFINITION OF *n*-FACTORIAL

For any positive integer n,

$$n! = n(n - 1)(n - 2) \cdots (3)(2)(1)$$

and

$$0! = 1.$$

DEFINITION OF THE BINOMIAL COEFFICIENT

For nonnegative integers n and r, with $r \leq n$, the symbol $\binom{n}{r}$ is defined as

$$\binom{n}{r} = \frac{n!}{r!(n - r)!}.$$

BINOMIAL THEOREM

For any positive integer n,

$$(x + y)^n = x^n + \binom{n}{n-1}x^{n-1}y + \binom{n}{n-2}x^{n-2}y^2 + \binom{n}{n-3}x^{n-3}y^3$$

$$+ \cdots + \binom{n}{n-r}x^{n-r}y^r + \cdots + \binom{n}{1}xy^{n-1} + y^n.$$

EXPRESSION FOR THE *r*TH TERM OF THE BINOMIAL EXPANSION

The rth term of the binomial expansion of $(x + y)^n$, where $n \geq r - 1$, is

$$\binom{n}{n-(r-1)}x^{n-(r-1)}y^{r-1}.$$

8.5 Mathematical
Induction

PROOF BY MATHEMATICAL INDUCTION

To prove a statement by mathematical induction, (1) prove that it is true for $n = 1$ and (2) show that if it is true for $n = k$, then it must also be true for $n = k + 1$.

8.6 Counting Theory

THE FUNDAMENTAL PRINCIPLE OF COUNTING

If n independent events occur, with

$$m_1 \text{ ways for event 1 to occur,}$$

$$m_2 \text{ ways for event 2 to occur,}$$

$$\cdot$$
$$\cdot$$
$$\cdot$$

and

$$m_n \text{ ways for event } n \text{ to occur,}$$

then there are

$$m_1 \cdot m_2 \cdot \ldots \cdot m_n$$

different ways for all n events to occur.

FORMULA FOR PERMUTATIONS OF *n* ELEMENTS TAKEN *r* AT A TIME

If $P(n, r)$ denotes the number of permutations of n elements taken r at a time, with $r \leq n$, then

$$P(n, r) = \frac{n!}{(n - r)!}.$$

FORMULA FOR COMBINATIONS OF *n* ELEMENTS TAKEN *r* AT A TIME

If $\binom{n}{r}$ or $C(n, r)$ represents the number of combinations of n things taken r at a time, with $r \leq n$, then

$$\binom{n}{r} = C(n, r) = \frac{n!}{(n - r)!\, r!}.$$

8.7 Probability

DEFINITIONS OF SAMPLE SPACE AND EVENT

The set S of all possible outcomes of a given experiment is called the sample space for the experiment. Any subset of S is called an event.

DEFINITION OF PROBABILITY OF EVENT *E*

In a sample space with equally likely outcomes, the probability of an event E, written $P(E)$, is the ratio of the number of outcomes in sample space S that belong to event E, $n(E)$, to the total number of outcomes in sample space S, $n(S)$. That is,

$$P(E) = \frac{n(E)}{n(S)}.$$

DEFINITION OF COMPLEMENT

The complement of an event E, written E', is the set of all outcomes not in E. Thus,

$$P(E) + P(E') = 1.$$

DEFINITION OF ODDS

The odds in favor of an event E are expressed as the ratio of $P(E)$ to $P(E')$ or as the fraction $P(E)/P(E')$.

FORMULA FOR THE PROBABILITY OF THE UNION OF TWO EVENTS

For any events E and F,

$$P(E \text{ or } F) = P(E \cup F) = P(E) + P(F) - P(E \cap F).$$

PROPERTIES OF PROBABILITY

For any events E and F,
1. $0 \leq P(E) \leq 1$
2. $P(\text{a certain event}) = 1$
3. $P(\text{an impossible event}) = 0$
4. $P(E') = 1 - P(E)$
5. $P(E \text{ or } F) = P(E \cup F) = P(E) + P(F) - P(E \cap F).$

BINOMIAL PROBABILITY

If an experiment consists of repeated independent trials with only two outcomes in each trial, success or failure, it is called a binomial experiment. Let the probability of success in one trial be p. Then the probability of failure is $1 - p$, and the probability of r successes in n trials is given by

$$\binom{n}{r} p^r (1 - p)^{n-r}.$$

CHAPTER 8   REVIEW EXERCISES

Write the first five terms for each sequence. State whether the sequence is arithmetic, geometric, or neither.

1. $a_n = \dfrac{n}{n+1}$ 　　　　**2.** $a_n = (-2)^n$ 　　　　**3.** $a_n = 2(n+3)$ 　　　　**4.** $a_n = n(n+1)$

5. $a_1 = 5$; for $n \geq 2, a_n = a_{n-1} - 3$

In Exercises 6–9, write the first five terms of the sequence described.

6. Arithmetic, $a_2 = 10, d = -2$ 　　　　　　**7.** Arithmetic, $a_3 = \pi, a_4 = 1$

8. Geometric, $a_1 = 6, r = 2$ 　　　　　　**9.** Geometric, $a_1 = -5, a_2 = -1$

10. An arithmetic sequence has $a_5 = -3$ and $a_{15} = 17$. Find a_1 and a_n.

11. A geometric sequence has $a_1 = -8$ and $a_7 = -1/8$. Find a_4 and a_n.

Find a_8 for each arithmetic sequence.

12. $a_1 = 6, d = 2$ 　　　　　　**13.** $a_1 = 6x - 9, a_2 = 5x + 1$

Find S_{12} for each arithmetic sequence.

14. $a_1 = 2, d = 3$ 　　　　　　**15.** $a_2 = 6, d = 10$

Find a_5 for each geometric sequence.

16. $a_1 = -2, r = 3$ 　　　　　　**17.** $a_3 = 4, r = \dfrac{1}{5}$

Find S_4 for each geometric sequence.

18. $a_1 = 3, r = 2$ 　　**19.** $a_1 = -1, r = 3$ 　　**20.** $\dfrac{3}{4}, -\dfrac{1}{2}, \dfrac{1}{3}, \cdots$

Evaluate the sums that exist.

21. $\displaystyle\sum_{i=1}^{7} (-1)^{i-1}$ 　　**22.** $\displaystyle\sum_{i=1}^{5} (i^2 + i)$ 　　**23.** $\displaystyle\sum_{i=1}^{4} \dfrac{i+1}{i}$ 　　**24.** $\displaystyle\sum_{j=1}^{10} (3j - 4)$

25. $\displaystyle\sum_{j=1}^{2500} j$ 　　**26.** $\displaystyle\sum_{i=1}^{5} 4 \cdot 2^i$ 　　**27.** $\displaystyle\sum_{i=1}^{\infty} \left(\dfrac{4}{7}\right)^i$ 　　**28.** $\displaystyle\sum_{i=1}^{\infty} -2\left(\dfrac{6}{5}\right)^i$

Evaluate the sums that converge. If the series diverges, say so.

29. $24 + 8 + \dfrac{8}{3} + \dfrac{8}{9} + \cdots$ 　　　　**30.** $-\dfrac{3}{4} + \dfrac{1}{2} - \dfrac{1}{3} + \dfrac{2}{9} - \cdots$

31. $\dfrac{1}{12} + \dfrac{1}{6} + \dfrac{1}{3} + \dfrac{2}{3} + \cdots$ 　　　　**32.** $.9 + .09 + .009 + .0009 + \cdots$

Evaluate the sum, where $x_1 = 0, x_2 = 1, x_3 = 2, x_4 = 3, x_5 = 4,$ and $x_6 = 5$.

33. $\displaystyle\sum_{i=1}^{4} (x_i^2 - 6)$ 　　　　**34.** $\displaystyle\sum_{i=1}^{6} f(x_i) \, \Delta x$; $f(x) = (x - 2)^3, \Delta x = .1$

Write each sum, using summation notation.

35. $4 - 1 - 6 - \cdots - 66$ 　　　　**36.** $10 + 14 + 18 + \cdots + 86$

37. $4 + 12 + 36 + \cdots + 972$ 　　　　**38.** $\dfrac{5}{6} + \dfrac{6}{7} + \dfrac{7}{8} + \cdots + \dfrac{12}{13}$

Use the binomial theorem to expand the following.

39. $(x + 2y)^4$ 　　**40.** $(3z - 5w)^3$ 　　**41.** $\left(3\sqrt{x} - \dfrac{1}{\sqrt{x}}\right)^5$ 　　**42.** $(m^3 - m^{-2})^4$

Find the indicated term or terms for each expansion.

43. Sixth term of $(4x - y)^8$

44. Seventh term of $(m - 3n)^{14}$

45. First four terms of $(x + 2)^{12}$

46. Last three terms of $(2a + 5b)^{16}$

47. What kinds of statements are proved by mathematical induction? Give examples.

48. Describe a proof by mathematical induction.

Use mathematical induction to prove that each statement is true for every positive integer n.

49. $1 + 3 + 5 + 7 + \cdots + (2n - 1) = n^2$

50. $2 + 6 + 10 + 14 + \cdots + (4n - 2) = 2n^2$

51. $2 + 2^2 + 2^3 + \cdots + 2^n = 2(2^n - 1)$

52. $1^3 + 3^3 + 5^3 + \cdots + (2n - 1)^3 = n^2(2n^2 - 1)$

53. Is a Social Security number an example of a permutation or a combination?

Find the value of each expression.

54. $P(9, 2)$ **55.** $P(6, 0)$ **56.** $\dbinom{8}{3}$ **57.** $9!$ **58.** $C(10, 5)$

Solve each problem.

59. *Wedding Plans* Two people are planning their wedding. They can select from 2 different chapels, 4 soloists, 3 organists, and 2 ministers. How many different wedding arrangements are possible?

60. *Couch Styles* Bob Schiffer, who is furnishing his apartment, wants to buy a new couch. He can select from 5 different styles, each available in 3 different fabrics, with 6 color choices. How many different couches are available?

61. *Summer Job Assignments* Four students are to be assigned to 4 different summer jobs. Each student is qualified for all 4 jobs. In how many ways can the jobs be assigned?

62. *Conference Delegations* A student body council consists of a president, vice-president, secretary/treasurer, and 3 representatives at large. Three members are to be selected to attend a conference.

    (a) How many different such delegations are possible?

    (b) How many are possible if the president must attend?

63. *Tournament Outcomes* Nine football teams are competing for first-, second-, and third-place titles in a statewide tournament. In how many ways can the winners be determined?

64. *License Plates* How many different license plates can be formed with a letter followed by 3 digits and then 3 letters? How many such license plates have no repeats?

65. *Drawing a Marble* A marble is drawn at random from a box containing 4 green, 5 black, and 6 white marbles. Find the following probabilities.

    (a) A green marble is drawn.

    (b) A marble that is not black is drawn.

    (c) A blue marble is drawn.

66. *Drawing a Marble* Refer to Exercise 65 and answer each question.

    (a) What are the odds in favor of drawing a green marble?

    (b) What are the odds against drawing a white marble?

    (c) What are the odds in favor of drawing a marble that is not white?

Drawing a Card A card is drawn from a standard deck of 52 cards. Find the probability that the following is drawn.

67. A black king

68. A face card or an ace

69. An ace or a diamond

70. A card that is not a diamond

Swimming Pool Filter Samples A sample shipment of 5 swimming pool filters is chosen. The probability of exactly 0, 1, 2, 3, 4, or 5 filters being defective is given in the following table.

Number Defective	0	1	2	3	4	5
Probability	.31	.25	.18	.12	.08	.06

Find the probability that the given number of filters are defective.

71. No more than 3

72. At least 2

73. More than 5

74. *Die Rolls* A die is rolled 12 times. Find the probability that exactly 2 of the rolls result in a five.

75. *Coin Tosses* A coin is tossed 10 times. Find the probability that exactly 4 of the tosses result in a tail.

CHAPTER 8   TEST

1. Write the first five terms for each sequence. State whether the sequence is arithmetic, geometric, or neither.

(a) $a_n = (-1)^n(n + 2)$ **(b)** $a_n = -3 \cdot \left(\frac{1}{2}\right)^n$

(c) $a_1 = 2, a_2 = 3;$ for $n \geq 3, a_n = a_{n-1} + 2a_{n-2}$

2. In each sequence described, find a_5.

(a) An arithmetic sequence with $a_1 = 1$ and $a_3 = 25$

(b) A geometric sequence with $a_1 = 81$ and $r = -2/3$

3. Find the sum of the first ten terms of the sequence described.

(a) Arithmetic, with $a_1 = -43$ and $d = 12$

(b) Geometric, with $a_1 = 5$ and $r = -2$

4. Evaluate each sum that exists.

(a) $\displaystyle\sum_{i=1}^{30} (5i + 2)$ **(b)** $\displaystyle\sum_{i=1}^{5} (-3 \cdot 2^i)$

(c) $\displaystyle\sum_{i=1}^{\infty} (2^i) \cdot 4$ **(d)** $\displaystyle\sum_{i=1}^{\infty} 54\left(\frac{2}{9}\right)^i$

5. (a) Use the binomial theorem to expand $(2x - 3y)^4$.

(b) Find the third term in the expansion of $(w - 2y)^6$.

6. Evaluate each of the following.

(a)
```
10 nCr 2
```

(b) $\dbinom{7}{3}$

(c)
```
7!
```

(d) $P(11, 3)$

7. Use mathematical induction to prove that for all natural numbers n,

$$8 + 14 + 20 + 26 + \cdots + (6n + 2) = 3n^2 + 5n.$$

8. Solve each problem involving counting theory.

(a) *Athletic Shoe Choices*   A sports-shoe manufacturer makes athletic shoes in 4 different styles. Each style comes in 3 different colors, and each color comes in 2 different shades. How many different types of shoes can be made?

(b) *Club Officer Choices*   A club with 20 members plans to elect a president, a secretary, and a treasurer from its membership. If a member can hold at most one office, in how many ways can the three offices be filled?

(c) *Club Officer Choices*   Refer to the problem in part (b). If there are 8 men and 12 women in the club, in how many different ways can 2 men and 3 women be chosen to attend a conference?

9. *Drawing a Card*   A single card is drawn from a standard deck of 52 cards.

(a) Find the probability of drawing a red three.

(b) Find the probability of drawing a card that is not a face card.

(c) Find the probability of drawing a king or a spade.

(d) What are the odds in favor of drawing a face card?

10. *Rolling a Die*   An experiment consists of rolling a die eight times. Find the probability of each of the following.

(a) Exactly three rolls result in a 4.

(b) All eight rolls result in a 6.

CHAPTER 8   PROJECT

USING EXPERIMENTAL PROBABILITIES TO SIMULATE THE MAKEUP OF A FAMILY

In Section 8.7, we calculated many theoretical probabilities. But even though the probability of obtaining a head on one toss of a coin is $\frac{1}{2}$, it does not mean that you will always get a head on exactly one-half of the tosses. In the experiments that follow, we will simulate the makeup of a family, and experimentally determine the probability that a family of 4 children will consist of 2 boys and 2 girls. Do you expect this probability to be $\frac{1}{2}$?

Probabilities can be calculated in two ways. One way is *theoretically*, as described in Section 8.7. The other is *experimentally*, by actually conducting the experiment and recording the successes compared with the total number of trials. The **Law of Large Numbers** says that the larger the number of trials in an experiment, the closer the experimental probability will be to the theoretical probability.

ACTIVITIES

1. Work with a partner. One member of the group should flip two coins simultaneously 20 times and the other should tally the results in a table similar to the one below. The possible events are 2 heads and 0 tails, 1 head and 1 tail, and 0 heads and 2 tails.

Event	Tally	Relative Frequency	Experimental Probability (as a decimal)
2 heads, 0 tails 1 head, 1 tail 0 heads, 2 tails		$\overline{20}$ $\overline{20}$ $\overline{20}$	

(a) List the sample space and calculate the *theoretical* probability of each event.
(b) Explain why the three outcomes are not equally likely.
(c) Your experimental probabilities probably do not match the theoretical probabilities. Why not? How can you obtain better results?

2. With your partner, repeat the experiment but this time flip four coins simultaneously 20 times and tally the results. The possible events are given in the table below.

Event	Tally	Relative Frequency	Experimental Probability (as a decimal)
4 heads, 0 tails 3 heads, 1 tail 2 heads, 2 tails 1 head, 3 tails 0 heads, 4 tails		$\overline{20}$ $\overline{20}$ $\overline{20}$ $\overline{20}$ $\overline{20}$	

(a) List the sample space and calculate the *theoretical* probability of each event.
(b) How frequently did you get 2 heads and 2 tails? Are you surprised?
(c) Your experimental probabilities probably do not match the theoretical probabilities. Why not? How can you obtain better results?

After working Activity 2, you probably agree that actual coin tossing is quite time-consuming. To simulate coin tosses, we can use a random integer generator on a graphing calculator, such as the TI-83. The first line in Figure A shows the syntax for generating a list of 4 integers on the closed interval [0, 1]; a toss of four coins can be simulated this way. Let us agree that 0 represents tails and 1 represents heads. See the six sample displays in Figure A.

By finding the sum of the entries in the list, we can determine the outcome. For example, a sum of 0 means that there are 4 tails, a sum of 1 means that there are 3 tails and 1 head, a sum of 2 means that there are 2 tails and 2 heads, and so on. The calculator can keep a tally of the sums, as shown in the short program on the home screen in Figure B.

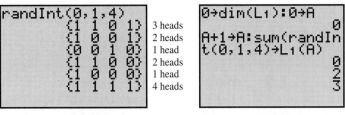

FIGURE A	**FIGURE B**

After running the program a rather large number of times, we can plot the results in a histogram, as seen in Figure C. The first display of Figure D shows how many times the experiment was run (150), the experimental probability for a sum of 2 (meaning 2 tails and 2 heads), and the theoretical probability for 2 heads and 2 tails.

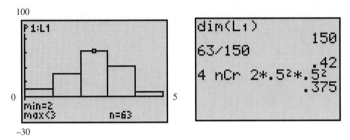

The sum 2 appears 63 times.

FIGURE C	**FIGURE D**

3. With your partner, duplicate the procedure just described to determine an experimental probability of having 2 girls and 2 boys in a family of 4 children. Let 0 represent the outcome of having a girl and 1 represent the outcome of having a boy. Use the calculator to obtain 200 outcomes. How close is your result to the theoretical probability? It should be much closer than the one you found in Activity 2. Why is this so?

R

Reference: Basic Algebraic Concepts

*I*n this reference, we present a review of topics that are usually studied in beginning and intermediate algebra courses. You may wish to refer to the various sections of this reference from time to time if you need to refresh your memory on these basic concepts.

R.1 REVIEW OF RULES FOR EXPONENTS AND OPERATIONS WITH POLYNOMIALS

Rules for Exponents ◆ Addition and Subtraction of Polynomials ◆ Multiplication of Polynomials

RULES FOR EXPONENTS

Work with exponents can be simplified by using the rules for exponents. By definition, the notation a^m (where m is a positive integer and a is a real number) means that a appears as a factor m times. In the same way, a^n (where n is a positive integer) means that a appears as a factor n times. In the product $a^m \cdot a^n$, the number a would appear $m + n$ times.

PRODUCT RULE

For all positive integers m and n and every real number a,

$$a^m \cdot a^n = a^{m+n}.$$

EXAMPLE 1   *Using the Product Rule*   Find the following products.

(a) $y^4 \cdot y^7$ **(b)** $(6z^5)(9z^3)(2z^2)$ **(c)** $2k^m(k^{1+m})$

SOLUTION

(a) $y^4 \cdot y^7 = y^{4+7} = y^{11}$

(b) $(6z^5)(9z^3)(2z^2) = (6 \cdot 9 \cdot 2) \cdot (z^5z^3z^2)$ Commutative and associative properties

$\qquad\qquad\qquad\quad = 108z^{5+3+2}$ Product rule

$\qquad\qquad\qquad\quad = 108z^{10}$

(c) $(2k^m)(k^{1+m}) = 2k^{m+(1+m)}$ Product rule (if *m* is a positive integer)

$\qquad\qquad\qquad = 2k^{1+2m}$ ◆

An exponent of zero is defined as follows. (The justification is found in Section R.4.)

DEFINITION OF a^0

For any nonzero real number *a*,
$$a^0 = 1.$$

EXAMPLE 2   *Using the Definition of a^0*   Evaluate each of the following.

(a) 3^0 **(b)** $(-4)^0$ **(c)** -4^0 **(d)** $-(-4)^0$ **(e)** $(7r)^0$,   if $r \neq 0$

SOLUTION

(a) $3^0 = 1$

(b) $(-4)^0 = 1$

Replace *a* with -4 in the definition.

(c) $-4^0 = -1$, since -4^0 means $-(4^0) = -1$.

(d) $-(-4)^0 = -(1) = -1$

(e) $(7r)^0 = 1$,   if $r \neq 0$ ◆

The expression $(2^5)^3$ can be written as
$$(2^5)^3 = 2^5 \cdot 2^5 \cdot 2^5.$$

By a generalization of the product rule for exponents, this product is
$$(2^5)^3 = 2^{5+5+5} = 2^{15}.$$

The same exponent could have been obtained by multiplying 3 and 5. This example suggests the first of the **power rules** given below. The others are found in a similar way.

POWER RULES

For all nonnegative integers *m* and *n* and all real numbers *a* and *b*,
$$(a^m)^n = a^{mn} \qquad (ab)^m = a^m b^m \qquad \left(\frac{a}{b}\right)^m = \frac{a^m}{b^m} \qquad (b \neq 0).$$

EXAMPLE 3   *Using the Power Rules*   Apply the power rules to simplify each expression.

(a) $(5^3)^2$ **(b)** $(3^4x^2)^3$ **(c)** $\left(\dfrac{2^5}{b^4}\right)^3$,   if $b \neq 0$

SOLUTION

(a) $(5^3)^2 = 5^{3(2)} = 5^6$

(b) $(3^4x^2)^3 = (3^4)^3(x^2)^3 = 3^{4(3)}x^{2(3)} = 3^{12}x^6$

(c) $\left(\dfrac{2^5}{b^4}\right)^3 = \dfrac{(2^5)^3}{(b^4)^3} = \dfrac{2^{15}}{b^{12}},$ if $b \neq 0$ ◆

The quotient rule for exponents is found in Section R.4 of this reference, after the discussion of negative exponents.

ADDITION AND SUBTRACTION OF POLYNOMIALS

Since the variables used in polynomials represent real numbers, a polynomial represents a real number. This means that all the properties of the real numbers hold for polynomials. In particular, the distributive property holds, so

$$3m^5 - 7m^5 = (3 - 7)m^5 = -4m^5.$$

Thus, polynomials are added by adding coefficients of like terms; polynomials are subtracted by subtracting coefficients of like terms.

EXAMPLE 4 *Adding and Subtracting Polynomials* Add or subtract, as indicated.

(a) $(2y^4 - 3y^2 + y) + (4y^4 + 7y^2 + 6y)$

(b) $(-3m^3 - 8m^2 + 4) - (m^3 + 7m^2 - 3)$

(c) $8m^4p^5 - 9m^3p^5 + (11m^4p^5 + 15m^3p^5)$

(d) $4(x^2 - 3x + 7) - 5(2x^2 - 8x - 4)$

SOLUTION

(a) $(2y^4 - 3y^2 + y) + (4y^4 + 7y^2 + 6y)$

$$= (2 + 4)y^4 + (-3 + 7)y^2 + (1 + 6)y$$

$$= 6y^4 + 4y^2 + 7y$$

(b) $(-3m^3 - 8m^2 + 4) - (m^3 + 7m^2 - 3)$

$$= (-3 - 1)m^3 + (-8 - 7)m^2 + [4 - (-3)]$$

$$= -4m^3 - 15m^2 + 7$$

(c) $8m^4p^5 - 9m^3p^5 + (11m^4p^5 + 15m^3p^5) = 19m^4p^5 + 6m^3p^5$

(d) $4(x^2 - 3x + 7) - 5(2x^2 - 8x - 4)$

$$= 4x^2 - 4(3x) + 4(7) - 5(2x^2) - 5(-8x) - 5(-4)$$

Distributive property

$$= 4x^2 - 12x + 28 - 10x^2 + 40x + 20$$

Associative property

$$= -6x^2 + 28x + 48$$

Add like terms.

◆

As shown in parts (a), (b), and (d) of Example 4, polynomials in one variable are often written with their terms in *descending order*, so the term of highest degree is first, the one with the next highest degree is next, and so on.

MULTIPLICATION OF POLYNOMIALS

The associative and distributive properties, together with the properties of exponents, can also be used to find the product of two polynomials. For example, to find the product of $3x - 4$ and $2x^2 - 3x + 5$, treat $3x - 4$ as a single expression and use the distributive property as follows.

$$(3x - 4)(2x^2 - 3x + 5) = (3x - 4)(2x^2) - (3x - 4)(3x) + (3x - 4)(5)$$

Now, use the distributive property three separate times on the right of the equals sign to get

$$(3x - 4)(2x^2 - 3x + 5)$$
$$= (3x)(2x^2) - 4(2x^2) - (3x)(3x) - (-4)(3x) + (3x)5 - 4(5)$$
$$= 6x^3 - 8x^2 - 9x^2 + 12x + 15x - 20$$
$$= 6x^3 - 17x^2 + 27x - 20.$$

It is sometimes more convenient to write such a product vertically, as follows.

$$
\begin{array}{r}
2x^2 - 3x + 5 \\
3x - 4 \\
\hline
-8x^2 + 12x - 20 \\
6x^3 - 9x^2 + 15x \\
\hline
6x^3 - 17x^2 + 27x - 20
\end{array}
\qquad \text{Add in columns.}
$$

EXAMPLE 5   *Multiplying Polynomials*   Multiply $(3p^2 - 4p + 1)(p^3 + 2p - 8)$.

SOLUTION   Multiply each term of the second polynomial by each term of the first and add these products. It is most efficient to work vertically with polynomials of more than two terms, so that like terms can be placed in columns.

$$
\begin{array}{r}
3p^2 - 4p + 1 \\
p^3 + 2p - 8 \\
\hline
-24p^2 + 32p - 8 \\
6p^3 - 8p^2 + 2p \\
3p^5 - 4p^4 + p^3 \\
\hline
3p^5 - 4p^4 + 7p^3 - 32p^2 + 34p - 8
\end{array}
\quad
\begin{array}{l}
\text{Multiply } 3p^2 - 4p + 1 \text{ by } -8. \\
\text{Multiply } 3p^2 - 4p + 1 \text{ by } 2p. \\
\text{Multiply } 3p^2 - 4p + 1 \text{ by } p^3. \\
\text{Add in columns.} \qquad \blacklozenge
\end{array}
$$

The FOIL method is a convenient way to find the product of two binomials. The memory aid FOIL (for First, Outside, Inside, Last) gives the pairs of terms to be multiplied to get the product, as shown in the next examples.

EXAMPLE 6   *Using FOIL to Multiply Two Binomials*   Find each product.

(a) $(6m + 1)(4m - 3)$ **(b)** $(2x + 7)(2x - 7)$ **(c)** $(2k^n - 5)(k^n + 3)$

SOLUTION

$$
\begin{array}{cccc}
\text{F} & \text{O} & \text{I} & \text{L}
\end{array}
$$

(a) $(6m + 1)(4m - 3) = (6m)(4m) + (6m)(-3) + 1(4m) + 1(-3)$
$$= 24m^2 - 14m - 3$$

(b) $(2x + 7)(2x - 7) = 4x^2 - 14x + 14x - 49$
$$= 4x^2 - 49$$

(c) $(2k^n - 5)(k^n + 3) = 2k^{2n} + 6k^n - 5k^n - 15$
$$= 2k^{2n} + k^n - 15 \qquad \blacklozenge$$

In parts (a) and (c) of Example 6, the product of two binomials was a trinomial, while in part (b) the product of two binomials was a binomial. The product of two binomials of the forms $x + y$ and $x - y$ is always a binomial. Check by multiplying that the following is true.

PRODUCT OF THE SUM AND DIFFERENCE OF TWO TERMS

$$(x + y)(x - y) = x^2 - y^2$$

This product is called the **difference of two squares.** Since products of this type occur frequently, it is important to *memorize this formula.*

EXAMPLE 7 *Multiplying the Sum and Difference of Two Terms* Find each product.

(a) $(3p + 11)(3p - 11)$ **(b)** $(5m^3 - 3)(5m^3 + 3)$

(c) $(9k - 11r^3)(9k + 11r^3)$

SOLUTION

(a) Using the pattern discussed above, replace x with $3p$ and y with 11.

$$(3p + 11)(3p - 11) = (3p)^2 - 11^2 = 9p^2 - 121$$

(b) $(5m^3 - 3)(5m^3 + 3) = (5m^3)^2 - 3^2 = 25m^6 - 9$

(c) $(9k - 11r^3)(9k + 11r^3) = (9k)^2 - (11r^3)^2 = 81k^2 - 121r^6$ ◆

The **squares of binomials** are also special products. The products $(x + y)^2$ and $(x - y)^2$ are shown in the following box.

SQUARES OF BINOMIALS

$$(x + y)^2 = x^2 + 2xy + y^2$$

$$(x - y)^2 = x^2 - 2xy + y^2$$

This formula is also one that occurs frequently and *should be memorized.* When factoring polynomials, you will need to be able to recognize these patterns.

EXAMPLE 8 *Using the Formulas for Squares of Binomials* Find each product.

(a) $(2m + 5)^2$ **(b)** $(3x - 7y^4)^2$

SOLUTION

(a) $(2m + 5)^2 = (2m)^2 + 2(2m)(5) + 5^2$

$$= 4m^2 + 20m + 25$$

(b) $(3x - 7y^4)^2 = (3x)^2 - 2(3x)(7y^4) + (7y^4)^2$

$$= 9x^2 - 42xy^4 + 49y^8$$ ◆

R.1 EXERCISES

Use the properties of exponents to simplify each exponential expression. Leave answers with exponents.

1. $(-4)^3 \cdot (-4)^2$ **2.** $(-5)^2 \cdot (-5)^6$ **3.** 2^0 **4.** -2^0

5. $(5m)^0 \quad (m \neq 0)$ **6.** $(-4z)^0 \quad (z \neq 0)$ **7.** $(2^2)^5$ **8.** $(6^4)^3$

9. $(2x^5y^4)^3$ **10.** $(-4m^3n^9)^2$ **11.** $-\left(\dfrac{p^4}{q}\right)^2$ **12.** $\left(\dfrac{r^8}{s^2}\right)^3$

Find each of the following sums and differences.

13. $(3x^2 - 4x + 5) + (-2x^2 + 3x - 2)$ **14.** $(4m^3 - 3m^2 + 5) + (-3m^3 - m^2 + 5)$

15. $(12y^2 - 8y + 6) - (3y^2 - 4y + 2)$ **16.** $(8p^2 - 5p) - (3p^2 - 2p + 4)$

17. $(6m^4 - 3m^2 + m) - (2m^3 + 5m^2 + 4m) + (m^2 - m)$ **18.** $-(8x^3 + x - 3) + (2x^3 + x^2) - (4x^2 + 3x - 1)$

Find each of the following products.

19. $(4r - 1)(7r + 2)$ **20.** $(5m - 6)(3m + 4)$ **21.** $\left(3x - \dfrac{2}{3}\right)\left(5x + \dfrac{1}{3}\right)$

22. $\left(2m - \dfrac{1}{4}\right)\left(3m + \dfrac{1}{2}\right)$ **23.** $4x^2(3x^3 + 2x^2 - 5x + 1)$ **24.** $2b^3(b^2 - 4b + 3)$

25. $(2z - 1)(-z^2 + 3z - 4)$ **26.** $(k + 2)(12k^3 - 3k^2 + k + 1)$ **27.** $(m - n + k)(m + 2n - 3k)$

28. $(r - 3s + t)(2r - s + t)$ **29.** $(a - b + 2c)^2$ **30.** $(k - y + 3m)^2$

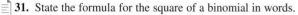 **31.** State the formula for the square of a binomial in words.

32. State the formula for the product of the sum and difference of two terms in words.

Use the special products formulas for each of the following.

33. $(2m + 3)(2m - 3)$ **34.** $(8s - 3t)(8s + 3t)$ **35.** $(4m + 2n)^2$

36. $(a - 6b)^2$ **37.** $(5r + 3t^2)^2$ **38.** $(2z^4 - 3y)^2$

39. $[(2p - 3) + q]^2$ **40.** $[(4y - 1) + z]^2$ **41.** $[(3q + 5) - p][(3q + 5) + p]$

42. $[(9r - s) + 2][(9r - s) - 2]$ **43.** $[a + (b + c)][a - (b + c)]$ **44.** $[(k + 2) + r][(k + 2) - r]$

45. $[(3a + b) - 1]^2$ **46.** $[(2m + 7) - n]^2$

Perform the indicated operations.

47. $(6p + 5q)(3p - 7q)$ **48.** $(2p - 1)(3p^2 - 4p + 5)$

49. $(3x - 4y)^3$ **50.** $(r^5 - r^3 + r) + (3r^5 - 4r^4 + r^3 + 2r)$

51. $(6k - 3)^2$ **52.** $(4x + 3y)(4x - 3y)$

53. $(p^3 - 4p^2 + p) - (3p^2 + 2p + 7)$ **54.** $(2z + y)(3z - 4y)$

55. $(7m + 2n)(7m - 2n)$ **56.** $(3p + 5)^2$

57. $-3(4q^2 - 3q + 2) + 2(-q^2 + q - 4)$ **58.** $2(3r^2 + 4r + 2) - 3(-r^2 + 4r - 5)$

59. $p(4p - 6) + 2(3p - 8)$ **60.** $m(5m - 2) + 9(5 - m)$

61. $-y(y^2 - 4) + 6y^2(2y - 3)$ **62.** $-z^3(9 - z) + 4z(2 + 3z)$

Find each of the following products. Assume that all variables used in exponents represent non-negative integers.

63. $(k^m + 2)(k^m - 2)$ **64.** $(y^x - 4)(y^x + 4)$ **65.** $(b^r + 3)(b^r - 2)$

66. $(q^y + 4)(q^y + 3)$ **67.** $(3p^x + 1)(p^x - 2)$ **68.** $(2^a + 5)(2^a + 3)$

69. $(m^x - 2)^2$ **70.** $(z^r + 5)^2$ **71.** $(q^p - 5p^q)^2$

72. $(3y^x - 2x^y)^2$ **73.** $(3k^a - 2)^3$ **74.** $(r^x - 4)^3$

Suppose one polynomial has degree m and another has degree n, where m and n are natural numbers with $n < m$. Find the degree of the following for the polynomials.

75. Sum

76. Difference

77. Product

78. What would be the degree of the square of the polynomial of degree m?

R.2 REVIEW OF FACTORING

Basic Concepts and Greatest Common Factors ◆ Factoring by Grouping ◆ Factoring Trinomials ◆ Factoring Special Products ◆ Factoring by Substitution

BASIC CONCEPTS AND GREATEST COMMON FACTORS

The process of finding polynomials whose product equals a given polynomial is called **factoring.** For example, since $4x + 12 = 4(x + 3)$, both 4 and $x + 3$ are called **factors** of $4x + 12$. Also, $4(x + 3)$ is called the **factored form** of $4x + 12$. A polynomial that cannot be written as a product of two polynomials with integer coefficients is a **prime** or **irreducible polynomial.** A polynomial is **factored completely** when it is written as a product of prime polynomials with integer coefficients.

Polynomials are factored by using the distributive property. For example, to factor $6x^2y^3 + 9xy^4 + 18y^5$, we look for a monomial that is the greatest common factor of each term. The terms of this polynomial have $3y^3$ as the greatest common factor. By the distributive property,

$$6x^2y^3 + 9xy^4 + 18y^5 = (3y^3)(2x^2) + (3y^3)(3xy) + (3y^3)(6y^2)$$
$$= 3y^3(2x^2 + 3xy + 6y^2).$$

EXAMPLE 1 *Factoring out the Greatest Common Factor* Factor out the greatest common factor from each polynomial.

(a) $9y^5 + y^2$

(b) $6x^2t + 8xt + 12t$

(c) $14m^4(m + 1) - 28m^3(m + 1) - 7m^2(m + 1)$

SOLUTION

(a) The greatest common factor is y^2.

$$9y^5 + y^2 = y^2 \cdot 9y^3 + y^2 \cdot 1$$
$$= y^2(9y^3 + 1)$$

(b) $6x^2t + 8xt + 12t = 2t(3x^2 + 4x + 6)$

(c) The greatest common factor is $7m^2(m + 1)$. Use the distributive property as follows.

$$14m^4(m + 1) - 28m^3(m + 1) - 7m^2(m + 1)$$
$$= [7m^2(m + 1)](2m^2 - 4m - 1)$$
$$= 7m^2(m + 1)(2m^2 - 4m - 1) \qquad ◆$$

FACTORING BY GROUPING

When a polynomial has more than three terms, it can sometimes be factored by a method called **factoring by grouping.** For example, to factor

$$ax + ay + 6x + 6y,$$

collect the terms into two groups so that each group has a common factor.

$$ax + ay + 6x + 6y = (ax + ay) + (6x + 6y)$$

Factor each group, getting

$$ax + ay + 6x + 6y = a(x + y) + 6(x + y).$$

The quantity $(x + y)$ is now a common factor, which can be factored out, producing

$$ax + ay + 6x + 6y = (x + y)(a + 6).$$

It is not always obvious which terms should be grouped. Experience and repeated trials are the most reliable tools for factoring by grouping.

EXAMPLE 2 *Factoring by Grouping* Factor by grouping.

(a) $mp^2 + 7m + 3p^2 + 21$ **(b)** $2y^2 - 2z - ay^2 + az$

SOLUTION

(a) Group the terms as follows.

$$mp^2 + 7m + 3p^2 + 21 = (mp^2 + 7m) + (3p^2 + 21)$$

Factor out the greatest common factor from each group.

$$(mp^2 + 7m) + (3p^2 + 21) = m(p^2 + 7) + 3(p^2 + 7)$$
$$= (p^2 + 7)(m + 3) \qquad p^2 + 7 \text{ is a common factor.}$$

(b) Grouping terms as above gives

$$2y^2 - 2z - ay^2 + az = (2y^2 - 2z) + (-ay^2 + az)$$
$$= 2(y^2 - z) + a(-y^2 + z).$$

The expression $-y^2 + z$ is the negative of $y^2 - z$, so the terms should be grouped as follows.

$$2y^2 - 2z - ay^2 + az = (2y^2 - 2z) - (ay^2 - az)$$
$$= 2(y^2 - z) - a(y^2 - z) \qquad \text{Factor each group.}$$
$$= (y^2 - z)(2 - a) \qquad \text{Factor out } y^2 - z.$$

Later, in Example 5(e), we show another way to factor by grouping three of the four terms.

FACTORING TRINOMIALS

Factoring is the opposite of multiplication. Since the product of two binomials is usually a trinomial, we can expect factorable trinomials (that have terms with no common factor) to have two binomial factors. Thus, factoring trinomials requires using FOIL backwards.

EXAMPLE 3   *Factoring Trinomials*   Factor each trinomial.

(a) $4y^2 - 11y + 6$ **(b)** $6p^2 - 7p - 5$

SOLUTION

(a)   To factor this polynomial, we must find integers a, b, c, and d such that

$$4y^2 - 11y + 6 = (ay + b)(cy + d)$$
$$= acy^2 + (ad + bc)y + bd.$$

By using FOIL, we see that $ac = 4$ and $bd = 6$. The positive factors of 4 are 4 and 1 or 2 and 2. Since the middle term is negative, we consider only negative factors of 6. The possibilities are -2 and -3 or -1 and -6. Now we try various arrangements of these factors until we find one that gives the correct coefficient of y.

$$(2y - 1)(2y - 6) = 4y^2 - 14y + 6 \qquad \text{Incorrect}$$
$$(2y - 2)(2y - 3) = 4y^2 - 10y + 6 \qquad \text{Incorrect}$$
$$(y - 2)(4y - 3) = 4y^2 - 11y + 6 \qquad \text{Correct}$$

The last trial gives the correct factorization.

(b)   Again, we try various possibilities. The positive factors of 6 could be 2 and 3 or 1 and 6. As factors of -5 we have only -1 and 5 or -5 and 1. Try different combinations of these factors until the correct one is found.

$$(2p - 5)(3p + 1) = 6p^2 - 13p - 5 \qquad \text{Incorrect}$$
$$(3p - 5)(2p + 1) = 6p^2 - 7p - 5 \qquad \text{Correct}$$

Finally, $6p^2 - 7p - 5$ factors as $(3p - 5)(2p + 1)$.   ◆

FACTORING SPECIAL PRODUCTS

Each of the special patterns of multiplication given earlier can be used in reverse to get a pattern for factoring. Perfect square trinomials can be factored as follows.

PERFECT SQUARE TRINOMIALS

$$x^2 + 2xy + y^2 = (x + y)^2$$
$$x^2 - 2xy + y^2 = (x - y)^2$$

These formulas should be memorized.

EXAMPLE 4   *Factoring Perfect Square Trinomials*   Factor each polynomial.

(a) $16p^2 - 40pq + 25q^2$ **(b)** $169x^2 + 104xy^2 + 16y^4$

SOLUTION

(a)   Since $16p^2 = (4p)^2$ and $25q^2 = (5q)^2$, use the second pattern shown above, with $4p$ replacing x and $5q$ replacing y to get

$$16p^2 - 40pq + 25q^2 = (4p)^2 - 2(4p)(5q) + (5q)^2$$
$$= (4p - 5q)^2.$$

Make sure that the middle term of the trinomial being factored, $-40pq$ here, is twice the product of the two terms in the binomial $4p - 5q$.

$$-40pq = 2(4p)(-5q)$$

(b) $169x^2 + 104xy^2 + 16y^4 = (13x + 4y^2)^2$, since $2(13x)(4y^2) = 104xy^2$.   ◆

The pattern for the product of the sum and difference of two terms gives the following factorization.

DIFFERENCE OF TWO SQUARES

$$x^2 - y^2 = (x + y)(x - y)$$

EXAMPLE 5   *Factoring a Difference of Squares*   Factor each of the following polynomials.

(a) $4m^2 - 9$ **(b)** $256k^4 - 625m^4$ **(c)** $(a + 2b)^2 - 4c^2$

(d) $y^{4q} - z^{2q}$ **(e)** $x^2 - 6x + 9 - y^4$

SOLUTION

(a)   First, recognize that $4m^2 - 9$ is the difference of two squares, since $4m^2 = (2m)^2$ and $9 = 3^2$. Use the pattern for the difference of two squares, with $2m$ replacing x and 3 replacing y. Doing this gives

$$4m^2 - 9 = (2m)^2 - 3^2$$
$$= (2m + 3)(2m - 3).$$

(b)   Use the difference of two squares pattern twice as follows:

$$256k^4 - 625m^4 = (16k^2)^2 - (25m^2)^2$$
$$= (16k^2 + 25m^2)(16k^2 - 25m^2)$$
$$= (16k^2 + 25m^2)(4k + 5m)(4k - 5m).$$

(c) $(a + 2b)^2 - 4c^2 = (a + 2b)^2 - (2c)^2$
$$= [(a + 2b) + 2c][(a + 2b) - 2c]$$
$$= (a + 2b + 2c)(a + 2b - 2c)$$

(d) $y^{4q} - z^{2q} = (y^{2q} + z^q)(y^{2q} - z^q)$

(e)   Group the first three terms to get a perfect square trinomial. Then, use the difference of squares pattern.

$$x^2 - 6x + 9 - y^4 = (x^2 - 6x + 9) - y^4$$
$$= (x - 3)^2 - (y^2)^2$$
$$= [(x - 3) + y^2][(x - 3) - y^2]$$
$$= (x - 3 + y^2)(x - 3 - y^2)$$ ◆

Two other special results of factoring are listed below. Each can be verified by multiplying on the right side of the equation.

DIFFERENCE AND SUM OF TWO CUBES

Difference of two cubes $x^3 - y^3 = (x - y)(x^2 + xy + y^2)$

Sum of two cubes $x^3 + y^3 = (x + y)(x^2 - xy + y^2)$

EXAMPLE 6   *Factoring the Sum or Difference of Cubes*   Factor each polynomial.

(a) $x^3 + 27$ **(b)** $m^3 - 64n^3$ **(c)** $8q^6 + 125p^9$

SOLUTION

(a)   Notice that $27 = 3^3$, so the expression is a sum of two cubes. Use the second pattern given above.

$$x^3 + 27 = x^3 + 3^3 = (x + 3)(x^2 - 3x + 9)$$

(b)   Since $64n^3 = (4n)^3$, the given polynomial is a difference of two cubes. To factor, use the first pattern in the box above, replacing x with m and y with $4n$.

$$m^3 - 64n^3 = m^3 - (4n)^3$$
$$= (m - 4n)[m^2 + m(4n) + (4n)^2]$$
$$= (m - 4n)(m^2 + 4mn + 16n^2)$$

(c)   Write $8q^6$ as $(2q^2)^3$ and $125p^9$ as $(5p^3)^3$, so that the given polynomial is a sum of two cubes.

$$8q^6 + 125p^9 = (2q^2)^3 + (5p^3)^3$$
$$= (2q^2 + 5p^3)[(2q^2)^2 - (2q^2)(5p^3) + (5p^3)^2]$$
$$= (2q^2 + 5p^3)(4q^4 - 10q^2p^3 + 25p^6)$$ ◆

FACTORING BY SUBSTITUTION

Sometimes a polynomial can be factored by substituting one expression for another. The next example shows this **method of substitution.**

EXAMPLE 7   *Factoring by Substitution*   Factor each polynomial.

(a) $6z^4 - 13z^2 - 5$ **(b)** $10(2a - 1)^2 - 19(2a - 1) - 15$
(c) $(2a - 1)^3 + 8$

SOLUTION

(a)   Replace z^2 with y, so that $y^2 = (z^2)^2 = z^4$. This replacement gives

$$6z^4 - 13z^2 - 5 = 6y^2 - 13y - 5.$$

Factor $6y^2 - 13y - 5$ as

$$6y^2 - 13y - 5 = (2y - 5)(3y + 1).$$

Replacing y with z^2 gives

$$6z^4 - 13z^2 - 5 = (2z^2 - 5)(3z^2 + 1).$$

(Some students prefer to factor this type of trinomial directly by using trial and error with FOIL.)

(b)   Replacing $2a - 1$ with m gives

$$10m^2 - 19m - 15 = (5m + 3)(2m - 5).$$

Now, replace m with $2a - 1$ in the factored form and simplify.

$$10(2a - 1)^2 - 19(2a - 1) - 15$$
$$= [5(2a - 1) + 3][2(2a - 1) - 5] \qquad \text{Let } m = 2a - 1.$$
$$= (10a - 5 + 3)(4a - 2 - 5) \qquad \text{Multiply.}$$
$$= (10a - 2)(4a - 7) \qquad \text{Add.}$$
$$= 2(5a - 1)(4a - 7) \qquad \text{Factor out the common factor.}$$

(c)   Let $2a - 1 = K$ to get

$$(2a - 1)^3 + 8 = K^3 + 8$$
$$= K^3 + 2^3$$
$$= (K + 2)(K^2 - 2K + 2^2).$$

Replacing K with $2a - 1$ gives

$$(2a - 1)^3 + 8 = (2a - 1 + 2)[(2a - 1)^2 - 2(2a - 1) + 4] \qquad \text{Let } K = 2a - 1.$$
$$= (2a + 1)(4a^2 - 4a + 1 - 4a + 2 + 4) \qquad \text{Multiply.}$$
$$= (2a + 1)(4a^2 - 8a + 7). \qquad \text{Combine terms.}$$

◆

R.2   EXERCISES

Factor the greatest common factor from each polynomial.

1. $4k^2m^3 + 8k^4m^3 - 12k^2m^4$

2. $28r^4s^2 + 7r^3s - 35r^4s^3$

3. $2(a + b) + 4m(a + b)$

4. $4(y - 2)^2 + 3(y - 2)$

5. $(2y - 3)(y + 2) + (y + 5)(y + 2)$

6. $(6a - 1)(a + 2) + (6a - 1)(3a - 1)$

7. $(5r - 6)(r + 3) - (2r - 1)(r + 3)$

8. $(3z + 2)(z + 4) - (z + 6)(z + 4)$

9. $2(m - 1) - 3(m - 1)^2 + 2(m - 1)^3$

10. $5(a + 3)^3 - 2(a + 3) + (a + 3)^2$

Factor each of the following by grouping.

11. $6st + 9t - 10s - 15$

12. $10ab - 6b + 35a - 21$

13. $rt^3 + rs^2 - pt^3 - ps^2$

14. $2m^4 + 6 - am^4 - 3a$

15. $6p^2 - 14p + 15p - 35$

16. $8r^2 - 10r + 12r - 15$

17. $20z^2 - 8zx - 45zx + 18x^2$

18. $16a^2 + 10ab - 24ab - 15b^2$

19. $15 - 5m^2 - 3r^2 + m^2r^2$

Factor each trinomial.

20. $6a^2 - 48a - 120$

21. $8h^2 - 24h - 320$

22. $3m^3 + 12m^2 + 9m$

23. $9y^4 - 54y^3 + 45y^2$

24. $6k^2 + 5kp - 6p^2$

25. $14m^2 + 11mr - 15r^2$

26. $5a^2 - 7ab - 6b^2$

27. $12s^2 + 11st - 5t^2$

28. $9x^2 - 6x^3 + x^4$

29. $30a^2 + am - m^2$

30. $24a^4 + 10a^3b - 4a^2b^2$

31. $18x^5 + 15x^4z - 75x^3z^2$

32. When asked to factor $6x^4 - 3x^2 - 3$ completely, a student gave the following result: $6x^4 - 3x^2 - 3 = (2x^2 + 1)(3x^2 - 3)$. Is this answer correct? Explain why or why not.

Factor each perfect square trinomial.

33. $9m^2 - 12m + 4$

34. $16p^2 - 40p + 25$

35. $32a^2 - 48ab + 18b^2$

36. $20p^2 - 100pq + 125q^2$

37. $4x^2y^2 + 28xy + 49$

38. $9m^2n^2 - 12mn + 4$

39. $(a - 3b)^2 - 6(a - 3b) + 9$

40. $(2p + q)^2 - 10(2p + q) + 25$

Factor each difference of two squares.

41. $9a^2 - 16$ **42.** $16q^2 - 25$ **43.** $25s^4 - 9t^2$ **44.** $36z^2 - 81y^4$

45. $(a + b)^2 - 16$ **46.** $(p - 2q)^2 - 100$ **47.** $p^4 - 625$ **48.** $m^4 - 81$

49. Which of the following is the correct complete factorization of $x^4 - 1$?

    (a) $(x^2 - 1)(x^2 + 1)$ **(b)** $(x^2 + 1)(x + 1)(x - 1)$ **(c)** $(x^2 - 1)^2$ **(d)** $(x - 1)^2(x + 1)^2$

50. Which of the following is the correct factorization of $x^3 + 8$?

    (a) $(x + 2)^3$ **(b)** $(x + 2)(x^2 + 2x + 4)$ **(c)** $(x + 2)(x^2 - 2x + 4)$ **(d)** $(x + 2)(x^2 - 4x + 4)$

Factor each sum or difference of cubes.

51. $8 - a^3$ **52.** $r^3 + 27$ **53.** $125x^3 - 27$ **54.** $8m^3 - 27n^3$

55. $27y^9 + 125z^6$ **56.** $27z^3 + 729y^3$ **57.** $(r + 6)^3 - 216$ **58.** $(b + 3)^3 - 27$

59. $27 - (m + 2n)^3$ **60.** $125 - (4a - b)^3$

61. Is the following factorization of $3a^4 + 14a^2 - 5$ correct? Explain. If it is incorrect, give the correct factors.

$$3a^4 + 14a^2 - 5 = 3u^2 + 14u - 5 \qquad \text{Let } u = a^2.$$
$$= (3u - 1)(u + 5)$$

Factor each of the following, using the method of substitution.

62. $m^4 - 3m^2 - 10$ **63.** $a^4 - 2a^2 - 48$ **64.** $7(3k - 1)^2 + 26(3k - 1) - 8$

65. $6(4z - 3)^2 + 7(4z - 3) - 3$ **66.** $9(a - 4)^2 + 30(a - 4) + 25$ **67.** $20(4 - p)^2 - 3(4 - p) - 2$

Factor by any method.

68. $a^3(r + s) + b^2(r + s)$ **69.** $4b^2 + 4bc + c^2 - 16$ **70.** $(2y - 1)^2 - 4(2y - 1) + 4$

71. $x^2 + xy - 5x - 5y$ **72.** $8r^2 - 3rs + 10s^2$ **73.** $p^4(m - 2n) + q(m - 2n)$

74. $36a^2 + 60a + 25$ **75.** $4z^2 + 28z + 49$ **76.** $6p^4 + 7p^2 - 3$

77. $1000x^3 + 343y^3$ **78.** $b^2 + 8b + 16 - a^2$ **79.** $125m^6 - 216$

80. $q^2 + 6q + 9 - p^2$ **81.** $12m^2 + 16mn - 35n^2$ **82.** $216p^3 + 125q^3$

83. $4p^2 + 3p - 1$ **84.** $100r^2 - 169s^2$ **85.** $144z^2 + 121$

86. $(3a + 5)^2 - 18(3a + 5) + 81$ **87.** $(x + y)^2 - (x - y)^2$ **88.** $4z^4 - 7z^2 - 15$

Factor each of the following polynomials. Assume that all variables used in exponents represent positive integers.

89. $r^2 + rs^q - 6s^{2q}$ **90.** $6z^{2a} - z^a x^b - x^{2b}$ **91.** $9a^{4k} - b^{8k}$

92. $16y^{2c} - 25x^{4c}$ **93.** $4y^{2a} - 12y^a + 9$ **94.** $25x^{4c} - 20x^{2c} + 4$

95. $6(m + p)^{2k} + (m + p)^k - 15$ **96.** $8(2k + q)^{4z} - 2(2k + q)^{2z} - 3$

Find a value of b or c that will make the following expressions perfect square trinomials.

97. $4z^2 + bz + 81$ **98.** $9p^2 + bp + 25$ **99.** $100r^2 - 60r + c$ **100.** $49x^2 + 70x + c$

R.3   REVIEW OF OPERATIONS WITH RATIONAL EXPRESSIONS

Rational Expressions in Lowest Terms ◆ Multiplication and Division ◆ Addition and Subtraction ◆ Complex Fractions

RATIONAL EXPRESSIONS IN LOWEST TERMS

An expression that is the quotient of two algebraic expressions (with denominator not 0) is called a **fractional expression.** The most common fractional expressions are those that are the quotients of two polynomials; these are called **rational expressions.** Since fractional expressions involve quotients, it is important to keep track of values of the variable that satisfy the requirement that no denominator be 0. For example, $x \neq -2$ in the rational expression

$$\frac{x + 6}{x + 2}$$

because replacing x with -2 makes the denominator equal 0. Similarly, in

$$\frac{(x + 6)(x + 4)}{(x + 2)(x + 4)},$$

$x \neq -2$ and $x \neq -4$.

Just as the fraction $\frac{6}{8}$ is written in lowest terms as $\frac{3}{4}$, rational expressions may also be written in lowest terms. This is done with the fundamental principle.

FUNDAMENTAL PRINCIPLE OF RATIONAL EXPRESSIONS

$$\frac{ac}{bc} = \frac{a}{b} \quad (b \neq 0, c \neq 0)$$

EXAMPLE 1 *Writing in Lowest Terms* Write each expression in lowest terms.

(a) $\dfrac{2p^2 + 7p - 4}{5p^2 + 20p}$ **(b)** $\dfrac{6 - 3k}{k^2 - 4}$

SOLUTION

(a) Factor the numerator and the denominator to get

$$\frac{2p^2 + 7p - 4}{5p^2 + 20p} = \frac{(2p - 1)(p + 4)}{5p(p + 4)}.$$

By the fundamental principle,

$$\frac{2p^2 + 7p - 4}{5p^2 + 20p} = \frac{2p - 1}{5p}.$$

In the original expression, p cannot be 0 or -4, because $5p^2 + 20p \neq 0$, so this result is valid only for values of p other than 0 and -4. In the examples that follow, we shall always assume such restrictions when reducing rational expressions.

(b) Factor to get

$$\frac{6 - 3k}{k^2 - 4} = \frac{3(2 - k)}{(k + 2)(k - 2)}.$$

The factors $2 - k$ and $k - 2$ have opposite signs. Because of this, multiply numerator and denominator by -1 as follows.

$$\frac{6 - 3k}{k^2 - 4} = \frac{3(2 - k)(-1)}{(k + 2)(k - 2)(-1)}$$

Since $(k - 2)(-1) = -k + 2$, or $2 - k$,

$$\frac{6 - 3k}{k^2 - 4} = \frac{3(2 - k)(-1)}{(k + 2)(2 - k)},$$

giving

$$\frac{6 - 3k}{k^2 - 4} = \frac{-3}{k + 2}.$$

Working in an alternative way would lead to the equivalent result

$$\frac{3}{-k-2}.$$ ◆

MULTIPLICATION AND DIVISION

Rational expressions are multiplied and divided by using definitions from earlier experience with fractions.

> **MULTIPLICATION AND DIVISION OF FRACTIONS**
>
> For fractions $\frac{a}{b}$ and $\frac{c}{d}$ ($b \neq 0$, $d \neq 0$),
>
> $$\frac{a}{b} \cdot \frac{c}{d} = \frac{ac}{bd}$$
>
> $$\frac{a}{b} \div \frac{c}{d} = \frac{a}{b} \cdot \frac{d}{c} \quad \left(\text{if } \frac{c}{d} \neq 0\right).$$

EXAMPLE 2 *Multiplying and Dividing Rational Expressions* Multiply or divide, as indicated.

(a) $\dfrac{2y^2}{9} \cdot \dfrac{27}{8y^5}$

(b) $\dfrac{3m^2 - 2m - 8}{3m^2 + 14m + 8} \cdot \dfrac{3m + 2}{3m + 4}$

(c) $\dfrac{5}{8m + 16} \div \dfrac{7}{12m + 24}$

(d) $\dfrac{3p^2 + 11p - 4}{24p^3 - 8p^2} \div \dfrac{9p + 36}{24p^4 - 36p^3}$

SOLUTION

(a) $\dfrac{2y^2}{9} \cdot \dfrac{27}{8y^5} = \dfrac{2y^2 \cdot 27}{9 \cdot 8y^5}$

$= \dfrac{2 \cdot 9 \cdot 3 \cdot y^2}{2 \cdot 9 \cdot 4 \cdot y^2 \cdot y^3}$ Factor.

$= \dfrac{3}{4y^3}$ Fundamental principle

The product was written in lowest terms in the last step.

(b) $\dfrac{3m^2 - 2m - 8}{3m^2 + 14m + 8} \cdot \dfrac{3m + 2}{3m + 4} = \dfrac{(m - 2)(3m + 4)}{(m + 4)(3m + 2)} \cdot \dfrac{3m + 2}{3m + 4}$ Factor.

$= \dfrac{(m - 2)(3m + 4)(3m + 2)}{(m + 4)(3m + 2)(3m + 4)}$ Multiply fractions.

$= \dfrac{m - 2}{m + 4}$ Fundamental principle

(c) $\dfrac{5}{8m + 16} \div \dfrac{7}{12m + 24} = \dfrac{5}{8(m + 2)} \div \dfrac{7}{12(m + 2)}$ Factor.

$\qquad = \dfrac{5}{8(m + 2)} \cdot \dfrac{12(m + 2)}{7}$ Definition of division

$\qquad = \dfrac{5 \cdot 12(m + 2)}{8 \cdot 7(m + 2)}$ Multiply.

$\qquad = \dfrac{15}{14}$ Fundamental principle

(d) $\dfrac{3p^2 + 11p - 4}{24p^3 - 8p^2} \div \dfrac{9p + 36}{24p^4 - 36p^3} = \dfrac{(p + 4)(3p - 1)}{8p^2(3p - 1)} \div \dfrac{9(p + 4)}{12p^3(2p - 3)}$

$\qquad = \dfrac{(p + 4)(3p - 1)(12p^3)(2p - 3)}{8p^2(3p - 1)(9)(p + 4)}$

$\qquad = \dfrac{12p^3(2p - 3)}{9 \cdot 8p^2} = \dfrac{p(2p - 3)}{6}$ ◆

ADDITION AND SUBTRACTION

Adding and subtracting rational expressions also depends on definitions from earlier experience with fractions.

ADDITION AND SUBTRACTION OF FRACTIONS

For fractions $\frac{a}{b}$ and $\frac{c}{d}$ $(b \neq 0, d \neq 0)$,

$$\frac{a}{b} + \frac{c}{d} = \frac{ad + bc}{bd}$$

$$\frac{a}{b} - \frac{c}{d} = \frac{ad - bc}{bd}.$$

In practice, rational expressions are normally added or subtracted after rewriting all the rational expressions with a common denominator found with the steps given below.

FINDING A COMMON DENOMINATOR

1. Write each denominator as a product of prime factors.
2. Form a product of all the different prime factors. Each factor should have as exponent the *greatest* exponent that appears on that factor.

EXAMPLE 3   *Adding or Subtracting Rational Expressions*   Add or subtract, as indicated.

(a) $\dfrac{5}{9x^2} + \dfrac{1}{6x}$ **(b)** $\dfrac{y + 2}{y^2 - y} - \dfrac{3y}{2y^2 - 4y + 2}$

(c) $\dfrac{3}{(x - 1)(x + 2)} - \dfrac{1}{(x + 3)(x - 4)}$

690 CHAPTER R Reference: Basic Algebraic Concepts

SOLUTION

(a) Write each denominator as a product of prime factors as follows.

$$9x^2 = 3^2 \cdot x^2$$
$$6x = 2^1 \cdot 3^1 \cdot x^1$$

For the common denominator, form the product of all the prime factors, with each factor having the greatest exponent that appears on it. Here the greatest exponent on 2 is 1, while both 3 and x have a greatest exponent of 2. The common denominator is

$$2^1 \cdot 3^2 \cdot x^2 = 18x^2.$$

Now, use the fundamental principle to write both of the given expressions with this denominator, then add.

$$\frac{5}{9x^2} + \frac{1}{6x} = \frac{5 \cdot 2}{9x^2 \cdot 2} + \frac{1 \cdot 3x}{6x \cdot 3x}$$

$$= \frac{10}{18x^2} + \frac{3x}{18x^2}$$

$$= \frac{10 + 3x}{18x^2}$$

Always check at this point to see that the answer is in lowest terms.

(b) Factor each denominator, giving

$$\frac{y+2}{y^2-y} - \frac{3y}{2y^2-4y+2} = \frac{y+2}{y(y-1)} - \frac{3y}{2(y-1)^2}.$$

The common denominator, by the method above, is $2y(y-1)^2$. Write each rational expression with this denominator and subtract as follows.

$$\frac{y+2}{y(y-1)} - \frac{3y}{2(y-1)^2} = \frac{(y+2) \cdot 2(y-1)}{y(y-1) \cdot 2(y-1)} - \frac{3y \cdot y}{2(y-1)^2 \cdot y}$$

$$= \frac{2(y^2+y-2)}{2y(y-1)^2} - \frac{3y^2}{2y(y-1)^2}$$

$$= \frac{2y^2+2y-4-3y^2}{2y(y-1)^2} \quad \text{Multiply and subtract.}$$

$$= \frac{-y^2+2y-4}{2y(y-1)^2} \quad \text{Combine terms.}$$

(c) The common denominator here is $(x-1)(x+2)(x+3)(x-4)$. Write each fraction with this common denominator, then perform the subtraction.

$$\frac{3}{(x-1)(x+2)} - \frac{1}{(x+3)(x-4)}$$

$$= \frac{3(x+3)(x-4)}{(x-1)(x+2)(x+3)(x-4)} - \frac{(x-1)(x+2)}{(x+3)(x-4)(x-1)(x+2)}$$

$$= \frac{3(x^2-x-12) - (x^2+x-2)}{(x-1)(x+2)(x+3)(x-4)}$$

$$= \frac{3x^2-3x-36-x^2-x+2}{(x-1)(x+2)(x+3)(x-4)}$$

$$= \frac{2x^2-4x-34}{(x-1)(x+2)(x+3)(x-4)} \quad \blacklozenge$$

COMPLEX FRACTIONS

Any quotient of two rational expressions is called a **complex fraction.** Complex fractions often can be simplified by the methods shown in the following example.

EXAMPLE 4   *Simplifying Complex Fractions*   Simplify each complex fraction.

(a) $\dfrac{6 - \dfrac{5}{k}}{1 + \dfrac{5}{k}}$ **(b)** $\dfrac{\dfrac{a}{a+1} + \dfrac{1}{a}}{\dfrac{1}{a} + \dfrac{1}{a+1}}$

SOLUTION

(a) Multiply both numerator and denominator by the least common denominator of all the fractions, k.

$$\frac{k\left(6 - \dfrac{5}{k}\right)}{k\left(1 + \dfrac{5}{k}\right)} = \frac{6k - k\left(\dfrac{5}{k}\right)}{k + k\left(\dfrac{5}{k}\right)} = \frac{6k - 5}{k + 5}$$

(b) Multiply both numerator and denominator by the least common denominator of all the fractions, in this case $a(a + 1)$.

$$\frac{\dfrac{a}{a+1} + \dfrac{1}{a}}{\dfrac{1}{a} + \dfrac{1}{a+1}} = \frac{\left(\dfrac{a}{a+1} + \dfrac{1}{a}\right)a(a+1)}{\left(\dfrac{1}{a} + \dfrac{1}{a+1}\right)a(a+1)}$$

$$= \frac{\dfrac{a}{a+1}(a)(a+1) + \dfrac{1}{a}(a)(a+1)}{\dfrac{1}{a}(a)(a+1) + \dfrac{1}{a+1}(a)(a+1)} \qquad \text{Distributive property}$$

$$= \frac{a^2 + (a+1)}{(a+1) + a}$$

$$= \frac{a^2 + a + 1}{2a + 1}$$

As an alternative method of solution, first perform the indicated additions in the numerator and denominator, and then divide.

$$\frac{\dfrac{a}{a+1} + \dfrac{1}{a}}{\dfrac{1}{a} + \dfrac{1}{a+1}} = \frac{\dfrac{a^2 + 1(a+1)}{a(a+1)}}{\dfrac{1(a+1) + 1(a)}{a(a+1)}} \qquad \begin{array}{l}\text{Get a common denominator;}\\ \text{add terms in numerator and}\\ \text{denominator.}\end{array}$$

$$= \frac{\dfrac{a^2 + a + 1}{a(a+1)}}{\dfrac{2a + 1}{a(a+1)}} \qquad \begin{array}{l}\text{Combine terms in numerator}\\ \text{and denominator.}\end{array}$$

$$= \frac{a^2 + a + 1}{a(a+1)} \cdot \frac{a(a+1)}{2a+1} \qquad \text{Definition of division}$$

$$= \frac{a^2 + a + 1}{2a + 1} \qquad \begin{array}{l}\text{Multiply fractions and write in}\\ \text{lowest terms.}\end{array} \quad \blacklozenge$$

R.3   EXERCISES

For each of the following, give the restrictions on the variable.

1. $\dfrac{x - 2}{x + 6}$ **2.** $\dfrac{x + 5}{x - 3}$ **3.** $\dfrac{2x}{5x - 3}$ **4.** $\dfrac{6x}{2x - 1}$ **5.** $\dfrac{-8}{x^2 + 1}$ **6.** $\dfrac{3x}{3x^2 + 7}$

7. Which one of the following expressions is equivalent to $\dfrac{x^2 + 4x + 3}{x + 1}$?

  (a) $x + 3$ **(b)** $x + 7$ **(c)** $5x + 3$ **(d)** $x^2 + 7$

8. Explain why $\dfrac{x - 1}{2 - x}$ is equivalent to $\dfrac{1 - x}{x - 2}$.

Write each of the following in lowest terms.

9. $\dfrac{25p^3}{10p^2}$ **10.** $\dfrac{14z^3}{6z^2}$ **11.** $\dfrac{8k + 16}{9k + 18}$ **12.** $\dfrac{20r + 10}{30r + 15}$

13. $\dfrac{3(t + 5)}{(t + 5)(t - 3)}$ **14.** $\dfrac{-8(y - 4)}{(y + 2)(y - 4)}$ **15.** $\dfrac{8x^2 + 16x}{4x^2}$ **16.** $\dfrac{36y^2 + 72y}{9y}$

17. $\dfrac{m^2 - 4m + 4}{m^2 + m - 6}$ **18.** $\dfrac{r^2 - r - 6}{r^2 + r - 12}$ **19.** $\dfrac{8m^2 + 6m - 9}{16m^2 - 9}$ **20.** $\dfrac{6y^2 + 11y + 4}{3y^2 + 7y + 4}$

Find each of the following products or quotients.

21. $\dfrac{15p^3}{9p^2} \div \dfrac{6p}{10p^2}$ **22.** $\dfrac{3r^2}{9r^3} \div \dfrac{8r^3}{6r}$

23. $\dfrac{2k + 8}{6} \div \dfrac{3k + 12}{2}$ **24.** $\dfrac{5m + 25}{10} \cdot \dfrac{12}{6m + 30}$

25. $\dfrac{x^2 + x}{5} \cdot \dfrac{25}{xy + y}$ **26.** $\dfrac{3m - 15}{4m - 20} \cdot \dfrac{m^2 - 10m + 25}{12m - 60}$

27. $\dfrac{4a + 12}{2a - 10} \div \dfrac{a^2 - 9}{a^2 - a - 20}$ **28.** $\dfrac{6r - 18}{9r^2 + 6r - 24} \cdot \dfrac{12r - 16}{4r - 12}$

29. $\dfrac{p^2 - p - 12}{p^2 - 2p - 15} \cdot \dfrac{p^2 - 9p + 20}{p^2 - 8p + 16}$ **30.** $\dfrac{x^2 + 2x - 15}{x^2 + 11x + 30} \cdot \dfrac{x^2 + 2x - 24}{x^2 - 8x + 15}$

31. $\dfrac{m^2 + 3m + 2}{m^2 + 5m + 4} \div \dfrac{m^2 + 5m + 6}{m^2 + 10m + 24}$ **32.** $\dfrac{y^2 + y - 2}{y^2 + 3y - 4} \div \dfrac{y^2 + 3y + 2}{y^2 + 4y + 3}$

33. $\dfrac{2m^2 - 5m - 12}{m^2 - 10m + 24} \div \dfrac{4m^2 - 9}{m^2 - 9m + 18}$ **34.** $\dfrac{6n^2 - 5n - 6}{6n^2 + 5n - 6} \cdot \dfrac{12n^2 - 17n + 6}{12n^2 - n - 6}$

35. $\left(1 + \dfrac{1}{x}\right)\left(1 - \dfrac{1}{x}\right)$ **36.** $\left(3 + \dfrac{2}{y}\right)\left(3 - \dfrac{2}{y}\right)$

37. $\dfrac{x^3 + y^3}{x^2 - y^2} \cdot \dfrac{x + y}{x^2 - xy + y^2}$ **38.** $\dfrac{8y^3 - 125}{4y^2 - 20y + 25} \cdot \dfrac{2y - 5}{y}$

39. $\dfrac{x^3 + y^3}{x^3 - y^3} \cdot \dfrac{x^2 - y^2}{x^2 + 2xy + y^2}$ **40.** $\dfrac{x^2 - y^2}{(x - y)^2} \cdot \dfrac{x^2 - xy + y^2}{x^2 - 2xy + y^2} \div \dfrac{x^3 + y^3}{(x - y)^4}$

41. Which of the following rational expressions equals -1?

  (a) $\dfrac{x - 4}{x + 4}$ **(b)** $\dfrac{-x - 4}{x + 4}$ **(c)** $\dfrac{x - 4}{4 - x}$ **(d)** $\dfrac{x - 4}{-x - 4}$

42. In your own words, explain how to find the least common denominator for two fractions.

Perform each of the following additions or subtractions.

43. $\dfrac{3}{2k} + \dfrac{5}{3k}$

44. $\dfrac{8}{5p} + \dfrac{3}{4p}$

45. $\dfrac{a+1}{2} - \dfrac{a-1}{2}$

46. $\dfrac{y+6}{5} - \dfrac{y-6}{5}$

47. $\dfrac{3}{p} + \dfrac{1}{2}$

48. $\dfrac{9}{r} - \dfrac{2}{3}$

49. $\dfrac{1}{6m} + \dfrac{2}{5m} + \dfrac{4}{m}$

50. $\dfrac{8}{3p} + \dfrac{5}{4p} + \dfrac{9}{2p}$

51. $\dfrac{1}{a+1} - \dfrac{1}{a-1}$

52. $\dfrac{1}{x+z} + \dfrac{1}{x-z}$

53. $\dfrac{m+1}{m-1} + \dfrac{m-1}{m+1}$

54. $\dfrac{2}{x-1} + \dfrac{1}{1-x}$

55. $\dfrac{3}{a-2} - \dfrac{1}{2-a}$

56. $\dfrac{q}{p-q} - \dfrac{q}{q-p}$

57. $\dfrac{x+y}{2x-y} - \dfrac{2x}{y-2x}$

58. $\dfrac{m-4}{3m-4} + \dfrac{3m+2}{4-3m}$

59. $\dfrac{1}{a^2-5a+6} - \dfrac{1}{a^2-4}$

60. $\dfrac{-3}{m^2-m-2} - \dfrac{1}{m^2+3m+2}$

61. $\dfrac{1}{x^2+x-12} - \dfrac{1}{x^2-7x+12} + \dfrac{1}{x^2-16}$

62. $\dfrac{2}{2p^2-9p-5} + \dfrac{p}{3p^2-17p+10} - \dfrac{2p}{6p^2-p-2}$

63. $\dfrac{3a}{a^2+5a-6} - \dfrac{2a}{a^2+7a+6}$

64. $\dfrac{2k}{k^2+4k+3} + \dfrac{3k}{k^2+5k+6}$

Perform the indicated operations.

65. $\dfrac{1+\frac{1}{x}}{1-\frac{1}{x}}$

66. $\dfrac{2-\frac{2}{y}}{2+\frac{2}{y}}$

67. $\dfrac{\frac{1}{x+1}-\frac{1}{x}}{\frac{1}{x}}$

68. $\dfrac{\frac{1}{y+3}-\frac{1}{y}}{\frac{1}{y}}$

69. $\dfrac{1+\frac{1}{1-b}}{1-\frac{1}{1+b}}$

70. $m - \dfrac{m}{m+\frac{1}{2}}$

71. $\dfrac{m-\frac{1}{m^2-4}}{\frac{1}{m+2}}$

72. $\dfrac{\frac{3}{p^2-16}+p}{\frac{1}{p-4}}$

73. $\left(\dfrac{3}{p-1} - \dfrac{2}{p+1}\right)\left(\dfrac{p-1}{p}\right)$

74. $\left(\dfrac{y}{y^2-1} - \dfrac{y}{y^2-2y+1}\right)\left(\dfrac{y-1}{y+1}\right)$

75. $\dfrac{\frac{1}{x+h}-\frac{1}{x}}{h}$

76. $\dfrac{1}{h}\left(\dfrac{1}{(x+h)^2+9} - \dfrac{1}{x^2+9}\right)$

R.4 REVIEW OF NEGATIVE AND RATIONAL EXPONENTS

Negative Exponents ◆ The Quotient Rule for Exponents ◆ Rational Exponents ◆ Complex Fractions That Involve Negative Exponents

NEGATIVE EXPONENTS

Earlier we introduced some rules for exponents: the product rule and the power rules. We now complete our review of exponential expressions, beginning with a rule for division.

In the product rule $a^m \cdot a^n = a^{m+n}$, the exponents are *added*. If $a \neq 0$,

$$\frac{a^3}{a^7} = \frac{a \cdot a \cdot a}{a \cdot a \cdot a \cdot a \cdot a \cdot a \cdot a} = \frac{1}{a \cdot a \cdot a \cdot a} = \frac{1}{a^4}.$$

This suggests that we should *subtract* exponents when dividing. Subtracting exponents gives

$$\frac{a^3}{a^7} = a^{3-7} = a^{-4}.$$

The only way to keep these results consistent is to define a^{-4} as $\frac{1}{a^4}$. This example suggests the following definition.

DEFINITION OF NEGATIVE EXPONENTS

If a is a nonzero real number and n is any integer, then

$$a^{-n} = \frac{1}{a^n}.$$

EXAMPLE 1   *Using the Definition of a Negative Exponent*   Evaluate each expression in parts (a)–(c). In parts (d) and (e), write the expression without negative exponents.

(a) 4^{-2} **(b)** $\left(\frac{2}{5}\right)^{-3}$ **(c)** -4^{-2} **(d)** x^{-4} **(e)** xy^{-3}

SOLUTION

(a) $4^{-2} = \frac{1}{4^2} = \frac{1}{16}$

(b) $\left(\frac{2}{5}\right)^{-3} = \frac{1}{\left(\frac{2}{5}\right)^3} = \frac{1}{\frac{8}{125}} = \frac{125}{8}$

(c) $-4^{-2} = -\frac{1}{4^2} = -\frac{1}{16}$

(d) $x^{-4} = \frac{1}{x^4}$ $(x \neq 0)$

(e) $xy^{-3} = x \cdot \frac{1}{y^3} = \frac{x}{y^3}$ $(y \neq 0)$ ◆

Part (b) of Example 1 showed that
$$\left(\frac{2}{5}\right)^{-3} = \frac{125}{8} = \left(\frac{5}{2}\right)^3.$$

This result can be generalized. If $a \neq 0$ and $b \neq 0$, then for any integer n,
$$\left(\frac{a}{b}\right)^{-n} = \left(\frac{b}{a}\right)^n.$$

THE QUOTIENT RULE FOR EXPONENTS

The quotient rule for exponents follows from the definition of exponents, as shown above.

QUOTIENT RULE

For all integers m and n and all nonzero real numbers a,

$$\frac{a^m}{a^n} = a^{m-n}.$$

By the quotient rule, if $a \neq 0$,

$$\frac{a^m}{a^m} = a^{m-m} = a^0.$$

On the other hand, any nonzero quantity divided by itself equals 1. This is why we defined $a^0 = 1$ earlier.

EXAMPLE 2 *Using the Quotient Rule* Use the quotient rule to simplify each expression. Assume that all variables represent nonzero real numbers.

(a) $\dfrac{12^5}{12^2}$ **(b)** $\dfrac{a^5}{a^{-8}}$ **(c)** $\dfrac{16m^{-9}}{12m^{11}}$ **(d)** $\dfrac{25r^7z^5}{10r^9z}$

(e) $\dfrac{x^{5y}}{x^{3y}}$, if y is an integer

SOLUTION

(a) $\dfrac{12^5}{12^2} = 12^{5-2} = 12^3$ **(b)** $\dfrac{a^5}{a^{-8}} = a^{5-(-8)} = a^{13}$

(c) $\dfrac{16m^{-9}}{12m^{11}} = \dfrac{16}{12} \cdot m^{-9-11} = \dfrac{4}{3}m^{-20} = \dfrac{4}{3} \cdot \dfrac{1}{m^{20}} = \dfrac{4}{3m^{20}}$

(d) $\dfrac{25r^7z^5}{10r^9z} = \dfrac{25}{10} \cdot \dfrac{r^7}{r^9} \cdot \dfrac{z^5}{z^1} = \dfrac{5}{2}r^{-2}z^4 = \dfrac{5z^4}{2r^2}$

(e) $\dfrac{x^{5y}}{x^{3y}} = x^{5y-3y} = x^{2y}$, if y is an integer ◆

The rules for exponents stated earlier in this reference also apply to negative exponents.

EXAMPLE 3 *Using the Rules for Exponents* Use the rules for exponents to simplify each expression. Write answers without negative exponents. Assume that all variables represent nonzero real numbers.

(a) $3x^{-2}(4^{-1}x^{-5})^2$ **(b)** $\dfrac{5m^{-3}}{10m^{-5}}$ **(c)** $\dfrac{12p^3q^{-1}}{8p^{-2}q}$

SOLUTION

(a) $3x^{-2}(4^{-1}x^{-5})^2 = 3x^{-2}(4^{-2}x^{-10})$ Power rule

$\qquad = 3 \cdot 4^{-2} \cdot x^{-2+(-10)}$ Rearrange factors; use the product rule.

$\qquad = 3 \cdot 4^{-2} \cdot x^{-12}$

$\qquad = \dfrac{3}{16x^{12}}$ Write with positive exponents.

(b) $\dfrac{5m^{-3}}{10m^{-5}} = \dfrac{5}{10}m^{-3-(-5)}$ Quotient rule

$\qquad = \dfrac{1}{2}m^2$ or $\dfrac{m^2}{2}$

(c) $\dfrac{12p^3q^{-1}}{8p^{-2}q} = \dfrac{12}{8} \cdot \dfrac{p^3}{p^{-2}} \cdot \dfrac{q^{-1}}{q^1}$

$= \dfrac{3}{2} \cdot p^{3-(-2)}q^{-1-1}$ Quotient rule

$= \dfrac{3}{2}\, p^5q^{-2}$

$= \dfrac{3p^5}{2q^2}$ Write with positive exponents.   ◆

RATIONAL EXPONENTS

The definition of a^n can be extended to rational values of n by defining $a^{1/n}$ to be the nth root of a. By one of the power rules of exponents (extended to a rational exponent),

$$(a^{1/n})^n = a^{(1/n)n} = a^1 = a,$$

suggesting that $a^{1/n}$ is a number whose nth power is a.

DEFINITION OF $a^{1/n}$, n EVEN $a^{1/n}$, n ODD

  i. If n is an *even* positive integer, and if $a > 0$, then $a^{1/n}$ is the positive real number whose nth power is a. That is, $(a^{1/n})^n = a$. In this case, $a^{1/n}$ is the principal nth root of a.
  ii. If n is an *odd* positive integer, and a *is any real number*, then $a^{1/n}$ is the positive or negative real number whose nth power is a. That is, $(a^{1/n})^n = a$.

EXAMPLE 4   *Using the Definition of $a^{1/n}$*   Evaluate each expression.

(a) $36^{1/2}$ **(b)** $-100^{1/2}$ **(c)** $-(225)^{1/2}$ **(d)** $625^{1/4}$ **(e)** $(-1296)^{1/4}$

(f) $(-27)^{1/3}$ **(g)** $-32^{1/5}$

SOLUTION

(a) $36^{1/2} = 6$ because $6^2 = 36$ **(b)** $-100^{1/2} = -10$

(c) $-(225)^{1/2} = -15$ **(d)** $625^{1/4} = 5$

(e) $(-1296)^{1/4}$ is not defined, but $-1296^{1/4} = -6$

(f) $(-27)^{1/3} = -3$

(g) $-32^{1/5} = -2$ ◆

We now discuss more general rational exponents. The notation $a^{m/n}$ should be defined so that all the past rules for exponents still hold. For the power rule to hold, $(a^{1/n})^m$ must equal $a^{m/n}$. Therefore, $a^{m/n}$ is defined as follows.

DEFINITION OF RATIONAL EXPONENTS

For all integers m, all positive integers n, and all real numbers a for which $a^{1/n}$ is defined,

$$a^{m/n} = (a^{1/n})^m.$$

EXAMPLE 5 *Using the Definition of $a^{m/n}$* Evaluate each expression.

(a) $125^{2/3}$ **(b)** $32^{7/5}$ **(c)** $-81^{3/2}$ **(d)** $(-4)^{5/2}$

(e) $(-27)^{2/3}$ **(f)** $16^{-3/4}$

SOLUTION

(a) $125^{2/3} = (125^{1/3})^2 = 5^2 = 25$

(b) $32^{7/5} = (32^{1/5})^7 = 2^7 = 128$

(c) $-81^{3/2} = -(81^{1/2})^3 = -9^3 = -729$

(d) $(-4)^{5/2}$ is not defined because $(-4)^{1/2}$ is not defined.

(e) $(-27)^{2/3} = [(-27)^{1/3}]^2 = (-3)^2 = 9$

(f) $16^{-3/4} = \dfrac{1}{16^{3/4}} = \dfrac{1}{(16^{1/4})^3} = \dfrac{1}{2^3} = \dfrac{1}{8}$ ◆

By starting with $(a^{1/n})^m$ and $(a^m)^{1/n}$, and raising each expression to the nth power, it can be shown that $(a^{1/n})^m$ is equal to $(a^m)^{1/n}$. This means that $a^{m/n}$ could be defined in either of the following ways.

For all real numbers a, integers m, and positive integers n for which $a^{1/n}$ is defined,

$$a^{m/n} = (a^{1/n})^m \qquad \text{or} \qquad a^{m/n} = (a^m)^{1/n}.$$

Now, $a^{m/n}$ can be evaluated in either of two ways: as $(a^{1/n})^m$ or as $(a^m)^{1/n}$. It is usually easier to find $(a^{1/n})^m$. For example, $27^{4/3}$ can be evaluated in either of two ways:

$$27^{4/3} = (27^{1/3})^4 = 3^4 = 81$$

$$27^{4/3} = (27^4)^{1/3} = 531,441^{1/3} = 81.$$

The form $(27^{1/3})^4$ is easier to evaluate.

It can be shown that all the earlier results concerning integer exponents also apply to rational exponents. These definitions and rules are summarized here.

RULES FOR EXPONENTS

Let r and s be rational numbers. The results below are valid for all positive numbers a and b.

$$a^r \cdot a^s = a^{r+s} \qquad (ab)^r = a^r \cdot b^r \qquad (a^r)^s = a^{rs}$$

$$\frac{a^r}{a^s} = a^{r-s} \qquad \left(\frac{a}{b}\right)^r = \frac{a^r}{b^r} \qquad a^{-r} = \frac{1}{a^r}$$

EXAMPLE 6 *Using the Definitions and Rules for Exponents* Use the definitions and rules for exponents to simplify each expression.

(a) $\dfrac{27^{1/3} \cdot 27^{5/3}}{27^3}$

(b) $81^{5/4} \cdot 4^{-3/2}$

(c) $6y^{2/3} \cdot 2y^{1/2}$, where $y \geq 0$

(d) $\left(\dfrac{3m^{5/6}}{y^{3/4}}\right)^2 \cdot \left(\dfrac{8y^3}{m^6}\right)^{2/3}$, where $m > 0,\ y > 0$

(e) $m^{2/3}(m^{7/3} + 2m^{1/3})$

SOLUTION

(a) $\dfrac{27^{1/3} \cdot 27^{5/3}}{27^3} = \dfrac{27^{1/3+5/3}}{27^3}$ Product rule

$$= \dfrac{27^2}{27^3} = 27^{2-3}$$ Quotient rule

$$= 27^{-1} = \dfrac{1}{27}$$

(b) $81^{5/4} \cdot 4^{-3/2} = (81^{1/4})^5 (4^{1/2})^{-3} = 3^5 \cdot 2^{-3} = \dfrac{3^5}{2^3}$ or $\dfrac{243}{8}$

(c) $6y^{2/3} \cdot 2y^{1/2} = 12y^{2/3+1/2} = 12y^{7/6}$,   where $y \geq 0$

(d) $\left(\dfrac{3m^{5/6}}{y^{3/4}}\right)^2 \cdot \left(\dfrac{8y^3}{m^6}\right)^{2/3} = \dfrac{9m^{5/3}}{y^{3/2}} \cdot \dfrac{4y^2}{m^4} = 36m^{5/3-4}y^{2-3/2}$

$$= \dfrac{36y^{1/2}}{m^{7/3}}   (m > 0, y > 0)$$

(e) $m^{2/3}(m^{7/3} + 2m^{1/3}) = (m^{2/3+7/3} + 2m^{2/3+1/3}) = m^3 + 2m$ ◆

The next example shows how to factor when negative or rational exponents are involved.

EXAMPLE 7   *Factoring an Expression with Negative or Rational Exponents*
Factor out the smallest power of the variable. Assume that all variables represent positive real numbers.

(a) $9x^{-2} - 6x^{-3}$ **(b)** $4m^{1/2} + 3m^{3/2}$ **(c)** $y^{-1/3} + y^{2/3}$

SOLUTION

(a)   The smallest exponent here is -3. Since 3 is a common numerical factor, factor out $3x^{-3}$.

$$9x^{-2} - 6x^{-3} = 3x^{-3}(3x - 2)$$

Check by multiplying on the right. The factored form can now be written without negative exponents as $\dfrac{3(3x - 2)}{x^3}$.

(b) $4m^{1/2} + 3m^{3/2} = m^{1/2}(4 + 3m)$. To check this result, multiply $m^{1/2}$ by $4 + 3m$.

(c) $y^{-1/3} + y^{2/3} = y^{-1/3}(1 + y)$. The factored form can be written with only positive exponents as $\dfrac{1 + y}{y^{1/3}}$.   ◆

COMPLEX FRACTIONS THAT INVOLVE NEGATIVE EXPONENTS

Negative exponents are sometimes used to write complex fractions. Recall that complex fractions are simplified either by first multiplying the numerator and denominator by the greatest common multiple of all the denominators, or by performing any indicated operations in the numerator and the denominator and then using the definition of division for fractions.

EXAMPLE 8 *Simplifying a Fraction with Negative Exponents* Simplify $\dfrac{(x+y)^{-1}}{x^{-1}+y^{-1}}$. Write the result with only positive exponents.

SOLUTION   Begin by using the definition of a negative integer exponent. Then, perform the indicated operations.

$$\frac{(x+y)^{-1}}{x^{-1}+y^{-1}} = \frac{\dfrac{1}{x+y}}{\dfrac{1}{x}+\dfrac{1}{y}}$$

$$= \frac{\dfrac{1}{x+y}}{\dfrac{y+x}{xy}}$$

$$= \frac{1}{x+y} \cdot \frac{xy}{x+y}$$

$$= \frac{xy}{(x+y)^2} \qquad \blacklozenge$$

R.4 EXERCISES

Simplify each of the following. Assume that all variables represent positive real numbers.

1. $(-4)^{-3}$ **2.** $(-5)^{-2}$ **3.** $\left(\dfrac{1}{2}\right)^{-3}$ **4.** $\left(\dfrac{2}{3}\right)^{-2}$ **5.** $-4^{1/2}$

6. $25^{1/2}$ **7.** $8^{2/3}$ **8.** $-81^{3/4}$ **9.** $27^{-2/3}$ **10.** $(-32)^{-4/5}$

11. $\left(-\dfrac{4}{9}\right)^{-3/2}$ **12.** $\left(\dfrac{1}{8}\right)^{-1/2}$ **13.** $\left(\dfrac{27}{64}\right)^{-4/3}$ **14.** $\left(\dfrac{121}{100}\right)^{-3/2}$ **15.** $\left(\dfrac{1}{4}\right)^{-5/2}$

16. $\left(\dfrac{1}{8}\right)^{-7/3}$ **17.** $(16p^4)^{1/2}$ **18.** $(36r^6)^{1/2}$ **19.** $(27x^6)^{2/3}$ **20.** $(64a^{12})^{5/6}$

21. Why is $a^{1/n}$ defined to be the nth root of a (with appropriate restrictions)?

22. Why is 0^{-3} not defined?

23. Which of the following expressions is equivalent to $(2x^{-3/2})^2$?

(a) $2x^{-3}$ (b) 2^{-3} (c) $2^2x^{-3/4}$ (d) $\dfrac{2^2}{x^3}$

24. Explain how you would evaluate $(-27)^{5/3}$.

Perform the indicated operations. Write your answers with only positive exponents. Assume that all variables represent positive real numbers.

25. $2^{-3} \cdot 2^{-4}$ **26.** $5^{-2} \cdot 5^{-6}$ **27.** $27^{-2} \cdot 27^{-1}$ **28.** $9^{-4} \cdot 9^{-1}$

29. $\dfrac{4^{-2} \cdot 4^{-1}}{4^{-3}}$ **30.** $\dfrac{3^{-1} \cdot 3^{-4}}{3^2 \cdot 3^{-2}}$ **31.** $(m^{2/3})(m^{5/3})$ **32.** $(x^{4/5})(x^{2/5})$

33. $(1 + n)^{1/2}(1 + n)^{3/4}$ **34.** $(m + 7)^{-1/6}(m + 7)^{-2/3}$ **35.** $(2y^{3/4}z)(3y^{-2}z^{-1/3})$ **36.** $(4a^{-1}b^{2/3})(a^{3/2}b^{-3})$

37. $(4a^{-2}b^7)^{1/2} \cdot (2a^{1/4}b^3)^5$ **38.** $(x^{-2}y^{1/3})^5 \cdot (8x^2y^{-2})^{-1/3}$ **39.** $\left(\dfrac{r^{-2}}{s^{-5}}\right)^{-3}$ **40.** $\left(\dfrac{p^{-1}}{q^{-5}}\right)^{-2}$

41. $\left(\dfrac{-a}{b^{-3}}\right)^{-1}$ **42.** $\dfrac{7^{-1/3}7r^{-3}}{7^{2/3}r^{-2}}$ **43.** $\dfrac{12^{5/4}y^{-2}}{12^{-1}y^{-3}}$ **44.** $\dfrac{6k^{-4}(3k^{-1})^{-2}}{2^3k^{1/2}}$

45. $\dfrac{8p^{-3}(4p^2)^{-2}}{p^{-5}}$ **46.** $\dfrac{k^{-3/5}h^{-1/3}t^{2/5}}{k^{-1/5}h^{-2/3}t^{1/5}}$ **47.** $\dfrac{m^{7/3}n^{-2/5}p^{3/8}}{m^{-2/3}n^{3/5}p^{-5/8}}$ **48.** $\dfrac{m^{2/5}m^{3/5}m^{-4/5}}{m^{1/5}m^{-6/5}}$

49. $\dfrac{-4a^{-1}a^{2/3}}{a^{-2}}$ **50.** $\dfrac{8y^{2/3}y^{-1}}{2^{-1}y^{3/4}y^{-1/6}}$ **51.** $\dfrac{(k + 5)^{1/2}(k + 5)^{-1/4}}{(k + 5)^{3/4}}$ **52.** $\dfrac{(x + y)^{-5/8}(x + y)^{3/8}}{(x + y)^{1/8}(x + y)^{-1/8}}$

53. $\left(\dfrac{x^4y^3z}{16x^{-6}yz^5}\right)^{-1/2}$ **54.** $\left(\dfrac{p^3r^9}{27p^{-3}r^{-6}}\right)^{-1/3}$

Perform the indicated operations. Write your answers without denominators. Assume that all variables used in denominators are not zero and all variables used as exponents represent rational numbers.

55. $(r^{3/p})^{2p}(r^{1/p})^{p^2}$ **56.** $(m^{2/x})^{x/3}(m^{x/4})^{2/x}$ **57.** $\dfrac{m^{1-a}m^a}{m^{-1/2}}$ **58.** $\dfrac{(y^{3-b})(y^{2b-1})}{y^{1/2}}$

59. $\dfrac{(x^{n/2})(x^{3n})^{1/2}}{x^{1/n}}$ **60.** $\dfrac{(a^{2/3})(a^{1/x})}{(a^{x/3})^{-2}}$ **61.** $\dfrac{(p^{1/n})(p^{1/m})}{p^{-m/n}}$ **62.** $\dfrac{(q^{2r/3})(q^r)^{-1/3}}{(q^{4/3})^{1/r}}$

Find each of the following products. Assume that all variables represent positive real numbers.

63. $y^{5/8}(y^{3/8} - 10y^{11/8})$ **64.** $p^{11/5}(3p^{4/5} + 9p^{19/5})$ **65.** $-4k(k^{7/3} - 6k^{1/3})$ **66.** $-5y(3y^{9/10} + 4y^{3/10})$

67. $(x + x^{1/2})(x - x^{1/2})$ **68.** $(2z^{1/2} + z)(z^{1/2} - z)$ **69.** $(r^{1/2} - r^{-1/2})^2$ **70.** $(p^{1/2} - p^{-1/2})(p^{1/2} + p^{-1/2})$

Factor, using the given common factor. Assume all variables represent positive real numbers.

71. $4k^{-1} + k^{-2}; \quad k^{-2}$ **72.** $y^{-5} - 3y^{-3}; \quad y^{-5}$ **73.** $9z^{-1/2} + 2z^{1/2}; \quad z^{-1/2}$

74. $3m^{2/3} - 4m^{-1/3}; \quad m^{-1/3}$ **75.** $p^{-3/4} - 2p^{-7/4}; \quad p^{-7/4}$ **76.** $6r^{-2/3} - 5r^{-5/3}; \quad r^{-5/3}$

77. $(p + 4)^{-3/2} + (p + 4)^{-1/2} + (p + 4)^{1/2}; \quad (p + 4)^{-3/2}$ **78.** $(3r + 1)^{-2/3} + (3r + 1)^{1/3} + (3r + 1)^{4/3}; \quad (3r + 1)^{-2/3}$

R.5   REVIEW OF RADICALS

Radical Notation ◆ Rules for Radicals ◆ Simplifying Radicals ◆ Operations with Radicals

RADICAL NOTATION

In Section R.4 of this reference chapter, the notation $a^{1/n}$ was used for the nth root of a for appropriate values of a and n. An alternative (and more familiar) notation for $a^{1/n}$ uses *radical notation.*

RADICAL NOTATION FOR $a^{1/n}$

If a is a real number, n is a positive integer, and $a^{1/n}$ is defined, then

$$\sqrt[n]{a} = a^{1/n}.$$

The symbol $\sqrt[n]{}$ is a **radical sign,** the number a is the **radicand,** and n is the **index** of the radical $\sqrt[n]{a}$. It is customary to use the familiar notation $\sqrt{a}$ instead of $\sqrt[2]{a}$ for the square root.

For even values of n (square roots, fourth roots, and so on) and $a > 0$, there are two nth roots, one positive and one negative. In such cases, the notation $\sqrt[n]{a}$ represents the positive root, the **principal nth root.** The negative root is written $-\sqrt[n]{a}$.

EXAMPLE 1 *Evaluating Roots*   Evaluate each root.

(a) $\sqrt[4]{16}$ (b) $-\sqrt[4]{16}$ (c) $\sqrt[4]{-16}$

(d) $\sqrt[5]{-32}$ (e) $\sqrt[3]{1000}$ (f) $\sqrt[6]{\dfrac{64}{729}}$

SOLUTION

(a) $\sqrt[4]{16} = 16^{1/4} = (2^4)^{1/4} = 2$ (b) $-\sqrt[4]{16} = -16^{1/4} = -(2^4)^{1/4} = -2$

(c) $\sqrt[4]{-16}$ is not defined.   (d) $\sqrt[5]{-32} = [(-2)^5]^{1/5} = -2$

(e) $\sqrt[3]{1000} = 10$ (f) $\sqrt[6]{\dfrac{64}{729}} = \left(\dfrac{2^6}{3^6}\right)^{1/6} = \dfrac{2}{3}$ ◆

With $a^{1/n}$ written as $\sqrt[n]{a}$, $a^{m/n}$ also can be written with radicals.

RADICAL NOTATION FOR $a^{m/n}$

If a is a real number, m is an integer, n is a positive integer, and $\sqrt[n]{a}$ is defined, then

$$a^{m/n} = \left(\sqrt[n]{a}\right)^m = \sqrt[n]{a^m}.$$

EXAMPLE 2 *Converting from Rational Exponents to Radicals*   Write in radical form and simplify.

(a) $8^{2/3}$ (b) $(-32)^{4/5}$ (c) $-16^{3/4}$ (d) $x^{5/6}, \quad x \ge 0$ (e) $3x^{2/3}$

(f) $2p^{-1/2}, \quad p > 0$ (g) $(3a + b)^{1/4}, \quad 3a + b \ge 0$

SOLUTION

(a) $8^{2/3} = \left(\sqrt[3]{8}\right)^2 = 2^2 = 4$

(b) $(-32)^{4/5} = \left(\sqrt[5]{-32}\right)^4 = (-2)^4 = 16$

(c) $-16^{3/4} = -\left(\sqrt[4]{16}\right)^3 = -(2)^3 = -8$

(d) $x^{5/6} = \sqrt[6]{x^5}, \quad x \ge 0$

(e) $3x^{2/3} = 3\sqrt[3]{x^2}$

(f) $2p^{-1/2} = \dfrac{2}{p^{1/2}} = \dfrac{2}{\sqrt{p}}, \quad p > 0$

(g) $(3a + b)^{1/4} = \sqrt[4]{3a + b}, \quad 3a + b \ge 0$ ◆

EXAMPLE 3 *Converting from Radicals to Rational Exponents*   Write in exponential form.

(a) $\sqrt[4]{x^5}, \quad x \ge 0$ (b) $\sqrt{3y}, \quad y \ge 0$ (c) $10\left(\sqrt[5]{z}\right)^2$ (d) $5\sqrt[3]{(2x^4)^7}$

(e) $\sqrt{p^2 + q}, \quad p^2 + q \ge 0$

SOLUTION

(a) $\sqrt[4]{x^5} = x^{5/4}, \quad x \ge 0$ (b) $\sqrt{3y} = (3y)^{1/2}, \quad y \ge 0$

(c) $10\left(\sqrt[5]{z}\right)^2 = 10z^{2/5}$ (d) $5\sqrt[3]{(2x^4)^7} = 5(2x^4)^{7/3} = 5 \cdot 2^{7/3}x^{28/3}$

(e) $\sqrt{p^2 + q} = (p^2 + q)^{1/2}$, $p^2 + q \geq 0$ ◆

By the definition of $\sqrt[n]{a}$, for any positive integer n, if $\sqrt[n]{a}$ is defined, then

$$\left(\sqrt[n]{a}\right)^n = a.$$

If a is positive, or if a is negative and n is an odd positive integer,

$$\sqrt[n]{a^n} = a.$$

Because of the conditions just given, we *cannot* simply write $\sqrt{x^2} = x$. For example, if $x = -5$,

$$\sqrt{x^2} = \sqrt{(-5)^2} = \sqrt{25} = 5 \neq x.$$

To take care of the fact that a negative value of x can produce a positive result, we use absolute value. For any real number a,

$$\sqrt{a^2} = |a|.$$

For example,

$$\sqrt{(-9)^2} = |-9| = 9, \quad \text{and} \quad \sqrt{13^2} = |13| = 13.$$

This result can be generalized to any even nth root.

$\sqrt[n]{a^n}$

If n is an even positive integer, $\sqrt[n]{a^n} = |a|$, and if n is an odd positive integer, $\sqrt[n]{a^n} = a$.

EXAMPLE 4   *Using Absolute Value to Simplify Roots*   Use absolute value as applicable to simplify the following expressions.

(a) $\sqrt{p^4}$ **(b)** $\sqrt[4]{p^4}$ **(c)** $\sqrt{16m^8 r^6}$ **(d)** $\sqrt[6]{(-2)^6}$ **(e)** $\sqrt[5]{m^5}$
(f) $\sqrt{(2k+3)^2}$ **(g)** $\sqrt{x^2 - 4x + 4}$

SOLUTION

(a) $\sqrt{p^4} = |p^2| = p^2$ **(b)** $\sqrt[4]{p^4} = |p|$ **(c)** $\sqrt{16m^8 r^6} = |4m^4 r^3| = 4m^4 |r^3|$
(d) $\sqrt[6]{(-2)^6} = |-2| = 2$ **(e)** $\sqrt[5]{m^5} = m$ **(f)** $\sqrt{(2k+3)^2} = |2k+3|$
(g) $\sqrt{x^2 - 4x + 4} = \sqrt{(x-2)^2} = |x - 2|$ ◆

For the remainder of this reference, to avoid difficulties when working with variable radicands, we usually will assume that all variables in radicands represent only nonnegative real numbers.

RULES FOR RADICALS

Three key rules for working with radicals are given below. These rules are just the power rules for exponents written in radical notation.

RULES FOR RADICALS

For all real numbers a and b, and positive integers m and n for which the indicated roots are defined,

$$\sqrt[n]{a} \cdot \sqrt[n]{b} = \sqrt[n]{ab} \qquad \sqrt[n]{\dfrac{a}{b}} = \dfrac{\sqrt[n]{a}}{\sqrt[n]{b}} \quad (b \neq 0) \qquad \sqrt[m]{\sqrt[n]{a}} = \sqrt[mn]{a}.$$

EXAMPLE 5 *Using the Rules for Radicals to Simplify Radical Expressions*
Apply the rules for radicals.

(a) $\sqrt{6} \cdot \sqrt{54}$ **(b)** $\sqrt[3]{m} \cdot \sqrt[3]{m^2}$ **(c)** $\sqrt{\dfrac{7}{64}}$ **(d)** $\sqrt[4]{\dfrac{a}{b^4}}$

(e) $\sqrt[7]{\sqrt[3]{2}}$ **(f)** $\sqrt[4]{\sqrt{3}}$

SOLUTION
(a) $\sqrt{6} \cdot \sqrt{54} = \sqrt{6 \cdot 54} = \sqrt{324} = 18$
(b) $\sqrt[3]{m} \cdot \sqrt[3]{m^2} = \sqrt[3]{m^3} = m$
(c) $\sqrt{\dfrac{7}{64}} = \dfrac{\sqrt{7}}{\sqrt{64}} = \dfrac{\sqrt{7}}{8}$
(d) $\sqrt[4]{\dfrac{a}{b^4}} = \dfrac{\sqrt[4]{a}}{\sqrt[4]{b^4}} = \dfrac{\sqrt[4]{a}}{b}, \quad a \geq 0,\ b > 0$
(e) $\sqrt[7]{\sqrt[3]{2}} = \sqrt[21]{2}$ Use the third rule for radicals.
(f) $\sqrt[4]{\sqrt{3}} = \sqrt[8]{3}$ ◆

SIMPLIFYING RADICALS

In working with fractions, it is customary to write a fraction in its simplest form. For example, $\frac{10}{2}$ is written as 5, $-\frac{9}{6}$ is written as $-\frac{3}{2}$, and $\frac{4}{16}$ is written as $\frac{1}{4}$. Similarly, expressions with radicals are often written in their simplest forms.

> ### SIMPLIFIED RADICALS
>
> An expression with radicals is simplified when all of the following conditions are satisfied.
>
> 1. The radicand has no factor raised to a power greater than or equal to the index.
> 2. The radicand has no fractions.
> 3. No denominator contains a radical.
> 4. Exponents in the radicand and the index of the radical have no common factor.
> 5. All indicated operations have been performed (if possible).

EXAMPLE 6 *Simplifying Radicals*   Simplify each of the following. Assume that all variables represent nonnegative real numbers.

(a) $\sqrt{175}$ **(b)** $-3\sqrt[5]{32}$ **(c)** $\sqrt{288m^5}$

(d) $\sqrt[3]{81x^5y^7z^6}$ **(e)** $\dfrac{4}{\sqrt{3}}$ **(f)** $\dfrac{2}{\sqrt[3]{x}}, x \neq 0$

SOLUTION
(a) $\sqrt{175} = \sqrt{25 \cdot 7} = \sqrt{25} \cdot \sqrt{7} = 5\sqrt{7}$
(b) $-3\sqrt[5]{32} = -3\sqrt[5]{2^5} = -3 \cdot 2 = -6$
(c) $\sqrt{288m^5} = \sqrt{144m^4 \cdot 2m}$
$= 12m^2\sqrt{2m}$

(d) $\sqrt[3]{81x^5y^7z^6} = \sqrt[3]{27 \cdot 3 \cdot x^3 \cdot x^2 \cdot y^6 \cdot y \cdot z^6}$ Factor.

$\qquad\qquad\quad = \sqrt[3]{27x^3y^6z^6(3x^2y)}$ Group all perfect cubes.

$\qquad\qquad\quad = 3xy^2z^2\sqrt[3]{3x^2y}$ Remove all pefect cubes from the radical.

(e) $\dfrac{4}{\sqrt{3}} = \dfrac{4}{\sqrt{3}} \cdot \dfrac{\sqrt{3}}{\sqrt{3}} = \dfrac{4\sqrt{3}}{3}$ To eliminate the radical denominator, we multiply by $\dfrac{\sqrt{3}}{\sqrt{3}} = 1$.

(f) $\dfrac{2}{\sqrt[3]{x}} = \dfrac{2}{\sqrt[3]{x}} \cdot \dfrac{\sqrt[3]{x^2}}{\sqrt[3]{x^2}}$ $\dfrac{\sqrt[3]{x^2}}{\sqrt[3]{x^2}} = 1$ and $\sqrt[3]{x} \cdot \sqrt[3]{x^2} = \sqrt[3]{x^3} = x$

$\qquad = \dfrac{2\sqrt[3]{x^2}}{\sqrt[3]{x^3}}$

$\qquad = \dfrac{2\sqrt[3]{x^2}}{x}$ ◆

If the index of the radical and an exponent in the radicand have a common factor, the radical can be simplified by writing it in exponential form, simplifying the rational exponent, and then writing the result as a radical again.

EXAMPLE 7   *Simplifying Radicals by Writing Them with Rational Exponents*
Simplify the following radicals by first rewriting with rational exponents.

(a) $\sqrt[6]{3^2}$ **(b)** $\sqrt[6]{x^{12}y^3}, \; y \geq 0$ **(c)** $\sqrt[9]{\sqrt{6^3}}$

SOLUTION

(a) $\sqrt[6]{3^2} = 3^{2/6} = 3^{1/3} = \sqrt[3]{3}$

(b) $\sqrt[6]{x^{12}y^3} = (x^{12}y^3)^{1/6} = x^2y^{3/6} = x^2y^{1/2} = x^2\sqrt{y}, \; y \geq 0$

(c) $\sqrt[9]{\sqrt{6^3}} = \sqrt[9]{6^{3/2}} = (6^{3/2})^{1/9} = 6^{1/6} = \sqrt[6]{6}$ ◆

In Example 7(a), we simplified $\sqrt[6]{3^2}$ as $\sqrt[3]{3}$. However, to simplify $\left(\sqrt[6]{x}\right)^2$, the variable x must be nonnegative. For example, suppose we write

$$(-8)^{2/6} = [(-8)^{1/6}]^2.$$

This result is not a real number, because $(-8)^{1/6}$ is not a real number. On the other hand,

$$(-8)^{1/3} = -2.$$

Here, even though $2/6 = 1/3$,

$$\left(\sqrt[6]{x}\right)^2 \neq \sqrt[3]{x}.$$

If a is nonnegative, then it is always true that $a^{m/n} = a^{mp/(np)}$. Reducing rational exponents on negative bases must be considered case by case.

OPERATIONS WITH RADICALS

Radicals with the same radicand and the same index, such as $3\sqrt[4]{11pq}$ and $-7\sqrt[4]{11pq}$, are called **like radicals.** Like radicals are added or subtracted by using the distributive property. Only like radicals can be combined. It is sometimes necessary to simplify radicals before adding or subtracting.

EXAMPLE 8   *Adding and Subtracting Like Radicals*   Add as indicated. Assume all variables represent positive real numbers.

(a) $3\sqrt[4]{11pq} + \left(-7\sqrt[4]{11pq}\right)$ **(b)** $7\sqrt{2} - 8\sqrt{18} + 4\sqrt{72}$

(c) $\sqrt{98x^3y} + 3x\sqrt{32xy}$

Quick Reference Guide for

A Graphical Approach to
College Algebra, Second Edition

A Graphical Approach to
College Algebra and Trigonometry, Second Edition

A Graphical Approach to
Precalculus, Second Edition

John Hornsby, *University of New Orleans*
Margaret L. Lial, *American River College*

*T*he final chapter of this text is a reference chapter—*Chapter R Reference: Basic Algebraic Concepts*. This chapter provides material for students wishing to review basic topics that are prerequisites for the study of college algebra, trigonometry, and precalculus. The titles of the sections of this reference chapter are as follows:

R.1   Review of Rules for Exponents and Operations with Polynomials
R.2   Review of Factoring
R.3   Review of Operations with Rational Expressions
R.4   Review of Negative and Rational Exponents
R.5   Review of Radicals

The following guide is included for students who wish to review material from the reference chapter before studying the main chapters of the text.

Text Chapter	Suggested Sections for Review
1 Rectangular Coordinates, Functions, and Analysis of Linear Functions	R.1, R.5
2 Analysis of Graphs of Functions	R.1, R.3, R.4

If your course includes topics from chapters involving trigonometry, you should have reviewed all five sections from Chapter R by the time you begin to study those chapters.

Available for purchase at your campus bookstore, from Addison Wesley Longman, or on the World Wide Web, these learning aids are designed to help you through this course.

Student's Solutions Manual
ISBN 0-321-03945-9

Graphing Calculator Manual
ISBN 0-321-03948-3

InterAct Math® Tutorial Software
Windows version ISBN 0-321-03547-X
Macintosh version ISBN 0-321-03548-8

Visit our Web site:
http://hepg.awl.com
Keyword: Hornsby

SOLUTION

(a) $3\sqrt[4]{11pq} + \left(-7\sqrt[4]{11pq}\right) = -4\sqrt[4]{11pq}$

(b) First, remove all perfect square factors from under the radical. Then, use the distributive property as follows.

$$7\sqrt{2} - 8\sqrt{18} + 4\sqrt{72}$$
$$= 7\sqrt{2} - 8\sqrt{9 \cdot 2} + 4\sqrt{36 \cdot 2}$$
$$= 7\sqrt{2} - 8 \cdot 3\sqrt{2} + 4 \cdot 6\sqrt{2}$$
$$= 7\sqrt{2} - 24\sqrt{2} + 24\sqrt{2}$$
$$= 7\sqrt{2}$$

(c) $\sqrt{98x^3y} + 3x\sqrt{32xy} = \sqrt{49 \cdot 2 \cdot x^2 \cdot x \cdot y} + 3x\sqrt{16 \cdot 2 \cdot x \cdot y}$
$$= 7x\sqrt{2xy} + (3x)(4)\sqrt{2xy}$$
$$= 7x\sqrt{2xy} + 12x\sqrt{2xy}$$
$$= 19x\sqrt{2xy} \qquad \text{Distributive property} \qquad \blacklozenge$$

Multiplying radical expressions is much like multiplying polynomials.

EXAMPLE 9 *Multiplying Radical Expressions* Multiply.

(a) $(\sqrt{2} + 3)(\sqrt{8} - 5)$ **(b)** $(\sqrt{7} - \sqrt{10})(\sqrt{7} + \sqrt{10})$

SOLUTION

(a) $(\sqrt{2} + 3)(\sqrt{8} - 5) = \sqrt{2}(\sqrt{8}) - \sqrt{2}(5) + 3\sqrt{8} - 3(5)$ FOIL
$$= \sqrt{16} - 5\sqrt{2} + 3(2\sqrt{2}) - 15 \qquad \text{Multiply.}$$
$$= 4 - 5\sqrt{2} + 6\sqrt{2} - 15$$
$$= -11 + \sqrt{2} \qquad \text{Combine terms.}$$

(b) $(\sqrt{7} - \sqrt{10})(\sqrt{7} + \sqrt{10}) = (\sqrt{7})^2 - (\sqrt{10})^2$ Product of the sum and difference of two terms
$$= 7 - 10$$
$$= -3 \qquad \blacklozenge$$

Condition 3 of the rules for simplifying radicals described above requires that no denominator contain a radical.

In Example 9(b), we saw that

$$(\sqrt{7} - \sqrt{10})(\sqrt{7} + \sqrt{10}) = -3,$$

a rational number. This suggests a way to rationalize a denominator that is a binomial in which one or both terms is a radical. The expressions $a\sqrt{m} + b\sqrt{n}$ and $a\sqrt{m} - b\sqrt{n}$ are **conjugates.**

EXAMPLE 10 *Rationalizing a Binomial Denominator* Rationalize the denominator of $\dfrac{1}{1 - \sqrt{2}}$.

SOLUTION As suggested above, the best approach here is to multiply both numerator and denominator by the conjugate of the denominator, in this case $1 + \sqrt{2}$.

$$\frac{1}{1 - \sqrt{2}} = \frac{1(1 + \sqrt{2})}{(1 - \sqrt{2})(1 + \sqrt{2})} = \frac{1 + \sqrt{2}}{1 - 2} = -1 - \sqrt{2} \qquad \blacklozenge$$

R.5   EXERCISES

Write in radical form. Assume all variables are positive real numbers.

1. $(-m)^{2/3}$ **2.** $p^{5/4}$ **3.** $5m^{4/5}$ **4.** $-2k^{1/6}$

5. $-4z^{-1/3}$ **6.** $10a^{-3/4}$ **7.** $(2m + p)^{2/3}$ **8.** $(5r + 3t)^{4/7}$

Write in exponential form. Assume all variables are nonnegative real numbers.

9. $\sqrt[5]{k^2}$ **10.** $-\sqrt[4]{z^5}$ **11.** $-\sqrt[3]{a^2}$ **12.** $2\sqrt{m^9}$

13. $-3\sqrt{5p^3}$ **14.** $m\sqrt{2y^5}$ **15.** $18\sqrt{m^2n^3p}$ **16.** $-12\sqrt[5]{x^3y^2z^4}$

 17. What is wrong with the statement $\sqrt[3]{4} \cdot \sqrt[3]{4} = 4$?

18. Which of the following expressions is *not* simplified? Give the simplified form.

     (a) $\sqrt[3]{2y}$ **(b)** $\dfrac{\sqrt{5}}{2}$ **(c)** $\sqrt[4]{m^3}$ **(d)** $\sqrt{\dfrac{3}{4}}$

19. Explain how to rationalize the denominator of $\sqrt[3]{\dfrac{3}{2}}$.

20. How can we multiply $\sqrt{2}$ and $\sqrt[3]{2}$?

Simplify each of the following. Assume that all variables represent positive real numbers.

21. $\sqrt[3]{125}$ **22.** $\sqrt[4]{81}$ **23.** $\sqrt[5]{-3125}$ **24.** $\sqrt[3]{343}$ **25.** $\sqrt{50}$

26. $\sqrt{45}$ **27.** $\sqrt[3]{81}$ **28.** $\sqrt[3]{250}$ **29.** $-\sqrt[4]{32}$ **30.** $-\sqrt[4]{243}$

31. $-\sqrt{\dfrac{9}{5}}$ **32.** $-\sqrt[3]{\dfrac{3}{2}}$ **33.** $-\sqrt[3]{\dfrac{4}{5}}$ **34.** $\sqrt[4]{\dfrac{3}{2}}$ **35.** $\sqrt[3]{16(-2)^4(2)^8}$

36. $\sqrt[3]{25(3)^4(5)^3}$ **37.** $\sqrt{8x^5z^8}$ **38.** $\sqrt{24m^6n^5}$ **39.** $\sqrt[3]{16z^5x^8y^4}$ **40.** $-\sqrt[6]{64a^{12}b^8}$

41. $\sqrt[4]{m^2n^7p^8}$ **42.** $\sqrt[4]{x^8y^7z^9}$ **43.** $\sqrt[4]{x^4 + y^4}$ **44.** $\sqrt[3]{27 + a^3}$ **45.** $\sqrt{\dfrac{2}{3x}}$

46. $\sqrt{\dfrac{5}{3p}}$ **47.** $\sqrt{\dfrac{x^5y^3}{z^2}}$ **48.** $\sqrt{\dfrac{g^3h^5}{r^3}}$ **49.** $\sqrt[3]{\dfrac{8}{x^2}}$ **50.** $\sqrt[3]{\dfrac{9}{16p^4}}$

51. $\sqrt[4]{\dfrac{g^3h^5}{9r^6}}$ **52.** $\sqrt[4]{\dfrac{32x^5}{y^5}}$ **53.** $\dfrac{\sqrt[3]{mn} \cdot \sqrt[3]{m^2}}{\sqrt[3]{n^2}}$ **54.** $\dfrac{\sqrt[3]{8m^2n^3} \cdot \sqrt[3]{2m^2}}{\sqrt[3]{32m^4n^3}}$ **55.** $\dfrac{\sqrt[4]{32x^5y} \cdot \sqrt[4]{2xy^4}}{\sqrt[4]{4x^3y^2}}$

56. $\dfrac{\sqrt[4]{rs^2t^3} \cdot \sqrt[4]{r^3s^2t}}{\sqrt[4]{r^2t^3}}$ **57.** $\sqrt[3]{\sqrt{4}}$ **58.** $\sqrt[4]{\sqrt[3]{2}}$ **59.** $\sqrt[6]{\sqrt[3]{x}}$ **60.** $\sqrt[8]{\sqrt[4]{y}}$

Simplify each of the following, assuming that all variables represent nonnegative numbers.

61. $4\sqrt{3} - 5\sqrt{12} + 3\sqrt{75}$ **62.** $2\sqrt{5} - 3\sqrt{20} + 2\sqrt{45}$ **63.** $3\sqrt{28p} - 4\sqrt{63p} + \sqrt{112p}$

64. $9\sqrt{8k} + 3\sqrt{18k} - \sqrt{32k}$ **65.** $2\sqrt[3]{3} + 4\sqrt[3]{24} - \sqrt[3]{81}$ **66.** $\sqrt[3]{32} - 5\sqrt[3]{4} + 2\sqrt[3]{108}$

67. $\dfrac{1}{\sqrt{3}} - \dfrac{2}{\sqrt{12}} + 2\sqrt{3}$ **68.** $\dfrac{1}{\sqrt{2}} + \dfrac{3}{\sqrt{8}} + \dfrac{1}{\sqrt{32}}$ **69.** $\dfrac{5}{\sqrt{2}} - \dfrac{2}{\sqrt[3]{16}} + \dfrac{1}{\sqrt[3]{54}}$

70. $\dfrac{-4}{\sqrt[3]{3}} + \dfrac{1}{\sqrt[3]{24}} - \dfrac{2}{\sqrt[3]{81}}$ **71.** $(\sqrt{2} + 3)(\sqrt{2} - 3)$ **72.** $(\sqrt{5} + \sqrt{2})(\sqrt{5} - \sqrt{2})$

73. $(\sqrt[3]{11} - 1)(\sqrt[3]{11^2} + \sqrt[3]{11} + 1)$ **74.** $(\sqrt[3]{7} + 3)(\sqrt[3]{7^2} - 3\sqrt[3]{7} + 9)$ **75.** $(\sqrt{3} + \sqrt{8})^2$

76. $(\sqrt{2} - 1)^2$ **77.** $(3\sqrt{2} + \sqrt{3})(2\sqrt{3} - \sqrt{2})$ **78.** $(4\sqrt{5} - 1)(3\sqrt{5} + 2)$

79. $(2\sqrt[3]{3} + 1)(\sqrt[3]{3} - 4)$ **80.** $(\sqrt[3]{4} + 3)(5\sqrt[3]{4} + 1)$

Rationalize the denominator of each of the following. Assume that all variables represent non-negative numbers and that no denominators are zero.

81. $\dfrac{\sqrt{3}}{\sqrt{5} + \sqrt{3}}$

82. $\dfrac{\sqrt{7}}{\sqrt{3} - \sqrt{7}}$

83. $\dfrac{1 + \sqrt{3}}{3\sqrt{5} + 2\sqrt{3}}$

84. $\dfrac{\sqrt{7} - 1}{2\sqrt{7} + 4\sqrt{2}}$

85. $\dfrac{p}{\sqrt{p} + 2}$

86. $\dfrac{\sqrt{r}}{3 - \sqrt{r}}$

87. $\dfrac{a}{\sqrt{a + b} - 1}$

88. $\dfrac{3m}{2 + \sqrt{m + n}}$

In advanced mathematics, it is sometimes useful to write a radical expression with a rational numerator. The procedure is similar to rationalizing the denominator. Rationalize the numerator of each of the following expressions.

89. $\dfrac{1 + \sqrt{2}}{2}$

90. $\dfrac{1 - \sqrt{3}}{3}$

91. $\dfrac{\sqrt{x}}{1 + \sqrt{x}}$

92. $\dfrac{\sqrt{p}}{1 - \sqrt{p}}$

93. $\dfrac{\sqrt{x} + \sqrt{x + 1}}{\sqrt{x} - \sqrt{x + 1}}$

94. $\dfrac{\sqrt{p} + \sqrt{p^2 - 1}}{\sqrt{p} - \sqrt{p^2 - 1}}$

Write the following without radicals. Use absolute value if necessary.

95. $\sqrt{(m + n)^2}$

96. $\sqrt[4]{(a + 2b)^4}$

97. $\sqrt{z^2 - 6zx + 9x^2}$

98. $\sqrt[3]{(r + 2s)(r^2 + 4rs + 4s^2)}$

Answers to Selected Exercises

In this section we provide the answers that we think most students will obtain when they work the exercises using the methods explained in the text. If your answer does not look exactly like the one given here, it is not necessarily wrong. In many cases there are equivalent forms of the answer that are correct. For example, if the answer section shows $\frac{3}{4}$ and your answer is .75, you have obtained the right answer but written it in a different (yet equivalent) form. Unless the directions specify otherwise, .75 is just as valid an answer as $\frac{3}{4}$. In general, if your answer does not agree with the one given, see whether it can be transformed into the other form. If it can, then it is the correct answer. If you still have doubts, talk with your instructor.

CHAPTER 1

Section 1.1 (page 9)

1. (a) 10 **(b)** 0, 10 **(c)** $-6, -\frac{12}{4}$ (or -3), 0, 10 **(d)** $-6, -\frac{12}{4}$ (or -3), $-\frac{5}{8}$, 0, .31, $.\overline{3}$, 10 **(e)** $-\sqrt{3}, 2\pi, \sqrt{17}$
(f) All are real numbers. **3. (a)** None are natural numbers. **(b)** None are whole numbers.
(c) $-\sqrt{100}$ (or -10), -1 **(d)** $-\sqrt{100}$ (or -10), $-\frac{13}{6}$, -1, 5.23, 9.14, 3.14, $\frac{22}{7}$ **(e)** None are irrational numbers.
(f) All are real numbers. **5.** 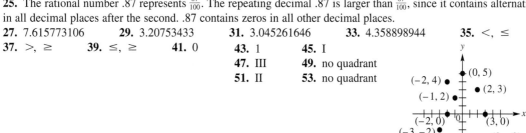 **7.** **9.** D

11. B **13.** C **15.** G **17.** $.8\overline{3}$ **19.** $-4.\overline{3}$ **21.** $.\overline{2}$ **23.** $.0\overline{81}$
25. The rational number .87 represents $\frac{87}{100}$. The repeating decimal $.\overline{87}$ is larger than $\frac{87}{100}$, since it contains alternating 8 and 7 in all decimal places after the second. .87 contains zeros in all other decimal places.
27. 7.615773106 **29.** 3.20753433 **31.** 3.045261646 **33.** 4.358898944 **35.** $<, \leq$
37. $>, \geq$ **39.** $\leq, \geq$ **41.** 0 **43.** 1 **45.** I
47. III **49.** no quadrant
51. II **53.** no quadrant

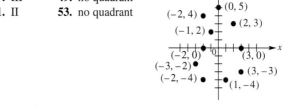

55. I or III **57.** II or IV **59.** It must lie on the y-axis. **61.** Answers will vary. For example, for the TI-83, use ZOOM 6. **63.** $[-5, 5]$ by $[-25, 25]$ **65.** $[-60, 60]$ by $[-100, 100]$ **67.** $[-500, 300]$ by $[-300, 500]$

69.

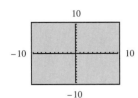

71.

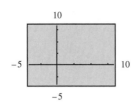

73.

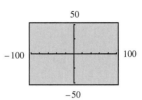

75. There are no tick marks. To set a screen with no tick marks on the axes, use Xscl = 0 and Yscl = 0.

Section 1.2 (page 22)

1. $(-1, 4)$

3. $(-\infty, 0)$

5. $[1, 2)$

7. $\{x \mid -4 < x < 3\}$ **9.** $\{x \mid x \le -1\}$ **11.** $\{x \mid -2 \le x < 6\}$

13. $\{x \mid x \le -4\}$ **15.** Use a parenthesis if the symbol is $<$ or $>$, and use a square bracket if the symbol is $\le$ or $\ge$.
17. domain: $\{5, 3, 4, 7\}$; range: $\{1, 2, 9, 6\}$; function   **19.** domain: $(-\infty, \infty)$; range: $(-\infty, \infty)$; function
21. domain: $[-4, 4]$; range: $[-3, 3]$; not a function   **23.** domain: $[2, \infty)$; range: $[0, \infty)$; function
25. domain: $[-9, \infty)$; range: $(-\infty, \infty)$; not a function   **27.** domain: $\{-5, -2, -1, -.5, 0, 1.75, 3.5\}$; range: $\{-1, 2, 3,$
$3.5, 4, 5.75, 7.5\}$; function   **29.** domain: $\{0, 10, 20, 30, 40, 50\}$; range: $\{100, 150, 175, 200, 250, 300, 400\}$; not a
function   **31.** domain: $\{2, 5, 11, 17, 3\}$; range: $\{1, 7, 20\}$; function   **33. (a)–(d)** Answers will vary. See the
definitions.   **35.** 1.5 (million pounds)   **37.** 1987 and 1990   **39.** 12 seconds (approximately)
41. 35 miles per hour (approximately)   **43.** 20   **45.** 2   **47.** 4   **49.** -5.5 **51.** 0   **53.** -5
55. 10   **57.** 31   **59.** 6   **61.** 11   **63.** 3   **65.** 7   **67.** 7   **69.** -4 **71.** -3.5
73. $3a - 1$ **75.** $2r^2 + r + 3$ **77.** $7p + 2s - 1$ **79.** $2xh + h^2$

Section 1.3 (page 39)

1. yes; $m = 3, b = 12$ **3.** yes; $m = -.3, b = .12$ **5.** no   **7.** no   **9.** no
11. (a) 4
(b) -4
(c) $(-\infty, \infty)$
(d) $(-\infty, \infty)$
(e) 1

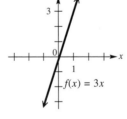

13. (a) 2
(b) -6
(c) $(-\infty, \infty)$
(d) $(-\infty, \infty)$
(e) 3

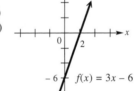

15. (a) 5
(b) 2
(c) $(-\infty, \infty)$
(d) $(-\infty, \infty)$
(e) $-\frac{2}{5}$

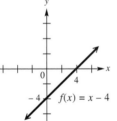

17. (a) 0
(b) 0
(c) $(-\infty, \infty)$
(d) $(-\infty, \infty)$
(e) 3

19. The point $(0, 0)$ must lie on the line.
21. (a) none
(b) -3
(c) $(-\infty, \infty)$
(d) $\{-3\}$
(e) 0

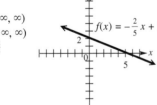

23. (a) -1.5
(b) none
(c) $\{-1.5\}$
(d) $(-\infty, \infty)$
(e) undefined

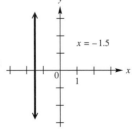

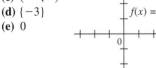

25. (a) 2
(b) none
(c) {2}
(d) $(-\infty, \infty)$
(e) undefined

$x = 2$

27. constant function   **29.** $y = 0$ **31.** Window B
33. Window B

35. $f(4.3) = -12.65$

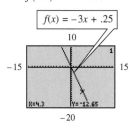

$f(x) = -3x + .25$

37. $f(-1.3) = 6.23$

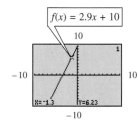

$f(x) = 2.9x + 10$

39. $\frac{1}{5}$ **41.** $\frac{7}{9}$ **43.** 0

45. horizontal

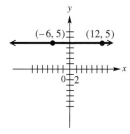

This is the graph of a function.

47. 4   **49.** -3.5 **51.** .75   **53.** $m = 4; b = 2; Y_1 = 4x + 2$

55. (a) The slope of $-.0221$ indicates that on the average from 1912 to 1992, the 5000 meter run is being run .0221 second faster every Olympic game. It is negative because the times are generally decreasing as time progresses.   **(b)** World War II (1939–1945) included the years 1940 and 1944.   **(c)** Yes. If it continues in a linear pattern, eventually the winning time is 0, which is unrealistic.   **57. (a)** A   **(b)** C   **(c)** D   **(d)** B

59.   (a)

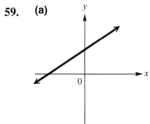

(b)

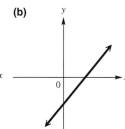

(c)

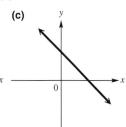

(d)

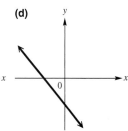

61. y_1; Both slopes are positive.   **63.** y_2; Slope of y_1 is positive and slope of y_2 is negative.
65. y_1; Slope of y_1 is negative and slope of y_2 is positive.

67.
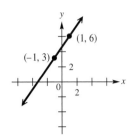

69.
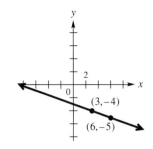

71.
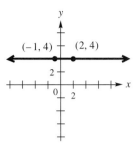

73.
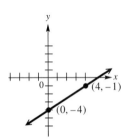

75. $y = \frac{3}{4}x - 4$ **77.** The line has a positive slope and a negative y-intercept.

Section 1.4 (page 54)

1. $y = -2x + 5$ **3.** $y = -1.5x - 3.5$ **5.** $y = -.5x - 3$ **7.** $y = 3x + 4$ **9.** $y = -1.5x + 6.5$
11. $y = 3x + 8$ **13.** $y = 8x + 12$ **15. (a)** $x = 2$ **(b)** $x = -5$ **(c)** $y = 8$ **(d)** $y = -4$
17. The y-coordinate is 0.
The graphs of the lines in Exercises 19–23 are the same as the graphs in Exercises 11–15 in Section 1.3. In the answers that follow, we give the x-intercept first and the y-intercept second. See the graphs given in Section 1.3.
19. 4; −4   **21.** 2; −6   **23.** 5; 2   **25.**

We used (0, 0) and (1, 3).

27.
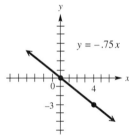

We used (0, 0) and (4, −3).

29. $y = -\frac{5}{3}x + 5$

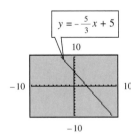

31. $y = \frac{2}{7}x + \frac{4}{7}$

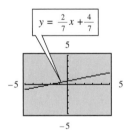

33. $y = -\frac{3}{4}x + \frac{25}{8}$

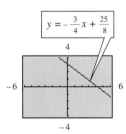

35. $c = 17$ **37.** $b = 84$ **39.** $c = \sqrt{89}$ **41.** $b = 4$ **43. (a)** $8\sqrt{2}$ **(b)** $(9, 3)$ **45. (a)** $\sqrt{34}$
(b) $(-5.5, -3.5)$ **47. (a)** $2\sqrt{10}$ **(b)** $(-1, 4)$ **49.** $7555   **51.** $3495   **53.** No. The midpoint of
the segment joining (1970, 3968) and (1980, 8414) is (1975, 6191). The actual cutoff for 1975 was $5500, not $6191.
55. $y = -\frac{1}{3}x + \frac{11}{3}$ **57.** $y = \frac{5}{3}x + \frac{13}{3}$ **59.** $x = -5$ (It is not possible to write this equation in slope-intercept
form.)   **61.** $y = -.2x + 7$ **63.** $y = .5x$ **65.** They are *not* parallel. Line y_1 has slope 2.3 and line y_2 has
slope 2.3001. Since $2.3 \neq 2.3001$, and nonvertical parallel lines must have *equal* slopes, they can't be parallel. They only
appear to be parallel in the figure because the slopes differ by .0001, a very small number.   **67. (a)** $a + b$; $a + b$;
$a^2 + 2ab + b^2$ **(b)** $\frac{1}{2}ab$; $4\left(\frac{1}{2}ab\right) = 2ab$; c^2 **(c)** $2ab + c^2$ **(d)** $a^2 + 2ab + b^2$; $2ab + c^2$ **(e)** $a^2 + b^2$; c^2

69. (a) the Pythagorean theorem and its converse   **(b)** $\sqrt{x_1^2 + m_1^2 x_1^2}$ **(c)** $\sqrt{x_2^2 + m_2^2 x_2^2}$
(d) $\sqrt{(x_2 - x_1)^2 + (m_2 x_2 - m_1 x_1)^2}$ **(f)** $-2x_1 x_2(m_1 m_2 + 1) = 0$ **(g)** Since $x_1 \neq 0$, $x_2 \neq 0$, $m_2 m_1 + 1 = 0$, this implies
$m_1 m_2 = -1$.   **(h)** The product of the slopes of perpendicular lines, neither of which is parallel to an axis, is -1.
71. 3   **72.** 3   **73.** equal   **74.** $\sqrt{10}$ **75.** $2\sqrt{10}$ **76.** $3\sqrt{10}$ **77.** The sum is $3\sqrt{10}$,
which is equal to the answer in Exercise 76.   **78.** *B; C; A; C* (The order of the last two may be reversed.)
79. The midpoint is (3, 3), which is the same as the middle entry in the table.   **80.** 7.5

Section 1.5 (page 68)

1. -4 **3.** 0   **5.** $-\frac{23}{19}$ **7.** $\frac{8}{3}$ **9.** $-\frac{29}{6}$ **11.** {10}   **13.** {−1.3}   **15.** {2}
17. The *y*-value is the function value (range value) for *both* y_1 and y_2 when the solution of the equation is substituted for *x*.
19. {−.8}   **21.** {−16}   **23.** ∅; This is called a contradiction.   **25.** {12}   **27.** {3}
29. {−2}   **31.** {7}   **33.** {75}   **35.** {−1}   **37.** {0}   **39.** $\{\frac{5}{4}\}$ **41.** {4}   **43.** {1.5}
45. {16.07}   **47.** {−1.46}   **49.** {−3.92}   **51.** {16.07}   **53.** {−1.46}   **55.** {−3.92}
57. identity   **59.** contradiction   **61.** contradiction   **63.** Graph $y_1 = .06x + .09(15 - x)$ and $y_2 = .07(15)$ to
find that the coordinates of the point of intersection are $x = 10$, $y = 1.05$. (This is the intersection-of-graphs method.) The
solution set contains this *x*-value, 10: {10}.   **65.** (a); -4.05 could be a solution. In the first view, $y > 0$, and in the
second, $y < 0$. The solution must therefore be between the two *x*-values shown. Choice (a) is the only such choice.
67. When subtracting the right-hand side, $4x - 12$, parentheses must be inserted around it. The correct solution would be
found by graphing $y_1 = 2x + 3 - (4x - 12)$.   **69.** {3.65}   **71.** {.90}

Section 1.6 (page 80)

1. [1965, 1976]   **3.** [1965, 1978]   **5.** [1965, 1990]   **7.** {3}   **9.** $(-\infty, 3)$ **11.** $(3, \infty)$
13. $[3, \infty)$ **15.** {3}   **17.** {3}   **19.** By subtracting $g(x)$ from both sides of $f(x) = g(x)$, we get an equivalent
equation, $f(x) - g(x) = 0$. It has the same solution set.   **21. (a)** $(20, \infty)$ **(b)** $(-\infty, 20)$ **(c)** $[20, \infty)$ **(d)** $(-\infty, 20]$
23. (a) $[.4, \infty)$ **(b)** $(-\infty, .4)$ **(c)** $(.4, \infty)$ **(d)** $(-\infty, .4]$ **25. (a)** {4} **(b)** $(4, \infty)$ **(c)** $(-\infty, 4)$
27. (a) $(6, \infty)$ **(b)** $[6, \infty)$ **(c)** $(-\infty, 6)$ **(d)** $(-\infty, 6]$ **29. (a)** $(-\infty, -3]$ **(b)** $(-3, \infty)$
31. (a) $\left(-\frac{3}{4}, \infty\right)$ **(b)** $\left(-\infty, -\frac{3}{4}\right]$ **33. (a)** $(-\infty, 15]$ **(b)** $(15, \infty)$ **35. (a)** $(-6, \infty)$ **(b)** $(-\infty, -6]$
37. (a) $[-8, \infty)$ **(b)** $(-\infty, -8)$ **39. (a)** $(25, \infty)$ **(b)** $(-\infty, 25]$ **41.** $(-\infty, 29.90)$ **43.** $[7.28, \infty)$
45. $(4.20, \infty)$ **47.** $(-\infty, 12)$ **49.** $[12, \infty)$ **51.** $(-2, 12)$ **53.** ∅   **55.** ∅   **57.** [1, 4]
59. $(-6, -4)$ **61.** $(-16, 19]$ **63.** $(-\infty, \infty)$ **65.** ∅   **67.** ∅   **69.** $(-\infty, \infty)$ **71. (a)** 6; {6}
(b) Every *y*-value is greater than 0 (i.e., positive); $(6, \infty)$ **(c)** Every *y*-value is less than 0 (i.e., negative); $(-\infty, 6)$

Section 1.7 (page 91)

1. a negative correlation close to -1 **3.** a positive correlation close to $+1$ **5.** no correlation
7. (a)   **(b)**

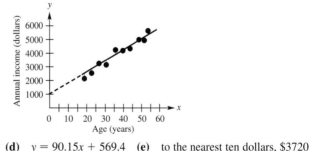

(c)

x	y	x · y	x²
19	2150	40,850	361
23	2550	58,650	529
27	3250	87,750	729
31	3150	97,650	961
36	4250	153,000	1296
40	4200	168,000	1600
44	4350	191,400	1936
49	5000	245,000	2401
52	4950	257,400	2704
54	5650	305,100	2916
Sums: 375	39,500	1,604,800	15,433

(d) $y = 90.15x + 569.4$ **(e)** to the nearest ten dollars, $3720
(f) $\Sigma y^2 = 167,660,000$; $r = .98$ (to two decimal places). Because
$r \approx 1$, there is a high positive correlation.

9. (a) $y = .556x - 17.8$ **(b)** $48.9°$ **(c)** $r = 1$ (The experimental points must lie perfectly along a line.)
11. (a) $y = 1.99x - 51.23$ **(b)** 78 **(c)** $r = .91$ **13. (a)**

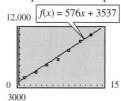

$f(x) = 576x + 3537$. The slope of 576 indicates that tuition and fees have increased roughly $576 per year.

(b) $f(10) = 9297$. This value, \$9297, is slightly less than the true value. **(c)** In 1970, $x = -10$ and $f(-10) = 576(-10) + 3537 = -2223$, which is clearly incorrect. The year 2010 is many years in the future. Many factors could affect the tuition and f would probably be inaccurate that far into the future. **(d)** $y = 587.66x + 3305.09$; $r = .997$
(e) \$15,646 (Use $x = 21$.)
15. (a)

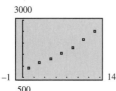

(b)

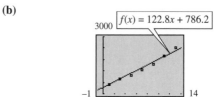

The data are not exactly linear. If they were linear the tuition increase would be equal between each two-year interval. The data do appear to be somewhat linear. A linear function might approximate the tuition.

$f(x) = 122.8x + 786.2$. The slope of 122.8 indicates that tuition and fees increased roughly $122.80 per year from 1981 to 1993.

(c) 1984: \$1277.40; This is slightly more than the true value. **(d)** $y = 130.27x + 710.55$; $r = .99$ **(e)** \$3446 (Use $x = 21$.)
17. $y = .0139x + 53,270$; $r = .528$; There is a positive correlation but it is not very strong, since it is not close to $+1$.
19. $y = .101x + 11.6$; $r = .909$; There is a fairly strong positive correlation, because .909 is close to $+1$.
21. $m = 2$ **22.** $y = 2x + 4$ **23.** $(4, 12)$ **24.** $y = 2(4) + 4 = 12$; yes **25.** $y = 2x + 4$; yes
26.

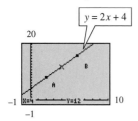

$y = 2x + 4$

27. $122°$ **29.** $32.\overline{2}°$

Section 1.8 (page 105)

1. $P = \dfrac{I}{RT}$ **3.** $W = \dfrac{P - 2L}{2}$ or $W = \dfrac{P}{2} - L$ **5.** $h = \dfrac{2A}{b_1 + b_2}$ **7.** $h = \dfrac{S - 2\pi r^2}{2\pi r}$ or $h = \dfrac{S}{2\pi r} - r$
9. $C = \frac{5}{9}(F - 32)$ **11.** 10 cm by 17 cm **13.** length: 20 inches; width: 12 inches **15.** 328.6 yards
17. $266\frac{2}{3}$ gallons **19.** 2 liters **21.** 2.4 liters **23.** 4 liters **25.** 35 pounds per square foot
27. $51\frac{3}{7}$ feet **29.** 12,500 fish **31. (a)** $C(x) = .02x + 200$ **(d)** For $x < 10,000$, a loss; for $x > 10,000$, a profit
(b) $R(x) = .04x$
(c) 10,000

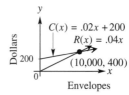

33. (a) $C(x) = 3.00x + 2300$ **(d)** For $x < 920$, a loss; for $x > 920$, a profit
(b) $R(x) = 5.50x$
(c) 920

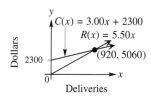

35. 96   **37.** 94 or greater   **39.** \$160   **41.** 65 miles per hour   **43. (a)** $y = 640x + 1100$
(b) \$17,100   **(c)** Locate the point $(25, 17,100)$ on the graph of $y = 640x + 1100$.   **45. (a)** $y = 2600x + 120,000$
(b) \$156,400; Locate the point $(14, 156,400)$.   **(c)** It represents the annual appreciation in value of the house (\$2600 per year).   **47. (a)** $V(x) = 900x$ **(b)** $\frac{3}{50}x$ **(c)** $A = 2.4$ ach   **(d)** It should be increased by $3.\overline{3}$ times. (Smoking areas require more than triple the normal ventilation.)   **49. (a)** approximately .000021 for each individual
(b) $C(x) = .000021x$ **(c)** approximately 2.1 cases   **(d)** approximately 413,000 deaths per year   **51. (a)** year 6, or 1988   **(b)** 1988; They do correspond very closely.   **(c)** 39.33%   **53.** 36 inches   **55.** \$7866
57. about 44%   **59.** about 17.678 million   **61. (a)** 500 cu cm   **(b)** 90°C   **(c)** -273°C

Chapter 1 Review Exercises (page 112)

1. $6\sqrt{17}$ **3.** -4 **5.** $-\frac{3}{4}$ **7.** 36   **9.** 11   **11. (a)** -3 **(b)** $y = -3x + 2$ **(c)** $(2.5, -5.5)$
(d) $\sqrt{10}$ **13.** C   **15.** A   **17.** E   **19.** domain: $[-6, 6]$; range: $[-6, 6]$ **21.** I   **23.** B
25. I   **27.** O   **29.** $\left\{\frac{46}{7}\right\}$ **31.** $\left(-\frac{10}{3}, \frac{46}{3}\right]$ **33.** $(2.5, \infty)$ **35.** $C(x) = 30x + 150$ **37.** 20
39. $f = \dfrac{AB(p + 1)}{24}$ **41. (a)** 41°F   **(b)** about 21,000 feet   **(c)** Graph $y = -3.52x + 58.6$. Then, find the coordinates of the point where $x = 5$ to support the answer in (a). Finally, find the coordinates of the point where $y = -15$ to support the answer in (b).   **43. (a)** $y = f(x) = 2.065x - .0456$ **(b)** The slope 2.065 indicates that the number of passengers at these airports is predicted to approximately double between 1992 and 2005.   **(c)** $y = 2.081x - .0931$
(d) $f(4.9) \approx 10.1$ million passengers; this agrees favorably with the FAA prediction of 10.3 million.   **45.** about 1999
47. $y = 4x + 120$; 4 feet   **49.** 4 feet   **51.** 1 hour, 8 minutes, 42.6 seconds   **53.** 150 kg per m²
55. The amount of ozone remaining after filtration is 79.8 ppb. Since $79.8 > 50$, it did not remove enough of the ozone.
57. $3\frac{3}{7}$ liters   **59.** $[300, \infty)$

Chapter 1 Test (page 117)

1. (a) (i) $(-\infty, \infty)$ **(ii)** $[2, \infty)$ **(iii)** no x-intercepts   **(iv)** 3   **(b) (i)** $(-\infty, \infty)$ **(ii)** $(-\infty, 0]$ **(iii)** 3   **(iv)** -3
(c) (i) $[-4, \infty)$ **(ii)** $[0, \infty)$ **(iii)** -4 **(iv)** 2   **2. (a)** $\{-4\}$ **(b)** $(-\infty, -4)$ **(c)** $[-4, \infty)$ **(d)** $\{-4\}$
3. (a) $\{5.5\}$ **(b)** $(-\infty, 5.5)$ **(c)** $(5.5, \infty)$ **(d)** $(-\infty, 5.5]$ **4. (a)** $\{-1\}$; the check leads to $-3 = -3$.
(b)

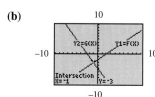

$(-1, \infty)$; the graph of $y_1 = f(x)$ is *above* the graph of $y_2 = g(x)$ for domain values greater than -1.   **(c)** $(-\infty, -1)$; the graph of $y_1 = f(x)$ is *below* the graph of $y_2 = g(x)$ for domain values less than -1.

5. (a) $\{-8\}$ **(b)** $[-8, \infty)$ **(c)**

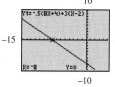

The x-intercept is -8, supporting the result of part (a). The graph of the linear function lies below or on the x-axis for domain values greater than or equal to -8, supporting the result of part (b).

6. (a) \$13.35   **(b)** $m = \frac{11}{12} \approx .92$; This means that during the years 1980 to 1992, the monthly rate increased on the average approximately 92¢ per year.   **7. (a)** $y = -2x - 1$ **(b)** $y = -\frac{1}{2}x + \frac{7}{2}$ **8.** $W = \dfrac{S - 2LH}{2H + 2L}$
9. (a) $f(x) = -.68x + 34.8$ **(b)** $f(40) = 7.6$ (degrees); This value is lower than the actual value.   **(c)** $y = -.65x + 32.7$

(d)   30.89

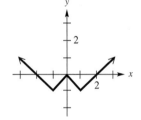

7.5   45.5

8.11

$r = -.98$; Using $y = -.65(40) + 32.7$, we obtain 6.7 (degrees).   **10.** .1 mm
(There is a discrepancy in the screen display due to roundoff.)

CHAPTER 2

Section 2.1 (page 133)

1. $(-\infty, \infty)$ **3.** $[0, \infty)$ **5.** $(-\infty, -3); (-3, \infty)$ **7.** While it seems to be continuous at first glance, there is a discontinuity at $x = -3$, as indicated by the calculator not giving a corresponding value of y for $x = -3$. This happens because -3 causes the denominator to equal zero, giving an undefined expression.   **9. (a)** $(-\infty, 0]$ **(b)** $[0, \infty)$
(c) none   **11. (a)** $[4, \infty)$ **(b)** $(-\infty, -1]$ **(c)** $[-1, 4]$ **13. (a)** none   **(b)** $(-\infty, \infty)$ **(c)** none
15. (a) none   **(b)** both intervals $(-\infty, -2]$ and $[3, \infty)$ **(c)** $(-2, 3)$ **17.** increasing   **19.** decreasing
21. increasing   **23. (a)** no   **(b)** yes   **(c)** no   **25. (a)** yes   **(b)** no   **(c)** no   **27. (a)** yes   **(b)** yes
(c) yes   **29. (a)** no   **(b)** no   **(c)** yes   **31. (a)** no   **(b)** no   **(c)** no   **33.** $(-1.625, 2.0352051)$
35. $(.5, -.84089642)$ **37.** $(-5.092687, .9285541)$
39. (a)

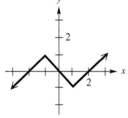

(b)

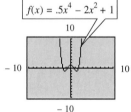

41. even   **43.** even

45. Show that $f(-x) = f(x)$.   **47.** Show that $f(-x) = -f(x)$.
49. symmetric with respect to the origin   **51.** symmetric with respect to the y-axis   **53.** symmetric with respect to neither the origin nor the y-axis

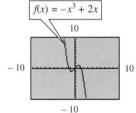

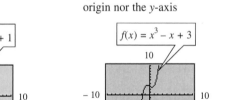

55. odd   **57.** even   **59.** neither even nor odd

In Exercises 61–63, we give a calculator graph of what the sketch should look like.
61.

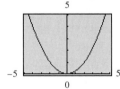

63.

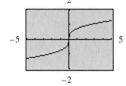

65.

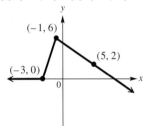

The slope of the tangent line is negative.

66.

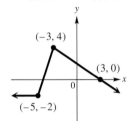

The slope of the tangent line is positive.

67.

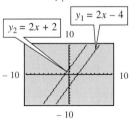

The slope of the tangent line is 0.
(Tangent line coincides with *x*-axis.)

68. If a tangent line is drawn at a point where the function is decreasing, the slope is negative. If a tangent line is drawn at a point where the function is increasing, the slope is positive. If a tangent line is drawn at a point where the function changes from decreasing to increasing, the slope is 0.

69. true   **71.** false   **73.** true   **75.** true   **77.** true

Section 2.2 (page 142)

1. D   **3.** C   **5.** B   **7.** A   **9.** C   **11.** B   **13.** C   **15.** A   **17.** E   **19.** B
21. D   **23.** E   **25.** A   **27.** (a) $(-\infty, \infty)$ (b) $[-3, \infty)$ **29.** (a) $[2, \infty)$ (b) $[-4, \infty)$
31. (a) $(-\infty, \infty)$ (b) $(-\infty, \infty)$ **33.** B   **35.** D   **37.** 4   **39.** -3
41.

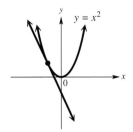

43.

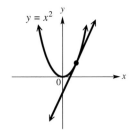

45.

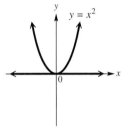

46. 2   **47.** $y_1 = 2x - 4$ **48.** $(1, 4)$ and $(3, 8)$ **49.** 2   **50.** $y_2 = 2x + 2$
51.

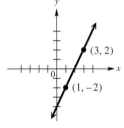

The graph of y_2 can be obtained by shifting the graph of y_1 upward 6 units. The constant, 6, comes from the 6 we added to each *y*-value in Exercise 48.

52. *c*; *c*; the same as; *c*; upward (or positive vertical)

53. B   **55.** A   **57.** $y = (x + 4)^2 + 3$ (a) $[-4, \infty)$ (b) $(-\infty, -4]$ **59.** $y = x^3 - 5$ (a) $(-\infty, \infty)$
(b) none   **61.** *h* can be any real number, while *k* must equal 38.   **63.** (a) $\{3, 4\}$ (b) $(-\infty, 3) \cup (4, \infty)$
(c) $(3, 4)$ **64.** (a) $\{\sqrt{2}\}$ (b) $(\sqrt{2}, \infty)$ (c) $(-\infty, \sqrt{2})$ **65.** (a) $\{-4, 5\}$ (b) $(-\infty, -4] \cup [5, \infty)$
(c) $[-4, 5]$ **66.** (a) $\emptyset$ (b) $[1, \infty)$ (c) $\emptyset$ **67.** $y = 588(x - 1980) + 3305$ **69.** (a) $y = 12.6x + 22.8$
(b) $y = 12.6(x - 1985) + 22.8$ **(c)** $y = 12.6x - 24{,}988.2$; Use the distributive property in part (b) to get this same equation.

Section 2.3 (page 155)

1.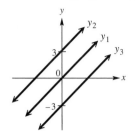

3.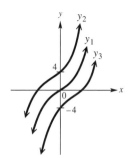

5.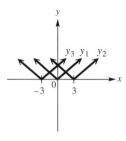

7.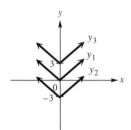

9.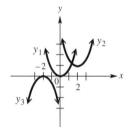

11.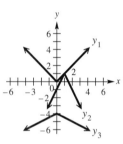

13.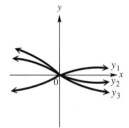

15.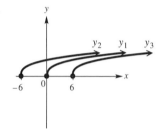

17.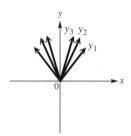

19. $4; x$ **21.** 2; left; $\frac{1}{4}$; x; 3; downward (or negative)   **23.** 3; right; 6 **25.** $y = \frac{1}{2}x^2 - 7$

27. $y = 4.5\sqrt{x - 3} - 6$ **29.** F   **31.** D   **33.** B   **35.** $g(x) = -(x - 5)^2 - 2$

37. (a) 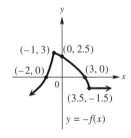   **(b)** 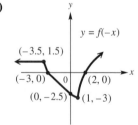   **(c)** 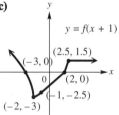   **(d)** $f(0) = 1$

39. (a)    **(b)** 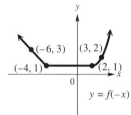   **(c)** 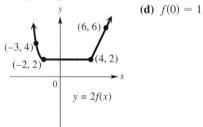   **(d)** -1 and 4

41. (a)

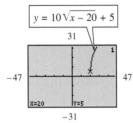

(b) The graph of $y = f(-x)$ is the same as that of $y = -f(x)$, shown in part (a).

(c)

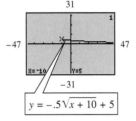

(d) symmetry with respect to the origin   **43. (a)** r is an x-intercept.   **(b)** $-r$ is an x-intercept.   **(c)** $-r$ is an x-intercept.

45. domain: $[20, \infty)$; range: $[5, \infty)$

46. domain: $[-15, \infty)$; range: $(-\infty, -18]$

47. domain: $[-10, \infty)$; range: $(-\infty, 5]$

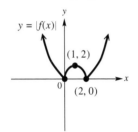

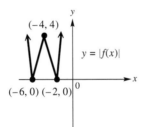

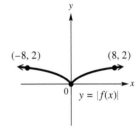

48. The domain is $[h, \infty)$ and the range is $[k, \infty)$.   **49. (a)** $[-1, 2]$ **(b)** $(-\infty, -1]$ **(c)** $[2, \infty)$ **51. (a)** $[1, \infty)$ **(b)** $[-2, 1]$ **(c)** $(-\infty, -2]$ **53.** $y_2 = 5(2) = 10$ when $x = 8$ **55.** $y_2 = 5(-2.154435) = -10.772175$ when $x = -10$ **57.** decreases   **59.** increases   **61. (a)** It is symmetric with respect to the y-axis.   **(b)** It is symmetric with respect to the y-axis.   **63. (a)** $\{1.89, 5.32\}$ **(b)** $(1.89, 5.32)$ **(c)** $(-\infty, 1.89) \cup (5.32, \infty)$
64. (a) $\{.47, .89, 2.30\}$ **(b)** $(-\infty, .47) \cup (.89, 2.30)$ **(c)** $(.47, .89) \cup (2.30, \infty)$

Section 2.4 (page 168)

1.

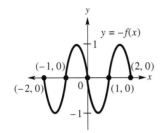

3.

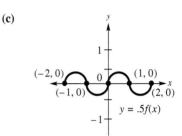

5.

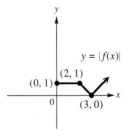

7. $y = |f(x)|$ has the same graph as $y = f(x)$.   **9.**

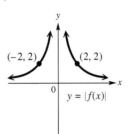

11.

13. We reflect the graph of $y = f(x)$ across the x-axis for all points for which $y < 0$. If $y \geq 0$, the graph remains unchanged.
15. A   **17.** B

19. (a)   **(b)**   **(c)**

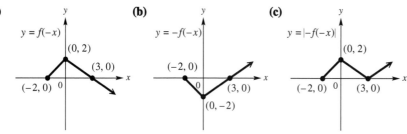

21. $[2, \infty)$ **23.** Figure (b) shows the graph of $y = f(x)$, while (a) shows the graph of $y = |f(x)|$.

25. Property 1: $|(-10)(-20)| = |-10| \cdot |-20|$
$$|200| = 10 \cdot 20$$
$$200 = 200 \checkmark$$

Property 2: $\left|\dfrac{-10}{-20}\right| = \dfrac{|-10|}{|-20|}$
$$\left|\dfrac{1}{2}\right| = \dfrac{10}{20}$$
$$\dfrac{1}{2} = \dfrac{1}{2} \checkmark$$

Property 3: $|-10| = |-(-10)|$
$$10 = |10|$$
$$10 = 10 \checkmark$$

Property 4: $|-10| + |-20| \geq |-10 + (-20)|$
$$10 + 20 \geq |-30|$$
$$30 \geq 30$$
True, because $30 = 30$.

27. (a) greater than   **(b)** equal to   **(c)** a straight line (segment)   **(d)** ">" holds when a and b have opposite signs. "=" holds when a and b have the same sign, or when a or b is equal to 0.   **29. (a)** $\{-2, 8\}$ **(b)** $(-\infty, -2) \cup (8, \infty)$
(c) $(-2, 8)$ **31. (a)** $\{1, 5\}$ **(b)** $(-\infty, 1] \cup [5, \infty)$ **(c)** $[1, 5]$ **33. (a)** $\left\{-\frac{7}{4}\right\}$ **(b)** $\left(-\infty, -\frac{7}{4}\right) \cup \left(-\frac{7}{4}, \infty\right)$
(c) $\emptyset$ **35. (a)** $\emptyset$ **(b)** $\emptyset$ **(c)** $(-\infty, \infty)$ **37. (a)** $\{-1, 6\}$ **(b)** $(-1, 6)$ **(c)** $(-\infty, -1) \cup (6, \infty)$
39. (a) $\{4\}$ **(b)** $\{4\}$ **(c)** $(-\infty, \infty)$ **41. (a)** $\left\{-8, \frac{6}{5}\right\}$ **(b)** $(-\infty, -8) \cup \left(\frac{6}{5}, \infty\right)$ **(c)** $\left(-8, \frac{6}{5}\right)$ **43. (a)** $\left\{-3, \frac{5}{3}\right\}$
(b) $(-\infty, -3] \cup \left[\frac{5}{3}, \infty\right)$ **(c)** $\left[-3, \frac{5}{3}\right]$ **45. (a)** $\left\{-\frac{1}{2}\right\}$ **(b)** $\left(-\infty, -\frac{1}{2}\right)$ **(c)** $\left(-\frac{1}{2}, \infty\right)$ **47. (a)** $\{-4, 2\}$
(b) $(-\infty, -4) \cup (2, \infty)$ **(c)** $(-4, 2)$ **49. (a)** $\{-20, 4\}$ **(b)** $(-\infty, -20] \cup [4, \infty)$ **(c)** $[-20, 4]$
50. The "V-shaped" graph is the graph of $f(x) = |.5x + 6|$, since this is the typical shape of such an absolute value function.
51. The straight-line graph is the graph of $g(x) = 3x - 14$, since this is a linear function.   **52.** $\{8\}$ **53.** $(-\infty, 8)$
54. $(8, \infty)$ **55.** $\{8\}$ **57.** $[2.75, 6.5]$ **59.** $(-\infty, \infty)$ **61.** $\emptyset$ **63.** $(2.66\overline{3}, 2.67)$
65. $(-1.60002, -1.59998)$ **67.** $94°F$ **69.** $98°F$ **71. (a)** $P_d = 9$ **(b)** 113 or 147   **73.** $[6.5, 9.5]$
75. $|x - 123| \leq 25$; $|x - 21| \leq 5$

Section 2.5 (page 179)

1. (a) -10 **(b)** -2 **(c)** -1 **(d)** 2   **(e)** 4   **3. (a)** -10 **(b)** 2   **(c)** 5   **(d)** 3   **(e)** 5
5.   **7.**

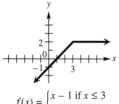

$$f(x) = \begin{cases} x - 1 & \text{if } x \leq 3 \\ 2 & \text{if } x > 3 \end{cases}$$

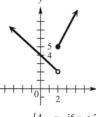

$$f(x) = \begin{cases} 4 - x & \text{if } x < 2 \\ 1 + 2x & \text{if } x \geq 2 \end{cases}$$

9.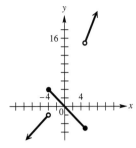

$$f(x) = \begin{cases} 2 + x & \text{if } x < -4 \\ -x & \text{if } -4 \le x \le 5 \\ 3x & \text{if } x > 5 \end{cases}$$

11.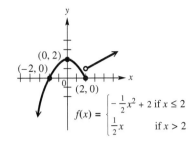

$$f(x) = \begin{cases} -\frac{1}{2}x^2 + 2 & \text{if } x \le 2 \\ \frac{1}{2}x & \text{if } x > 2 \end{cases}$$

13.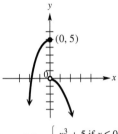

$$f(x) = \begin{cases} x^3 + 5 & \text{if } x \le 0 \\ -x^2 & \text{if } x > 0 \end{cases}$$

15.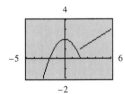

$$f(x) = \begin{cases} x - 1 & \text{if } x \le 3 \\ 2 & \text{if } x > 3 \end{cases}$$

17.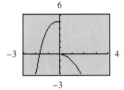

$$f(x) = \begin{cases} 4 - x & \text{if } x < 2 \\ 1 + 2x & \text{if } x \ge 2 \end{cases}$$

19.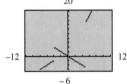

$$f(x) = \begin{cases} 2 + x & \text{if } x < -4 \\ -x & \text{if } -4 \le x \le 5 \\ 3x & \text{if } x > 5 \end{cases}$$

21.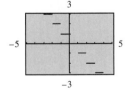

$$f(x) = \begin{cases} -\frac{1}{2}x^2 + 2 & \text{if } x \le 2 \\ \frac{1}{2}x & \text{if } x > 2 \end{cases}$$

23.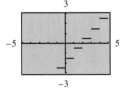

$$f(x) = \begin{cases} x^3 + 5 & \text{if } x \le 0 \\ -x^2 & \text{if } x > 0 \end{cases}$$

24. $[3, \infty)$ **25.** $[-5, 3)$ **26.** $(-\infty, -5)$ **27.** There is an "overlap" of intervals, since the number 4 satisfies both conditions. To truly be a function, every x-value is used only once.

Although there are many possible ways to describe the functions graphed in Exercises 29–33, we give only the most basic rule in these answers.

29. $f(x) = \begin{cases} 2 & \text{if } x \le 0 \\ -1 & \text{if } x > 1 \end{cases}$
domain: $(-\infty, 0] \cup (1, \infty)$
range: $\{-1, 2\}$

31. $f(x) = \begin{cases} x & \text{if } x \le 0 \\ 2 & \text{if } x > 0 \end{cases}$
domain: $(-\infty, \infty)$
range: $(-\infty, 0] \cup \{2\}$

33. $f(x) = \begin{cases} \sqrt[3]{x} & \text{if } x < 1 \\ x + 1 & \text{if } x \ge 1 \end{cases}$
domain: $(-\infty, \infty)$
range: $(-\infty, 1) \cup [2, \infty)$

35. Because of the choice of window, the graph of $y = [\![x]\!]$ looks like the graph of $y = x$. If one does not know how $[\![x]\!]$ is defined, one may incorrectly conclude that $[\![x]\!] = x$ for all x, but this is true only when x is an integer.

37.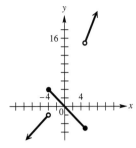

$y = [\![x]\!] - 1.5$

39.

$y = -[\![x]\!]$

41. The graph of $y = [\![x]\!]$ is shifted 1.5 units downward.   **42.** The graph of $y = [\![-x]\!]$ is reflected across the y-axis.
43. The graph of $y = [\![x]\!]$ is reflected across the x-axis.   **44.** The graph of $y = [\![x]\!]$ is shifted 2 units to the left.

45.

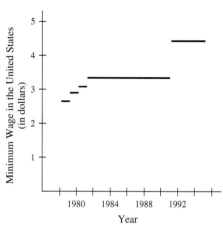

47. 2-pound package: $13; 2.5-pound package: $16; 5.8-pound package: $25

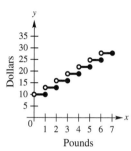

The range is {10, 13, 16, 19, 22, 25, 28}.

49. (a) $.30   **(b)** $.57   **(c)** $1.11   **(d)**

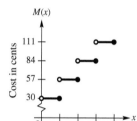

(e) domain: $(0, \infty)$
range: {30, 57, 84, 111, ...}

51.

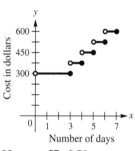

53. (a) 140   **(b)** 220   **(c)** 220   **(d)** 220
(e) 220   **(f)** 60   **(g)** 60   **(h)**

55. $.32   **57.** $.78

Section 2.6 (page 190)

1. (a) $(f + g)(x) = 10x + 2$; $(f - g)(x) = -2x - 4$; $(fg)(x) = 24x^2 + 6x - 3$ **(b)** domain is $(-\infty, \infty)$ in all cases
(c) $\left(\dfrac{f}{g}\right)(x) = \dfrac{4x - 1}{6x + 3}$; its domain is $\left\{x \mid x \neq -\dfrac{1}{2}\right\}$ **(d)** $(f \circ g)(x) = 24x + 11$; its domain is $(-\infty, \infty)$
(e) $(g \circ f)(x) = 24x - 3$; its domain is $(-\infty, \infty)$ **3. (a)** $(f + g)(x) = |x + 3| + 2x$; $(f - g)(x) = |x + 3| - 2x$;
$(fg)(x) = |x + 3|(2x)$ **(b)** domain is $(-\infty, \infty)$ in all cases   **(c)** $\left(\dfrac{f}{g}\right)(x) = \dfrac{|x + 3|}{2x}$; its domain is $\{x \mid x \neq 0\}$
(d) $(f \circ g)(x) = |2x + 3|$; its domain is $(-\infty, \infty)$ **(e)** $(g \circ f)(x) = 2|x + 3|$; its domain is $(-\infty, \infty)$
5. (a) $(f + g)(x) = \sqrt[3]{x + 4} + x^3 + 5$; $(f - g)(x) = \sqrt[3]{x + 4} - x^3 - 5$; $(fg)(x) = \left(\sqrt[3]{x + 4}\right)(x^3 + 5)$ **(b)** domain is
$(-\infty, \infty)$ in all cases   **(c)** $\left(\dfrac{f}{g}\right)(x) = \dfrac{\sqrt[3]{x + 4}}{x^3 + 5}$; its domain is $\left\{x \mid x \neq \sqrt[3]{-5}\right\}$ **(d)** $(f \circ g)(x) = \sqrt[3]{x^3 + 9}$; its domain
is $(-\infty, \infty)$ **(e)** $(g \circ f)(x) = x + 9$; its domain is $(-\infty, \infty)$ **7. (a)** $(f + g)(x) = \sqrt{x^2 + 3} + x + 1$;
$(f - g)(x) = \sqrt{x^2 + 3} - x - 1$; $(fg)(x) = \left(\sqrt{x^2 + 3}\right)(x + 1)$ **(b)** domain is $(-\infty, \infty)$ in all cases

(c) $\left(\dfrac{f}{g}\right)(x) = \dfrac{\sqrt{x^2 + 3}}{x + 1}$; its domain is $\{x \mid x \neq -1\}$ **(d)** $(f \circ g)(x) = \sqrt{x^2 + 2x + 4}$; its domain is $(-\infty, \infty)$ (*Note:* To see that this is the domain, graph $y = x^2 + 2x + 4$ and note that $y > 0$ for all x.)   **(e)** $(g \circ f)(x) = \sqrt{x^2 + 3} + 1$; its domain is $(-\infty, \infty)$ **9.** $256x^2 + 48x + 2$ **11.** 2450   **13.** 55   **15.** 1848   **17.** $-\frac{6}{7}$ **19.** 1122   **21.** 97   **23.** 5   **25.** 0   **27.** 3   **29.** 2   **31.** 1   **33.** 1   **35.** 9   **37.** 1

39. We are not given enough information to determine $f(g(1))$, which is $f(9)$.   **41.** -5 **43.** 3

49. The graph of y_2 can be obtained by *reflecting* the graph of y_1 across the line $y_3 = x$.

50. The graph of y_2 can be obtained by *reflecting* the graph of y_1 across the line $y_3 = x$.

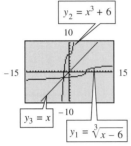

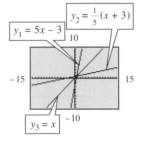

51. 4   **53.** $-12x - 1 - 6h$ **55.** $3x^2 + 3xh + h^2$

We give only one of many possible correct pairs of functions f and g in Exercises 57–61.

57. $f(x) = x^2, g(x) = 6x - 2$ **59.** $f(x) = \sqrt{x}, g(x) = x^2 - 1$ **61.** $f(x) = \sqrt{x} + 12, g(x) = 6x$

63. (a) $C(x) = 10x + 500$ **(b)** $R(x) = 35x$ **(c)** $P(x) = 35x - (10x + 500)$ or $P(x) = 25x - 500$ **(d)** 21 items
(e) The smallest whole number value for which $P(x) > 0$ is 21. Use a window of $[0, 30]$ by $[-1000, 500]$ for example.
65. (a) $C(x) = 100x + 2700$ **(b)** $R(x) = 280x$ **(c)** $P(x) = 280x - (100x + 2700)$ or $P(x) = 180x - 2700$
(d) 16 items   **(e)** The smallest whole number value for which $P(x) > 0$ is 16. Use a window of $[0, 30]$ by $[-3000, 500]$ for example.

67. (a) $V(r) = \frac{4}{3}\pi(r + 3)^3 - \frac{4}{3}\pi r^3$ **(b)**

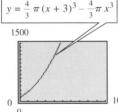

This appears to be a portion of a parabola, formed by translating the squaring function.

(c) 1168.67 cubic inches
(d) $V(4) = \frac{4}{3}\pi(7)^3 - \frac{4}{3}\pi(4^3) \approx 1168.67$

69. (a) $P = 6x$; $P(x) = 6x$; This is a linear function.   **(b)** x represents the width of the rectangle and y represents the perimeter.

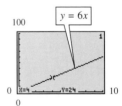

(c) (See graph for part (b).)
Perimeter $= 24$. This is the y-value shown on the screen for the integer x-value 4.

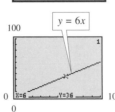

(d) (Answers may vary.)
If the perimeter y of a rectangle satisfying the given conditions is 36, then the width x is 6.

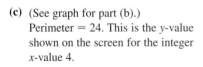

71. (a) $A(2x) = \sqrt{3}x^2$ **(b)** $A(16) = 64\sqrt{3}$ **(c)** On the graph of $y = \dfrac{\sqrt{3}}{4}x^2$, locate the point where $x = 16$ to find $y \approx 110.85$, an approximation for $64\sqrt{3}$.

73. (a) $(A \circ r)(t) = 4\pi t^2$ **(b)** $(A \circ r)(t)$ is a composite function that expresses the area of the circular region covered by the pollutants as a function of time t (in hours). **(c)** (Use $t = 4$.) 64π square miles **(d)** Graph $y = 4\pi x^2$ and show that for $x = 4$, $y \approx 201$ (an approximation for 64π).

Chapter 2 Review Exercises (page 196)

1. F   **3.** G   **5.** B   **7.** E   **9.** $[0, \infty)$ **11.** $(-\infty, \infty)$ **13.** $(-\infty, \infty)$ **15.** $[0, \infty)$
17. $Y_1 = \sqrt{x + 4}$ and $Y_2 = -\sqrt{x + 4}$ **19. (a)** $(-\infty, -2), [-2, 1], (1, \infty)$ **(b)** $[-2, 1]$ **(c)** $(-\infty, -2)$ **(d)** $(1, \infty)$
(e) $(-\infty, \infty)$ **(f)** $\{-2\} \cup [-1, 1] \cup (2, \infty)$ **21.** x-axis symmetry, y-axis symmetry, origin symmetry; not a function
23. y-axis symmetry; even function   **25.** x-axis symmetry; not a function   **27.** Start with the graph of $y = x^2$.
Shift it 4 units to the left, stretch vertically by a factor of 3, reflect across the x-axis, and shift 8 units downward.
29.   **31.** $\{-6, 1\}$ **33.** $(-\infty, -6] \cup [1, \infty)$

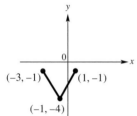

35. (a) $\{-3, 5\}$; When $x = -3$, we get $8 = 8$, and when $x = 5$, we also get $8 = 8$. **(b) (i)** $(-3, 5)$
(ii) $(-\infty, -3) \cup (5, \infty)$ **37.** $(-\infty, 0) \cup (2, \infty)$ **39.** $[2, \infty)$ **41.** $\varnothing$
43.   **45.**   **47.**

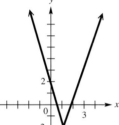

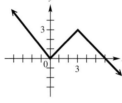

$f(x) = \begin{cases} |x| & \text{if } x < 3 \\ 6 - x & \text{if } x \geq 3 \end{cases}$

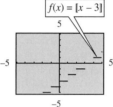

$f(x) = \begin{cases} -4x + 2 & \text{if } x \leq 1 \\ 3x - 5 & \text{if } x > 1 \end{cases}$

49. $f(x) = \begin{cases} \frac{3}{250}x + \frac{1}{20} & \text{if } 0 \leq x \leq 15 \\ .23 & \text{if } 15 < x \leq 20 \end{cases}$ **51.** $4x^2 - 3x - 8$ **53.** 44   **55.** $16k^2 - 6k - 8$
57. undefined   **59.** $(-\infty, -1) \cup (-1, 4) \cup (4, \infty)$ **61.** 2
In Exercise 63, we give only one of many possible choices for f and g.
63. $f(x) = x^2$, $g(x) = x^3 - 3x$ **65.** $(P \circ f)(a) = 18a^2 + 24a + 9$
67. $P(x) = x + (x + 2) + x + (x + 2)$ or $2x + 2(x + 2) = 4x + 4$; linear function

Chapter 2 Test (page 199)

1. (a) D **(b)** D **(c)** C **(d)** B **(e)** C **(f)** C **(g)** C **(h)** D **(i)** D **(j)** C
2. (a)   **(b)**   **(c)**

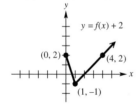

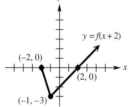

(d)

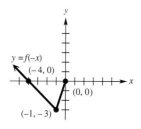

(e)

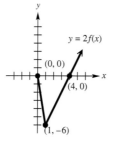

(f)

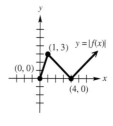

3. (a) $(-3, 6)$ **(b)** $(-3, -6)$ **(c)**

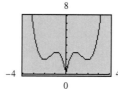

(We give an actual screen here. The drawing should resemble it.)

4. (a) Shift the graph of $y = \sqrt[3]{x}$ 2 units to the left, stretch by a factor of 4, and shift 5 units down.   **(b)**

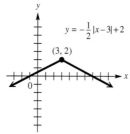

domain: $(-\infty, \infty)$; range: $(-\infty, 2]$

5. (a) $(-\infty, -3)$ **(b)** $(4, \infty)$ **(c)** $[-3, 4]$ **(d)** $(-\infty, -3), [-3, 4], (4, \infty)$ **(e)** $(-\infty, \infty)$ **(f)** $(-\infty, 2)$

6.

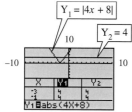

(a) $\{-3, -1\}$; The x-coordinates of the points of intersection of the graphs of y_1 and y_2 are -3 and -1.

(b) $(-3, -1)$; The graph of y_1 lies below the graph of y_2 for x-values between -3 and -1.

(c) $(-\infty, -3) \cup (-1, \infty)$; The graph of y_1 lies above the graph of y_2 for x-values less than -3 or for x-values greater than -1.

7. (a) $2x^2 - x + 1$ **(b)** $\dfrac{2x^2 - 3x + 2}{-2x + 1}$ **(c)** $\left(-\infty, \frac{1}{2}\right) \cup \left(\frac{1}{2}, \infty\right)$ **(d)** $8x^2 - 2x + 1$ **(e)** $4x + 2h - 3$

8. (a)

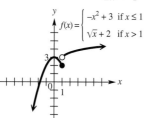

(b)

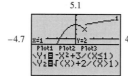

$y_1 = (-x^2 + 3)/(x \le 1)$;

$y_2 = \left(\sqrt{x} + 2\right)/(x > 1)$

9. (a), (b)

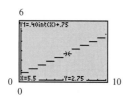

$2.75 is the cost for a 5.5-minute call.   **(c)**

X	Y1
.5	.75
1	1.15
1.5	1.15
2	1.55
2.5	1.55
3	1.95
3.5	1.95

$Y_1\blacksquare.40int(X)+.75$

10. (a) $C(x) = 3300 + 4.50x$
(b) $R(x) = 10.50x$
(c) $P(x) = R(x) - C(x) = 6.00x - 3300$
(d) 551

(e)

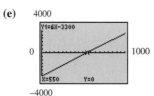

The first integer x-value for which $P(x) > 0$ is 551.

CHAPTER 3

Section 3.1 (page 209)

1. (a) 0 **(b)** -9 **(c)** imaginary   **3. (a)** π **(b)** 0 **(c)** real   **5. (a)** 0 **(b)** $\sqrt{6}$ **(c)** imaginary
7. (a) 2 **(b)** 5 **(c)** imaginary   **9.** $10i$ **11.** $-20i$ **13.** $-i\sqrt{39}$ **15.** $5 + 2i$ **17.** -2.87
19. 6.4   **21.** $16 + 24i$ **23.** A real number a may be written as $a + 0i$, and thus it is a complex number.
However, a complex number with a nonzero imaginary part, such as $2 + 3i$, is not a real number.   **31.** $7 - i$
33. 2   **35.** $1 - 10i$ **37.** $-14 + 2i$ **39.** $5 - 12i$ **41.** 13   **43.** 7   **45.** $25i$ **47.** $2 + 12i$
49. $-70 + 140i$ **51.** Verify that $(-3)^3 - (-3)^2 - 7(-3) + 15$ is equal to 0.   **52.** Verify that
$(2 - i)^3 - (2 - i)^2 - 7(2 - i) + 15$ is equal to 0.   **53.** Verify that $(2 + i)^3 - (2 + i)^2 - 7(2 + i) + 15$ is equal to 0.
54. They are complex conjugates.   **55.** $-5 - i$ **57.** $-1 - 2i$ **59.** $2i$ **61.** $\frac{3}{5} - \frac{4}{5}i$ **63.** $3i$
65. We are multiplying by 1, the identity element for multiplication.   **67.** $-i$ **69.** i **71.** -1 **73.** 1

Section 3.2 (page 220)

1. (a) domain: $(-\infty, \infty)$; range: $[0, \infty)$ **(b)** $(2, 0)$ **(c)** $x = 2$ **(d)** $[2, \infty)$ **(e)** $(-\infty, 2]$ **(f)** minimum point; 0
(g) concave up   **3. (a)** domain: $(-\infty, \infty)$; range: $[-4, \infty)$ **(b)** $(-3, -4)$ **(c)** $x = -3$ **(d)** $[-3, \infty)$
(e) $(-\infty, -3]$ **(f)** minimum point; -4 **(g)** concave up   **5. (a)** domain: $(-\infty, \infty)$; range: $(-\infty, 2]$ **(b)** $(-3, 2)$
(c) $x = -3$ **(d)** $(-\infty, -3]$ **(e)** $[-3, \infty)$ **(f)** maximum point; 2 **(g)** concave down   **7. (a)** domain: $(-\infty, \infty)$;
range: $(-\infty, -3]$ **(b)** $(-1, -3)$ **(c)** $x = -1$ **(d)** $(-\infty, -1]$ **(e)** $[-1, \infty)$ **(f)** maximum point; -3 **(g)** concave
down   **9. (a)** $(.5, -24.5)$ **(b)** -3 and 4 **(c)** -24 **(d)** Use $[-10, 10]$ by $[-30, 10]$ for a good view.
11. (a) $(1, -16)$ **(b)** -3 and 5 **(c)** -15 **(d)** Use $[-10, 10]$ by $[-20, 10]$ for a good view.   **13. (a)** $(1.5, 4.5)$
(b) 0 and 3 **(c)** 0 **(d)** Use $[-10, 10]$ by $[-10, 10]$ for a good view.   **15. (a)** $(2.75, -42.25)$ **(b)** $-\frac{1}{2}$ and 6
(c) -12 **(d)** Use $[-5, 10]$ by $[-50, 10]$ for a good view.   **17.** B   **19.** D   **21. (a)** $(2.71, 5.20)$
(b) -1.33 and 6.74   **23. (a)** $(1.12, .56)$ **(b)** There are no x-intercepts.   **25. (a)** $(-.52, -5.00)$
(b) -1.77 and .74   **27.** C and F   **29.** A and D   **31.** $c > 0$ for A, B, and C; $c < 0$ for D, E, and F; $c = 0$
for none of these   **33.** A   **35.** ∪   **37.** ∩   **39.** ∩   **41.** ∪   **43. (a)** $\left\{-3, \frac{1}{2}\right\}$
(b) $\left(-3, \frac{1}{2}\right)$ **(c)** $(-\infty, -3) \cup \left(\frac{1}{2}, \infty\right)$ **44. (a)** $\{-.62, 1.62\}$ **(b)** $(-\infty, -.62] \cup [1.62, \infty)$ **(c)** $[-.62, 1.62]$
45. (a) $(4, -12)$ **(b)** minimum point **(c)** -12 is the minimum value. **(d)** $[-12, \infty)$ **(e)** ∪   **47. (a)** $(1.5, 2)$
(b) maximum point **(c)** 2 is the maximum value. **(d)** $(-\infty, 2]$ **(e)** ∩   **49.** $P(x) = 3x^2 + 6x - 1$
51. $P(x) = .5x^2 - 8x + 35$ **53.** $P(x) = -\frac{2}{3}x^2 - \frac{16}{3}x - \frac{38}{3}$
55. $P(x) = -2(x - 1)^2 + 4$ or
$P(x) = -2x^2 + 4x + 2$

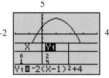

Section 3.3 (page 235)

1. (d); $\left\{-\frac{1}{3}, 7\right\}$ **3.** (c); $\{-4, 3\}$
To support graphically the results in Exercises 5–11, graph the parabola defined by the left side, the line defined by the right
side, and locate graphically the points of intersection. The x-values support the real solutions.
5. $\{\pm 4\}$ **7.** $\{\pm 3\}$ **9.** $\{\pm 4i\}$ **11.** $\left\{\pm 3i\sqrt{2}\right\}$ **13.** $\left\{1 \pm \sqrt{5}\right\}$ **15.** $\left\{-\frac{1}{2} \pm \frac{1}{2}i\right\}$

17. $\left\{\dfrac{1 \pm \sqrt{5}}{2}\right\}$ **19.** $\{3 \pm \sqrt{2}\}$ **21.** $\left\{\dfrac{3}{2} \pm \dfrac{\sqrt{2}}{2}i\right\}$ **23.** 0; one real solution (a double solution), rational

25. 84; two real solutions, both irrational   **27.** -23; no real solutions   **29.** E   **31.** D   **33.** C
35. $\{2, 4\}$ **37.** $(-\infty, 2) \cup (4, \infty)$ **39.** $(-\infty, 3) \cup (3, \infty)$ **41.** $(-\infty, \infty)$ **43.** There are no real
solutions. It has two complex solutions, both imaginary.   **45.** 3   **46.** $x = 3$ **47.** yes; negative
48. positive   **49.** $y = f(x)$ **50.** $y = g(x)$ **51.** $\{-.5, 7\}$
53. $\{-.4509456768, 1.267442258\}$

To support your answers graphically in Exercises 55–59, consider the polynomial on the left to be $P(x)$. Graph $y = P(x)$, and then determine the x-values for which the graph intersects, is above, or is below the x-axis. These values correspond to the solutions of $P(x) = 0$, $P(x) > 0$, and $P(x) < 0$, respectively.
55. (a) $(-\infty, -3] \cup [-1, \infty)$ **(b)** the interval $(-3, -1)$ **57. (a)** $\left(-\infty, \frac{1}{2}\right) \cup (4, \infty)$ **(b)** $\left[\frac{1}{2}, 4\right]$
59. (a) $\left[1 - \sqrt{2}, 1 + \sqrt{2}\right]$ **(b)** $\left(-\infty, 1 - \sqrt{2}\right) \cup \left(1 + \sqrt{2}, \infty\right)$ **61.** -4 and 2   **62.** the interval $(-4, 2)$
63. The graph of y_2 is obtained by reflecting the graph of y_1 across the x-axis.   **64.** the interval $(-4, 2)$
65. They are the same.   **66.** Because multiplying a function f by -1 causes the graph to reflect across the x-axis, the original solution set of $f(x) < 0$ is the same as the solution set of $-f(x) > 0$. The same holds true for the solution sets of $f(x) > 0$ and $-f(x) < 0$.   **67. (a)** $\{-2, 3\}$ **(b)** the interval $(-2, 3)$ **(c)** $(-\infty, -2) \cup (3, \infty)$
69. (a) $\{-.5, 2.5\}$ **(b)** $(-\infty, -.5] \cup [2.5, \infty)$ **(c)** $[-.5, 2.5]$ **71. (a)** $\{-1.12, .92\}$ **(b)** $(-\infty, -1.12) \cup (.92, \infty)$
(c) the interval $(-1.12, .92)$ **73. (a)** $\{.30, 2.82\}$ **(b)** the interval $(.30, 2.82)$ **(c)** $(-\infty, .30) \cup (2.82, \infty)$
75. $t = \dfrac{\pm \sqrt{2sg}}{g}$ **77.** $v = \dfrac{\pm\sqrt[4]{Frk^3M^3}}{kM}$ **79.** $R = \dfrac{E^2 - 2Pr \pm E\sqrt{E^2 - 4Pr}}{2P}$

Section 3.4 (page 246)

1. (a)   **3. (a)** $30 - x$ **(b)** $0 < x < 30$ **(c)** $P(x) = x(30 - x)$ or $P(x) = -x^2 + 30x$ **(d)** 15 and 15; The maximum product is 225. Locate the point $(15, 225)$ on the graph. It is the vertex.   **5. (a)** $640 - 2x$ **(b)** $0 < x < 320$
(c) $A(x) = x(640 - 2x)$ or $A(x) = -2x^2 + 640x$ **(d)** Choose x to be between approximately 57.04 feet and 85.17 feet or between approximately 234.83 feet and 262.96 feet.   **(e)** 160 feet by 320 feet; The maximum area is 51,200 square feet. Locate the point $(160, 51{,}200)$ on the graph. It is the vertex.   **7. (a)** $2x$ **(b)** length: $2x - 4$; width: $x - 4$; $x > 4$
(c) $V(x) = 2(2x - 4)(x - 4)$ or $V(x) = 4x^2 - 24x + 32$ **(d)** 8 inches by 20 inches; Locate the point $(12, 320)$ on the graph.   **(e)** 13.0 to 14.2 inches   **9. (a)** The value of t cannot be negative since t represents time elapsed from the throw.   **(b)** Since the rock was thrown from ground level, s_0, the original height of the rock, is 0.
(c) $s(t) = -16t^2 + 90t$ **(d)** 99 feet   **(e)** After 2.8125 seconds, the maximum height, 126.5625 feet, is attained. Locate the vertex $(2.8125, 126.5625)$.   **(f)** 5.625 seconds; Locate the point $(5.625, 0)$.   **11. (a)** The ball will not reach 355 feet, because the graph of $y_1 = -16x^2 + 150x$ does not intersect the graph of $y_2 = 355$.   **(b)** It reaches a height of 355 feet at $t \approx 1.4$ and $t \approx 14.2$ seconds.   **13.** 128   **15. (a)** $80 - x$ **(b)** $400 + 20x$ **(c)** $R(x) = (80 - x)(400 + 20x)$ or $R(x) = 32{,}000 + 1200x - 20x^2$ **(d)** 5 or 55   **(e)** \$1000   **17. (a)** 3.5 feet   **(b)** approximately .2 second and 2.3 seconds   **(c)** 1.25 feet   **(d)** approximately 3.78 feet   **19. (a)** approximately 19.2 hours   **(b)** 84.3 ppm (109.8 is not in the interval $[50, 100]$.)   **21.** When $x \approx 2.16$; Since 2.16 occurs during year 2, which is 1987, we would expect the minimum number of degrees to be in 1987. This number was approximately 590,000.
23. (a)   258,092   **(b)** $f(x) = 1817.95(x - 2)^2 + 620$

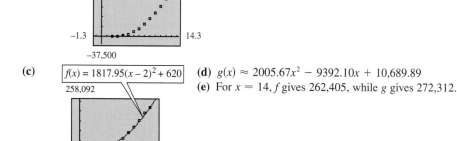

(c) $\boxed{f(x) = 1817.95(x-2)^2 + 620}$ **(d)** $g(x) \approx 2005.67x^2 - 9392.10x + 10{,}689.89$
258,092   **(e)** For $x = 14$, f gives 262,405, while g gives 272,312.

There is a very good fit.

25. (a)

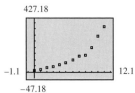

(b) $f(x) \approx 2.926x^2 + 13$ **(c)** $g(x) \approx 3.60x^2 - 11.35x + 32.71$

(d)

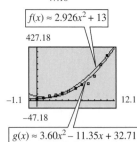

(e) For $x = 12$, f gives 434, while g gives 415. In this case, f is the better predictor.

Section 3.5 (page 264)

1. A   **3.** one   **5.** B and D   **7.** one   **9.** B   **11.** 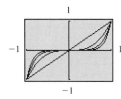   **13.** 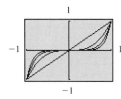   **15.**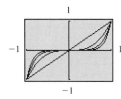
17. 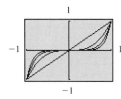   **19.** D   **21.** B
23.

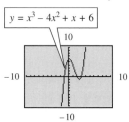

As the exponent n gets larger (but remains odd), the graph "flattens out" in the window $[-1, 1]$ by $[-1, 1]$. The graph of $y = x^7$ will lie between the x-axis and the graph of $y = x^5$ in this window.

$y = x^n$ for $n = 1, 3, 5, 7$

25. They all take the shape 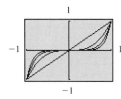 . As n gets larger, the graphs get "steeper."

27.

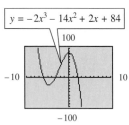

$y = x^3 - 4x^2 + x + 6$

(a) $(-\infty, \infty)$ **(b)** $(2.54, -.88)$; not an absolute minimum   **(c)** $(.13, 6.06)$; not an absolute maximum   **(d)** $(-\infty, \infty)$ **(e)** x-intercepts: $-1, 2, 3$; y-intercept: 6
(f) $(-\infty, .13]$; $[2.54, \infty)$ **(g)** $[.13, 2.54]$

29.

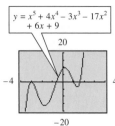

$y = -2x^3 - 14x^2 + 2x + 84$

(a) $(-\infty, \infty)$ **(b)** $(-4.74, -27.03)$; not an absolute minimum   **(c)** $(.07, 84.07)$; not an absolute maximum   **(d)** $(-\infty, \infty)$ **(e)** x-intercepts: $-6, -3.19, 2.19$; y-intercept: 84
(f) $[-4.74, .07]$ **(g)** $(-\infty, -4.74]$; $[.07, \infty)$

31.

$y = x^5 + 4x^4 - 3x^3 - 17x^2$
$+ 6x + 9$

(a) $(-\infty, \infty)$ **(b)** $(-1.73, -16.39)$; $(1.35, -3.49)$; neither is an absolute minimum
(c) $(-3, 0)$; $(.17, 9.52)$; neither is an absolute maximum   **(d)** $(-\infty, \infty)$ **(e)** x-intercepts: $-3, -.62, 1, 1.62$; y-intercept: 9   **(f)** $(-\infty, -3]$; $[-1.73, .17]$; $[1.35, \infty)$
(g) $[-3, -1.73]$; $[.17, 1.35]$

33.

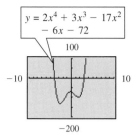

$y = 2x^4 + 3x^3 - 17x^2 - 6x - 72$

(a) $(-\infty, \infty)$ **(b)** $(-2.63, -132.69)$ is an absolute minimum; $(1.68, -99.90)$
(c) $(-.17, -71.48)$; no absolute maximum   **(d)** $[-132.69, \infty)$ **(e)** x-intercepts: -4, 3;
y-intercept: -72 **(f)** $[-2.63, -.17]$; $[1.68, \infty)$ **(g)** $(-\infty, -2.63]$; $[-.17, 1.68]$

35.

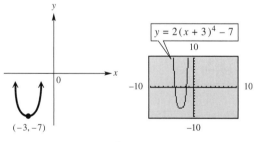

$y = -x^6 + 24x^4 - 144x^2 + 256$

(a) $(-\infty, \infty)$ **(b)** $(-2, 0)$; $(2, 0)$; neither is an absolute minimum   **(c)** $(-3.46, 256)$,
$(0, 256)$, and $(3.46, 256)$ all are absolute maximum points   **(d)** $(-\infty, 256]$
(e) x-intercepts: -4, -2, 2, 4; y-intercept: 256   **(f)** $(-\infty, -3.46]$; $[-2, 0]$; $[2, 3.46]$
(g) $[-3.46, -2]$; $[0, 2]$; $[3.46, \infty)$

37. When n is even, $P(x) = x^n$ is always concave up.   **38.** When n is odd, $P(x) = x^n$ is concave down for negative values of
x, and concave up for positive values of x. (When $x = 0$, there is a point of inflection—see Section 2.1.)
39. The graph is concave up for $x > 0$, and for $x < 0$ the graph is concave down. The point $(0, 0)$ is an inflection point.
40. The end behavior must be $\curvearrowright$, indicating downward concavity as $|x| \to \infty$.
There are many possible valid windows in Exercises 41–47. We give only one in each case.
41. $[-10, 10]$ by $[-40, 10]$ **43.** $[-10, 20]$ by $[-1500, 500]$ **45.** $[-10, 10]$ by $[-20, 500]$
47. local maximum at $(2, 3.67)$; local minimum at $(3, 3.5)$ **49.** local maximum at $(-3.33, -1.85)$; local minimum at
$(-4, -2)$ **51.** two; 2.10 and 2.15   **53.** none
55. Shift the graph of $y = x^4$ three units to the left,
stretch vertically by a factor of 2, and then shift
downward 7 units.

56. Shift the graph of $y = x^4$ one unit to the left,
stretch vertically by a factor of 3, reflect across
the x-axis, and shift upward 12 units.

57. Shift the graph of $y = x^3$ one unit to the right,
stretch vertically by a factor of 3, reflect across
the x-axis, and shift upward 12 units.

58. Shift the graph of $y = x^5$ one unit to the right,
shrink vertically by a factor of .5, and shift 13
units upward.

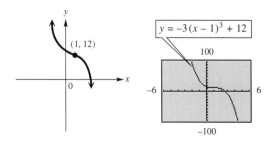

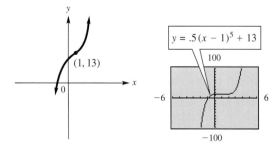

59. false   **61.** true   **63.** true   **65.** false
67. (a) Show that $P(-x) = P(x)$.   **(c)** Show that $P(-3) = 0$.
    (b) The single negative x-intercept is -3.   **(d)** The single positive x-intercept is 3.

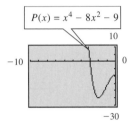

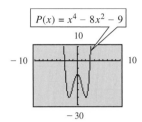

68. (a)

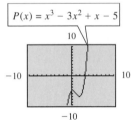

(b) It will be the same curve, but shifted 4 units to the right.
(c)

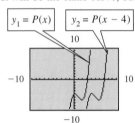

69. (a) $(-\infty, -.26]$; $[1.93, \infty)$ **(b)** $[-.26, 1.93]$
70. (a)

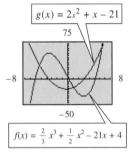

(b) $g(x) = 0$ for $x = -3.5$ and $x = 3$. The x-coordinate of the local maximum of f is -3.5, and the x-coordinate of the local minimum of f is 3.
(c) f is increasing and g is nonnegative on the intervals $(-\infty, -3.5]$ and $[3, \infty)$.
(d) f is decreasing and g is nonpositive on the interval $[-3.5, 3]$.

Section 3.6 (page 277)

1. $P(1) = -5$ and $P(2) = 2$ differ in sign, so there must be a zero of P between 1 and 2. It is approximately 1.79.
3. $P(2) = 2$ and $P(2.5) = -.25$ differ in sign, so there must be a zero of P between 2 and 2.5. It is approximately 2.39.
5. $P(2) = 16$ and $P(1.5) = -.375$ differ in sign, so there must be zero of P between 2 and 1.5. It is approximately 1.52.
7. There is a zero between 2 and 2.5.   **9.** $Q(x) = x^2 - 3x - 2$; $R = 0$ **11.** $Q(x) = 3x^2 + 4x$; $R = 3$
13. $Q(x) = x^3 - x^2 - 6x$; $R = 0$ **15.** 2   **17.** -25 **19.** -5 **21.** yes   **23.** no
25. yes   **27.** $x + 3, x - 1, x - 4$ **28.** $-3, 1, 4$ **29.** $-3, 1, 4$
30. -10; -10 **31.** $(-3, -1) \cup (4, \infty)$ **32.** $(-\infty, -3) \cup (1, 4)$ **33.** $\dfrac{-1 + \sqrt{5}}{2}, \dfrac{-1 - \sqrt{5}}{2}$
35. $\dfrac{1 + \sqrt{13}}{6}, \dfrac{1 - \sqrt{13}}{6}$ **37.** $P(x) = (x - 2)(2x - 5)(x + 3)$ **39.** $P(x) = (x + 4)(3x - 1)(2x + 1)$
41. Possible: 0 or 2 positive real zeros, 1 negative real zero; Actual: 0 positive, 1 negative
43. Possible: 1 positive real zero, 1 negative real zero; Actual: 1 positive, 1 negative
45. Possible: 0 or 2 positive real zeros, 1 or 3 negative real zeros; Actual: 0 positive, 1 negative

Section 3.7 (page 286)

Answers may vary in Exercises 1–5. We give only one possible answer.
1. $P(x) = x^3 - 8x^2 + 21x - 20$ **3.** $P(x) = x^3 - 5x^2 + x - 5$ **5.** $P(x) = x^3 - 6x^2 + 10x$
7. $P(x) = -\frac{1}{6}x^3 + \frac{13}{6}x + 2$ **9.** $P(x) = -\frac{1}{2}x^3 - \frac{1}{2}x^2 + x$ **11.** $P(x) = -x^3 + 6x^2 - 10x + 8$
13. $-1 + i, -1 - i$ **15.** $-1 + \sqrt{2}, -1 - \sqrt{2}$ **17.** $-3i, \dfrac{1}{2} + \dfrac{\sqrt{3}}{2}i, \dfrac{1}{2} - \dfrac{\sqrt{3}}{2}i$

Answers may vary in Exercises 19–23. We give only one possible answer.

19. $P(x) = x^2 - x - 20$ **21.** $P(x) = x^4 + x^3 - 5x^2 + x - 6$ **23.** $P(x) = x^4 + 2x^3 - 10x^2 - 6x + 45$

25. Use synthetic division twice, with $k = -2$.   **27.** 1, 3, 5

$$
\begin{array}{r}
-2)\overline{1 \quad 2 \quad -7 \quad -20 \quad -12} \\
\underline{-2 \quad 0 \quad 14 \quad 12} \\
-2)\overline{1 \quad 0 \quad -7 \quad -6 \quad 0} \\
\underline{-2 \quad 4 \quad 6} \\
1 \quad -2 \quad -3 \quad 0
\end{array}
$$

$x^2 - 2x - 3 = (x - 3)(x + 1)$

Zeros are $-2, 3, -1$.

$P(x) = (x + 2)^2(x - 3)(x + 1)$

29. The polynomial must also have a zero of $1 - i$, which would make it impossible for it to be of degree 3.

31.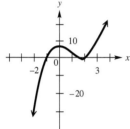

$P(x) = 2x^3 - 5x^2 - x + 6$

33.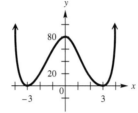

$P(x) = x^4 - 18x^2 + 81$

35.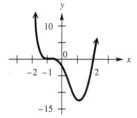

$P(x) = 2x^4 + x^3 - 6x^2 - 7x - 2$

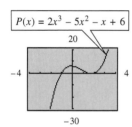

$P(x) = 2x^3 - 5x^2 - x + 6$

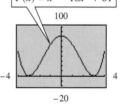

$P(x) = x^4 - 18x^2 + 81$

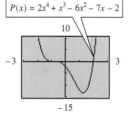

$P(x) = 2x^4 + x^3 - 6x^2 - 7x - 2$

37. **(a)** $\pm 1, \pm 2, \pm 5, \pm 10$ **(b)** Eliminate values less than -2 or greater than 5.   **(c)** $-2, -1, 5$
(d) $(x + 2)(x + 1)(x - 5)$ **39.** **(a)** $\pm 1, \pm 2, \pm 3, \pm 5, \pm 6, \pm 10, \pm 15, \pm 30$ **(b)** Eliminate values less than -5 or
greater than 2.   **(c)** $-5, -3, 2$ **(d)** $(x + 5)(x + 3)(x - 2)$ **41.** **(a)** $\pm 1, \pm 2, \pm 3, \pm 4, \pm 6, \pm 12, \pm\frac{1}{2}, \pm\frac{1}{3}, \pm\frac{1}{6}, \pm\frac{2}{3}, \pm\frac{3}{2}, \pm\frac{4}{3}$
(b) Eliminate values less than -4 or greater than $\frac{3}{2}$.   **(c)** $-4, -\frac{1}{3}, \frac{3}{2}$ **(d)** $(x + 4)(3x + 1)(2x - 3)$ **43.** **(a)** The

possible rational zeros are of the form $\dfrac{p}{q}$, where $p \in \{\pm 1, \pm 2, \pm 3, \pm 6\}$ and $q \in \{\pm 1, \pm 2, \pm 3, \pm 4, \pm 6, \pm 12\}$.   **(b)** Eliminate

values less than $-\frac{3}{2}$ or greater than $\frac{1}{2}$.   **(c)** $-\frac{3}{2}, -\frac{2}{3}, \frac{1}{2}$ **(d)** $(3x + 2)(2x + 3)(2x - 1)$ **45.** $-2, -1, \frac{5}{2}$ **47.** $\frac{3}{2}, 4$
49. $P(k) = (k - k) \cdot Q(k) + R$
$\qquad\quad = 0 \cdot Q(k) + R$
$\qquad\quad = 0 + R$
$\qquad\quad = R$
Therefore, when $P(x)$ is divided by $x - k$, the remainder R is equal to $P(k)$.

Section 3.8 (page 298)

1. $\{-2, -1.5, 1.5, 2\}$ **3.** $\{-4, 4, -i, i\}$ **5.** $\{-8, 1, 8\}$ **7.** $\{-1.5, 0, 1\}$ **9.** $\left\{-\sqrt{7}, -1, \sqrt{7}\right\}$; Use

$\pm\sqrt{7} \approx \pm 2.65$ to support these two solutions.   **11.** $\left\{\dfrac{-1 - \sqrt{73}}{6}, 0, \dfrac{-1 + \sqrt{73}}{6}\right\}$; Use $\dfrac{-1 - \sqrt{73}}{6} \approx -1.59$ and

$\dfrac{-1 + \sqrt{73}}{6} \approx 1.26$ to support these two solutions.   **13.** $\left\{0, -\dfrac{1}{2} - i\dfrac{\sqrt{3}}{2}, -\dfrac{1}{2} + i\dfrac{\sqrt{3}}{2}\right\}$ **15.** $\{-4i, -i, i, 4i\}$

17. $\left\{2, -3, -1 \pm i\sqrt{3}, \frac{3}{2} \pm \frac{3}{2}i\sqrt{3}\right\}$ **19.** $\left\{\pm\sqrt{\dfrac{6 + \sqrt{33}}{3}}, \pm\sqrt{\dfrac{6 - \sqrt{33}}{3}}\right\}$; Approximations are $\pm.29$ and ±1.98.

21. It is not a comprehensive graph, because end behavior as $x \to \infty$ is not shown, and not all x-intercepts are shown.
22. symmetry with respect to the y-axis   **23.** $\left\{-5, -\sqrt{3}, \sqrt{3}, 5\right\}$ **24.** Because it is of degree 4 and there are four real solutions, there can be no imaginary solutions.   **25. (a)** $\{-2, 1, 4\}$ **(b)** $(-\infty, -2) \cup (1, 4)$
(c) $(-2, 1) \cup (4, \infty)$ **27. (a)** $\{-2.5, 1, 3 \text{ (mult. 2)}\}$ **(b)** $(-2.5, 1)$ **(c)** $(-\infty, -2.5) \cup (1, 3) \cup (3, \infty)$
29. (a) $\{-3 \text{ (mult. 2)}, 0, 2\}$ **(b)** $\{-3\} \cup [0, 2]$ **(c)** $(-\infty, 0] \cup [2, \infty)$
31. The given root is approximately 2.8473221.   **33.** The given root is approximately .22102254.

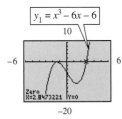

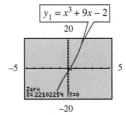

35. (a) 2   **(b)**   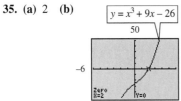   **(c)** $-1 \pm 2i\sqrt{3}$ **37.** $\{-.88, 2.12, 4.86\}$ **39.** $\{1.52\}$

41. $\{-.40, 2.02\}$ **43.** $\{-i, i\}$ **45.** $\left\{-1, \dfrac{1}{2} - \dfrac{\sqrt{3}}{2}i, \dfrac{1}{2} + \dfrac{\sqrt{3}}{2}i\right\}$ **47.** $\left\{3, -\dfrac{3}{2} - \dfrac{3\sqrt{3}}{2}i, -\dfrac{3}{2} + \dfrac{3\sqrt{3}}{2}i\right\}$

49. $\{-2, 2, -2i, 2i\}$ **51.** $\left\{-1, 1, -\dfrac{1}{2} \pm \dfrac{\sqrt{3}}{2}i, \dfrac{1}{2} \pm \dfrac{\sqrt{3}}{2}i\right\}$ **53. (a)** $0 < x < 6$

(b) $V(x) = x(12 - 2x)(18 - 2x)$ or $V(x) = 4x^3 - 60x^2 + 216x$ **(c)** $x \approx 2.35$; maximum volume ≈ 228.16 cubic inches

(d) between .42 and 5   **55.** approximately 2.61 inches   **57. (a)** $x - 1$ **(b)** $\sqrt{x^2 - (x - 1)^2}$
(c) $2x^3 - 5x^2 + 4x - 28{,}225 = 0$ **(d)** hypotenuse: 25 inches; legs: 24 inches and 7 inches   **59.** The maximum volume is approximately 66.15 cubic inches, when $x \approx 1.59$ inches.
61. $y = 13{,}333.\overline{3}x^3 - 87{,}000x^2 + 122{,}666.\overline{6}x + 118{,}000$
(a)   **(b)** 167,000   **(c)** There is quite a large discrepancy.

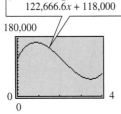

63. (a)

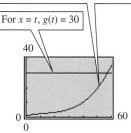

For $x = t$,
$f(t) = 2.8 \times 10^{-4}t^3 - .011t^2 + .23t + .93$

For $x = t$, $g(t) = 30$

(b) The graphs intersect at $x = t \approx 56.9$. Since $t = 0$ corresponds to 1930, this would be during 1986.
(c) An increasing percentage of females have smoked during this time period. Smoking has been shown to increase the likelihood of lung cancer.

65. (a) $y \approx 13.36x^2 + 41.56x + 150.8$ **(b)** $y \approx -5.75x^3 + 65.11x^2 - 94.14x + 247.4$

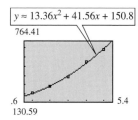

$y \approx 13.36x^2 + 41.56x + 150.8$

764.41

.6   5.4
130.59

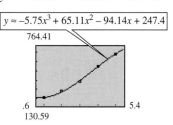

$y \approx -5.75x^3 + 65.11x^2 - 94.14x + 247.4$

764.41

.6   5.4
130.59

(c) $y \approx -4.71x^4 + 50.75x^3 - 168.29x^2 + 289.25x + 44$ **(d)** The quadratic function gives 881, the cubic gives 785, and the quartic gives 579. The quadratic function is the best predictor in this case.

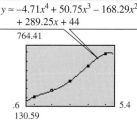

$y \approx -4.71x^4 + 50.75x^3 - 168.29x^2 + 289.25x + 44$

764.41

.6   5.4
130.59

67. (a) If the length L of the pendulum increases, so does the period of oscillation T.   **(b)** There are a number of ways. One way is to realize that $k = \dfrac{L}{T^n}$ for some integer n. The ratio should be the constant k for each data point when the correct n is found.   **(c)** $k \approx .81; n = 2$ **(d)** 2.48 seconds   **(e)** T increases by a factor of $\sqrt{2} \approx 1.414$.

Chapter 3 Review Exercises (page 307)

1. $18 - 4i$ **3.** $14 - 52i$ **5.** $\frac{1}{10} + \frac{3}{10}i$ **7.** $(-\infty, \infty)$ **9.** $\cup$ **11.** -8 **13.** $\left[\frac{3}{2}, \infty\right); \left(-\infty, \frac{3}{2}\right]$
15. The graph intersects the x-axis at -1 and 4, supporting the answer in (a). It lies above the x-axis when $x < -1$ or $x > 4$, supporting the answer in (b). It lies on or below the x-axis when x is between -1 and 4 (inclusive), supporting the answer in (c).   **17.** The graph is concave up for all values in the domain.   **19.** $\{-.52, 2.59\}$ **21.** $(1.04, 6.37)$
23. (a) 3 **(b)** -3 **(c)** none **(d)** two   **25.** 25 seconds   **27.** 6.3 seconds and 43.7 seconds
29. (a) $V(x) = 4(3x - 8)(x - 8)$ **(b)** 20 inches by 60 inches   **(c)** One way is to show that the graphs of $y_1 = 2496$ and $y_2 = V(x)$ intersect at the point where $x = 20$.   **31.** $Q(x) = x^2 + 4x + 1; R = -7$ **33.** -1 **35.** 28
37. $7 - 2i$

In Exercises 39 and 41, other answers are possible.
39. $P(x) = x^3 - 13x^2 + 46x - 48$ **41.** $P(x) = x^4 + 5x^3 + x^2 - 9x + 2$ **43.** yes
45. $P(x) = -2x^3 + 6x^2 + 12x - 16$

47. Any polynomial that can be factored as $a(x - b)^3$ works. One example is $P(x) = 2(x - 1)^3$.

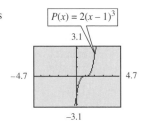

49. $1, -\frac{1}{2}, \pm 2i$ **51.** It has three real solutions. The one that is an integer is 3.   **53.** $\dfrac{-1 + \sqrt{5}}{2}, \dfrac{-1 - \sqrt{5}}{2}$

55. **(a)** $\left(\dfrac{-1 - \sqrt{5}}{2}, \dfrac{-1 + \sqrt{5}}{2} \right) \cup (3, \infty)$ **(b)** $\left(-\infty, \dfrac{-1 - \sqrt{5}}{2} \right] \cup \left[\dfrac{-1 + \sqrt{5}}{2}, 3 \right]$

57. $(x + 2)(x - 1)(x - 3)^2$ **59.** odd   **61.** none   **63.** the open interval (d, h) **65.** $r - pi$

67. A factor is $x + 4$. The zeros are $-4, \dfrac{5 + \sqrt{21}}{2}$, and $\dfrac{5 - \sqrt{21}}{2}$. The factored form is

$(x + 4)\left(x - \left(\dfrac{5 + \sqrt{21}}{2} \right) \right)\left(x - \left(\dfrac{5 - \sqrt{21}}{2} \right) \right)$.   **69.** false   **71.** true   **73.** true   **75.** two

77. $Q(x) = -2x^4 + 5x^3 + 4x^2 - 12x$ **79.** $(-.97, -54.15)$ **81.** **(a)** $\left[-\sqrt{7}, -\frac{2}{3} \right] \cup \left[\sqrt{7}, \infty \right)$

(b) $\left(-\infty, -\sqrt{7} \right) \cup \left(-\frac{2}{3}, \sqrt{7} \right)$ **83.** 4 inches by 4 inches by 4 inches

85. approximately 2044 thousand (i.e., 2,044,000)

Chapter 3 Test (page 312)

1. **(a)** $20 - 9i$ **(b)** $4 - i$ **(c)** i **(d)** $12 + 16i$ **2.** **(a)** $(-1, 8)$

(b)

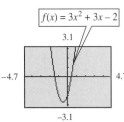

(c) $-3, 1$

(d) 6
(e) domain: $(-\infty, \infty)$; range: $(-\infty, 8]$
(f) increasing: $(-\infty, -1]$; decreasing: $[-1, \infty)$

3. **(a)** $\left\{ \dfrac{-3 \pm \sqrt{33}}{6} \right\}$ **(b)**

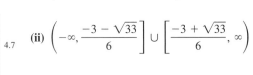

(i) $\left(\dfrac{-3 - \sqrt{33}}{6}, \dfrac{-3 + \sqrt{33}}{6} \right)$

(ii) $\left(-\infty, \dfrac{-3 - \sqrt{33}}{6} \right] \cup \left[\dfrac{-3 + \sqrt{33}}{6}, \infty \right)$

4. **(a)**

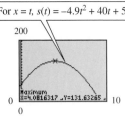

(b) approximately 4.08 seconds
(c) approximately 131.63 meters
(d) $-4.9t^2 + 40t + 50 = 0$; approximately 9.26 seconds

5. (a)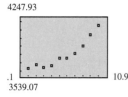

(b) $f(x) \approx 6.531(x - 1)^2 + 3629$

(c) $y \approx 7.996x^2 - 33.098x + 3679.58$

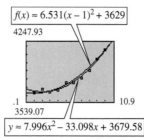

6. (a) $-1 + i\sqrt{3}, -1 - i\sqrt{3}$ **(b)** ⌣

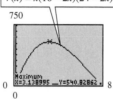

7. (a) $\left\{-\frac{5}{2}, \frac{5}{2}, -i, i\right\}$ **(b)**

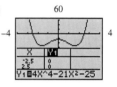

(c) It is symmetric with respect to the y-axis. **(d) (i)** $\left(-\infty, -\frac{5}{2}\right] \cup \left[\frac{5}{2}, \infty\right)$ **(ii)** $\left(-\frac{5}{2}, \frac{5}{2}\right)$

8. (a) approximately .189, 1, approximately 3.633 **(b)** two   **9. (a)** $V(x) = x(16 - 2x)(24 - 2x), 0 < x < 8$

(b) approximately 3.14 inches   **(c)** approximately 540.83 cubic inches

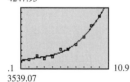

10. (a) $y \approx .661x^3 - 2.904x^2 + 17.177x + 3622.9$ **(b)** $y \approx -.0427x^4 + 1.600x^3 - 9.777x^2 + 35.959x + 3608.25$

(c)

(d) the quartic function in part (b)

Both functions are
graphed over the
data points.

CHAPTER 4

Section 4.1 (page 329)

1. A, B, C   **3.** A   **5.** A   **7.** A, C, D

9. To obtain the graph of f, stretch the graph of $y = \frac{1}{x}$ by a factor of 2.

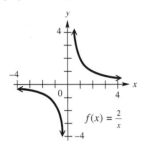

$f(x) = \frac{2}{x}$

11. To obtain the graph of f, shift the graph of $y = \frac{1}{x}$ 2 units to the left.

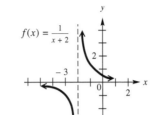

$f(x) = \frac{1}{x+2}$

$x = -2$

13. To obtain the graph of f, shift the graph of $y = \frac{1}{x}$ 1 unit up.

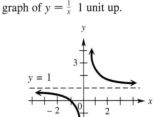

$y = 1$

$f(x) = \frac{1}{x} + 1$

15. To obtain the graph of f, reflect the graph of $y = \frac{1}{x^2}$ across the x-axis and stretch vertically by a factor of 2.

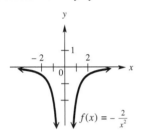

$f(x) = -\frac{2}{x^2}$

17. To obtain the graph of f, shift the graph of $y = \frac{1}{x^2}$ 3 units to the right.

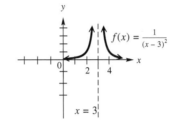

$f(x) = \frac{1}{(x-3)^2}$

$x = 3$

19. C   **21.** B

In Exercises 23–29, V.A. represents vertical asymptote, H.A. represents horizontal asymptote, and O.A. represents oblique asymptote.

23. V.A.: $x = 5$; H.A.: $y = 0$ **25.** V.A.: $x = -\frac{1}{2}$; H.A.: $y = -\frac{3}{2}$ **27.** V.A.: $x = -3$; O.A.: $y = x - 3$

29. V.A.: $x = -2$, $x = \frac{5}{2}$; H.A.: $y = \frac{1}{2}$ **31.** (a)   **33.** Set the denominator, $x^2 + x + 4$, equal to 0 and solve. Any real solutions give the x-values of the vertical asymptotes.   **34.** It has no real solutions if the discriminant is negative.

35. The complex solutions are $x = -\frac{1}{2} \pm \frac{i\sqrt{15}}{2}$. There are no real solutions. The function f has no vertical asymptotes.

36. The connected mode is acceptable because this rational function is continuous on its entire domain.

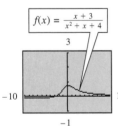

$f(x) = \frac{x+3}{x^2+x+4}$

37.

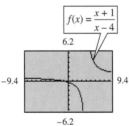

$f(x) = \frac{x+1}{x-4}$

39.

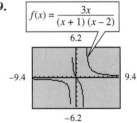

$f(x) = \frac{3x}{(x+1)(x-2)}$

41.

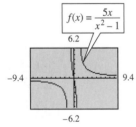

$f(x) = \frac{5x}{x^2-1}$

43.

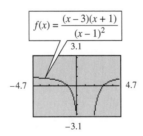

$f(x) = \frac{(x-3)(x+1)}{(x-1)^2}$

45.

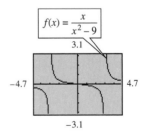

$f(x) = \frac{x}{x^2-9}$

47.
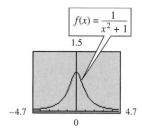
$$f(x) = \frac{1}{x^2 + 1}$$

49.
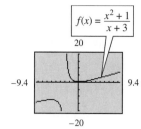
$$f(x) = \frac{x^2 + 1}{x + 3}$$

51.
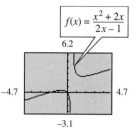
$$f(x) = \frac{x^2 + 2x}{2x - 1}$$

53.
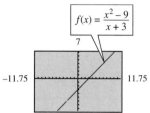
$$f(x) = \frac{x^2 - 9}{x + 3}$$

55. $p = 4, q = 2$ **57.** $p = -2, q = -1$

59. $f(x) = \dfrac{(x - 3)(x + 2)}{(x - 2)(x + 2)}$ or $f(x) = \dfrac{x^2 - x - 6}{x^2 - 4}$

61. $f(x) = \dfrac{x - 2}{x(x - 4)}$ or $f(x) = \dfrac{x - 2}{x^2 - 4x}$

63.
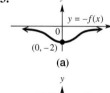
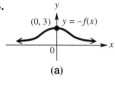

64.
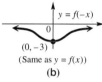

65.
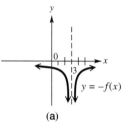

66.
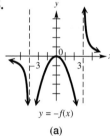
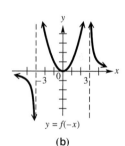

67. $y = -2x - 8$
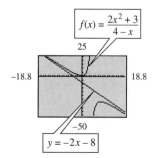
$$f(x) = \frac{2x^2 + 3}{4 - x}$$
$y = -2x - 8$

69. $y = -x + 3$
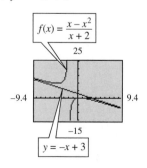
$$f(x) = \frac{x - x^2}{x + 2}$$
$y = -x + 3$

71. $y = x - 5$; the graph of f intersects it at $(.5, -4.5)$.
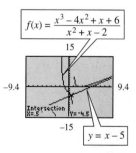
$$f(x) = \frac{x^3 - 4x^2 + x + 6}{x^2 + x - 2}$$
$y = x - 5$

73. (a) $g(x) = x - 3$ **(b)** f is undefined at $x = -3$, as indicated by the error message. **(c)** The "hole" is at $(-3, -6)$.
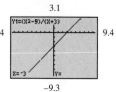

75. (a) The horizontal asymptote has equation $y = 0$ (the x-axis).   **(b)** The horizontal asymptote has equation $y = \dfrac{a}{b}$, where a is the leading coefficient of $p(x)$, and b is the leading coefficient of $q(x)$.   **(c)** The oblique asymptote has equation $y = ax + b$, where $ax + b$ is the quotient (with remainder disregarded) found by dividing $p(x)$ by $q(x)$.
77. As $|x| \to \infty$, $Y_1 \to 3$.

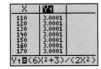

Section 4.2 (page 342)

1. (a) $\emptyset$ **(b)** $(-\infty, -2)$ **(c)** $(-2, \infty)$ **3. (a)** $\{-1\}$ **(b)** $(-1, 0)$ **(c)** $(-\infty, -1) \cup (0, \infty)$ **5. (a)** $\{0\}$
(b) $(-2, 0) \cup (2, \infty)$ **(c)** $(-\infty, -2) \cup (0, 2)$ **7. (a)** $\{2.5\}$ **(b)** $(2.5, 3)$ **(c)** $(-\infty, 2.5) \cup (3, \infty)$ **9. (a)** $\{3\}$
(b) $(-5, 3]$ **(c)** $(-\infty, -5) \cup [3, \infty)$ **11. (a)** $\emptyset$ **(b)** $(-\infty, -2)$ **(c)** $(-2, \infty)$ **13. (a)** $\left\{\frac{9}{5}\right\}$
(b) $(-\infty, 1) \cup \left(\frac{9}{5}, \infty\right)$ **(c)** $\left(1, \frac{9}{5}\right)$ **15. (a)** $\{-2\}$ **(b)** $(-\infty, -2] \cup (1, 2)$ **(c)** $[-2, 1) \cup (2, \infty)$ **17. (a)** $\emptyset$
(b) $\emptyset$ **(c)** $(-\infty, 2) \cup (2, \infty)$ **19. (a)** $\emptyset$ **(b)** $(-\infty, -1)$ **(c)** $(-1, \infty)$ **21.** If an appropriate window is not used, there may be parts of the graph that are not visible. This is particularly true for rational functions. For example, in this window, the U-shaped portion of the graph that lies in the interval $(1, 2)$ is not visible at all. We must know how to solve analytically because incorrect conclusions may be drawn from windows that do not show comprehensive graphs.

23. $\{4, 9\}$ **25.** $\left\{\dfrac{-3 - \sqrt{29}}{2}, \dfrac{-3 + \sqrt{29}}{2}\right\}$ -4.193 and 1.1926 are approximations for these exact values.

In Exercises 27–35, real solutions are verified in the same manner as in Exercises 23–25.

27. $\left\{\dfrac{3 - \sqrt{3}}{3}, \dfrac{3 + \sqrt{3}}{3}\right\}$ **29.** $\left\{\pm\frac{1}{2}, \pm i\right\}$ **31.** $\emptyset$ **33.** $\{-10\}$ **35.** $\left\{\frac{27}{56}\right\}$ **37.** $(3, \infty)$
38. $(-6, \infty)$ **39.** $(-1, 1]$ **40.** $\left(-5, \frac{1}{2}\right]$ **41.** $\emptyset$ **42.** $\emptyset$ **43. (a)** $\{-3.54\}$
(b) $(-\infty, -3.54) \cup (1.20, \infty)$ **(c)** $(-3.54, 1.20)$ **45.** $-1.30, 1, 2.30$

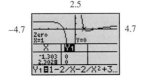

46. $(-\infty, 0) \cup (0, \infty)$ **47.** The equation is $x^2 - x - 3 = 0$, and its solutions are $\dfrac{1 \pm \sqrt{13}}{2}$.

48. $\dfrac{1 - \sqrt{13}}{2} \approx -1.30$ and $\dfrac{1 + \sqrt{13}}{2} \approx 2.30$ **49.** His solution is not correct. If $x + 2 < 0$, it is necessary to reverse the direction of the inequality symbol. He should use a sign graph or the method explained in the exposition preceding Exercise 37 in this section.

51. (a) 26 per minute   **(b)** 5 parking attendants   **53. (a)** 1   **(b)** One possibility is that there are more possible causes of death as an individual gets older.

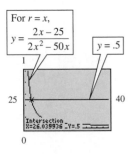

55. The graphs of the function and rational approximation are given on the same screen. From the graphs, we can see that all the rational approximations give excellent results on the interval [1, 15].
(a) choice (iii), $r_3(x)$ **(b)** choice (ii), $r_2(x)$ **(c)** choice (iv), $r_4(x)$ **(d)** choice (i), $r_1(x)$

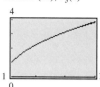   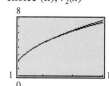   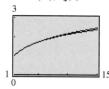   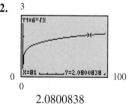

57. 5 times   **59.** inversely; height; 24   **61.** $\frac{32}{15}$ **63.** $\frac{18}{125}$
65. increases; decreases   **67.** .0444 ohm   **69.** 4.3 vibrations per second   **71.** 26.67 days
73. 7500 pounds   **75.** 4   **77.** .004 second   **79.** 5.1   **81.** -1.4

Section 4.3 (page 357)

1. 13   **3.** -2 **5.** 729   **7.** $\frac{1}{25}$ **9.** 100
The number of displayed digits may vary in Exercises 11–27.
11. (a) 6.2449979984   **(b)** 11.9916637711   **(c)** 95.2417975471   **13.** 1.44224957 (rational approximation)
15. 2.65 (exact value)   **17.** -2.571281591 (rational approximation)   **19.** 1.464591888 (rational approximation)
21. 1.174618943 (rational approximation)   **23.** 1.267463962 (rational approximation)   **25.** .0322515344
(rational approximation)   **27.** 1.181352075 (rational approximation)   **29. (a)** .125; $\frac{1}{8}$
(b) $\left(\sqrt[4]{16}\right)^{-3} = \sqrt[4]{16^{-3}}$ (There are others.) Each is equal to .125.   **(c)** Show that $.125 = \frac{1}{8}$.
31. Because $a^{1/2} = \sqrt{a}$, $\sqrt{\sqrt{a}} = (a^{1/2})^{1/2} = a^{1/4} = \sqrt[4]{a}$, and $\sqrt{\sqrt{\sqrt{a}}} = [(a^{1/2})^{1/2}]^{1/2} = a^{1/8} = \sqrt[8]{a}$. Therefore,
$\sqrt[16]{65{,}536} = \sqrt{\sqrt{\sqrt{\sqrt{65{,}536}}}} = 65{,}536^{1/16} = 2$.

33. $2.3^{1/3} = \sqrt[3]{2.3} \approx 1.3200061$ **35.** $4^{-1/2} = \sqrt{4^{-1}} = .5$ **37.** $7^{3/5} = \sqrt[5]{7^3} \approx 3.2140958$
39.

2.080083823

40.

2.080083823

41.

In this table, $Y_1 = \sqrt[6]{x}$.
2.08008382305

42. 3

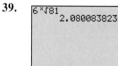

2.0800838

43. No, π is not exactly equal to $\sqrt[4]{\frac{2143}{22}}$. The two decimal *approximations* agree to eight decimal places.
45. $\left[-\frac{5}{4}, \infty\right)$ **47.** $(-\infty, 6]$ **49.** $(-\infty, \infty)$ **51.** $[-7, 7]$ **53.** $[-1, 0] \cup [1, \infty)$
55. $(-\infty, -1] \cup [7, \infty)$
56. $(-\infty, -1] \cup [7, \infty)$

57. When $f(x) \geq 0$, $g(x)$ is a real number; or, the solution set of $f(x) \geq 0$ is the domain of g.

58. The domain is $(-\infty, \infty)$, because $x^2 + 10$ is always positive.

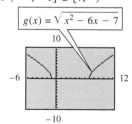

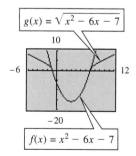

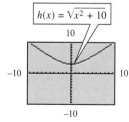

59. Because $h(-x) = h(x)$, the graph is symmetric with respect to the y-axis.   **60. (a)** $\emptyset$ **(b)** $(-\infty, \infty)$ **(c)** $\emptyset$
In Exercises 61–67, we do not give the graphs, only the answers to parts (a)–(d).
61. (a) $[0, \infty)$ **(b)** $\left[-\frac{5}{4}, \infty\right)$ **(c)** none   **(d)** $\{-1.25\}$ **63. (a)** $(-\infty, 0]$ **(b)** $(-\infty, 6]$ **(c)** none   **(d)** $\{6\}$
65. (a) $(-\infty, \infty)$ **(b)** $(-\infty, \infty)$ **(c)** none   **(d)** $\{3\}$ **67. (a)** $[0, 7]$ **(b)** $[-7, 0]$ **(c)** $[0, 7]$ **(d)** $\{-7, 7\}$

69. Shift the graph of $y = \sqrt{x}$ 3 units to the left, and stretch vertically by a factor of 3.   **71.** Shift the graph of
$y = \sqrt{x}$ 4 units to the left, stretch vertically by a factor of $\sqrt{7}$, and shift 4 units up.   **73.** Shift the graph of $y = \sqrt[3]{x}$
2 units to the left, stretch vertically by a factor of 3, and shift 5 units down.

75. The graph is a circle.
$$y_1 = \sqrt{100 - x^2};\ y_2 = -\sqrt{100 - x^2}$$

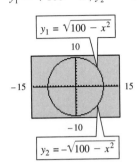

77. The graph is a (shifted) circle.
$$y_1 = \sqrt{9 - (x - 2)^2};\ y_2 = -\sqrt{9 - (x - 2)^2}$$

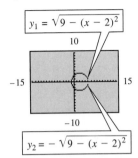

79. The graph is a horizontal parabola.
$$y_1 = -3 + \sqrt{x};\ y_2 = -3 - \sqrt{x}$$

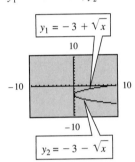

81. The graph is a horizontal parabola.
$$y_1 = -2 + \sqrt{.5x + 3.5};\ y_2 = -2 - \sqrt{.5x + 3.5}$$

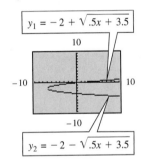

Section 4.4 (page 369)

1. (a) $\{4\}$ (b) $[4, \infty)$ (c) $[-5, 4]$ **3.** (a) $\{-3\}$ (b) $(-\infty, -3)$ (c) $(-3, \infty)$ **5.** (a) $\{-1\}$
(b) $\left[-\frac{7}{3}, -1\right)$ (c) $(-1, \infty)$ **7.** (a) $\{4\}$ (b) $(4, \infty)$ (c) $\left[-\frac{4}{3}, 4\right)$ **9.** (a) $\left\{\frac{5}{4}\right\}$ (b) $\left[1, \frac{5}{4}\right]$ (c) $\left[\frac{5}{4}, \infty\right)$
11. (a) $\{4, 20\}$ (b) $[2, 4) \cup (20, \infty)$ (c) $(4, 20)$ **13.** (a) $\left\{\frac{1}{4}, 1\right\}$ (b) $\left(-\infty, \frac{1}{4}\right) \cup (1, \infty)$ (c) $\left(\frac{1}{4}, 1\right)$
15. (a) $\{31\}$ (b) $(31, \infty)$ (c) $[15, 31)$ **17.** (a) $\{-3, 1\}$ (b) $(-\infty, -3) \cup (1, \infty)$ (c) $(-3, -2] \cup [0, 1)$
19. (a) $\left\{\frac{1}{4}, 1\right\}$ (b) $\left(-\infty, \frac{1}{4}\right) \cup (1, \infty)$ (c) $\left(\frac{1}{4}, 1\right)$ **21.** (a) $\emptyset$ (b) $\emptyset$ (c) $\left[-\frac{2}{3}, 1\right]$

23. Solution set: $\left\{\dfrac{7 - \sqrt{13}}{2}\right\}$ **25.** Solution set: $\emptyset$ **27.** Solution set: $\{0, 1\}$

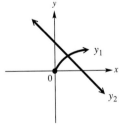

One point of intersection

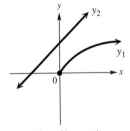

No points of intersection

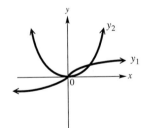

Two points of intersection

29. The graphs of $y_1 = \sqrt{x}$ and $y_2 = -x - 5$ have no points of intersection. Thus, the equation $y_1 = y_2$ (or $\sqrt{x} = -x - 5$) has no solution.

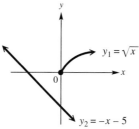

31. $(4x - 4)^{1/3} = (x + 1)^{1/2}$

32. 6   **33.** $(4x - 4)^2 = (x + 1)^3$ **34.** $16x^2 - 32x + 16 = x^3 + 3x^2 + 3x + 1$, and, thus, $x^3 - 13x^2 + 35x - 15 = 0$.

35. The equation has three real roots.

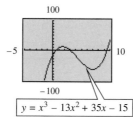

36.
$$3\overline{)1 \quad -13 \quad 35 \quad -15}$$
$$ 3 \quad -30 \quad 15$$
$$\overline{1 \quad -10 \quad 5 \quad 0}$$
$$P(3) = 0$$

37. $P(x) = (x - 3)(x^2 - 10x + 5)$ **38.** $\left\{5 \pm 2\sqrt{5}\right\}$ **39.** $3, 5 + 2\sqrt{5}, 5 - 2\sqrt{5}$

40. The original equation has two real solutions.   **41.** $\left\{3, 5 + 2\sqrt{5}\right\}$; for the calculator solution, $5 + 2\sqrt{5} \approx 9.47$

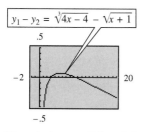

42. The solution set of the original equation is a subset of the solution set of the equation in Exercise 34. The extraneous solution $5 - 2\sqrt{5}$ was obtained when both sides of the original equation were raised to the sixth power.

43. 5.35 feet   **45. (a)** 173 miles **(b)** 211 miles   **47. (a)** about 133 species **(b)** about 327 species

49. (a) $-60.9°F$ **(b)** $-64.4°F$ **50. (a)** $-63.4°F$ **(b)** $-46.7°F$ **51. (a)** $-63°F$ **(b)** $-47°F$

52. The formula in Exercise 50 provides a better model.   **53. (a)** $20 - x$ **(b)** x must be between 0 and 20.

(c) $AP = \sqrt{x^2 + 12^2}$; $BP = \sqrt{(20 - x)^2 + 16^2}$ **(d)** $f(x) = \sqrt{x^2 + 12^2} + \sqrt{(20 - x)^2 + 16^2}, 0 < x < 20$

(e)

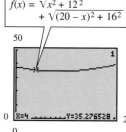

$f(4) \approx 35.28$. This means that when the stake is 4 feet from the base of the 12-foot pole, approximately 35.28 feet of wire will be required.

(f) When $x \approx 8.57$ feet, $f(x)$ is a minimum (approximately 34.41 feet).

(g) This problem has examined how the total amount of wire used can be expressed in terms of the distance from the stake at P to the base of the 12-foot pole. We find that the amount of wire used can be minimized when the stake is approximately 8.57 feet from the 12-foot pole.

55. Since $x \approx 1.31$, the hunter must travel $8 - x \approx 8 - 1.31 = 6.69$ miles along the river.   **57.** At about 1:20 P.M., they will be approximately 33.28 miles apart, their minimum distance from each other.   **59.** Values of x less than 15 will give a negative number under a square root radical. The calculator will give only real number values for Y_1.

61. $\{4\}$ **63.** $\{-.4542187292\}$

Section 4.5 (page 381)

1. The screens suggest they are inverse functions.   **3.** They are not inverse functions.   **5.** one-to-one
7. x; $(g \circ f)(x)$ **9.** (b, a) **11.** $y = x$ **13.** does not; it is not one-to-one   **15.** one-to-one
17. not one-to-one   **19.** one-to-one   **21.** not one-to-one   **23.** one-to-one   **25.** one-to-one
27. not one-to-one   **29.** not one-to-one   **31.** one-to-one   **33.** Because the end behavior will be either ⤴
or ⤵ , the graph will fail the horizontal line test.   **35.** untying your shoelaces
37. leaving a room   **39.** yes   **41.** no   **43.** yes   **45.** no   **47.** yes

49.
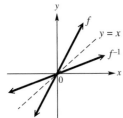
51.
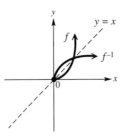
53. $f^{-1}(x) = \dfrac{x + 4}{3}$
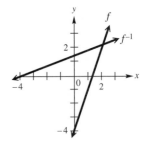

55. $f^{-1}(x) = 3x$

57. $f^{-1}(x) = \sqrt[3]{x - 1}$
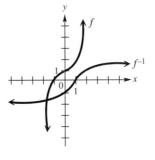

59. It is not one-to-one. For example, $f(1) = 1$ and $f(-1) = 1$, $f(3) = 81$ and $f(-3) = 81$, and so on.
61. $f^{-1}(x) = \frac{1}{2}x + 4$

62.
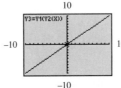
This is the identity function.

63.
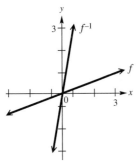
For each input x, the output is also x.
64.
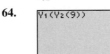
For example, suppose the name is GARBARINO. Then, $k = 9$, and the result is 9. This will happen for *any* value of k; the output is also k.

GARBARINO has 9 letters.

In Exercises 65–69, we give only one of several possible choices for domain restriction.
65. $[0, \infty)$ **67.** $[6, \infty)$ **69.** $[0, \infty)$

Answers to Exercises 71 and 73 are based on the restrictions chosen in Exercises 65 and 67.

71. $f^{-1}(x) = \sqrt{4 - x}$ **73.** $f^{-1}(x) = x + 6 \ (x \geq 0)$ **75.** $f^{-1}(x) = \dfrac{x + 2}{3}$; the message reads MIGUEL HAS

ARRIVED.   **77.** 6858   124   2743   63   511   124   1727   4095; $f^{-1}(x) = \sqrt[3]{x + 1}$

Chapter 4 Review Exercises (page 388)

1. Shift the graph of $y = \dfrac{1}{x}$ three units to the left, stretch vertically by a factor of 2, and reflect across the x-axis.

3. Stretch the graph of $y = \dfrac{1}{x}$ by a factor of 4, and shift 3 units downward.

5.

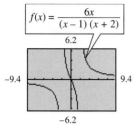

$f(x) = \dfrac{6x}{(x-1)(x+2)}$

6.2

−9.4　　9.4

−6.2

H.A.: $y = 0$;
V.A.: $x = 1, x = -2$

7.

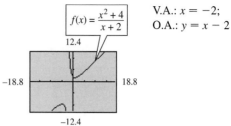

$f(x) = \dfrac{x^2+4}{x+2}$

12.4

−18.8　　18.8

−12.4

V.A.: $x = -2$;
O.A.: $y = x - 2$

9.

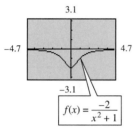

3.1

−4.7　　4.7

−3.1

$f(x) = \dfrac{-2}{x^2+1}$

H.A.: $y = 0$

11. The degree of the numerator will be exactly 1 more than the degree of the denominator.

13. (a) $\left\{\frac{2}{3}\right\}$　**(b)** $\left(-1, \frac{2}{3}\right)$　**(c)** $(-\infty, -1) \cup \left(\frac{2}{3}, \infty\right)$

15. (a) $\{0\}$　**(b)** $(-\infty, -1) \cup [0, 2)$　**(c)** $(-1, 0] \cup (2, \infty)$

17. $\{-2\}$　　**19.** $(-2, -1)$　　**21. (a)**

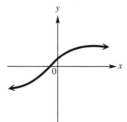

$C(x) = \dfrac{6.7x}{100-x}$

150

0

0　　100

X=95　Y=127.3

(b) approximately 127.3 thousand dollars (See the graph in part (a).)

23. 847　　**25.** 71.11 kilograms　　**27.**

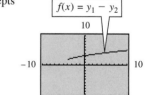

29.

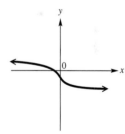

31. 2　　**33.** -100　　**35.** 8　　**37.** .5 (exact)　　**39.** .0625 (exact)　　**41.** $(-\infty, 0]$　　**43.** $[2, \infty)$

45. $y_1 = -4 + \sqrt{25 - x^2}, y_2 = -4 - \sqrt{25 - x^2}$　　**47.** $\{-2\}$　　**49.** $\{2, 5\}$　　**51.** $(-\infty, 2) \cup (5, \infty)$

53. $(-\infty, 2) \cup (5, \infty)$　　**55. (a)** $\{2\}$　**(b)** $[-2.5, 2)$　**(c)** $(2, \infty)$　　**57. (a)** $\{-1\}$　**(b)** $[-1, \infty)$　**(c)** $(-\infty, -1]$

59. There are no solutions to $y_1 = y_2$.

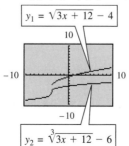

$y_1 = \sqrt{3x + 12} - 4$

10

−10　　10

−10

$y_2 = \sqrt[3]{3x + 12} - 6$

60. There are no x-intercepts for $f(x) = y_1 - y_2$.

$f(x) = y_1 - y_2$

10

−10　　10

−10

61. The graph of $-f(x) = y_2 - y_1$ is the reflection of the graph of $f(x) = y_1 - y_2$ across the x-axis.

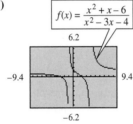

62. The graph of $y = f(-x)$ is the reflection of the graph of $y = f(x)$ across the y-axis.

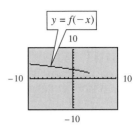

63. 3.1 seconds   **65.** not one-to-one   **67.** not one-to-one   **69.** $(-\infty, \infty)$ **71.** $f^{-1}(x) = \dfrac{x^3 + 7}{2}$

75. AWESOME DUDE

Chapter 4 Test (page 391)

1. (a) $x = -1, x = 4$ **(f)**
   (b) $y = 1$
   (c) 1.5
   (d) $-3, 2$
   (e) $(.5, 1)$

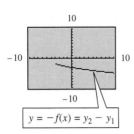

2. (a) -4 **(b)**

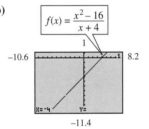

3. $y = 2x + 5$

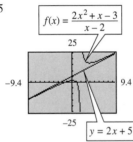

4. (a) $\{5\}$ **(b)** $(-\infty, -2) \cup (2, 5]$ **5.** 92; undernourished

6. approximately 8.6 centimeters; approximately 1394.9 square centimeters

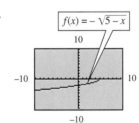

7.

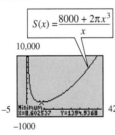

   (a) $(-\infty, 5]$
   (b) $(-\infty, 0]$
   (c) increases
   (d) $\{5\}$
   (e) $(-\infty, 5)$

8. (a) $\{0\}$
   (b) $(-\infty, 0)$
   (c) $[0, 4]$

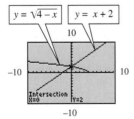

9. The cable should be laid underwater from P to a point S which is on the bank 400 yards away from Q, in the direction of R.

10. (a) For all $a \neq b$, $f(a) \neq f(b)$. Furthermore, it passes the horizontal line test and is an increasing function.
   (b) $f^{-1}(x) = 2x + 6$ **(c)** $(f \circ f^{-1})(x) = \frac{1}{2}(2x + 6) - 3 = x$; $(f^{-1} \circ f)(x) = 2(\frac{1}{2}x - 3) + 6 = x$

(d)

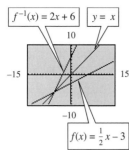

$f^{-1}(x) = 2x + 6$ $y = x$

$f(x) = \frac{1}{2}x - 3$

The graph of f^{-1} is a reflection of the graph of f across the line $y = x$.

CHAPTER 5

Section 5.1 (page 406)

1. 8.952419619   **3.** .3752142272   **5.** .0868214883   **7.** 13.1207791   **9.** Show that the point $\left(\sqrt{10}, 8.9524196\right)$ lies on the graph of $y = 2^x$.   **11.** Show that the point $\left(\sqrt{2}, .37521423\right)$ lies on the graph of $y = \left(\frac{1}{2}\right)^x$.
13. 2.3   **15.** .75   **17.** .31   **19.** 4   **21.** $\frac{1}{125}$
23. (a) $a > 1$
(b) domain: $(-\infty, \infty)$; range: $(0, \infty)$; asymptote: $y = 0$
(c)   **(d)** domain: $(-\infty, \infty)$; range: $(-\infty, 0)$; asymptote: $y = 0$
(e)   **(f)** domain: $(-\infty, \infty)$; range: $(0, \infty)$; asymptote: $y = 0$

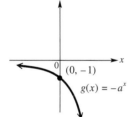

$g(x) = -a^x$ $(0, -1)$

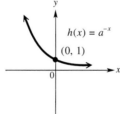

$h(x) = a^{-x}$ $(0, 1)$

25. Shift the graph of $y = 2^x$ five units to the left and 3 units down.   **26.** Shift the graph of $y = 2^x$ one unit to the right and 4 units up.   **27.** Reflect the graph of $y = 2^x$ across the y-axis, and shift 1 unit up.   **28.** Reflect the graph of $y = 2^x$ across the y-axis, and shift 6 units down.   **29.** Stretch the graph of $y = 2^x$ by a factor of 3, and reflect across the x-axis.   **30.** Reflect the graph of $y = 2^x$ across the y-axis, stretch by a factor of 4, and reflect across the x-axis.
31. (a) $\left\{\frac{1}{3}\right\}$ **(b)** $\left(\frac{1}{3}, \infty\right)$ **(c)** $\left(-\infty, \frac{1}{3}\right)$ **33. (a)** $\{-2\}$ **(b)** $(-\infty, -2]$ **(c)** $[-2, \infty)$ **35. (a)** $\{0\}$ **(b)** $(-\infty, 0)$
(c) $(0, \infty)$ **37. (a)** $\left\{\frac{1}{5}\right\}$ **(b)** $\left(-\infty, \frac{1}{5}\right)$ **(c)** $\left(\frac{1}{5}, \infty\right)$ **39.** 22.19795128; Show that (3.1, 22.197951) lies on the graph of $y = e^x$.
(In Exercises 41 and 43, graphical support is done in a manner similar to Exercise 39.)
41. .7788007831   **43.** 4.113250379   **45. (a)** \$22,510.18 **(b)** \$22,529.85   **47. (a)** \$33,504.35
(b) \$33,504.71   **49.** Plan A is better, because it pays \$102.65 more in interest.   **51.** \$1.16; \$2.44; \$7.08; \$18.12;
\$59.34; \$145.80; \$318.43   **53.**   **55.**

$f(x) = \dfrac{e^x - e^{-x}}{2}$

$f(x) = x \cdot 2^x$

In Exercises 57–63, we give the approximate solutions to the nearest hundredth.
57. $\{1.28\}$ **59.** $\{3.58\}$ **61.** $\{2.08\}$ **63.** $\{.90\}$

Section 5.2 (page 417)

1. $\log_3 81 = 4$ **3.** $\log_{1/2} 16 = -4$ **5.** $\log_{10} .0001 = -4$ or $\log .0001 = -4$ **7.** $\log_e 1 = 0$ or $\ln 1 = 0$
9. $6^2 = 36$ **11.** $\sqrt{3^8} = 81$ **13.** $10^{-3} = .001$ **15.** $10^5 = \sqrt{10}$

17. The expression $\log_a x$ represents the exponent to which a must be raised in order to obtain x.   **19.** 3   **21.** -3
23. 24   **25.** $-\frac{1}{6}$ **27.** $-\frac{11}{2}$ **29. (a)** 19 **(b)** 17 **(c)** $\frac{1}{3}$ **(d)** $\frac{1}{2}$ **31.** 1.633468456
33. $-.1062382379$ **35.** 4.341474094   **37.** 3.761200116   **39.** $-.244622583$ **41.** 9.996613531
43. (a) .3741982579, 1.374198258, 2.374198258, 3.374198258 **(b)** $2.367 \times 10^0, 2.367 \times 10^1, 2.367 \times 10^2, 2.367 \times 10^3$
(c) In each case, the decimal digits are the same. The difference is that the whole number part corresponds to the exponent
on 10 in scientific notation.
45.

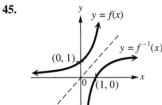

46. $f^{-1}(x) = \log_a x$ **47.** $f^{-1}(x) = \ln x$ **48.** $f^{-1}(x) = \log_{10} x$ or
$f^{-1}(x) = \log x$

49. 3.2   **51.** 8.4   **53.** 2×10^{-3} **55.** 1.6×10^{-5} **57.** 4.9 years   **59.** 13.9 years
61. $\log_3 2 - \log_3 5$ **63.** $\log_2 6 + \log_2 x - \log_2 y$ **65.** $1 + \frac{1}{2}\log_5 7 - \log_5 3$ **67.** no change possible
69. $\log_k p + 2\log_k q - \log_k m$ **71.** $\frac{1}{2}(\log_m 5 + 3\log_m r - 5\log_m z)$ **73.** $\log_a \dfrac{xy}{m}$ **75.** $\log_m \dfrac{a^2}{b^6}$
77. $\log_a[(z-1)^2(3z+2)]$ **79.** $\log_5 \dfrac{5^{1/3}}{m^{1/3}}$ **81.** 1.430676558   **83.** .5943161289   **85.** .9595390462
87. 1.892441722   **89.** Reflect the graph of $y = 3^x$ across the x-axis and shift 7 units upward.
90.

    91. 1.7712437   **92.** $\{\log_3 7\}$ **93.** $\dfrac{\log 7}{\log 3} = \dfrac{\ln 7}{\ln 3} \approx 1.771243749$

94. The approximations are close enough to support the conclusion that the x-intercept is equal to $\log_3 7$.

Section 5.3 (page 428)

1.

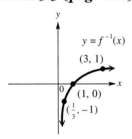

domain: $(0, \infty)$;
range: $(-\infty, \infty)$;
f^{-1} increases on $(0, \infty)$;
V.A.: $x = 0$

3.

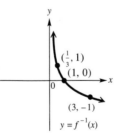

domain: $(0, \infty)$;
range: $(-\infty, \infty)$;
f^{-1} decreases on $(0, \infty)$;
V.A.: $x = 0$

5.

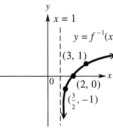

domain: $(1, \infty)$;
range: $(-\infty, \infty)$;
f^{-1} increases on $(1, \infty)$;
V.A.: $x = 1$

7. logarithmic   **9.** $(0, \infty)$ **11.** $\left(-\frac{7}{3}, \infty\right)$ **13.** $(-\infty, \infty)$ **15.** $(-\infty, -3) \cup (7, \infty)$
17. $\left(-\infty, \dfrac{-1 - \sqrt{5}}{2}\right) \cup \left(\dfrac{-1 + \sqrt{5}}{2}, \infty\right)$ **19.** $(-1, 0) \cup (1, \infty)$ **21.** B   **23.** D   **25.** A   **27.** C

In Exercises 29–33, other descriptions may be possible.
29. The graph of $y = \log_2 x$ is shifted 4 units to the left.   **31.** The graph of $y = \log_2 x$ is stretched vertically by a factor
of 3 and is shifted 1 unit up.   **33.** The graph of $y = \log_2 x$ is reflected across the y-axis and is shifted 1 unit up.

35. $(-\infty, 0) \cup \left(\frac{1}{2}, \infty\right)$ **36.** $y = f(x) = \dfrac{\log (2x^2 - x)}{\log 4}$ **37.** $-\frac{1}{2}, 1$ **38.** $x = 0, x = \frac{1}{2}$

39. In general, to find the y-intercept for the graph of $y = f(x)$, we evaluate $f(0)$. Because 0 is not in the domain of f, $f(0)$ is not defined, and there is no y-intercept.

40. 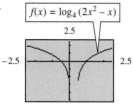   Solution set of $f(x) = 0$: $\left\{-\frac{1}{2}, 1\right\}$

Solution set of $f(x) < 0$: $\left(-\frac{1}{2}, 0\right) \cup \left(\frac{1}{2}, 1\right)$

Solution set of $f(x) > 0$: $\left(-\infty, -\frac{1}{2}\right) \cup (1, \infty)$

41. 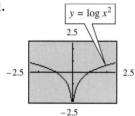   The graphs are not the same because the domain of $y = \log x^2$ is $(-\infty, 0) \cup (0, \infty)$, while the domain of $y = 2 \log x$ is $(0, \infty)$. The power rule does not apply if the argument is nonpositive.

43. (a) $\dfrac{3}{2}$ **(b)** $\log_9 27 = \dfrac{\log 27}{\log 9} = 1.5 = \dfrac{3}{2}$ **45. (a)** $-\dfrac{3}{4}$ **(b)** $\log_{16}\left(\dfrac{1}{8}\right) = \dfrac{\log\left(\dfrac{1}{8}\right)}{\log 16} = -.75 = -\dfrac{3}{4}$

(c) 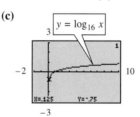   **(c)**

In Exercises 47–51, graph $y = \log x$ or $y = \ln x$ as appropriate. Let x equal the value given and show that y is equal to the approximation given in these answers. Check, using the log or ln key.

47. .84509804   **49.** 1.9459101   **51.** 1.1673173

53.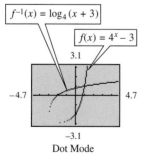

$f^{-1}(x) = \log_4(x + 3)$. The domain of f is equal to the range of f^{-1}, and vice versa. The x- and y-intercepts are reversed. The horizontal asymptote for f, $y = -3$, is transformed into the vertical asymptote for f^{-1}, $x = -3$.

55.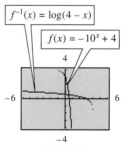

$f^{-1}(x) = \log(4 - x)$. The domain of f is equal to the range of f^{-1}, and vice versa. The x- and y-intercepts are reversed. The horizontal asymptote for f, $y = 4$, is transformed into the vertical asymptote for f^{-1}, $x = 4$.

57. $\log 5.0118723 \approx .7$ **59.** Values of x greater than or equal to 4 lead to the logarithm of a nonpositive number, which is not a real number.

Section 5.4 (page 438)

1. (a) $\{1.4036775\}$ (b) $(1.4036775, \infty)$ **3.** (a) $\{-1\}$ (b) $(-1, \infty)$

There are other ways to express the answers in Exercises 5 and 7.

5. (a) $\left\{\dfrac{\log 10}{\log 3}\right\}$ (b) $\left(\dfrac{\log 10}{\log 3}, \infty\right)$ (c) $\left(-\infty, \dfrac{\log 10}{\log 3}\right)$ **7.** (a) $\left\{\dfrac{\log\left(\frac{1}{8}\right)}{\log\left(\frac{2}{5}\right)}\right\}$ (b) $\left(-\infty, \dfrac{\log\left(\frac{1}{8}\right)}{\log\left(\frac{2}{5}\right)}\right]$ (c) $\left[\dfrac{\log\left(\frac{1}{8}\right)}{\log\left(\frac{2}{5}\right)}, \infty\right)$

9. (a) $\{5\}$ (b) $(3, 5)$ (c) $(5, \infty)$ **11.** $\{2.5\}$ **13.** $\{3\}$ **15.** (a) $(0, 4)$ (b) $(4, \infty)$

17. $\log(3x + 2) + \log(x - 1) = \log 10$; $\left\{\dfrac{1 + \sqrt{145}}{6}\right\}$ **19.** (a) $\{-.535\}$ (b) $(-.535, \infty)$ (c) $(-\infty, -.535)$

21. (a) $\{-.123\}$ (b) $(-\infty, -.123)$ (c) $(-.123, \infty)$ **23.** (a) $\{1.236\}$ (b) $(0, 1.236]$ (c) $[1.236, \infty)$
25. The graph of $y = \ln x - \ln(x + 1) - \ln 5$ does not intersect the x-axis.   **27.** $\{17.106\}$

The answers in Exercises 29–37 may have alternative equivalent forms.

29. $t = e^{(p-r)/k}$ **31.** $t = -\dfrac{1}{k}\log\left(\dfrac{T - T_0}{T_1 - T_0}\right)$ **33.** $k = \dfrac{\ln\left(\dfrac{A - T_0}{C}\right)}{-t}$ **35.** $x = \dfrac{\ln\left(\dfrac{y - A - B}{-B}\right)}{-C}$

37. $A = \dfrac{B}{x^C}$ **39.** Because $(a^m)^n = a^{mn}$, $(e^x)^2 = e^{x \cdot 2} = e^{2x}$.   **40.** $(e^x - 3)(e^x - 1) = 0$ **41.** $\{0, \ln 3\}$

42.

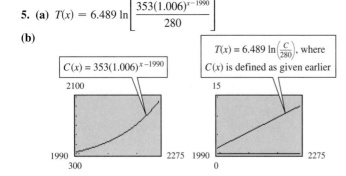

$y = e^{2x} - 4e^x + 3$

The graph intersects the x-axis at 0 and $\ln 3 \approx 1.099$.   **43.** $(-\infty, 0) \cup (\ln 3, \infty)$

44. $(0, \ln 3)$ **45.** $\{-.767, 2, 4\}$ **47.** $\{2.454, 5.659\}$ **49.** $\{-.443\}$ **51.** $\{17.475\}$
53. $\{-2, 2\}$ **55.** $\{1, 10\}$ **57.** $\{-3\}$ **59.** $\{\ln 2, \ln 4\}$

Section 5.5 (page 451)

1. (a)

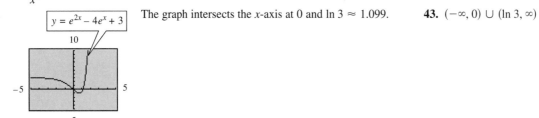

Yes, the levels appear to grow exponentially.
(b) Levels will double by 2065 and triple by 2132.

3. (a) linear (because $R = k \cdot \ln 2$ and $\ln 2$ is a constant) (b) $T(k) = 1.03k \ln\left(\dfrac{C}{C_0}\right)$ (c) between 7°F and 11°F

5. (a) $T(x) = 6.489 \ln\left[\dfrac{353(1.006)^{x-1990}}{280}\right]$

(b)

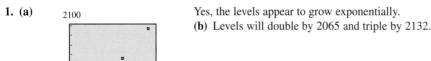

$C(x) = 353(1.006)^{x-1990}$

$T(x) = 6.489 \ln\left(\dfrac{C}{280}\right)$, where $C(x)$ is defined as given earlier

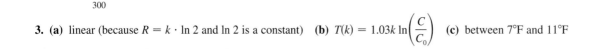

C is an exponential function and *T* is a linear function over the same time period. While the carbon dioxide levels in the atmosphere increase at an exponential rate, the average global temperature will rise at a linear rate.
(c) The slope is .0388. This means that the temperature is expected to rise at an average rate of .04°F per year from 1990 to 2275.
(d) $x \approx 2208.9$; $C \approx 1308$ ppm (about 4.67 times above preindustrial carbon dioxide levels)

7. (a) 404 grams **(b)** 327 grams **(c)** 173 grams **(d)** 13.08 years **(e)**

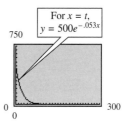

9. (a) 21 **(b)** 70 **(c)** 91 **(d)** 120 **(e)** 140 **11. (a)** about $4{,}000{,}000I_0$ **(b)** about $3{,}200{,}000I_0$ **(c)** It is about 1.25 times as great. **13.** Magnitude 1 is about 6.3 times as great as magnitude 3. **15.** 59,800 feet
17. (a) $P(T) = 1 - e^{-.0034 - .0053T}$ **(b)**

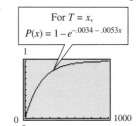

(c) $P(60) \approx .275$ or 27.5%. The reduction in carbon emissions from a tax of \$60 per ton of carbon is 27.5%.
(d) $T = \$130.14$

19. about 4200 years old **21.** about 13,000 years old **23.** 71.7°C **25.** about 66 million; We must assume that the rate of increase continues to be logarithmic. **27. (a)** about 46.2 years **(b)** about 46.0 years
29. about 1.8 years **31.** 7% compounded quarterly; \$800.31 **33.** about 6.14% **35.** \$5583.95
37. 12% **39. (a)** \$205.52 **(b)** \$1364.96 **41. (a)** \$471.58 **(b)** \$29,884.40 **43. (a)** 19 years, 39 days
(b) 11 years, 166 days **45. (a)** about 2,700,000 **(b)** about 3,000,000 **(c)** about 9,500,000 **47.** 23 days
49. about 59 mg **51.** about 1503 thousand, or 1.503 million **53. (a)** $(0, 1.6), (60, 39); 1.6 = a(b^0), 39 = a(b^{60})$
(b) $a = 1.6, b = 1.0547; f(x) = 1.6(1.0547)^x$ **(c)** 10.9 million users; This is close to the correct answer of 10.3 million users. **(d)** May 1994 **55.** 2016 **57.** about 375 billion dollars

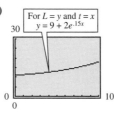

59. (a) 11 **(b)** 12.6 **(c)** 18.0 **(d)**

61. (a)

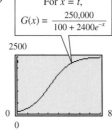

(b) 590; 589 **(c)** 2.8; 2.7726

Chapter 5 Review Exercises (page 460)

1. C   **3.** D   **5.** $0 < a < 1$ **7.** $(0, \infty)$ **9.**

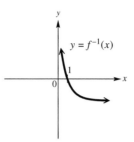

11. (a) $\left\{\tfrac{3}{5}\right\}$
   (b) $\left[\tfrac{3}{5}, \infty\right)$
   (c) $\left(-\infty, \tfrac{3}{5}\right]$

13. (a) $\left\{-\tfrac{2}{3}\right\}$ **(b)** $\left(-\infty, -\tfrac{2}{3}\right)$ **(c)** $\left(-\tfrac{2}{3}, \infty\right)$ **15.** $(-.7666647, .58777476)$ **17.** 1.765668555

19. 4.065602093 **21.** 0   **23.** 12   **25.** 1.584962501 **27.** E   **29.** B   **31.** F   **33.** 3

35. $\log_3 m + \log_3 n - \log_3 5 - \log_3 r$ **37.** $2 \log_5 x + 4 \log_5 y + \tfrac{3}{5} \log_5 m + \tfrac{1}{5} \log_5 p$ **39. (a)** $\{2\}$ **(b)** $(2, \infty)$

(c) $(0, 2)$ **41. (a)** $\{1.4\}$ **(b)** $(1, 1.4]$ **(c)** $[1.4, \infty)$ **43.** $\left\{\tfrac{3}{2}\right\}$ **45.** $\{1.303\}$ **47.** $\{2\}$ **49.** $\left\{\tfrac{1}{2}\right\}$

51. $c = de^{(N-a)/b}$ **53.** $\{1.87\}$ **55.** 9.6% **57.** $\$93,761.31$ **59. (a)** $\$2322.37$ **(b)** $\$2323.67$

(c) 36.6 years   **61. (a)** .0054 gram per liter **(b)** .00073 gram per liter **(c)** .000013 gram per liter **(d)** .75 mile

63. (a) 329.3 grams **(b)** 13.9 days   **65. (a)**

There appears to be a linear relationship.
(b) By taking logarithms on both sides, the function can be written as $\ln P = kx + \ln C$, which is a linear function in the form $\ln P = ax + b$. Here, $a = k$ and $b = \ln C$, with $y = \ln P$.

Chapter 5 Test (page 463)

1. (a) B
   (b) A
   (c) C
   (d) D

2. (a)

$f(x) = -2^{x-1} + 8$

(b) Shift the graph of $y = 2^x$ one unit to the right, reflect it across the x-axis, and shift 8 units upward.
(c) domain: $(-\infty, \infty)$; range: $(-\infty, 8)$
(d) Yes, it has a horizontal asymptote whose equation is $y = 8$.
(e) x-intercept: 4; y-intercept: 7.5   Use a calculator to find $f(0)$ to support the y-intercept, and find the x-intercept, using the root-locating capability.

3. (a) $\{.5\}$ **(b)** $(-\infty, .5)$ **(c)** $(.5, \infty)$ **4. (a)** $\$12,442.11$ **(b)** $\$12,460.77$ **5.** The expression $\log_5 27$ is the exponent to which 5 must be raised in order to obtain 27. To find an approximation with a calculator, we use the change-of-base rule.   **6. (a)** 1.659 **(b)** 6.153 **(c)** 6.049   **7.** $3 \log m + \log n - \tfrac{1}{2} \log y$

8. (a) $\{2\}$; The extraneous solution is -4.   **(b)** $y_1 = \dfrac{\log x}{\log 2} + \dfrac{\log(x + 2)}{\log 2} - 3$

$y_1 = \log_2 x + \log_2 (x + 2) - 3$

(c) $(2, \infty)$

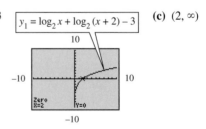

9. (a) $\left\{\dfrac{\log 18}{\log 48}\right\}$ **(b)** $\{.747\}$ **10. (a)** B **(b)** D **(c)** C **(d)** A

11. (a)   For $t = x$, $A(x) = x^2 - x + 350$ **(b)**   For $t = x$, $A(x) = 350 \log (x + 1)$

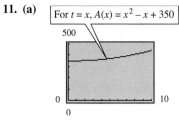

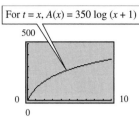

(c)   For $t = x$, $A(x) = 350(.75)^x$ **(d)**   For $t = x$, $A(x) = 100(.95)^x$ Function (c) best describes $A(t)$.

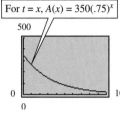

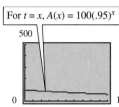

12. (a)   6,000,000   **(b)** Function (iii) best describes the data because it increases at a faster rate as x increases.

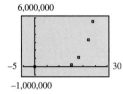

(c) $f(x) = 2300e^{.3241x}$ is one answer. Answers may vary somewhat.   **(d)** about 28,000,000

$f(x) = 2300e^{.3241x}$

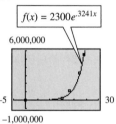

CHAPTER 6

Section 6.1 (page 477)

1. C   **3.** F   **5.** G   **7.** J   **9.** B   **11.** $(x - 1)^2 + (y - 4)^2 = 9$ **13.** $x^2 + y^2 = 1$
15. $\left(x - \frac{2}{3}\right)^2 + \left(y + \frac{4}{5}\right)^2 = \frac{9}{49}$ **17.** $(x + 1)^2 + (y - 2)^2 = 25$ **19.** $(x + 3)^2 + (y + 2)^2 = 4$ **21.** $(2, -3)$
22. $3\sqrt{5}$ **23.** $(x - 2)^2 + (y + 3)^2 = 45$ **24.**   10   **25.** $(x + 2)^2 + (y + 1)^2 = 41$

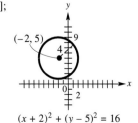

27. domain: $[-6, 6]$; y
    range: $[-6, 6]$

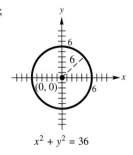

$x^2 + y^2 = 36$

29. domain: $[-6, 2]$; y
    range: $[1, 9]$

$(x + 2)^2 + (y - 5)^2 = 16$

31.

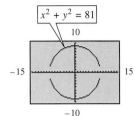

$x^2 + y^2 = 81$

33.

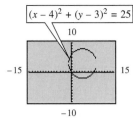

$(x - 4)^2 + (y - 3)^2 = 25$

35. It is the single point (3, 3).

37. domain: $[-6, 6]$; points of intersection: $(-6, 0)$, $(6, 0)$ **39.** domain: $[1, 9]$; points of intersection: $(1, 2)$, $(9, 2)$

41. domain: $[2, \infty)$; point of intersection: $(2, 1)$ **43.** center: $(-3, -4)$; radius: 4   **45.** center: $(6, -5)$; radius: 6

47. center: $(0, 1)$; radius: 7   **49.** The point $(-3, 4)$ lies on the circles $(x - 1)^2 + (y - 4)^2 = 16$, $(x + 6)^2 + y^2 = 25$, and $(x - 5)^2 + (y + 2)^2 = 100$.   **51.** $x^2 + y^2 = 1296$; $x^2 + y^2 = 576$; $x^2 + y^2 = 144$ **53.** B   **55.** A

57. H   **59.** G   **61. (a)** III   **(b)** II   **(c)** IV   **(d)** I   **63.** focus: $(0, 1)$; directrix: $y = -1$; axis: y-axis

65. focus: $\left(0, \frac{1}{36}\right)$; directrix: $y = -\frac{1}{36}$; axis: y-axis   **67.** focus: $\left(-\frac{1}{128}, 0\right)$; directrix: $x = \frac{1}{128}$; axis: x-axis

69. focus: $(-1, 0)$; directrix: $x = 1$; axis: x-axis   **71.** $x = \frac{1}{20} y^2$ **73.** $y = x^2$ **75.** $y = x^2$

77. $x = -\frac{1}{3} y^2$ **79.** $x = \frac{3}{4} y^2$

In the answers for Exercises 81–95, we provide traditional graphs.

81.

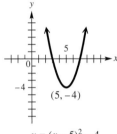

$y = (x - 5)^2 - 4$

vertex: $(5, -4)$;
axis: $x = 5$;
domain: $(-\infty, \infty)$;
range: $[-4, \infty)$

83.

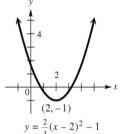

$y = \frac{2}{3}(x - 2)^2 - 1$

vertex: $(2, -1)$;
axis: $x = 2$;
domain: $(-\infty, \infty)$;
range: $[-1, \infty)$

85.

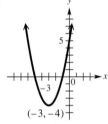

$y = x^2 + 6x + 5$

vertex: $(-3, -4)$;
axis: $x = -3$;
domain: $(-\infty, \infty)$;
range: $[-4, \infty)$

87.

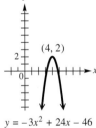

$y = -3x^2 + 24x - 46$

vertex: $(4, 2)$; axis: $x = 4$; domain: $(-\infty, \infty)$; range: $(-\infty, 2]$

To obtain calculator-generated graphs in Exercises 89–95, solve for y_1 and y_2 in terms of x, and use a square window.

89.

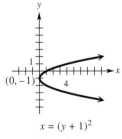

$x = (y + 1)^2$

vertex: $(0, -1)$;
axis: $y = -1$;
domain: $[0, \infty)$;
range: $(-\infty, \infty)$

91.

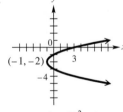

$x = (y + 2)^2 - 1$

vertex: $(-1, -2)$;
axis: $y = -2$;
domain: $[-1, \infty)$;
range: $(-\infty, \infty)$

93.

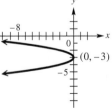

$x = -2(y + 3)^2$

vertex: $(0, -3)$;
axis: $y = -3$;
domain: $(-\infty, 0]$;
range: $(-\infty, \infty)$

95.

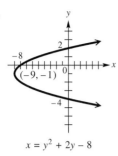

$x = y^2 + 2y - 8$

vertex: $(-9, -1)$;
axis: $y = -1$;
domain: $[-9, \infty)$;
range: $(-\infty, \infty)$

97. (a)

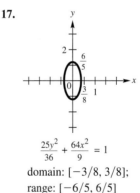

Mars
$Y_2 = x - \frac{6.3}{961}x^2$

Earth
$Y_1 = x - \frac{16.1}{961}x^2$

120

0   180
0

Earth: $y = x - \frac{16.1}{961} x^2$; Mars: $y = x - \frac{6.3}{961} x^2$
(b) approximately 93 feet

99. approximately 43.8 feet   **101.** 60 feet
103. Let P be (x, y). The focus is $F(c, 0)$, and the directrix has equation $x = -c$. If $D(-c, y)$ is on the directrix, then $d(P, F) = d(P, D)$. So $\sqrt{(x - c)^2 + (y - 0)^2} = \sqrt{(x + c)^2 + (y - y)^2}$. This gives $\sqrt{(x - c)^2 + y^2} = \sqrt{(x + c)^2}$. Squaring both sides and expanding gives $x^2 - 2xc + c^2 + y^2 = x^2 + 2xc + c^2$. Subtracting x^2 and c^2 and rearranging terms gives $y^2 = 4cx$, or $x = \dfrac{1}{4c} y^2$.

Section 6.2 (page 489)

1. E   **3.** H   **5.** A   **7.** I   **9.** D   **11.** A circle can be interpreted as an ellipse whose two foci actually have the same coordinates; the "coinciding foci" give the center of the circle.

In the answers for Exercises 13–21, we provide traditional graphs. To obtain calculator-generated graphs, solve for y_1 and y_2 in terms of x, and use a square window.

13.

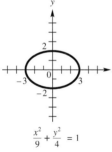

$\dfrac{x^2}{9} + \dfrac{y^2}{4} = 1$

domain: $[-3, 3]$;
range: $[-2, 2]$

15.

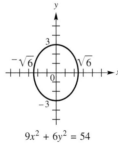

$9x^2 + 6y^2 = 54$

domain: $\left[-\sqrt{6}, \sqrt{6}\right]$;
range: $[-3, 3]$

17.

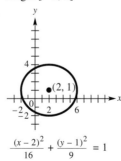

$\dfrac{25y^2}{36} + \dfrac{64x^2}{9} = 1$

domain: $[-3/8, 3/8]$;
range: $[-6/5, 6/5]$

19.

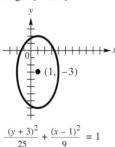

$(1, -3)$

$\dfrac{(y + 3)^2}{25} + \dfrac{(x - 1)^2}{9} = 1$

domain: $[-2, 4]$;
range: $[-8, 2]$

21.

$(2, 1)$

$\dfrac{(x - 2)^2}{16} + \dfrac{(y - 1)^2}{9} = 1$

domain: $[-2, 6]$;
range: $[-2, 4]$

23. A hyperbola centered at the origin is symmetric with respect to both axes, as well as symmetric with respect to the origin.

25. $16 - \dfrac{16(x-2)^2}{9} \geq 0$ **26.** Its graph is a parabola.   **27.**

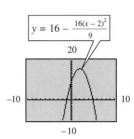

28. The graph of $y = 16 - \dfrac{16(x-2)^2}{9}$ lies above or on the x-axis in the interval $[-1, 5]$.   **29.** In Figure 22, we see

that the domain is $[-1, 5]$. This corresponds to the solution set found graphically in Exercise 28.   **30.** $[-1, 5]$

In the answers for Exercises 31–39, we provide traditional graphs. To obtain calculator-generated graphs, solve for y_1 and y_2 in terms of x, and use a square window.

31.

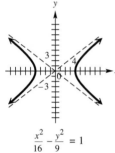

$$\dfrac{x^2}{16} - \dfrac{y^2}{9} = 1$$
domain: $(-\infty, -4] \cup [4, \infty)$;
range: $(-\infty, \infty)$

33.

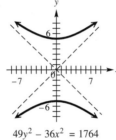

$49y^2 - 36x^2 = 1764$
domain: $(-\infty, \infty)$;
range: $(-\infty, -6] \cup [6, \infty)$

35.

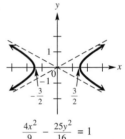

$$\dfrac{4x^2}{9} - \dfrac{25y^2}{16} = 1$$
domain: $(-\infty, -3/2] \cup [3/2, \infty)$;
range: $(-\infty, \infty)$

37.

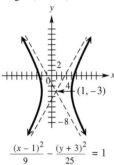

$$\dfrac{(x-1)^2}{9} - \dfrac{(y+3)^2}{25} = 1$$
domain: $(-\infty, -2] \cup [4, \infty)$;
range: $(-\infty, \infty)$

39.

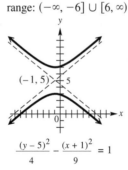

$$\dfrac{(y-5)^2}{4} - \dfrac{(x+1)^2}{9} = 1$$
domain: $(-\infty, \infty)$;
range: $(-\infty, 3] \cup [7, \infty)$

41.

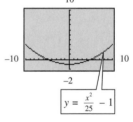
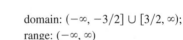

The graph of $y = \dfrac{x^2}{25} - 1$ lies above or on the x-axis in $(-\infty, -5] \cup [5, \infty)$. This set is the same as the domain of the given hyperbola.

42. (a) The graph of $y = 9 + 4x^2$ can be obtained by vertically stretching the graph of $y = x^2$ by a factor of 4, and shifting 9 units upward.   **(b)** A   **(c)** The domain of the graph in A, that of $y = 9 + 4x^2$, is $(-\infty, \infty)$, which is the domain of the original hyperbola.   **43.** The domain of the ellipse is $[-2, 2]$, while the domain of the hyperbola is $(-\infty, -2] \cup [2, \infty)$.

45. $\dfrac{x^2}{16} + \dfrac{y^2}{12} = 1$ **47.** $\dfrac{x^2}{36} + \dfrac{y^2}{20} = 1$ **49.** $\dfrac{(y+2)^2}{25} + \dfrac{(x-3)^2}{16} = 1$ **51.** $\dfrac{x^2}{9} - \dfrac{y^2}{7} = 1$ **53.** $\dfrac{y^2}{9} - \dfrac{x^2}{25} = 1$

55. $(-2, 0), (2, 0)$ **56.**

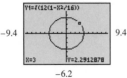

In addition to $(3, 2.2912878)$ shown on the screen, other points are $(0, 3.4641016)$ and $(-3, -2.291288)$.

57. The points satisfy the equation.　　**58.**

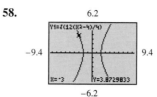

$(-4, 0)$, $(4, 0)$; In addition to the point shown on the screen, other points are $(-2, 0)$, $(2, 0)$, and $(4, 6)$.

59. They satisfy the equation.　　**60.** Exercise 57 demonstrates that the points on the graph satisfy the definition of the ellipse for that particular ellipse. Exercise 59 demonstrates similarly that the definition of the hyperbola is satisfied for that hyperbola.　　**61.** One such ellipse would have an equation of $\dfrac{x^2}{100} + \dfrac{y^2}{64} = 1$.　　**63.** 348.2 feet　　**65.** It must be just under 12 feet tall.

67. (a)

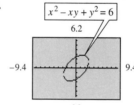

(b) minimum: approximately 341 miles; maximum: approximately 669 miles

Section 6.3 (page 503)

1. circle　　**3.** parabola　　**5.** parabola　　**7.** ellipse　　**9.** hyperbola　　**11.** hyperbola　　**13.** ellipse
15. no graph　　**17.** circle　　**19.** parabola　　**21.** point　　**23.** no graph　　**25.** circle
27. parabola　　**29.** hyperbola　　**31.** ellipse　　**33.** no graph
35.

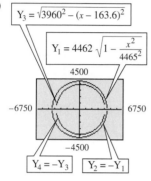

37.

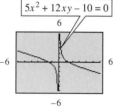

39. line　　**41.** $\frac{1}{2}$　　**43.** $\sqrt{2}$

45. $\dfrac{\sqrt{21}}{7}$　　**47.** $\dfrac{\sqrt{10}}{3}$　　**49.** $y = \frac{1}{32}x^2$　　**51.** $\dfrac{x^2}{36} + \dfrac{y^2}{27} = 1$　　**53.** $\dfrac{x^2}{36} - \dfrac{y^2}{108} = 1$　　**55.** $y = -\frac{1}{4}x^2$

57. $\dfrac{y^2}{9} + \dfrac{25x^2}{81} = 1$　　**59.** C, A, B, D

61. (a) Neptune: $\dfrac{(x - .2709)^2}{30.1^2} + \dfrac{y^2}{30.1^2} = 1$;　**(b)**

Pluto: $\dfrac{(x - 9.8106)^2}{39.4^2} + \dfrac{y^2}{38.16^2} = 1$

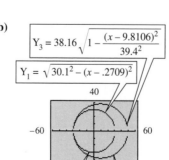

63. approximately 55 million miles

Section 6.4 (page 508)

1.
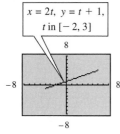

$x = 2t,\ y = t + 1,$
t in $[-2, 3]$

$y = \frac{1}{2}x + 1$, for x in $[-4, 6]$

3.
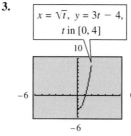

$x = \sqrt{t},\ y = 3t - 4,$
t in $[0, 4]$

$y = 3x^2 - 4$, for x in $[0, 2]$

5.
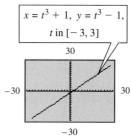

$x = t^3 + 1,\ y = t^3 - 1,$
t in $[-3, 3]$

$y = x - 2$, for x in $[-26, 28]$

7.
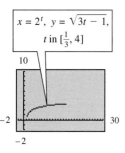

$x = 2^t,\ y = \sqrt{3t - 1},$
t in $\left[\frac{1}{3}, 4\right]$

$x = 2^{(y^2+1)/3}$, for y in $\left[0, \sqrt{11}\right]$

or $y^2 = \dfrac{3 \ln x}{\ln 2} - 1$, for x in $\left[\sqrt[3]{2}, 16\right]$

9.
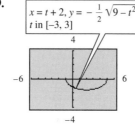

$x = t + 2,\ y = -\frac{1}{2}\sqrt{9 - t^2}$
t in $[-3, 3]$

$y = -\frac{1}{2}\sqrt{9 - (x - 2)^2}$,

for x in $[-1, 5]$

11.
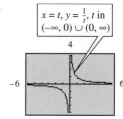

$x = t,\ y = \frac{1}{t},\ t$ in
$(-\infty, 0) \cup (0, \infty)$

$y = \dfrac{1}{x}$, for x in

$(-\infty, 0) \cup (0, \infty)$

13. $x = 3\left(\dfrac{y}{4}\right)^{2/3}$, for y in $(-\infty, \infty)$ **15.** $y = \sqrt{x^2 + 2}$, for x in $(-\infty, \infty)$ **17.** $y = \dfrac{1}{x}$, for x in $(0, \infty)$

19. $y = 1 - 2x^2$, for x in $(0, \infty)$ **21.** $y = \dfrac{1}{x}$, for $x \neq 0$ **23.** $y = \ln x$, for x in $(0, \infty)$

Other answers are possible for Exercises 25 and 27.

25. $x = \dfrac{1}{2}t,\ y = t + 3;\ x = \dfrac{t + 3}{2},\ y = t + 6$ **27.** $x = \dfrac{1}{3}t,\ y = \sqrt{t + 2},\ t$ in $[-2, \infty);\ x = \dfrac{t - 2}{3},\ y = \sqrt{t},\ t$ in $[0, \infty)$

29. (a) 17.7 seconds   **(b)** 5000 feet   **(c)** 1250 feet

33.
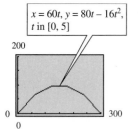

$x = 60t,\ y = 80t - 16t^2,$
t in $[0, 5]$

$y = \frac{4}{3}x - \frac{1}{225}x^2$

35. Many answers are possible, two of which are $x = t$,
$y - y_1 = m(t - x_1)$ and $t = x - x_1,\ y = mt + y_1$.

Chapter 6 Review Exercises (page 511)

1. $(x + 2)^2 + (y - 3)^2 = 25$ **3.** $(x + 8)^2 + (y - 1)^2 = 289$ **5.** $(2, -3);\ r = 1$ **7.** $\left(-\frac{7}{2}, -\frac{3}{2}\right);\ r = \dfrac{3\sqrt{6}}{2}$

9. It is the point (4, 5).   **11.** 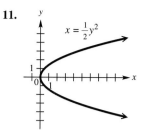   **13.**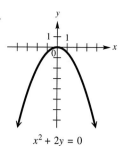

$x = \frac{1}{2}y^2$

$x^2 + 2y = 0$

$\left(\frac{1}{2}, 0\right); x = -\frac{1}{2}; x\text{-axis}$ $\left(0, -\frac{1}{2}\right); y = \frac{1}{2}; y\text{-axis}$

15. $x = \frac{2}{25}y^2$ **17.** $x + 5 = \frac{1}{28}(y - 6)^2$ **19.** 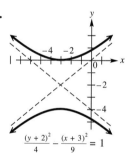   **21.**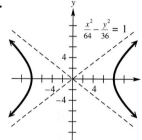

$\frac{y^2}{9} + \frac{x^2}{5} = 1$

$\frac{x^2}{64} - \frac{y^2}{36} = 1$

$(0, -3), (0, 3)$

$(-8, 0), (8, 0)$

23. 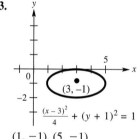   **25.**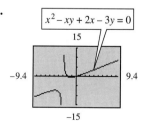

$(3, -1)$

$\frac{(x - 3)^2}{4} + (y + 1)^2 = 1$

$(1, -1), (5, -1)$

$\frac{(y + 2)^2}{4} - \frac{(x + 3)^2}{9} = 1$

$(-3, 0), (-3, -4)$

27. $\frac{y^2}{16} + \frac{x^2}{12} = 1$ **29.** $\frac{y^2}{16} - \frac{x^2}{9} = 1$ **31.** $\frac{4y^2}{81} + \frac{4x^2}{45} = 1$ **33. (a)** $(-1, -3)$ **(b)** 5

(c) $y_1 = -3 + \sqrt{24 - 2x - x^2}; y_2 = -3 - \sqrt{24 - 2x - x^2}$ **35.** C   **37.** E   **39.** D   **41.** $\frac{\sqrt{5}}{3}$

43. 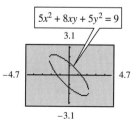   **45.** 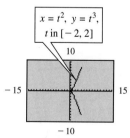   **47.**

$5x^2 + 8xy + 5y^2 = 9$

$x^2 - xy + 2x - 3y = 0$

$x = t^2, y = t^3, t \text{ in } [-2, 2]$

49. a portion of a parabola   **51.** $y = \sqrt{x^2 + 1}$, for x in $[0, \infty)$ **53.** 66.8 and 67.7 million miles

55. elliptic   **57.** The required increase in velocity is less when D is larger.

59. For example, the point (1, 3.354102) is on the ellipse. The distance between (2, 0) and (1, 3.354102) is 3.500000032. The distance between (1, 3.354102) and the line $x = 8$ is $8 - 1 = 7$. Since $(1/2)7 = 3.5$, the equation is satisfied.

Chapter 6 Test (page 513)

1. (a) B
   (b) E
   (c) F
   (d) A
   (e) C
   (f) D

2. $\left(\frac{1}{32}, 0\right)$; $x = -\frac{1}{32}$

3.

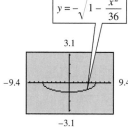

yes

4. $y_1 = 7\sqrt{\dfrac{x^2}{25} - 1}$; $y_2 = -7\sqrt{\dfrac{x^2}{25} - 1}$

5. hyperbola; center: (0, 0); vertices: (0, 2) and (0, −2); foci: $\left(0, \sqrt{13}\right)$ and $\left(0, -\sqrt{13}\right)$ **6.** the point (−1, 2)

7. semicircle; radius: 3; center: (0, 4)   **8.** parabola; vertex: (3, 4); focus: $\left(\frac{7}{2}, 4\right)$ **9. (a)** $y = -\frac{1}{8}x^2$

(b) $\dfrac{x^2}{11/4} + \dfrac{y^2}{9} = 1$ or $\dfrac{4x^2}{11} + \dfrac{y^2}{9} = 1$ **10.** $\dfrac{x^2}{400} + \dfrac{y^2}{144} = 1$; approximately 10.39 feet

CHAPTER 7

Section 7.1 (page 529)

1. In the first equation, 3 = 3, and in the second equation, 1 = 1.   **3.** In the first equation, 3 = 3, and in the second equation, 0 = 0.   **5.** In the first equation, 2 = 2, and in the second equation, 34 = 34.   **7.** In the first equation, 2 = 2, and in the second equation, 11 = 11.   **9.** single solution   **11.** infinitely many solutions   **13.** infinitely many solutions   **15.** {(6, 15)}   **17.** {(2, −3)}   **19.** Multiply equation (4) by 3 and equation (5) by 4.   **21.** $\left\{\left(\dfrac{y + 9}{4}, y\right)\right\}$ **23.** {(4, −2)}   **25.** {(12, 6)}   **27.** $5t + 15u = 16$; $5t + 4u = 5$

28. $t = \frac{1}{5}$; $u = 1$ **29.** $x = 5$, $y = 1$ **30.** $y = \dfrac{-15x}{5 - 16x}$ **31.** $y = \dfrac{-4x}{5 - 5x}$

32.

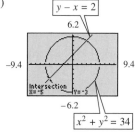

Dot Mode

33. {(2, 2)}   **35.** {(−8.71, 15.67)}   **37.** (−5, −3)

39.

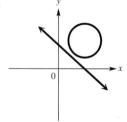

41.

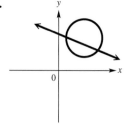

43.

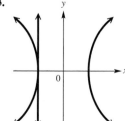

45.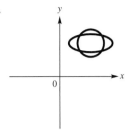

47.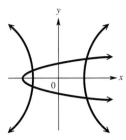

49. $\{(1, 1), (-2, -2)\}$

51. $\left\{\left(-\frac{3}{5}, \frac{7}{5}\right), (-1, 1)\right\}$ **53.** $\{(3, 1), (3, -1), (-3, 1), (-3, -1)\}$ **55.** $\{(-3, 0), (3, 0)\}$
57. $\left\{\left(x, \pm\sqrt{10 - x^2}\right)\right\}$ **59.** $\left\{\left(\sqrt{5}, 0\right), \left(-\sqrt{5}, 0\right), \left(\sqrt{5}, \sqrt{5}\right), \left(-\sqrt{5}, -\sqrt{5}\right)\right\}$
61. $\left\{\left(\dfrac{1 + \sqrt{13}}{2}, \dfrac{-1 + \sqrt{13}}{2}\right), \left(\dfrac{-1 - \sqrt{21}}{2}, \dfrac{3 + \sqrt{21}}{2}\right)\right\}$ **63.** $\{(-.79, .62), (.88, .77)\}$ **65.** $\{(.06, 2.88)\}$
67. \$12,000 at 7%; \$4000 at 8%   **69.** $13\frac{1}{3}$ gallons of 98-octane; $26\frac{2}{3}$ gallons of 92-octane   **71.** \$18,000 at 4.5%;
\$12,000 at 5%   **73.** $x = 2100$; $R = C = 52{,}000$ dollars   **75.** $p = 32$ dollars; $q = 80$
77. (a) 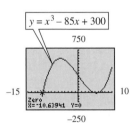   **(b)** 36   **(c)** 54   **79.** $y = 4x - 4$ **81.** $5, -5$

83. Find the coordinates of the two points of intersection, using a system, and then use one of any of several methods discussed so far in this book to find the equation of the line joining the two points.
85. $R \approx 187$; $R_t \approx 645$ **87.**

$$5)\overline{\begin{array}{rrrr} 1 & 0 & -85 & 300 \\ & 5 & 25 & -300 \\ \hline 1 & 5 & -60 & 0 \end{array}} \leftarrow 5 \text{ is a solution.}$$

88. $x^3 - 85x + 300 = (x - 5)(x^2 + 5x - 60)$ **89.** Use the quadratic formula with $a = 1$, $b = 5$, and $c = -60$.
90. 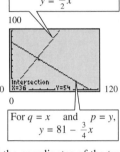   **91.** $\dfrac{-5 - \sqrt{265}}{2}$ **92.** Because x represents a length, it must be a positive number.

Section 7.2 (page 539)

1. In the first equation, we get $-1 = -1$. In the second, we get $-6 = -6$, and in the third, we get $19 = 19$.   **3.** In the first equation, we get $-.4 = -.4$. In the second, we get $1.8 = 1.8$, and in the third, we get $-.7 = -.7$.   **5.** $\{(1, 2, -1)\}$
7. $\{(2, 0, 3)\}$ **9.** $\emptyset$ **11.** $\{(1, 2, 3)\}$ **13.** $\{(-1, 2, 1)\}$ **15.** $\{(4, 1, 2)\}$ **17.** $\left\{\left(\frac{1}{2}, \frac{2}{3}, -1\right)\right\}$
19. $\{(2, 4, 2)\}$ **21.** $\left\{\left(-1, 1, \frac{1}{3}\right)\right\}$ **23. (a)** As an example, $x + 2y + z = 5$, $2x - y + 3z = 4$. (There are others.)
(b) As an example, $x + y + z = 5$, $2x - y + 3z = 4$. (There are others.)   **(c)** As an example, $2x + 2y + 2z = 8$,
$2x - y + 3z = 4$. (There are others.)   **25.** For example, the ceiling and two perpendicular walls intersect in a point.

27. $\left\{\left(\dfrac{4 - z}{3}, \dfrac{4z - 7}{3}, z\right)\right\}$ **29.** $\left\{\left(\dfrac{69 - z}{9}, \dfrac{z + 66}{9}, z\right)\right\}$ **31.** $\{(3z - 24, z - 13, z)\}$ **33.** 12 cents;
13 nickels; 4 quarters   **35.** 30 barrels each of \$150 and \$190 glue; 210 barrels of \$120 glue   **37.** 28 inches;
17 inches; 14 inches   **39.** \$50,000 at 5%; \$10,000 at 4.5%; \$40,000 at 3.75%   **41.** either 10 buffets, 5 chairs, and
no tables; or 11 buffets, 1 chair, and 1 table   **43.** $y = .75x^2 + .25x - .5$ **45.** $y = x^2 + 2x + 1$
47. (a) $C = \frac{17}{450}t^2 + \frac{1}{10}t + 315$ **(b)** 2048   **49.** $x^2 + y^2 + x - 7y = 0$ **51.** $a = -2, b = 20, c = 5; s(8) = 37$

Section 7.3 (page 550)

1. $\begin{bmatrix} -4 & -8 \\ 4 & 7 \end{bmatrix}$ **3.** $\begin{bmatrix} 1 & 5 & 6 \\ -1 & 8 & 5 \\ 4 & 7 & 0 \end{bmatrix}$ **5.** $\begin{bmatrix} -3 & 1 & -4 \\ 2 & 1 & 3 \\ -17 & 0 & -13 \end{bmatrix}$ **7.** $\left[\begin{array}{cc|c} 2 & 3 & 11 \\ 1 & 2 & 8 \end{array}\right]$ **9.** $\left[\begin{array}{cc|c} 1 & 5 & 6 \\ 1 & 2 & 8 \end{array}\right]$

11. $\left[\begin{array}{ccc|c} 2 & 1 & 1 & 3 \\ 3 & -4 & 2 & -7 \\ 1 & 1 & 1 & 2 \end{array}\right]$ **13.** $\left[\begin{array}{ccc|c} 1 & 1 & 0 & 2 \\ 0 & 2 & 1 & -4 \\ 0 & 0 & 1 & 2 \end{array}\right]$ **15.** $\begin{aligned} 2x + y &= 1 \\ 3x - 2y &= -9 \end{aligned}$ **17.** $\begin{aligned} x &= 2 \\ y &= 3 \\ z &= -2 \end{aligned}$

19. $\begin{aligned} 3x + 2y + z &= 1 \\ 2y + 4z &= 22 \\ -x - 2y + 3z &= 15 \end{aligned}$ **21.** $\{(2, 3)\}$ **23.** $\{(-3, 0)\}$ **25.** $\left\{\left(\frac{7}{2}, -1\right)\right\}$ **27.** $\emptyset$ **29.** $\left\{\left(\dfrac{3y + 1}{6}, y\right)\right\}$

31. $\{(-2, 1, 3)\}$ **33.** $\{(-1, 23, 16)\}$ **35.** $\{(3, 2, -4)\}$ **37.** $\{(2, 1, -1)\}$ **39.** In both cases, we
simply write the coefficients and do not write the variables. This is possible because we agree beforehand on the order in which
the variables appear.   **41.** $\left\{\left(\dfrac{5z + 14}{5}, \dfrac{5z - 12}{5}, z\right)\right\}$ **43.** $\left\{\left(\dfrac{12 - z}{7}, \dfrac{4z - 6}{7}, z\right)\right\}$ **45.** $\emptyset$

47. wife: 40 days; husband: 32 days   **49.** model 201: 5 bicycles; model 301: 8 bicycles   **51.** \$10,000 at 8%;
\$7000 at 10%; \$8000 at 9%   **53.** $\{(0, 2, -2, 1)\}$ **55.** $2 = a + b + c$ **56.** $-13 = 4a - 2b + c$
57. $-8 = 9a + 3b + c$ **58.** a must be negative because the parabola opens downward.   **59.** c must be positive
because the graph intersects the y-axis above the origin. (The value of c is the y-intercept.)
60. $a = -2, b = 3, c = 1$ **61.** $f(-1.5) = -8$
62.

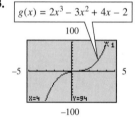

Show that the point $(-1.5, -8)$ lies on the
graph of the function $f(x) = -2x^2 + 3x + 1$.

63.

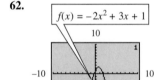

$a = 2, b = -3, c = 4, d = -2; g(4) = 94$; The point $(4, 94)$
lies on the graph of $g(x) = 2x^3 - 3x^2 + 4x - 2$.

65. $\left[\begin{array}{ccc|c} 1 & 0 & \frac{7}{3} & -\frac{4}{3} \\ 0 & 1 & -\frac{8}{3} & \frac{14}{3} \\ 0 & 0 & 1 & -1 \end{array}\right]$ **66.** $\left[\begin{array}{ccc|c} 1 & 0 & 0 & 1 \\ 0 & 1 & -\frac{8}{3} & \frac{14}{3} \\ 0 & 0 & 1 & -1 \end{array}\right]$ **67.** $\left[\begin{array}{ccc|c} 1 & 0 & 0 & 1 \\ 0 & 1 & 0 & 2 \\ 0 & 0 & 1 & -1 \end{array}\right]$ **68.** $\begin{aligned} x &= 1 \\ y &= 2 \\ z &= -1 \end{aligned}$

69. $(1, 2, -1)$; yes   **70.**

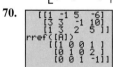

71. (a) $a + 871b + 11.5c + 3d = 239$
$a + 847b + 12.2c + 2d = 234$
$a + 685b + 10.6c + 5d = 192$
$a + 969b + 14.2c + 1d = 343$

(b) $\left[\begin{array}{cccc|c} 1 & 871 & 11.5 & 3 & 239 \\ 1 & 847 & 12.2 & 2 & 234 \\ 1 & 685 & 10.6 & 5 & 192 \\ 1 & 969 & 14.2 & 1 & 343 \end{array}\right]$ The solution is $a \approx -715.457, b \approx .34756, c \approx 48.6585$, and $d \approx 30.71951$.

(c) $F = -715.457 + .34756A + 48.6585P + 30.71951W$ **(d)** approximately 323, which is very close to 320

Section 7.4 (page 562)

1. Answers will vary.   **3.** 2×2; square   **5.** 3×4 **7.** 2×1; column   **9.** 3×3; square
11. $x = -15; y = 5; k = 3$ **13.** $z = 18; r = 3; s = 3; p = 3; a = \frac{3}{4}$
15. Only matrices of the same dimensions may be added. Simply add corresponding elements.
17. $\begin{bmatrix} 8 & -43 & -18 \\ 26 & 29 & 6 \\ -2 & 10 & 43 \end{bmatrix}$ **19.** It is not possible to find the sum.   **21.** $\begin{bmatrix} 13 & 3 & 0 & -2 \\ 9 & -12 & 4 & 8 \\ 12 & -11 & -1 & 9 \end{bmatrix}$

23. $\begin{bmatrix} -12x + 8y & -x + y \\ x & 8x - y \end{bmatrix}$ **25.** $\begin{bmatrix} 7 & 2 \\ 9 & 0 \\ 8 & 6 \end{bmatrix}; \begin{bmatrix} 7 & 9 & 8 \\ 2 & 0 & 6 \end{bmatrix}$ **27.** $\begin{bmatrix} 5411 & 11,352 \\ 9371 & 15,956 \end{bmatrix}; \begin{bmatrix} 5411 & 9371 \\ 11,352 & 15,956 \end{bmatrix}$

29. $\begin{bmatrix} -4 & 8 \\ 0 & 6 \end{bmatrix}$ **31.** $\begin{bmatrix} 8 & -16 \\ 0 & -12 \end{bmatrix}$ **33.** $\begin{bmatrix} 2 & 6 \\ -4 & 6 \end{bmatrix}$ **35.** $\begin{bmatrix} 60 & -10 \\ -44 & 9 \end{bmatrix}$ **37.** $\begin{bmatrix} 2 & 6 & 5 \\ -4 & -7 & 9 \end{bmatrix}$
39. $4 \times 4; 2 \times 2$ **41.** 3×2; BA is not defined.   **43.** Neither AB nor BA is defined.
45. columns; rows   **47.** (corresponding) first row; (corresponding) second column; adding
49. $\begin{bmatrix} pa + qb & pc + qd \\ ra + sb & rc + sd \end{bmatrix}$ **51.** $\begin{bmatrix} -17 \\ -1 \end{bmatrix}$ **53.** $\begin{bmatrix} 17 & -10 \\ 1 & 2 \end{bmatrix}$ **55.** $\begin{bmatrix} -2 & 10 \\ 0 & 8 \end{bmatrix}$
57. $\begin{bmatrix} -2 & 5 & 0 \\ 6 & 6 & 1 \\ 12 & 2 & -3 \end{bmatrix}$ **59.** $[2 \quad 7 \quad -4]$ **61.** $\begin{bmatrix} 20 & 0 & 8 \\ 8 & 0 & 4 \end{bmatrix}$ **63.** The product is not defined.

65. $[-8]$ **67.** $\begin{bmatrix} -6 & 12 & 18 \\ 4 & -8 & -12 \\ -2 & 4 & 6 \end{bmatrix}$ **69.** $\begin{bmatrix} 16 & 22 \\ 7 & 19 \end{bmatrix}$ **71.** $\begin{bmatrix} -10 & 16 & 26 \\ 5 & 7 & 22 \end{bmatrix}$

73. no; The answers are different (CA is not defined); no

75. (a) $\begin{bmatrix} 50 & 100 & 30 \\ 10 & 90 & 50 \\ 60 & 120 & 40 \end{bmatrix}$ **(b)** $\begin{bmatrix} 12 \\ 10 \\ 15 \end{bmatrix}$ or $[12 \quad 10 \quad 15]$ **(c)** $\begin{bmatrix} 2050 \\ 1770 \\ 2520 \end{bmatrix}$ or $[2050 \quad 1770 \quad 2520]$ **(d)** $6340

77. always true   **79.** always true

Section 7.5 (page 572)

1. -36 **3.** 7 **5.** 0 **7.** -26 **9.** 8 **11.** 17 **13.** 166 **15.** 0 **17.** 0
19. -88 **21.** If we expand about that particular row or column, the product of each element and its cofactor will be zero, and the sum of all these zeros will be zero. Therefore, the determinant will be zero.   **22.** $5x^2 + 2x - 6$

23. $5x^2 + 2x - 6 = 45$; quadratic   **24.** $\{3, -3.4\}$ **25.** $\det\begin{bmatrix} 3 & 2 & 1 \\ -1 & 3 & 4 \\ -2 & 0 & 5 \end{bmatrix} = 45$ and

$\det\begin{bmatrix} -3.4 & 2 & 1 \\ -1 & -3.4 & 4 \\ -2 & 0 & 5 \end{bmatrix} = 45$ are both true.   **27.** $\{0\}$ **29.** $\left\{-\frac{1}{2}, 6\right\}$ **31.** $\{-2\}$ **33.** $\left\{-\frac{2}{3}\right\}$ **35.** 0

37. -1; The first and second rows of the matrix were interchanged.   **39.** 0   **41.** $x + 3y - 11 = 0$
42. The equation of the line is $x + 3y - 11 = 0$, which is equivalent to the equation in Exercise 41.

43. $y - y_1 = \dfrac{y_2 - y_1}{x_2 - x_1}(x - x_1)$ **44.** When the determinant is expanded, the equation is the same as the answer to

Exercise 43.   **45.** $\{(2, 2)\}$ **47.** $\{(2, -5)\}$ **49.** $\left\{\left(\dfrac{4 - 2y}{3}, y\right)\right\}$ (Use another method.)   **51.** $\{(2, 3)\}$

53. $\{(-1, 2, 1)\}$ **55.** $\{(-3, 4, 2)\}$ **57.** $\{(0, 0, -1)\}$ **59.** $\emptyset$ (Use another method.)
61. $\{(-1, 2, 5, 1)\}$ **63.** If $D = 0$, Cramer's rule *cannot* be applied, because there is no *unique* solution.
65. (a) 1 (b) 5/2 (c) 19/2   **66.** (a) 7 (b) 33/2 (c) 7/2

Section 7.6 (page 581)

1. yes   **3.** no   **5.** yes   **7.** no   **9.** The inverse will not exist if the determinant is equal to 0.
11. $\begin{bmatrix} 0 & \frac{1}{2} \\ -1 & \frac{1}{2} \end{bmatrix}$ **13.** The inverse does not exist.   **15.** $\begin{bmatrix} 2 & 1 \\ -\frac{3}{2} & -\frac{1}{2} \end{bmatrix}$ **17.** $\begin{bmatrix} -2.5 & 5 \\ 12.5 & -15 \end{bmatrix}$

19. $\begin{bmatrix} 1 & 0 & 0 \\ 0 & -1 & 0 \\ -1 & 0 & 1 \end{bmatrix}$ **21.** $\begin{bmatrix} 15 & 4 & -5 \\ -12 & -3 & 4 \\ -4 & -1 & 1 \end{bmatrix}$ **23.** $\begin{bmatrix} -\frac{10}{3} & \frac{5}{9} & -\frac{10}{9} \\ \frac{20}{3} & \frac{5}{9} & \frac{80}{9} \\ -5 & \frac{5}{6} & -\frac{20}{3} \end{bmatrix}$ **25.** The inverse does not exist.

27. $\begin{bmatrix} 2x + 4z & 2y + 4w \\ x - z & y - w \end{bmatrix}$ **28.** $2x + 4z = 1,\ 2y + 4w = 0,\ x - z = 0,\ y - w = 1$

29. $x = 1/6,\ z = 1/6,\ y = 2/3,\ w = -1/3$ **30.** $A^{-1} = \begin{bmatrix} \frac{1}{6} & \frac{2}{3} \\ \frac{1}{6} & -\frac{1}{3} \end{bmatrix}$ **31.** $\begin{bmatrix} 1 & 0 & \frac{1}{6} & \frac{2}{3} \\ 0 & 1 & \frac{1}{6} & -\frac{1}{3} \end{bmatrix}$

32. Matrix B is the same as matrix A^{-1}. Perform row transformations on $[A \mid I_2]$ to change it to $[I_2 \mid B]$. Matrix B is A^{-1}.

33. $A = \begin{bmatrix} 1 & 1 \\ 2 & -1 \end{bmatrix}$; $X = \begin{bmatrix} x \\ y \end{bmatrix}$; $B = \begin{bmatrix} 8 \\ 4 \end{bmatrix}$ **35.** $A = \begin{bmatrix} 4 & 5 \\ 2 & 3 \end{bmatrix}$; $X = \begin{bmatrix} x \\ y \end{bmatrix}$; $B = \begin{bmatrix} 7 \\ 5 \end{bmatrix}$ **37.** $\{(-2, 4)\}$

39. $\{(4, -6)\}$ **41.** $\left\{\left(\frac{5}{2}, -1\right)\right\}$ **43.** $\{(3, 0, 2)\}$ **45.** $\left\{\left(12, -\frac{15}{11}, -\frac{65}{11}\right)\right\}$ **47.** $\{(0, 2, -2, 1)\}$
49. $P(x) = -2x^3 + 5x^2 - 4x + 3$ **51.** $P(x) = x^4 + 2x^3 + 3x^2 - x - 1$
53. (a) $602.7 = a + 5.543b + 37.14c$ (b) $a \approx -490.547, b \approx -89, c = 42.71875$
$656.7 = a + 6.933b + 41.30c$
$778.5 = a + 7.638b + 45.62c$
(c) $S = -490.547 - 89A + 42.71875B$ (d) approximately 843.5
(e) $S \approx 1547.5$; Using only three consecutive years to forecast six years into the future is probably not very accurate.

Section 7.7 (page 590)

1. **3.** **5.**

7.

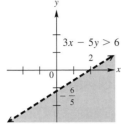

$3x - 5y > 6$

9.

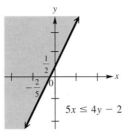

$5x \leq 4y - 2$

11.

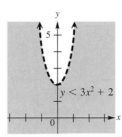

$y < 3x^2 + 2$

13.

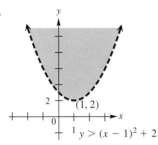

$y > (x - 1)^2 + 2$

15.

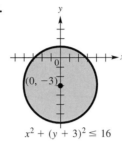

$x^2 + (y + 3)^2 \leq 16$

17.

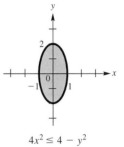

$4x^2 \leq 4 - y^2$

19.

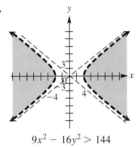

$9x^2 - 16y^2 > 144$

21. (b)
23. (d)
25. C
27. A

29.

$x + y \leq 4$
$x - 2y \geq 6$

31.

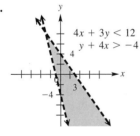

$4x + 3y < 12$
$y + 4x > -4$

33.

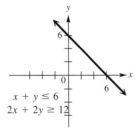

$x + y \leq 6$
$2x + 2y \geq 12$

Only the points of the
line are included.

35.

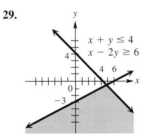

$x + 2y \leq 4$
$y \geq x^2 - 1$

37.

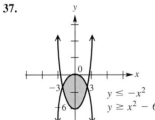

$y \leq -x^2$
$y \geq x^2 - 6$

39.

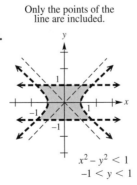

$x^2 - y^2 < 1$
$-1 < y < 1$

41.

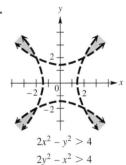

$2x^2 - y^2 > 4$
$2y^2 - x^2 > 4$

43.

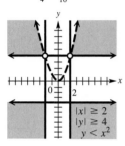

$$\frac{x^2}{16} + \frac{y^2}{9} \le 1$$
$$\frac{x^2}{4} - \frac{y^2}{16} \ge 1$$

45.

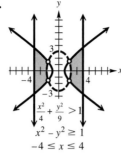

$$\frac{x^2}{4} + \frac{y^2}{9} > 1$$
$$x^2 - y^2 \ge 1$$
$$-4 \le x \le 4$$

47.

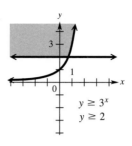

$$y \ge 3^x$$
$$y \ge 2$$

49.

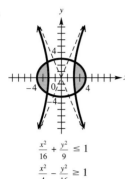

$$|x| \ge 2$$
$$|y| \ge 4$$
$$y < x^2$$

51.

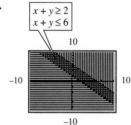

$$y \le |x + 2|$$
$$\frac{x^2}{16} - \frac{y^2}{9} \le 1$$

53. (d)   **55.** A   **57.** B

59.

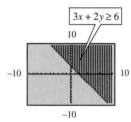

$3x + 2y \ge 6$

61.

$x + y \ge 2$
$x + y \le 6$

63.

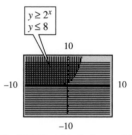

$y \ge 2^x$
$y \le 8$

65.

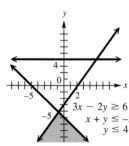

$3x - 2y \ge 6$
$x + y \le -5$
$y \le 4$

67. maximum value: 65 at (5, 10)
minimum value: 8 at (1, 1)

69. \$1120, with 4 pigs and 12 geese

71. 3.75 servings of A and 1.875 servings of B, for a minimum cost of \$1.69   **73.** 1600 Type 1 and 0 Type 2 for a maximum revenue of \$160   **75.** 0 medical kits and 4000 containers of water

Section 7.8 (page 598)

1. $\dfrac{5}{3x} + \dfrac{-10}{3(2x + 1)}$ **3.** $\dfrac{6}{5(x + 2)} + \dfrac{8}{5(2x - 1)}$ **5.** $\dfrac{5}{6(x + 5)} + \dfrac{1}{6(x - 1)}$ **7.** $\dfrac{-2}{x + 1} + \dfrac{2}{x + 2} + \dfrac{4}{(x + 2)^2}$

9. $\dfrac{4}{x} + \dfrac{4}{1 - x}$ **11.** $\dfrac{15}{x} + \dfrac{-5}{x + 1} + \dfrac{-6}{x - 1}$ **13.** $1 + \dfrac{-2}{x + 1} + \dfrac{1}{(x + 1)^2}$

15. $x^3 - x^2 + \dfrac{-1}{3(2x + 1)} + \dfrac{2}{3(x + 2)}$ **17.** $\dfrac{1}{9} + \dfrac{-1}{x} + \dfrac{25}{18(3x + 2)} + \dfrac{29}{18(3x - 2)}$ **19.** $\dfrac{-3}{5x^2} + \dfrac{3}{5(x^2 + 5)}$

21. $\dfrac{-2}{7(x + 4)} + \dfrac{6x - 3}{7(3x^2 + 1)}$ **23.** $\dfrac{1}{4x} + \dfrac{-8}{19(2x + 1)} + \dfrac{-(9x + 24)}{76(3x^2 + 4)}$ **25.** $\dfrac{-1}{x} + \dfrac{2x}{2x^2 + 1} + \dfrac{2x + 3}{(2x^2 + 1)^2}$

27. $\dfrac{-1}{x + 2} + \dfrac{3}{(x^2 + 4)^2}$ **29.** $5x^2 + \dfrac{3}{x} + \dfrac{-1}{x + 3} + \dfrac{2}{x - 1}$ **31.** correct   **33.** incorrect

Chapter 7 Review Exercises (page 601)

1. $\{(5, 7)\}$ **3.** $\{(5, -3)\}$ **5.** $\left\{(1, 1), \left(\frac{7}{5}, -\frac{1}{5}\right)\right\}$ **7.** $\left\{\left(3\sqrt{3}, \sqrt{2}\right), \left(-3\sqrt{3}, \sqrt{2}\right), \left(3\sqrt{3}, -\sqrt{2}\right), \left(-3\sqrt{3}, -\sqrt{2}\right)\right\}$
9. $\{(-2, 0), (1, 1)\}$ **11.** It cannot, because two distinct lines will intersect in *at most* one point. If the lines coincide, there are infinitely many solutions.   **13.** $\{(-1, 2, 3)\}$ **15.** $\{(1, 3, -1)\}$ **17.** No, because two distinct planes intersect in a line, so there are infinitely many solutions. If the planes coincide, there are infinitely many solutions, also. If the planes are parallel, there are no solutions.   **19.** $\{(-3, 2)\}$ **21.** $\{(0, 1, 0)\}$

23. $\begin{bmatrix} -4 \\ 6 \\ 1 \end{bmatrix}$ **25.** $\begin{bmatrix} -4 & -4 \\ -7 & 16 \end{bmatrix}$ **27.** $\begin{bmatrix} 14x + 28y & 42x + 22y \\ 18x - 46y & 70x + 18y \end{bmatrix}$ **29.** $\begin{bmatrix} 11 & 20 \\ 14 & 40 \end{bmatrix}$ **31.** $\begin{bmatrix} -3 \\ 10 \end{bmatrix}$

33. $\begin{bmatrix} -2 & 22 & 31 \\ 25 & 12 & 9 \end{bmatrix}$ **35.** (a)   **37.** yes   **39.** no   **41.** $\begin{bmatrix} 3 & -1 \\ -5 & 2 \end{bmatrix}$ **43.** $\begin{bmatrix} \frac{1}{2} & 0 \\ \frac{1}{10} & \frac{1}{5} \end{bmatrix}$

45. $\begin{bmatrix} \frac{2}{3} & 0 & -\frac{1}{3} \\ \frac{1}{3} & 0 & -\frac{2}{3} \\ -\frac{2}{3} & 1 & \frac{1}{3} \end{bmatrix}$ **47.** $\{(2, 2)\}$ **49.** $\{(2, 1)\}$

51. dependent equations; $\left\{\left(\dfrac{6z + 5}{11}, \dfrac{31z + 2}{11}, z\right)\right\}$ **53.** $\{(-1, 0, 2)\}$ **55.** -25 **57.** -44
59. $\left\{-\frac{7}{3}\right\}$ **61.** {all real numbers}   **63.** (a) 5   (b) 30   (c) -50 (d) $x = 6, y = -10$
65. If $D = 0$, then there would be division by zero, which is undefined. When $D = 0$, the system either has no solution or infinitely many solutions.   **67.** $\{(-4, 2)\}$ **69.** $\{(-4, 6, 2)\}$ **71.** $\left\{\left(\frac{172}{67}, -\frac{14}{67}, -\frac{87}{67}\right)\right\}$
73. 80 $3\frac{1}{2}''$ diskettes; 20 $5\frac{1}{4}''$ diskettes   **75.** 11 ml of 5%; 3 ml of 15%; 6 ml of 10%

77.

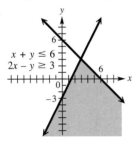

79.

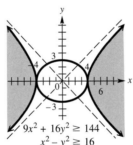

81. minimum of 40 at $(0, 20)$ or $\left(\frac{10}{3}, \frac{40}{3}\right)$ or at any point between

83. $\dfrac{3}{x + 2} + \dfrac{2}{x - 2}$

Chapter 7 Test (page 604)

1. (a) The first is the equation of a hyperbola; the second is the equation of a line.   (b) 0, 1, or 2   (c) $\left\{\left(-\frac{11}{35}, \frac{68}{35}\right), (1, -2)\right\}$
(d) There are two intersection points.   **2.** (a) $\{(2, 0, -1)\}$ (b) The same steps are used. The difference is that with the Gaussian method, we work with the rows of a matrix (with no variables), while with the echelon method we work with the equations of a system.   **3.** (a) $\begin{bmatrix} 8 & 3 \\ 0 & -11 \\ 15 & 19 \end{bmatrix}$ (b) not possible   (c) $\begin{bmatrix} -5 & 16 \\ 19 & 2 \end{bmatrix}$ **4.** (a) yes   (b) yes
(c) not necessarily, because matrix multiplication is not commutative   (d) Only BA can be found.   **5.** (a) -58

(b) -844 **6.** $\{(-6, 7)\}$ **7. (a)** $A = \begin{bmatrix} 1 & 1 & -1 \\ 2 & -3 & -1 \\ 1 & 2 & 2 \end{bmatrix}$, $X = \begin{bmatrix} x \\ y \\ z \end{bmatrix}$, $B = \begin{bmatrix} -4 \\ 5 \\ 3 \end{bmatrix}$ **(b)** $A^{-1} = \begin{bmatrix} \frac{1}{4} & \frac{1}{4} & \frac{1}{4} \\ \frac{5}{16} & -\frac{3}{16} & \frac{1}{16} \\ -\frac{7}{16} & \frac{1}{16} & \frac{5}{16} \end{bmatrix}$

(c) $\{(1, -2, 3)\}$ **(d)** With the new equation replacing the first equation, the determinant of the coefficient matrix A is 0, so A^{-1} does not exist.   **8.** (b)   **9. (a)** a quadratic function $y = ax^2 + bx + c$

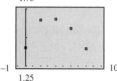

(b) $1.41 = c$, $1.66 = 16a + 4b + c$, $1.40 = 64a + 8b + c$
(c) $f(x) = -.0159375x^2 + .12625x + 1.41$

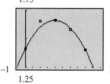

(d) approximately 1.08 million vehicle thefts

10. (a) $x_1 + x_4 = 1000$, $x_1 + x_2 = 1100$, $x_2 + x_3 = 700$, $x_3 + x_4 = 600$

(b) $\begin{bmatrix} 1 & 0 & 0 & 1 & | & 1000 \\ 1 & 1 & 0 & 0 & | & 1100 \\ 0 & 1 & 1 & 0 & | & 700 \\ 0 & 0 & 1 & 1 & | & 600 \end{bmatrix}$; $\begin{bmatrix} 1 & 0 & 0 & 1 & | & 1000 \\ 0 & 1 & 0 & -1 & | & 100 \\ 0 & 0 & 1 & 1 & | & 600 \\ 0 & 0 & 0 & 0 & | & 0 \end{bmatrix}$ **(c)** $x_1 \le 1000$, $x_2 \le 700$, $x_3 \le 600$, $x_4 \le 600$

(d) No. One solution is $x_1 = 900$, $x_2 = 200$, $x_3 = 500$, $x_4 = 100$. Many solutions are possible.

CHAPTER 8

Section 8.1 (page 616)

1. 14, 18, 22, 26, 30   **3.** 1, 2, 4, 8, 16   **5.** $0, \frac{1}{9}, \frac{2}{27}, \frac{1}{27}, \frac{4}{243}$ **7.** $-2, 4, -6, 8, -10$ **9.** $1, \frac{7}{6}, 1, \frac{5}{6}, \frac{19}{27}$
11. The nth term is the term that is in position n. For example, in the sequence 2, 4, 6, 8, . . . , the *third* term is 6, which is in position 3. The nth term here is given by $2n$.   **13.** finite   **15.** finite   **17.** infinite   **19.** finite
21. $-2, 1, 4, 7$ **23.** 1, 1, 2, 3   **25.** 35   **27.** $\frac{25}{12}$ **29.** 288   **31.** 3   **33.** $-1 + 1 + 3 + 5 + 7$
35. $-10 - 4 + 0$ **37.** $0 + \frac{1}{2} + \frac{2}{3} + \frac{3}{4}$ **39.** $-3.5 + .5 + 4.5 + 8.5$ **41.** $0 + 4 + 16 + 36$

43. $-1 - \frac{1}{3} - \frac{1}{5} - \frac{1}{7}$ **45.** $\sum_{j=3}^{7} (12 - 3j)$ **47.** $\sum_{j=0}^{9} 2(3)^{j+1}$ **49.** $\sum_{j=0}^{10} (j^2 - 4j + 3)$ **51.** 90

53. 220   **55.** 304   **57.** converges to $\frac{1}{2}$ **59.** diverges   **61.** converges to $e \approx 2.71828$
63. (a) $N_{j+1} = 2N_j$ for $j \ge 1$ **(b)** 1840   **(c)** 15,000

(d) The growth is very rapid. Since there is a doubling of the bacteria at equal intervals, their growth is exponential.

Section 8.2 (page 624)

1. 3   **3.** -5 **5.** $x + 2y$ **7.** 8, 14, 20, 26, 32   **9.** 5, 3, 1, -1, -3 **11.** 14, 12, 10, 8, 6
13. $a_8 = 19$; $a_n = 3 + 2n$ **15.** $a_8 = 7$; $a_n = n - 1$ **17.** $a_8 = -6$; $a_n = 10 - 2n$ **19.** $a_8 = -3$;

$a_n = -39 + \dfrac{9n}{2}$ **21.** $a_8 = x + 21$; $a_n = x + 3n - 3$ **23.** 3   **25.** 5   **27.** 215   **29.** 125

31. 230   **33.** 77.5   **35.** $a_1 = 7$; $d = 5$ **37.** $a_1 = 1$; $d = -\frac{20}{11}$ **39.** 18   **41.** 140
43. -684 **45.** 500,500   **47.** $f(1) = m + b$; $f(2) = 2m + b$; $f(3) = 3m + b$ **48.** yes   **49.** m
50. $a_n = mn + b$ **51.** 328.3   **53.** 172.884   **55.** 1281   **57.** 4680   **59.** 54,800
61. 713 inches   **63.** All terms are the same constant.   **65.** yes

Section 8.3 (page 632)

1. $\frac{5}{3}$, 5, 15, 45   **3.** $\frac{5}{8}$, $\frac{5}{4}$, $\frac{5}{2}$, 5, 10   **5.** $a_5 = 80$; $a_n = 5(-2)^{n-1}$ **7.** $a_5 = -108$; $a_n = \left(-\frac{4}{3}\right) \cdot 3^{n-1}$
9. $a_5 = -729$; $a_n = (-9)(-3)^{n-1}$ **11.** $a_5 = -324$; $a_n = -4 \cdot 3^{n-1}$ **13.** $a_5 = \frac{125}{4}$; $a_n = \left(\frac{4}{5}\right)\left(\frac{5}{2}\right)^{n-1}$
15. $x - 5 = .6 - x$ **16.** The solution is 2.8; 5, 2.8, .6   **17.** $\dfrac{x}{5} = \dfrac{.6}{x}$ **18.** The solution is $\sqrt{3}$; 5, $\sqrt{3}$, .6
19. $a_1 = 125$; $r = \frac{1}{5}$ **21.** $a_1 = -2$; $r = \frac{1}{2}$ **23.** 682   **25.** $\frac{99}{8}$ **27.** 860.95   **29.** 363   **31.** $\frac{189}{4}$
33. 2032   **35.** The sum exists if $|r| < 1$.   **37.** 2; does not converge   **39.** $\frac{1}{2}$ **41.** $\frac{128}{7}$ **43.** $\frac{1000}{9}$
45. $\frac{8}{3}$ **47.** 4   **49.** $\frac{3}{7}$ **51.** $g(1) = ab$; $g(2) = ab^2$; $g(3) = ab^3$ **52.** yes; The common ratio is b.
53. $a_n = ab^n$ **54.** The independent variable is in the exponent.   **55.** 97.739   **57.** .212
59. $12,487.56   **61.** $32,029.33   **63. (a)** $a_n = a_1 \cdot 2^{n-1}$ **(b)** 15 (rounded from 14.28)   **(c)** 560 minutes, or
9 hours and 20 minutes   **65.** approximately 13.4%   **67.** $26,214.40   **69.** 200 centimeters
71. 62; 2046   **73.** $\frac{1}{64}$ meter

Section 8.4 (page 640)

1. 20   **3.** 35   **5.** 56   **7.** 45   **9.** 1   **11.** n **13.** 9
15. $x^6 + 6x^5y + 15x^4y^2 + 20x^3y^3 + 15x^2y^4 + 6xy^5 + y^6$ **17.** $p^5 - 5p^4q + 10p^3q^2 - 10p^2q^3 + 5pq^4 - q^5$
19. $r^{10} + 5r^8s + 10r^6s^2 + 10r^4s^3 + 5r^2s^4 + s^5$ **21.** $p^4 + 8p^3q + 24p^2q^2 + 32pq^3 + 16q^4$
23. $2401p^4 + 2744p^3q + 1176p^2q^2 + 224pq^3 + 16q^4$
25. $729x^6 - 2916x^5y + 4860x^4y^2 - 4320x^3y^3 + 2160x^2y^4 - 576xy^5 + 64y^6$
27. $\dfrac{m^6}{64} - \dfrac{3m^5}{16} + \dfrac{15m^4}{16} - \dfrac{5m^3}{2} + \dfrac{15m^2}{4} - 3m + 1$ **29.** $4r^4 + \dfrac{8\sqrt{2}r^3}{m} + \dfrac{12r^2}{m^2} + \dfrac{4\sqrt{2}r}{m^3} + \dfrac{1}{m^4}$ **31.** $-3584h^3j^5$
33. $319,770a^{16}b^{14}$ **35.** $38,760x^6y^{42}$ **37.** $90,720x^{28}y^{12}$ **39.** 11   **41.** exact: 3,628,800; approximate:
3,598,695.619   **42.** $\approx .830\%$ **43.** exact: 479,001,600; approximate: 475,687,486.5; $\approx .692\%$
44. exact: 6,227,020,800; approximate: 6,187,239,475; $\approx .639\%$; As n gets larger, the percent error decreases.
45. .942   **47.** 1.015

Section 8.5 (page 647)

1. natural numbers   **3.** $n = 1$ and $n = 2$ **25.** $x^2 + xy + y^2 = \dfrac{x^3 - y^3}{x - y}$
26. $x^3 - y^3 = (x - y)(x^2 + xy + y^2)$ **27.** $x^{2n} - x^{2n-1}y + \ldots - xy^{2n-1} + y^{2n} = \dfrac{x^{2n+1} + y^{2n+1}}{x + y}$
28. power; exponents   **29.** power; exponents

Section 8.6 (page 654)

1. 19,958,400   **3.** 72   **5.** 5   **7.** 6   **9.** 1   **11.** 495   **13. (a)** permutation   **(b)** permutation
(c) combination   **(d)** combination   **(e)** permutation   **(f)** combination   **(g)** permutation   **15.** 30   **17. (a)** 27,600
(b) 35,152   **(c)** 1104   **19.** 15   **21. (a)** 17,576,000   **(b)** 17,576,000   **(c)** 456,976,000   **23.** 720   **25.** 120
27. 2730   **29.** 120; 30,240   **31.** 27,405   **33.** 20   **35.** 10   **37.** 28   **39. (a)** 84   **(b)** 10
(c) 40   **(d)** 28   **41.** 1680   **43.** 15   **45.** 479,001,600   **47. (a)** 56   **(b)** 462   **(c)** 3080   **(d)** 8526
49. 15,120   **59. (a)** $3.04140932 \times 10^{64}$ **(b)** $8.320987113 \times 10^{81}$ **(c)** $8.247650592 \times 10^{90}$

Section 8.7 (page 663)

1. Let h = heads and t = tails. $S = \{h\}$ **3.** $S = \{(h, h, h), (h, h, t), (h, t, h), (t, h, h), (h, t, t), (t, h, t), (t, t, h), (t, t, t)\}$
5. Let c = correct, w = wrong. $S = \{(c, c, c), (c, c, w), (c, w, c), (w, c, c), (w, w, c), (w, c, w), (c, w, w), (w, w, w)\}$
7. (a) $\{h\}$; 1   **(b)** $\emptyset$; 0   **9. (a)** $\{(c, c, c)\}$; $\frac{1}{8}$ **(b)** $\{(w, w, w)\}$; $\frac{1}{8}$ **(c)** $\{(c, c, w), (c, w, c), (w, c, c)\}$; $\frac{3}{8}$
(d) $\{(c, w, w), (w, c, w), (w, w, c), (c, c, w), (c, w, c), (w, c, c), (c, c, c)\}$; $\frac{7}{8}$ **11.** A probability will always be a number
between 0 and 1, inclusive of both. $\frac{6}{5} > 1$, and so this answer must be incorrect.   **13. (a)** $\frac{1}{5}$ **(b)** 0   **(c)** $\frac{7}{15}$ **(d)** 1 to 4
(e) 7 to 8   **15.** 1 to 4   **17.** $\frac{2}{5}$ **19. (a)** $\frac{1}{2}$ **(b)** $\frac{7}{10}$ **(c)** $\frac{2}{5}$ **21. (a)** .62   **(b)** .27   **(c)** .11   **(d)** .89
23. $\frac{48}{28,561} \approx .001681$ **25. (a)** .72   **(b)** .70   **(c)** .79   **27.** approximately .313
29. approximately .031   **31.** .5   **33.** approximately 4.6×10^{-10} **35.** approximately .875
37. The probabilities, in order, are .125, .375, .375, and .125.   **39. (a)** approximately 40.4%   **(b)** approximately
4.7%   **(c)** approximately .2%; This means that in a large family or group of people, it is highly unlikely that everyone will

become sick even though the disease is highly infectious.   **41. (a)** $P_{00} = 1$; $P_{01} = 0$; $P_{02} = 0$; $P_{10} = \frac{1}{6}$; $P_{11} = \frac{2}{3}$;

$P_{12} = \frac{1}{6}$; $P_{20} = 0$; $P_{21} = 0$; $P_{22} = 1$ **(b)** $P = \begin{bmatrix} P_{00} & P_{01} & P_{02} \\ P_{10} & P_{11} & P_{12} \\ P_{20} & P_{21} & P_{22} \end{bmatrix} = \begin{bmatrix} 1 & 0 & 0 \\ 1/6 & 2/3 & 1/6 \\ 0 & 0 & 1 \end{bmatrix}$

(c) The matrix exhibits symmetry. The sum of the probabilities in each row is equal to 1. The greatest probabilities lie along the diagonal. This means that a mother cell is most likely to produce a daughter cell like itself.

Chapter 8 Review Exercises (page 669)

1. $\frac{1}{2}, \frac{2}{3}, \frac{3}{4}, \frac{4}{5}, \frac{5}{6}$; neither   **3.** 8, 10, 12, 14, 16; arithmetic   **5.** 5, 2, -1, -4, -7; arithmetic
7. $3\pi - 2, 2\pi - 1, \pi, 1, -\pi + 2$ **9.** $-5, -1, -\frac{1}{5}, -\frac{1}{25}, -\frac{1}{125}$ **11.** $a_4 = -1$; $a_n = -8\left(\frac{1}{2}\right)^{n-1}$ or $a_4 = 1$;
$a_n = -8\left(-\frac{1}{2}\right)^{n-1}$ (There are other ways to express a_n.)   **13.** $-x + 61$ **15.** 612   **17.** $\frac{4}{25}$ **19.** -40
21. 1   **23.** $\frac{73}{12}$ **25.** 3,126,250   **27.** $\frac{4}{3}$ **29.** 36   **31.** diverges   **33.** -10
35. $\sum\limits_{i=1}^{15} (-5i + 9)$ **37.** $\sum\limits_{i=1}^{6} 4(3)^{i-1}$ **39.** $x^4 + 8x^3y + 24x^2y^2 + 32xy^3 + 16y^4$
41. $243x^{5/2} - 405x^{3/2} + 270x^{1/2} - 90x^{-1/2} + 15x^{-3/2} - x^{-5/2}$ **43.** $-3584x^3y^5$
45. $x^{12} + 24x^{11} + 264x^{10} + 1760x^9$ **47.** The statement has as its domain the set of natural numbers. For example, if n
is a natural number, $1 + 2 + 3 + \ldots + n = \dfrac{n(n+1)}{2}$ and $1 + 3 + 5 + \ldots + (2n - 1) = n^2$.   **53.** permutation
55. 1   **57.** 362,880   **59.** 48   **61.** 24   **63.** 504   **65. (a)** $\frac{4}{15}$ **(b)** $\frac{2}{3}$ **(c)** 0   **67.** $\frac{1}{26}$
69. $\frac{4}{13}$ **71.** .86   **73.** 0   **75.** approximately .205

Chapter 8 Test (page 671)

1. (a) $-3, 4, -5, 6, -7$; neither   **(b)** $-\frac{3}{2}, -\frac{3}{4}, -\frac{3}{8}, -\frac{3}{16}, -\frac{3}{32}$; geometric   **(c)** 2, 3, 7, 13, 27; neither
2. (a) 49 **(b)** 16   **3. (a)** 110 **(b)** -1705 **4. (a)** 2385 **(b)** -186 **(c)** does not exist **(d)** $\frac{108}{7}$
5. (a) $16x^4 - 96x^3y + 216x^2y^2 - 216xy^3 + 81y^4$ **(b)** $60w^4y^2$ **6. (a)** 45 **(b)** 35 **(c)** 5040 **(d)** 990
8. (a) 24 **(b)** 6840 **(c)** 6160   **9. (a)** $\frac{1}{26}$ **(b)** $\frac{10}{13}$ **(c)** $\frac{4}{13}$ **(d)** 3 to 10   **10. (a)** approximately .104
(b) approximately .000000595

CHAPTER R

Section R.1 (page 679)

1. $(-4)^5$ **3.** 1   **5.** 1   **7.** 2^{10} **9.** $2^3x^{15}y^{12}$ **11.** $-\dfrac{p^8}{q^2}$ **13.** $x^2 - x + 3$
15. $9y^2 - 4y + 4$ **17.** $6m^4 - 2m^3 - 7m^2 - 4m$ **19.** $28r^2 + r - 2$ **21.** $15x^2 - \frac{7}{3}x - \frac{2}{9}$
23. $12x^5 + 8x^4 - 20x^3 + 4x^2$ **25.** $-2z^3 + 7z^2 - 11z + 4$ **27.** $m^2 + mn - 2n^2 - 2km + 5kn - 3k^2$
29. $a^2 - 2ab + b^2 + 4ac - 4bc + 4c^2$ **31.** Find the sum of the square of the first term, twice the product of the two
terms, and the square of the last term.   **33.** $4m^2 - 9$ **35.** $16m^2 + 16mn + 4n^2$ **37.** $25r^2 + 30rt^2 + 9t^4$
39. $4p^2 - 12p + 9 + 4pq - 6q + q^2$ **41.** $9q^2 + 30q + 25 - p^2$ **43.** $a^2 - b^2 - 2bc - c^2$
45. $9a^2 + 6ab + b^2 - 6a - 2b + 1$ **47.** $18p^2 - 27pq - 35q^2$ **49.** $27x^3 - 108x^2y + 144xy^2 - 64y^3$
51. $36k^2 - 36k + 9$ **53.** $p^3 - 7p^2 - p - 7$ **55.** $49m^2 - 4n^2$ **57.** $-14q^2 + 11q - 14$ **59.** $4p^2 - 16$
61. $11y^3 - 18y^2 + 4y$ **63.** $k^{2m} - 4$ **65.** $b^{2r} + b^r - 6$ **67.** $3p^{2x} - 5p^x - 2$ **69.** $m^{2x} - 4m^x + 4$
71. $q^{2p} - 10q^pp^q + 25p^{2q}$ **73.** $27k^{3a} - 54k^{2a} + 36k^a - 8$ **75.** m **77.** $m + n$

Section R.2 (page 685)

1. $4k^2m^3(1 + 2k^2 - 3m)$ **3.** $2(a + b)(1 + 2m)$ **5.** $(y + 2)(3y + 2)$ **7.** $(r + 3)(3r - 5)$
9. $(m - 1)(2m^2 - 7m + 7)$ **11.** $(2s + 3)(3t - 5)$ **13.** $(t^3 + s^2)(r - p)$ **15.** $(3p - 7)(2p + 5)$
17. $(5z - 2x)(4z - 9x)$ **19.** $(3 - m^2)(5 - r^2)$ **21.** $8(h - 8)(h + 5)$ **23.** $9y^2(y - 5)(y - 1)$
25. $(7m - 5r)(2m + 3r)$ **27.** $(4s + 5t)(3s - t)$ **29.** $(5a + m)(6a - m)$ **31.** $3x^3(3x - 5z)(2x + 5z)$
33. $(3m - 2)^2$ **35.** $2(4a - 3b)^2$ **37.** $(2xy + 7)^2$ **39.** $(a - 3b - 3)^2$ **41.** $(3a + 4)(3a - 4)$
43. $(5s^2 + 3t)(5s^2 - 3t)$ **45.** $(a + b + 4)(a + b - 4)$ **47.** $(p^2 + 25)(p + 5)(p - 5)$ **49.** (b)
51. $(2 - a)(4 + 2a + a^2)$ **53.** $(5x - 3)(25x^2 + 15x + 9)$ **55.** $(3y^3 + 5z^2)(9y^6 - 15y^3z^2 + 25z^4)$

57. $r(r^2 + 18r + 108)$ **59.** $(3 - m - 2n)(9 + 3m + 6n + m^2 + 4mn + 4n^2)$ **61.** It is incomplete, because a^2 has not been substituted back for u.   **63.** $(a^2 - 8)(a^2 + 6)$ **65.** $2(8z - 3)(6z - 5)$
67. $(18 - 5p)(17 - 4p)$ **69.** $(2b + c + 4)(2b + c - 4)$ **71.** $(x + y)(x - 5)$ **73.** $(m - 2n)(p^4 + q)$
75. $(2z + 7)^2$ **77.** $(10x + 7y)(100x^2 - 70xy + 49y^2)$ **79.** $(5m^2 - 6)(25m^4 + 30m^2 + 36)$
81. $(6m - 7n)(2m + 5n)$ **83.** $(4p - 1)(p + 1)$ **85.** prime   **87.** $4xy$ **89.** $(r + 3s^q)(r - 2s^q)$
91. $(3a^{2k} + b^{4k})(3a^{2k} - b^{4k})$ **93.** $(2y^a - 3)^2$ **95.** $[3(m + p)^k + 5][2(m + p)^k - 3]$ **97.** ± 36 **99.** 9

Section R.3 (page 692)

1. $x \neq -6$ **3.** $x \neq \frac{3}{5}$ **5.** no restrictions   **7.** (a)   **9.** $\dfrac{5p}{2}$ **11.** $\frac{8}{9}$ **13.** $\dfrac{3}{t - 3}$

15. $\dfrac{2x + 4}{x}$ **17.** $\dfrac{m - 2}{m + 3}$ **19.** $\dfrac{2m + 3}{4m + 3}$ **21.** $\dfrac{25p^2}{9}$ **23.** $\frac{2}{9}$ **25.** $\dfrac{5x}{y}$ **27.** $\dfrac{2(a + 4)}{a - 3}$

29. 1 **31.** $\dfrac{m + 6}{m + 3}$ **33.** $\dfrac{m - 3}{2m - 3}$ **35.** $\dfrac{x^2 - 1}{x^2}$ **37.** $\dfrac{x + y}{x - y}$ **39.** $\dfrac{x^2 - xy + y^2}{x^2 + xy + y^2}$

41. (b) and (c)   **43.** $\dfrac{19}{6k}$ **45.** 1 **47.** $\dfrac{6 + p}{2p}$ **49.** $\dfrac{137}{30m}$ **51.** $\dfrac{-2}{(a + 1)(a - 1)}$

53. $\dfrac{2m^2 + 2}{(m - 1)(m + 1)}$ **55.** $\dfrac{4}{a - 2}$ or $\dfrac{-4}{2 - a}$ **57.** $\dfrac{3x + y}{2x - y}$ or $\dfrac{-3x - y}{y - 2x}$ **59.** $\dfrac{5}{(a - 2)(a - 3)(a + 2)}$

61. $\dfrac{x - 11}{(x + 4)(x - 4)(x - 3)}$ **63.** $\dfrac{a^2 + 5a}{(a + 6)(a + 1)(a - 1)}$ **65.** $\dfrac{x + 1}{x - 1}$ **67.** $\dfrac{-1}{x + 1}$ **69.** $\dfrac{(2 - b)(1 + b)}{b(1 - b)}$

71. $\dfrac{m^3 - 4m - 1}{m - 2}$ **73.** $\dfrac{p + 5}{p(p + 1)}$ **75.** $\dfrac{-1}{x(x + h)}$

Section R.4 (page 699)

1. $-\frac{1}{64}$ **3.** 8 **5.** -2 **7.** 4 **9.** $\frac{1}{9}$ **11.** not defined   **13.** $\frac{256}{81}$ **15.** 32 **17.** $4p^2$

19. $9x^4$ **21.** It is the expression which, when raised to the power n, gives a, just like $\sqrt[n]{a}$.   **23.** (d)   **25.** $\dfrac{1}{2^7}$

27. $\dfrac{1}{27^3}$ **29.** 1 **31.** $m^{7/3}$ **33.** $(1 + n)^{5/4}$ **35.** $\dfrac{6z^{2/3}}{y^{5/4}}$ **37.** $2^6 a^{1/4} b^{37/2}$ **39.** $\dfrac{r^6}{s^{15}}$

41. $-\dfrac{1}{ab^3}$ **43.** $12^{9/4} y$ **45.** $\dfrac{1}{2p^2}$ **47.** $\dfrac{m^3 p}{n}$ **49.** $-4a^{5/3}$ **51.** $\dfrac{1}{(k + 5)^{1/2}}$ **53.** $\dfrac{4z^2}{x^5 y}$

55. $r^{6 + p}$ **57.** $m^{3/2}$ **59.** $x^{(2n^2 - 1)/n}$ **61.** $p^{(m + n + m^2)/(mn)}$ **63.** $y - 10y^2$ **65.** $-4k^{10/3} + 24k^{4/3}$

67. $x^2 - x$ **69.** $r - 2 + r^{-1}$ or $r - 2 + \dfrac{1}{r}$ **71.** $k^{-2}(4k + 1)$ or $\dfrac{4k + 1}{k^2}$ **73.** $z^{-1/2}(9 + 2z)$ or $\dfrac{9 + 2z}{z^{1/2}}$

75. $p^{-7/4}(p - 2)$ or $\dfrac{p - 2}{p^{7/4}}$ **77.** $(p + 4)^{-3/2}(p^2 + 9p + 21)$ or $\dfrac{p^2 + 9p + 21}{(p + 4)^{3/2}}$

Section R.5 (page 706)

1. $\sqrt[3]{(-m)^2}$ or $\left(\sqrt[3]{-m}\right)^2$ **3.** $5\sqrt[5]{m^4}$ or $5\left(\sqrt[5]{m}\right)^4$ **5.** $\dfrac{-4}{\sqrt[3]{z}}$ **7.** $\sqrt[3]{(2m + p)^2}$ or $\left(\sqrt[3]{2m + p}\right)^2$

9. $k^{2/5}$ **11.** $-a^{2/3}$ **13.** $-3 \cdot 5^{1/2} p^{3/2}$ **15.** $18mn^{3/2} p^{1/2}$ **17.** The product is $\sqrt[3]{4^2} = \sqrt[3]{16}$, not 4.

19. Rewrite as $\dfrac{\sqrt[3]{3}}{\sqrt[3]{2}}$ and multiply both the numerator and the denominator by $\sqrt[3]{4}$.   **21.** 5 **23.** -5

25. $5\sqrt{2}$ **27.** $3\sqrt[3]{3}$ **29.** $-2\sqrt[4]{2}$ **31.** $-\dfrac{3\sqrt{5}}{5}$ **33.** $-\dfrac{\sqrt[3]{100}}{5}$ **35.** $32\sqrt[3]{2}$

37. $2x^2 z^4 \sqrt{2x}$ **39.** $2zx^2 y \sqrt[3]{2z^2 x^2 y}$ **41.** $np^2 \sqrt[4]{m^2 n^3}$ **43.** cannot be simplified further   **45.** $\dfrac{\sqrt{6x}}{3x}$

47. $\dfrac{x^2 y \sqrt{xy}}{z}$ **49.** $\dfrac{2\sqrt[3]{x}}{x}$ **51.** $\dfrac{h\sqrt[4]{9g^3 hr^2}}{3r^2}$ **53.** $\dfrac{m\sqrt[3]{n^2}}{n}$ **55.** $2\sqrt[4]{x^3 y^3}$ **57.** $\sqrt[3]{2}$

59. $\sqrt[18]{x}$ **61.** $9\sqrt{3}$ **63.** $-2\sqrt{7p}$ **65.** $7\sqrt[3]{3}$ **67.** $2\sqrt{3}$ **69.** $\dfrac{13\sqrt[3]{4}}{6}$ **71.** -7 **73.** 10

75. $11 + 4\sqrt{6}$ **77.** $5\sqrt{6}$ **79.** $2\sqrt[3]{9} - 7\sqrt[3]{3} - 4$ **81.** $\dfrac{\sqrt{15} - 3}{2}$ **83.** $\dfrac{3\sqrt{5} + 3\sqrt{15} - 2\sqrt{3} - 6}{33}$

85. $\dfrac{p(\sqrt{p} - 2)}{p - 4}$ **87.** $\dfrac{a(\sqrt{a + b} + 1)}{a + b - 1}$ **89.** $\dfrac{-1}{2(1 - \sqrt{2})}$ **91.** $\dfrac{x}{\sqrt{x} + x}$ **93.** $\dfrac{-1}{2x - 2\sqrt{x(x + 1)} + 1}$

95. $|m + n|$ **97.** $|z - 3x|$

Index of Applications

Index

GEOMETRIC FORMULAS

Figure	Formulas	Examples
Square	Perimeter: $P = 4s$ Area: $A = s^2$	
Rectangle	Perimeter: $P = 2L + 2W$ Area: $A = LW$	
Triangle	Perimeter: $P = a + b + c$ Area: $A = \frac{1}{2}bh$	
Pythagorean Theorem (for Right Triangles)	$c^2 = a^2 + b^2$	
Sum of the Angles of a Triangle	$A + B + C = 180°$	
Circle	Diameter: $d = 2r$ Circumference: $\quad C = 2\pi r = \pi d$ Area: $A = \pi r^2$	
Parallelogram	Area: $A = bh$ Perimeter: $P = 2a + 2b$	